Lecture Notes in Mathematics 2017

Editors:
J.-M. Morel, Cachan
B. Teissier, Paris

Lars Diening · Petteri Harjulehto
Peter Hästö · Michael Růžička

Lebesgue and Sobolev Spaces
with Variable Exponents

 Springer

Lars Diening
LMU Munich
Institute of Mathematics
Theresienstr. 39
80333 Munich
Germany
lars@diening.de

Petteri Harjulehto
University of Helsinki
Department of Mathematics and Statistics
Helsinki
Finland
petteri.harjulehto@helsinki.fi

Peter Hästö
University of Oulu
Department of Mathematical Sciences
Oulu
Finland
peter.hasto@helsinki.fi

Michael Růžička
University of Freiburg
Institute of Mathematics
Eckerstr. 1
79104 Freiburg
Germany
rose@mathematik.uni-freiburg.de

ISBN 978-3-642-18362-1   e-ISBN 978-3-642-18363-8
DOI 10.1007/978-3-642-18363-8
Springer Heidelberg Dordrecht London New York

Lecture Notes in Mathematics ISSN print edition: 0075-8434
                              ISSN electronic edition: 1617-9692

Library of Congress Control Number: 2011925547

Mathematics Subject Classification (2011): 46E30, 46E35, 26D10, 31B15, 35J60, 35Q35, 76W05,
                              76A05, 35B65, 35D05

Cover design: deblik, Berlin

Printed on acid-free paper

Springer is part of Springer Science+Business Media (www.springer.com)

Preface

In the past few years the subject of variable exponent spaces has undergone a vast development. Nevertheless, the standard reference is still the article by Kováčik and Rákosník from 1991. This paper covers only basic properties, such as reflexivity, separability, duality and first results concerning embeddings and density of smooth functions. In particular, the boundedness of the maximal operator, proved by Diening in 2002, and its consequences are missing.

Naturally, progress on more advanced properties is scattered in a large number of articles. The need to introduce students and colleagues to the main results led around 2005 to some short survey articles. Moreover, Diening gave lectures at the University of Freiburg in 2005 and Růžička gave a course in 2006 at the Spring School NAFSA 8 in Prague. The usefulness of a more comprehensive treatment was clear, and so we decided in the summer of 2006 to write a book containing both basic and advanced properties, with improved assumptions. Two further lecture courses were given by Hästö based on our material in progress (2008 in Oulu and 2009 at the Spring School in Paseky); another summary is Diening's 2007 habilitation thesis.

It has been our goal to make the book accessible to graduate students as well as a valuable resource for researchers. We present the basic and advanced theory of function spaces with variable exponents and applications to partial differential equations. Not only do we summarize much of the existing literature but we also present new results of our most recent research, including unifying approaches generated while writing the book.

Writing such a book would not have been possible without various sources of support. We thank our universities for their hospitality and the Academy of Finland and the DFG research unit "Nonlinear Partial Differential Equations: Theoretical and Numerical Analysis" for financial support. We also wish to express our appreciation of our fellow researchers whose results are presented and ask for understanding for the lapses, omissions and misattributions that may have entered the text. Thanks are also in order to Springer Verlag for their cooperation and assistance in publishing the book.

We thank our friends, colleagues and especially our families for their continuous support and patience during the preparation of this book.

Finally, we hope that you find this book useful in your journey into the world of variable exponent Lebesgue and Sobolev spaces.

Munich, Germany   *Lars Diening*
Helsinki, Finland   *Petteri Harjulehto*
Oulu, Finland   *Peter Hästö*
Freiburg, Germany   *Michael Růžička*
November 2010

Contents

1   **Introduction**.. 1
     1.1   History of Variable Exponent Spaces 2
     1.2   Structure of the Book 4
     1.3   Summary of Central Results................................... 7
     1.4   Notation and Background...................................... 10

Part I   Lebesgue Spaces

2   **A Framework for Function Spaces** 21
     2.1   Basic Properties of Semimodular Spaces...................... 21
     2.2   Conjugate Modulars and Dual Semimodular Spaces......... 29
     2.3   Musielak–Orlicz Spaces: Basic Properties.................... 34
     2.4   Uniform Convexity 42
     2.5   Separability 48
     2.6   Conjugate Φ-Functions 52
     2.7   Associate Spaces and Dual Spaces 57
     2.8   Embeddings and Operators 66

3   **Variable Exponent Lebesgue Spaces** 69
     3.1   The Lebesgue Space Φ-Function.............................. 69
     3.2   Basic Properties 73
     3.3   Embeddings 82
     3.4   Properties for Restricted Exponents 87
     3.5   Limit of Exponents 92
     3.6   Convolution* 94

4   **The Maximal Operator**.. 99
     4.1   Logarithmic Hölder Continuity 100
     4.2   Point-Wise Estimates 104
     4.3   The Boundedness of the Maximal Operator.................. 110
     4.4   Weak-Type Estimates and Averaging Operators.............. 115
     4.5   Norms of Characteristic Functions 122
     4.6   Mollification and Convolution 127

4.7   Necessary Conditions for Boundedness* 131
4.8   Preimage of the Maximal Operator* 138

5   **The Generalized Muckenhoupt Condition*** 143
5.1   Non Sufficiency of log-Hölder Continuity* 143
5.2   Class $\mathcal{A}^*$ 149
5.3   Class $\mathcal{A}$ for Variable Exponent Lebesgue Spaces* 157
5.4   Class $\mathcal{A}_\infty^*$ 159
5.5   A Sufficient Condition for the Boundedness of M * 168
5.6   Characterization of (Strong-)Domination* 175
5.7   The Case of Lebesgue Spaces with Variable Exponents* 180
5.8   Weighted Variable Exponent Lebesgue Spaces* 192

6   **Classical Operators** 199
6.1   Riesz Potentials 199
6.2   The Sharp Operator $M^\sharp f$ 206
6.3   Calderón–Zygmund Operators 208

7   **Transfer Techniques** 213
7.1   Complex Interpolation 213
7.2   Extrapolation Results 218
7.3   Local-to-Global Results 222
7.4   Ball/Cubes-to-John 237

Part II   Sobolev Spaces

8   **Introduction to Sobolev Spaces** 247
8.1   Basic Properties 247
8.2   Poincaré Inequalities 252
8.3   Sobolev-Poincaré Inequalities and Embeddings 265
8.4   Compact Embeddings 272
8.5   Extension Operator 275
8.6   Limiting Cases of Sobolev Embeddings* 283

9   **Density of Regular Functions** 289
9.1   Basic Results on Density 290
9.2   Density with Continuous Exponents 293
9.3   Density with Discontinuous Exponents* 299
9.4   Density of Continuous Functions* 305
9.5   The Lipschitz Truncation Method* 310

10   **Capacities** 315
10.1   Sobolev Capacity 315
10.2   Relative Capacity 322
10.3   The Relationship Between the Capacities 331
10.4   Sobolev Capacity and Hausdorff Measure 334

11 Fine Properties of Sobolev Functions 339
  11.1 Quasicontinuity 339
  11.2 Sobolev Spaces with Zero Boundary Values 345
  11.3 Exceptional Sets in Variable Exponent Sobolev Spaces 350
  11.4 Lebesgue Points 352
  11.5 Failure of Existence of Lebesgue Points* 360

12 Other Spaces of Differentiable Functions 367
  12.1 Trace Spaces 368
  12.2 Homogeneous Sobolev Spaces 378
  12.3 Sobolev Spaces with Negative Smoothness 383
  12.4 Bessel Potential Spaces* 388
  12.5 Besov and Triebel–Lizorkin Spaces* 392

Part III Applications to Partial Differential Equations

13 Dirichlet Energy Integral and Laplace Equation 401
  13.1 The One Dimensional Case 402
  13.2 Minimizers 412
  13.3 Harmonic and Superharmonic Functions 417
  13.4 Harnack's Inequality for A-harmonic Functions 421

14 PDEs and Fluid Dynamics 437
  14.1 Poisson Problem 437
  14.2 Stokes Problem 446
  14.3 Divergence Equation and Consequences 459
  14.4 Electrorheological Fluids 470

References 483

List of Symbols 501

Index 505

Chapter 1
Introduction

The field of variable exponent function spaces has witnessed an explosive growth in recent years. For instance, a search for "variable exponent" in Mathematical Reviews yields 15 articles before 2000, 31 articles between 2000 and 2004, and 267 articles between 2005 and 2010. This is a crude measure with some misclassifications, but it is nevertheless quite telling.

The standard reference for basic properties has been the article [258] by Kováčik and Rákosník from 1991. (The same properties were derived by different methods by Fan and Zhao [149] 10 years later.) Some surveys of the field exist, e.g. [99, 345], but they are already quite dated. When we started writing this book, in 2006, it seemed possible to derive a more coherent foundation for the field with simpler and better proofs. This turned out to be somewhat more challenging than we had anticipated, but it is fair to say that the understanding of the basics of the field has now, in 2010, reached a certain stability and maturity. Thus we have tried to write a usable, self-contained monograph collecting all the basic properties of variable exponent Lebesgue and Sobolev spaces, which fills the need of having a readily available reference with unified notation and terminology.

Since most of the results contained in this book are no more than ten years old, we have generally credited the original authors of results mid-text, often noting also previous contributions. Our selection of topics is based to some extent on our personal interests, but we have tried to include all the most important and general results, and make note of several other ones along with references to sources for further information.

Many of the very early contributions are largely superseded by more recent results, and so we include here a brief history of the field from its inception in 1931 to approximately 2000 in the next section. The second section of this chapter provides an outline of the rest of the book.

In Sect. 1.3 we summarize the most important basic properties of variable exponent spaces from the book, as well as some properties which do not hold. We also provide a diagram which shows the connections between different central assumptions on the exponent. This section is meant as a reference for locating the results one needs, and is not self-contained.

L. Diening et al., *Lebesgue and Sobolev Spaces with Variable Exponents,*
Lecture Notes in Mathematics 2017, DOI 10.1007/978-3-642-18363-8_1,
© Springer-Verlag Berlin Heidelberg 2011

Finally, in Sect. 1.4 we introduce some notation and conventions used throughout the book; we also recall many well-known definitions and results from real and functional analysis, topology and measure theory which are needed later on. No proofs are included for these standard results, but references are provided and they can be consulted if necessary. The results from this section are used in many places later on; we also introduce some standard results later in the book if they are needed only in a single proof or section.

1.1 History of Variable Exponent Spaces

Variable exponent Lebesgue spaces appeared in the literature for the first time already in a 1931 article by Orlicz [319]. In this article the following question is considered: let (p_i) (with $p_i > 1$) and (x_i) be sequences of real numbers such that $\sum_i x_i^{p_i}$ converges. What are the necessary and sufficient conditions on (y_i) for $\sum_i x_i y_i$ to converge? It turns out that the answer is that $\sum_i (\lambda y_i)^{p_i'}$ should converge for some $\lambda > 0$ and $p_i' = p_i/(p_i - 1)$. This is essentially Hölder's inequality in the space $\ell^{p(\cdot)}$. Orlicz also considered the variable exponent function space $L^{p(\cdot)}$ on the real line, and proved the Hölder inequality in this setting.

However, after this one paper, Orlicz abandoned the study of variable exponent spaces, to concentrate on the theory of the function spaces that now bear his name (but see also [308]). In the theory of Orlicz spaces, one defines the space L^φ to consist of those measurable functions $u\colon \Omega \to \mathbb{R}$ for which

$$\varrho(\lambda u) = \int_\Omega \varphi(\lambda |u(x)|)\, dx < \infty$$

for some $\lambda > 0$ (φ has to satisfy certain conditions, see Example 2.3.12 (b)). Abstracting certain central properties of ϱ, we are led to a more general class of so-called modular function spaces which were first systematically studied by Nakano [309, 310]. In the appendix [p. 284] of the first of these books, Nakano mentions explicitly variable exponent Lebesgue spaces as an example of the more general spaces he considers. The duality property mentioned above is again observed.

Following the work of Nakano, modular spaces were investigated by several people, most importantly by groups at Sapporo (Japan), Voronezh (USSR), and Leiden (Netherlands). Somewhat later, a more explicit version of these spaces, modular function spaces, were investigated by Polish mathematicians, for instance Hudzik, Kamińska and Musielak. For a comprehensive presentation of modular function spaces, see the monograph [307] by Musielak.

Variable exponent Lebesgue spaces on the real line have been independently developed by Russian researchers, notably Sharapudinov. These investigations originated in a paper by Tsenov from 1961 [366], and were briefly touched on by Portnov [325, 326]. The question raised by Tsenov and solved by Sharapudinov [351–353] is the minimization of

$$\int_a^b |u(x) - v(x)|^{p(x)} dx,$$

where u is a fixed function and v varies over a finite dimensional subspace of $L^{p(\cdot)}([a, b])$. In [351] Sharapudinov also introduced the Luxemburg norm for the Lebesgue space and showed that this space is reflexive if the exponent satisfies $1 < p^- \leqslant p^+ < \infty$. In the mid-1980s Zhikov [392] started a new line of investigation, that was to become intimately related to the study of variable exponent spaces, considering variational integrals with non-standard growth conditions. Another early PDE paper is [257] by Kováčik, but this paper appears to have had little influence on later developments.

The next major step in the investigation of variable exponent spaces was the paper by Kováčik and Rákosník in the early 1990s [258]. This paper established many of the basic properties of Lebesgue and Sobolev spaces in $\mathbb{R}^n$. During the following ten years there were many scattered efforts to understand these spaces.

At the turn of the millennium various developments lead to the start of a period of systematic intense study of variable exponent spaces: First, the connection was made between variable exponent spaces and variational integrals with non-standard growth and coercivity conditions (e.g., [4, 393]). It was also observed that these non-standard variational problems are related to modeling of so-called electrorheological fluids, see [328, 329, 337]. Moreover, progress in physics and engineering over the past ten year have made the study of fluid mechanical properties of these fluids an important issue, see [90, 337, 369]. (Later on, other applications have emerged in thermorheological fluids [34] and image processing [1, 53, 70, 269].)

Even more important from the point of view of the present book is the fact that the "correct" condition for regularity of variable exponents was found. This condition, which we call log-Hölder continuity, was used by Diening [91] to show that the maximal operator is bounded on $L^{p(\cdot)}(\Omega)$ when Ω is bounded. He also showed that the boundedness holds in $L^{p(\cdot)}(\mathbb{R}^n)$ if the exponent is constant outside a compact set. The case of unbounded domains was soon improved by Cruz-Uribe, Fiorenza and Neugebauer [84] and, independently, Nekvinda [314] so that a decay condition replaces the constancy at infinity. The boundedness of the maximal operator opens up the door for treating a plethora of other operators. For instance one can then consider the Riesz potential operator and thus prove Sobolev embeddings. Such results indeed followed in quick succession starting from the middle of the 00s.

The boundedness of the maximal operator and other operators is a subtle question and improvements on these initial results have been made since then in many papers. In this book we present mature versions of these results as well as more recent advances. In particular, we would like to emphasize the efforts to remove spurious bounds on the exponents from previous results which were the consequence of technical rather than substantial issues. In particular, we have made a point of replacing the assumptions $1 < p^-$ and $p^+ < \infty$ by $1 \leqslant p^-$ and $p^+ \leqslant \infty$ whenever possible.

1.2 Structure of the Book

This book is divided into three parts. The first part deals with variable exponent Lebesgue spaces, and the second one deals with variable exponent Sobolev spaces. These form the main content of the book. In the third part we give a selection of applications of these results to partial differential equations. Some sections and one chapter are marked by an asterisk. These we consider more advanced content which may be omitted on first reading.

Figure 1.1 illustrates the main dependencies among the chapters. As indicated by the triple line, Chaps. 3, 4, 6 and 8 form the core of the book. They

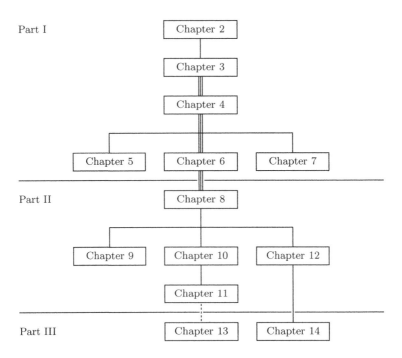

Fig. 1.1 The main dependencies among chapters

deal with basic properties of Lebesgue spaces, the maximal operator, other operators, and Sobolev spaces, respectively. Any course based on the book would likely include at least these chapters, although Chap. 8 could be omitted if one is not interested in differentiability. Results from Chap. 5 and 7 are used in Chap. 8 and later to some extent, but it is not unreasonable to skip these at first reading if one is interested mainly in Sobolev spaces.

Chapter 2 is not properly about variable exponent spaces but rather introduces the more general frameworks of semimodular spaces and Musielak–Orlicz spaces. Since these topics have not been treated in this generality in widely available sources we have included this preliminary chapter for completeness. It can be skipped by readers mostly interested in more advanced properties of variable exponent spaces. On the other hand, many basic properties, including completeness, reflexivity, separability and uniform convexity, follow in the variable exponent setting directly from the more general case. It should also be stressed that the study of semimodular spaces, rather than modular spaces, allows us to treat variable exponent spaces with unbounded exponents in a uniform manner, in contrast to many previous investigations which have used a more *ad hoc* approach (cf. Remark 3.2.3).

Chapter 3 relies heavily on Chap. 2: we directly obtain completeness, reflexivity, separability and uniform convexity. The more complicated general conditions translate into simple (and optimal) assumptions on the variable exponent (see Sect. 1.3). Another important topic in these sections is the norm dual formula, which we derive in the framework of associate spaces; this is another component which allows us to avoid earlier restrictions on the variable exponent that follow from dual space considerations.

Chapter 4 introduces a slate of new techniques to deal with the maximal function and averaging operators. These are the central advances of the past few years which have made possible the rapid expansion of the field. In contrast to previous investigations, our general techniques allow for the systematic inclusion of unbounded exponents. After introducing the logarithmic Hölder continuity condition, we derive "Hölder"-type inequalities

$$\left(\fint_Q |f| \, dx \right)^{p(y)} \leqslant \fint_Q |f|^{p(x)} \, dx + \text{error}$$

where $y \in Q$ and "error" denotes an appropriate error term. This estimate suffices for the boundedness of the maximal operator in unbounded domains and for unbounded exponents. If we use the boundedness of the maximal operator we always incur the restriction $p^- > 1$, which is in fact necessary by Theorem 4.7.1. Therefore we also study two tools without this shortcoming: weak-type estimates and averaging operators. We prove, for instance, that

$$\|1\|_{L^{p(\cdot)}(Q)} \approx |Q|^{\frac{1}{p_Q}}$$

for all cubes Q, where $\frac{1}{p}$ is log-Hölder continuous but p is possibly unbounded, and p_Q is the harmonic average of p on Q. Also convolution is shown to work without bounds on the exponent.

Chapter 6 consists of a fairly straightforward application of the methods from Chap. 4 to other operators such as the Riesz potential, the sharp operator and singular integrals.

The first part includes two optional chapters. Chapter 5 contains a more abstract treatment of the boundedness of the maximal operator in terms of the so-called class $\mathcal{A}$. This class consists of those exponents for which a suitable collection of averaging operators are bounded. It provides the right context for a necessary and sufficient condition of the boundedness of the maximal operator similar to the Muckenhoupt classes for weighted Lebesgue spaces. Working with averaging rather than maximal operators allows us to remove superfluous restrictions on the exponent from below which had appeared in various previous results. This is the case for instance for the Poincaré inequality, which is considered in Chap. 8.

Chapter 7 is a collection of methods which we call "transfer techniques". The idea is that we start with a result in one setting and obtain it in another setting "for free". The best known example of such a technique is interpolation, which has played an important unifying role in the development of the theory of constant exponent spaces. Unfortunately, it is not possible to interpolate from constant exponents to variable exponents. Therefore other techniques are also included, namely, extrapolation and a result for generalizing statements for balls to statements in (possible unbounded) John domains.

The first chapter in the second part, Chap. 8, relies substantially on the results from the first part. First we "translate" the results from Chap. 3 to results for Sobolev spaces. Hence we prove completeness, reflexivity, separability and uniform convexity, again under optimal assumptions on the exponent. More sophisticated results like Sobolev embeddings and Poincaré inequalities are proved by recourse to results on the maximal and other operators. We also include a short section on compact embeddings and present a recent extension result. Again, several results are presented for the first time including the cases $p^- = 1$ and/or $p^+ = \infty$.

After the first chapter, the second part of the book splits into three relatively independent strands. Chapter 9 deals with the density of smooth and continuous functions in Sobolev spaces, which turns out to be an elusive and difficult question which is not fully understood yet. We present several sufficient conditions for density, as well as examples when density does not hold.

Chapter 10 introduces a Sobolev and a relative capacity, which measure set size on a finer scale than the Lebesgue measure. We study their relationship with each other and with the Hausdorff measure. The capacities are used in Chap. 11 to the study of fine properties of Sobolev functions, such as quasicontinuity, removability, Lebesgue points and function with zero boundary value.

The third strand, in Chap. 12, deals with other spaces of "Sobolev type", i.e. spaces of functions with at least some (possibly fractional) smoothness. In particular, trace, homogeneous Sobolev, Bessel potential, Besov, and Triebel–Lizorkin spaces are considered.

In the third part, we consider applications to partial differential equations of the theory developed in the first two parts. The third part consists of two chapters. In Chap. 13, we consider PDE of non-standard growth, i.e. differential equations where the main term is of the form $-\operatorname{div}(|\nabla u|^{p(\cdot)-2}\nabla u)$. In this case $W^{1,p(\cdot)}(\Omega)$ is the natural space in which to look for solutions. The approach of the chapter continues the minimal assumptions-theme of previous chapters. In particular, we add continuity assumptions on the exponent only as necessary. This part is based on capacity methods and fine properties of the functions from Chaps. 10 and 11. Chapter 14 is the culmination of the other strand in Part II: here we use traces and homogeneous spaces from Chap. 12, Calderón–Zygmund operators (Sect. 6.3), as well as the Lipschitz truncation method (Sect. 9.5), and the transfer technique from Sect. 7.4. We first treat classical linear PDE with data in variable exponent spaces, namely the Poisson and Stokes problems and the divergence equation. The latter leads to generalizations of further classical results to variable exponent spaces. Finally these results and the theory of pseudomonotone operators are applied in Sect. 14.4 to prove the existence of solutions to the steady equations for the motion of electrorheological fluids, which is again a PDE with a version of the variable $p(\cdot)$-Laplacian as a main elliptic term.

1.3 Summary of Central Results

In this section we highlight the similarities and differences between constant exponent and variable exponent spaces; we also emphasize the assumptions on the exponent needed for the properties. First we list properties which do not require any regularity of the exponent. The second section features a diagram which illustrates the quite complex relationship between the different conditions used when dealing with more advanced properties such as boundedness of various operators. In the final section we list some properties which essentially never hold in the variable exponent context.

Elementary Properties

Here we collect the most important properties of variable exponent Lebesgue and Sobolev spaces which hold without advanced conditions on the exponent.

For Any Measurable Exponent p

- $L^{p(\cdot)}$ and $W^{1,p(\cdot)}$ are Banach spaces (Theorem 3.2.7, Theorem 8.1.6).
- The modular $\varrho_{p(\cdot)}$ and the norm $\|\cdot\|_{p(\cdot)}$ are lower semicontinuous with respect to (sequential) weak convergence and almost everywhere convergence (Theorem 3.2.9, Lemma 3.2.8, Lemma 3.2.10).
- Hölder's inequality holds (Lemma 3.2.20).
- $L^{p(\cdot)}$ is a Banach function space (Theorem 3.2.13).
- $(L^{p(\cdot)})' \cong L^{p'(\cdot)}$ and the norm conjugate formula holds (Theorem 3.2.13, Corollary 3.2.14).

For Any Measurable Bounded Exponent p

- $L^{p(\cdot)}$ and $W^{1,p(\cdot)}$ are separable spaces (Lemma 3.4.4, Theorem 8.1.6).
- The Δ_2-condition holds (Theorem 3.4.1).
- Bounded functions are dense in $L^{p(\cdot)}$ and $W^{1,p(\cdot)}$ (Corollary 3.4.10, Lemma 9.1.1).
- C_0^∞ is dense in $L^{p(\cdot)}$ (Theorem 3.4.12).

For Any Measurable Exponent p with $1 < p^- \leqslant p^+ < \infty$

- $L^{p(\cdot)}$ and $W^{1,p(\cdot)}$ are reflexive (Theorem 3.4.7, Theorem 8.1.6).
- $L^{p(\cdot)}$ and $W^{1,p(\cdot)}$ are uniformly convex (Theorem 3.4.9, Theorem 8.1.6).

The log-Hölder and Other Conditions

The diagram in Fig. 1.2 illustrates the relationship between more advanced conditions imposed on the exponent p. Arrows represent implications, with the relevant theorem or lemma number quoted. The three bullets are conjunctions of the conditions, e.g., $p^- > 1$ and $\mathcal{P}^{\log}$ together imply that the maximal operator M is bounded, by Theorem 4.3.8.

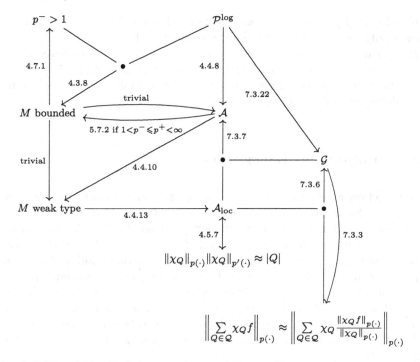

Fig. 1.2 The relationship between various conditions on the exponent

Warnings!

Here are some results and techniques from constant exponent spaces which do not hold in the variable exponent setting even when the exponent is very regular, e.g., $p \in \mathcal{P}^{\log}$ or $p \in C^{\infty}(\overline{\Omega})$ with $1 < p^- \leqslant p^+ < \infty$.

- The space $L^{p(\cdot)}$ is not rearrangement invariant; the translation operator $T_h : L^{p(\cdot)} \to L^{p(\cdot)}$, $T_h f(x) := f(x+h)$ is not bounded; Young's convolution inequality $\|f * g\|_{p(\cdot)} \leqslant c \|f\|_1 \|g\|_{p(\cdot)}$ does not hold (Sect. 3.6).
- The formula

$$\int_{\Omega} |f(x)|^p \, dx = p \int_0^{\infty} t^{p-1} |\{x \in \Omega : |f(x)| > t\}| \, dt$$

has no variable exponent analogue.

- Maximal, Poincaré, Sobolev, etc., inequalities do not hold in a modular form. For instance, Lerner showed that

$$\int_{\mathbb{R}^n} |Mf|^{p(x)} \, dx \leqslant c \int_{\mathbb{R}^n} |f|^{p(x)} \, dx$$

if and only if $p \in (1, \infty]$ is constant [267, Theorem 1.1]. For the Poincaré inequality see Example 8.2.7 and the discussion after it.

- Interpolation is not so useful, since variable exponent spaces never result as an interpolant of constant exponent spaces (see Sect. 7.1).
- Solutions of the $p(\cdot)$-Laplace equation are not scalable, i.e. λu need not be a solution even if u is (Example 13.1.9).

1.4   Notation and Background

In this section we clarify the basic notation used in the book. Moreover we give precise formulations of some basic results which are frequently used.

We use the symbol $:=$ to define the left-hand side by the right-hand side. For constants we use the letters $c, c_1, c_2, C, C_1, C_2, \ldots$, or other letters specifically mentioned to be constants. The symbol c without index stands for a generic constant which may vary from line to line. In theorems, propositions and lemmas we give precise dependencies of the constants on the involved other quantities. We use $x \approx y$ if there exist constants c_1, c_2 such that $c_1 x \leqslant y \leqslant c_2 x$. The Euler constant is denoted by e and the imaginary unit is denoted by i. For sets A and B the notation $A \subset B$ includes also the case $A = B$.

By $\mathbb{R}^n$ we denote the n-dimensional Euclidean space, and $n \in \mathbb{N}$ always stands for the dimension of the space. By U and V we denote open sets and by F closed sets of the topological space under consideration, usually $\mathbb{R}^n$. A compact set will usually be denoted by K. For $A, E \subset X$ we use the notation $A \subset\subset E$ if the closure $\overline{A}$ is compact and $\overline{A} \subset E$. By Ω we always denote an open subset of $\mathbb{R}^n$. If the set has additional properties it will be stated explicitly. A *domain* $\Omega \subset \mathbb{R}^n$ is a connected open set. We will also use domains with specific conditions on the boundary, such as John domains (cf. Definition 7.4.1).

Balls will be denoted by B. The ball with radius r and center $x_0 \in \mathbb{R}^n$ will be denoted by $B(x_0, r)$. We usually denote cubes in $\mathbb{R}^n$ by Q, and by a cube we always mean a *non-degenerate cube with faces parallel to the coordinate axes*. However, in many places Q stands for cubes or balls, since the statements hold for both of them, but this will be mentioned explicitly. For a ball B we will denote the ball with α times the radius and the same center by αB. Similarly, for a cube Q we will denote by αQ the cube with α times the diameter and same center as Q. For *half-spaces* of $\mathbb{R}^n$ we use the notation $\mathbb{R}^n_> := \{x \in \mathbb{R}^n : x_n > 0\}$, $\mathbb{R}^n_\geqslant := \{x \in \mathbb{R}^n : x_n \geqslant 0\}$, $\mathbb{R}^n_< := \{x \in \mathbb{R}^n : x_n < 0\}$, and $\mathbb{R}^n_\leqslant := \{x \in \mathbb{R}^n : x_n \leqslant 0\}$, where $x = (x_1, \ldots, x_n)$. For $a, b \in \mathbb{R}^n$ we use (a, b) and $[a, b]$ to denote the *open and closed segment*, respectively, connecting a and b.

Functional Analysis

A *Banach space* $(X, \|\cdot\|_X)$ is a normed vector space over the field of real numbers $\mathbb{R}$ or the field of complex numbers $\mathbb{C}$, which is complete with respect the norm $\|\cdot\|_X$. The Cartesian product $X : \prod_{j=1}^{N} X_j$ of Banach spaces $(X_j, \|\cdot\|_{X_j})$ consists of points $(x_1, \dots, x_N)$ and is equipped with any of the equivalent norms $\|x\|_X = \|x\|_{X,r} := (\sum_{j=1}^{N} \|x_j\|_{X_j}^{r})^{\frac{1}{r}}$, $1 \leqslant r < \infty$, and an obvious modification for $r = \infty$. If $X_j = Y$ for $j = 1, \dots, N$, we write $X = Y^N$. Sometimes it is useful to equip a vector space with a *quasinorm* instead of a norm. A quasinorm satisfies all properties of a norm except the triangle inequality which is replaced by $\|x + y\| \leqslant c (\|x\| + \|y\|)$ for some $c > 0$.

Let X and Y be normed vector spaces. The mapping $F \colon X \to Y$ is *bounded* if $\|F(a)\|_Y \leqslant C \|a\|_X$ for all $a \in X$. It is an *isomorphism* if F and F^{-1} are bijective, linear and continuous. Clearly, a linear mapping is bounded if and only if it is continuous.

Let X and Y be normed spaces, both subsets of a Hausdorff space Z (i.e. distinct points possess disjoint neighborhoods). Then the intersection $X \cap Y$ equipped with the norm $\|z\|_{X \cap Y} = \max \{\|z\|_X, \|z\|_Y\}$ and the sum $X + Y := \{x + y : x \in X, y \in Y\}$ equipped with the norm

$$\|z\|_{X+Y} = \inf \{\|x\|_X + \|y\|_Y : x \in X, y \in Y, z = x + y\}$$

are normed spaces. If X and Y are Banach spaces, then $X \cap Y$ and $X + Y$ are Banach spaces as well.

The *dual space* X^* of a Banach space X consists of all bounded, linear functionals $F \colon X \to \mathbb{R}$ (or $\mathbb{C}$). The *duality pairing* between X^* and X is defined by $\langle F, X \rangle_{X^*, X} = \langle F, X \rangle := F(x)$ for $F \in X^*$, $x \in X$. The dual space is equipped with the *dual norm* $\|F\|_{X^*} := \sup_{\|x\|_X \leqslant 1} \langle F, x \rangle$, which makes X^* a Banach space. We have the following versions and consequences of the Hahn–Banach theorem (cf. [58, Corollary I.2, Theorem I.7, Corollary I.4], [335]).

Theorem 1.4.1. *Let X be a Banach space and let $Y \subset X$ be a closed, linear subspace. Every bounded, linear functional $F \in Y^*$ can be extended to a bounded, linear functional $\widetilde{F} \in X^*$ satisfying*

$$\|\widetilde{F}\|_{X^*} = \|F\|_{Y^*}.$$

Here we mean by extension that $\langle F, y \rangle_{Y^*, Y} = \langle \widetilde{F}, y \rangle_{X^*, X}$ for all $y \in Y$.

Theorem 1.4.2. *Let X be a Banach space and let E and K be convex, disjoint non-empty subsets of X. If E is closed and K compact, then there exists a closed hyperplane which strictly separates E and K, i.e. there exists $F \in X^*$, $\alpha \in \mathbb{R}$ and $\varepsilon > 0$ such that $\operatorname{Re} F(x) + \varepsilon \leqslant \alpha \leqslant \operatorname{Re} F(y) - \varepsilon$ for all $x \in E$, $y \in K$.*

Corollary 1.4.3. *Let X be a Banach space. Then we have for all $x \in X$*

$$\|x\|_X = \sup_{\|F\|_{X^*} \leqslant 1} |\langle F, x \rangle| .$$

A Banach space is called *separable* if it contains a dense, countable subset. We denote the bidual space by $X^{**} := (X^*)^*$. A Banach space X is called *reflexive* if the *natural injection* $\iota\colon X \to X^{**}$, given by $\langle \iota x, F \rangle_{X^{**}, X^*} := \langle F, x \rangle_{X^*, X}$, is surjective. A norm $\| \cdot \|$ on a Banach space X is called *uniformly convex* if for every $\varepsilon > 0$ there exists $\delta(\varepsilon) > 0$ such that for all $x, y \in X$ satisfying $\|x\|, \|y\| \leqslant 1$, the inequality $\|x - y\| > \varepsilon$ implies $\|(x + y)/2\| < 1 - \delta(\varepsilon)$. A Banach space X is called *uniformly convex*, if there exists a uniformly convex norm $\| \cdot \|'$, which is equivalent to the original norm of X. These properties are inherited to closed linear subspaces and Cartesian products. More precisely we have (cf. [11, Chap. I]):

Proposition 1.4.4. *Let X be a Banach space and let Y denote either a closed subset of X or a Cartesian product X^N. Then:*

(i) *Y is a Banach space.*
(ii) *If X is reflexive, then Y is reflexive.*
(iii) *If X is separable, then Y is separable.*
(iv) *If X is uniformly convex, then X is reflexive.*
(v) *If X is uniformly convex, then Y is uniformly convex.*

We say that a Banach space X is *continuously embedded* into a Banach space Y, $X \hookrightarrow Y$, if $X \subset Y$ and there exists a constant $c > 0$ such that $\|x\|_Y \leqslant c \|x\|_X$ for all $x \in X$. The embedding of X into Y is called *compact*, $X \hookrightarrow\hookrightarrow Y$, if $X \hookrightarrow Y$ and bounded sets in X are precompact in Y. A sequence $(x_k)_{k \in \mathbb{N}} \subset X$ is called *(strongly) convergent* to $x \in X$, if $\lim_{k \to \infty} \|x_k - x\|_X = 0$. It is called *weakly convergent* if $\lim_{k \to \infty} \langle F, x_k \rangle = 0$ for all $F \in X^*$. An embedding $X \hookrightarrow Y$ is compact if and only if weakly convergent sequences in X are mapped to strongly convergent sequences in Y. Note that each set which is closed with respect the weak topology (convergence) is also closed with respect to the strong topology (convergence). The converse implication is in general false. However it holds for convex sets (cf. [58, Theorem III.7]).

Let $(X, \|\cdot\|_X)$ be a Banach space and $A \subset X$ a set. The *closure of A with respect to the norm* $\|\cdot\|_X$, $\overline{A}^{\|\cdot\|_X}$, is the smallest closed set Y that contains A. The closure of a set A is denoted by $\overline{A}$ when the space is clear from the context.

Measures and Covering Theorems

We denote by (A, Σ, μ) a *measure space* (cf. [184]). If not stated otherwise μ will always be a *σ-finite, complete measure* on Σ with $\mu(A) > 0$. If there is no danger of confusion we omit Σ from the notation. We use the usual convention of identifying two μ-measurable functions on A if they agree *almost everywhere*, i.e. if they agree up to a set of μ-measure zero. The characteristic function of a set $E \subset A$ will be denoted by χ_E.

A measure μ is called *doubling* if balls have finite and positive measure and there is a constant $c \geqslant 1$ such that

$$\mu(2B) \leqslant c\,\mu(B) \qquad \text{for all balls } B.$$

A measure μ is called *atom-less* if for any measurable set A with $\mu(A) > 0$ there exists a measurable subset A' of A such that $\mu(A) > \mu(A') > 0$. For a sequence (A_k) of sets we write $A_k \nearrow A$ if $A_k \subset A_{k+1}$ for $k \in \mathbb{N}$ and $A = \bigcup_{k=1}^{\infty} A_k$. We write $A_k \searrow A$ if $A_k \supset A_{k+1}$ for $k \in \mathbb{N}$ and $A = \bigcap_{k=1}^{\infty} A_k$.

The Lebesgue integral of a Lebesgue-measurable function $f \colon A \to \mathbb{K}$, where $\mathbb{K}$ is either $\mathbb{R}$ or $\mathbb{C}$, is defined in the standard way (cf. [334, Chap. 1]) and denoted by $\int_A f \, d\mu$. If there is no danger of confusion we will write "measurable" instead of "μ-measurable", "almost everywhere" instead of "μ-almost everywhere", etc. The most prominent example for our purposes are: $A = \Omega$ is an open subset of $\mathbb{R}^n$, μ is the n-dimensional Lebesgue measure, and Σ is the σ-algebra of Lebesgue-measurable subsets of Ω; or $A = \mathbb{Z}^n$, μ is the counting measure, and Σ is the power set of $\mathbb{Z}^n$. In the former case the Lebesgue integral will be denoted by $\int_\Omega f \, dx$ and the measure of a measurable subset $E \subset \Omega$ will be denoted by $|E|$.

We need some covering theorems. We state the basic covering theorem in metric measure spaces. The stronger Besicovitch covering theorem does not hold in general in metric measure spaces and hence it is stated in $\mathbb{R}^n$. For the proof for the basic covering theorem see for example [217, Theorem 1.2] or [129, Theorem 1, p. 27] and for the Besicovitch covering theorem [129, Theorem 2, p. 30] or [288, Theorem 2.7, p. 30].

Theorem 1.4.5 (Basic covering theorem). *Let $\mathcal{F}$ be any collection of balls in a metric space with*

$$\sup\{\mathrm{diam}(B) : B \in \mathcal{F}\} < \infty.$$

Then there exists a countable subcollection $\mathcal{G}$ of pair-wise disjoint balls in $\mathcal{F}$ such that

$$\bigcup_{B \in \mathcal{F}} B \subset \bigcup_{B \in \mathcal{G}} 5B.$$

Theorem 1.4.6 (Besicovitch covering theorem). *Let A be a bounded set in $\mathbb{R}^n$. For each $x \in A$ a cube (or ball) $Q_x \subset \mathbb{R}^n$ centered at x is given. Then one can choose, from among the given sets $\{Q_x\}_{x\in A}$ a sequence $\{Q_j\}_{j\in\mathbb{N}}$ (possibly finite) such that:*

(a) *The set A is covered by the sequence, $A \subset \cup_{j\in\mathbb{N}}Q_j$.*
(b) *No point of $\mathbb{R}^n$ is in more than θ_n sets of the sequence $\{Q_j\}_{j\in\mathbb{N}}$, i.e.*

$$\sum_{j=1}^{\infty} \chi_{Q_j} \leqslant \theta_n.$$

(c) *The sequence $\{Q_j\}_{j\in\mathbb{N}}$ can be divided in ξ_n families of disjoint sets.*

The numbers θ_n and ξ_n depend only on the dimension n.

A proof for the existence of a partition of unity can be found for example in [11, Theorem 3.14, p. 51] or in [280, Theorem 1.44, p. 25].

Theorem 1.4.7 (Partition of unity). *Let $\mathcal{U}$ be a family of open sets which cover a closed set $E \subset \mathbb{R}^n$. Then there exists a family $\mathcal{F}$ of functions in $C_0^\infty(\mathbb{R}^n)$ with values in $[0,1]$ such that:*

(a) $\sum_{f\in\mathcal{F}} f(x) = 1$ *for every $x \in E$.*
(b) *For each function $f \in \mathcal{F}$, there exists $U \in \mathcal{U}$ such that spt $f \subset U$.*
(c) *If $K \subset E$ is compact, then spt $f \cap K \neq \emptyset$ for only finitely many $f \in \mathcal{F}$.*

The family $\mathcal{F}$ is said to be a partition of unity of E subordinate to the open covering $\mathcal{U}$.

Integration

Let $\Omega \subset \mathbb{R}^n$ be an open set equipped with the n-dimensional Lebesgue measure. For $s \in [1,\infty]$ and $k \in \mathbb{N}$, we denote by $L^s(\Omega)$ and $W^{k,s}(\Omega)$ the classical *Lebesgue* and *Sobolev spaces*, respectively (cf. [11]). If there is no danger of confusion we omit the set Ω and abbreviate L^s and $W^{k,s}$. The *gradient* of a *Sobolev function u*, i.e. a function belonging to some Sobolev space, is given by $\nabla u := (\partial_1 u, \ldots, \partial_n u)$, where $\partial_i u$, $i = 1, \ldots, n$, are the *weak partial derivatives* of u. We also use the notation $\frac{\partial u}{\partial x_i}$ for $\partial_i u$. If $u \in W^{k,s}(\Omega)$ we denote higher order weak derivatives by $\partial_\alpha u := \frac{\partial^{\alpha_1 + \ldots + \alpha_n} u}{\partial^{\alpha_1} x_1 \cdots \partial^{\alpha_n} x_n}$, where α is a multi-index with $|\alpha| \leqslant k$. By $\nabla^k u$ we denote the tensor consisting of all weak derivatives of u of order k, i.e. $\nabla^k u := (\partial_\alpha u)_{|\alpha|=k}$. In most cases we do not distinguish in the notation of the function spaces and the corresponding norms between scalar, vector-valued or tensor-valued functions.

By $L^1_{\text{loc}}(\Omega)$ we denote the space of all locally integrable functions f, i.e. $f \in L^1(U)$ for all open $U\subset\subset\Omega$. We equip L^1_{loc} by the *initial topology* of those embeddings, i.e. the coarsest topology such that $L^1_{\mathrm{loc}}(\Omega) \hookrightarrow L^1(U)$ for every open $U \subset\subset \Omega$. Analogously, we define $W^{k,1}_{\mathrm{loc}}(\Omega)$ for $k \in \mathbb{N}$ to consist of the functions such that $f \in W^{k,1}(U)$ for all open $U \subset\subset \Omega$. We equip $W^{k,1}_{\mathrm{loc}}(\Omega)$ with the initial topology induced by the embeddings $W^{k,1}_{\mathrm{loc}}(\Omega) \hookrightarrow W^{k,1}(U)$ for all open $U \subset\subset \Omega$. A function $f \colon \Omega \to \mathbb{K}$ has *compact support* if there exists a compact set $K \subset \Omega$ such that $f = f\chi_K$. For an exponent $s \in [1, \infty]$ the *dual exponent* s' is defined by $\frac{1}{s} + \frac{1}{s'} = 1$, with the usual convention $\frac{1}{\infty} = 0$.

We denote by $C(\overline{\Omega})$ the space of uniformly continuous functions equipped with the supremum norm $\|f\|_\infty = \sup_{x \in \overline{\Omega}} |f(x)|$. By $C^k(\overline{\Omega})$, $k \in \mathbb{N}$, we denote the space of functions f, such $\partial_\alpha f \in C(\overline{\Omega})$ for all $|\alpha| \leqslant k$. The space is equipped with the norm $\sup_{|\alpha| \leqslant k} \|\partial_\alpha f\|_\infty$. The set of *smooth functions* in Ω is denoted by $C^\infty(\Omega)$—it consists of functions which are continuously differentiable arbitrarily many times. The set $C_0^\infty(\Omega)$ is the subset of $C^\infty(\Omega)$ of functions which have compact support. We equip $C_0^\infty(\Omega)$ with the initial topology of the embeddings $C_0^\infty(\Omega) \hookrightarrow C^k(\overline{U})$ for all $k \in \mathbb{N}_0$ and open $U \subset\subset \Omega$.

A *standard mollifier* is a non-negative, radially symmetric and radially decreasing function $\psi \in C_0^\infty(B(0,1))$ with $\int_{B(0,1)} \psi\,dx = 1$. We call $\{\psi_\varepsilon\}$ a *standard mollifier family (on $\mathbb{R}^n$)* if ψ is a standard mollifier and $\psi_\varepsilon(\xi) := \varepsilon^{-n}\psi(\xi/\varepsilon)$. By a *modulus of continuity* we mean an increasing, continuous, concave function $\omega \colon \mathbb{R}_\geqslant \to \mathbb{R}_\geqslant$ with $\omega(0) = 0$. A real- or complex-valued function f has modulus of continuity ω if $|f(x) - f(y)| \leqslant \omega(|x - y|)$ for all x and y in the domain of f.

In the case of a measure space (A, μ) we denote by $L^s(A, \mu)$, $s \in [1, \infty]$, the corresponding Lebesgue space. A sequence (f_i) of μ-measurable functions is said to *converge in measure* to the function f if to every $\varepsilon > 0$ there exists a natural number N such that

$$\mu\bigl\{ x \in A : |f_i(x) - f(x)| \geqslant \varepsilon \bigr\} \leqslant \varepsilon$$

for all $i \geqslant N$. For $\mu(A) < \infty$, it is well known that if $f_i \to f$ μ-almost everywhere, then $f_i \to f$ in measure; on the other hand $\lim_{i \to \infty} \int_A |f_i - f|\,d\mu = 0$ does not imply convergence in measure unless we pass to an appropriate subsequence.

Theorem 1.4.8. *Let (A, Σ, μ) be a σ-finite, complete measure space. Assume that f and (f_i) are in $L^1(A, \mu)$, $\mu(A) < \infty$ and*

$$\lim_{i \to \infty} \int_A |f_i - f|\,d\mu = 0.$$

Then there exists a subsequence (f_{i_k}) such that $f_{i_k} \to f$ μ-almost everywhere.

We write $a_k \nearrow a$ if (a_k) is a sequence increasing to a. We frequently use the following convergence results, which can be found e.g. in [11, 184]:

Theorem 1.4.9 (Monotone convergence). *Let (A, Σ, μ) be a σ-finite, complete measure space. Let (f_k) be a sequence of μ-measurable functions with $f_k \nearrow f$ μ-almost everywhere and $\int_A f_1 \, d\mu > -\infty$. Then*

$$\lim_{n \to \infty} \int_A f_k \, d\mu = \int_A f \, d\mu.$$

Theorem 1.4.10 (Dominated convergence). *Let (A, Σ, μ) be a σ-finite, complete measure space. Let (f_k) be a sequence of μ-measurable functions with $f_k \to f$ μ-almost everywhere. If there exists a function $h \in L^1(A, \mu)$ such that $|f_k| \leqslant h$ μ-almost everywhere for all $k \in \mathbb{N}$, then $f \in L^1(A, \mu)$ and*

$$\lim_{n \to \infty} \int_A f_k \, d\mu = \int_A f \, d\mu.$$

Lemma 1.4.11 (Fatou). *Let (A, Σ, μ) be a σ-finite, complete measure space. Let (f_k) be a sequence of μ-measurable functions and let $g \in L^1(A, \mu)$. If $f_k \geqslant g$ μ-almost everywhere for all $n \in \mathbb{N}$, then*

$$\int_A \liminf_{n \to \infty} f_k \, d\mu \leqslant \liminf_{n \to \infty} \int_A f_k \, d\mu.$$

Let $(f_n) \subset L^s(A, \mu)$ and set $\lambda(E) := \limsup_{n \to \infty} \int_E |f_n|^s \, d\mu$, for $E \subset A$ measurable. We say that (f_n) is *equi-integrable* if:

1. For all $\varepsilon > 0$ there exists a $\delta > 0$ with $\lambda(E) < \varepsilon$ for all measurable E with $\mu(E) < \delta$.
2. For all $\varepsilon > 0$ there exists a measurable set A_0 with $\mu(A_0) < \infty$ and $\lambda(A \setminus A_0) < \varepsilon$.

Theorem 1.4.12 (Vitali). *Let (A, Σ, μ) be a σ-finite, complete measure space. Let $1 \leqslant s < \infty$. Let the sequence $(f_n) \subset L^s(A, \mu)$ converge μ-almost everywhere to a μ-measurable function f. Then $f \in L^s(A, \mu)$ and $\|f_n - f\|_{L^s(A, \mu)} \to 0$ if and only if (f_n) is equi-integrable.*

Recall that ν is *absolutely continuous* with respect to μ if $\nu(E) = 0$ for all $E \in \Sigma$ with $\mu(E) = 0$. Another important result is the following (cf. [184]):

Theorem 1.4.13 (Radon–Nikodym). *Let (A, Σ, μ) be a finite measure space and ν be a finite, signed measure on (A, Σ). If ν is absolutely continuous with respect to μ, then there exists a unique function $g \in L^1(A, \mu)$ such that*

$$\nu(E) = \int_E g \, d\mu \quad \text{for all } E \in \Sigma.$$

Theorem 1.4.14 (Jensen's inequality). *Let (A, Σ, μ) be a finite measure space with $\mu(A) = 1$. If f is a real function in $L^1(A, \mu)$, $a < f(x) < b$ for all $x \in A$, and if φ is convex on (a, b), then*

$$\varphi\left(\int_A f \, d\mu\right) \leqslant \int_A \varphi \circ f \, d\mu.$$

The classical *sequence spaces* $l^s(\mathbb{Z}^n)$, $s \in [1, \infty]$, are defined as the Lebesgue spaces $L^s(\mathbb{Z}^n, \mu)$ with μ being the counting measure. It is well known that for $s, q \in [1, \infty]$ the embedding $l^s(\mathbb{Z}^n) \hookrightarrow l^q(\mathbb{Z}^n)$ holds if and only if $s \leqslant q$.

The space of *distributions* $\mathcal{D}'(\Omega)$ is a superset of the space $L^1_{\mathrm{loc}}(\Omega)$ of locally integrable functions. To state the definition, we first equip the space $\mathcal{D}(\Omega) := C_0^\infty(\Omega)$ with such a topology that $f_k \to f$ if and only if

$$\bigcup_k \mathrm{spt}\, f_k \subset\subset \Omega \qquad \text{and} \qquad \lim_{k \to \infty} \sup_\Omega \left|\partial_\alpha(f_k - f)\right| = 0$$

for every multi-index α. Then $\mathcal{D}'(\Omega)$ is the dual of $\mathcal{D}(\Omega)$, i.e. it consists of all bounded linear functionals on $\mathcal{D}(\Omega)$. If $f \in L^1_{\mathrm{loc}}(\Omega)$, then $T_f \in \mathcal{D}'(\Omega)$, where

$$\langle T_f, \varphi \rangle := \int_\Omega f(x)\varphi(x) \, dx$$

for $\varphi \in \mathcal{D}(\Omega)$. For simplicity, one may denote this by $\langle f, \varphi \rangle$ and write $L^1_{\mathrm{loc}}(\Omega) \subset \mathcal{D}'(\Omega)$. If $f \in \mathcal{D}'(\Omega)$, then its *distributional derivative* is the distribution g satisfying

$$\langle g, \varphi \rangle = -\langle f, \varphi' \rangle$$

for all $\varphi \in \mathcal{D}(\Omega)$. Note that this corresponds to partial integration if $f \in C^1(\Omega)$.

Part I
Lebesgue Spaces

Chapter 2
A Framework for Function Spaces

In this chapter we study *modular spaces* and *Musielak–Orlicz spaces* which provide the framework for a variety of different function spaces, including classical (weighted) Lebesgue and Orlicz spaces and variable exponent Lebesgue spaces. Although our aim mainly is to study the latter, it is important to see the connections between all of these spaces. Many of the results in this chapter can be found in a similar form in [307], but we include them to make this exposition self-contained. Research in the field of Musielak–Orlicz functions is still active and we refer to [69] for newer results and references.

Our first two sections deal with the more general case of semimodular spaces. Then we move to basic properties of Musielak–Orlicz spaces in Sect. 2.3. Sections 2.4 and 2.5 deal with the uniform convexity and the separability of the Musielak–Orlicz spaces. In Sects. 2.6 and 2.7 we study dual spaces, and a related concept, associate spaces. Finally, we consider embeddings in Sect. 2.8.

2.1 Basic Properties of Semimodular Spaces

For the investigation of weighted Lebesgue spaces it is enough to stay in the framework of Banach spaces. In particular, the space and its topology is described in terms of a norm. However, in the context of Orlicz spaces this is not the best way. Instead, it is better to start with the so-called modular which then induces a norm. In the case of classical Lebesgue spaces the modular is $\int |f(x)|^p \, dx$ compared to the norm $(\int |f(x)|^p \, dx)^{\frac{1}{p}}$. In some cases the modular has certain advantages compared to the norm, since it inherits all the good properties of the integral. The *modular spaces* defined below capture this advantage.

We are mainly interested in vector spaces defined over $\mathbb{R}$. However, there is no big difference in the definition of real valued and complex valued modular spaces. To avoid a double definition we let $\mathbb{K}$ be either $\mathbb{R}$ or $\mathbb{C}$.

The function ϱ is said to be *left-continuous* if the mapping $\lambda \mapsto \varrho(\lambda x)$ is left-continuous on $[0, \infty)$ for every $x \in X$, i.e. $\lim_{\lambda \to 1^-} \varrho(\lambda x) = \varrho(x)$. Here

L. Diening et al., *Lebesgue and Sobolev Spaces with Variable Exponents*,
Lecture Notes in Mathematics 2017, DOI 10.1007/978-3-642-18363-8_2,
© Springer-Verlag Berlin Heidelberg 2011

$a \to b^-$ means that a tends to b from below, i.e. $a < b$ and $a \to b$; $a \to b^+$ is defined analogously.

Definition 2.1.1. Let X be a $\mathbb{K}$-vector space. A function $\varrho \colon X \to [0, \infty]$ is called a *semimodular* on X if the following properties hold.

(a) $\varrho(0) = 0$.
(b) $\varrho(\lambda x) = \varrho(x)$ for all $x \in X, \lambda \in \mathbb{K}$ with $|\lambda| = 1$.
(c) ϱ is convex.
(d) ϱ is left-continuous.
(e) $\varrho(\lambda x) = 0$ for all $\lambda > 0$ implies $x = 0$.

A semimodular ϱ is called a *modular* if

(f) $\varrho(x) = 0$ implies $x = 0$.

A semimodular ϱ is called *continuous* if

(g) the mapping $\lambda \mapsto \varrho(\lambda x)$ is continuous on $[0, \infty)$ for every $x \in X$.

Remark 2.1.2. Note that our semimodulars are always convex, in contrast to some other sources.

Before we proceed let us provide a few examples.

Definition 2.1.3. Let (A, Σ, μ) be a σ-finite, complete measure space. Then by $L^0(A, \mu)$ we denote the space of all $\mathbb{K}$-valued, μ-measurable functions on A. Two functions are identical, if they agree almost everywhere.

In the special case that μ is the n-dimensional Lebesgue measure, Ω is a μ-measurable subset of $\mathbb{R}^n$, and Σ is the σ-algebra of μ-measurable subsets of Ω we abbreviate $L^0(\Omega) := L^0(\Omega, \mu)$.

Example 2.1.4.

(a) If $1 \leqslant p < \infty$, then

$$\varrho_p(f) := \int_\Omega |f(x)|^p \, dx$$

defines a continuous modular on $L^0(\Omega)$.

(b) Let $\varphi_\infty(t) := \infty \cdot \chi_{(1,\infty)}(t)$ for $t \geqslant 0$, i.e. $\varphi_\infty(t) = 0$ for $t \in [0, 1]$ and $\varphi_\infty(t) = \infty$ for $t \in (1, \infty)$. Then

$$\varrho_\infty(f) := \int_\Omega \varphi_\infty(|f(x)|) \, dx$$

defines a semimodular on $L^0(\Omega)$ which is not continuous.

(c) Let $\omega \in L^1_{\text{loc}}(\Omega)$ with $\omega > 0$ almost everywhere and $1 \leqslant p < \infty$. Then

$$\varrho(f) := \int\limits_\Omega |f(x)|^p \omega(x)\, dx$$

defines a continuous modular on $L^0(\Omega)$.

(d) The integral expression

$$\varrho(f) := \int\limits_\Omega \exp(|f(x)|) - 1\, dx$$

defines a modular on $L^0(\Omega)$. It is not continuous: if $f \in L^2(\Omega)$ is such that $|f| > 2$ and $|f| \notin L^p(\Omega)$ for any $p > 2$, then $\varrho(\lambda \log |f|) = \infty$ for $\lambda > 2$ but $\varrho(2 \log |f|) < \infty$.

(e) If $1 \leqslant p < \infty$, then

$$\varrho_p((x_j)) := \sum_{j=0}^{\infty} |x_j|^p\, dx$$

defines a continuous modular on $\mathbb{R}^\mathbb{N}$.

(f) For $f \in L^0(\Omega)$ we define the decreasing rearrangement, $f^* \colon [0, \infty) \to [0, \infty)$ by the formula $f^*(s) := \sup\{t \colon |\,|f| > t| > s\}$. For $1 \leqslant q \leqslant p < \infty$ the expression

$$\varrho(f) := \int\limits_0^\infty |f^*(s^{p/q})|^q\, ds$$

defines a continuous modular on $L^0(\Omega)$.

Let ϱ be a semimodular on X. Then by convexity and non-negative of ϱ and $\varrho(0) = 0$ it follows that $\lambda \mapsto \varrho(\lambda x)$ is non-decreasing on $[0, \infty)$ for every $x \in X$. Moreover,

$$\begin{aligned}
\varrho(\lambda x) = \varrho(|\lambda|\, x) \leqslant |\lambda|\, \varrho(x) \qquad &\text{for all } |\lambda| \leqslant 1, \\
\varrho(\lambda x) = \varrho(|\lambda|\, x) \geqslant |\lambda|\, \varrho(x) \qquad &\text{for all } |\lambda| \geqslant 1.
\end{aligned} \tag{2.1.5}$$

In the definition of a semimodular or modular the set X is usually chosen to be larger than necessary. The idea behind this is to choose the same large set X for different modulars like in our Examples 2.1.4(a), (b), (c), (d) and (f). Then depending on the modular we pick interesting subsets from this set X.

Definition 2.1.6. If ϱ be a semimodular or modular on X, then

$$X_\varrho := \left\{ x \in X \colon \lim_{\lambda \to 0} \varrho(\lambda x) = 0 \right\}$$

is called a *semimodular space* or *modular space*, respectively. The limit $\lambda \to 0$ takes place in $\mathbb{K}$.

Since $\varrho(\lambda x) = \varrho(|\lambda| \, x)$ it is enough to require $\lim_{\lambda \to 0} \varrho(\lambda x)$ with $\lambda \in (0, \infty)$. Due to (2.1.5) we can alternatively define X_ϱ by

$$X_\varrho := \{x \in X \colon \varrho(\lambda x) < \infty \text{ for some } \lambda > 0\},$$

since for $\lambda' < \lambda$ we have by (2.1.5) that

$$\varrho(\lambda' x) = \varrho\Big(\frac{\lambda'}{\lambda} \lambda x\Big) \leqslant \frac{\lambda'}{\lambda} \varrho(\lambda x) \to 0$$

as $\lambda' \to 0$.

In the next theorem, like elsewhere, the infimum of the empty set is by definition infinity.

Theorem 2.1.7. *Let ϱ be a semimodular on X. Then X_ϱ is a normed $\mathbb{K}$-vector space. The norm, called the* Luxemburg norm, *is defined by*

$$\|x\|_\varrho := \inf \Big\{\lambda > 0 \colon \varrho\Big(\frac{1}{\lambda} x\Big) \leqslant 1\Big\}.$$

Proof. We begin with the vector space property of X_ϱ. Let $x, y \in X_\varrho$ and $\alpha \in \mathbb{K} \setminus \{0\}$. From the definition of X_ϱ and $\varrho(\alpha x) = \varrho(|\alpha| \, x)$ it is clear that $\alpha x \in X_\varrho$. By the convexity of ϱ we estimate

$$0 \leqslant \varrho\big(\lambda(x + y)\big) \leqslant \tfrac{1}{2}\varrho(2\lambda x) + \tfrac{1}{2}\varrho(2\lambda y) \xrightarrow{\lambda \to 0} 0.$$

Hence, $x + y \in X_\varrho$. It is clear that $0 \in X_\varrho$. This proves that X_ϱ is a $\mathbb{K}$-vector space.

It is clear that $\|x\|_\varrho < \infty$ for all $x \in X_\varrho$ and $\|0\|_\varrho = 0$. For $\alpha \in \mathbb{K}$ we have

$$\|\alpha x\|_\varrho = \inf \Big\{\lambda > 0 \colon \varrho\Big(\frac{\alpha x}{\lambda}\Big) \leqslant 1\Big\} = |\alpha| \inf \Big\{\lambda > 0 \colon \varrho\Big(\frac{1}{\lambda} x\Big) \leqslant 1\Big\}$$
$$= |\alpha| \, \|x\|_\varrho.$$

Let $x, y \in X$ and $u > \|x\|_\varrho$ and $v > \|y\|_\varrho$. Then $\varrho(x/u) \leqslant 1$ and $\varrho(y/v) \leqslant 1$, hence, by the convexity of ϱ,

$$\varrho\Big(\frac{x + y}{u + v}\Big) = \varrho\Big(\frac{u}{u + v} \frac{x}{u} + \frac{v}{u + v} \frac{y}{v}\Big) \leqslant \frac{u}{u + v}\varrho\Big(\frac{x}{u}\Big) + \frac{v}{u + v}\varrho\Big(\frac{y}{v}\Big) \leqslant 1.$$

Thus $\|x + y\|_\varrho \leqslant u + v$, and we obtain $\|x + y\|_\varrho \leqslant \|x\|_\varrho + \|y\|_\varrho$.

If $\|x\|_\varrho = 0$, then $\varrho(\alpha x) \leqslant 1$ for all $\alpha > 0$. Therefore,

$$\varrho(\lambda x) \leqslant \beta \varrho\Big(\frac{\lambda x}{\beta}\Big) \leqslant \beta$$

for all $\lambda > 0$ and $\beta \in (0, 1]$, where we have used (2.1.5). This implies $\varrho(\lambda x) = 0$ for all $\lambda > 0$. Thus $x = 0$.   □

The norm in the previous theorem is more generally known as the *Minkowski functional* of the set $\{x \in X : \varrho(x) \leqslant 1\}$, see Remark 2.1.16. The Minkowski functional was first introduced by Kolmogorov in [253] long before the appearance of the Luxemburg norm. Nevertheless, we use the name "Luxemburg norm" as it is customary in the theory of Orlicz spaces.

In the following example we use the notation of Example 2.1.4.

Example 2.1.8 (Classical Lebesgue spaces). Let $1 \leqslant p < \infty$. Then the corresponding modular space $(L^0(\Omega))_{\varrho_p}$ coincides with the classical Lebesgue space L^p, i.e.

$$\|f\|_p := \|f\|_{\varrho_p} = \left(\int_\Omega |f(x)|^p \, dx \right)^{\frac{1}{p}}.$$

Similarly, the corresponding semimodular space $(L^0(\Omega))_{\varrho_\infty}$ coincides with the classical Lebesgue space L^∞, i.e.

$$\|f\|_\infty := \|f\|_{\varrho_\infty} = \operatorname*{ess\,sup}_{x \in \Omega} |f(x)|.$$

The norm $\|\cdot\|_\varrho$ defines our standard topology on X_ϱ. So for $x_k, x \in X_\varrho$ we say that x_k *converges strongly* or *in norm* to x if $\|x_k - x\|_\varrho \to 0$. In this case we write $x_k \to x$. The next lemma characterizes this topology in terms of the semimodular. Here it suffices to study null-sequences.

Lemma 2.1.9. *Let ϱ be a semimodular on X and $x_k \in X_\varrho$. Then $x_k \to 0$ for $k \to \infty$ if and only if $\lim_{k\to\infty} \varrho(\lambda x_k) = 0$ for all $\lambda > 0$.*

Proof. Assume that $\|x_k\|_\varrho \to 0$ and $\lambda > 0$. Then $\|K \lambda x_k\|_\varrho < 1$ for all $K > 1$ and large k. Thus $\varrho(K \lambda x_k) \leqslant 1$ for large k, hence

$$\varrho(\lambda x_k) \leqslant \frac{1}{K} \varrho(K \lambda x_k) \leqslant \frac{1}{K}$$

for large k, by (2.1.5). This implies $\varrho(\lambda x_k) \to 0$.

Assume now that $\varrho(\lambda x_k) \to 0$ for all $\lambda > 0$. Then $\varrho(\lambda x_k) \leqslant 1$ for large k. In particular, $\|x_k\|_\varrho \leqslant 1/\lambda$ for large k. Since $\lambda > 0$ was arbitrary, we get $\|x_k\|_\varrho \to 0$. In other words $x_k \to 0$.   □

Apart from our standard topology on X_ϱ, which was induced by the norm, it is possible to define another type of convergence by means of the semimodular.

Definition 2.1.10. Let ϱ be a semimodular on X and $x_k, x \in X_\varrho$. Then we say that x_k is modular convergent (ϱ-convergent) to x if there exists $\lambda > 0$ such that $\varrho(\lambda(x_k - x)) \to 0$. We denote this by $x_k \overset{\varrho}{\to} x$.

It is clear from Lemma 2.1.9 that modular convergence is weaker than norm convergence. Indeed, for norm convergence we have $\lim_{k \to \infty} \varrho(\lambda(x_k - y)) = 0$ for all $\lambda > 0$, while for modular convergence this only has to hold for some $\lambda > 0$.

For some semimodular spaces modular convergence and norm convergence coincide and for others they differ:

Lemma 2.1.11. *Let X_ϱ be a semimodular space. Then modular convergence and norm convergence are equivalent if and only if $\varrho(x_k) \to 0$ implies $\varrho(2x_k) \to 0$.*

Proof. "$\Rightarrow$": Let modular convergence and norm convergence be equivalent and let $\varrho(x_k) \to 0$ with $x_k \in X_\varrho$. Then $x_k \to 0$ and by Lemma 2.1.9 it follows that $\varrho(2x_k) \to 0$.

"$\Leftarrow$": Let $x_k \in X_\varrho$ with $\varrho(x_k) \to 0$. We have to show that $\varrho(\lambda x_k) \to 0$ for all $\lambda > 0$. For fixed $\lambda > 0$ choose $m \in \mathbb{N}$ such that $2^m \geqslant \lambda$. Then by repeated application of the assumption we get $\lim_{k \to \infty} \varrho(2^m x_k) = 0$. Then $0 \leqslant \lim_{k \to \infty} \varrho(\lambda x_k) \leqslant \lambda 2^{-m} \lim_{k \to \infty} \varrho(2^m x_k) = 0$ by (2.1.5). This proves that $x_k \to 0$. $\square$

If either of the equivalent conditions in the previous lemma hold, then we say that the semimodular satisfies the *weak Δ_2-condition*.

If ϱ is a semimodular that satisfies the weak Δ_2-condition, then ϱ is already a modular. Indeed, if $\varrho(x) = 0$, then the constant sequence x is modular convergent to 0 and therefore convergent to 0 with respect to the norm, but this implies $x = 0$.

Lemma 2.1.12. *Let be a semimodular on X that satisfies the weak Δ_2-condition. Then for every $\varepsilon > 0$ there exists $\delta > 0$ such that $\varrho(f) \leqslant \delta$ implies $\|f\|_\varrho \leqslant \varepsilon$.*

Proof. This is an immediate consequence of the equivalence of modular and norm convergence. $\square$

Example 2.1.13. The weak Δ_2-condition of modulars is satisfied in Examples 2.1.4 (a) and (c). Examples 2.1.4 (b) and (d) do not satisfy this condition.

Let us study the closed and open unit ball of X_ϱ. The left-continuity of ϱ is of special significance. The following lemma is of great technical importance. We will invoke it by mentioning the *unit ball property*, or, when more clarity is needed, the *norm-modular unit ball property*.

Lemma 2.1.14 (Norm-modular unit ball property). *Let ϱ be a semi-modular on X. Then $\|x\|_\varrho \leqslant 1$ and $\varrho(x) \leqslant 1$ are equivalent. If ϱ is continuous, then also $\|x\|_\varrho < 1$ and $\varrho(x) < 1$ are equivalent, as are $\|x\|_\varrho = 1$ and $\varrho(x) = 1$.*

Proof. If $\varrho(x) \leqslant 1$, then $\|x\|_\varrho \leqslant 1$ by definition of $\|\cdot\|_\varrho$. If on the other hand $\|x\|_\varrho \leqslant 1$, then $\varrho(x/\lambda) \leqslant 1$ for all $\lambda > 1$. Since ϱ is left-continuous it follows that $\varrho(x) \leqslant 1$.

Let ϱ be continuous. If $\|x\|_\varrho < 1$, then there exists $\lambda < 1$ with $\varrho(x/\lambda) \leqslant 1$. Hence by (2.1.5) it follows that $\varrho(x) \leqslant \lambda\varrho(x/\lambda) \leqslant \lambda < 1$. If on the other hand $\varrho(x) < 1$, then by the continuity of ϱ there exists $\gamma > 1$ with $\varrho(\gamma x) < 1$. Hence $\|\gamma x\|_\varrho \leqslant 1$ and $\|x\|_\varrho \leqslant 1/\gamma < 1$. The equivalence of $\|x\|_\varrho = 1$ and $\varrho(x) = 1$ now follows immediately from the cases "$\leqslant 1$" and "< 1". □

A simple example of a semimodular which is left-continuous but not continuous is given by $\varrho_\infty(t) = \infty \cdot \chi_{(1,\infty)}(t)$ on $X = \mathbb{R}$. This is a semimodular on $\mathbb{R}$ and $\|x\|_{\varrho_\infty} = |x|$.

Corollary 2.1.15. *Let ϱ be a semimodular on X and $x \in X_\varrho$.*

(a) *If $\|x\|_\varrho \leqslant 1$, then $\varrho(x) \leqslant \|x\|_\varrho$.*
(b) *If $1 < \|x\|_\varrho$, then $\|x\|_\varrho \leqslant \varrho(x)$.*
(c) *$\|x\|_\varrho \leqslant \varrho(x) + 1$.*

Proof. (a) The claim is obvious for $x = 0$, so let us assume that $0 < \|x\|_\varrho \leqslant 1$. By the unit ball property (Lemma 2.1.14) and $\|x/\|x\|_\varrho\|_\varrho = 1$ it follows that $\varrho(x/\|x\|_\varrho) \leqslant 1$. Since $\|x\|_\varrho \leqslant 1$, it follows from (2.1.5) that $\varrho(x)/\|x\|_\varrho \leqslant 1$.
(b) Assume that $\|x\|_\varrho > 1$. Then $\varrho(x/\lambda) > 1$ for $1 < \lambda < \|x\|_\varrho$ and by (2.1.5) it follows that $1 < \varrho(x)/\lambda$. Since λ was arbitrary, $\varrho(x) \geqslant \|x\|_\varrho$.
(c) This follows immediately from (b). □

Remark 2.1.16. Let $K := \{x \in X_\varrho \colon \varrho(x) \leqslant 1\}$. Then the unit ball property states that $K = \overline{B(0,1)}$, the closed unit ball with respect to the norm. This provides an alternative proof of the fact that $\|\cdot\|_\varrho$ is a norm. Indeed, K is a balanced, i.e. $\lambda K := \{\lambda x \colon x \in K\} \subset K$ for all $|\lambda| \leqslant 1$, convex set. Moreover, by definition of X_ϱ the set K is absorbing for X_ϱ, i.e. $\bigcup_{\lambda > 0}(\lambda K) = X_\varrho$. Therefore, the *Minkowski functional* of K, namely $x \mapsto \inf\{\lambda > 0 \colon \frac{1}{\lambda}x \in K\}$, defines a norm on X_ϱ. But this functional is exactly $\|\cdot\|_\varrho$ which is therefore a norm on X_ϱ.

We have seen in Remark 2.1.16 that $\{x \in X_\varrho \colon \varrho(x) \leqslant 1\}$ is closed. This raises the question whether $\{x \in X \colon \varrho(x) \leqslant \alpha\}$ is closed for every $\alpha \in [0, \infty)$. This is equivalent to the lower semicontinuity of ϱ on X_ϱ, hence the next theorem gives a positive answer.

Theorem 2.1.17. *Let ϱ be a semimodular on X. Then ϱ is lower semicontinuous on X_ϱ, i.e.*

$$\varrho(x) \leqslant \liminf_{k \to \infty} \varrho(x_k)$$

for all $x_k, x \in X_\varrho$ with $x_k \to x$ (in norm) for $k \to \infty$.

Proof. Let $x_k, x \in X_\varrho$ with $x_k \to x$ for $k \to \infty$. We begin with the case $\varrho(x) < \infty$. By Lemma 2.1.9, $\lim_{k \to \infty} \varrho(\gamma(x - x_k)) = 0$ for all $\gamma > 0$. Let $\varepsilon \in (0, \frac{1}{2})$. Then, by convexity of ϱ,

$$\varrho((1 - \varepsilon)x) = \varrho\left(\frac{1}{2}x + \frac{1 - 2\varepsilon}{2}(x - x_k) + \frac{1 - 2\varepsilon}{2}x_k\right)$$

$$\leqslant \frac{1}{2}\varrho(x) + \frac{1}{2}\varrho\left((1 - 2\varepsilon)(x - x_k) + (1 - 2\varepsilon)x_k\right)$$

$$\leqslant \frac{1}{2}\varrho(x) + \frac{2\varepsilon}{2}\varrho\left(\frac{1 - 2\varepsilon}{2\varepsilon}(x - x_k)\right) + \frac{1 - 2\varepsilon}{2}\varrho(x_k).$$

We pass to the limit $k \to \infty$:

$$\varrho((1 - \varepsilon)x) \leqslant \frac{1}{2}\varrho(x) + \frac{1 - 2\varepsilon}{2}\liminf_{k \to \infty} \varrho(x_k).$$

Now letting $\varepsilon \to 0^+$ and using the left-continuity of ϱ, we get

$$\varrho(x) \leqslant \frac{1}{2}\varrho(x) + \frac{1}{2}\liminf_{k \to \infty} \varrho(x_k).$$

Since $\varrho(x) < \infty$, we get $\varrho(x) \leqslant \liminf_{k \to \infty} \varrho(x_k)$. This completes the proof in the case $\varrho(x) < \infty$.

Assume now that $x \in X_\varrho$ with $\varrho(x) = \infty$. If $\liminf_{k \to \infty} \varrho(x_k) = \infty$, then there is nothing to show. So we can assume $\liminf_{k \to \infty} \varrho(x_k) < \infty$. Let $\lambda_0 := \sup\{\lambda > 0 \colon \varrho(\lambda x) < \infty\}$. Since $x \in X_\varrho$, we have $\lambda_0 > 0$. Moreover, $\varrho(x) = \infty$ implies $\lambda_0 \leqslant 1$. For all $\lambda \in (0, \lambda_0)$ the inequality $\varrho(\lambda x) < \infty$ holds, so

$$\varrho(\lambda x) \leqslant \liminf_{k \to \infty} \varrho(\lambda x_k) \leqslant \liminf_{k \to \infty} \varrho(x_k)$$

for all $\lambda \in (0, \lambda_0)$ by the first part of the proof. The left-continuity of ϱ implies that

$$\varrho(\lambda_0 x) \leqslant \liminf_{k \to \infty} \varrho(x_k).$$

If $\lambda_0 = 1$, then the proof is finished. Finally we show, by contradiction, that $\lambda_0 \notin (0, 1)$. So let $\lambda_0 \in (0, 1)$. Choose $\lambda_1 \in (\lambda_0, 1)$ and $\alpha \in (0, 1)$ such that

$$\frac{\lambda_1 - \lambda_0}{\lambda_0} + \alpha + \lambda_0 = 1.$$

The convexity of ϱ implies

$$
\varrho(\lambda_1 x) = \varrho\bigg((\lambda_1 - \lambda_0)x + \lambda_0(x - x_k) + \lambda_0 x_k \bigg)
$$

$$
\leqslant \frac{\lambda_1 - \lambda_0}{\lambda_0} \varrho(\lambda_0 x) + \alpha\varrho\bigg(\frac{\lambda_0}{\alpha}(x - x_k) \bigg) + \lambda_0 \varrho(x_k).
$$

We pass to the limit $k \to \infty$:

$$
\varrho(\lambda_1 x) \leqslant \frac{\lambda_1 - \lambda_0}{\lambda_0} \varrho(\lambda_0 x) + \lambda_0 \liminf_{k\to\infty} \varrho(x_k) \leqslant (1 - \alpha) \liminf_{k\to\infty} \varrho(x_k).
$$

Since $\liminf_k \varrho(x_k) < \infty$, we get $\varrho(\lambda_1 x) < \infty$. But this and $\lambda_1 > \lambda_0$ contradict the definition of λ_0.   □

Remark 2.1.18. It follows from Theorem 2.1.17 that the sets $\{x \in X \colon \varrho(x) \leqslant \alpha\}$ are closed for every $\alpha \in [0, \infty)$. Since these sets are convex, it follows that they are also closed with respect to the weak topology of X_ϱ (cf. Sect. 1.4, Functional analysis).

Remark 2.1.19. Let ϱ be a semimodular on X. Then

$$
\|\|x\|\|_\varrho := \inf_{\lambda > 0} \lambda\bigg(1 + \varrho\Big(\frac{1}{\lambda}x\Big) \bigg)
$$

defines a norm on X_ϱ and

$$
\|x\|_\varrho \leqslant \|\|x\|\|_\varrho \leqslant 2\|x\|_\varrho.
$$

This norm is called the *Amemiya norm*. For a proof see [307].

2.2   Conjugate Modulars and Dual Semimodular Spaces

The dual space of a normed space X is the set of all linear, bounded functionals from X to $\mathbb{K}$. It is denoted by X^*. It is well known that X^* equipped with the norm

$$
\|x^*\|_{X^*} := \sup_{\|x\|_X \leqslant 1} |\langle x^*, x\rangle|
$$

is a Banach space. Here we use the notation $\langle x^*, x\rangle := x^*(x)$. The study of the dual of X is a standard tool to get a better understanding of the space X itself. In this section we examine the dual space of X_ϱ.

Lemma 2.2.1. *Let ϱ be a semimodular on X. A linear functional x^* on X_ϱ is bounded with respect to $\|\cdot\|_\varrho$ if and only if there exists $\gamma > 0$ such that for every $x \in X_\varrho$*

$$|\langle x^*, x \rangle| \leqslant \gamma(\varrho(x) + 1).$$

Proof. If $x^* \in X_\varrho^*$ and $x \in X_\varrho$, then $\langle x^*, x \rangle \leqslant \|x^*\|_{X_\varrho^*} \|x\|_{X_\varrho} \leqslant \|x^*\|_{X_\varrho^*}(1 + \varrho(x))$ by Corollary 2.1.15. Assume conversely that the inequality holds. Then

$$\left|\left\langle x^*, \frac{x}{\|x\|_\varrho + \varepsilon} \right\rangle\right| \leqslant \gamma\left(\varrho\left(\frac{x}{\|x\|_\varrho + \varepsilon}\right) + 1\right) \leqslant 2\gamma$$

for every $\varepsilon > 0$, hence $\|x^*\|_{X_\varrho^*} \leqslant 2\gamma$.   □

Definition 2.2.2. Let ϱ be a semimodular on X. Then by X_ϱ^* we denote the dual space of $(X_\varrho, \|\cdot\|_\varrho)$. Furthermore, we define $\varrho^* \colon X_\varrho^* \to [0, \infty]$ by

$$\varrho^*(x^*) := \sup_{x \in X_\varrho} \left(|\langle x^*, x \rangle| - \varrho(x)\right).$$

We call ϱ^* the *conjugate semimodular* of ϱ.

Note the difference between the spaces X_ϱ^* and X_{ϱ^*}: the former is the dual space of X_ϱ, whereas the latter is the semimodular space defined by ϱ^*.

By definition of the functional ϱ^* we have

$$|\langle x^*, x \rangle| \leqslant \varrho(x) + \varrho^*(x^*) \tag{2.2.3}$$

for all $x \in X_\varrho$ and $x^* \in X_\varrho^*$. This inequality is a generalized version of the classical Young inequality.

Theorem 2.2.4. *Let ϱ be a semimodular on X. Then ϱ^* is a semimodular on X_ϱ^*.*

Proof. It is easily seen that $\varrho^*(0) = 0$, $\varrho^*(\lambda x^*) = \varrho^*(x^*)$ for $|\lambda| = 1$, and $\varrho^*(x^*) \geqslant 0$ for every $x^* \in X_\varrho^*$. Let $x_0^*, x_1^* \in X_\varrho^*$ and $\theta \in (0, 1)$. Then

$$\begin{aligned}
\varrho^*\left((1 - \theta)x_0^* + \theta x_1^*\right) &= \sup_{x \in X} \left(\left|\langle (1 - \theta)x_0^* + \theta x_1^*, x \rangle\right| - \varrho(x)\right) \\
&\leqslant (1 - \theta) \sup_{x \in X} \left(|\langle x_0^*, x \rangle| - \varrho(x)\right) \\
&\quad + \theta \sup_{x \in X} \left(|\langle x_1^*, x \rangle| - \varrho(x)\right) \\
&= (1 - \theta)\varrho^*(x_0^*) + \theta \varrho^*(x_1^*).
\end{aligned}$$

Finally, let $\varrho^*(\lambda x^*) = 0$ for every $\lambda > 0$. For $x \in X_\varrho$ choose $\eta > 0$ such that $\varrho(\eta x) < \infty$. Then by (2.2.3)

$$\lambda \eta \, |\langle x^*, x \rangle| \leqslant \varrho(\eta x) + \varrho^*(\lambda x^*) = \varrho(\eta x).$$

Taking $\lambda \to \infty$ we obtain $|\langle x^*, x \rangle| = 0$. Hence $x^* = 0$. It remains to show that ϱ^* is left-continuous. For $\lambda \to 1^-$ and $x^* \in X_\varrho^*$ we have

$$\lim_{\lambda \to 1^-} \varrho^*(\lambda x^*) = \lim_{\lambda \to 1^-} \sup_{x \in X} \left(|\langle \lambda x^*, x \rangle| - \varrho(x) \right)$$
$$= \sup_{0 < \lambda < 1} \sup_{x \in X} \left(|\lambda| \, |\langle x^*, x \rangle| - \varrho(x) \right)$$
$$= \sup_{x \in X} \left(|\langle x^*, x \rangle| - \varrho(x) \right) = \varrho^*(x).$$

Thus ϱ^* is left-continuous.   □

For a semimodular ϱ on X we have defined the conjugate semimodular ϱ^* on X_ϱ^*. By duality we can proceed further and define ϱ^{**} the conjugate semimodular of ϱ^* on the bidual $X_\varrho^{**} := (X_\varrho^*)^*$. The functional ϱ^{**} is called the *biconjugate semimodular of ϱ on X_ϱ^{**}*. Using the natural injection ι of X_ϱ into its bidual X_ϱ^{**}, the mapping $x \mapsto \varrho^{**}(\iota x)$ defines a semimodular on X_ϱ, which we call the *biconjugate semimodular of ϱ on X_ϱ*. For simplicity of notation it is also denoted by ϱ^{**} neglecting the extra injection ι. In particular, we have

$$\varrho^{**}(x) = \sup_{x^* \in X_\varrho^*} \left(|\langle x^*, x \rangle| - \varrho^*(x^*) \right) \tag{2.2.5}$$

for all $x \in X_\varrho$. Certainly the formula is also valid for all $x \in X_\varrho^{**}$, by the definition of ϱ^{**} on X_ϱ^{**}, if we interpret $\langle x^*, x \rangle$ as $\langle x, x^* \rangle_{X_\varrho^{**} \times X_\varrho^*}$.

Analogously to the fact that $\iota : X_\varrho \to X_\varrho^{**}$ is an isometry, it turns out that the biconjugate ϱ^{**} and ϱ coincide on X_ϱ.

Theorem 2.2.6. *Let ϱ be a semimodular on X. Then $\varrho^{**} = \varrho$ on X_ϱ.*

Proof. Exactly as in the proof of Theorem 2.2.4 we can prove that ϱ^{**} is a semimodular on X_ϱ. By definition of ϱ^{**} and (2.2.3) we get for $x \in X_\varrho$

$$\varrho^{**}(x) = \sup_{x^* \in X_\varrho^*} \left(|\langle x^*, x \rangle| - \varrho^*(x^*) \right)$$
$$= \sup_{x^* \in X_\varrho^*, \varrho^*(x^*) < \infty} \left(|\langle x^*, x \rangle| - \varrho^*(x^*) \right)$$
$$\leqslant \sup_{x^* \in X_\varrho^*, \varrho^*(x^*) < \infty} \left(\varrho(x) + \varrho^*(x^*) - \varrho^*(x^*) \right)$$
$$= \varrho(x).$$

It remains to show $\varrho^{**}(x) \geqslant \varrho(x)$. We prove this by contradiction. Assume to the contrary that there exists $x_0 \in X_\varrho$ with $\varrho^{**}(x_0) < \varrho(x_0)$. In particular, $\varrho^{**}(x_0) < \infty$. We define the epigraph of ϱ by

$$\mathrm{epi}(\varrho) := \bigcup_{\lambda \in \mathbb{R}} \left\{ (x, \gamma) \in X_\varrho \times \mathbb{R} \colon \gamma \geqslant \varrho(x) \right\}.$$

Since ϱ is convex and lower semicontinuous (Theorem 2.1.17), the set $\mathrm{epi}(\varrho)$ is convex and closed (cf. [58, Sect. I.3]). Moreover, due to $\varrho^{**}(x_0) < \varrho(x_0)$ the point $(x_0, \varrho^{**}(x_0))$ is not contained in $\mathrm{epi}(\varrho)$. So by the Hahn–Banach Theorem 1.4.2 there exists a functional on $X_\varrho \times \mathbb{R}$ which strictly separates $\mathrm{epi}(\varrho)$ from $(x_0, \varrho^{**}(x_0))$. So there exist $\alpha, \beta \in \mathbb{R}$ and $x^* \in X_\varrho^*$ with

$$\langle x^*, x \rangle - \beta \varrho(x) < \alpha < \langle x^*, x_0 \rangle - \beta \varrho^{**}(x_0)$$

for all $x \in X_\varrho$. The choice $x = x_0$ and the estimate $\varrho^{**}(x_0) < \varrho(x_0)$ imply $\beta > 0$. We multiply by $\frac{1}{\beta}$ and get

$$\left\langle \frac{x^*}{\beta}, x \right\rangle - \varrho(x) < \frac{\alpha}{\beta} < \left\langle \frac{x^*}{\beta}, x_0 \right\rangle - \varrho^{**}(x_0)$$

for all $x \in X_\varrho$. Due to (2.2.5) the right-hand side is bounded by $\varrho^*(\frac{x^*}{\beta})$. Now, taking the supremum on the left-hand side over $x \in X_\varrho$ implies

$$\varrho^* \left(\frac{x^*}{\beta} \right) \leqslant \frac{\alpha}{\beta} < \varrho^* \left(\frac{x^*}{\beta} \right).$$

This is the desired contradiction.   □

For two semimodulars ϱ, κ on X we write $\varrho \leqslant \kappa$ if $\varrho(f) \leqslant \kappa(f)$ for every $f \in X$.

Corollary 2.2.7. *Let ϱ, κ be semimodulars on X. Then $\varrho \leqslant \kappa$ if and only if $\kappa^* \leqslant \varrho^*$.*

Proof. If $\varrho \leqslant \kappa$, then by definition of the conjugate semimodular follows easily $\kappa^* \leqslant \varrho^*$. If however $\kappa^* \leqslant \varrho^*$, then $\varrho^{**} \leqslant \kappa^{**}$ and by Theorem 2.2.6 follows $\varrho \leqslant \kappa$.   □

From Theorem 2.1.17 we already know that the modular ϱ is lower semicontinuous on X_ϱ with respect to convergence in norm. This raises the question of whether ϱ is also lower semicontinuous on X_ϱ with respect to weak convergence. Let $f_k, f \in X_\varrho$. As usual we say that f_k *converges weakly to f if $\langle g^*, f_k \rangle \to \langle g^*, f \rangle$ for all $g^* \in X_\varrho^*$. In this case we write $f_k \rightharpoonup f$.

Theorem 2.2.8. *Let ϱ be a semimodular on X, then the semimodular ϱ is weakly (sequentially) lower semicontinuous, i.e. if $f_k \rightharpoonup f$ weakly in X_ϱ, then $\varrho(f) \leqslant \liminf_{k \to \infty} \varrho(f_k)$.*

Proof. Let $f_k, f \in X_\varrho$ with $f_k \rightharpoonup f$. Then, by Theorem 2.2.6, $\varrho = \varrho^{**}$, which implies

$$\varrho(f) = \varrho^{**}(f) = \sup_{g^* \in X_\varrho^*} \left(|\langle g^*, f \rangle| - \varrho^*(g^*) \right)$$

$$= \sup_{g^* \in X_\varrho^*} \left(\lim_{k \to \infty} |\langle g^*, f_k \rangle| - \varrho^*(g^*) \right)$$

$$\leqslant \liminf_{k \to \infty} \left(\sup_{g^* \in X_\varrho^*} \left(|\langle g^*, f_k \rangle| - \varrho^*(g^*) \right) \right)$$

$$= \liminf_{k \to \infty} \varrho^{**}(f_k)$$

$$= \liminf_{k \to \infty} \varrho(f_k). \qquad \qquad \square$$

In the definition of ϱ^* the supremum is taken over all $x \in X_\varrho$. However, it is possible to restrict this to the closed unit ball of X_ϱ.

Lemma 2.2.9. *If ϱ is a semimodular on X, then*

$$\varrho^*(x^*) = \sup_{x \in X_\varrho, \|x\|_\varrho \leqslant 1} \left(|\langle x^*, x \rangle| - \varrho(x) \right) = \sup_{x \in X_\varrho, \varrho(x) \leqslant 1} \left(|\langle x^*, x \rangle| - \varrho(x) \right)$$

for $x^* \in X_\varrho^*$ with $\|x^*\|_{X_\varrho^*} \leqslant 1$.

Proof. The equivalence of the suprema follows from the unit ball property (Lemma 2.1.14). Let $\|x^*\|_{X_\varrho^*} \leqslant 1$. By the definition of the dual norm we have

$$\sup_{\|x\|_\varrho > 1} \left(|\langle x^*, x \rangle| - \varrho(x) \right) \leqslant \sup_{\|x\|_\varrho > 1} \left(\|x^*\|_{X_\varrho^*} \|x\|_\varrho - \varrho(x) \right)$$

$$\leqslant \sup_{\|x\|_\varrho > 1} \left(\|x\|_\varrho - \varrho(x) \right).$$

If $\|x\|_\varrho > 1$, then $\varrho(x) \geqslant \|x\|_\varrho$ by Corollary 2.1.15, and so the right-hand side of the previous inequality is non-positive. Since ϱ^* is defined as a supremum, and is always non-negative, we see that the points with $\|x\|_\varrho > 1$ do not affect the supremum, and so the claim follows. $\qquad \square$

Since ϱ^* is a semimodular on X_ϱ^*, it defines another norm $\|\cdot\|_{\varrho^*}$ on X_ϱ^*. We next want to compare it with the norm $\|\cdot\|_{X_\varrho^*}$.

Theorem 2.2.10. *If ϱ be a semimodular on X, then for every $x^* \in X_\varrho^*$*

$$\|x^*\|_{\varrho^*} \leqslant \|x^*\|_{X_\varrho^*} \leqslant 2\|x^*\|_{\varrho^*}.$$

Proof. We first prove the second inequality. By the unit ball property (Lemma 2.1.14) the inequalities $\|x\|_\varrho \leqslant 1$ and $\varrho(x) \leqslant 1$ are equivalent. Hence,

$$\|x^*\|_{X_\varrho^*} = \sup_{\|x\|_\varrho \leqslant 1} |\langle x^*, x \rangle| \leqslant \sup_{\varrho(x) \leqslant 1} \left(\varrho^*(x^*) + \varrho(x) \right) \leqslant \varrho^*(x^*) + 1.$$

If $\|x^*\|_{\varrho^*} \leqslant 1$, then $\varrho^*(x^*) \leqslant 1$ by the unit ball property and we conclude that $\|x^*\|_{X_\varrho^*} \leqslant 2$. The conclusion follows from this by a *scaling argument*: if $\|x^*\|_{\varrho^*} > 0$, then set $y^* := x^*/\|x^*\|_{\varrho^*}$. Since $\|y^*\|_{\varrho^*} = 1$, we conclude that $\|y^*\|_{X_\varrho^*} \leqslant 2\|y^*\|_{\varrho^*}$. Multiplying by $\|x^*\|_{\varrho^*}$ gives the result.

Assume now that $\|x^*\|_{X_\varrho^*} \leqslant 1$. Then by Lemma 2.2.9 and Corollary 2.1.15 (c)

$$\varrho^*(x^*) = \sup_{x \in X_\varrho, \varrho(x) \leqslant 1} \left(|\langle x^*, x\rangle| - \varrho(x)\right) \leqslant \sup_{x \in X_\varrho, \varrho(x) \leqslant 1} \left(\|x\|_\varrho - \varrho(x)\right) \leqslant 1.$$

Hence, $\|x^*\|_{\varrho^*} \leqslant 1$. The scaling argument gives $\|x^*\|_{\varrho^*} \leqslant \|x^*\|_{X_\varrho^*}$ □

Note the scaling argument technique used in the previous proof. It is one of the central methods for dealing with these kind of spaces, and it will be used often in what follows.

With the help of the conjugate semimodular ϱ^* it is also possible to define yet another norm on X_ϱ by means of duality. Luckily this norm is equivalent to the norm $\|\cdot\|_\varrho$.

Theorem 2.2.11. *Let ϱ be a semimodular on X. Then*

$$\|x\|_\varrho' := \sup \left\{|\langle x^*, x\rangle| : x^* \in X_\varrho^*, \|x^*\|_{\varrho^*} \leqslant 1\right\}$$
$$= \sup \left\{|\langle x^*, x\rangle| : x^* \in X_\varrho^*, \varrho^*(x^*) \leqslant 1\right\}$$

defines a norm on X_ϱ. This norm is called the Orlicz norm. *For all $x \in X_\varrho$ we have $\|x\|_\varrho \leqslant \|x\|_\varrho' \leqslant 2\|x\|_\varrho$.*

Proof. By the unit ball property (Lemma 2.1.14) the two suprema are equal. If $\|x\|_\varrho \leqslant 1$ and $\|x^*\|_{\varrho^*} \leqslant 1$, then $\varrho(x) \leqslant 1$ and $\varrho^*(x^*) \leqslant 1$. Hence, $|\langle x, x^*\rangle| \leqslant \varrho(x) + \varrho^*(x^*) \leqslant 2$. Therefore $\|x\|_\varrho' \leqslant 2$. A scaling argument proves $\|x\|_\varrho' \leqslant 2\|x\|_\varrho$.

If $\|x\|_\varrho' \leqslant 1$, then $|\langle x^*, x\rangle| \leqslant 1$ for all $x^* \in X_\varrho^*$ with $\|x^*\|_{\varrho^*} \leqslant 1$. In particular, by Theorem 2.2.10 we have $|\langle x^*, x\rangle| \leqslant 1$ for all $x^* \in X_\varrho^*$ with $\|x^*\|_{X_\varrho^*} \leqslant 1$. Hence, Corollary 1.4.3 implies $\|x\|_\varrho \leqslant 1$. We have thus shown that $\|x\|_\varrho \leqslant \|x\|_\varrho'$.   □

2.3 Musielak–Orlicz Spaces: Basic Properties

In this section we start our journey towards more concrete spaces. Instead of general semimodular spaces, we will consider spaces where the modular is given by the integral of a real-valued function.

Definition 2.3.1. A convex, left-continuous function $\varphi : [0, \infty) \to [0, \infty]$ with $\varphi(0) = 0$, $\lim_{t \to 0^+} \varphi(t) = 0$, and $\lim_{t \to \infty} \varphi(t) = \infty$ is called a Φ-*function*. It is called *positive* if $\varphi(t) > 0$ for all $t > 0$.

In fact, there is a very close relationship between Φ-functions and semi-modulars on $\mathbb{R}$.

Lemma 2.3.2. *Let $\varphi\colon [0,\infty) \to [0,\infty]$ and let ϱ denote its even extension to $\mathbb{R}$, i.e. $\varrho(t) := \varphi(|t|)$ for all $t \in \mathbb{R}$. Then φ is a Φ-function if and only if ϱ is a semimodular on $\mathbb{R}$ with $X_\varrho = \mathbb{R}$. Moreover, φ is a positive Φ-function if and only if ϱ is a modular on $\mathbb{R}$ with $X_\varrho = \mathbb{R}$.*

Proof. "$\Rightarrow$": Let φ be a Φ-function. Since $\lim_{t\to 0+} \varphi(t) = 0$, we have $X_\varrho = \mathbb{R}$. To prove that ϱ is a semimodular on $\mathbb{R}$ it remains to prove that $\varrho(\lambda t_0) = 0$ for all $\lambda > 0$ implies $t_0 = 0$. So assume that $\varrho(\lambda t_0) = 0$ for all $\lambda > 0$. Since $\lim_{t\to\infty} \varphi(t) = \infty$, there exists $t_1 > 0$ with $\varphi(t_1) > 0$. Thus there exists no $\lambda > 0$ such that $t_1 = \lambda t_0$, which implies that $t_0 = 0$. Hence ϱ is a semimodular. Assume that φ is additionally positive. If $\varrho(s) = 0$, then $\varphi(|s|) = 0$ and therefore $s = 0$. This proves that ϱ is a modular.

"$\Leftarrow$": Let ϱ be a semimodular on $\mathbb{R}$ with $X_\varrho = \mathbb{R}$. Since $X_\varrho = \mathbb{R}$, there exists $t_2 > 0$ such that $\varrho(t_2) < \infty$. From (2.1.5) follows that $0 \leqslant \varphi(t) \leqslant t/t_2 \varphi(t_2)$ for all $t \in [0,t_2]$, which implies that $\lim_{t\to 0+} \varphi(t) = 0$. Since $1 \neq 0$, there exists $\lambda > 0$ such that $\varrho(\lambda \cdot 1) \neq 0$. In particular there exists $t_3 > 0$ with $\varphi(t_3) > 0$ and $\varphi(kt_3) \geqslant k\varphi(t_3) > 0$ by (2.1.5) for all $k \in \mathbb{N}$. Since k is arbitrary, we get $\lim_{t\to\infty} \varphi(t) = \infty$. We have proved that φ is a Φ-function. Assume additionally that ϱ is a modular. In particular $\varrho(t) = \varphi(|t|) = 0$ implies $t = 0$. Hence by negation we get that $t > 0$ implies $\varphi(t) > 0$, so φ is positive. $\square$

Let us remark that if φ is a Φ-function then on the set $\{t \geqslant 0\colon \varphi(t) < \infty\}$ it has the form

$$\varphi(t) = \int_0^t a(\tau)\, d\tau, \qquad (2.3.3)$$

where $a(t)$ is the right-derivative of $\varphi(t)$ (see [330], Theorem 1.3.1). Moreover, the function $a(t)$ is non-increasing and right-continuous.

The following lemma is an easy consequence of the left-continuity, convexity, and monotonicity of φ. However, it is also possible to use Lemma 2.3.2 and Theorem 2.1.17 to prove this.

Lemma 2.3.4. *Every Φ-function is lower semicontinuous.*

Example 2.3.5. Let $1 \leqslant p < \infty$. Define

$$\varphi_p(t) := \frac{1}{p}t^p,$$

$$\varphi_\infty(t) := \infty \cdot \chi_{(1,\infty)}(t)$$

for all $t \geqslant 0$. Then φ_p and φ_∞ are Φ-functions. Moreover, φ_p is continuous and positive, while φ_∞ is only left-continuous and lower semicontinuous but not positive.

Remark 2.3.6. Let φ be a Φ-function. As a lower semicontinuous function φ satisfies

$$\varphi(\inf A) \leqslant \inf \varphi(A)$$

for every non-empty set $A \subset [0, \infty)$. The reverse estimate might fail as the example φ_∞ with $A := (1, \infty)$ shows. However, for every $\lambda > 1$ we have

$$\inf \varphi(A) \leqslant \varphi(\lambda \inf A).$$

Indeed, if $\inf A = 0$, then the claim follows by $\lim_{t \to 0^+} \varphi(t) = 0$. If $\inf A > 0$, then we can find $t \in A$ such that $\inf A \leqslant t \leqslant \lambda \inf A$. Now, the monotonicity of φ implies $\inf \varphi(A) \leqslant \varphi(t) \leqslant \varphi(\lambda \inf A)$.

Remark 2.3.7. Let φ be a Φ-function. Then, as a convex function, φ is continuous if and only if φ is finite on $[0, \infty)$.

The following properties of Φ-functions are very useful:

$$\begin{aligned} \varphi(rt) &\leqslant r\varphi(t), \\ \varphi(st) &\geqslant s\varphi(t), \end{aligned} \tag{2.3.8}$$

for any $r \in [0, 1]$, $s \in [1, \infty)$ and $t \geqslant 0$ (compare with (2.1.5)). This is a simple consequence of the convexity of φ and $\varphi(0) = 0$. Inequality (2.3.8) further implies that $\varphi(a) + \varphi(b) \leqslant \frac{a}{a+b}\varphi(a+b) + \frac{b}{a+b}\varphi(a+b) = \varphi(a+b)$ for $a + b > 0$ for all $a, b \geqslant 0$ which combined with convexity yields

$$\varphi(a) + \varphi(b) \leqslant \varphi(a+b) \leqslant \tfrac{1}{2}(\varphi(2a) + \varphi(2b)).$$

Although it is possible to define function spaces using Φ-functions, these are not sufficiently general for our needs. In the case of variable exponent Lebesgue spaces (see Chap. 3) we need our function φ to depend also on the location in the space. So we need to generalize Φ-functions in such a way that they may depend on the space variable.

Definition 2.3.9. Let (A, Σ, μ) be a σ-finite, complete measure space. A real function $\varphi \colon A \times [0, \infty) \to [0, \infty]$ is said to be a *generalized Φ-function* on (A, Σ, μ) if:

(a) $\varphi(y, \cdot)$ is a Φ-function for every $y \in A$.
(b) $y \mapsto \varphi(y, t)$ is measurable for every $t \geqslant 0$.

If φ is a generalized Φ-function on (A, Σ, μ), we write $\varphi \in \Phi(A, \mu)$. If Ω is an open subset of $\mathbb{R}^n$ and μ is the n-dimensional Lebesgue measure we abbreviate this as $\varphi \in \Phi(\Omega)$ or say that φ is a generalized Φ-function on Ω.

In what follows we always make the natural assumption that our measure μ is not identically zero.

Certainly every Φ-function is a generalized Φ-function if we set $\varphi(y,t) := \varphi(t)$ for $y \in A$ and $t \in [0,\infty)$. Also, from (2.3.8) and Lemma 2.3.4 we see that $\varphi(y,\cdot)$ is non-decreasing and lower semicontinuous on $[0,\infty)$ for every $y \in A$.

We say that a function is *simple* if it is the linear combination of characteristic functions of measurable sets with finite measure, $\sum_{i=1}^{k} s_i \chi_{A_i}(x)$ with $\mu(A_1), \ldots, \mu(A_k) < \infty$, $s_1, \ldots, s_k \in \mathbb{K}$. We denote the set of simple functions by $S(A,\mu)$. If Ω is an open subset of $\mathbb{R}^n$ and μ is the n-dimensional Lebesgue measure we abbreviate this by $S(\Omega)$.

We next show that every generalized Φ-function generates a semimodular on $L^0(A,\mu)$.

Lemma 2.3.10. *If $\varphi \in \Phi(A,\mu)$ and $f \in L^0(A,\mu)$, then $y \mapsto \varphi(y,|f(y)|)$ is μ-measurable and*

$$\varrho_\varphi(f) := \int_A \varphi(y,|f(y)|)\, d\mu(y)$$

is a semimodular on $L^0(A,\mu)$. If φ is positive, then ϱ_φ is a modular. We call ϱ_φ the semimodular induced by φ.

Proof. By splitting the function into its positive and negative (real and imaginary) part it suffices to consider the case $f \geqslant 0$. Let $f_k \nearrow f$ point-wise where f_k are non-negative simple functions. Then

$$\varphi(y,|f_k(y)|) = \sum_j \varphi(y,\alpha_j^k) \cdot \chi_{A_j^k}(y),$$

which is measurable and $\varphi(y,f_k(y)) \nearrow \varphi(y,f(y))$. Thus $\varphi(\cdot,f(\cdot))$ is measurable.

Obviously, $\varrho_\varphi(0) = 0$ and $\varrho_\varphi(\lambda x) = \varrho_\varphi(x)$ for $|\lambda| = 1$. The convexity of ϱ_φ is a direct consequence of the convexity of φ. Let us show the left-continuity of ϱ_φ: if $\lambda_k \to 1^-$ and $y \in A$, then $0 \leqslant \varphi(y,\lambda_k f(y)) \to \varphi(y,f(y))$ by the left-continuity and monotonicity of $\varphi(y,\cdot)$. Hence $\varrho_\varphi(\lambda_k f) \to \varrho_\varphi(f)$, by the theorem of monotone convergence. So ϱ_φ is left-continuous in the sense of Definition 2.1.1 (d).

Assume now that $f \in L^0(A,\mu)$ is such that $\varrho_\varphi(\lambda f) = 0$ for all $\lambda > 0$. So for any $k \in \mathbb{N}$ we have $\varphi(y,kf(y)) = 0$ for almost all $y \in A$. Since $\mathbb{N}$ is countable we deduce that $\varphi(y,kf(y)) = 0$ for almost all $y \in A$ and all $k \in \mathbb{N}$. The convexity of φ and $\varphi(y,0) = 0$ imply that $\varphi(y,\lambda f(y)) = 0$ for almost all $y \in A$ and all $\lambda > 0$. Since $\lim_{t\to\infty} \varphi(y,t) = \infty$ for all $y \in A$, this implies that $|f(y)| = 0$ for almost all $y \in A$, hence $f = 0$. So ϱ_φ is a semimodular on $L^0(A,\mu)$.

Assume now that φ is positive and that $\varrho_\varphi(f) = 0$. Then $\varphi(y, f(y)) = 0$ for almost all $y \in A$. Since φ is positive, $f(y) = 0$ for almost all $y \in A$, thus $f = 0$. This proves that ϱ_φ is a modular on $L^0(A, \mu)$.   □

Since every $\varphi \in \Phi(A, \mu)$ generates a semimodular it is natural to study the corresponding semimodular space.

Definition 2.3.11. Let $\varphi \in \Phi(A, \mu)$ and let ϱ_φ be given by

$$\varrho_\varphi(f) := \int_A \varphi(y, |f(y)|) \, d\mu(y)$$

for all $f \in L^0(A, \mu)$. Then the semimodular space

$$
\begin{aligned}
(L^0(A, \mu))_{\varrho_\varphi} &= \{f \in L^0(A, \mu): \lim_{\lambda \to 0} \varrho_\varphi(\lambda f) = 0\} \\
&= \{f \in L^0(A, \mu) : \varrho_\varphi(\lambda f) < \infty \text{ for some } \lambda > 0\}
\end{aligned}
$$

will be called *Musielak–Orlicz space* and denoted by $L^\varphi(A, \mu)$ or L^φ, for short. The norm $\|\cdot\|_{\varrho_\varphi}$ is denoted by $\|\cdot\|_\varphi$, thus

$$\|f\|_\varphi = \inf\left\{\lambda > 0: \varrho_\varphi\left(\frac{x}{\lambda}\right) \leqslant 1\right\}.$$

The Musielak–Orlicz spaces are also called *generalized Orlicz spaces*. They provide a good framework for many function spaces. Here are some examples.

Example 2.3.12. Let (A, Σ, μ) be a σ-finite, complete measure space.

(a) The (semi)modulars given in Example 2.1.4 (a)–(c) give rise to (weighted) Lebesgue spaces.
(b) Let φ be a Φ-function. Then

$$\varrho_\varphi(f) = \int_A \varphi(|f(y)|) \, d\mu(y)$$

is a semimodular on $L^0(A, \mu)$. If φ is positive, then ϱ is a modular on $L^0(A, \mu)$ and the space $L^\varphi(A, \mu)$ is called an *Orlicz space*.
With suitable choices of φ, A and μ, this includes all modulars in Example 2.1.4 except (f).
(c) Example 2.1.4 (f) is not a Musielak–Orlicz space.

As a semimodular space, $L^\varphi = (L^\varphi, \|\cdot\|_\varphi)$ is a normed space, which, in fact, is complete.

Theorem 2.3.13. *Let* $\varphi \in \Phi(A, \mu)$. *Then* $L^\varphi(A, \mu)$ *is a Banach space.*

Before we get to the proof of Theorem 2.3.13 we need to prove two useful lemmas.

Lemma 2.3.14. *Let $\varphi \in \Phi(A, \mu)$ and $\mu(A) < \infty$. Then every $\|\cdot\|_\varphi$-Cauchy sequence is also a Cauchy sequence with respect to convergence in measure.*

Proof. Fix $\varepsilon > 0$ and let $V_t := \{y \in A \colon \varphi(y, t) = 0\}$ for $t > 0$. Then V_t is measurable. For all $y \in A$ the function $t \mapsto \varphi(y, t)$ is non-decreasing and $\lim_{t \to \infty} \varphi(y, t) = \infty$, so $V_t \searrow \emptyset$ as $t \to \infty$. Therefore, $\lim_{k \to \infty} \mu(V_k) = \mu(\emptyset) = 0$, where we have used that $\mu(A) < \infty$. Thus, there exists $K \in \mathbb{N}$ such that $\mu(V_K) < \varepsilon$. Note that if φ is positive then $V_t = \emptyset$ for all $t > 0$ and we do not need this step in the proof.

For a μ-measurable set $E \subset A$ define

$$\nu_K(E) := \varrho_\varphi(K \chi_E) = \int_E \varphi(y, K) \, d\mu(y).$$

If E is μ-measurable with $\nu_K(E) = 0$, then $\varphi(y, K) = 0$ for μ-almost every $y \in E$. Thus $\mu(E \setminus V_K) = 0$ by the definition of V_K. Hence, E is a $\mu|_{A \setminus V_K}$-null set, which means that the measure $\mu|_{A \setminus V_K}$ is absolutely continuous with respect to ν_K.

Since $\mu(A \setminus V_K) \leqslant \mu(A) < \infty$ and $\mu|_{A \setminus V_K}$ is absolutely continuous with respect to ν_K, there exists $\delta \in (0, 1)$ such that $\nu_K(E) \leqslant \delta$ implies $\mu(E \setminus V_K) \leqslant \varepsilon$ (cf. [184, Theorem 30.B]). Since f_k is a $\|\cdot\|_\varphi$-Cauchy sequence, there exists $k_0 \in \mathbb{N}$ such that $\|K \varepsilon^{-1} \delta^{-1}(f_m - f_k)\|_\varphi \leqslant 1$ for all $m, k \geqslant k_0$. Assume in the following $m, k \geqslant k_0$, then by (2.1.5) and the norm-modular unit ball property (Lemma 2.1.14)

$$\varrho_\varphi\big(K \varepsilon^{-1}(f_m - f_k)\big) \leqslant \delta \varrho_\varphi\big(K \varepsilon^{-1} \delta^{-1}(f_m - f_k)\big) \leqslant \delta.$$

Let us write $E_{m,k,\varepsilon} := \{y \in A \colon |f_m(y) - f_k(y)| \geqslant \varepsilon\}$. Then

$$\nu_K(E_{m,k,\varepsilon}) = \int_{E_{m,k,\varepsilon}} \varphi(y, K) \, d\mu(y) \leqslant \varrho_\varphi\big(K \varepsilon^{-1}(f_m - f_k)\big) \leqslant \delta.$$

By the choice of δ, this implies that $\mu(E_{m,k,\varepsilon} \setminus V_K) \leqslant \varepsilon$. With $\mu(V_K) < \varepsilon$ we have $\mu(E_{m,k,\varepsilon}) \leqslant 2\varepsilon$. Since $\varepsilon > 0$ was arbitrary, this proves that f_k is a Cauchy sequence with respect to convergence in measure.

If $\|f_k\|_\varphi \to 0$, then as above there exists $K \in \mathbb{N}$ such that $\mu(\{|f_k| \geqslant \varepsilon\}) \leqslant 2\varepsilon$ for all $k \geqslant K$. This proves $f_k \to 0$ in measure. $\qquad\square$

Lemma 2.3.15. *Let $\varphi \in \Phi(A, \mu)$. Then every $\|\cdot\|_\varphi$-Cauchy sequence $(f_k) \subset L^\varphi$ has a subsequence which converges μ-almost everywhere to a measurable function f.*

Proof. Recall that μ is σ-finite. Let $A = \bigcup_{i=1}^{\infty} A_i$ with A_i pairwise disjoint and $\mu(A_i) < \infty$ for all $i \in \mathbb{N}$. Then, by Lemma 2.3.14, (f_k) is a Cauchy sequence with respect to convergence in measure on A_1. Therefore there exists a measurable function $f \colon A_1 \to \mathbb{K}$ and a subsequence of f_k which converges to f μ-almost everywhere. Repeating this argument for every A_i and passing to the diagonal sequence we get a subsequence (f_{k_j}) and a μ-measurable function $f \colon A \to \mathbb{K}$ such that $f_{k_j} \to f$ μ-almost everywhere.   □

Let us now get to the proof of the completeness of L^φ.

Proof of Theorem 2.3.13. Let (f_k) be a Cauchy sequence. By Lemma 2.3.15 there exists a subsequence f_{k_j} and a μ-measurable function $f \colon A \to \mathbb{K}$ such that $f_{k_j} \to f$ for μ-almost every $y \in A$. This implies $\varphi(y, |f_{k_j}(y) - f(y)|) \to 0$ μ-almost everywhere. Let $\lambda > 0$ and $0 < \varepsilon < 1$. Since (f_k) is a Cauchy sequence, there exists $K = K(\lambda, \varepsilon) \in \mathbb{N}$ such that $\|\lambda(f_m - f_k)\|_\varphi < \varepsilon$ for all $m, k \geqslant N$, which implies $\varrho_\varphi(\lambda(f_m - f_k)) \leqslant \varepsilon$ by Corollary 2.1.15. Therefore by Fatou's lemma

$$
\varrho_\varphi\big(\lambda(f_m - f)\big) = \int_A \lim_{j \to \infty} \varphi\big(y, \lambda|f_m(y) - f_{k_j}(y)|\big)\, d\mu(y)
$$

$$
\leqslant \liminf_{j \to \infty} \int_A \varphi\big(y, \lambda|f_m(y) - f_{k_j}(y)|\big)\, d\mu(y)
$$

$$
= \liminf_{j \to \infty} \varrho_\varphi\big(\lambda(f_m - f_{k_j})\big)
$$

$$
\leqslant \varepsilon.
$$

So $\varrho_\varphi(\lambda(f_m - f)) \to 0$ for $m \to \infty$ and all $\lambda > 0$ and $\|f_k - f\|_\varphi \to 0$ by Lemma 2.1.9. Thus every Cauchy sequence converges in L^φ, as was to be shown.   □

The next lemma collects analogues of the classical Lebesgue integral convergence results.

Lemma 2.3.16. *Let* $\varphi \in \Phi(A, \mu)$ *and* $f_k, f, g \in L^0(A, \mu)$.

(a) *If* $f_k \to f$ μ-*almost everywhere, then* $\varrho_\varphi(f) \leqslant \liminf_{k \to \infty} \varrho_\varphi(f_k)$.
(b) *If* $|f_k| \nearrow |f|$ μ-*almost everywhere, then* $\varrho_\varphi(f) = \lim_{k \to \infty} \varrho_\varphi(f_k)$.
(c) *If* $f_k \to f$ μ-*almost everywhere and* $|f_k| \leqslant |g|$ μ-*almost everywhere, and* $\varrho_\varphi(\lambda g) < \infty$ *for every* $\lambda > 0$, *then* $f_k \to f$ *in* L^φ.

These properties are called Fatou's lemma (for the modular), monotone convergence *and* dominated convergence, *respectively.*

Proof. By Lemma 2.3.4 the mappings $\varphi(y, \cdot)$ are lower semicontinuous. Thus Fatou's lemma implies

$$\varrho_\varphi(f) = \int\limits_A \varphi(y, \lim_{k\to\infty} |f_k(y)|) \, d\mu(y)$$

$$\leqslant \int\limits_A \liminf_{k\to\infty} \varphi(y, |f_k(y)|) \, d\mu(y)$$

$$\leqslant \liminf_{k\to\infty} \int\limits_A \varphi(y, |f_k(y)|) \, d\mu(y)$$

$$= \liminf_{k\to\infty} \varrho_\varphi(f_k).$$

This proves (a).

To prove (b) let $|f_k| \nearrow |f|$. Then by the left-continuity and monotonicity of $\varphi(y, \cdot)$, we have $0 \leqslant \varphi(\cdot, |f_k(\cdot)|) \nearrow \varphi(\cdot, |f(\cdot)|)$ almost everywhere. So, the theorem of monotone convergence gives

$$\varrho_\varphi(f) = \int\limits_A \varphi(y, \lim_{k\to\infty} |f_k(y)|) \, d\mu(y)$$

$$= \int\limits_A \lim_{k\to\infty} \varphi(y, |f_k(y)|) \, d\mu(y)$$

$$= \lim_{k\to\infty} \int\limits_A \varphi(y, |f_k(y)|) \, d\mu(y)$$

$$= \lim_{k\to\infty} \varrho_\varphi(f_k).$$

To prove (c) assume that $f_k \to f$ almost everywhere, $|f_k| \leqslant |g|$, and $\varrho(\lambda g) < \infty$ for every $\lambda > 0$. Then $|f_k - f| \to 0$ almost everywhere, $|f| \leqslant |g|$ and $|f_k - f| \leqslant 2|g|$. Since $\varrho_\varphi(2\lambda g) < \infty$, we can use the theorem of dominated convergence to conclude that

$$\lim_{k\to\infty} \varrho_\varphi(\lambda|f - f_k|) = \int\limits_A \varphi\left(y, \lim_{k\to\infty} \lambda|f(y) - f_k(y)|\right) d\mu(y) = 0.$$

Since $\lambda > 0$ was arbitrary, Lemma 2.1.9 implies that $f_k \to f$ in L^φ. □

Let us summarize a few additional properties of L^φ. Properties (a), (b), (c) and (d) of the next theorem are known as *circularity, solidity, Fatou's lemma (for the norm)*, and the *Fatou property*, respectively.

Theorem 2.3.17. *Let $\varphi \in \Phi(A, \mu)$. Then the following hold.*

(a) $\|f\|_\varphi = \| |f| \|_\varphi$ *for all $f \in L^\varphi$.*

(b) *If $f \in L^\varphi$, $g \in L^0(A, \mu)$, and $0 \leqslant |g| \leqslant |f|$ μ-almost everywhere, then $g \in L^\varphi$ and $\|g\|_\varphi \leqslant \|f\|_\varphi$.*

(c) *If $f_k \to f$ almost everywhere, then $\|f\|_\varphi \leqslant \liminf_{k\to\infty} \|f_k\|_\varphi$.*
(d) *If $|f_k| \nearrow |f|$ μ-almost everywhere with $f_k \in L^\varphi(A, \mu)$ and $\sup_k \|f_k\|_\varphi < \infty$, then $f \in L^\varphi(A, \mu)$ and $\|f_k\|_\varphi \nearrow \|f\|_\varphi$.*

Proof. The properties (a) and (b) are obvious. Let us now prove (c). So let $f_k \to f$ μ-almost everywhere. There is nothing to prove for $\liminf_{k\to\infty} \|f_k\|_\varphi = \infty$. Let $\lambda > \liminf_{k\to\infty} \|f_k\|_\varphi$. Then $\|f_k\|_\varphi < \lambda$ for large k. Thus by the unit ball property $\varrho_\varphi(f_k/\lambda) \leqslant 1$ for large k. Now Fatou's lemma for the modular (Lemma 2.3.16) implies $\varrho_\varphi(f/\lambda) \leqslant 1$. So $\|f\|_\varphi \leqslant \lambda$ again by the unit ball property, which implies $\|f\|_\varphi \leqslant \liminf_{k\to\infty} \|f_k\|_\varphi$.

It remains to prove (d). So let $|f_k| \nearrow |f|$ μ-almost everywhere with $\sup_k \|f_k\|_\varphi < \infty$. From (a) and (c) follows $\|f\|_\varphi \leqslant \liminf_{k\to\infty} \|f_k\|_\varphi \leqslant \sup_k \|f_k\|_\varphi < \infty$, which proves $f \in L^\varphi$. On the other hand $|f_k| \nearrow |f|$ and (b) implies that $\|f_k\|_\varphi \nearrow \limsup_{k\to\infty} \|f_k\|_\varphi \leqslant \|f\|_\varphi$. Thus $\lim_{k\to\infty} \|f_k\|_\varphi = \|f\|_\varphi$ and $\|f_k\|_\varphi \nearrow \|f\|_\varphi$. $\square$

2.4 Uniform Convexity

In this section we study sufficient conditions for the uniform convexity of a modular space X_ϱ and the Musielak–Orlicz space L^φ. We first show that the uniform convexity of the Φ-function implies that of the modular; and that the uniform convexity of the semimodular combined with the Δ_2-condition implies the uniform convexity of the norm. The section is concluded by some further properties of uniformly convex modulars. Let us start with the Δ_2-condition of the Φ-function and some implications.

Definition 2.4.1. We say that $\varphi \in \Phi(A, \mu)$ satisfies the Δ_2-*condition* if there exists $K \geqslant 2$ such that

$$\varphi(y, 2t) \leqslant K\varphi(y, t)$$

for all $y \in A$ and all $t \geqslant 0$. The smallest such K is called the Δ_2-*constant of φ.*

Analogously, we say that a semimodular ϱ on X satisfies the Δ_2-*condition* if there exists $K \geqslant 2$ such that $\varrho(2f) \leqslant K \varrho(f)$ for all $f \in X_\varrho$. Again, the smallest such K is called the Δ_2-*constant of ϱ.*

If $\varphi \in \Phi(A, \mu)$ satisfy the Δ_2-condition, then ϱ_φ satisfies the Δ_2-condition with the same constant. Moreover, ϱ_φ satisfies the weak Δ_2-condition for modulars, so by Lemma 2.1.11 modular convergence and norm convergence are equivalent; and $E \subset L^\varphi(\Omega, \mu)$ is bounded with respect to the norm if and only if it is bounded with respect to the modular, i.e. $\sup_{f \in E} \|f\| < \infty$ if and only if $\sup_{f \in E} \varrho_\varphi(f) < \infty$.

Corollary 2.1.15 shows that a small norm implies a small modular. The following result shows the reverse implication.

Lemma 2.4.2. *Let ϱ be a semimodular on X that satisfies the Δ_2-condition. Let K be the Δ_2-constant of ϱ. Then for every $\varepsilon > 0$ there exists $\delta = \delta(\varepsilon, K) > 0$ such that $\varrho(f) \leqslant \delta$ implies $\|f\|_\varrho \leqslant \varepsilon$.*

Proof. For $\varepsilon > 0$ choose $j \in \mathbb{N}$ with $2^{-j} \leqslant \varepsilon$. Let $\delta := K^j$ and $\varrho(f) \leqslant \delta$. Then $\varrho(2^j f) \leqslant K^j \varrho(f) \leqslant 1$ and the unit ball property yields $\|f\|_\varrho \leqslant 2^{-j} \leqslant \varepsilon$. $\square$

Lemma 2.4.3. *Let ϱ be a semimodular on X that satisfies the Δ_2-condition with constant K. Then ϱ is a continuous modular and for every $\varepsilon > 0$ there exists $\delta = \delta(\varepsilon, K) > 0$ such that $\varrho(f) \leqslant 1 - \varepsilon$ implies $\|f\|_\varrho \leqslant 1 - \delta$ for $f \in X_\varrho$.*

Proof. If $\varrho(f) = 0$, then $\varrho(2^m f) \leqslant K^m \varrho(f) = 0$, where K is the Δ_2-constant of φ. This proves $f = 0$, so ϱ is a modular. We already know that ϱ is left-continuous, so it suffices to show $\varrho(x) = \lim_{\lambda \to 1+} \varrho(\lambda x)$. By monotonicity we have $\varrho(x) \leqslant \liminf_{\lambda \to 1+} \varrho(\lambda x)$. It follows by convexity of ϱ that

$$\varrho(af) \leqslant (2 - a)\varrho(f) + (a - 1)\varrho(2f) \leqslant \big((2 - a) + K(a - 1)\big)\varrho(f)$$
$$\leqslant \big(1 + (K - 1)(a - 1)\big)\varrho(f)$$

for every $a \in [1, 2]$. Hence $\varrho(x) \geqslant \liminf_{\lambda \to 1+} \varrho(\lambda x)$, which completes the proof of continuity.

Let $\varepsilon > 0$ and $f \in X_\varrho$ with $\varrho(f) \leqslant 1 - \varepsilon$. Fix $a = a(K, \varepsilon) \in (1, 2)$ such that the right-hand side of the previous inequality is bounded by one. Then $\varrho(af) \leqslant 1$ and the unit ball property implies $\|af\|_\varrho \leqslant 1$. The claim follows with $1 - \delta := \frac{1}{a}$. $\square$

In the previous sections we worked with general $\varphi \in \Phi(A, \mu)$. The corresponding Musielak–Orlicz spaces include the classical spaces L^p with $1 \leqslant p \leqslant \infty$, see Example 2.1.8. Sometimes, however, it is better to work with a subclass of $\Phi(A, \mu)$, called N-functions. These functions will have better properties (N stands for *nice*) but the special cases $p = 1$ and $p = \infty$ are excluded. This corresponds to the experience that also in the classical case the "borderline" cases $p = 1$ and $p = \infty$ are often treated differently.

Definition 2.4.4. A Φ-function φ is said to be an *N-function* if it is continuous and positive and satisfies $\lim_{t \to 0} \frac{\varphi(t)}{t} = 0$ and $\lim_{t \to \infty} \frac{\varphi(t)}{t} = \infty$.

A function $\varphi \in \Phi(A, \mu)$ is said to be a *generalized N-function* if $\varphi(y, \cdot)$ is for every $y \in \Omega$ an N-function.

If φ is a generalized N-function on (A, μ), we write $\varphi \in N(A, \mu)$ for short. If Ω is an open subset of $\mathbb{R}^n$ and μ is the n-dimensional Lebesgue measure we abbreviate $\varphi \in N(\Omega)$.

Definition 2.4.5. A function $\varphi \in N(A, \mu)$ is called *uniformly convex* if for every $\varepsilon > 0$ there exists $\delta > 0$ such that

$$|u - v| \leqslant \varepsilon \max\{u, v\} \quad \text{or} \quad \varphi\Big(y, \frac{u + v}{2}\Big) \leqslant (1 - \delta)\frac{\varphi(y, u) + \varphi(y, v)}{2}$$

for all $u, v \geqslant 0$ and every $y \in A$.

Remark 2.4.6. If $\varphi(x,t) = t^q$ with $q \in (1, \infty)$, then φ is uniformly convex. To prove this, we have to show that for $u, v \geqslant 0$ the estimate $|u - v| > \varepsilon \max\{v, u\}$ implies $(\frac{u+v}{2})^q \leqslant (1 - \delta(\varepsilon))\frac{1}{2}(u^q + v^q)$ with $\delta(\varepsilon) > 0$ for every $\varepsilon > 0$. Without loss of generality we can assume $\varepsilon \in (0, \frac{1}{2})$. By homogeneity it suffices to consider the case $v = 1$ and $0 \leqslant u \leqslant 1$. So we have to show that $u \in [0, 1-\varepsilon)$ implies $(\frac{1+u}{2})^q \leqslant (1-\delta(\varepsilon))\frac{1}{2}(1+u^q)$. Define $f(\tau) := 2^{1-q}\frac{(1+u)^q}{(1+u^q)}$. Then f is continuous on $[0,1]$ and has its maximum at 1. This proves as desired $f(u) \leqslant \delta(\varepsilon)$ for all $u \in [0, 1 - \varepsilon)$.

It follows by division with q that $\varphi(x,t) = \frac{1}{q}t^q$ with $1 < q < \infty$ is also uniformly convex.

Definition 2.4.5 is formulated for $u, v \geqslant 0$. However, the following lemma shows that this can be relaxed to values in $\mathbb{K}$.

Lemma 2.4.7. Let $\varphi \in N(A, \mu)$ be uniformly convex. Then for every $\varepsilon_2 > 0$ there exists $\delta_2 > 0$ such that

$$|a - b| \leqslant \varepsilon_2 \max\{|a|, |b|\} \quad or \quad \varphi\left(y, \left|\frac{a+b}{2}\right|\right) \leqslant (1 - \delta_2)\frac{\varphi(y, |a|) + \varphi(y, |b|)}{2}.$$

for all $a, b \in \mathbb{K}$ and every $y \in A$.

Proof. Fix $\varepsilon_2 > 0$. For $\varepsilon := \varepsilon_2/2$ let $\delta > 0$ be as in Definition 2.4.5. Let $|a - b| > \varepsilon_2 \max\{|a|, |b|\}$. If $||a| - |b|| > \varepsilon \max\{|a|, |b|\}$, then the claim follows by $|a + b| \leqslant |a| + |b|$ and choice of δ with $\delta_2 = \delta$. So assume in the following $||a| - |b|| \leqslant \varepsilon \max\{|a|, |b|\}$. Then

$$|a - b| > \varepsilon_2 \max\{|a|, |b|\} = 2\varepsilon \max\{|a|, |b|\} \geqslant 2||a| - |b||.$$

Therefore,

$$\left|\frac{a+b}{2}\right|^2 = \frac{|a|^2}{2} + \frac{|b|^2}{2} - \left|\frac{a-b}{2}\right|^2$$
$$\leqslant \frac{|a|^2}{2} + \frac{|b|^2}{2} - \frac{3}{4}\left|\frac{a-b}{2}\right|^2 - \left(\frac{|a| - |b|}{2}\right)^2$$
$$= \left(\frac{|a| + |b|}{2}\right)^2 - \frac{3}{4}\left|\frac{a-b}{2}\right|^2.$$

Since $|a - b| > \varepsilon_2 \max\{|a|, |b|\} \geqslant \varepsilon_2(|a| + |b|)/2$, it follows that

$$\left|\frac{a+b}{2}\right|^2 \leqslant \left(1 - \frac{3\varepsilon_2^2}{16}\right)\left(\frac{|a| + |b|}{2}\right)^2.$$

Let $\delta_2 := 1 - \sqrt{1 - \frac{3\varepsilon_2^2}{16}} > 0$, then $|\frac{a+b}{2}| \leqslant (1 - \delta_2)\frac{|a|+|b|}{2}$. This, (2.1.5) and the convexity of φ imply

$$\varphi\left(y, \left|\frac{a+b}{2}\right|\right) \leqslant (1-\delta_2)\varphi\left(y, \frac{|a|+|b|}{2}\right) \leqslant (1-\delta_2)\frac{\varphi(y,|a|)+\varphi(y,|b|)}{2}. \qquad \square$$

Remark 2.4.8. If $u, v \in \mathbb{K}$ satisfies $|a-b| \leqslant \varepsilon_2 \max\{|a|,|b|\}$ with $\varepsilon_2 \in (0,1)$, then $\frac{|a-b|}{2} \leqslant \varepsilon_2 \frac{|a|+|b|}{2}$ and by the convexity of φ follows

$$\varphi\left(y, \frac{|a-b|}{2}\right) \leqslant \varepsilon_2 \frac{\varphi(y,|a|)+\varphi(y,|b|)}{2}. \tag{2.4.9}$$

Therefore, we can replace the first alternative in Lemma 2.4.7 by the weaker version (2.4.9).

We need the following concept of uniform convexity for the semimodular.

Definition 2.4.10. A semimodular ϱ on X is called *uniformly convex* if for every $\varepsilon > 0$ there exists $\delta > 0$ such that

$$\varrho\left(\frac{f-g}{2}\right) \leqslant \varepsilon \frac{\varrho(f)+\varrho(g)}{2} \quad \text{or} \quad \varrho\left(\frac{f+g}{2}\right) \leqslant (1-\delta)\frac{\varrho(f)+\varrho(g)}{2}$$

for all $f, g \in X_\varrho$.

Theorem 2.4.11. *Let $\varphi \in N(A,\mu)$ be uniformly convex. Then ϱ_φ is uniformly convex.*

Proof. Let $\varepsilon_2, \delta_2 > 0$ be as in Lemma 2.4.7 and let $\varepsilon := 2\varepsilon_2$. There is nothing to show if $\varrho_\varphi(f) = \infty$ or $\varrho_\varphi(g) = \infty$. So in the following let $\varrho_\varphi(f), \varrho_\varphi(g) < \infty$, which implies by convexity $\varrho(\frac{f+g}{2}), \varrho(\frac{f-g}{2}) < \infty$.

Assume that $\varrho_\varphi(\frac{f-g}{2}) > \varepsilon \frac{\varrho_\varphi(f)+\varrho_\varphi(g)}{2}$. We show that

$$\varrho_\varphi\left(\frac{f+g}{2}\right) \leqslant \left(1 - \frac{\delta_2 \varepsilon}{2}\right)\frac{\varrho_\varphi(f)+\varrho_\varphi(g)}{2},$$

which proves that ϱ_φ is uniformly convex. Define

$$E := \left\{ y \in A \colon |f(y)-g(y)| > \frac{\varepsilon}{2} \max\{|f(y)|, |g(y)|\} \right\}.$$

It follows from Remark 2.4.8 that (2.4.9) holds for almost all $y \in A \setminus E$. In particular,

$$\varrho_\varphi\left(\chi_{A\setminus E}\frac{f-g}{2}\right) \leqslant \frac{\varepsilon}{2}\frac{\varrho_\varphi(\chi_{A\setminus E}f)+\varrho_\varphi(\chi_{A\setminus E}g)}{2} \leqslant \frac{\varepsilon}{2}\frac{\varrho_\varphi(f)+\varrho_\varphi(g)}{2}.$$

This and $\varrho_\varphi(\frac{f-g}{2}) > \varepsilon \frac{\varrho_\varphi(f) + \varrho_\varphi(g)}{2}$ imply

$$\varrho_\varphi\left(\chi_E \frac{f-g}{2}\right) = \varrho_\varphi\left(\frac{f-g}{2}\right) - \varrho_\varphi\left(\chi_{A\setminus E}\frac{f-g}{2}\right) > \frac{\varepsilon}{2}\frac{\varrho_\varphi(f) + \varrho_\varphi(g)}{2}.$$
(2.4.12)

On the other hand it follows by the definition of E and the choice of δ_2 in Lemma 2.4.7 that

$$\varrho_\varphi\left(\chi_E \frac{f+g}{2}\right) \leqslant (1-\delta_2)\frac{\varrho_\varphi(\chi_E f) + \varrho_\varphi(\chi_E g)}{2}.$$
(2.4.13)

We estimate

$$\frac{\varrho_\varphi(f) + \varrho_\varphi(g)}{2} - \varrho_\varphi\left(\frac{f+g}{2}\right) \geqslant \frac{\varrho_\varphi(\chi_E f) + \varrho_\varphi(\chi_E g)}{2} - \varrho_\varphi\left(\chi_E\frac{f+g}{2}\right),$$

where we have split the domain of the involved integrals into the sets E and $A \setminus E$ and have used $\frac{1}{2}(\varphi(f) + \varphi(g)) - \varphi(\frac{f+g}{2}) \geqslant 0$ on $A \setminus E$. This, (2.4.13), the convexity and (2.4.12) imply

$$\begin{aligned}
\frac{\varrho_\varphi(f) + \varrho_\varphi(g)}{2} - \varrho_\varphi\left(\frac{f+g}{2}\right) &\geqslant \delta_2 \frac{\varrho_\varphi(\chi_E f) + \varrho_\varphi(\chi_E g)}{2} \\
&\geqslant \delta_2 \varrho_\varphi\left(\chi_E \frac{f-g}{2}\right) \\
&\geqslant \frac{\delta_2 \varepsilon}{2}\frac{\varrho_\varphi(f) + \varrho_\varphi(g)}{2}. \qquad \square
\end{aligned}$$

The question arises if uniform convexity of the semimodular ϱ implies the uniform convexity of X_ϱ. This turns out to be true under the Δ_2-condition.

Theorem 2.4.14. *Let ϱ be a uniformly convex semimodular on X that satisfies the Δ_2-condition. Then the norm $\|\cdot\|_\varrho$ on X_ϱ is uniformly convex. Hence, X_ϱ is uniformly convex.*

Proof. Fix $\varepsilon > 0$. Let $x, y \in X$ with $\|x\|_\varrho, \|y\|_\varrho \leqslant 1$ and $\|x - y\|_\varrho > \varepsilon$. Then $\|\frac{x-y}{2}\| > \frac{\varepsilon}{2}$ and by Lemma 2.4.2 there exists $\alpha = \alpha(\varepsilon) > 0$ such that $\varrho(\frac{x-y}{2}) > \alpha$. By the unit ball property we have $\varrho(x), \varrho(y) \leqslant 1$, so $\varrho(\frac{x-y}{2}) > \alpha \frac{\varrho(x) + \varrho(y)}{2}$. Since ϱ is uniformly convex, there exists $\beta = \beta(\alpha) > 0$ such that $\varrho(\frac{x+y}{2}) \leqslant (1 - \beta)\frac{\varrho(x) + \varrho(y)}{2} \leqslant 1 - \beta$. Now Lemma 2.4.3 implies the existence of $\delta = \delta(K, \beta) > 0$ with $\|\frac{x+y}{2}\|_\varrho \leqslant 1 - \delta$. This proves the uniform convexity of $\|\cdot\|_\varrho$. $\square$

Remark 2.4.15. If $\varphi \in N(A, \mu)$ is uniformly convex and satisfies the Δ_2-condition, then it follows by the combination of Theorems 2.4.11 and 2.4.14 that the norm $\|\cdot\|_\varphi$ of $L^\varphi(A, \mu)$ is uniformly convex. Hence, $L^\varphi(A, \mu)$ is also uniformly convex.

We will later need that the sum of uniformly convex semimodulars is again uniformly convex.

Lemma 2.4.16. *If* ϱ_1, ϱ_2 *are uniformly convex semimodulars on* X, *then* $\varrho := \varrho_1 + \varrho_2$ *is uniformly convex.*

Proof. If $\varepsilon > 0$, then there exists $\delta > 0$ such that

$$\varrho_j\left(\frac{f-g}{2}\right) \leqslant \varepsilon \frac{\varrho_j(f) + \varrho_j(g)}{2} \quad \text{or} \quad \varrho_j\left(\frac{f+g}{2}\right) \leqslant (1-\delta)\frac{\varrho_j(f) + \varrho_j(g)}{2}$$

for $j = 1, 2$. We show that

$$\varrho\left(\frac{f-g}{2}\right) \leqslant 2\varepsilon \frac{\varrho(f) + \varrho(g)}{2} \quad \text{or} \quad \varrho\left(\frac{f+g}{2}\right) \leqslant (1-\delta\varepsilon)\frac{\varrho(f) + \varrho(g)}{2},$$

since this proves the uniform convexity of ϱ. Fix f and g and assume that $\varrho(\frac{f-g}{2}) > 2\varepsilon\frac{\varrho(f)+\varrho(g)}{2}$. Without loss of generality, we can assume that $\varrho_1(\frac{f-g}{2})$ $\geqslant \varrho_2(\frac{f-g}{2})$ for this specific choice of f and g. Therefore, $\varrho_1(\frac{f-g}{2}) > \varepsilon\frac{\varrho(f)+\varrho(g)}{2} \geqslant$ $\varepsilon\frac{\varrho_1(f)+\varrho_1(g)}{2}$. So the choice of δ implies

$$\varrho_1\left(\frac{f+g}{2}\right) \leqslant (1-\delta)\frac{\varrho_1(f) + \varrho_1(g)}{2}.$$

Taking into account the convexity of ϱ_2, we obtain

$$\varrho\left(\frac{f+g}{2}\right) \leqslant \frac{\varrho(f) + \varrho(g)}{2} - \delta\frac{\varrho_1(f) + \varrho_1(g)}{2}.$$

Since $\frac{\varrho_1(f)+\varrho_1(g)}{2} \geqslant \varrho_1(\frac{f-g}{2}) > \varepsilon\frac{\varrho(f)+\varrho(g)}{2}$, this implies

$$\varrho\left(\frac{f+g}{2}\right) \leqslant (1-\delta\varepsilon)\frac{\varrho(f) + \varrho(g)}{2}. \qquad \square$$

It is well known that on uniformly convex spaces weak convergence $x_k \rightharpoonup x$ combined with convergence of the norms $\|x_k\| \to \|x\|$ implies strong convergence $x_k \to x$. The following lemma is in this spirit.

Lemma 2.4.17. *Let* ϱ *be a uniformly convex semimodular on* X. *Let* $x_k, x \in X_\varrho$ *such that* $x_k \rightharpoonup x$, $\varrho(x_k) \to \varrho(x)$ *and* $\varrho(x) < \infty$. *Then*

$$\varrho\left(\frac{x_k - x}{2}\right) \to 0.$$

Proof. We proceed by contradiction. Assume that the claim is wrong and there exists $\varepsilon > 0$ and a subsequence x_{k_j} such that

$$\varrho\left(\frac{x_{k_j} - x}{2}\right) > \varepsilon \tag{2.4.18}$$

for all $j \in \mathbb{N}$. Since ϱ is uniformly continuous, there exists $\delta > 0$ such that

$$\varrho\left(\frac{x_k - x}{2}\right) \leqslant \varepsilon \quad \text{or} \quad \varrho\left(\frac{x_k + x}{2}\right) \leqslant (1 - \delta)\frac{\varrho(x_k) + \varrho(x)}{2}.$$

In particular, our subsequence always satisfies the second alternative. Together with $\frac{1}{2}(x_k + x) \rightharpoonup x$, the weak lower semicontinuity of ϱ (Theorem 2.2.8) and $\varrho(x_k) \to \varrho(x)$ implies that

$$\varrho(x) \leqslant \liminf_{j \to \infty} \varrho\left(\frac{x_{k_j} + x}{2}\right) \leqslant (1 - \delta)\liminf_{j \to \infty} \frac{\varrho(x_{k_j}) + \varrho(x)}{2} = (1 - \delta)\varrho(x).$$

Using $\varrho(x) < \infty$ we get $\varrho(x) = 0$. It follows by convexity and $\varrho(x_k) \to \varrho(x)$ that

$$\varrho\left(\frac{x_k - x}{2}\right) \leqslant \frac{\varrho(x_k) + \varrho(x)}{2} \to \varrho(x) = 0$$

for $n \to \infty$. This contradicts (2.4.18).   □

Remark 2.4.19. If ϱ satisfies the (weak) Δ_2-condition, then under the conditions of the previous lemma, $\varrho(\lambda(x_k - x)) \to 0$ for all $\lambda > 0$ and $x_k \to x$ in X_ϱ by Lemma 2.1.11.

2.5 Separability

We next prove basic properties of Musielak–Orlicz spaces that require some additional structure. Since these properties do not even hold for the full range $p \in [1, \infty]$ of classical Lebesgue spaces, it is clear that some restrictions are necessary. In this section we consider separability.

We first define some function classes related to L^φ. The set E^φ of finite elements will be later important in the approximability by simple functions, see Theorem 2.5.9.

Definition 2.5.1. Let $\varphi \in \Phi(A, \mu)$. The set

$$L^\varphi_{OC} := L^\varphi_{OC}(A, \mu) := \{f \in L^\varphi : \varrho_\varphi(f) < \infty\} \tag{2.5.2}$$

is called the *Musielak–Orlicz class*. Let

$$E^\varphi := E^\varphi(A, \mu) := \{f \in L^\varphi : \varrho_\varphi(\lambda f) < \infty \text{ for all } \lambda > 0\}. \qquad (2.5.3)$$

The elements of $E^\varphi(A, \mu)$ are called *finite*.

Let us start with a few examples:

(a) Let $\varphi(y, t) = t^p$ with $1 \leqslant p < \infty$. Then $E^\varphi = L^\varphi_{OC} = L^\varphi = L^p$.
(b) Let $\varphi(y, t) = \infty \cdot \chi_{(1,\infty)}(t)$. Then

$$E^\varphi = \{0\},$$
$$L^\varphi_{OC} = \{f : |f| \leqslant 1 \text{ almost everywhere}\},$$
$$L^\varphi = L^\infty.$$

(c) Let $\varphi(y, t) = \exp(t) - 1$ and $\Omega = (0, 1)$. Then $\varphi \in \Phi(\Omega)$ is positive and
    continuous but $E^\varphi \neq L^\varphi_{OC} \neq L^\varphi$. Indeed, if $f := \sum_{k=1}^\infty \frac{k}{2} \chi_{(2^{-k}, 2^{-k+1})}$,
    then $f \in L^\varphi_{OC} \setminus E^\varphi$ and $2f \in L^\varphi \setminus L^\varphi_{OC}$.

By definition of E^φ, L^φ_{OC}, and L^φ it is clear that $E^\varphi \subset L^\varphi_{OC} \subset L^\varphi$.
Moreover, by convexity of φ the set L^φ_{OC} is convex and the sets E^φ and L^φ
are linear subspaces of L^0. There is a special relation of E^φ and L^φ to L^φ_{OC}:
E^φ is the biggest vector space in L^φ_{OC} and L^φ is the smallest vector space
in L^0 containing L^φ_{OC}.

In some cases the inclusions $E^\varphi \subset L^\varphi_{OC} \subset L^\varphi$ are strict and in other cases
equality holds. In fact, it is easily seen that $E^\varphi = L^\varphi_{OC} = L^\varphi$ is equivalent
to the implication $f \in L^\varphi_{OC} \Rightarrow 2f \in L^\varphi_{OC}$. The Δ_2-condition (see Definition 2.4.1) implies that $\varrho_\varphi(2^m f) \leqslant K^m \varrho_\varphi(f)$, where K is the Δ_2-constant,
from which we conclude that

$$E^\varphi(A, \mu) = L^\varphi_{OC}(A, \mu) = L^\varphi(A, \mu).$$

Remark 2.5.4. The set E^φ is a closed subset of L^φ. Indeed, let $f_k \to f$
in L^φ with $f_k \in E^\varphi$. For $\lambda > 0$ we have $\varrho_\varphi(2\lambda(f_k - f)) \to 0$ as $k \to \infty$.
In particular, $\varrho_\varphi(2\lambda(f_{k_\lambda} - f)) \leqslant 1$ for some k_λ. By convexity $\varrho_\varphi(\lambda f) \leqslant$
$\frac{1}{2}\varrho_\varphi(2\lambda(f_{k_\lambda} - f)) + \frac{1}{2}\varrho_\varphi(2\lambda f_{k_\lambda}) \leqslant \frac{1}{2} + \frac{1}{2}\varrho_\varphi(2\lambda f_{k_\lambda}) < \infty$, which shows that
$f \in E^\varphi$.

In the approximation of measurable functions it is very useful to work with
simple functions. To be able to approximate a function f by simple functions
we have to assume an additional property of φ:

Definition 2.5.5. A function $\varphi \in \Phi(A, \mu)$ is called *locally integrable* on A if
$\varrho_\varphi(t\chi_E) < \infty$ for all $t \geqslant 0$ and all μ-measurable $E \subset A$ with $\mu(E) < \infty$.

Note that local integrability in the previous definition differs from the one
used in L^1_{loc}, where we assume integrability over compact subsets.

If $\varphi \in \Phi(A, \mu)$ is locally integrable, then the set of simple functions $S(A, \mu)$
is contained in E^φ. Actually, the property $S(A, \mu) \subset E^\varphi$ is equivalent to the
local integrability of φ.

Example 2.5.6. Let $\varphi \in \Phi(A, \mu)$ with $\varphi(y, t) = \psi(t)$ where ψ is a continuous Φ-function. Then φ is locally integrable. Indeed, due to the continuity we know that $t \mapsto \psi(t)$ is everywhere finite on $[0, \infty)$. Therefore, $\varrho_\varphi(t\chi_E) = \mu(E)\psi(t) < \infty$ for all $t \geqslant 0$ and $\mu(E) < \infty$.

Proposition 2.5.7. *Let* $\varphi \in \Phi(A, \mu)$ *be locally integrable. Then for every* $\lambda > 0$ *and* $\varepsilon > 0$ *there exists* $\delta > 0$ *such that* $\mu(E) \leqslant \delta$ *implies* $\varrho_\varphi(\lambda\chi_E) \leqslant \varepsilon$ *and* $\|\chi_E\|_\varphi \leqslant \frac{1}{\lambda}$.

Proof. We begin with the proof of $\varrho_\varphi(\lambda\chi_E) \leqslant \varepsilon$ by contradiction. Assume to the contrary that there exist $\lambda > 0$ and $\varepsilon > 0$ and a sequence (E_k) such that $\mu(E_k) \leqslant 2^{-k}$ and $\varrho_\varphi(\lambda\chi_{E_k}) > \varepsilon$. Let $G_k := \bigcup_{m=k}^\infty E_m$, and note that $\mu(G_k) \leqslant \sum_{m=k}^\infty 2^{-m} = 2^{1-k} \to 0$ as $k \to \infty$. Since φ is locally integrable and $\mu(G_1) \leqslant 1$, we have $\varrho_\varphi(\lambda\chi_{G_1}) < \infty$. Moreover, $\lambda\chi_{G_k} \leqslant \lambda\chi_{G_1}$ and $\lambda\chi_{G_k} \to 0$ almost everywhere. Thus, we conclude by dominated convergence that $\varrho_\varphi(\lambda\chi_{G_k}) \to 0$. This contradicts $\varrho_\varphi(\lambda\chi_{G_k}) \geqslant \varrho_\varphi(\lambda\chi_{E_k}) > \varepsilon$ for every k.

The claim $\|\chi_E\|_\varphi \leqslant \frac{1}{\lambda}$ follows from $\varrho_\varphi(\lambda\chi_E) \leqslant \varepsilon$ by the choice $\varepsilon = 1$ and the unit ball property. $\square$

Remark 2.5.8. If $f \in L^\varphi$ has the property that $\|\chi_{E_k} f\|_\varphi \to 0$ if $E_k \searrow \emptyset$, then we say that f *has absolutely continuous norm*. If follows easily by dominated convergence (Lemma 2.3.16) that every $f \in E^\varphi$ has absolutely continuous norm. .

Theorem 2.5.9. *Let* $\varphi \in \Phi(A, \mu)$ *be locally integrable and let* $S := S(A, \mu)$ *be the set of simple functions. Then* $\overline{S}^{\|\cdot\|_\varphi} = E^\varphi(A, \mu)$.

Proof. The local integrability implies that $S \subset E^\varphi$. Since E^φ is closed by Remark 2.5.4, it suffices to show that every $f \in E^\varphi$ is in the closure of S. Let $f \in E^\varphi$ with $f \geqslant 0$. Since $f \in L^0(A)$, there exist $f_k \in S$ with $0 \leqslant f_k \nearrow f$ almost everywhere. So $f_k \to f$ in L^φ by the theorem of dominated convergence (Lemma 2.3.16). Thus, f is in the closure of S. If we drop the assumption $f \geqslant 0$, then we split x into positive and negative parts (real and imaginary parts) which belong again to E^φ. $\square$

We now investigate the problem of separability of E^φ. Let (A, Σ, μ) be a σ-finite, complete measure space. Here, we need the notion of separable measures: recall that a measure μ is called *separable* if there exists a sequence $(E_k) \subset \Sigma$ with the following properties:

(a) $\mu(E_k) < \infty$ for all $k \in \mathbb{N}$.
(b) For every $E \in \Sigma$ with $\mu(E) < \infty$ and every $\varepsilon > 0$ there exists an index k such that $\mu(E \triangle E_k) < \varepsilon$, where $\triangle$ denotes the symmetric difference defined through $E \triangle E_k := (E \setminus E_k) \cup (E_k \setminus E)$.

For instance the Lebesgue measure on $\mathbb{R}^n$ and the counting measure on $\mathbb{Z}^n$ are separable. Recall that a Banach space is separable if it contains a dense, countable subset.

Theorem 2.5.10. *Let $\varphi \in \Phi(A, \mu)$ be locally integrable and let μ be separable. Then $E^{\varphi}(A, \mu)$ is separable.*

Proof. Let S_0 be the set of all simple functions g of the form $g = \sum_{i=1}^{k} a_i \chi_{E_i}$ with $a_i \in \mathbb{Q}$ and E_i is as in the definition of a separable measure. By Theorem 2.5.9 it suffices to prove that S_0 is dense in S. Let $f \in S$. Then we can write f in the form $f = \sum_{i=1}^{k} b_i \chi_{B_i}$ with $b_i \in \mathbb{R}$, $B_i \in \Sigma$ pairwise disjoint and $\mu(B_i) < \infty$. Let $\lambda > 0$ be arbitrary and define $b := \max_{1 \leqslant i \leqslant k} |b_i|$. Since φ is locally integrable, we know by Proposition 2.5.7 that the integral of $y \mapsto \varphi(y, 4k\lambda b)$ is small over small sets. Hence, by the separability of μ we find measurable sets $E_{j_1}, \ldots, E_{j_k}$ of finite measure such that

$$\int_{E_{j_i} \triangle B_i} \varphi(y, 4k\lambda b)\, d\mu(y) \leqslant 1.$$

Let $B := \bigcup_{i=1}^{k} B_i$. Then $\int_B \varphi(y, 2\lambda \eta)\, d\mu(y) \to 0$ for $\eta \to 0$, since $\mu(B) < \infty$ and φ is locally integrable. Let $\delta > 0$ be such that $\int_B \varphi(y, 2\lambda \delta)\, d\mu(y) \leqslant 1$. Choose rational numbers $a_1, \ldots, a_k$ such that $|b_i - a_i| < \delta$ and $|a_i| \leqslant 2b$ for $i = 1, \ldots, k$. Let $g := \sum_{i=1}^{k} a_i \, \chi_{E_{j_i}}$. Then

$$|f - g| = \left| \sum_{i=1}^{k} (b_i - a_i) \chi_{B_i} \right| + \left| \sum_{i=1}^{k} a_i \left(\chi_{B_i} - \chi_{E_{j_i}} \right) \right|$$

$$\leqslant \sum_{i=1}^{k} |b_i - a_i| \chi_{B_i} + \sum_{i=1}^{k} |a_i| \, \chi_{E_{j_i} \triangle B_i}$$

$$\leqslant \delta \chi_B + \sum_{i=1}^{k} 2b \, \chi_{E_i \triangle B_i}.$$

Hence, by the previous estimate and convexity,

$$\varrho_\varphi\big(\lambda(f - g)\big) \leqslant \frac{1}{2} \varrho_\varphi\big(2\lambda\delta \, \chi_B\big) + \frac{1}{2k} \sum_{i=1}^{k} \varrho_\varphi\left(4k\lambda b \, \chi_{E_i \triangle B_i}\right)$$

$$= \frac{1}{2} \int_B \varphi(y, 2\lambda\delta)\, d\mu(y) + \frac{1}{2k} \sum_{i=1}^{k} \int_{E_i \triangle B_i} \varphi(y, 4k\lambda b)\, d\mu(y).$$

The right-hand side of the previous estimate is at most 1 and so $\|f - g\|_\varphi \leqslant \frac{1}{\lambda}$ by the unit ball property. Since $\lambda > 0$ was arbitrary, this completes the proof. $\qquad \square$

2.6 Conjugate Φ-Functions

In this section we specialize the results from Sect. 2.2 Conjugate modulars and dual semimodular spaces to Φ-functions and generalized Φ-functions. Apart from the general results, we are also able to prove stronger results in this special case.

By Lemma 2.3.2 we know that every Φ-function defines (by even extension) a semimodular on $\mathbb{R}$. This motivates to transfer the definition of a conjugate semimodular in a point-wise sense to generalized Φ-functions:

Definition 2.6.1. Let $\varphi \in \Phi(A, \mu)$. Then for any $y \in A$ we denote by $\varphi^*(y, \cdot)$ the *conjugate function* of $\varphi(y, \cdot)$ which is defined by

$$\varphi^*(y, u) = \sup_{t \geq 0} \big(tu - \varphi(y, t)\big)$$

for all $u \geq 0$ and $y \in \Omega$.

This definition applies in particular in the case when φ is a (non-generalized) Φ-function, in which case

$$\varphi^*(u) = \sup_{t \geq 0} \big(tu - \varphi(t)\big)$$

concurs with the *Legendre transformation of* φ. By definition of φ^*,

$$tu \leqslant \varphi(t) + \varphi^*(u) \tag{2.6.2}$$

for every $t, u \geqslant 0$. This inequality is called *Young's inequality*. If φ is a Φ-function and $\varrho(t) := \varphi(|t|)$ is its even extension to $\mathbb{R}$, then $\varrho^*(t) = \varphi^*(|t|)$ for all $t \in \mathbb{R}$.

As a special case of Theorem 2.2.6 we have

Corollary 2.6.3. *Let* $\varphi \in \Phi(A, \mu)$. *Then* $(\varphi^*)^* = \varphi$. *In particular,*

$$\varphi(y, t) = \sup_{u \geqslant 0} \big(tu - \varphi^*(y, u)\big)$$

for all $y \in \Omega$ *and all* $t \geqslant 0$.

Lemma 2.6.4. *Let* φ, ψ *be* Φ-*functions.*

(a) *The estimate* $\varphi(t) \leqslant \psi(t)$ *holds for all* $t \geqslant 0$ *if and only if* $\psi^*(u) \leqslant \varphi^*(u)$ *for all* $u \geqslant 0$.
(b) *Let* $a, b > 0$. *If* $\psi(t) = a\varphi(bt)$ *for all* $t \geqslant 0$, *then* $\psi^*(u) = a\varphi^*(\frac{u}{ab})$ *for all* $u \geqslant 0$.

Proof. We begin with the proof of (a). Let $\varphi(t) \leqslant \psi(t)$ for all $t \geqslant 0$. Then

$$\psi^*(u) = \sup_{t \geqslant 0} \big(tu - \psi(t)\big) \leqslant \sup_{t \geqslant 0} \big(tu - \varphi(t)\big) = \varphi^*(u)$$

for all $u \geqslant 0$. The reverse claim follows using $\varphi^{**} = \varphi$ and $\psi^{**} = \psi$ by Corollary 2.6.3. Let us now prove (b). Let $a, b > 0$ and $\psi(t) = a\varphi(bt)$ for all $t \geqslant 0$. Then

$$\psi^*(u) = \sup_{t \geqslant 0} \big(tu - \psi(t)\big) = \sup_{t \geqslant 0} \big(tu - a\varphi(bt)\big) = \sup_{t \geqslant 0} a\Big(t\frac{u}{ab} - \varphi(t)\Big)$$

$$= a\psi^*\Big(\frac{u}{ab}\Big)$$

for all $u \geqslant 0$. $\qquad\square$

The following result is the generalization of the classical Hölder inequality $\int |f||g|\,d\mu \leqslant \|f\|_q \|g\|_{q'}$ to the Musielak–Orlicz spaces. The extra constant 2 cannot be omitted.

Lemma 2.6.5 (Hölder's inequality). *Let $\varphi \in \Phi(A, \mu)$. Then*

$$\int\limits_A |f|\,|g|\,d\mu(y) \leqslant 2\|f\|_\varphi \|g\|_{\varphi^*}$$

for all $f \in L^\varphi(A, \mu)$ and $g \in L^{\varphi^*}(A, \mu)$.

Proof. Let $f \in L^\varphi$ and $g \in L^{\varphi^*}$. The claim is obvious for $f = 0$ or $g = 0$, so we can assume $f \neq 0$ and $g \neq 0$. Due to the unit ball property, $\varrho_\varphi(f/\|f\|_\varphi) \leqslant 1$ and $\varrho_{\varphi^*}(g/\|g\|_{\varphi^*}) \leqslant 1$. Thus, using Young's inequality (2.6.2), we obtain

$$\int\limits_A \frac{|f(y)|}{\|f\|_\varphi} \frac{|g(y)|}{\|g\|_{\varphi^*}} \, d\mu(y) \leqslant \int\limits_A \varphi\Big(y, \frac{|f(y)|}{\|f\|_\varphi}\Big) + \varphi^*\Big(y, \frac{|g(y)|}{\|g\|_{\varphi^*}}\Big) \, d\mu(y)$$

$$= \varrho_\varphi(f/\|f\|_\varphi) + \varrho_{\varphi^*}(g/\|g\|_{\varphi^*})$$

$$\leqslant 2.$$

Multiplying through by $\|f\|_\varphi \|g\|_{\varphi^*}$ yields the claim. $\qquad\square$

Let us recall the definitions of N-function and generalized N-function from Definition 2.4.4. A Φ-function φ is said to be an N-function if it is continuous and positive and satisfies $\lim_{t \to 0} \frac{\varphi(t)}{t} = 0$ and $\lim_{t \to \infty} \frac{\varphi(t)}{t} = \infty$. A function $\varphi \in \Phi(A, \mu)$ is said to be a generalized N-function if $\varphi(y, \cdot)$ is for every $y \in \Omega$ an N-function.

Note that by continuity N-functions only take values in $[0, \infty)$. Let $\varphi \in N(A, \mu)$ be an N-function. As was noted in (2.3.3), the function has a right-derivative, denoted by $\varphi'(y, \cdot)$, and

$$\varphi(y,t) = \int_0^t \varphi'(y,\tau)\, d\tau$$

for all $y \in A$ and all $t \geqslant 0$. The right-derivative $\varphi'(y, \cdot)$ is non-decreasing and right-continuous.

Lemma 2.6.6. *Let φ be an N-function. Then*

$$\frac{t}{2}\varphi'\left(\frac{t}{2}\right) \leqslant \varphi(t) \leqslant t\varphi'(t)$$

for all $t \geqslant 0$

Proof. Using the monotonicity of φ' we get

$$\varphi(t) = \int_0^t \varphi'(\tau)\, d\tau \leqslant \int_0^t \varphi'(t)\, d\tau = t\varphi'(t),$$

$$\varphi(t) = \int_0^t \varphi'(\tau)\, d\tau \geqslant \int_{t/2}^t \varphi'(t/2)\, d\tau = \frac{t}{2}\varphi'\left(\frac{t}{2}\right)$$

for all $t \geqslant 0$. □

Remark 2.6.7. If φ is a generalized N-function, which satisfies the Δ_2-condition (Definition 2.4.1), then Lemma 2.6.6 implies $\varphi(y,t) \approx \varphi'(y,t)t$ uniformly in $y \in A$ and $t \geqslant 0$.

Let $\varphi \in N(A, \mu)$. Then we already know that $\varphi'(y, \cdot)$ is for any $y \in A$ non-decreasing right-continuous, $\varphi'(y,0) = 0$, and $\lim_{t\to\infty} \varphi'(y,t) = \infty$. Define

$$b(y,u) := \inf\{t \geqslant 0 \colon \varphi'(y,t) > u\}.$$

Then $b(y, \cdot)$ has the same properties, i.e. $b(y, \cdot)$ is for any $y \in A$ non-decreasing, right-continuous, $b(y,0) = 0$, and $\lim_{t\to\infty} b(y,t) = \infty$. The function $b(y, \cdot)$ is the *right-continuous inverse* of $\varphi'(y, \cdot)$ and we therefore denote it by $(\varphi')^{-1}(y,u)$. It is easy to see that the right-continuous inverse of $(\varphi')^{-1}$ is again φ', i.e. $((\varphi')^{-1})^{-1} = \varphi'$. The function $(\varphi')^{-1}$ is important, since we can use it to represent the conjugate function φ^*.

Theorem 2.6.8. *If $\varphi \in N(A, \mu)$, then $\varphi^* \in N(A, \mu)$ and $(\varphi^*)' = (\varphi')^{-1}$. In particular,*

$$\varphi^*(y,t) = \int_0^t (\varphi')^{-1}(y,\tau)\, d\tau$$

for all $y \in A$ and $t \geqslant 0$.

Proof. It suffices to prove the claim point-wise, and thus we may assume without loss of generality that $\varphi(y, t)$ is independent of y, i.e. an N-function.

It is easy to see that φ' is non-decreasing, right-continuous and satisfies $(\varphi')^{-1}(0) = 0$, $(\varphi')^{-1}(t) > 0$ for $t > 0$, and $\lim_{t \to \infty}(\varphi')^{-1}(t) = \infty$. Thus,

$$\psi(t) := \int_0^t (\varphi')^{-1}(\tau)\, d\tau$$

for $t \geqslant 0$ defines an N-function. In particular, φ and ψ are finite.

Note that $\sigma < \varphi'(\tau)$ is equivalent to $(\varphi')^{-1}(\sigma) < \tau$. Hence, the sets

$$\{(\tau, \sigma) \in [0, \infty) \times [0, \infty) : \sigma < \varphi'(\tau)\}$$
$$\{(\tau, \sigma) \in [0, \infty) \times [0, \infty) : (\varphi')^{-1}(\sigma) \geqslant \tau\}$$

are complementary with respect to $[0, \infty) \times [0, \infty)$. Therefore, we can estimate with the help of the theorem of Fubini

$$0 \leqslant tu = \int_0^t \int_0^u d\sigma\, d\tau$$

$$= \iint_{\{0 \leqslant \tau \leqslant t, \sigma \leqslant u:\, 0 \leqslant \sigma < \varphi'(\tau)\}} d\sigma\, d\tau + \iint_{\{0 \leqslant \tau \leqslant t, 0 \leqslant \sigma \leqslant u:\, (\varphi')^{-1}(\sigma) \geqslant \tau\}} d\sigma\, d\tau$$

$$= \int_0^t \int_0^{\min\{u, \varphi'(\tau)\}} d\sigma\, d\tau + \int_0^u \int_0^{\min\{t, (\varphi')^{-1}(\sigma)\}} d\tau\, d\sigma$$

$$\leqslant \int_0^t \varphi'(\tau)\, d\tau + \int_0^u (\varphi')^{-1}(\sigma)\, d\sigma$$

$$= \varphi(t) + \psi(u).$$

If $u = \varphi'(t)$ or $t = (\varphi')^{-1}(u)$, then $\min\{u, \varphi'(\tau)\} = \varphi'(\tau)$ and $\min\{t, (\varphi')^{-1}(\sigma)\} = (\varphi')^{-1}(\sigma)$ in the integrals of the third line. So in this case we have equality in the penultimate step. Since $\varphi^*(u) = \sup_t(ut - \varphi(t))$ it follows that $\varphi^* = \psi$. $\qquad\square$

Remark 2.6.9. Let φ be an N-function. Then it follows from the previous proof that the right-derivative $(\varphi^*)'$ of φ^* satisfies $(\varphi^*)' = (\varphi')^{-1}$ for all $t \geqslant 0$. Young's inequality $tu \leqslant \varphi(t) + \varphi^*(u)$ holds with equality if $u = \varphi'(t)$ or $t = (\varphi')^{-1}(u)$.

Theorem 2.6.8 enables us to calculate the conjugate function of N-functions. Let us present three examples:

(a) Let $\varphi(t) = e^t - t - 1$. Then $\varphi'(t) = e^t - 1$ and $(\varphi^*)'(u) = (\varphi')^{-1}(u) = \log(1 + u)$. Integration over u implies $\varphi^*(u) = (1 + t)\log(1 + t) - t$.

(b) Let $\varphi(t) = \frac{1}{p}t^p$ for $1 < p < \infty$. Then $\varphi'(t) = t^{p-1}$ and $(\varphi^*)'(u) = (\varphi')^{-1}(u) = u^{\frac{1}{p-1}} = u^{p'-1}$ with $\frac{1}{p} + \frac{1}{p'} = 1$. Integration over u implies $\varphi^*(u) = \frac{1}{p'}u^{p'}$.

(c) Let $\varphi(t) = t^p$ for $1 < p < \infty$. Then $\varphi'(t) = pt^{p-1}$ and $(\varphi^*)'(u) = (\varphi')^{-1}(u) = (u/p)^{\frac{1}{p-1}} = p^{\frac{1}{1-p}}u^{p'-1}$ with $\frac{1}{p} + \frac{1}{p'} = 1$. Integration over u implies $\varphi^*(u) = p^{\frac{1}{1-p}}\frac{1}{p'}u^{p'} = p^{-p'}(p-1)u^{p'}$.

Remark 2.6.10. We have seen that the supremum in Remark 2.6.9 is attained for any N-function φ. However, this is not the case if φ is only a Φ-function. Indeed, if $\varphi(t) = t$, then $\varphi^*(u) = \infty \cdot \chi_{\{u>1\}}(u)$. However, $tu = \varphi_1(t) + (\varphi_1)^*(u)$ only holds if $u = 1$ and $t \geqslant 0$ or if $u \in [0, 1]$ and $t = 0$.

There are a lot of nice estimates for N-functions. Let us collect a few.

Lemma 2.6.11. *Let φ be an N-function. Then for all $t \geqslant 0$ and all $\varepsilon > 0$*

$$t \leqslant \varphi^{-1}(t)(\varphi^*)^{-1}(t) \leqslant 2t, \tag{2.6.12}$$

$$(\varphi^*)'\big(\varphi'(t) - \varepsilon\big) \leqslant t \leqslant (\varphi^*)'\big(\varphi'(t)\big), \tag{2.6.13}$$

$$\varphi'\big((\varphi^*)'(t) - \varepsilon\big) \leqslant t \leqslant \varphi'\big((\varphi^*)'(t)\big), \tag{2.6.14}$$

$$\varphi^*\big(\varphi'(t)\big) \leqslant t\varphi'(t), \tag{2.6.15}$$

$$\varphi^*\Big(\frac{\varphi(t)}{t}\Big) \leqslant \varphi(t), \tag{2.6.16}$$

where we assumed $t > 0$ in (2.6.16).

Proof. We first note that $(\varphi^*)' = (\varphi')^{-1}$ by Remark 2.6.9. Let $t \geqslant 0$ and $\varepsilon > 0$. The first part of (2.6.13) follows from

$$(\varphi^*)'\big(\varphi'(t) - \varepsilon\big) = \inf\{a \geqslant 0 \colon \varphi'(a) > \varphi'(t) - \varepsilon\} \leqslant t.$$

The second part of (2.6.13) follows from

$$\varphi'\big((\varphi^*)'(t)\big) = \varphi'\big(\inf\{a \geqslant 0 \colon \varphi'(a) > t\}\big)$$
$$= \inf\{\varphi'(a) \geqslant 0 \colon \varphi'(a) > t\} \geqslant t,$$

where we have used that φ' is right-continuous and non-decreasing.

Now, (2.6.14) is a consequence of (2.6.13) using $(\varphi^*)^* = \varphi$. By Young's inequality (2.6.2) we estimate

$$\varphi^{-1}(t)(\varphi^*)^{-1}(t) \leqslant t + t = 2t.$$

With Lemma 2.6.6 for φ and φ^* and (2.6.13) we deduce

$$\varphi^*(\varphi'(t)) \leqslant (\varphi^*)'(\varphi'(t))\varphi'(t) \leqslant t\varphi'(t),$$

$$\varphi^*\left(\frac{\varphi(t)}{t} - \varepsilon\right) \leqslant \left(\frac{\varphi(t)}{t} - \varepsilon\right)(\varphi^*)'\left(\frac{\varphi(t)}{t} - \varepsilon\right) \leqslant \frac{\varphi(t)}{t}(\varphi^*)'(\varphi'(t) - \varepsilon) \leqslant \varphi(t).$$

Letting $\varepsilon \to 0$ in the latter inequality yields (2.6.16). Setting $t = \varphi^{-1}(u)$ in (2.6.16) gives

$$\varphi^*\left(\frac{u}{\varphi^{-1}(u)}\right) \leqslant u.$$

From this it follows that $u \leqslant \varphi^{-1}(u)(\varphi^*)^{-1}(u)$. $\square$

Note that if φ and φ^* satisfy the Δ_2-condition (Definition 2.4.1), than all the "$\leqslant$"-signs in Lemma 2.6.11 can be replaced by "$\approx$"-signs.

2.7 Associate Spaces and Dual Spaces

In the case of classical Lebesgue spaces it is well known that there is a natural embedding of $L^{q'}$ into $(L^q)^*$ for $1 \leqslant q \leqslant \infty$ and $\frac{1}{q} + \frac{1}{q'} = 1$. In particular, for every $g \in L^{q'}$ the mapping $J_g \colon f \mapsto \int fg\,d\mu$ is an element of $(L^q)^*$. Even more, if $1 \leqslant q < \infty$, then the mapping $g \mapsto J_g$ is an isometry from $L^{q'}$ to $(L^q)^*$. Besides the nice characterization of the dual space, this has the consequence that

$$\|f\|_q = \sup_{\|g\|_{q'} \leqslant 1} \int fg\,d\mu$$

for every $1 \leqslant q \leqslant \infty$. This formula is often called the *norm conjugate formula*. The cases $q = 1$ and $q = \infty$ need special attention, since $(L^1)^* = L^\infty$ but $(L^\infty)^* \neq L^1$. However, the isometry $(L^1)^* = L^\infty$ suffices for the proof of the formula when $q = 1$ and $q = \infty$.

In the case of Musielak–Orlicz spaces we have a similar situation. We will see that L^{φ^*} can be naturally embedded into $(L^\varphi)^*$. Moreover, the mapping $g \mapsto J_g$ is an isomorphism under certain assumptions on φ, which exclude for example the case $L^\varphi = L^\infty$. The mapping is not an isometry but its operator norm lies in the interval $[1, 2]$.

The norm conjugate formula above requires more attention in the case of Musielak–Orlicz spaces. Certainly, we cannot expect equality but only equivalence up to a factor of 2. Since the space L^φ can partly behave like L^1 and partly like L^∞, there are cases, where $(L^\varphi)^* \neq L^{\varphi^*}$ and $L^\varphi \neq (L^{\varphi^*})^*$. This is in particular the case for our generalized Lebesgue spaces $L^{p(\cdot)}$ (see Chap. 3) when p take the values 1 and ∞ on some subsets. To derive an equivalent of the norm conjugate formula for L^φ, we need to study the associate space, which is a closed subspaces of $(L^\varphi)^*$ generated by measurable functions.

Definition 2.7.1. Let $\varphi \in \Phi(A, \mu)$. Then

$$\left(L^\varphi(A, \mu)\right)' := \{g \in L^0(A, \mu) \,:\, \|g\|_{(L^\varphi(A,\mu))'} < \infty\}$$

with norm

$$\|g\|_{(L^\varphi(A,\mu))'} := \sup_{f \in L^\varphi \,:\, \|f\|_\varphi \leqslant 1} \int_A |f|\,|g|\,d\mu,$$

will be called the *associate space of* $L^\varphi(A, \mu)$ or $(L^\varphi)'$ for short.

In the definition of the norm of the associate space $(L^\varphi)'$ it suffices to take the supremum over simple function from L^φ:

Lemma 2.7.2. *Let* $\varphi \in \Phi(A, \mu)$. *Then*

$$\|g\|_{(L^\varphi)'} = \sup_{f \in S \cap L^\varphi \,:\, \|f\|_\varphi \leqslant 1} \int_A |f|\,|g|\,d\mu$$

for all $g \in (L^\varphi(A, \mu))'$.

Proof. For $g \in (L^\varphi)'$ let $|||g|||$ in this proof denote the right-hand side of the expression in the lemma. It is obvious that $|||g|||_\varphi \leqslant \|g\|_{(L^\varphi)'}$. To prove the reverse let $f \in L^\varphi$ with $\|f\|_\varphi \leqslant 1$. We have to prove that $\int |f|\,|g|\,d\mu \leqslant |||g|||$. Let (f_k) be a sequence of simple functions such that $0 \leqslant f_k \nearrow |f|$ almost everywhere. In particular, $f_k \in S(A, \mu) \cap L^\varphi$ and $\|f_k\|_\varphi \leqslant \|f\|_\varphi \leqslant 1$, since L^φ is solid (Theorem 2.3.17 (b)). Since $0 \leqslant f_k|g| \nearrow |f||g|$, we can conclude by the theorem of monotone convergence and the definition of $|||g|||$ that

$$\int |f||g|\,d\mu = \lim_{k \to \infty} \int f_k|g|\,d\mu \leqslant |||g|||.$$

The claim follows by taking the supremum over all possible f.   □

As an immediate consequence of Hölder's inequality (Lemma 2.6.5) we have
$L^{\varphi^*}(A, \mu) \hookrightarrow (L^\varphi(A, \mu))'$ and

$$\|g\|_{(L^\varphi)'} \leqslant 2 \|g\|_{\varphi^*}$$

for every $g \in L^{\varphi^*}(A, \mu)$.

If $g \in (L^\varphi)'$ and $f \in L^\varphi$, then $fg \in L^1$ by definition of the associate space. In particular, the integral $\int fg\,d\mu$ is well defined and

$$\left| \int fg\,d\mu \right| \leqslant \|g\|_{(L^\varphi)'} \|f\|_\varphi.$$

Thus $f \mapsto \int fg\,d\mu$ defines an element of the dual space $(L^\varphi)^*$ with $\|g\|_{(L^\varphi)^*} \leqslant \|g\|_{(L^\varphi)'}$. Therefore, for every $g \in (L^\varphi)'$ we can define an element $J_g \in (L^\varphi)^*$ by

$$J_g : f \mapsto \int fg\,d\mu \tag{2.7.3}$$

and we have $\|J_g\|_{(L^\varphi)^*} \leqslant \|g\|_{(L^\varphi)'}$. Since L^φ is circular (Theorem 2.3.17 (a)), we even have

$$\|J_g\|_{(L^\varphi)^*} = \sup_{f \in L^\varphi : \|f\|_\varphi \leqslant 1} \left| \int fg\,d\mu \right|$$

$$= \sup_{f \in L^\varphi : \|f\|_\varphi \leqslant 1} \int |f|\,|g|\,d\mu = \|g\|_{(L^\varphi)'}$$

for every $g \in (L^\varphi)'$. Obviously, $g \mapsto J_g$ is linear. Hence, $g \mapsto J_g$ defines an isometric, *natural embedding* of $(L^\varphi)' \hookrightarrow (L^\varphi)^*$. So the associate space $(L^\varphi)'$ is isometrically isomorphic to a closed subset of the dual space $(L^\varphi)^*$ and therefore itself a Banach space. It is easy to see that $(L^\varphi)'$ is circular and solid. We have the following inclusions of Banach spaces

$$L^{\varphi^*} \hookrightarrow (L^\varphi)' \hookrightarrow (L^\varphi)^*.$$

Under rather few assumptions on φ, we will see that the first inclusion is surjective and therefore an isomorphism even if L^{φ^*} is not isomorphic to the dual space $(L^\varphi)^*$. Therefore, the notion of the associate space is more flexible than that of the dual space.

The mapping $g \mapsto J_g$ can also be used to define natural embeddings

$$L^\varphi = L^{\varphi^{**}} \hookrightarrow (L^{\varphi^*})' \hookrightarrow (L^{\varphi^*})^*.$$

if we replace above φ by φ^* and use $\varphi = \varphi^{**}$ (Corollary 2.6.3).

Since $L^{\varphi^*} \hookrightarrow (L^\varphi)' \hookrightarrow (L^\varphi)^*$ via the embedding $g \mapsto J_g$, we can evaluate the conjugate semimodular $(\varrho_\varphi)^*$ at J_g for every $g \in L^{\varphi^*}$. As a direct consequence of Young's inequality (2.6.2) we have

$$(\varrho_\varphi)^*(J_g) = \sup_{f \in L^\varphi} \left(J_g(f) - \varrho_\varphi(f) \right) \leqslant \varrho_{\varphi^*}(g).$$

Theorem 2.7.4. *Let* $\varphi \in \Phi(A, \mu)$ *be such that* $S(A, \mu) \subset L^\varphi(A, \mu)$. *Then* $L^{\varphi^*}(A, \mu) = (L^\varphi(A, \mu))'$, $\varrho_{\varphi^*}(g) = (\varrho_\varphi)^*(J_g)$ *and*

$$\|g\|_{\varphi^*} \leqslant \|g\|_{(L^\varphi)'} = \|J_g\|_{(L^\varphi)^*} \leqslant 2\,\|g\|_{\varphi^*}$$

for every $g \in L^{\varphi^*}(A, \mu)$, *where* $J_g : f \mapsto \int_A fg\,d\mu$. (*or complex-valued functions, the constant* 2 *should be replaced by* 4.)

Proof. For the sake of simplicity we assume $\mathbb{K} = \mathbb{R}$. In the case $\mathbb{K} = \mathbb{C}$ we can proceed analogously by splitting g into its real and imaginary part.

We already know that $L^{\varphi^*} \subset (L^\varphi)'$, $\|g\|_{(L^\varphi)'} = \|J_g\|_{(L^\varphi)^*} \leqslant 2\,\|g\|_{\varphi^*}$, and $(\varrho_\varphi)^*(J_g) \leqslant \varrho_{\varphi^*}(g)$ for every $g \in L^{\varphi^*}$. Fix $g \in (L^\varphi)'$. We claim that $g \in L^{\varphi^*}$ and $\varrho_{\varphi^*}(g) \leqslant (\varrho_\varphi)^*(J_g)$.

Since μ is σ-finite, we find measurable sets $A_k \subset A$ with $\mu(A_k) < \infty$ and $A_1 \subset A_2 \subset \ldots$ such that $A = \bigcup_{k=1}^\infty A_k$. Let $\{q_1, q_2, \ldots\}$ be a countable, dense subset of $[0, \infty)$ with $q_j \neq q_k$ for $j \neq k$ and $q_1 = 0$. For $k \in \mathbb{N}$ and $y \in A$ define

$$r_k(y) := \chi_{A_k}(y) \max_{j=1,\ldots,k} \big(q_j\,|g(y)| - \varphi(y, q_j)\big).$$

The special choice $q_1 = 0$ implies $r_k(y) \geqslant 0$ for all $y \geqslant 0$. Since $\{q_1, q_2, \ldots\}$ is dense in $[0, \infty)$ and $\varphi(y, \cdot)$ is left-continuous, $r_k(y) \nearrow \varphi^*(y, |g(y)|)$ for any $y \in A$ as $k \to \infty$. For every $k \in \mathbb{N}$ there exists a simple function f_k with $f_k(A) \subset \{q_1, \ldots, q_k\}$ and $f_k(y) = 0$ for all $y \in A \setminus A_k$ such that

$$r_k(y) = f_k(y)\,|g(y)| - \varphi(y, f_k(y))$$

for all $y \in A$. As a simple function, f_k belongs by assumption to $L^\varphi(A, \mu)$. Define $h_k(y) := f_k(y)\,\mathrm{sgn}(g(y))$ for $y \in A$, where $\mathrm{sgn}(a)$ denotes the sign of a. Then also h_k is a simple function (here we use $\mathbb{K} = \mathbb{R}$) and therefore

$$(\varrho_\varphi)^*(J_g) \geqslant J_g(h_k) - \varrho_\varphi(h_k) = \int_A g(y)h_k(y) - \varphi(y, |h_k(y)|)\,d\mu(y).$$

By the definition of h_k it follows that

$$(\varrho_\varphi)^*(J_g) \geqslant \int_A |g(y)|\,f_k(y) - \varphi(y, |f_k(y)|)\,d\mu(y) = \int_A r_k(y)\,d\mu(y).$$

Since $r_k \geqslant 0$ and $r_k(y) \nearrow \varphi^*(y, |g(y)|)$, we get by the theorem of monotone convergence that

$$(\varrho_\varphi)^*(J_g) \geqslant \limsup_{k \to \infty} \int_A r_k(y)\,d\mu(y) = \int_A \varphi^*(y, |g(y)|)\,d\mu(y) = \varrho_{\varphi^*}(g).$$

Together with $(\varrho_\varphi)^*(J_g) \leqslant \varrho_{\varphi^*}(g)$ we get $(\varrho_\varphi)^*(J_g) = \varrho_{\varphi^*}(g)$.

Since $g \mapsto J_g$ is linear, it follows that $(\varrho_\varphi)^*(\lambda J_g) = \varrho_{\varphi^*}(\lambda g)$ for every $\lambda > 0$ and therefore $\|g\|_{\varphi^*} = \|J_g\|_{(\varrho_\varphi)^*} \leqslant \|J_g\|_{(L^\varphi)^*} = \|g\|_{(L^\varphi)'}$ using in the second step Theorem 2.2.10. $\square$

Theorem 2.7.4 allows us to generalize the norm conjugate formula to L^φ.

Corollary 2.7.5 (Norm conjugate formula). *Let $\varphi \in \Phi(A, \mu)$. If $S(A, \mu) \subset L^{\varphi^*}(A, \mu)$, then*

$$\|f\|_\varphi \leqslant \sup_{g \in L^{\varphi^*} \, : \, \|g\|_{\varphi^*} \leqslant 1} \int |f| \, |g| \, d\mu \leqslant 2 \, \|f\|_\varphi$$

for every $f \in L^0(A, \mu)$. The supremum is unchanged if we replace the condition $g \in L^{\varphi^*}$ by $g \in S(A, \mu)$.

Proof. Applying Theorem 2.7.4 to φ^* and taking into account that $\varphi^{**} = \varphi$, we have

$$\|f\|_\varphi \leqslant \|f\|_{(L^{\varphi^*})'} \leqslant 2 \, \|f\|_\varphi$$

for $f \in L^\varphi$. That the supremum does not change for $g \in S(A, \mu)$ follows by Lemma 2.7.2. The claim also follows in the case $f \in L^0 \setminus L^{\varphi^*} = L^0 \setminus (L^\varphi)'$, since both sides of the formula are infinite. $\square$

Remark 2.7.6. Since μ is σ-finite it suffices in Theorem 2.7.4 and Corollary 2.7.5 to assume $S(A_k, \mu) \subset L^\varphi(A, \mu)$, where (A_k) is a sequence with $A_k \nearrow A$ and $\mu(A_k) < \infty$ for all k. This is important for example in weighted Lebesgue spaces $L^q_\omega(\mathbb{R}^n)$ with Muckenhoupt weights.

Definition 2.7.7. A normed space $(Y, \|\cdot\|_Y)$ with $Y \subset L^0(A, \mu)$ is called a *Banach function space*, if

(a) $(Y, \|\cdot\|_Y)$ is circular, solid and satisfies the Fatou property.
(b) If $\mu(E) < \infty$, then $\chi_E \in Y$.
(c) If $\mu(E) < \infty$, then $\chi_E \in Y'$, i.e. $\int_E |f| \, d\mu \leqslant c(E)\|f\|_Y$ for all $f \in Y$.

From Theorem 2.3.17 we know that L^φ satisfies (a) for every $\varphi \in \Phi(A, \mu)$ so one need only check (b) and (c). These properties are equivalent to $S \subset L^\varphi$ and $S \subset (L^\varphi)'$, where S is the set of simple functions. These inclusions may or may not hold, depending on the function φ.

Definition 2.7.8. A generalized Φ-function $\varphi \in \Phi(A, \mu)$ is called *proper* if the set of simple functions $S(A, \mu)$ satisfies $S(A, \mu) \subset L^\varphi(A, \mu) \cap (L^\varphi(A, \mu))'$.

So φ is proper if and only if L^φ is a Banach function space. Moreover, if φ is proper then the norm conjugate formula for L^φ and L^{φ^*} holds (Corollary 2.7.5) and $L^{\varphi^*} = (L^\varphi)'$.

Corollary 2.7.9. *Let $\varphi \in \Phi(A, \mu)$. Then the following are equivalent:*

(a) φ *is proper.*
(b) φ^* *is proper.*
(c) $S(A, \mu) \subset L^\varphi(A, \mu) \cap L^{\varphi^*}(A, \mu)$.

Proof. If (a) or (c) holds, then $S \subset L^\varphi$. Hence $(L^\varphi)' = L^{\varphi^*}$ by Theorem 2.7.4, which obviously implies the equivalence of (a) and (c).

Applying this equivalence for the function φ^*, and taking into account that $\varphi^{**} = \varphi$, yields the equivalence of (b) and (c). $\qquad\square$

Remark 2.7.10. The conditions $\chi_E \in L^\varphi$ and $\chi_E \in (L^\varphi)'$ for $\mu(E) < \infty$ in Definition 2.7.7 can be interpreted in terms of embeddings. Indeed, $\chi_E \in L^\varphi$ implies $L^{\varphi^*} \hookrightarrow L^1(E)$. The condition $\chi_E \in (L^\varphi)'$ is equivalent to $L^\varphi(E) \hookrightarrow L^1(E)$. In particular, if φ is proper, then $L^\varphi(\Omega) \hookrightarrow L^1_{\mathrm{loc}}(\Omega)$ and $L^{\varphi^*}(\Omega) \hookrightarrow L^1_{\mathrm{loc}}(\Omega)$.

Remark 2.7.11. Let $\varphi \in \Phi$ be proper; so L^φ is a Banach function space. It has been shown in [43, Proposition 3.6] that $f \in L^\varphi$ has absolutely continuous norm (see Remark 2.5.8) if and only if f has the following property: If g_k, $g \in L^0$ with $|g_k| \leqslant |f|$ and $g_k \to g$ almost everywhere, then $g_k \to g$ in L^φ. Thus, f acts as a majorant in the theorem of dominated convergence.

It has been shown by Lorentz and Luxemburg that the second associate space X'' of a Banach function space coincides with X with equality of norms, see [43, Theorem 2.7]. In particular, $(L^\varphi)'' = L^\varphi$ with equality of norms if φ is proper. For the sake of completeness we include a proof of this result in our setting.

Theorem 2.7.12. *Let $\varphi \in \Phi(A, \mu)$ be proper. Then $L^{\varphi^*}(A, \mu) = (L^\varphi(A, \mu))'$ and $(L^{\varphi^*}(A, \mu))' = L^\varphi(A, \mu)$. Moreover, $(L^\varphi(A, \mu))'' = L^\varphi(A, \mu)$ with equality of norms, i.e. $\|f\|_\varphi = \|f\|_{(L^\varphi)''}$ for all $f \in L^\varphi(A, \mu)$.*

Proof. The equalities $L^{\varphi^*} = (L^\varphi)'$ and $(L^{\varphi^*})' = L^\varphi$ follow by Theorem 2.7.4 and as a consequence $(L^\varphi)'' = (L^{\varphi^*})' = L^{\varphi^{**}} = L^\varphi$ using $\varphi^{**} = \varphi$. It only remains to prove the equality of norms. Let $f \in L^\varphi$, then

$$\|f\|_{(L^\varphi)''} = \sup_{g \in (L^\varphi)' \, : \, \|g\|_{(L^\varphi)'} \leqslant 1} \int |f| \, |g| \, d\mu \leqslant \|f\|_\varphi.$$

We now prove $\|f\|_\varphi \leqslant \|f\|_{(L^\varphi)''}$. We begin with the case $\mu(A) < \infty$. If $f = 0$, there is nothing to show, so assume $f \neq 0$. Let B denote the unit ball of L^φ. Due to Remark 2.7.10 and $\mu(A) < \infty$, we have $L^\varphi(A) \hookrightarrow L^1(A)$, so $B \subset L^1(A)$. Moreover, B is a closed, convex subset of $L^1(A)$. Indeed, if $u_k \in B$ with $u_k \to u$ in $L^1(A)$, then $u_k \to u$ μ-almost everywhere for a subsequence, so Fatou's lemma for the norm (Theorem 2.3.17) implies $u \in B$.

Let $h := \lambda f/\|f\|_\varphi$ with $\lambda > 1$, then $h \notin B$, so by the Hahn–Banach Theorem 1.4.2 there exists a functional on $(L^1(A))^*$ separating B and f. In other words, there exists a function $g \in L^\infty(A)$ and $\gamma \in \mathbb{R}$ such that

$$\mathrm{Re}\left(\int vg\,d\mu\right) \leqslant \gamma < \mathrm{Re}\left(\int hg\,d\mu\right)$$

for all $v \in B$, where we have used the representation of $(L^1(A))^*$ by $L^\infty(A)$. From $g \in L^\infty(A)$ and $\chi_A \in (L^\varphi)'$ it follows by solidity of $(L^\varphi)'$ that $g \in (L^\varphi)'$. Moreover, the circularity of L^φ implies that

$$\int |v||g|\,d\mu \leqslant \gamma < \int |h||g|\,d\mu = \frac{\lambda}{\|f\|_\varphi}\int |f||g|\,d\mu \leqslant \frac{\lambda\|f\|_{(L^\varphi)''}\|g\|_{(L^\varphi)'}}{\|f\|_\varphi}$$

for all $v \in B$. In other words,

$$\|g\|_{(L^\varphi)'} \leqslant \frac{\lambda\|f\|_{(L^\varphi)''}\|g\|_{(L^\varphi)'}}{\|f\|_\varphi}.$$

Using $\|g\|_{(L^\varphi)'} < \infty$ we get $\|f\|_\varphi \leqslant \lambda\|f\|_{(L^\varphi)''}$. This proves $\|f\|_\varphi \leqslant \|f\|_{(L^\varphi)''}$ and therefore $\|f\|_\varphi = \|f\|_{(L^\varphi)''}$.

It remains to consider the case $\mu(A) = \infty$. Choose $A_k \subset A$ with $\mu(A_k) < \infty$, $A_1 \subset A_2 \subset \ldots$, and $A = \bigcup_{k=1}^\infty A_k$. Then $\|f\chi_{A_k}\|_\varphi = \|f\|_{L^\varphi(A_k)} = \|f\|_{(L^\varphi(A_k))''}$ $= \|f\chi_{A_k}\|_{(L^\varphi(A))''}$ by the first part. Now, with the Fatou property of L^φ and $(L^\varphi)''$ we conclude $\|f\|_\varphi = \|f\|_{(L^\varphi)''}$. $\square$

Remark 2.7.13. Let $\varphi \in \Phi(A,\mu)$ be proper. Then we can use Theorem 2.7.12 Hölder's inequality to derive the formula

$$\frac{1}{2}\|f\|_\varphi \leqslant \sup_{h \in L^{\varphi^*}\,:\,\|h\|_{\varphi^*} \leqslant 1} \int |f|\,|h|\,d\mu \leqslant 2\,\|f\|_\varphi.$$

for all $f \in L^0(A,\mu)$. This is a weaker version of the norm conjugate formula in Corollary 2.7.5, with an extra factor $\frac{1}{2}$ on the left-hand side.

We are now able to characterize the dual space of L^φ.

Theorem 2.7.14. *Let $\varphi \in \Phi(A,\mu)$ be proper and locally integrable, and suppose that $E^\varphi = L^\varphi$. Then $V : g \mapsto J_g$ is an isomorphism from $L^{\varphi^*}(A,\mu)$ to $(L^\varphi(A,\mu))^*$.*

Proof. By Theorem 2.7.4 V is an isomorphism from L^{φ^*} onto its image $\mathrm{Im}(V) \subset (L^\varphi)^*$. In particular, $\mathrm{Im}(V)$ is a closed subspace of $(L^\varphi)^*$. Since φ is locally integrable $\overline{S} = E^\varphi$ by Theorem 2.5.9, so that $\overline{S} = E^\varphi = L^\varphi$.

We have to show that V is surjective. We begin with the case $\mu(A) < \infty$. Let $J \in (L^\varphi)^*$. For any measurable set $E \subset A$ we define $\tau(E) := J(\chi_E)$, which is well defined since $S \subset L^\varphi$. We claim that τ is a signed, finite measure on A. Obviously, τ is a set function with $\tau(E_1 \cup E_2) = \tau(E_1) + \tau(E_2)$ for E_1, E_2 disjoint measurable sets. Let (E_j) be sequence of pairwise disjoint, measurable sets. Let $E := \bigcup_{j=1}^\infty E_j$. Then $\sum_{j=1}^k \chi_{E_j} \to \chi_E$ almost everywhere and by dominated convergence (Lemma 2.3.16) using $\chi_E \in L^\varphi = E^\varphi$ we find that $\sum_{j=1}^k \chi_{E_j} \to \chi_E$ in L^φ. This and the continuity of J imply

$$\sum_{j=1}^\infty \tau(E_j) = \sum_{j=1}^\infty J(\chi_{E_j}) = J(\chi_E) = \tau(E),$$

which proves that τ is σ-additive. The estimate

$$|\tau(E)| = |J(\chi_E)| \leqslant \|J\|_{(L^\varphi)^*} \|\chi_E\|_\varphi \leqslant \|J\|_{(L^\varphi)^*} \|\chi_A\|_\varphi$$

for all measurable E, proves that τ is a signed, finite measure. If $\mu(E) = 0$, then $\tau(E) = J(\chi_E) = 0$, so τ is absolutely continuous with respect to μ. Thus by the Radon–Nikodym Theorem 1.4.13 there exists a function $g \in L^1(A)$ such that

$$J(f) = \int_A f g \, d\mu \tag{2.7.15}$$

for all $f = \chi_E$ with E measurable and therefore by linearity for all $f \in S$. We claim that $\|g\|_{(L^\varphi)'} \leqslant \|J\|_{(L^\varphi)^*}$. Due to Lemma 2.7.2 it suffices to show that $\int |f| |g| \, d\mu \leqslant \|J\|_{(L^\varphi)^*}$ for every $f \in S = S \cap L^\varphi$ with $\|f\|_\varphi \leqslant 1$. Fix such an f. If $\mathbb{K} = \mathbb{R}$, then $\operatorname{sgn} g$ is a simple function. However, to include the case $\mathbb{K} = \mathbb{C}$, we need to approximate $\operatorname{sgn} g$ by simple function as follows. Since $\operatorname{sgn} g \in L^\infty$, we find a sequence (h_k) of simple functions with $h_k \to \operatorname{sgn} g$ almost everywhere and $|h_k| \leqslant 1$. Since $|f| h_k \in S$ and $\| |f| h_k\|_\varphi \leqslant \|f\|_\varphi \leqslant 1$, we estimate $\int |f| h_k g \, dx = J(|f| h_k) \leqslant \|J\|_{(L^\varphi)^*}$ using (2.7.15). We have $|f| h_k g \to |f| |g|$ almost everywhere and $|f h_k g| \leqslant |f| |g| \in L^1$, since $g \in L^1$ and $f \in L^\infty$ as a simple function. Therefore, by the theorem of dominated convergence we conclude $\int |f| |g| \, dx = \lim_{k \to \infty} \int |f| h_k g \, dx \leqslant \|J\|_{(L^\varphi)^*}$. This yields $\|g\|_{(L^\varphi)'} \leqslant \|J\|_{(L^\varphi)^*}$. Then $g \in L^{\varphi^*}$ follows from $(L^\varphi)' = L^{\varphi^*}$ by Theorem 2.7.4. By (2.7.3) and (2.7.15) the functionals J_g and J agree on the set S. So the continuity of J and J_g and $\overline{S} = L^\varphi$ imply $J = J_g$ proving the surjectivity of $g \mapsto J_g$ in the case $\mu(A) < \infty$.

It remains to prove the surjectivity for μ σ-finite. Choose $A_k \subset A$ with $\mu(A_k) < \infty$, $A_1 \subset A_2 \subset \ldots$, and $A = \bigcup_{k=1}^\infty A_k$. By restriction we see that $J \in (L^\varphi(A_k))^*$ for each $J \in (L^\varphi(A))^*$. Since $\mu(A_k) < \infty$, there exists $g_k \in L^{\varphi^*}(A_k)$ such that $J(f) = J_{g_k}(f)$ for any $f \in L^\varphi(A_k)$ and $\|g_k\|_{\varphi^*} \leqslant \|J\|_{(L^\varphi)^*}$. The injectivity of $g \mapsto J_g$ implies $g_j = g_k$ on A_j for all $k \geqslant j$. So $g := g_k$ on A_k is well defined and $J(f) = J_g(f)$ for all $f \in L^\varphi(A_k)$ and every

k. Since $|g_k| \nearrow |g|$ almost everywhere and $\sup_k \|g_k\|_{\varphi^*} \leqslant \|J\|_{(L^\varphi)^*}$, it follows by the Fatou property of L^{φ^*} that $\|g\|_{\varphi^*} \leqslant \|J\|_{(L^\varphi)^*}$.

It remains to prove $J = J_g$. Let $f \in L^\varphi$. Then by Fatou's lemma (Lemma 2.3.16), $f \chi_{A_k} \to f$ in L^φ. Hence, the continuity of J and J_g and $J(f \chi_{A_k}) = J_g(f \chi_{A_k})$ yields $J(f) = J_g(f)$ as desired.   □

Remark 2.7.16. (a) If φ is proper and locally integrable, then the condition $L^\varphi = E^\varphi$ is equivalent to the density of the set S of simple functions in L^φ, see Theorem 2.5.9.

(b) If μ is atom-free, then the assumptions "locally integrable" and "$E^\varphi = L^\varphi$" are also necessary for $V : g \mapsto J_g$ from $L^{\varphi^*} = (L^\varphi)'$ to $(L^\varphi)^*$ to be an isomorphism. Indeed, if V is an isomorphism, then it has been shown in [43, Theorem 4.1] that every function $f \in L^\varphi$ has absolutely continuous norm (see Remark 2.5.8). In particular, every χ_E with $\mu(E) < \infty$ has absolutely continuous norm. We prove that φ is locally integrable by contradiction, so assume that there exists a measurable set E and $\lambda > 0$ such that $\mu(E) < \infty$ and $\varrho_\varphi(\lambda \chi_E) = \infty$. Since μ is atom-free there exists a sequence (E_k) of pairwise disjoint, measurable sets such that $E_k \searrow \emptyset$ and $\varrho_\varphi(\lambda \chi_{E_k}) = \infty$. In particular, $\|\chi_{E_k}\|_\varphi \geqslant \frac{1}{\lambda}$. However, since χ_E has absolutely continuous norm, we should have $\|\chi_{E_k}\|_\varphi = \|\chi_E \chi_{E_k}\|_\varphi \to 0$, which gives the desired contradiction. Thus, φ is locally integrable. If follows from Theorem 2.5.9 that $E^\varphi = \overline{S}$, where S are the simple functions. Moreover, since V is an isomorphism, by the norm conjugate formula in Lemma 2.7.2 it follows that $S^\circ = \{0\}$, where S° is the annihilator of S. This implies $E^\varphi = \overline{S} = S^{\circ\circ} = L^{\varphi^*}$.

The reflexivity of L^φ can be reduced to the characterization of $(L^\varphi)^*$ and $(L^{\varphi^*})^*$.

Lemma 2.7.17. *Let* $\varphi \in \Phi(A, \mu)$ *be proper. Then* L^φ *is reflexive, if and only if the natural embeddings* $V \colon L^{\varphi^*} \to (L^\varphi)^*$ *and* $U \colon L^\varphi \to (L^{\varphi^*})^*$ *are isomorphisms.*

Proof. Let ι denote the natural injection of L^φ into its bidual $(L^\varphi)^{**}$. It is easy to see that $V^* \circ \iota = U$. Indeed,

$$\langle V^* \iota f, g \rangle = \langle \iota f, V g \rangle = \langle V g, f \rangle = \int f(x) g(x) \, d\mu = \langle U f, g \rangle$$

for $f \in L^\varphi$ and $g \in L^{\varphi^*}$. If V and U are isomorphisms, then $\iota = (V^*)^{-1} \circ U$ must be an isomorphism and L^φ is reflexive.

Assume now that L^φ is reflexive. We have to show that U and V are isomorphisms. We already know from Theorem 2.7.4 (since φ is proper) that U and V are isomorphisms from L^φ and L^{φ^*} to their images $\mathrm{Im}(U)$ and $\mathrm{Im}(V)$, respectively. In particular, V is a closed operator and as a consequence $\mathrm{Im}(V^*) = (\ker(V))^\circ$. The injectivity of V implies that V^* is surjective. So

$U = V^* \circ \iota$ is surjective as well. This proves that U is an isomorphism. The formula $U = V^* \circ \iota$ implies that V^* is also an isomorphism. Since V is a closed operator, we have $\mathrm{Im}(V) = (\ker(V^*))^\circ$. The injectivity of V^* proves that V is surjective and therefore an isomorphism. $\qquad\square$

By Theorem 2.7.14 and Lemma 2.7.17 we immediately get the reflexivity of L^φ.

Corollary 2.7.18. *Let* $\varphi \in \Phi(A, \mu)$ *be proper. If* φ *and* φ^* *are locally integrable,* $E^\varphi = L^\varphi$ *and* $E^{\varphi^*} = L^{\varphi^*}$, *then* L^φ *is reflexive.*

2.8   Embeddings and Operators

In this section we characterize bounded, linear operators from one Musielak–Orlicz space to another. Recall that the operator S is said to be bounded from L^φ to L^ψ if $\|Sf\|_\varphi \leqslant C\|f\|_\psi$. We want to characterize this in terms of the modular. The study of embeddings is especially important to us, i.e. we want to know when the identity is a bounded operator. Such embeddings, which are denoted by $L^\varphi \hookrightarrow L^\psi$, can be characterized by comparing φ pointwise with ψ.

Let us begin with a characterization of bounded, sub-linear operators. Let $\varphi, \psi \in \Phi(A, \mu)$ and let $S \colon L^\varphi(A, \mu) \to L^\psi(A, \mu)$ be sub-linear. By the norm-modular unit ball property, S is bounded if and only if there exist $c > 0$ such that

$$\varrho_\varphi(f) \leqslant 1 \implies \varrho_\psi(Sf/c) \leqslant 1.$$

If φ and ψ satisfy the Δ_2-condition, then this is equivalent to the existence of $c_1, c_2 > 0$ such that

$$\varrho_\varphi(f) \leqslant c_1 \implies \varrho_\psi(Sf) \leqslant c_2$$

(since the Δ_2-condition allows us to move constants out of the modular).

Theorem 2.8.1. *Let* $\varphi, \psi \in \Phi(A, \mu)$ *and let the measure* μ *be atom-less. Then* $L^\varphi(A, \mu) \hookrightarrow L^\psi(A, \mu)$ *if and only if there exists* $c' > 0$ *and* $h \in L^1(A, \mu)$ *with* $\|h\|_1 \leqslant 1$ *such that*

$$\psi\left(y, \frac{t}{c'}\right) \leqslant \varphi(y, t) + h(y)$$

for almost all $y \in A$ *and all* $t \geqslant 0$.

Moreover, c' *is bounded by the embedding constant, whereas the embedding constant is bounded by* $2c'$.

Proof. Let us start by showing that the inequality implies the embedding. Let $\|f\|_\varphi \leqslant 1$, which yields by the unit ball property that $\varrho_\varphi(f) \leqslant 1$. Then

$$\varrho_\psi\left(\frac{f}{2c'}\right) \leqslant \frac{1}{2}\varrho_\psi\left(\frac{f}{c'}\right) \leqslant \frac{1}{2}\varrho_\varphi(f) + \frac{1}{2}\int\limits_A h(y)\,dy \leqslant 1.$$

This and the unit ball property yield $\|f/(2c')\|_\psi \leqslant 1$. Then the embedding follows by the scaling argument.

Assume next that the embedding holds with embedding constant c_1. For $y \in A$ and $t \geqslant 0$ define

$$\alpha(y,t) := \begin{cases} \psi(y,\frac{t}{c_1}) - \varphi(y,t) & \text{if } \varphi(y,t) < \infty, \\ 0 & \text{if } \varphi(y,t) = \infty. \end{cases}$$

Since $\varphi(y,\cdot)$ and $\psi(y,\cdot)$ are left-continuous for all $y \in A$, also $\alpha(y,\cdot)$ is left-continuous for all $y \in A$. Let (r_k) be a sequence of distinct numbers with $\{r_k : k \in \mathbb{N}\} = \mathbb{Q} \cap [0,\infty)$ and $r_1 = 0$. Then

$$\psi(y,\tfrac{r_k}{c_1}) \leqslant \varphi(y,r_k) + \alpha(y,r_k)$$

for all $k \in \mathbb{N}$ and $y \in A$. Define

$$b_k(y) := \max_{1 \leqslant j \leqslant k} \alpha(y,r_j).$$

Since $r_1 = 0$ and $\alpha(y,0) = 0$, we have $b_k \geqslant 0$. Moreover, the functions b_k are measurable and nondecreasing in k. The function $b := \sup_k b_k$ is measurable, non-negative, and satisfies

$$b(y) = \sup_{t \geqslant 0} \alpha(y,t),$$
$$\psi(y,\tfrac{t}{c_1}) \leqslant \varphi(y,t) + b(y)$$

for all $y \in A$ and all $t \geqslant 0$, where we have used that $\alpha(y,\cdot)$ is left-continuous and the density of $\{r_k : k \in \mathbb{N}\}$ in $[0,\infty)$.

We now show that $b \in L^1(A,\mu)$ with $\|b\|_1 \leqslant 1$. We consider first the case $|b| < \infty$ a.e., and assume to the contrary that there exists $\varepsilon > 0$ such that

$$\int\limits_A b(y)\,d\mu(y) \geqslant 1 + 2\varepsilon.$$

Define

$$V_k := \{y \in A : \alpha(y,r_k) > \tfrac{1}{1+\varepsilon}b(y)\},$$
$$W_{k+1} := V_{k+1} \setminus (V_1 \cup \cdots \cup V_k)$$

for all $k \in \mathbb{N}$. Note that $V_1 = \emptyset$ due to the special choice $r_1 = 0$. Since $\{r_k : k \in \mathbb{N}\}$ is dense in $[0, \infty)$ and $\alpha(y, \cdot)$ is left-continuous for every $y \in A$, we have $\bigcup_{k=1}^{\infty} V_k = \bigcup_{k=2}^{\infty} W_k = \{y \in A : b(y) > 0\}$.

Let $f := \sum_{k=2}^{\infty} r_k \chi_{W_k}$. For every $y \in W_k$ we have $\alpha(y, r_k) > 0$ and therefore $\varphi(y, r_k) < \infty$. If y is outside of $\bigcup_{k=2}^{\infty} W_k$, then $\varphi(y, |f(y)|) = 0$. This implies that $\varphi(y, |f(y)|)$ is everywhere finite. Moreover, by the definition of W_k and α we get

$$\psi\left(y, \frac{|f(y)|}{c_1}\right) \geqslant \varphi(y, |f(y)|) + \frac{1}{1+\varepsilon} b(y) \tag{2.8.2}$$

for all $y \in A$.

If $\varrho_\varphi(f) \leqslant 1$, then $\varrho_\psi(\frac{f}{c_1}) \leqslant 1$ by the unit ball property since c_1 is the embedding constant. However, this contradicts

$$\varrho_\psi(\tfrac{f}{c_1}) \geqslant \varrho_\varphi(f) + \frac{1}{1+\varepsilon} \int_A b(y)\, d\mu(y) \geqslant \frac{1+2\varepsilon}{1+\varepsilon} > 1,$$

where we have used (2.8.2) and $\bigcup_{k=2}^{\infty} W_k = \{y \in A : b(y) > 0\}$. So we can assume that $\varrho_\varphi(f) > 1$. Since μ is atom-less and $\varphi(y, |f(y)|)$ is almost everywhere finite, there exists $U \subset A$ with $\varrho_\varphi(f\chi_U) = 1$. Thus

$$\varrho_\psi(\tfrac{f}{c_1} \chi_U) \geqslant \varrho_\varphi(f\,\chi_U) + \tfrac{1}{1+\varepsilon} \int_U b(y)\, d\mu(y)$$
$$= 1 + \tfrac{1}{1+\varepsilon} \int_U b(y)\, d\mu(y). \tag{2.8.3}$$

Now, $\varrho_\varphi(f\chi_U) = 1$ implies that $\mu(U \cap \{f \neq 0\}) > 0$. Since $\{f \neq 0\} = \bigcup_{k=2}^{\infty} W_k = \{y \in A : b(y) > 0\}$ we get $\mu(U \cap \{y \in A : b(y) > 0\}) > 0$ and

$$\int_U b(y)\, d\mu(y) > 0.$$

This and (2.8.3) imply that

$$\varrho_\psi(f/c_1 \chi_U) > 1.$$

which contradicts $\varrho_\psi(f/c_1) \leqslant 1$. Thus the case where $|b| < \infty$ a.e. is complete.

If we assume that there exists $E \subset A$ with $b|_E = \infty$ and $\mu(E) > 0$, then a similar argument with $V_k := \{y \in E : \alpha(y, r_k) \geqslant \frac{2}{\mu(E)}\}$ yields a contradiction. Hence this case cannot occur, and the proof is complete by what was shown previously. $\square$

Chapter 3
Variable Exponent Lebesgue Spaces

In this chapter we define Lebesgue spaces with variable exponents, $L^{p(\cdot)}$. They differ from classical L^p spaces in that the exponent p is not constant but a function from Ω to $[1, \infty]$. The spaces $L^{p(\cdot)}$ fit into the framework of Musielak–Orlicz spaces and are therefore also semimodular spaces.

We first define the appropriate Φ-function for variable exponent spaces in Sect. 3.1 and study its properties. Then we are in a position to apply the results of general Musielak–Orlicz spaces to our case in Sect. 3.2. Section 3.3 deals with embeddings between spaces with different exponents. In Sect. 3.4 we have collected properties which are more restrictive in the sense that they hold only for exponents bounded away from 1 and/or ∞. The final two sections are more technical. First we develop tools for dealing with unbounded exponents in Sect. 3.5 and then we investigate failure of convolution in Sect. 3.6. The latter is a major topic also of Chap. 4.

3.1 The Lebesgue Space Φ-Function

For the definition of the variable exponent Lebesgue spaces it is necessary to introduce the kind of variable exponents that we are interested in.

Let us also mention that many results on the basic properties on $L^{p(\cdot)}$ from this chapter were proved first by Kováčik and Rákosník in [258]. These results were later reproved by Fan and Zhao in [149].

Definition 3.1.1. Let (A, Σ, μ) be a σ-finite, complete measure space. We define $\mathcal{P}(A, \mu)$ to be the set of all μ-measurable functions $p \colon A \to [1, \infty]$. Functions $p \in \mathcal{P}(A, \mu)$ are called *variable exponents on* A. We define $p^- := p_A^- := \operatorname{ess\,inf}_{y \in A} p(y)$ and $p^+ := p_A^+ := \operatorname{ess\,sup}_{y \in A} p(y)$. If $p^+ < \infty$, then we call p a *bounded variable exponent*.

If $p \in \mathcal{P}(A, \mu)$, then we define $p' \in \mathcal{P}(A, \mu)$ by $\frac{1}{p(y)} + \frac{1}{p'(y)} = 1$, where $\frac{1}{\infty} := 0$. The function p' is called the *dual variable exponent* of p.

In the special case that μ is the n-dimensional Lebesgue measure and Ω is an open subset of $\mathbb{R}^n$, we abbreviate $\mathcal{P}(\Omega) := \mathcal{P}(\Omega, \mu)$.

L. Diening et al., *Lebesgue and Sobolev Spaces with Variable Exponents*, Lecture Notes in Mathematics 2017, DOI 10.1007/978-3-642-18363-8_3, © Springer-Verlag Berlin Heidelberg 2011

For the definition of the space $L^{p(\cdot)}$ we need the corresponding generalized Φ-function. Interestingly, there are two natural choices. However, we will see that both generate the same space up to isomorphism.

Definition 3.1.2. For $t \geqslant 0$ and $1 \leqslant p < \infty$ we define

$$\widetilde{\varphi}_p(t) := \frac{1}{p}t^p,$$

$$\bar{\varphi}_p(t) := t^p.$$

Moreover we set

$$\bar{\varphi}_\infty(t) := \widetilde{\varphi}_\infty(t) := \infty \cdot \chi_{(1,\infty)}(t) = \begin{cases} 0 & \text{if } t \in [0,1], \\ \infty & \text{if } t \in (1,\infty). \end{cases}$$

For variable exponent $p \in \mathcal{P}(A,\mu)$ we define for $y \in A$ and $t \geqslant 0$

$$\widetilde{\varphi}_{p(\cdot)}(y,t) := \widetilde{\varphi}_{p(y)}(t) \qquad \text{and} \qquad \bar{\varphi}_{p(\cdot)}(y,t) := \bar{\varphi}_{p(y)}(t).$$

It is easy to see that both $\widetilde{\varphi}_q$ and $\bar{\varphi}_q$ are Φ-functions if $q \in [1,\infty]$. So $\widetilde{\varphi}_{p(\cdot)}$ and $\bar{\varphi}_{p(\cdot)}$ are generalized Φ-functions if $p \in \mathcal{P}(A,\mu)$. Even more, if $q \in (1,\infty)$ and $p \in \mathcal{P}(A,\mu)$ with $1 < p^- \leqslant p^+ < \infty$, then $\widetilde{\varphi}_q$ and $\bar{\varphi}_q$ are N-functions and $\widetilde{\varphi}_{p(\cdot)}$ and $\bar{\varphi}_{p(\cdot)}$ are generalized N-functions. If $q \in [1,\infty)$, then $\widetilde{\varphi}_q$ and $\bar{\varphi}_q$ are continuous and positive. The function $\widetilde{\varphi}_\infty = \bar{\varphi}_\infty$ is only left-continuous and it is not positive.

Both $\widetilde{\varphi}_p$ and $\bar{\varphi}_p$ have their advantages. The advantage of $\bar{\varphi}_p$ is that the corresponding Musielak–Orlicz space $L^{\bar{\varphi}_p}$ agrees for constant $p \in [1,\infty]$ exactly with the classical L^p spaces, see Example 2.1.8. In particular, for $f \in L^p(\Omega)$ we have $\|f\|_p = \|f\|_{\bar{\varphi}_p}$. Additionally, the generalized Φ-function $\bar{\varphi}_{p(\cdot)}$ has been used in the vast majority of papers on variable exponent function spaces.

The advantages of $\widetilde{\varphi}_p$ are its nice properties regarding conjugation, continuity, and convexity with respect to the exponent p. First, for all $t \geqslant 0$ the mapping $p \mapsto \widetilde{\varphi}_p(t)$ is continuous with respect to $p \in [1,\infty]$. In particular,

$$\widetilde{\varphi}_\infty(t) = \lim_{p \to \infty} \widetilde{\varphi}_p(t)$$

for all $t \geqslant 0$. This suggests that the expression $\frac{1}{p}t^p$ has for $p = \infty$ a natural interpretation, namely $\widetilde{\varphi}_\infty(t) = \infty \cdot \chi_{(1,\infty)}(t)$. Therefore, we sometimes will just write $\widetilde{\varphi}_p(t) = \frac{1}{p}t^p$ including the case $p = \infty$.

Second, $\widetilde{\varphi}_p$ acts nicely with respect to conjugation. For future reference we also need the corresponding result for $(\bar{\varphi}_{p(\cdot)})^*$.

Lemma 3.1.3. If $1 \leqslant q \leqslant \infty$, then $(\widetilde{\varphi}_q)^* = \widetilde{\varphi}_{q'}$ and

$$(\bar{\varphi}_q)^*(t) \leqslant \bar{\varphi}_{q'}(t) \leqslant (\bar{\varphi}_q)^*(2t)$$

for all $t \geqslant 0$.

Proof. We first show that $(\widetilde{\varphi}_q)^* = \overline{\varphi}_{q'}$ for $q \in [1, \infty]$. If $q \in (1, \infty)$, then the claim follows directly from Theorem 2.6.8 and example (b) thereafter. Moreover,

$$(\widetilde{\varphi}_1)^*(u) = \sup_{t \geqslant 0}(tu - t) = \sup_{t \geqslant 0}\big(t(u-1)\big) = \infty \cdot \chi_{(1,\infty)}(u) = \widetilde{\varphi}_\infty(u)$$

for all $u \geqslant 0$. Thus, $\widetilde{\varphi}_1(t) = (\widetilde{\varphi}_1)^{**}(t) = (\widetilde{\varphi}_\infty)^*(t)$ for all $t \geqslant 0$, where we have used Corollary 2.6.3.

Since $\widetilde{\varphi}_1 = \overline{\varphi}_1$, $\widetilde{\varphi}_\infty = \overline{\varphi}_\infty$, and $(\widetilde{\varphi}_q)^* = \overline{\varphi}_{q'}$ for all $q \in [1, \infty]$ by the previous case, it suffices to consider the case $1 < q < \infty$. The estimates

$$\frac{\overline{\varphi}_{q'}(t)}{(\overline{\varphi}_q)^*(t)} = q'q^{q'-1} \geqslant 1,$$

$$\frac{\overline{\varphi}_{q'}(t)}{(\overline{\varphi}_q)^*(2t)} = q'q^{q'-1}2^{-q'} \leqslant 1,$$

valid for all $t > 0$, yield the last assertion.   □

Third, $\widetilde{\varphi}_p$ has a certain convexity property with respect to p, which will turn out to be quite useful:

Lemma 3.1.4. *The mapping $a \mapsto \widetilde{\varphi}_{1/a}(t)$ is continuous and convex on $[0,1]$ for each $t \geqslant 0$, with the convention $\frac{1}{\infty} := 0$.*

Proof. The claim is obvious for $t = 0$, so assume $t > 0$. Define $g(a) := at^{\frac{1}{a}}$ for $a \in [0,1]$. Then $g(a) = \widetilde{\varphi}_{1/a}(t)$. We have to show that g is convex on $[0,1]$. An easy calculation shows that g is continuous on $[0,1]$ and

$$g''(a) = t^{\frac{1}{a}}\frac{(\log t)^2}{a^3} \geqslant 0$$

for $a \in (0,1]$. Thus, g is convex.   □

Remark 3.1.5. Let $q_0, q_1 \in [1, \infty]$. For $\theta \in [0,1]$ let $q_\theta \in [q_0, q_1]$ be defined through $\frac{1}{q_\theta} := \frac{1-\theta}{q_0} + \frac{\theta}{q_1}$. Then

$$\widetilde{\varphi}_{q_\theta}(t) \leqslant (1-\theta)\widetilde{\varphi}_{q_0}(t) + \theta\widetilde{\varphi}_{q_1}(t),$$

$$\min\big\{\overline{\varphi}_{q_0}(t), \overline{\varphi}_{q_1}(t)\big\} \leqslant \overline{\varphi}_{q_\theta}(t) \leqslant \max\big\{\overline{\varphi}_{q_0}(t), \overline{\varphi}_{q_1}(t)\big\}$$

for all $t \geqslant 0$. The estimate for $\widetilde{\varphi}_{q_\theta}$ follows by convexity (Lemma 3.1.4) and the estimate for $\overline{\varphi}_{q_\theta}(t)$ follows by direct calculation.

The two Φ-functions $\widetilde{\varphi}_p$ and $\overline{\varphi}_p$ are related in the following way.

Lemma 3.1.6. *Let $1 \leqslant q \leqslant \infty$. Then*

$$\widetilde{\varphi}_q(t) \leqslant \overline{\varphi}_q(t) \leqslant \widetilde{\varphi}_q(2t) \quad \text{for all } t \geqslant 0.$$

Proof. Since $\widetilde{\varphi}_\infty = \bar{\varphi}_\infty$ it suffices to consider the case $1 \leqslant q < \infty$. The case $t = 0$ follows from $\widetilde{\varphi}_q(0) = 0 = \bar{\varphi}_q(0)$. For $t > 0$ we have

$$\frac{\widetilde{\varphi}_q(t)}{\bar{\varphi}_q(t)} = \frac{1}{q} \leqslant 1,$$

$$\frac{\widetilde{\varphi}_q(2t)}{\bar{\varphi}_q(t)} = \frac{2^q}{q} \geqslant e \log 2 \geqslant 1.$$

This proves the claim.   □

Remark 3.1.7. It is also possible to show that for every $\lambda > 1$ there exists $c_\lambda \geqslant 1$ such that $\widetilde{\varphi}_q(t) \leqslant \bar{\varphi}_q(t) \leqslant c_\lambda \widetilde{\varphi}_q(\lambda t)$ for every $q \in [1, \infty]$ and $t \geqslant 0$.

In the following we will need the left-continuous inverse of a Φ-function.

Definition 3.1.8. For a Φ-function φ we define $\varphi^{-1} : [0, \infty) \to [0, \infty)$ by

$$\varphi^{-1}(t) := \inf \{\tau \geqslant 0 : \varphi(\tau) \geqslant t\}$$

for all $t \geqslant 0$. We call φ^{-1} the *left-continuous inverse of* φ.

For a generalized Φ-function $\varphi \in \Phi(A, \mu)$ the *left-continuous inverse* is defined pointwise in y, i.e. for all $y \in A$ let $\varphi^{-1}(y, \cdot) = (\varphi(y, \cdot))^{-1}$.

Let us collect a few properties of the left-continuous inverse, which follow from the properties of φ. Let φ be a Φ-function. Then φ^{-1} is non-decreasing and left-continuous on $[0, \infty)$. Moreover, $\varphi^{-1}(0) = 0$ and

$$\varphi\big(\varphi^{-1}(t)\big) \leqslant t \tag{3.1.9}$$

for all $t \geqslant 0$. We also have

$$t \leqslant \varphi^{-1}\big(\varphi(t)\big) \tag{3.1.10}$$

for all $t \geqslant 0$ with $\varphi(t) < \infty$.

Lemma 3.1.11. *If* $q \in [1, \infty)$, *then* $\widetilde{\varphi}_q^{-1}(t) = (q\,t)^{\frac{1}{q}}$, $\bar{\varphi}_q^{-1}(t) = t^{\frac{1}{q}}$ *and* $\widetilde{\varphi}_\infty^{-1}(t) = \bar{\varphi}_\infty^{-1}(t) = \chi_{(0,\infty)}(t)$ *for all* $t \geqslant 0$.
If $q \in [1, \infty]$ *and* $\frac{1}{q} + \frac{1}{q'} = 1$, *then*

$$t \leqslant \varphi_q^{-1}(t)\,\varphi_{q'}^{-1}(t) \leqslant 2t,$$

for all $t \geqslant 0$.

Proof. The formulas for $\widetilde{\varphi}_q^{-1}$, $\bar{\varphi}_q^{-1}$, and $\widetilde{\varphi}_\infty^{-1}$ follow easily by definition of the left-continuous inverse. The second claim is clear for $\bar{\varphi}$, since $\bar{\varphi}_q^{-1}(t)\bar{\varphi}_{q'}^{-1}(t) = t$ for all $q \in [1, \infty]$ and $t \geqslant 0$. For $\widetilde{\varphi}$ and $q \in (1, \infty)$, the claim follows from

$(\widetilde{\varphi}_q)^* = \widetilde{\varphi}_{q'}$ and Lemma 2.6.11. If $p = 1$, then $(\widetilde{\varphi}_1)^{-1}(t) = t$ and $(\widetilde{\varphi}_\infty)^{-1}(t) = \chi_{(0,\infty)}(t)$ for all $t \geqslant 0$. Thus, $\widetilde{\varphi}_1^{-1}(t)\,\widetilde{\varphi}_\infty^{-1}(t) = t$ for all $t \geqslant 0$. □

Lemma 3.1.12. *If $q \in [1, \infty]$, then*

$$\frac{1}{2}\varphi_q^{-1}(t) \leqslant \varphi_q^{-1}\left(\frac{t}{2}\right) \leqslant \bar{\varphi}_q^{-1}(t) \leqslant \widetilde{\varphi}_q^{-1}(t) \leqslant 2\,\bar{\varphi}_q^{-1}(t)$$

for all $t \geqslant 0$. Moreover,

$$\bar{\varphi}_q^{-1}(t)\bar{\varphi}_{q'}^{-1}\left(\frac{1}{t}\right) = 1 \qquad and \qquad 1 \leqslant \widetilde{\varphi}_q^{-1}(t)\,\widetilde{\varphi}_{q'}^{-1}\left(\frac{1}{t}\right) \leqslant 3$$

for all $t > 0$.

Proof. If $q \in [1, \infty)$, then $\varphi_q^{-1}(\frac{t}{2}) = 2^{-\frac{1}{q}}\varphi_p^{-1}(t)$ and the first claim follows from $\frac{1}{2} \leqslant 2^{-\frac{1}{q}} \leqslant 1$ and Lemma 3.1.6. The case $q = \infty$ follows from $\varphi_\infty^{-1}(t) = \chi_{(1,\infty)}$. The second claim follows from Lemma 3.1.11 and $1 \leqslant q^{2/q} \leqslant e^{2/e} < 3$. □

For $a \in (0, 1]$, $\bar{\varphi}_{1/a}^{-1}(t) = t^a$. Thus we immediately obtain

Lemma 3.1.13. *The mapping $a \mapsto \bar{\varphi}_{1/a}^{-1}(t)$ is convex on $(0, 1]$ for all $t \geqslant 0$.*

3.2 Basic Properties

We are now ready to define the variable exponents Lebesgue space.

Definition 3.2.1. Let $p \in \mathcal{P}(A, \mu)$ and let either $\varphi_{p(\cdot)} := \widetilde{\varphi}_{p(\cdot)}$ or $\varphi_{p(\cdot)} := \bar{\varphi}_{p(\cdot)}$. Hence we obtain a semimodular:

$$\varrho_{L^{p(\cdot)}(A)}(f) = \int_A \varphi_{p(x)}(|f(x)|)\,dx.$$

We define the *variable exponent Lebesgue space* $L^{p(\cdot)}(A, \mu)$ as the Musielak–Orlicz space $L^{\varphi_{p(\cdot)}}(A, \mu)$ with the norm $\|\cdot\|_{L^{p(\cdot)}(A,\mu)} := \|\cdot\|_{L^{\varphi_{p(\cdot)}}(A,\mu)}$.

In particular, the variable exponent Lebesgue space $L^{p(\cdot)}(A, \mu)$ is

$$L^{p(\cdot)}(A, \mu) = \left\{ f \in L^0(A, \mu) : \lim_{\lambda \to 0} \varrho_{L^{p(\cdot)}(A)}(\lambda f) = 0 \right\}$$

or equivalently

$$L^{p(\cdot)}(A, \mu) = \left\{ f \in L^0(A, \mu) : \varrho_{L^{p(\cdot)}(A)}(\lambda f) < \infty \text{ for some } \lambda > 0 \right\}$$

equipped with the norm

$$\|f\|_{L^{p(\cdot)}(A,\mu)} = \inf \left\{ \lambda > 0 : \varrho_{L^{p(\cdot)}(A)} \left(\frac{f}{\lambda} \right) \leqslant 1 \right\}.$$

Note that $\varrho_{L^{p(\cdot)}(A)}$ is a modular if p is finite everywhere. We abbreviate $\varrho_{L^{p(\cdot)}(A)}$ to $\varrho_{p(\cdot)}$ and $\|\cdot\|_{L^{p(\cdot)}(A,\mu)}$ to $\|\cdot\|_{p(\cdot)}$ if the set and the measure are clear from the context. Moreover, if $\Omega \subset \mathbb{R}^n$ and μ is the Lebesgue measure we simply write $L^{p(\cdot)}(\Omega)$ and if μ is the counting measure on $\mathbb{Z}^n$, then we write $l^{p(\cdot)}(\mathbb{Z}^n)$.

This definition seems ambiguous, since either $\varphi_{p(\cdot)} = \widetilde{\varphi}_{p(\cdot)}$ or $\varphi_{p(\cdot)} = \bar{\varphi}_{p(\cdot)}$. However, due to Lemma 3.1.6 it is clear that $L^{\widetilde{\varphi}_{p(\cdot)}} = L^{\bar{\varphi}_{p(\cdot)}}$ and

$$\|f\|_{\widetilde{\varphi}_{p(\cdot)}} \leqslant \|f\|_{\bar{\varphi}_{p(\cdot)}} \leqslant 2\|f\|_{\widetilde{\varphi}_{p(\cdot)}}. \tag{3.2.2}$$

Thus, the two definitions agree up to equivalence of norms with constant at most 2.

Recall that we have two relevant Φ-functions, $\widetilde{\varphi}_{p(\cdot)}$ and $\bar{\varphi}_{p(\cdot)}$. Usually, the exact norm of $L^{p(\cdot)}$ is not important, so we just work with $\varphi_{p(\cdot)}$ without specifying whether $\varphi_{p(\cdot)} = \widetilde{\varphi}_{p(\cdot)}$ or $\varphi_{p(\cdot)} = \bar{\varphi}_{p(\cdot)}$. If there is a difference in the choice of $\varphi_{p(\cdot)}$, then the specific choice for $\varphi_{p(\cdot)}$ will be specified.

Remark 3.2.3. Originally, the spaces $L^{p(\cdot)}$ have been introduced by Orlicz [319] in 1931 with $\varphi_{p(\cdot)} = \bar{\varphi}_{p(\cdot)}$ in the case $1 \leqslant p^- \leqslant p^+ < \infty$. The first definition of $L^{p(\cdot)}$ including the case $p^+ = \infty$ was given by Sharpudinov [351] and then, in the higher dimensional case, by Kováčik and Rákosník [258]. For measurable f they define

$$\varrho_{\mathrm{KR}}(f) := \bar{\varrho}_{p(\cdot)}(f \chi_{\{p \neq \infty\}}) + \|f \chi_{\{p=\infty\}}\|_\infty.$$

If is easy to see that ϱ_{KR} is a modular on $L^0(\Omega)$, the set of measurable functions. We denote the corresponding Luxemburg norm by

$$\|f\|_{\mathrm{KR}} = \inf \left\{ \lambda > 0 : \varrho_{\mathrm{KR}} \left(\frac{1}{\lambda} f \right) \leqslant 1 \right\}.$$

If $\mu(\{p = \infty\}) = 0$, then $\varrho_{\mathrm{KR}} = \bar{\varrho}_{p(\cdot)}$. But if $\mu(\{p = \infty\}) > 0$, then $\varrho_{\mathrm{KR}} \neq \bar{\varrho}_{p(\cdot)}$. Note that ϱ_{KR} is a modular, while our $\varrho_{p(\cdot)}$ is a only semimodular. In particular, $\varrho_{\mathrm{KR}}(f) = 0$ implies $f = 0$. For $\bar{\varrho}_{p(\cdot)}$ we only have that $\bar{\varrho}_{p(\cdot)}(\lambda f) = 0$ for all $\lambda > 0$ implies $f = 0$. This is due to the fact that $\bar{\varphi}_\infty$ is not a positive Φ-function. Since we developed most of the theory in Chap. 2 for semimodular spaces, we do not have to treat the set $\{p = \infty\}$ differently and we can work directly with $\varrho_{p(\cdot)}(f)$. This includes the case $p^+ = \infty$ in a more natural way.

Although ϱ_{KR} and $\bar{\varrho}_{p(\cdot)}$ differ if $p^+ = \infty$, they produce the same space up to isomorphism. Let us prove this: let $f \in L^0(\Omega)$ with $\|f\|_{\varrho_{\mathrm{KR}}} \leqslant 1$, so

$\varrho_{\mathrm{KR}}(f) \leqslant 1$ by the norm-modular unit ball property (Lemma 2.1.14). In particular, $\|f \chi_{\{p=\infty\}}\|_\infty \leqslant 1$. Thus $\bar{\varrho}_{p(\cdot)}(f \chi_{\{p=\infty\}}) = 0$, since $\bar{\varphi}_\infty(t) = 0$ for all $t \in [0,1]$. This proves that

$$\bar{\varrho}_{p(\cdot)}(f) = \bar{\varrho}_{p(\cdot)}(f \chi_{\{p\neq\infty\}}) + \bar{\varrho}_{p(\cdot)}(f \chi_{\{p=\infty\}}) = \bar{\varrho}_{p(\cdot)}(f \chi_{\{p\neq\infty\}}) \leqslant \bar{\varrho}_{\mathrm{KR}}(f) \leqslant 1.$$

So it follows that $\|f\|_{\bar{\varrho}_{p(\cdot)}} \leqslant 1$. The scaling argument shows that $\|f\|_{\bar{\varrho}_{p(\cdot)}} \leqslant \|f\|_{\mathrm{KR}}$.

Assume now that $\|f\|_{\bar{\varrho}_{p(\cdot)}} \leqslant 1$, so $\bar{\varrho}_{p(\cdot)}(f) \leqslant 1$ by the unit ball property. In particular, $\varrho_\infty(f \chi_{\{p=\infty\}}) = \bar{\varrho}_{p(\cdot)}(f \chi_{\{p=\infty\}}) \leqslant 1$ and therefore $|f| \leqslant 1$ almost everywhere on $\{p = \infty\}$. This proves that

$$\varrho_{\mathrm{KR}}(f) = \bar{\varrho}_{p(\cdot)}(f \chi_{\{p\neq\infty\}}) + \|f \chi_{\{p=\infty\}}\|_\infty \leqslant \bar{\varrho}_{p(\cdot)}(f) + 1 \leqslant 2.$$

This implies $\varrho_{\mathrm{KR}}(\frac{1}{2}f) \leqslant \frac{1}{2}\varrho_{\mathrm{KR}}(f) \leqslant 1$, so $\|f\|_{\mathrm{KR}} \leqslant 2$ by the norm-modular unit ball property. The scaling argument shows $\|f\|_{\mathrm{KR}} \leqslant 2\|f\|_{\bar{\varrho}_{p(\cdot)}}$.

Overall, we have shown that

$$\|f\|_{\bar{\varrho}_{p(\cdot)}} \leqslant \|f\|_{\mathrm{KR}} \leqslant 2\|f\|_{\bar{\varrho}_{p(\cdot)}},$$

for all $f \in L^0(\Omega)$. Thus ϱ_{KR} and $\varrho_{\bar{\varphi}_{p(\cdot)}}$ define the same space $L^{p(\cdot)}(\Omega)$, up to equivalence of norms.

For a constant exponent the relation between the modular and the norm is clear. For a variable exponent some more work is needed. We will invoke it by mentioning the *unit ball property*, or, when more clarity is needed, the *norm-modular unit ball property*.

Lemma 3.2.4 (Norm-modular unit ball property). *If $p \in \mathcal{P}(\Omega)$, then $\|f\|_{p(\cdot)} \leqslant 1$ and $\varrho_{p(\cdot)}(f) \leqslant 1$ are equivalent. For $f \in L^{p(\cdot)}(\Omega)$ we have*

(a) *If $\|f\|_{p(\cdot)} \leqslant 1$, then $\varrho_{p(\cdot)}(f) \leqslant \|f\|_{p(\cdot)}$.*
(b) *If $1 < \|f\|_{p(\cdot)}$, then $\|f\|_{p(\cdot)} \leqslant \varrho_{p(\cdot)}(f)$.*

This lemma follows directly from Lemma 2.1.14 and Corollary 2.1.15. The next lemma is a variant which is specific to the variable exponent context.

Lemma 3.2.5. *Let $p \in \mathcal{P}(\Omega)$ with $p^- < \infty$. If $\bar{\varrho}_{p(\cdot)}(f) > 0$ or $p^+ < \infty$, then*

$$\min\left\{\bar{\varrho}_{p(\cdot)}(f)^{\frac{1}{p^-}}, \bar{\varrho}_{p(\cdot)}(f)^{\frac{1}{p^+}}\right\} \leqslant \|f\|_{\bar{\varphi}_{p(\cdot)}} \leqslant \max\left\{\bar{\varrho}_{p(\cdot)}(f)^{\frac{1}{p^-}}, \bar{\varrho}_{p(\cdot)}(f)^{\frac{1}{p^+}}\right\}.$$

Proof. Suppose that $p^+ < \infty$. If $\bar{\varrho}_{p(\cdot)}(f) \leqslant 1$, then we need to prove that

$$\bar{\varrho}_{p(\cdot)}(f)^{\frac{1}{p^-}} \leqslant \|f\|_{p(\cdot)} \leqslant \bar{\varrho}_{p(\cdot)}(f)^{\frac{1}{p^+}}.$$

By homogeneity, the latter inequality is equivalent to $\|f/\overline{\varrho}_{p(\cdot)}(f)^{\frac{1}{p^+}}\|_{p(\cdot)} \leqslant 1$, which by the unit ball property is equivalent to

$$\int_\Omega \left(\frac{|f(x)|}{\overline{\varrho}_{p(\cdot)}(f)^{\frac{1}{p^+}}}\right)^{p(x)} dx \leqslant 1.$$

But since $\overline{\varrho}_{p(\cdot)}(f)^{-\frac{p(x)}{p^+}} \leqslant \overline{\varrho}_{p(\cdot)}(f)^{-1}$, this is clear. The other inequality and the case $\overline{\varrho}_{p(\cdot)}(f) \geqslant 1$ are similar.

Consider now $p^+ = \infty$ and $\overline{\varrho}_{p(\cdot)}(f) > 0$. In this case the upper inequality becomes $\|f\|_{p(\cdot)} \leqslant \max\{\overline{\varrho}_{p(\cdot)}(f)^{1/p^-}, 1\}$. If $\overline{\varrho}_{p(\cdot)}(f) \leqslant 1$, then $\|f\|_{p(\cdot)} \leqslant 1$, so the inequality holds. If $\overline{\varrho}_{p(\cdot)}(f) > 1$, then we need to show that

$$\int_\Omega \left(\frac{|f(x)|}{\overline{\varrho}_{p(\cdot)}(f)^{1/p^-}}\right)^{p(x)} dx \leqslant 1.$$

Since $\overline{\varrho}_{p(\cdot)}(f)^{-1} < 1$, we conclude that

$$\overline{\varrho}_{p(\cdot)}(f)^{\frac{-p(x)}{p^-}} \leqslant \begin{cases} 0, & \text{if } p(x) = \infty, \\ \overline{\varrho}_{p(\cdot)}(f)^{-1}, & \text{if } p(x) < \infty. \end{cases}$$

Hence

$$\int_\Omega \left(\frac{|f(x)|}{\overline{\varrho}_{p(\cdot)}(f)^{1/p^-}}\right)^{p(x)} dx \leqslant \int_\Omega \frac{|f(x)|^{p(x)}}{\varrho_{p(\cdot)}(f)} dx = 1.$$

The proof of the lower inequality is analogous.   □

Lemma 3.2.6. *Let $p \in \mathcal{P}(\mathbb{R}^n)$ and $s > 0$ be such that $sp^- \geqslant 1$. Then* $\||f|^s\|_{\overline{\varphi}_{p(\cdot)}} = \|f\|^s_{\overline{\varphi}_{sp(\cdot)}}.$

Proof. This follows from $\overline{\varphi}_{sp}(t) = \overline{\varphi}_p(t^s)$ and

$$\|f\|^s_{\overline{\varphi}_{sp(\cdot)}} = \left(\inf\{\lambda > 0 : \overline{\varrho}_{sp(\cdot)}(f/\lambda) \leqslant 1\}\right)^s$$
$$= \inf\{\lambda^s > 0 : \overline{\varrho}_{p(\cdot)}(|f|^s/\lambda^s) \leqslant 1\} = \||f|^s\|_{\overline{\varphi}_{p(\cdot)}}. \quad □$$

Let us begin with those properties of $L^{p(\cdot)}$ which can be derived directly by applying the results of Chap. 2. From Theorem 2.3.13 we immediately derive:

Theorem 3.2.7. *If $p \in \mathcal{P}(A, \mu)$, then $L^{p(\cdot)}(A, \mu)$ is a Banach space.*

Next we collect the continuity and lower semicontinuity results of Chap. 2. Recall that $E^{p(\cdot)}(A, \mu)$ denotes the set of finite elements of $L^{p(\cdot)}(A, \mu)$, see Definition 2.3.11. From Lemma 2.3.16 we deduce.

Lemma 3.2.8. *Let* $p \in \mathcal{P}(A, \mu)$ *and* $f_k, f, g \in L^0(A, \mu)$.

(a) *If* $f_k \to f$ μ-*almost everywhere, then* $\varrho_{p(\cdot)}(f) \leqslant \liminf_{k \to \infty} \varrho_{p(\cdot)}(f_k)$.
(b) *If* $|f_k| \nearrow |f|$ μ-*almost everywhere, then* $\varrho_{p(\cdot)}(f) = \lim_{k \to \infty} \varrho_{p(\cdot)}(f_k)$.
(c) *If* $f_k \to f$ μ-*almost everywhere,* $|f_k| \leqslant |g|$ μ-*almost everywhere and* $g \in E^{p(\cdot)}$, *then* $f_k \to f$ *in* $L^{p(\cdot)}$.

In analogy with the properties for the integral, the claims of the previous lemma will be called *Fatou's lemma (for the modular)*, *monotone convergence* and *dominated convergence*, respectively. From Theorem 2.2.8 we obtain.

Theorem 3.2.9. *If* $p \in \mathcal{P}(A, \mu)$, *then the modular is weakly (sequentially) lower semicontinuous, i.e.* $\varrho_{p(\cdot)}(f) \leqslant \liminf_{k \to \infty} \varrho_{p(\cdot)}(f_k)$ *if* $f_k \rightharpoonup f$ *weakly in* $L^{p(\cdot)}(A, \mu)$.

Since strong convergence implies weak convergence, the conclusion of the previous theorem holds also if $f_k \to f$ in $L^{p(\cdot)}(A, \mu)$. From Lemmas 2.3.14 and 2.3.15 we deduce.

Lemma 3.2.10. *Let* $p \in \mathcal{P}(A, \mu)$ *and let* $f_k \in L^{p(\cdot)}(A, \mu)$.

(a) *If* f_k *is a Cauchy sequence, then there exists a subsequence of* f_k *which converges* μ-*almost everywhere to a measurable function* f.
(b) *If* $\mu(A) < \infty$ *and* $\|f_k\|_{p(\cdot)} \to 0$, *then* $f_k \to 0$ *in measure.*

Theorem 2.3.17 implies that $L^{p(\cdot)}(A, \mu)$ is *circular*, *solid*, satisfies *Fatou's lemma (for the norm)* and has the *Fatou property*, i.e.

- $\|f\|_{p(\cdot)} = \big\| |f| \big\|_{p(\cdot)}$ for all $f \in L^{p(\cdot)}(A, \mu)$.
- If $f \in L^{p(\cdot)}(A, \mu)$, $g \in L^0(A, \mu)$ and $0 \leqslant |g| \leqslant |f|$ μ-almost everywhere, then $g \in L^{p(\cdot)}(A, \mu)$ and $\|g\|_{p(\cdot)} \leqslant \|f\|_{p(\cdot)}$.
- If $f_k \to f$ μ-almost everywhere, then $\|f\|_{p(\cdot)} \leqslant \liminf_{k \to \infty} \|f_k\|_{p(\cdot)}$.
- If $|f_k| \nearrow |f|$ μ-almost everywhere with $f_k \in L^{p(\cdot)}(A, \mu)$ and $\sup_k \|f_k\|_{p(\cdot)} < \infty$. Then $f \in L^{p(\cdot)}(A, \mu)$ and $\|f_k\|_{p(\cdot)} \nearrow \|f\|_{p(\cdot)}$, respectively.

In Definition 2.7.7 we introduced the notion of a Banach function space. In addition to being circular, solid and having the Fatou property, a *Banach function space* X has the property that all characteristic functions of μ-finite sets are elements of X and its associate space X'. In particular, all simple functions should be contained in X and X'. See Sect. 2.7 for the definition of the associate space X'.

Lemma 3.2.11. *Let* $p \in \mathcal{P}(A, \mu)$. *Then the set of simple functions* $S(A, \mu)$ *is contained in* $L^{p(\cdot)}(A, \mu)$ *and*

$$\min\{1, \mu(E)\} \leqslant \|\chi_E\|_{\bar{\varphi}_{p(\cdot)}} \leqslant \max\{1, \mu(E)\},$$

for every measurable set $E \subset A$.

Proof. Let $E \subset A$ be measurable with $\mu(E) < \infty$. Then

$$\overline{\varrho}_{p(\cdot)}\left(\frac{\chi_E}{\max\{1, \mu(E)\}}\right) = \int\limits_E \frac{1}{(\max\{1, \mu(E)\})^{p(x)}}\, dx$$

$$\leqslant \int\limits_E \frac{1}{\max\{1, \mu(E)\}}\, dx \leqslant 1.$$

Hence, by the unit ball property $\|\chi_E\|_{p(\cdot)} \leqslant \max\{1, \mu(E)\}$. Since simple functions are finite linear combinations of characteristic functions, we get $S(A, \mu) \subset L^{p(\cdot)}(A, \mu)$. Now, let $\lambda > 1$, then

$$\overline{\varrho}_{p(\cdot)}\left(\frac{\lambda\chi_E}{\min\{1, \mu(E)\}}\right) = \int\limits_E \frac{\lambda^{p(x)}}{(\min\{1, \mu(E)\})^{p(x)}}\, dx$$

$$\geqslant \int\limits_E \frac{\lambda}{\min\{1, \mu(E)\}}\, dx \geqslant \lambda > 1.$$

Hence, $\|\lambda\chi_E\|_{\overline{\varphi}_{p(\cdot)}} > \min\{1, \mu(E)\}$ for every $\lambda > 1$ (by the unit ball property), which proves $\|\chi_E\|_{\overline{\varphi}_{p(\cdot)}} \geqslant \min\{1, \mu(E)\}$. $\qquad\square$

The following lemma is an improved version of Lemma 3.2.11, which is especially useful if $\frac{1}{s^-} - \frac{1}{s^+}$ is small.

Lemma 3.2.12. *Let $s \in \mathcal{P}(A, \mu)$. Then*

$$\frac{1}{2}\min\left\{\mu(A)^{\frac{1}{s^+}}, \mu(A)^{\frac{1}{s^-}}\right\} \leqslant \|1\|_{L^{s(\cdot)}(A,\mu)} \leqslant 2\max\left\{\mu(A)^{\frac{1}{s^+}}, \mu(A)^{\frac{1}{s^-}}\right\}$$

for every measurable set A with $\mu(A) > 0$. If $\varphi_p = \overline{\varphi}_p$, then we can omit the factors $\frac{1}{2}$ and 2.

Proof. The case $\varphi_{p(\cdot)} = \overline{\varphi}_{p(\cdot)}$ follows from Lemma 3.2.5. The case $\varphi_{p(\cdot)} = \widetilde{\varphi}_{p(\cdot)}$ then follows by (3.2.2). $\qquad\square$

Let us apply the results of Sect. 2.7 to the spaces $L^{p(\cdot)}$.

Theorem 3.2.13. *Let $p \in \mathcal{P}(A, \mu)$. Then $\varphi_{p(\cdot)}$ is proper and $L^{p(\cdot)}(A, \mu)$ is a Banach function space. Its associate space satisfies $(L^{p(\cdot)}(A, \mu))' = L^{p'(\cdot)}(A, \mu)$ and*

$$\|g\|_{p'(\cdot)} \leqslant \|g\|_{(L^{p(\cdot)})'} \leqslant 2\|g\|_{p'(\cdot)} \qquad \text{if } \varphi_{p(\cdot)} = \widetilde{\varphi}_{p(\cdot)},$$

$$\frac{1}{2}\|g\|_{p'(\cdot)} \leqslant \|g\|_{(L^{p(\cdot)})'} \leqslant 2\|g\|_{p'(\cdot)} \qquad \text{if } \varphi_{p(\cdot)} = \overline{\varphi}_{p(\cdot)}$$

for every $g \in L^0(A, \mu)$.

Proof. If follows from Lemma 3.2.11 that simple functions are contained in $L^{p(\cdot)}$ and $L^{p'(\cdot)}$. Thus $\varphi_{p(\cdot)}$ is proper by Corollary 2.7.9 and therefore $L^{p(\cdot)}$ is a Banach function space, see also Sect. 2.7. We can apply Theorem 2.7.4 to $\varphi_{p(\cdot)}$ to get $(L^{p(\cdot)}(A,\mu))' = L^{p'(\cdot)}(A,\mu)$ and

$$\|g\|_{(\varphi_{p(\cdot)})^*} \leqslant \|g\|_{(L^{p(\cdot)})'} \leqslant 2\,\|g\|_{(\varphi_{p(\cdot)})^*},$$

which is the first estimate of the claim if $\varphi_{p(\cdot)} = \widetilde{\varphi}_{p(\cdot)}$, since $(\widetilde{\varphi}_{p(\cdot)})^* = \widetilde{\varphi}_{p'(\cdot)}$. From Lemma 3.1.3 we deduce

$$\|g\|_{(\widetilde{\varphi}_{p(\cdot)})^*} \leqslant \|g\|_{\widetilde{\varphi}_{p'(\cdot)}} \leqslant 2\,\|g\|_{(\widetilde{\varphi}_{p(\cdot)})^*},$$

which in combination with the previous estimate proves the second estimate of the claim.   □

Similar to Corollary 2.7.5 we derive from Theorem 3.2.13 the following norm conjugate formula of $L^{p(\cdot)}$.

Corollary 3.2.14 (Norm conjugate formula). *Let $p \in \mathcal{P}(A,\mu)$. Then*

$$\frac{1}{2}\|f\|_{p(\cdot)} \leqslant \sup_{g \in L^{p'(\cdot)}\,:\,\|g\|_{p'(\cdot)} \leqslant 1} \int |f|\,|g|\,d\mu \leqslant 2\,\|f\|_{p(\cdot)}$$

for all $f \in L^0(A,\mu)$. The factor $\frac{1}{2}$ can be omitted if $\varphi_{p(\cdot)} = \widetilde{\varphi}_{p(\cdot)}$.

The supremum is unchanged if we replace the condition $g \in L^{p'(\cdot)}(A,\mu)$ by $g \in S(A,\mu)$ or even $g \in S_c(\Omega)$ when $p \in \mathcal{P}(\Omega)$, where $S_c(\Omega)$ is the set of simple functions with compact support in Ω.

Proof. The proof of the formula is exactly the same as in Corollary 2.7.5 if we additionally use the estimates of Theorem 3.2.13. That the supremum does not change for $g \in S(A,\mu)$ follows by Lemma 2.7.2. The case $g \in S_c(\Omega)$ requires a simple straightforward modification of Lemma 2.7.2.   □

Since the norm conjugate formula can also be used for $f \in L^0$ (just measurable), it can be used to verify if a function belongs to $L^{p(\cdot)}$.

A critical property which holds for classical and variable exponent Lebesgue spaces, is Hölder's inequality, which we prove next. As usual, we start with Young's inequality.

Lemma 3.2.15 (Young's inequality). *Let $p, q, s \in [1, \infty]$ with*

$$\frac{1}{s} = \frac{1}{p} + \frac{1}{q}.$$

Then for all $a, b \geqslant 0$

$$\varphi_s(ab) \leqslant \varphi_p(a) + \varphi_q(b), \tag{3.2.16}$$

$$\bar\varphi_s(ab) \leqslant \frac{s}{p}\bar\varphi_p(a) + \frac{s}{q}\bar\varphi_q(b), \tag{3.2.17}$$

where we use the convention $\frac{s}{p} = \frac{s}{q} = 1$ for $s = p = q = \infty$. Moreover, if $1 \leqslant s < \infty$ then for all $a \geqslant 0$

$$\widetilde\varphi_p(a) = \sup_{b \geqslant 0} \left(\widetilde\varphi_s(ab) - \widetilde\varphi_q(b) \right). \tag{3.2.18}$$

Proof. Assume first that $s = \infty$. Then necessarily $p = q = \infty$. There is nothing to show for $a, b \in [0, 1]$, since in this case $\varphi_\infty(ab) = 0$. If $a > 1$ or $b > 1$, then $\varphi_p(a) = \infty$ or $\varphi_q(b) = \infty$, respectively. Thus the claim holds in this case also.

Assume then that $1 \leqslant s < \infty$. In order to prove (3.2.16) for $\widetilde\varphi$ it suffices to prove (3.2.18). If $s = 1$, then $p = q'$ and $\widetilde\varphi_p = (\widetilde\varphi_q)^*$ by Lemma 3.1.3. Thus,

$$\widetilde\varphi_p(a) = (\widetilde\varphi_q)^*(a) = \sup_{b \geqslant 0} \left(ab - \widetilde\varphi_q(b) \right) = \sup_{b \geqslant 0} \left(\widetilde\varphi_1(ab) - \widetilde\varphi_q(b) \right)$$

for all $a, b \geqslant 0$. If $1 < s < \infty$, then

$$1 = \frac{1}{p/s} + \frac{1}{q/s},$$

so by Lemma 3.1.3 $(\widetilde\varphi_{p/s})^* = \widetilde\varphi_{q/s}$. Using the case $s = 1$ we deduce

$$\widetilde\varphi_p(a) = \frac{1}{s}\widetilde\varphi_{p/s}(a^s) = \frac{1}{s}\sup_{b \geqslant 0} \left(a^s b^s - \widetilde\varphi_{q/s}(b^s) \right) = \sup_{b \geqslant 0} \left(\widetilde\varphi_s(ab) - \widetilde\varphi_q(b) \right)$$

for all $a, b \geqslant 0$.

It remains to prove (3.2.17), since this inequality is stronger than (3.2.16) for $\varphi = \bar\varphi$. If $s = \infty$, then $s = p = q = \infty$ and (3.2.17) follows from (3.2.16), since $\bar\varphi_\infty = \widetilde\varphi_\infty$. So in the following let $1 \leqslant s < \infty$. Now $s \leqslant p$ and $s \leqslant q$ and we obtain using the previous case that

$$\bar\varphi_s(ab) \leqslant s\widetilde\varphi_s(ab) \leqslant s\left(\widetilde\varphi_p(a) + \widetilde\varphi_q(b)\right) = \frac{s}{p}\bar\varphi_p(a) + \frac{s}{q}\bar\varphi_q(b) \leqslant \bar\varphi_p(a) + \bar\varphi_q(b)$$

for all $a, b \geqslant 0$. It remains to prove (3.2.16) with $\varphi = \bar{\varphi}$ for $1 \leqslant s < \infty$. Now $s \leqslant p$ and $s \leqslant q$ and we obtain using the previous case that

$$\bar{\varphi}_s(ab) \leqslant s\widetilde{\varphi}_s(ab) \leqslant s\big(\widetilde{\varphi}_p(a) + \widetilde{\varphi}_q(b)\big) = \frac{s}{p}\bar{\varphi}_p(a) + \frac{s}{q}\bar{\varphi}_q(b)$$

for all $a, b \geqslant 0$. $\square$

Remark 3.2.19. Note that (3.2.16) holds for both $\widetilde{\varphi}$ and $\bar{\varphi}$, but it is sharp only for $\widetilde{\varphi}$ as is shown by (3.2.18) and this example: if $s = 1$ and $p = q = 2$, then $\sup_{b \geqslant 0} \big(\bar{\varphi}_1(ab) - \bar{\varphi}_2(b)\big) = \frac{1}{4}a^2 \neq a^2 = \bar{\varphi}_2(a)$ for $a > 0$.

Lemma 3.2.20 (Hölder's inequality). *Let $p, q, s \in \mathcal{P}(A, \mu)$ be such that*

$$\frac{1}{s(y)} = \frac{1}{p(y)} + \frac{1}{q(y)}$$

for μ-almost every $y \in A$. Then

$$\varrho_{s(\cdot)}(fg) \leqslant \varrho_{p(\cdot)}(f) + \varrho_{q(\cdot)}(g), \tag{3.2.21}$$

$$\|fg\|_{s(\cdot)} \leqslant 2\|f\|_{p(\cdot)}\|g\|_{q(\cdot)}, \tag{3.2.22}$$

$$\|fg\|_{\bar{\varphi}_{s(\cdot)}} \leqslant \left(\left(\frac{s}{p}\right)^+ + \left(\frac{s}{q}\right)^+\right)\|f\|_{\bar{\varphi}_{p(\cdot)}}\|g\|_{\bar{\varphi}_{q(\cdot)}}, \tag{3.2.23}$$

for all $f \in L^{p(\cdot)}(A, \mu)$ and $g \in L^{q(\cdot)}(A, \mu)$, where in the case $s = p = q = \infty$ we use the convention $\frac{s}{p} = \frac{s}{q} = 1$.

In particular, $fg \in L^{s(\cdot)}(A, \mu)$. If additionally $f \in E^{p(\cdot)}(A, \mu)$ or $g \in E^{q(\cdot)}(A, \mu)$, then $fg \in E^{s(\cdot)}(A, \mu)$.

Proof. Let $f \in L^{p(\cdot)}$ and $g \in L^{q(\cdot)}$. Since f and g are measurable, also fg is measurable. Then (3.2.21) follows from (3.2.16) by integration over $y \in A$.

The following argument applies to both $\widetilde{\varphi}_{p(\cdot)}$ and $\bar{\varphi}_{p(\cdot)}$. If $\|f\|_{p(\cdot)} \leqslant 1$ and $\|g\|_{q(\cdot)} \leqslant 1$, then $\varrho_{p(\cdot)}(f) \leqslant 1$ and $\varrho_{q(\cdot)}(g) \leqslant 1$ by the unit ball property. Using (3.2.21) we estimate

$$\varrho_{s(\cdot)}\big(\tfrac{1}{2}fg\big) \leqslant \frac{1}{2}\varrho_{s(\cdot)}(fg) \leqslant \frac{1}{2}\big(\varrho_{p(\cdot)}(f) + \varrho_{q(\cdot)}(g)\big) \leqslant 1.$$

This implies $\|fg\|_{s(\cdot)} \leqslant 2$ by the unit ball property. The scaling argument proves (3.2.22).

Now let $\|f\|_{\bar{\varphi}_{p(\cdot)}} \leqslant 1$ and $\|g\|_{\bar{\varphi}_{q(\cdot)}} \leqslant 1$, then by the unit ball property $\bar{\varrho}_{p(\cdot)}(f) \leqslant 1$ and $\bar{\varrho}_{q(\cdot)}(g) \leqslant 1$. Using (3.2.17) integrated over $y \in A$ we get

$$\bar{\varrho}_{s(\cdot)}(ab) \leqslant \left(\frac{s}{p}\right)^+ \bar{\varrho}_{p(\cdot)}(f) + \left(\frac{s}{q}\right)^+ \bar{\varrho}_{q(\cdot)}(g) \leqslant \left(\frac{s}{p}\right)^+ + \left(\frac{s}{q}\right)^+.$$

This implies $\|fg\|_{\bar{\varphi}_{s(\cdot)}} \leqslant \operatorname{ess\,sup} \frac{s}{p} + \operatorname{ess\,sup} \frac{s}{q}$ by the unit ball property. The scaling argument proves (3.2.23).

Assume now that additionally $f \in E^{p(\cdot)}$, i.e. $\varrho_{p(\cdot)}(\lambda f) < \infty$ for every $\lambda > 0$. Let $\gamma > 0$ be such that $\varrho_{q(\cdot)}(g/\gamma) < \infty$. Then for every $\lambda > 0$

$$\varrho_{r(\cdot)}(\lambda fg) \leqslant \varrho_{p(\cdot)}(\lambda \gamma f) + \varrho_{q(\cdot)}(g/\gamma) < \infty.$$

Since $\lambda > 0$ was arbitrary, this proves $fg \in E^{r(\cdot)}$. The case $g \in E^{q(\cdot)}$ follows by symmetry. $\qquad\square$

The case $s = 1$ in Lemma 3.2.20 is of special interest:

$$\int_A |f|\,|g|\,d\mu \leqslant \varrho_{p(\cdot)}(f) + \varrho_{p'(\cdot)}(g),$$

$$\int_A |f|\,|g|\,d\mu \leqslant 2\|f\|_{p(\cdot)}\|g\|_{p'(\cdot)},$$

$$\int_A |f|\,|g|\,d\mu \leqslant \left(1 + \frac{1}{p^-} - \frac{1}{p^+}\right)\|f\|_{\bar{\varphi}_{p(\cdot)}}\|g\|_{\bar{\varphi}_{p'(\cdot)}}$$

for all $f \in L^{p(\cdot)}(A,\mu)$ and $g \in L^{p'(\cdot)}(A,\mu)$.

3.3   Embeddings

It is well known from the theory of classical Lebesgue spaces that $L^p(A)$ is a subspace of $L^q(A)$ with $p, q \in [1,\infty]$ if and only if $p \geqslant q$ and $|A| < \infty$. This suggests that a similar condition characterizes the embedding $L^{p(\cdot)}(A) \hookrightarrow L^{q(\cdot)}(A)$ for $p, q \in \mathcal{P}(A)$. Naturally, this question is closely related with the generalized Hölder inequality. We do not consider the case with different measures on the two sides of the embedding, for some result on this see [40].

We use the results of Sect. 2.8 to characterize the embeddings of variable exponent Lebesgue spaces. Recall that the norm of the embedding $L^{p(\cdot)}(A) \hookrightarrow L^{q(\cdot)}(A)$ is the smallest constant $K > 0$ for which $\|f\|_{q(\cdot)} \leqslant K\|f\|_{p(\cdot)}$.

Theorem 3.3.1. *Let $p, q \in \mathcal{P}(A,\mu)$. Define the exponent $r \in \mathcal{P}(A,\mu)$ by $\frac{1}{r(y)} := \max\left\{\frac{1}{q(y)} - \frac{1}{p(y)}, 0\right\}$ for all $y \in A$.*

(a) *If $q \leqslant p$ μ-almost everywhere and $1 \in L^{r(\cdot)}(A,\mu)$, then $L^{p(\cdot)}(A,\mu) \hookrightarrow L^{q(\cdot)}(A,\mu)$ with norm at most $2\|1\|_{L^{r(\cdot)}(A)}$.*

(b) *If the measure μ is atom-less and $L^{p(\cdot)}(A,\mu) \hookrightarrow L^{q(\cdot)}(A,\mu)$ with norm $K > 0$, then $q \leqslant p$ μ-almost everywhere and $\|1\|_{L^{r(\cdot)}(A)} \leqslant 4\,K$.*

Proof. We begin with the proof of (a). Since $q \leqslant p$ almost everywhere and $\frac{1}{r} + \frac{1}{p} = \frac{1}{q}$, we can apply Hölder's inequality, Lemma 3.2.20, to get

$$\|f\|_{q(\cdot)} \leqslant 2\|1\|_{r(\cdot)}\|f\|_{p(\cdot)}.$$

Let us now prove (b). We begin with the case $\varphi_{p(\cdot)} = \widetilde{\varphi}_{p(\cdot)}$. Assume that $L^{p(\cdot)}(A) \hookrightarrow L^{q(\cdot)}(A)$. Then by Theorem 2.8.1 there exists $h \in L^1(A, \mu)$ with $h \geqslant 0$ and $\|h\|_1 \leqslant 1$ such that

$$\widetilde{\varphi}_{q(y)}(t/K) \leqslant \widetilde{\varphi}_{p(y)}(t) + h(y) \tag{3.3.2}$$

for almost all $y \in A$ and all $t \geqslant 0$. The limit $t \to \infty$ implies that $q \leqslant p$ almost everywhere. If $q(y) < \infty$, then (3.2.18) and (3.3.2) imply that

$$\begin{aligned}
\widetilde{\varphi}_{r(y)}(1/K) &= \sup_{t \geqslant 0} \left(\widetilde{\varphi}_{q(y)}(t/K) - \widetilde{\varphi}_{p(y)}(t)\right) \\
&\leqslant \sup_{t \geqslant 0} \left(\widetilde{\varphi}_{p(y)}(t) - \widetilde{\varphi}_{p(y)}(t) + h(y)\right) \tag{3.3.3} \\
&= h(y).
\end{aligned}$$

If the set $E := \{q = \infty\}$ has measure zero, then we can integrate this inequality over $y \in A$ and get $\widetilde{\varrho}_{r(\cdot)}(1/K) \leqslant \|h\|_1 \leqslant 1$, so $1 \in L^{r(\cdot)}(\Omega)$ and $\|1\|_{r(\cdot)} \leqslant K$.

If $\mu(E) > 0$, then it follows from (3.3.2) with $t = 1$ that $\widetilde{\varphi}_\infty(1/K) \leqslant h(y)$ for almost all $y \in E$. Since h is a.e. finite on E and $\mu(E) > 0$, this implies $K \geqslant 1$. Since $r = \infty$ on the set E, we get

$$\widetilde{\varphi}_{r(y)}(1/K) = \varphi_\infty(1/K) = 0 \leqslant h(y)$$

for almost every $y \in E$. So (3.3.3) also holds on the set E. Thus we can proceed exactly as in the previous case to conclude $\|1\|_{r(\cdot)} \leqslant K$.

The case $\varphi_{p(\cdot)} = \bar{\varphi}_{p(\cdot)}$ follows from this using (3.2.2). $\qquad\square$

If $\mu(A) < \infty$, then by Lemma 3.2.11 and/or Lemma 3.2.12 the condition $1 \in L^{r(\cdot)}(A)$ of the last theorem is always satisfied. Hence,

Corollary 3.3.4. *Let $p, q \in \mathcal{P}(A, \mu)$ and let the measure μ be atom-less with $\mu(A) < \infty$. Then $L^{p(\cdot)}(A, \mu) \hookrightarrow L^{q(\cdot)}(A, \mu)$ if and only if $q \leqslant p$ μ-almost everywhere in A. The embedding constant is less or equal to $2(1 + \mu(A))$ and $2\max\left\{\mu(A)^{(\frac{1}{q} - \frac{1}{p})^+, (\frac{1}{q} - \frac{1}{p})^-}\right\}$.*

However, the condition $\mu(A) < \infty$ is not needed for $\|1\|_{r(\cdot)} < \infty$. See Proposition 4.1.8 for examples with $A = \mathbb{R}^n$, which are closely related to the following embedding result.

Lemma 3.3.5. *Let $p \in \mathcal{P}(\mathbb{R}^n)$ and $p_\infty \in [1, \infty]$. Define $s \in \mathcal{P}(\mathbb{R}^n)$ by*

$$\frac{1}{s(x)} := \left| \frac{1}{p(x)} - \frac{1}{p_\infty} \right|.$$

Then $1 \in L^{s(\cdot)}(\mathbb{R}^n)$ if and only if

$$L^{\max\{p(\cdot),p_\infty\}}(\mathbb{R}^n) \hookrightarrow L^{p(\cdot)}(\mathbb{R}^n) \hookrightarrow L^{\min\{p(\cdot),p_\infty\}}(\mathbb{R}^n).$$

Proof. If the embeddings hold, then Theorem 3.3.1 implies that $1 \in L^{s(\cdot)}$.

Assume now that $1 \in L^{s(\cdot)}(\mathbb{R}^n)$. Let $\gamma \in (0, 1)$ such that $\overline{\varrho}_{s(\cdot)}(\gamma) < \infty$. Define $r_1, r_2 \in \mathcal{P}(\mathbb{R}^n)$ by

$$\frac{1}{r_1(x)} := \min\left\{0, \frac{1}{p_\infty} - \frac{1}{p(x)}\right\} = \frac{1}{\max\{p(x), p_\infty\}} - \frac{1}{p(x)},$$

$$\frac{1}{r_2(x)} := \min\left\{0, \frac{1}{p(x)} - \frac{1}{p_\infty}\right\} = \frac{1}{p(x)} - \frac{1}{\max\{p(x), p_\infty\}}$$

for all $x \in \mathbb{R}^n$. Then $s \leqslant r_1$ and $s \leqslant r_2$ almost everywhere. Thus it follows from the definition that $\overline{\varrho}_{r_1(\cdot)}(\gamma) \leqslant \overline{\varrho}_{s(\cdot)}(\gamma) < \infty$ and $\overline{\varrho}_{r_2(\cdot)}(\gamma) \leqslant \overline{\varrho}_{s(\cdot)}(\gamma) < \infty$ for $\gamma \in (0, 1)$. In particular, $1 \in L^{r_1(\cdot)}$ and $1 \in L^{r_2(\cdot)}$. This and Theorem 3.3.1 prove the embeddings. □

The situation changes if the measure is not atom-less. In particular, the *variable exponent Lebesgue sequence space* $l^{p(\cdot)}(\mathbb{Z}^n)$ counting measure represents this kind of situation. It is well known that for $p, q \in [1, \infty]$ the classical Lebesgue sequence space $l^p(\mathbb{Z}^n)$ is a subset of $l^q(\mathbb{Z}^n)$ if and only if $p \leqslant q$. This condition generalizes to the cases of variable exponents.

Lemma 3.3.6. *Let $p, q \in \mathcal{P}(\mathbb{Z}^n)$ with $p \leqslant q$ on $\mathbb{Z}^n$. Then $l^{p(\cdot)}(\mathbb{Z}^n) \hookrightarrow l^{q(\cdot)}(\mathbb{Z}^n)$ and $\|f\|_{l^{q(\cdot)}(\mathbb{Z}^n)} \leqslant 2 \|f\|_{l^{p(\cdot)}(\mathbb{Z}^n)}$.*

Proof. Let $\varphi_{p(\cdot)} = \overline{\varphi}_{p(\cdot)}$ and let $f \in l^{p(\cdot)}(\mathbb{Z}^n)$ with $\|f\|_{l^{p(\cdot)}(\mathbb{Z}^n)} \leqslant 1$. Then by the unit ball property we have $\overline{\varrho}_{p(\cdot)}(f) \leqslant 1$. Since $\overline{\varrho}_{p(\cdot)}(f) = \sum_{k\in\mathbb{Z}^n} \overline{\varphi}_{p(k)}(|f(k)|)$, this implies $\overline{\varphi}_{p(k)}(|f(k)|) \leqslant 1$ for all $k \in \mathbb{N}$ and therefore $|f(k)| \leqslant 1$ for all $k \in \mathbb{N}$. Since $q \leqslant p$, we get

$$\overline{\varrho}_{q(\cdot)}(f) = \sum_{k\in\mathbb{Z}^n} \overline{\varphi}_{q(k)}(|f(k)|) \leqslant \sum_{k\in\mathbb{Z}^n} \overline{\varphi}_{p(k)}(|f(k)|) = \overline{\varrho}_{p(\cdot)}(f) \leqslant 1.$$

Therefore $\|f\|_{l^{q(\cdot)}(\mathbb{Z}^n)} \leqslant 1$. The claim follows by the scaling argument. The case $\varphi_{p(\cdot)} = \widetilde{\varphi}_{p(\cdot)}$ follows with the help of (3.2.2). □

We can combine Theorem 3.3.1 and Lemma 3.3.6 in a more general result for sequence spaces.

Theorem 3.3.7. *Let* $p, q, r \in \mathcal{P}(\mathbb{Z}^n)$ *with* $\frac{1}{r} = \max\{0, \frac{1}{q} - \frac{1}{p}\}$ *and* $1 \in l^{r(\cdot)}(\mathbb{Z}^n)$. *Then* $l^{p(\cdot)}(\mathbb{Z}^n) \hookrightarrow l^{q(\cdot)}(\mathbb{Z}^n)$.

Proof. Define $s \in \mathcal{P}(\mathbb{Z}^n)$ by $s := \max\{p, q\}$. Then $s \geqslant p$, $s \geqslant q$ and $\frac{1}{r} = \frac{1}{q} - \frac{1}{s}$. Thus by Lemma 3.3.6 and Theorem 3.3.1 it follows that $l^{p(\cdot)}(\mathbb{Z}^n) \hookrightarrow l^{s(\cdot)}(\mathbb{Z}^n) \hookrightarrow l^{q(\cdot)}(\mathbb{Z}^n)$. $\square$

We next characterize the embeddings of the sum and the intersection of variable exponent Lebesgue spaces. Let us introduce the usual notation. For two normed spaces X and Y (which are both embedded into a Hausdorff topological vector spaces Z) we equip the *intersection* $X \cap Y := \{f : f \in X, f \in Y\}$ and the *sum* $X + Y := \{g + h : g \in X, h \in Y\}$ with the norms

$$\|f\|_{X \cap Y} := \max\{\|f\|_X, \|f\|_Y\},$$
$$\|f\|_{X+Y} := \inf_{f=g+h, g \in X, h \in Y} (\|g\|_X + \|h\|_Y).$$

In the following let $1 \leqslant p \leqslant q \leqslant r \leqslant \infty$ be constants. We need estimates relating φ_q from above and below in terms of φ_p and φ_r. Since we can find $\theta \in [0, 1]$ such that $\frac{1}{q} = \frac{1-\theta}{q} + \frac{\theta}{r}$, it follows from the estimates in Remark 3.1.5 that

$$\varphi_q(t) \leqslant \varphi_p(t) + \varphi_r(t) \tag{3.3.8}$$

for all $t \geqslant 0$. Moreover, we have the lower estimate $\min\{\bar{\varphi}_p(t), \bar{\varphi}_r(t)\} \leqslant \bar{\varphi}_q(t)$. Although this lower estimate is sufficient for our purpose in the case of $\bar{\varphi}_q$, it does not hold with $\bar{\varphi}$ replaced by $\tilde{\varphi}$. Instead we need the estimates

$$\varphi_p(\max\{t-1, 0\}) \leqslant \varphi_q(t),$$
$$\varphi_r(\min\{t, 1\}) \leqslant \varphi_q(t) \tag{3.3.9}$$

for all $t \geqslant 0$. We begin with the first part of (3.3.9). It $t \in [0, 1]$, then $\varphi_p(\max\{t-1, 0\}) = 0$, so let us assume $t > 1$. If $q = \infty$, then $\varphi_q(t) = \infty$, so let us also assume $q < \infty$. As a consequence, also $p < \infty$. Define $a := t-1 > 0$. We estimate

$$\tilde{\varphi}_q(t) = \frac{1}{q}(1+a)^q = \frac{1}{q}\left((1+a)^{\frac{q}{p}}\right)^p \geqslant \frac{1}{q}\left(\frac{q}{p}a\right)^p \geqslant \frac{1}{p}a^p = \tilde{\varphi}_p(t)$$

with a similar estimate for $\bar{\varphi}_q$, which proves the first part of (3.3.9). We turn to the second part of (3.3.9). If $t > 1$, then the inequality is clear. For all $t \in [0, 1]$ we estimate

$$\tilde{\varphi}_r(t) = \frac{1}{r}t^r \leqslant \frac{1}{q}t^q = \tilde{\varphi}_q(t) \tag{3.3.10}$$

with a similar estimate for $\bar{\varphi}_q$. This concludes the proof of (3.3.9).

Theorem 3.3.11. *Let $p, q, r \in \mathcal{P}(A, \mu)$ with $p \leqslant q \leqslant r$ μ-almost everywhere in A. Then*

$$L^{p(\cdot)}(A, \mu) \cap L^{r(\cdot)}(A, \mu) \hookrightarrow L^{q(\cdot)}(A, \mu) \hookrightarrow L^{p(\cdot)}(A, \mu) + L^{r(\cdot)}(A, \mu).$$

The embedding constants are at most 2. More precisely, for $g \in L^{q(\cdot)}(A, \mu)$ the functions $g_0 := \operatorname{sgn} g \max\{|g| - 1, 0\}$ and $g_1 := \operatorname{sgn} g \min\{|g|, 1\}$ satisfy $g = g_0 + g_1$, $|g_0|, |g_1| \leqslant |g|$, $\|g_0\|_{p(\cdot)} \leqslant 1$ and $\|g_1\|_{r(\cdot)} \leqslant 1$.

Proof. Let $f \in L^{p(\cdot)} \cap L^{r(\cdot)}$ with $\max\{\|f\|_{p(\cdot)}, \|f\|_{r(\cdot)}\} \leqslant 1$. Then it follows by the norm-modular unit ball property that $\varrho_{p(\cdot)}(f) \leqslant 1$ and $\varrho_{r(\cdot)}(f) \leqslant 1$. From (3.3.8) it follows that $\varrho_{q(\cdot)}(f) \leqslant \varrho_{p(\cdot)}(f) + \varrho_{r(\cdot)}(f) \leqslant 2$. This yields $\varrho_{q(\cdot)}(f/2) \leqslant \frac{1}{2}\varrho_{q(\cdot)}(f) \leqslant 1$ using sub-linearity, (2.1.5), so $\|f\|_{q(\cdot)} \leqslant 2$ by the unit ball property. The scaling argument proves that $\|f\|_{q(\cdot)} \leqslant 2\|f\|_{L^{p(\cdot)} \cap L^{r(\cdot)}}$.

Now, let $g \in L^{q(\cdot)}(\Omega)$ with $\|g\|_{q(\cdot)} \leqslant 1$ so that $\varrho_{p(\cdot)}(g) \leqslant 1$ by the unit ball property. Define $g_0 := \operatorname{sgn} g \max\{|g| - 1, 0\}$ and $g_1 := \operatorname{sgn} g \min\{|g|, 1\}$. Then $g = g_0 + g_1$ and by (3.3.9) it follows that $\varrho_{r(\cdot)}(g_0) \leqslant \varrho_{q(\cdot)}(g) \leqslant 1$ and $\varrho_{p(\cdot)}(g_1) \leqslant \varrho_{q(\cdot)}(g) \leqslant 1$. The unit ball property implies $\|g_0\|_{r(\cdot)} \leqslant 1$ and $\|g_1\|_{p(\cdot)} \leqslant 1$. In particular, $\|g\|_{L^{p(\cdot)} + L^{r(\cdot)}} \leqslant 2$. The scaling argument proves $\|g\|_{L^{p(\cdot)} + L^{r(\cdot)}} \leqslant 2\|g\|_{q(\cdot)}$. $\square$

The following result is needed later in Theorem 3.6.5 in the study of the convolution operator.

Lemma 3.3.12. *Let $p \in \mathcal{P}(\mathbb{R}^n)$ and $p_\infty \in [1, \infty]$. Assume that $1 \in L^{s(\cdot)}(\mathbb{R}^n)$, where $s \in \mathcal{P}(\mathbb{R}^n)$ is defined by $\frac{1}{s(x)} := |\frac{1}{p(x)} - \frac{1}{p_\infty}|$. Then*

$$L^{p(\cdot)}(\mathbb{R}^n) \cap L^{p^+}(\mathbb{R}^n) \cong L^{p_\infty}(\mathbb{R}^n) \cap L^{p^+}(\mathbb{R}^n),$$

$$L^{p(\cdot)}(\mathbb{R}^n) \hookrightarrow L^{p_\infty}(\mathbb{R}^n) + L^{p^-}(\mathbb{R}^n).$$

Proof. Using Lemma 3.3.5, Theorem 3.3.11 twice, and then Lemma 3.3.5 again we deduce

$$L^{p(\cdot)} \cap L^{p^+} \hookrightarrow L^{\min\{p(\cdot), p_\infty\}} \cap L^{p^+}$$

$$\hookrightarrow L^{p_\infty} \cap L^{p^+}$$

$$\hookrightarrow L^{\max\{p(\cdot), p_\infty\}} \cap L^{p^+}$$

$$\hookrightarrow L^{p(\cdot)} \cap L^{p^+}.$$

This proves the first assertion. Analogously, by Lemma 3.3.5 and Theorem 3.3.11

$$L^{p(\cdot)} \hookrightarrow L^{\min\{p(\cdot), p_\infty\}} \hookrightarrow L^{p_\infty} + L^{p^-}.$$

This proves the second assertion. $\square$

3.4 Properties for Restricted Exponents

In this section we consider basic properties of variable exponent Lebesgue spaces that hold only under some additional conditions, namely when $p^+ < \infty$ and/or $p^- > 1$. Recall that p^- and p^+ denote the essential infimum and supremum of p, respectively.

The following theorem shows that the condition $p^+ < \infty$ plays an important role for the properties of $L^{p(\cdot)}$. Indeed, it shows that $p^+ < \infty$ is equivalent to $E^{p(\cdot)} = L^{p(\cdot)}$ which is needed for example for the characterization of the dual space $(L^{p(\cdot)})^*$. Recall that $L_{OC}^{\varphi_{p(\cdot)}}(A, \sigma, \mu)$ is the Musielak–Orlicz class of the modular defined by $\varphi_{p(\cdot)}$, see Definition 2.5.1.

Theorem 3.4.1. *Let $p \in \mathcal{P}(A, \mu)$. Then the following conditions are equivalent:*

(a) $E^{p(\cdot)}(A, \mu) = L_{OC}^{\varphi_{p(\cdot)}}(A, \mu)$.
(b) $L_{OC}^{\varphi_{p(\cdot)}}(A, \mu) = L^{p(\cdot)}(A, \mu)$.
(c) $E^{p(\cdot)}(A, \mu) = L^{p(\cdot)}(A, \mu)$.
(d) $\varphi_{p(\cdot)}$ *satisfies the Δ_2-condition with constant 2^{p^+}.*
(e) $p^+ < \infty$.
(f) $\varrho_{p(\cdot)}$ *satisfies the weak Δ_2-condition for modulars, i.e. modular convergence and norm convergence are the same.*
(g) $\varrho_{p(\cdot)}$ *is a continuous modular.*

Proof. (e) $\Rightarrow$ (d): This follows from $2^{p(y)} \leqslant 2^{p^+}$ for all $y \in A$.

(d) $\Rightarrow$ (c) and (f): This is a consequence of $\varrho_\varphi(2^k f) \leqslant 2^{kp^+} \varrho_\varphi(f)$.

(c) $\Rightarrow$ (b) and (a): Follows from $E^{p(\cdot)} \subset L_{OC}^{\varphi_{p(\cdot)}} \subset L^{p(\cdot)}$.

(d) $\Rightarrow$ (g): Follows from Lemma 2.4.3.

(a) or (b) or (g) or (f) $\Rightarrow$ (e): We prove the claim by contradiction: so let $p^+ = \infty$. We begin with the case $\mu(\{p = \infty\}) > 0$. Let $f := \chi_{\{p=\infty\}}$, then $\varrho_{p(\cdot)}(f) = 0$ and $\varrho_{p(\cdot)}(\lambda f) = \infty$ for $\lambda > 1$. This proves $f \in L_{OC}^{\varphi_{p(\cdot)}} \setminus E^{p(\cdot)}$ and $2f \in L^{p(\cdot)} \setminus L_{OC}^{\varphi_{p(\cdot)}}$, which contradicts (a) and (b), respectively. Moreover, $\lim_{\lambda \to 1^+} \varrho_{p(\cdot)}(\lambda f) = \infty \neq 0 = \varrho_{p(\cdot)}(f)$, which contradicts (g). If $f_k := f$, then $\varrho_{p(\cdot)}(f_k) = 0 \to 0$ and $\varrho_{p(\cdot)}(2f_k) = \infty \not\to 0$, which contradicts (f).

Assume now that $\mu(\{p = \infty\}) = 0$. Since $p^+ = \infty$, there exists a sequence $q_k \in [1, \infty)$ with $q_k \nearrow \infty$ and $q_k \geqslant k$ and pairwise disjoint sets E_k with $0 < \mu(E_k) < \infty$ and $E_k \subset \{y : q_k \leqslant p(y) < q_{k+1}\}$. Since p is bounded on the set E_k and $0 < \mu(E_k) < \infty$, the mapping $t \mapsto \varrho_{p(\cdot)}(t\chi_{E_k})$ is continuous for $t \geqslant 0$ with image $[0, \infty)$. Hence, there exists t_k with $\varrho_{p(\cdot)}(t_k \chi_{E_k}) = \frac{1}{2k^2}$. Let $f_k := t_k \chi_{E_k}$ then $\varrho_{p(\cdot)}(f_k) = \frac{1}{2k^2}$ and $\varrho_{p(\cdot)}(f_k) \to 0$ as $k \to \infty$. On the other hand $p \geqslant q_k \geqslant k$ on E_k implies for $\lambda > 1$ using (2.1.5)

$$\varrho_{p(\cdot)}(\lambda f_k) \geqslant \lambda^k \varrho_{p(\cdot)}(f_k) = \frac{\lambda^k}{2k^2} \to \infty.$$

We have found a sequence f_k with $\varrho_{p(\cdot)}(f_k) \to 0$ and $\varrho_{p(\cdot)}(2f_k) \to \infty$, which contradicts (f). Define

$$g_k := \sum_{j=1}^{k} f_j = \sum_{j=1}^{k} t_j \, \chi_{E_j}, \qquad g := \sum_{j=1}^{\infty} f_j = \sum_{j=1}^{\infty} t_j \, \chi_{E_j}.$$

Since $t_k \geqslant 0$, we have $0 \leqslant g_k \nearrow g$. Therefore monotone convergence, Lemma 3.2.8, implies that for $\lambda > 1$

$$\varrho_{p(\cdot)}(g) = \sum_{j=1}^{\infty} \varrho_{p(\cdot)}(f_j) = \sum_{j=1}^{\infty} \frac{1}{2j^2} \leqslant 1,$$

$$\varrho_{p(\cdot)}(\lambda g) = \sum_{j=1}^{\infty} \varrho_{p(\cdot)}(\lambda f_j) \geqslant \sum_{j=1}^{\infty} \frac{\lambda^{j-1}}{2j^2} = \infty.$$

This proves $g \in L_{OC}^{\varphi_{p(\cdot)}} \setminus E^{p(\cdot)}$ and $2g \in L^{p(\cdot)} \setminus L_{OC}^{\varphi_{p(\cdot)}}$, which contradicts (a) and (b), respectively. Moreover, $\lim_{\lambda \to 1^+} \varrho_{p(\cdot)}(\lambda g) = \infty$ and $\varrho_{p(\cdot)}(g) \leqslant 1$, which contradicts (g).   □

With the aid of the previous result we can extend the unit ball property (cf. Lemma 2.1.14):

Lemma 3.4.2 (Norm-modular unit ball property). *If $p \in \mathcal{P}(\Omega)$ is bounded, then $\|f\|_{p(\cdot)} \leqslant 1$ and $\varrho_{p(\cdot)}(f) \leqslant 1$ are equivalent, as are $\|f\|_{p(\cdot)} < 1$ and $\varrho_{p(\cdot)}(f) < 1$, and $\|f\|_{p(\cdot)} = 1$ and $\varrho_{p(\cdot)}(f) = 1$.*

Remark 3.4.3. Let $p \in \mathcal{P}(A, \mu)$ be a bounded exponent. Then $\varphi_{p(\cdot)}$ is locally integrable, since

$$\int_E \varphi_{p(y)}(\lambda) \, d\mu(y) \leqslant \mu(E) \max \left\{ \lambda^{p^-}, \lambda^{p^+} \right\}$$

for every measurable $E \subset A$ with $\mu(E) < \infty$ and every $\lambda > 0$. However, the local integrability of $\varphi_{p(\cdot)}$ does not imply that $p^+ < \infty$. Indeed, let $A := \mathbb{R}$ and let $E_k \subset \mathbb{R}$ be pairwise disjoint with $|E_k| = \exp(\exp(-k))$. Now, define $p(x) := k$ for $x \in E_k$ and $k \in \mathbb{N}$ and $p(x) = 1$ for $x \in \mathbb{R} \setminus \bigcup_{k=1}^{\infty} E_k$. Then for every $\lambda > 0$ and every $E \subset \mathbb{R}$ with $|E| < \infty$ we have

$$\bar{\varphi}_{p(\cdot)}(\lambda E) \leqslant \lambda |E| + \sum_{k=1}^{\infty} \lambda^k \exp \left(\exp(-k) \right) < \infty.$$

Thus $\bar{\varphi}_{p(\cdot)}$ is locally integrable but $p^+ = \infty$.

The boundedness of the exponent also suffices for separability:

Lemma 3.4.4. *Let $p \in \mathcal{P}(A, \mu)$ be a bounded exponent and let μ be separable. Then $L^{p(\cdot)}(A, \mu)$ is separable.*

Proof. Since $\varphi_{p(\cdot)}$ is locally integrable by Remark 3.4.3, we can apply Theorem 2.5.10 to show that $E^{p(\cdot)}$ is separable. Since $p^+ < \infty$, we further have $E^{p(\cdot)} = L^{p(\cdot)}$ (Theorem 3.4.1). □

We can directly apply this lemma to $L^{p(\cdot)}(\Omega)$ with $\Omega \subset \mathbb{R}^n$ and $l^{p(\cdot)}(\mathbb{Z}^n)$:

Corollary 3.4.5. *If $p \in \mathcal{P}(\Omega)$ and $q \in \mathcal{P}(\mathbb{Z}^n)$ are bounded exponents, then $L^{p(\cdot)}(\Omega)$ and $l^{q(\cdot)}(\mathbb{Z}^n)$ are separable.*

Recall that for all $g \in L^{p'(\cdot)}$ the mapping J_g is defined as $J_g(f) = \int fg\,d\mu$, where $f \in L^{p(\cdot)}$, and belongs to $(L^{p(\cdot)})^*$ (cf. (2.7.3), Theorem 3.2.13).

Theorem 3.4.6. *Let $p \in \mathcal{P}(A, \mu)$ be a bounded exponent, then $V : g \mapsto J_g$ is an isomorphism from $L^{p'(\cdot)}(A, \mu)$ to $(L^{p(\cdot)}(A, \mu))^*$.*

Proof. From $p^+ < \infty$ it follows by Theorem 3.2.13, Remark 3.4.3 and Theorem 3.4.1 that $\varphi_{p(\cdot)}$ is proper and locally integrable and $E^{p(\cdot)} = L^{p(\cdot)}$. Now the claim follows by Theorem 2.7.14. □

Reflexivity and uniform convexity require even stronger assumptions on the exponent. Note that Dinca and Matei [111] studied uniform convexity in the case $p \geqslant 2$.

Theorem 3.4.7. *Let $p \in \mathcal{P}(A, \mu)$ with $1 < p^- \leqslant p^+ < \infty$. Then $L^{p(\cdot)}(A, \mu)$ is reflexive.*

Proof. Let $1 < p^- \leqslant p^+ < \infty$. Then it follows from Remark 3.4.3 that $\varphi_{p(\cdot)}$ and $(\varphi_{p(\cdot)})^*$ are locally integrable. Moreover, by Theorem 3.4.1 it follows that $E^{p(\cdot)} = L^{p(\cdot)}$ and $E^{p'(\cdot)} = L^{p'(\cdot)}$. Thus Corollary 2.7.18 shows that $L^{p(\cdot)}$ is reflexive. □

Remark 3.4.8. The condition $1 < p^- \leqslant p^+ < \infty$ in Theorem 3.4.7 is sharp if μ is atom-free. This has been proved first by Kováčik and Rákosník [258, Corollary 2.7] for $L^{p(\cdot)}(\Omega)$, i.e. in the case of the Lebesgue measure. Indeed, if $L^{p(\cdot)}(A, \mu)$ is reflexive and μ is atom free, then by Remark 2.7.16 (b) follows that $E^{p(\cdot)} = L^{p(\cdot)}$ and $E^{p'(\cdot)} = L^{p'(\cdot)}$. Thus, Theorem 3.4.1 implies $1 < p^- \leqslant p^+ < \infty$.

Theorem 3.4.9. *Let $p \in \mathcal{P}(\Omega)$ with $1 < p^- \leqslant p^+ < \infty$. Then $\varphi_{p(\cdot)}$ is a uniformly convex N-function, $\varrho_{p(\cdot)}$ is a uniformly convex semimodular and $\|\cdot\|_{p(\cdot)}$ is a uniformly convex norm. Hence, $L^{p(\cdot)}(\Omega)$ is uniformly convex.*

Proof. Note that $\varphi_{p(\cdot)}$ satisfies the Δ_2-condition since $p^+ < \infty$. In order to apply Theorems 2.4.11 and 2.4.14 we have to show that $\varphi_{p(\cdot)}$ is uniformly convex. In principle we have to show this for both $\bar{\varphi}_{p(\cdot)}$ and $\widetilde{\varphi}_{p(\cdot)}$, since the equivalence of norms does not transfer the uniform convexity. However, since $\bar{\varphi}_{p(y)}$ and $\widetilde{\varphi}_{p(y)}$ only differ for every $y \in \Omega$ by the multiplicative constant

$\frac{1}{p(y)}$, the uniform convexity of $\bar{\varphi}_{p(\cdot)}$ is equivalent to the uniform convexity of $\widetilde{\varphi}_{p(\cdot)}$. Thus it suffices to consider the case $\bar{\varphi}_{p(\cdot)}$.

Fix $\varepsilon > 0$. Let $u, v \geqslant 0$ be such that $|u - v| > \varepsilon \max\{u, v\}$. It follows from Remark 2.4.6 that the mapping $t \mapsto t^{p^-}$ is uniformly convex, since $p^- > 1$. Thus there exists $\delta = \delta(\varepsilon, p^-) > 0$ such that

$$\left(\frac{u + v}{2}\right)^{p^-} \leqslant (1 - \delta)\frac{u^{p^-} + v^{p^-}}{2}.$$

This and the convexity of $t \mapsto t^{\frac{p(y)}{p^-}}$ for $y \in \Omega$ imply

$$\left(\frac{u + v}{2}\right)^{p(y)} \leqslant \left((1 - \delta)\frac{u^{p^-} + v^{p^-}}{2}\right)^{\frac{p(y)}{p^-}} \leqslant (1 - \delta)\frac{u^{p(y)} + v^{p(y)}}{2}.$$

This proves that $\bar{\varphi}_{p(\cdot)}$ is uniformly convex. The semimodular $\varrho_{p(\cdot)}$ is uniformly convex by Theorem 2.4.11 and the norm $\|\cdot\|_{p(\cdot)}$ is uniformly convex by Theorem 2.4.14.   □

It is often the case that results are easier to prove for nice functions and then by density the results carry over to the general case. It is therefore of interest to find nice subsets of $L^{p(\cdot)}$ which are dense in $L^{p(\cdot)}$. If the exponent is bounded, then by Theorems 2.5.9 and 3.4.1 we immediately get the following density result.

Corollary 3.4.10. *If $p \in \mathcal{P}(\Omega)$ with $p^+ < \infty$, then simple functions are dense in $L^{p(\cdot)}(\Omega)$.*

Remark 3.4.11. Since simple function are a subset of $L^\infty(\Omega) \cap L^{p(\cdot)}(\Omega)$ it follows from Corollary 3.4.10 that $L^\infty(\Omega) \cap L^{p(\cdot)}(\Omega)$ is also dense is $L^{p(\cdot)}(\Omega)$ if $p^+ < \infty$. This fact was first shown by Kováčik and Rákosník in [258] for the case $p^+ < \infty$. Later Kalyabin [227] has proved that the condition $p^+ < \infty$ is necessary and sufficient for the density of $L^\infty(\Omega) \cap L^{p(\cdot)}(\Omega)$ in $L^{p(\cdot)}(\Omega)$ if the variable exponent is finite almost everywhere. We can use this result to characterize the density of $L^\infty(\Omega) \cap L^{p(\cdot)}(\Omega)$ in $L^{p(\cdot)}(\Omega)$ for general $p \in \mathcal{P}(\Omega)$:

Let $p \in \mathcal{P}(\Omega)$ and $\Omega_0 := \{y \in \Omega : p(y) < \infty\}$. Then $L^\infty(\Omega) \cap L^{p(\cdot)}(\Omega)$ is dense in $L^{p(\cdot)}(\Omega)$ if and only if ess $\sup_{y \in \Omega_0} p(y) < \infty$.

Hence, p might be unbounded when $L^\infty(\Omega) \cap L^{p(\cdot)}(\Omega)$ is dense in $L^{p(\cdot)}(\Omega)$, but on the subset Ω_0, where p is finite, it must be bounded.

For an open set $\Omega \subset \mathbb{R}^n$ let $C_0^\infty(\Omega)$ denote the set of smooth functions with compact support in Ω.

Theorem 3.4.12. *If $p \in \mathcal{P}(\Omega)$ with $p^+ < \infty$, then $C_0^\infty(\Omega)$ is dense in $L^{p(\cdot)}(\Omega)$.*

Proof. Since $p^+ < \infty$, simple functions are dense in $L^{p(\cdot)}(\Omega)$ (Corollary 3.4.10). Since a simple function belongs to $L^{p^-}(\Omega) \cap L^{p^+}(\Omega)$, it can be approximated by a sequence of $C_0^\infty(\Omega)$ functions in the same space, which yields the claim since $L^{p^+}(\Omega) \cap L^{p^-}(\Omega) \hookrightarrow L^{p(\cdot)}(\Omega)$ by Theorem 3.3.11. $\qquad\square$

As a consequence $C_0^\infty(\Omega)$ is dense in $L^{p'(\cdot)}(\Omega)$ if $p^- > 1$ and therefore the norm conjugate formula in Corollary 3.2.14 is unchanged if we replace the condition $g \in L^{p'(\cdot)}(\Omega)$ by $g \in C_0^\infty(\Omega)$.

Corollary 3.4.13 (Norm conjugate formula). *Let $p \in \mathcal{P}(\Omega)$ with $p^- > 1$. Then*

$$\frac{1}{2}\|f\|_{p(\cdot)} \leqslant \sup_{g \in C_0^\infty(\Omega)\,:\,\|g\|_{p'(\cdot)} \leqslant 1} \int |f|\,|g|\,dx \leqslant 2\,\|f\|_{p(\cdot)}$$

for all $f \in L^0(\Omega)$. The factor $\frac{1}{2}$ can be omitted if $\varphi_{p(\cdot)} = \tilde{\varphi}_{p(\cdot)}$.

In Corollary 4.6.6 we prove the norm conjugate formula without the assumption $p^- > 1$, however, there we require other regularity of the space.

Sometimes it is necessary to consider the subspace of $L^{p(\cdot)}(\Omega)$ consisting of functions with a vanishing integral. For domains with $|\Omega| < \infty$ we denote the space of such functions by

$$L_0^{p(\cdot)}(\Omega) := \left\{ f \in L^{p(\cdot)}(\Omega) : \int_\Omega f(x)\,dx = 0 \right\}.$$

(In contrast to the definition of C_0^∞, the index 0 in $L_0^{p(\cdot)}$ does not indicate compact support. However, in both cases the only constant within the space is zero.) In the case that $|\Omega| = \infty$ we set $L_0^{p(\cdot)}(\Omega) := L^{p(\cdot)}(\Omega)$. We will see that for a large class of exponents this is sensible. The space of compactly supported smooth functions with vanishing integral we denote by $C_{0,0}^\infty(\Omega)$.

Proposition 3.4.14. *Let Ω be a domain and let $p \in \mathcal{P}(\Omega)$ be a bounded exponent. If $|\Omega| < \infty$ or $p^- > 1$, then $C_{0,0}^\infty(\Omega)$ is dense in $L_0^{p(\cdot)}(\Omega)$.*

Proof. Let us first consider the case $|\Omega| < \infty$. Choose $\psi \in C_0^\infty(\Omega)$ satisfying $\int_\Omega \psi\,dx = 1$. For $f \in L_0^{p(\cdot)}(\Omega)$ Theorem 3.4.12 implies that there exists a sequence $(\tilde{f}_k) \subset C_0^\infty(\Omega)$ such that $\tilde{f}_k \to f$ in $L^{p(\cdot)}(\Omega)$. Since $|\Omega| < \infty$ we get by Hölder's inequality $\|f - \tilde{f}_k\|_1 \leqslant 2\,\|\chi_\Omega\|_{p'(\cdot)}\|f - \tilde{f}_k\|_{p(\cdot)}$. Consequently, we have $\tilde{f}_k \to f$ in $L^1(\Omega)$ and $\int_\Omega \tilde{f}_k\,dx \to \int_\Omega f\,dx = 0$. Setting $f_k := \tilde{f}_k - \psi \int_\Omega \tilde{f}_k\,dx$ we see that $f_k \in C_{0,0}^\infty(\Omega)$ and that

$$\|f - f_k\|_{p(\cdot)} \leqslant \|f - \tilde{f}_k\|_{p(\cdot)} + \|\psi\|_{p(\cdot)} \left| \int_\Omega \tilde{f}_k\,dx \right|,$$

which tends to zero for $k \to \infty$ in view of the above.

Let now Ω satisfy $|\Omega| = \infty$ and $p^- > 1$. Choose an increasing sequence of bounded domains $\Omega_j \subset\subset \Omega$ with $\bigcup_{j=1}^{\infty} \Omega_j = \Omega$, $|\Omega_j| \geqslant 1$ and non-negative functions $\psi_j \in C_0^{\infty}(\Omega_j)$ which satisfy $\int_{\Omega_j} \psi_j \, dx = 1$, $\psi_j \leqslant c |\Omega_j|^{-1} \chi_{\Omega_j}$. This is possible since one can take a mollification of $|\Omega_j|^{-1} \chi_{\Omega_j}$. From $\psi_j \leqslant c |\Omega_j|^{-1} \chi_{\Omega_j}$ and Theorem 3.3.11 it follows that

$$
\begin{aligned}
\|\psi_j\|_{p(\cdot)} &\leqslant c |\Omega_j|^{-1} \|\chi_{\Omega_j}\|_{L^{p^-} + L^{p^+}} \\
&\leqslant c \max \left\{ |\Omega_j|^{-1 + \frac{1}{p^-}}, |\Omega_j|^{-1 + \frac{1}{p^+}} \right\} \to 0
\end{aligned}
\tag{3.4.15}
$$

for $j \to \infty$. For $f \in L_0^{p(\cdot)}(\Omega)$ Theorem 3.4.12 implies that there exists a sequence $(\tilde{f}_k) \subset C_0^{\infty}(\Omega)$ such that $\tilde{f}_k \to f$ in $L^{p(\cdot)}(\Omega)$ and $\|\tilde{f}_k\|_{p(\cdot)} \leqslant \|f\|_{p(\cdot)} + 1$. We set $f_k := \tilde{f}_k - \psi_{j_k} \int_{\Omega_k} \tilde{f}_k \, dx$, where j_k is an increasing sequence in $\mathbb{N}$, which will be chosen below. By definition of f_k we have $f_k \in C_{0,0}^{\infty}(\Omega)$. With Hölder's inequality we estimate

$$
\begin{aligned}
\|f - f_k\|_{p(\cdot)} &\leqslant \|f - \tilde{f}_k\|_{p(\cdot)} + \|\psi_{j_k}\|_{p(\cdot)} 2 \|\chi_{\Omega_k}\|_{p'(\cdot)} \|\tilde{f}_k\|_{p(\cdot)} \\
&\leqslant \|f - \tilde{f}_k\|_{p(\cdot)} + \|\psi_{j_k}\|_{p(\cdot)} 2 \|\chi_{\Omega_k}\|_{p'(\cdot)} \left(\|f\|_{p(\cdot)} + 1 \right).
\end{aligned}
$$

The first term converges to zero for $k \to \infty$. Since $|\Omega_k| < \infty$, we have $\chi_{\Omega_k} \in L^{p'(\cdot)}(\Omega)$ as simple functions are contained in $L^{p'(\cdot)}(\Omega)$ by Lemma 3.2.11. According to (3.4.15), we can choose j_k such that $\|\psi_{j_k}\|_{p(\cdot)} \|\chi_{\Omega_k}\|_{p'(\cdot)} \leqslant 2^{-k}$. With this choice also the second term in the previous estimate converges to zero for $k \to \infty$. In particular, we have $f_k \to f$ in $L^{p(\cdot)}(\Omega)$. $\qquad\square$

3.5 Limit of Exponents

In this section we collect some continuity results with respect to convergence of the exponent. In particular, we examine the behavior of the semimodular $\varrho_{p_k(\cdot)}(f)$ and the norm $\|f\|_{p_k(\cdot)}$ if the exponent p_k converges pointwise to an exponent p. Let us mention that some other properties of the norm in the case $p^+ = \infty$ were studied by Edmunds, Lang and Nekvinda [116].

We begin with the continuity property of φ_q with respect to q. If $q_k, q \in [1, \infty]$ with $q_k \to q$, then it is easily checked that

$$
\begin{aligned}
\lim_{k \to \infty} \tilde{\varphi}_{q_k}(t) &= \tilde{\varphi}_q(t), \\
\lim_{k \to \infty} \bar{\varphi}_{q_k}(t) &= \bar{\varphi}_q(t) \qquad \text{if } q < \infty
\end{aligned}
\tag{3.5.1}
$$

for every $t \geqslant 0$.

Remark 3.5.2. If $q_k \to q = \infty$, then $\lim_{k\to\infty} \bar{\varphi}_{q_k}(t) = 0 = \bar{\varphi}_\infty(t)$ for $t \in [0,1)$ and $\lim_{k\to\infty} \bar{\varphi}_{q_k}(t) = \infty = \bar{\varphi}_\infty(t)$ for $t > 1$. However, $\lim_{k\to\infty} \bar{\varphi}_{q_k}(1) = 1 \neq 0 = \bar{\varphi}_\infty(1)$. This is the reason, why we had to exclude the case $q = \infty$ for $\bar{\varphi}$ in (3.5.1). Nevertheless, we have

$$\lim_{k\to\infty} \bar{\varphi}_{q_k}(\lambda t) \leqslant \bar{\varphi}_q(t) \leqslant \lim_{k\to\infty} \bar{\varphi}_{q_k}(t)$$

for all $t \geqslant 0$ and all $\lambda \in [0,1)$.

Remark 3.5.3. By Lemma 3.1.11 we have $\widetilde{\varphi}_q^{-1}(t) = (qt)^{\frac{1}{q}}$, $\bar{\varphi}_q^{-1}(t) = t^{\frac{1}{q}}$ for $1 \leqslant q < \infty$ and $\widetilde{\varphi}_\infty^{-1}(t) = \bar{\varphi}_\infty^{-1}(t) = \chi_{(0,\infty)}(t)$ for all $t \geqslant 0$. If follows easily that $\varphi_{q_n}^{-1}(t) \to \varphi_q(t)$ for all $t \geqslant 0$ and $q_n \to q$. For the case $\widetilde{\varphi}_q$ and $q = \infty$, we use $\lim_{q\to\infty} q^{\frac{1}{q}} = \exp(\lim_{q\to\infty} \frac{\log q}{q}) = 1$.

We deduce the following lower semicontinuity results for the semimodular and the norm.

Corollary 3.5.4. If $p_k, p \in \mathcal{P}(A,\mu)$ with $p_k \to p$ μ-almost everywhere, then $\varrho_{p(\cdot)}(f) \leqslant \liminf_{k\to\infty} \varrho_{p_k(\cdot)}(f)$ and $\|f\|_{p(\cdot)} \leqslant \liminf_{k\to\infty} \|f\|_{p_k(\cdot)}$ for all $f \in L^0(A,\mu)$.

Proof. The estimate $\varrho_{p(\cdot)}(f) \leqslant \liminf_{k\to\infty} \varrho_{p_k(\cdot)}(f)$ follows from (3.5.1) and Fatou's lemma in L^1. In the case $\varphi_{p(\cdot)} = \bar{\varphi}_{p(\cdot)}$ and $\mu(\{p = \infty\}) > 0$, we also need $\bar{\varphi}_q(t) \leqslant \lim_{k\to\infty} \bar{\varphi}_{q_k}(t)$ from Remark 3.5.2.

Now, let $\alpha := \liminf_{k\to\infty} \|f\|_{p_k(\cdot)}$. There is nothing to prove for $\alpha = \infty$, so let us assume that $\alpha < \infty$. For every $\lambda > \alpha$ we have $\|f\|_{p_k(\cdot)} \leqslant \lambda$ for large k and therefore by the unit ball property $\varrho_{p_k(\cdot)}(f/\lambda) \leqslant 1$ for large k. The first part of the corollary implies $\varrho_{p(\cdot)}(f/\lambda) \leqslant 1$ and hence $\|f\|_{p(\cdot)} \leqslant \lambda$ by the unit ball property. Since $\lambda > \alpha$ was arbitrary the claim follows. □

Under certain integrability conditions on f, the modular is also continuous with respect to pointwise convergence of the exponent.

Lemma 3.5.5. Let $r, s, p_k, p \in \mathcal{P}(A,\mu)$ with $r \leqslant p_k \leqslant s$ and $p_k \to p$ μ-almost everywhere. Let $f \in L^0(A,\mu)$ with $\varrho_{r(\cdot)}(f), \varrho_{s(\cdot)}(f) < \infty$. Then $\lim_{k\to\infty} \widetilde{\varrho}_{p_k(\cdot)}(f) = \widetilde{\varrho}_{p(\cdot)}(f)$.
If additionally $\mu(\{s = \infty\}) = 0$, then $\lim_{k\to\infty} \bar{\varrho}_{p_k(\cdot)}(f) = \bar{\varrho}_{p(\cdot)}(f)$.

Proof. By (3.3.8) we have $\varphi_{p_k(\cdot)}(f) \leqslant \varphi_{r(\cdot)}(f) + \varphi_{s(\cdot)}(f)$ pointwise, where the left-hand side converges pointwise by (3.5.1). Thus, the claim follows by the theorem of dominated convergence. □

Remark 3.5.6. If we drop the condition $\mu(\{s = \infty\}) = 0$, then it follows by Remark 3.5.2 that $\lim_{k\to\infty} \bar{\varrho}_{p_k(\cdot)}(\lambda f) \leqslant \bar{\varrho}_{p(\cdot)}(f) \leqslant \lim_{k\to\infty} \bar{\varrho}_{p_k(\cdot)}(f)$ for all $\lambda \in [0,1)$.

Theorem 3.5.7. *Let $p_k, p \in \mathcal{P}(A, \mu)$ with $p_k \nearrow p$ μ-almost everywhere and suppose that $\mu(A) < \infty$. Then for all $f \in L^{p(\cdot)}(A, \mu)$ holds*

$$\lim_{k \to \infty} \|f\|_{p_k(\cdot)} = \|f\|_{p(\cdot)}.$$

Proof. We know from Corollary 3.5.4 that $\|f\|_{p(\cdot)} \leqslant \liminf_{k \to \infty} \|f\|_{p_k(\cdot)}$, so it suffices to prove $\limsup_{k \to \infty} \|f\|_{p_k(\cdot)} \leqslant \|f\|_{p(\cdot)}$.

We begin with the proof for $\varphi_{p(\cdot)} = \widetilde{\varphi}_{p(\cdot)}$. The case $f = 0$ is obvious, so we can assume $\|f\|_{p(\cdot)} > 0$. Since $\mu(A) < \infty$, it follows by Lemma 3.2.20 (with $s = p_k$ and $q = 1$) and the unit ball property that for all $\lambda \in (0, 1)$

$$\widetilde{\varrho}_{p_1(\cdot)}\left(\frac{\lambda f}{\|f\|_{p(\cdot)}}\right) \leqslant \widetilde{\varrho}_{p(\cdot)}\left(\frac{\lambda f}{\|f\|_{p(\cdot)}}\right) + \widetilde{\varrho}_1(1) \leqslant 1 + \mu(A) < \infty$$

Therefore, we can apply Lemma 3.5.5 to get

$$\lim_{k \to \infty} \widetilde{\varrho}_{p_k(\cdot)}\left(\frac{\lambda f}{\|f\|_{p(\cdot)}}\right) = \widetilde{\varrho}_{p(\cdot)}\left(\frac{\lambda f}{\|f\|_{p(\cdot)}}\right) \leqslant \lambda \widetilde{\varrho}_{p(\cdot)}\left(\frac{f}{\|f\|_{p(\cdot)}}\right) \leqslant \lambda < 1$$

for all $\lambda \in (0, 1)$. Thus $\widetilde{\varrho}_{p_k(\cdot)}(\lambda f / \|f\|_{p(\cdot)}) \leqslant 1$ for large k, which implies that $\|f\|_{p_k(\cdot)} \leqslant \|f\|_{p(\cdot)}/\lambda$ for large k. Since $\lambda \in (0, 1)$ was arbitrary, this proves $\limsup_{k \to \infty} \|f\|_{p_k(\cdot)} \leqslant \|f\|_{p(\cdot)}$.

The proof of the case $\varphi_{p(\cdot)} = \overline{\varphi}_{p(\cdot)}$ is similar if we start our estimates with $\lim_{k \to \infty} \overline{\varrho}_{p_k(\cdot)}(\lambda^2 f / \|f\|_{p(\cdot)}) \leqslant \overline{\varrho}_{p(\cdot)}(\lambda f / \|f\|_{p(\cdot)})$ using Remark 3.5.6.   □

3.6 Convolution*

For two measurable functions f and g, we define the *convolution* by

$$f * g(z) := \int_{\mathbb{R}^n} f(z - y)g(y)\, dy = \int_{\mathbb{R}^n} f(y)g(z - y)\, dy$$

for every $z \in \mathbb{R}^n$ provided this formula makes sense. If the functions f and g are only defined on a subset Ω, then we extend them by zero outside of Ω before applying the convolution.

The operation of convolution on classical Lebesgue spaces is described by *Young's inequality* for convolution. It states that

$$\|f * g\|_r \leqslant \|f\|_p \|g\|_q.$$

for $f \in L^p(\mathbb{R}^n)$ and $g \in L^q(\mathbb{R}^n)$ when $p, q, r \in [1, \infty]$ with $\frac{1}{r} + 1 = \frac{1}{p} + \frac{1}{q}$. The case $q = 1$,

$$\|f * g\|_p \leqslant \|f\|_p \|g\|_1,$$

is of special interest.

Unfortunately these inequalities cannot be generalized to the spaces $L^{p(\cdot)}$ for non-constant p. This is a consequence of the fact that our spaces are not translation invariant. In fact, with the help of Theorem 3.3.1 we show that translations are bounded on $L^{p(\cdot)}(\Omega)$ if and only if the variable exponent p is constant, i.e. if we are in the setting of classical Lebesgue spaces.

Proposition 3.6.1. *Let $p \in \mathcal{P}(\mathbb{R}^n)$ and define the translation operator by $(\tau_h f)(y) := f(y - h)$. Then τ_h maps $L^{p(\cdot)}(\mathbb{R}^n)$ to $L^{p(\cdot)}(\mathbb{R}^n)$ for every $h \in \mathbb{R}^n$ if and only if p is constant.*

Proof. Suppose first that τ_h is bounded on $L^{p(\cdot)}(\mathbb{R}^n)$ for every $h \in \mathbb{R}^n$. Since $\|\tau_h f\|_{p(\cdot)} = \|f\|_{\tau_{-h}p(\cdot)}$, this implies that $L^{p(\cdot)}(\mathbb{R}^n) \hookrightarrow L^{\tau_{-h}p(\cdot)}(\mathbb{R}^n)$. From Theorem 3.3.1 (b) we deduce that $p \geqslant \tau_h p$ almost everywhere. Replacing h by $-h$ we see that $p \geqslant \tau_h p \geqslant p$ almost everywhere. Since h is arbitrary, p has to be constant. The opposite implication is immediate. $\square$

If $p \in \mathcal{P}(\mathbb{R}^n)$ is a non-constant exponent, then we can construct a single function $f \in L^{p(\cdot)}(\mathbb{R}^n)$ with $\tau_h f \notin L^{p(\cdot)}(\mathbb{R}^n)$ by a standard procedure. Namely, let $h \in \mathbb{R}^n \setminus \{0\}$ be such that τ_h is not bounded from $L^{p(\cdot)}(\mathbb{R}^n)$ to $L^{p(\cdot)}(\mathbb{R}^n)$. Choose $f_j \in L^{p(\cdot)}(\mathbb{R}^n)$ with $f_j \geqslant 0$, $\|f_j\|_{p(\cdot)} \leqslant 2^{-j}$ and $\|\tau_h f_j\|_{p(\cdot)} \geqslant 2^j$ and set $f := \sum_{j=1}^{\infty} f_j$. Then $\|f\|_{p(\cdot)} \leqslant \sum_{j=1}^{\infty} \|f_j\|_{p(\cdot)} \leqslant 1$ and $\|\tau_h f\|_{p(\cdot)} \geqslant \lim \|\tau_h f_j\|_{p(\cdot)} = \infty$.

Remark 3.6.2. The previous proposition also holds if we replace $\mathbb{R}^n$ by some open, non-empty set $\Omega \subset \mathbb{R}^n$. Arguing as in the proof of Proposition 3.6.1 we deduce that $p \geqslant \tau_h p \geqslant p$ on the set $(\Omega - h) \cap \Omega$. Since h is arbitrary this implies again that p is constant on all of Ω.

Theorem 3.6.3. *Let Ω be bounded and $p, r \in \mathcal{P}(\mathbb{R}^n)$ with $1 < p^- \leqslant p^+ < \infty$ and $1 < r^- \leqslant r^+ < \infty$. Then the convolution $* : (f, g) \mapsto f * g$ is bounded as a mapping from $L^{p(\cdot)}(\Omega) \times L^1(\mathbb{R}^n)$ to $L^{r(\cdot)}(\Omega)$ if and only if $p^- \geqslant r^+$.*

Proof. "$\Leftarrow$": Since $p^- \geqslant r^+$, Corollary 3.3.4 implies $L^{p(\cdot)}(\Omega) \hookrightarrow L^{r^+}(\Omega) \hookrightarrow L^{r(\cdot)}(\Omega)$. By Young's convolution inequality, $* : L^{r^+}(\Omega) \times L^1(\mathbb{R}^n) \to L^{r^+}(\Omega)$. Combining these, we obtain the claim.

"$\Rightarrow$": We proceed by contradiction and assume $p^- < r^+$. So there exists $h \in \mathbb{R}^n$ such that $p \geqslant \tau_{-h} r$ does not hold almost everywhere on $\Omega \cap (\Omega - h)$. Hence, it follows from Theorem 3.3.1 (b) and $\|\tau_h f\|_{r(\cdot)} = \|f\|_{\tau_{-h}r(\cdot)}$ that τ_h does not map $L^{p(\cdot)}(\Omega)$ continuously to $L^{r(\cdot)}(\Omega)$. By Proposition 3.6.1 and Remark 3.6.2 there exists $f \in L^{p(\cdot)}(\Omega)$ and $h \in \mathbb{R}^n \setminus \{0\}$ such that

$\tau_h f \notin L^{r(\cdot)}(\Omega)$. For $\psi \in C_0^\infty(\mathbb{R}^n)$, $\psi \geqslant 0$ and $\int \psi \, dx = 1$ define ψ_ε by $\psi_\varepsilon(y) := \varepsilon^{-n} \psi((y-h)/\varepsilon)$, and note that $f * \psi_\varepsilon \to \tau_h f$ in $L^1_{\mathrm{loc}}(\mathbb{R}^n)$. By assumption on the convolution $\|f * \psi_\varepsilon\|_{r(\cdot)} \leqslant \|f\|_{p(\cdot)} \|\psi\|_1 \leqslant c$. Since $L^{r(\cdot)}(\Omega)$ is reflexive, there exists a subsequence converging weakly in $L^{r(\cdot)}(\Omega)$ to a function $g \in L^{r(\cdot)}(\Omega)$ as $\varepsilon \to 0$. Since $L^{r(\cdot)}(\Omega) \hookrightarrow L^1_{\mathrm{loc}}(\Omega)$, we have $g = \tau_h f$. In particular the subsequence converges weakly in $L^{r(\cdot)}(\Omega)$ to $\tau_h f$. This contradicts $\tau_h f \notin L^{r(\cdot)}(\Omega)$.   □

This theorem has the following undesired consequence:

Corollary 3.6.4. *Let $p \in \mathcal{P}(\mathbb{R}^n)$ with $1 < p^- \leqslant p^+ < \infty$. Then*

$$\|f * g\|_{p(\cdot)} \leqslant c \|f\|_{p(\cdot)} \|g\|_1$$

for some $c > 0$ and all $f \in L^{p(\cdot)}(\mathbb{R}^n)$ and all $g \in L^1(\mathbb{R}^n)$ if and only if p is constant.

Proof. If the inequality holds, then by Theorem 3.6.3 we have $p_\Omega^- \geqslant p_\Omega^+$ for all bounded, open subsets $\Omega \subset \mathbb{R}^n$. Thus $p^- \geqslant p^+$ and p has to be constant. If on the other hand p is constant, then the inequality is a consequence of Young's inequality for convolution, which was stated in the beginning of the section.   □

Let $f \in L^{p(\cdot)}(\mathbb{R}^n)$. Then the preceding corollary shows that for $f * g$ to belong to $L^{p(\cdot)}(\mathbb{R}^n)$ it is in general not enough to assume $g \in L^1(\mathbb{R}^n)$. However, we can solve this problem by assuming more regularity for g. This will be useful for instance when dealing with the Bessel potential in Sect. 12.4.

Theorem 3.6.5. *Let $p, q \in \mathcal{P}(\mathbb{R}^n)$ and let $p_\infty, q_\infty \in [1, \infty]$ satisfy $p^- \leqslant p_\infty \leqslant q_\infty \leqslant q^+$. Assume that $1 \in L^{s(\cdot)}(\mathbb{R}^n)$, where $s \in \mathcal{P}(\mathbb{R}^n)$ is defined by $\frac{1}{s(x)} := \left| \frac{1}{p(x)} - \frac{1}{p_\infty} \right|$. Let $r_0, r_1 \in [1, \infty]$ be defined by*

$$\frac{1}{r_0} = 1 - \frac{1}{p_\infty} + \frac{1}{q_\infty} \quad \text{and} \quad \frac{1}{r_1} = 1 - \frac{1}{p^-} + \frac{1}{q^+}.$$

Let $*$ denote the convolution operator. Then the bilinear mapping

$$* : L^{p(\cdot)}(\mathbb{R}^n) \times (L^{r_0}(\mathbb{R}^n) \cap L^{r_1}(\mathbb{R}^n)) \to L^{q(\cdot)}(\mathbb{R}^n) \cap L^{q^+}(\mathbb{R}^n)$$

is bounded.

Proof. Note that $p^- \leqslant p_\infty \leqslant q_\infty \leqslant q^+$ ensures that $r_0, r_1 \in [1, \infty]$ are well defined and $r_0 \leqslant r_1$. Define $r_2, r_3 \in [1, \infty]$ by

$$\frac{1}{r_2} = 1 - \frac{1}{p^-} + \frac{1}{q_\infty}, \qquad \frac{1}{r_3} = 1 - \frac{1}{p_\infty} + \frac{1}{q^+}.$$

Young's inequality for convolution for constant exponents implies the bound-edness of the following bilinear mappings

$$* : L^{p^-} \times L^{r_1} \to L^{q^+},$$
$$* : L^{p^-} \times L^{r_2} \to L^{q\infty},$$
$$* : L^{p\infty} \times L^{r_3} \to L^{q^+},$$
$$* : L^{p\infty} \times L^{r_0} \to L^{q\infty}.$$

Therefore,

$$* : L^{p^-} \times (L^{r_2} \cap L^{r_1}) \to L^{q\infty} \cap L^{q^+},$$
$$* : L^{p\infty} \times (L^{r_0} \cap L^{r_3}) \to L^{q\infty} \cap L^{q^+}.$$

From $r_0 \leqslant \min\{r_2, r_3\} \leqslant \max\{r_2, r_3\} \leqslant r_1$ and Theorem 3.3.11 we deduce

$$L^{r_0} \cap L^{r_1} \hookrightarrow L^{r_2} \cap L^{r_1} \qquad \text{and} \qquad L^{r_0} \cap L^{r_1} \hookrightarrow L^{r_0} \cap L^{r_3}.$$

Combining these embeddings with our previous result implies that

$$* : (L^{p^-} + L^{p\infty}) \times (L^{r_0} \cap L^{r_1}) \to L^{q\infty} \cap L^{q^+}.$$

By Lemma 3.3.12, $L^{p(\cdot)} \hookrightarrow L^{p^-} + L^{p\infty}$, and $L^{q\infty} \cap L^{q^+} \hookrightarrow L^{q(\cdot)} \cap L^{q^+}$. Combining this with the previous formula concludes the proof.   □

Chapter 4
The Maximal Operator

In the previous chapters we studied the spaces $L^{p(\cdot)}$ with general variable exponent p. We have seen that many results hold for fairly wild exponents, including discontinuous ones, in this general setting. We studied completeness, separability, reflexivity, and uniform convexity. However, these are only basic properties of $L^{p(\cdot)}$. For the study of partial differential equations it is necessary to develop more advanced tools for the $L^{p(\cdot)}$ spaces: we are interested in mollification, the Riesz potential, singular integrals, and the Hardy–Littlewood maximal operator. For general variable exponents p it is not possible to transfer these tools to $L^{p(\cdot)}$, as our counterexample in Sect. 4.7 shows. It turns out that a certain regularity has to be assumed on p: the so-called log-Hölder continuity of p. We will see that this regularity is in some sense optimal and cannot be improved.

In Corollary 3.6.4 we saw that the inequality $\|f * g\|_{p(\cdot)} \leqslant c \|f\|_{p(\cdot)} \|g\|_1$ does not hold for non-constant $p \in \mathcal{P}(\Omega)$. This seems like a strong drawback for the theory of $L^{p(\cdot)}$-spaces, since the version for constant exponents is used in many applications. For example the technique of mollification or approximate identities relies on this fact and a failure of this technique would have drastic consequences. But Corollary 3.6.4 only states that we have no control of the convolution of an $L^{p(\cdot)}$ function (p non-constant) with an arbitrary L^1 function. The proof relied on the fact that we could approximate the translation operator by the convolution with a sequence of L^1 functions, i.e. we used a shifted version of an approximate identity. So we used functions ψ_ε which concentrated in the limit $\varepsilon \to 0$ in some point h, i.e. $\psi_\varepsilon \to \delta_h$ in the sense of distributions, where δ_h is the δ-distribution at h. The technique of mollification or approximate identities is however restricted to the case $\psi_\varepsilon \to \delta_0$ for $\varepsilon \to 0$. Since $f * \delta_0 = f$, this does not contradict the discontinuity of translations. Indeed, we will see below that for certain variable exponents p we have $f * \psi_\varepsilon \to f$ in $L^{p(\cdot)}$ for $\varepsilon \to 0$ if ψ_ε is an approximate identity.

L. Diening et al., *Lebesgue and Sobolev Spaces with Variable Exponents*, Lecture Notes in Mathematics 2017, DOI 10.1007/978-3-642-18363-8_4,

4.1 Logarithmic Hölder Continuity

In this section we introduce the most important condition on the exponent
in the study of variable exponent spaces, the log-Hölder continuity condition.

Definition 4.1.1. We say that a function $\alpha \colon \Omega \to \mathbb{R}$ is *locally* log-*Hölder
continuous* on Ω if there exists $c_1 > 0$ such that

$$|\alpha(x) - \alpha(y)| \leqslant \frac{c_1}{\log(e + 1/|x - y|)}$$

for all $x, y \in \Omega$. We say that α satisfies the log-*Hölder decay condition* if there
exist $\alpha_\infty \in \mathbb{R}$ and a constant $c_2 > 0$ such that

$$|\alpha(x) - \alpha_\infty| \leqslant \frac{c_2}{\log(e + |x|)}$$

for all $x \in \Omega$. We say that α is *globally* log-*Hölder continuous* in Ω if it is
locally log-Hölder continuous and satisfies the log-Hölder decay condition.
The constants c_1 and c_2 are called the *local* log-*Hölder constant* and the log-
Hölder decay constant, respectively. The maximum $\max\{c_1, c_2\}$ is just called
the log-*Hölder constant* of α.

The local log-Hölder condition was first used in the variable exponent
context by Zhikov [392]. Various authors have used different names for this
condition, e.g. weak Lipschitz, Dini–Lipschitz, and 0-Hölder. However, we
think these terms are ambiguous and prefer the name log-Hölder. Before
appearing in the variable exponent context, the same condition was used
with variable order Hölder spaces [177, 229, 332]. It is unclear to what extent
these studies were known to researchers of variable exponent spaces, however.

If α is globally log-Hölder continuous on an unbounded domain, e.g. $\mathbb{R}^n$,
then the constant α_∞ in Definition 4.1.1 is unique. Note that any globally
log-Hölder continuous function is bounded.

Remark 4.1.2. We define the chordal metric $d : \overline{\mathbb{R}^n} \times \overline{\mathbb{R}^n} \to \mathbb{R}$ by

$$d(x, y) = \frac{|x - y|}{\sqrt{1 + |x|^2}\sqrt{1 + |y|^2}} \quad \text{and} \quad d(x, \infty) = \frac{1}{\sqrt{1 + |x|^2}}$$

for $x, y \in \mathbb{R}^n$. The motivation for the term "global log-Hölder continuity"
comes from the fact that $\alpha \colon \overline{\mathbb{R}^n} \to \mathbb{R}$ is globally log-Hölder continuous if and
only if

$$|\alpha(x) - \alpha(y)| \leqslant \frac{c}{\log(e + 1/d(x, y))} \tag{4.1.3}$$

for all $x, y \in \overline{\mathbb{R}^n}$. Since $d(x, y) \leqslant |x - y|$ and $|x| \leqslant 1/d(x, \infty)$, it is clear that
(4.1.3) implies log-Hölder continuity. The other implication follows from the
inequality

$$\frac{1}{d(x,y)} \leqslant 2\frac{1+|x|^2}{|x-y|} + 2\sqrt{1+|x|^2} \leqslant c\left(1+|x|^4 + \frac{1}{|x-y|^2}\right),$$

which in turn follows since $\sqrt{1+|y|^2} \leqslant 2\sqrt{1+|x|^2} + 2|x-y|$. The details are left to the reader.

Definition 4.1.4. We define the following class of variable exponents

$$\mathcal{P}^{\log}(\Omega) := \left\{p \in \mathcal{P}(\Omega) \colon \tfrac{1}{p} \text{ is globally log-Hölder continuous}\right\}.$$

By $c_{\log}(p)$ or $c_{\log}$ we denote the log-Hölder constant of $\frac{1}{p}$. If Ω is unbounded, then we define p_∞ by $\frac{1}{p_\infty} := \lim_{|x|\to\infty} \frac{1}{p(x)}$. As usual we use the convention $\frac{1}{\infty} := 0$.

Note that although $\frac{1}{p}$ is bounded, the variable exponent p itself can be unbounded. We would also like to remark that the definition of p_∞ "commutes with duality", i.e. $p \in \mathcal{P}^{\log}(\Omega)$ if and only if $p' \in \mathcal{P}^{\log}(\Omega)$ and

$$(p_\infty)' = (p')_\infty.$$

Hence we do not have to distinguish between $(p_\infty)'$ and $(p')_\infty$, and write p'_∞ for short.

Remark 4.1.5. If $p \in \mathcal{P}(\Omega)$ with $p^+ < \infty$, then $p \in \mathcal{P}^{\log}(\Omega)$ if and only if p is globally log-Hölder continuous. This is due to the fact that $p \mapsto \frac{1}{p}$ is a bilipschitz mapping from $[p^-, p^+]$ to $[\frac{1}{p^+}, \frac{1}{p^-}]$.

The following lemma provides a characterization of local log-Hölder continuity. Recall the notation $\alpha_A^\pm$ for the supremum and infimum of α over a set A.

Lemma 4.1.6. *Let* $\alpha : \mathbb{R}^n \to \mathbb{R}$ *be continuous and bounded, i.e.* $-\infty < \alpha^- \leqslant \alpha^+ < \infty$. *The following conditions are equivalent:*

(a) *α is locally* log-*Hölder continuous.*
(b) *For all balls B we have $|B|^{\alpha_B^- - \alpha_B^+} \leqslant c$.*
(c) *For all balls B and all $x \in B$ we have $|B|^{\alpha_B^- - \alpha(x)} \leqslant c$.*
(d) *For all balls B and all $x \in B$ we have $|B|^{\alpha(x) - \alpha_B^+} \leqslant c$.*

Instead of balls it is also possible to use cubes.

Proof. (a) $\Rightarrow$ (b): Since $\alpha_B^- - \alpha_B^+$ is non-positive, the claim is clear for balls of radius greater than $\frac{1}{4}$. If B is a ball with radius less than this, we use the local log-Hölder condition:

$$|\alpha_B^- - \alpha_B^+| \log \frac{1}{|B|} \leqslant \frac{c_1 \log(1/|B|)}{\log(e + 1/\operatorname{diam}(B))} \leqslant \frac{c_1 n \log(1/|B|)}{\log(c/|B|)} \leqslant c.$$

(b) $\Rightarrow$ (a): Fix $x, y \in \mathbb{R}^n$ and choose a ball B_r with radius r such that $x, y \in B_r$ and $\frac{|x-y|}{2} < r < |x - y|$. Since $|B_r| \leqslant (2r)^n$,

$$\left(2|x - y|\right)^{-|\alpha(x)-\alpha(y)|} \leqslant (2r)^{-|\alpha(x)-\alpha(y)|} \leqslant |B_r|^{\frac{-|\alpha(x)-\alpha(y)|}{n}} \leqslant |B_r|^{\frac{\alpha_B^- - \alpha_B^+}{n}} \leqslant c_1^{\frac{1}{n}}.$$

Since $|\alpha^+ - \alpha^-| < \infty$, this proves $|x - y|^{-|\alpha(x)-\alpha(y)|} \leqslant c$ for some $c > 1$. We take the logarithm of this inequality to deduce $|\alpha(x) - \alpha(y)| \leqslant \frac{\log c}{|\log|x-y||}$. This takes care of the claim when $|x - y| < \frac{1}{2}$; on the other hand, there the claim is obvious when $|x - y| \geqslant \frac{1}{2}$, since α is bounded by assumption.

The equivalence of (b), (c) and (d) is clear by the continuity of α. $\square$

Many results below are stated for variable exponents p which are defined on the whole space $\mathbb{R}^n$. However, sometimes initially the variable exponent is only given on a subset $\Omega \subset \mathbb{R}^n$, i.e. $q \in \mathcal{P}^{\log}(\Omega)$. The following result ensures that such a variable exponent q can always be extended to $\mathbb{R}^n$ without changing the fundamental properties.

Proposition 4.1.7. *If $p \in \mathcal{P}^{\log}(\Omega)$, then it has an extension $q \in \mathcal{P}^{\log}(\mathbb{R}^n)$ with $c_{\log}(q) = c_{\log}(p)$, $q^- = p^-$, and $q^+ = p^+$. If Ω is unbounded, then additionally $q_\infty = p_\infty$.*

Proof. Let $c_1 > 0$ and $p_\infty \geqslant 1$ be such that

$$\left|\frac{1}{p(x)} - \frac{1}{p(y)}\right| \leqslant \frac{c_1}{\log(e + 1/|x - y|)} \quad \text{and} \quad |p(x) - p_\infty| \leqslant \frac{c_1}{\log(e + |x|)}.$$

for all points $x, y \in \Omega$. Since $t \mapsto 1/\log(e + 1/t)$ is a modulus of continuity, we can use the extension of McShane-type [289] to extend $\frac{1}{p}$ to $\mathbb{R}^n$ with the same modulus of continuity and lower and upper bound. More precisely, we define $a \in C(\mathbb{R}^n)$ by

$$\frac{1}{a(y)} := \sup_{z \in \Omega} \left(\frac{1}{p(z)} - \frac{c_1}{\log(e + 1/|z - y|)}\right)$$

for $y \in \mathbb{R}^n$. In particular, a is locally log-Hölder continuous with local log-Hölder constant less or equal to c_1, and $a(y) = 1/p(y)$ for all $y \in \Omega$.

In order to ensure that our extension satisfies also the log-Hölder decay condition and has the same lower and upper bound as p we define q by truncation

$$\frac{1}{q(y)} := \min\left\{\max\left\{\frac{1}{a(y)}, \frac{1}{p_\infty} - \frac{c_2}{\log(e + |x|)}, \frac{1}{p_\Omega^+}\right\}, \frac{1}{p_\infty} + \frac{c_2}{\log(e + |x|)}, \frac{1}{p_\Omega^-}\right\}$$

for all $y \in \mathbb{R}^n$. Since $x \mapsto \frac{c_2}{\log(e+|x|)}$ is globally log-Hölder continuous with constant c_2, we see the log-Hölder constant of $\frac{1}{q}$ does not exceed $\max\{c_1, c_2\} = c_{\log}(p)$. The decay condition of $\frac{1}{p}$ ensures that $\frac{1}{q(y)} = a(y) = \frac{1}{p(y)}$ for all $y \in \Omega$. Therefore q is the variable exponent we are looking for. $\qquad\square$

Proposition 4.1.7 was first proved by Diening and Růžička [103] under the additional assumption that q is constant outside a large ball. For a general variable exponent $q \in \mathcal{P}^{\log}(\Omega)$ the result was first proved by Cruz-Uribe, Fiorenza, Martell and Pérez [83, Lemma 4.3] by means of the Whitney decomposition. The proof that we included is simpler and originates from Diening and Hästö [96, Proposition 3.7].

Let $p \in \mathcal{P}(\mathbb{R}^n)$ and $\frac{1}{s} := |\frac{1}{p} - \frac{1}{p_\infty}|$. We saw in Lemma 3.3.5 that the condition $1 \in L^{s(\cdot)}$ is important for embeddings; we now show that this condition follows from log-Hölder continuity. The condition turns out also to be important for the boundedness of maximal operators.

Proposition 4.1.8. *Let $p, q \in \mathcal{P}^{\log}(\mathbb{R}^n)$ with $p_\infty = q_\infty$. If $s \in \mathcal{P}(\mathbb{R}^n)$ is given by $\frac{1}{s} := |\frac{1}{p} - \frac{1}{q}|$, then $1 \in L^{s(\cdot)}(\mathbb{R}^n)$ and for every $m > 0$ there exists $\gamma \in (0,1)$ only depending on $c_{\log}(p)$ and n such that*

$$\varphi_{s(y)}(\gamma) \leqslant (e + |y|)^{-m}$$

for all $y \in \mathbb{R}^n$. Moreover,

$$L^{\max\{p(\cdot), q(\cdot)\}}(\mathbb{R}^n) \hookrightarrow L^{q(\cdot)}(\mathbb{R}^n) \hookrightarrow L^{\min\{p(\cdot), q(\cdot)\}}(\mathbb{R}^n)$$

Proof. We begin with the estimate for $\varphi_{s(y)}(\gamma)$. If $s(y) = \infty$, then $\varphi_\infty(\gamma) = 0$. So let us assume $s(y) < \infty$. Since $p \in \mathcal{P}^{\log}(\mathbb{R}^n)$, we have

$$\left| \frac{1}{p(y)} - \frac{1}{p_\infty} \right| \leqslant \frac{c_{\log}(p)}{\log(e + |y|)}$$

for all $y \in \mathbb{R}^n$. Let $\gamma := \exp(-m\, c_{\log}(p))$. Since $s(y) < \infty$, $\bar{\varphi}_{s(y)}(\gamma) = \gamma^{s(y)}$, and so we estimate

$$\bar{\varphi}_{s(y)}(\gamma) \leqslant \exp\left(\frac{-m\, c_{\log}(p)}{|\frac{1}{p(y)} - \frac{1}{p_\infty}|} \right) \leqslant \exp\left(-m \log(e + |y|) \right) = (e + |y|)^{-m}.$$

This proves the estimate for $\bar{\varphi}_{s(y)}(\gamma)$. Since $\widetilde{\varphi}_{p(\cdot)} \leqslant \bar{\varphi}_{p(\cdot)}$, the conclusion holds also for $\widetilde{\varphi}_{p(\cdot)}$. If $m > n$, then $\varphi_{s(\cdot)}(\gamma) \leqslant (e + |\cdot|)^{-m} \in L^1(\mathbb{R}^n)$; hence $1 \in L^{s(\cdot)}(\mathbb{R}^n)$.

Define $s_1, s_2 \in \mathcal{P}^{\log}(\mathbb{R}^n)$ by $\frac{1}{s_1} = \max\{\frac{1}{q} - \frac{1}{p}, 0\}$ and $\frac{1}{s_2} = \max\{\frac{1}{p} - \frac{1}{q}, 0\}$. If we use the already shown claims with p replaced by $\max\{p(\cdot), q(\cdot)\}$ and $\min\{p(\cdot), q(\cdot)\}$, then we get $1 \in L^{s_1(\cdot)}(\mathbb{R}^n) \cap L^{s_2(\cdot)}(\mathbb{R}^n)$, respectively. The embeddings now follow directly from Theorem 3.3.1. $\qquad\square$

Combining the previous proposition with Lemma 3.3.12 we see that

$$L^{p(\cdot)}(\mathbb{R}^n) \cap L^{p^+}(\mathbb{R}^n) \cong L^{p_\infty}(\mathbb{R}^n) \cap L^{p^+}(\mathbb{R}^n)$$

when $p \in \mathcal{P}^{\log}$. From Theorem 3.3.7 we get an interesting corollary, see [313, 319].

Corollary 4.1.9. *If $p, q \in \mathcal{P}(\mathbb{Z}^n)$ satisfy the log-Hölder decay condition and $p_\infty = q_\infty$, then $l^{p(\cdot)}(\mathbb{Z}^n) \cong l^{q(\cdot)}(\mathbb{Z}^n)$.*

4.2 Point-Wise Estimates

Recall that $\varphi_{p_Q^-}^{-1}$ is the left-continuous inverse as defined in Definition 3.1.8.

Lemma 4.2.1. *Let $p \in \mathcal{P}^{\log}(\mathbb{R}^n)$. Then there exists $\beta \in (0,1)$ which only depends on $c_{\log}(p)$ such that*

$$\varphi_{p(x)}\left(\beta\varphi_{p_Q^-}^{-1}\big(\lambda|Q|^{-1}\big)\right) \leqslant \lambda|Q|^{-1},$$

for all $\lambda \in [0,1]$, any cube (or ball) $Q \subset \mathbb{R}^n$ and any $x \in Q$.

Proof. If $\lambda = 0$, then the claim follows from $\varphi_{p_Q^-}^{-1}(0) = 0$ and $\varphi_{p(x)}(0) = 0$. So let us assume in the following that $\lambda > 0$. If $p_Q^- = \infty$, then, by continuity of $\frac{1}{p}$, $p(x) = \infty$ for all $x \in \mathbb{R}^n$ and $\bar\varphi_\infty\big(\frac{1}{2}\bar\varphi_\infty^{-1}(\lambda|Q|^{-1})\big) = \bar\varphi_\infty\big(\frac{1}{2}\big) = 0$. Assume now that $p_Q^- < \infty$ and $p(x) < \infty$. By Lemma 4.1.6 there exists $\beta \in (0,1)$ such that

$$\beta|Q|^{\frac{1}{p(x)} - \frac{1}{p_Q^-}} \leqslant 1.$$

Now, multiply this by $|Q|^{-\frac{1}{p(x)}}$ and raise the result to the power of $p(x)$ to prove the claim for $\lambda = 1$ and $\varphi_{p(\cdot)} = \bar\varphi_{p(\cdot)}$. The case $0 \leqslant \lambda < 1$ follows from this and

$$\bar\varphi_{p(x)}\left(\beta\bar\varphi_{p_Q^-}^{-1}\big(\lambda|Q|^{-1}\big)\right) = \lambda^{\frac{p(x)}{p_Q^-}} \bar\varphi_{p(x)}\left(\beta\bar\varphi_{p_Q^-}^{-1}\big(|Q|^{-1}\big)\right) \leqslant \lambda|Q|^{-1},$$

It remains to consider the case $p(x) = \infty$ and $p_Q^- < \infty$. Since $\frac{1}{p}$ is continuous and $p_Q^- < \infty$ we can choose a sequence (x_k) from Q which tend to x' with $p(x_k) < \infty$ for all $k \in \mathbb{N}$ and $p(x') = \infty$. Then by Remark 3.5.2 $\bar\varphi_{p(x)}(t) = \bar\varphi_{p(x')}(t) \leqslant \lim_{k\to\infty} \bar\varphi_{p(x_k)}(t)$ for all $t \geqslant 0$. Hence, this case can be reduced to the previous case. This proves the claim for $\varphi_{p(\cdot)} = \bar\varphi_{p(\cdot)}$. By Lemmas 3.1.6 and 3.1.12,

$$\tilde{\varphi}_{p(x)}\left(\frac{\beta}{2}\tilde{\varphi}_{p_Q}^{-1}(\lambda|Q|^{-1})\right) \leqslant \tilde{\varphi}_{p(x)}\left(\beta\tilde{\varphi}_{p_Q}^{-1}(\lambda|Q|^{-1})\right) \leqslant \lambda|Q|^{-1}.$$

This proves the claim for $\varphi_{p(\cdot)} = \tilde{\varphi}_{p(\cdot)}$ with β replaced by $\beta/2$. □

We now derive a generalized version of Jensen's inequality for $\varphi_{p(\cdot)}$. For constant $q \in [1, \infty]$, $f \in L^q(Q)$ and a cube $Q \subset \mathbb{R}^n$ we have by Jensen's inequality

$$\varphi_q\left(\fint_Q |f(y)|\, dy\right) \leqslant \fint_Q \varphi_q(|f(y)|)\, dy.$$

However, this inequality only holds if the exponent is constant. The following lemma shows that it is possible to generalize the constant exponent case to the setting of variable exponents $p \in \mathcal{P}^{\log}(\mathbb{R}^n)$. The price to pay is a multiplicative constant on the left-hand side and an extra additive term on the right-hand side, which is independent of f as long as f is from the unit ball of $L^{p(\cdot)} + L^\infty$.

Lemma 4.2.2. *Let $p \in \mathcal{P}(\mathbb{R}^n)$ and let $\frac{1}{p}$ be locally log-Hölder continuous. Define $q \in \mathcal{P}^{\log}(\mathbb{R}^n \times \mathbb{R}^n)$ by*

$$\frac{1}{q(x,y)} := \max\left\{\frac{1}{p(x)} - \frac{1}{p(y)}, 0\right\}.$$

Then for any $\gamma \in (0,1)$ there exists $\beta \in (0,1)$ only depending on γ and $c_{\log}(p)$ such that

$$\varphi_{p(x)}\left(\beta\fint_Q |f(y)|\, dy\right) \leqslant \fint_Q \varphi_{p(y)}(|f(y)|)\, dy + \fint_Q \varphi_{q(x,y)}(\gamma)\,\chi_{\{0<|f(y)|\leqslant 1\}}\, dy$$

for every cube (or ball) $Q \subset \mathbb{R}^n$, $x \in Q$, and $f \in L^{p(\cdot)}(\mathbb{R}^n) + L^\infty(\mathbb{R}^n)$ with $\|f\|_{L^{p(\cdot)}(\mathbb{R}^n)+L^\infty(\mathbb{R}^n)} \leqslant 1$.

Proof. We prove the claim for $\varphi_{p(\cdot)} = \bar{\varphi}_{p(\cdot)}$. The case $\varphi_{p(\cdot)} = \tilde{\varphi}_{p(\cdot)}$ then follows easily by Lemma 3.1.6.

By convexity of $\bar{\varphi}_{p(y)}$ it suffices to prove the claim separately for $\|f\|_{p(\cdot)} \leqslant 1$ and $\|f\|_\infty \leqslant 1$. Let $Q \subset \mathbb{R}^n$ be a cube and $x \in Q$.

If $p_Q^- = \infty$, then $p(y) = \infty$ for all $y \in Q$ and the claim is just Jensen's inequality for the convex function $\bar{\varphi}_\infty$ with an extra positive term on the right-hand side. So we assume in the following $p_Q^- < \infty$.

Let $\beta > 0$ be as in Lemma 4.2.1. We can assume that $\beta \leqslant \gamma$. We split f into three parts

$$f_1(y) := f(y)\,\chi_{\{y\in Q:\,|f(y)|>1\}},$$
$$f_2(y) := f(y)\,\chi_{\{y\in Q:\,|f(y)|\leqslant 1,\,p(y)\leqslant p(x)\}},$$
$$f_3(y) := f(y)\,\chi_{\{y\in Q:\,|f(y)|\leqslant 1,\,p(y)>p(x)\}}.$$

Then $f = f_1 + f_2 + f_3$ and $|f_j| \leqslant |f|$, so $\overline{\varrho}_{p(\cdot)}(f_j) \leqslant \overline{\varrho}_{p(\cdot)}(f) \leqslant 1$, $j = 1, 2, 3$. By convexity of $\bar{\varphi}_{p(x)}$

$$\bar{\varphi}_{p(x)}\left(\frac{\beta}{3}\fint_Q |f(y)|\,dy\right) \leqslant \frac{1}{3}\sum_{j=1}^{3}\bar{\varphi}_{p(x)}\left(\beta\fint_Q |f_j(y)|\,dy\right) =: \frac{1}{3}(I_1 + I_2 + I_3).$$

So it suffices to consider the functions f_1, f_2, and f_3 independently. We start with f_1. The convexity of $\bar{\varphi}_{p_Q^-}$ and Jensen's inequality imply that

$$I_1 \leqslant \bar{\varphi}_{p(x)}\left(\beta\bar{\varphi}_{p_Q^-}^{-1}\left(\fint_Q \bar{\varphi}_{p_Q^-}(|f_1(y)|)\,dy\right)\right),$$

where we have used that $\bar{\varphi}_{p(x)}$ and $\bar{\varphi}_{p_Q^-}^{-1}$ are non-decreasing. Since $|f_1(y)| > 1$ or $|f_1(y)| = 0$ and $p_Q^- \leqslant p(y)$, we have $\bar{\varphi}_{p_Q^-}(|f_1(y)|) \leqslant \bar{\varphi}_{p(y)}(|f_1(y)|)$ by Lemma 3.1.6 and thus

$$I_1 \leqslant \bar{\varphi}_{p(x)}\left(\beta\bar{\varphi}_{p_Q^-}^{-1}\left(\fint_Q \bar{\varphi}_{p(y)}(|f_1(y)|)\,dy\right)\right).$$

If $\|f\|_\infty \leqslant 1$, then $f_1 = 0$ and $I_1 = 0$. If on the other hand $\|f\|_{p(\cdot)} \leqslant 1$, then $\overline{\varrho}_{p(\cdot)}(f) \leqslant 1$ and $\int_Q \bar{\varphi}_{p(y)}(|f_1(y)|)\,dy \leqslant 1$. So by Lemma 4.2.1 it follows with $\lambda = \int_Q \bar{\varphi}_{p(y)}(|f(y)|)\,dy$ that

$$I_1 \leqslant \fint_Q \bar{\varphi}_{p(y)}(|f_1(y)|)\,dy \leqslant \fint_Q \bar{\varphi}_{p(y)}(|f(y)|)\,dy.$$

Jensen's inequality implies that

$$I_2 \leqslant \fint_Q \bar{\varphi}_{p(x)}(\beta|f_2(y)|)\,dy.$$

Since $\beta|f_2(y)| \leqslant |f_2(y)| \leqslant 1$ and $\bar{\varphi}_{p(x)}(t) \leqslant \bar{\varphi}_{p(y)}(t)$ for all $t \in [0,1]$ when $p(y) \leqslant p(x)$ (see Lemma 3.1.6), we find that

$$I_2 \leqslant \fint_Q \bar{\varphi}_{p(y)}(\beta|f_2(y)|)\,dy \leqslant \fint_Q \bar{\varphi}_{p(y)}(|f_2(y)|)\,dy \leqslant \fint_Q \bar{\varphi}_{p(y)}(|f(y)|)\,dy.$$

Finally, for I_3 we get with Jensen's inequality

$$I_3 \leqslant \fint_Q \bar{\varphi}_{p(x)}\big(\beta|f(y)|\big)\,\chi_{\{y\in Q:\; 0<|f(y)|\leqslant 1, p(y)>p(x)\}}\,dy.$$

Now, Young's inequality (Lemma 3.2.15) and $\beta \leqslant \gamma$ give that

$$I_3 \leqslant \fint_Q \left(\bar{\varphi}_{p(y)}\left(\beta\frac{|f(y)|}{\gamma}\right) + \bar{\varphi}_{q(x,y)}(\gamma)\right)\chi_{\{y\in Q:\; 0<|f(y)|\leqslant 1, p(y)>p(x)\}}\,dy$$

$$\leqslant \fint_Q \bar{\varphi}_{p(y)}\big(|f(y)|\big)\,dy + \fint_Q \bar{\varphi}_{q(x,y)}(\gamma)\,\chi_{\{0<|f(y)|\leqslant 1, p(y)>p(x)\}}\,dy.$$

This proves the lemma. $\qquad\square$

In the case where the limit $\frac{1}{p_\infty} = \lim_{|x|\to\infty}\frac{1}{p(x)}$ exists, it is useful to split the second integral in the previous estimate into two parts by means of the following lemma:

Lemma 4.2.3. *Let $p \in \mathcal{P}^{\log}(\mathbb{R}^n)$. Let q be as in Lemma 4.2.2 and define $s \in \mathcal{P}(\mathbb{R}^n)$ by $\frac{1}{s(x)} := \left|\frac{1}{p(x)} - \frac{1}{p_\infty}\right|$. Then*

$$\widetilde{\varphi}_{q(x,y)}(t) \leqslant \widetilde{\varphi}_{s(x)}\big(t^{\frac{1}{2}}\big) + \widetilde{\varphi}_{s(y)}\big(t^{\frac{1}{2}}\big)$$

for every $t \in [0,1]$.

Proof. Let $t \in [0,1]$. For all $x, y \in \mathbb{R}^n$

$$0 \leqslant \frac{1}{q(x,y)} = \max\left\{0,\; \frac{1}{p(x)} - \frac{1}{p(y)}\right\} \leqslant \frac{1}{s(x)} + \frac{1}{s(y)} =: \frac{1}{q_{x,y}}.$$

Using (3.3.10) and the convexity of $a \mapsto \widetilde{\varphi}_{1/a}(t)$ (Lemma 3.1.4) we estimate

$$\widetilde{\varphi}_{q(x,y)}(t) \leqslant \widetilde{\varphi}_{q_{x,y}}(t) \leqslant \frac{1}{2}\widetilde{\varphi}_{\frac{s(x)}{2}}(t) + \frac{1}{2}\widetilde{\varphi}_{\frac{s(y)}{2}}(t) = \widetilde{\varphi}_{s(x)}\big(t^{\frac{1}{2}}\big) + \widetilde{\varphi}_{s(y)}\big(t^{\frac{1}{2}}\big).\qquad\square$$

The following theorem plays a central role in later proofs of strong and weak type estimates, as well as estimates of convolutions.

Theorem 4.2.4 (Key estimate). *Let $p \in \mathcal{P}^{\log}(\mathbb{R}^n)$. Then for every $m > 0$ there exists $\beta \in (0,1)$ only depending on m and $c_{\log}(p)$ such that*

$$\varphi_{p(x)}\left(\beta \fint_Q |f(y)|\,dy\right)$$

$$\leq \begin{cases} \fint_Q \varphi_{p(y)}(|f(y)|)dy + \dfrac{1}{2}\left(\fint_Q \left((e+|x|)^{-m}+(e+|y|)^{-m}\right)\chi_{\{0<|f(y)|\leq 1\}}dy\right)^{p^-} \\[2ex] \fint_Q \varphi_{p(y)}(|f(y)|)dy + \dfrac{1}{2}\fint_Q \left((e+|x|)^{-m}+(e+|y|)^{-m}\right)\chi_{\{0<|f(y)|\leq 1\}}dy \end{cases}$$

for every cube (or ball) $Q \subset \mathbb{R}^n$, all $x \in Q$, and all $f \in L^{p(\cdot)}(\mathbb{R}^n) + L^\infty(\mathbb{R}^n)$ with $\|f\|_{L^{p(\cdot)}(\mathbb{R}^n)+L^\infty(\mathbb{R}^n)} \leq 1$.

Note that if $p^+ < \infty$ then in the previous lemmas and theorem we can take the constant β out from $\varphi_{p(\cdot)}$. For example, in the later case of the previous theorem we obtain

$$\varphi_{p(x)}\left(\fint_Q |f(y)|\,dy\right)$$

$$\leq c\fint_Q \varphi_{p(y)}(|f(y)|)\,dy + c\fint_Q \left((e+|x|)^{-m}+(e+|y|)^{-m}\right)\chi_{\{0<|f(y)|\leq 1\}}\,dy,$$

where the constant c depends only on m, $c_{\log}(p)$ and p^+.

Proof of Theorem 4.2.4. Define $q := \frac{p}{p^-}$. As an immediate consequence of Lemma 4.2.2, Lemma 4.2.3 and Proposition 4.1.8 with exponent q, we obtain

$$\varphi_{q(x)}\left(\beta \fint_Q |f(y)|\,dy\right) \leq \fint_Q \varphi_{q(y)}(|f(y)|)\,dy$$

$$+ c\fint_Q \left((e+|x|)^{-m}+(e+|y|)^{-m}\right)\chi_{\{0<|f(y)|\leq 1\}}\,dy$$

for suitable $\beta \in (0,1)$. Raising both sides to the power of p^- and using Jensen's inequality on the first integral on the right-hand side yields the first inequality. By Jensen's inequality, p^- can be taken into the second integral and absorbed into $m' := mp^-$, which gives the second inequality. $\square$

If we then integrate the estimate in Theorem 4.2.4 over a cube (or ball) Q, then we get the following result.

Corollary 4.2.5. *Let $p \in \mathcal{P}^{\log}(\mathbb{R}^n)$. Then for every $m > 0$ there exists $\beta \in (0,1)$ only depending on m and $c_{\log}(p)$ such that*

$$\int_Q \varphi_{p(x)}\left(\beta \fint_Q |f(y)|\,dy\right) dx \leqslant \int_Q \varphi_{p(y)}(|f(y)|)\,dy + \int_Q (e + |y|)^{-m}\,dy,$$

$$\int_Q \varphi_{p(x)}\left(\beta \fint_Q |f(y)|\,dy\right) dx \leqslant \int_Q \varphi_{p(y)}(|f(y)|)\,dy + |\{y \in Q : 0 < |f(y)| \leqslant 1\}|$$

for every cube (or ball) $Q \subset \mathbb{R}^n$ and all $f \in L^{p(\cdot)}(\mathbb{R}^n) + L^\infty(\mathbb{R}^n)$ with $\|f\|_{L^{p(\cdot)}(\mathbb{R}^n)+L^\infty(\mathbb{R}^n)} \leqslant 1$.

For later use we also record the following modification of Lemma 4.2.1. Note that p_Q^- from Lemma 4.2.1 is replaced in Lemma 4.2.7 by p_Q, the harmonic mean:

Definition 4.2.6. Let $p \in \mathcal{P}(\mathbb{R}^n)$. For measurable $E \subset \mathbb{R}^n$ with $|E| \in (0, \infty)$ we define the *harmonic mean* $p_E \in [1, \infty]$ by

$$\frac{1}{p_E} = \fint_E \frac{1}{p(y)}\,dy.$$

We derive further properties of the harmonic mean p_Q in Sect. 4.5.

Lemma 4.2.7. *Let $p \in \mathcal{P}^{\log}(\mathbb{R}^n)$. Then for any $m > 0$ there exists $\beta \in (0,1)$, which only depends on the local log-Hölder continuity constant of $\frac{1}{p}$, such that*

$$\varphi_{p(x)}\left(\beta \varphi_{p_Q}^{-1}(|Q|^{-1})\right) \leqslant |Q|^{-1} + \frac{1}{2}(e + |x|)^{-mp^-} + \frac{1}{2}\left(\fint_Q (e + |y|)^{-m}\,dy\right)^{p^-},$$

for any cube (or ball) $Q \subset \mathbb{R}^n$ and any $x \in Q$.

Proof. We prove the claim for $\varphi_{p(\cdot)} = \bar{\varphi}_{p(\cdot)}$. The case $\varphi_{p(\cdot)} = \tilde{\varphi}_{p(\cdot)}$ follows easily with the help of Lemmas 3.1.6 and 3.1.12.

Define $f := \chi_Q \bar{\varphi}_{p(\cdot)}^{-1}(|Q|^{-1})$. Since $\bar{\varphi}_{p(\cdot)}^{-1}$ is the left-continuous inverse, we find that $\bar{\varphi}_{p(\cdot)}(f) = \chi_Q \bar{\varphi}_{p(\cdot)}(\bar{\varphi}_{p(\cdot)}^{-1}(|Q|^{-1})) \leqslant \chi_Q |Q|^{-1}$. Hence $\bar{\varrho}_{p(\cdot)}(f) \leqslant 1$ and $\|f\|_{p(\cdot)} \leqslant 1$ by the unit ball property. The convexity of the mapping $q \mapsto \bar{\varphi}_{1/q}$ and Jensen's inequality imply that

$$\bar{\varphi}_{p_Q}^{-1}(|Q|^{-1}) \leqslant \fint_Q \bar{\varphi}_{p(y)}^{-1}(|Q|^{-1})\,dy.$$

By Theorem 4.2.4 there exists $\beta > 0$ such that

$$\bar{\varphi}_{p(x)}\left(\beta \fint_Q |f(y)|\, dy\right) \leqslant \fint_Q \bar{\varphi}_{p(y)}(|f(y)|)\, dy + \frac{1}{2}(\mathrm{e} + |x|)^{-mp^-}$$

$$+ \frac{1}{2}\left(\fint_Q (\mathrm{e} + |y|)^{-m}\, dy\right)^{p^-}.$$

Since $\bar{\varrho}_{p(\cdot)}(f) \leqslant 1$, the first term on the right-hand side is less than or equal to $|Q|^{-1}$, which completes the proof. $\qquad\square$

Remark 4.2.8. The decay condition can be slightly weakened. Assume that p is locally log-Hölder continuous and satisfies $1 \in L^{s(\cdot)}$, where s is given by

$$\frac{1}{s(x)} := \left| \frac{1}{p(x)} - \frac{1}{p_\infty} \right|.$$

Then Lemma 4.2.2 holds, since it only requires the local log-Hölder continuity of $\frac{1}{p}$. Moreover, Theorem 4.2.4 remains true. We only have to replace $(\mathrm{e} + |x|)^{-m}$ in the proof by $\varphi_{s(x)}(\gamma)$ with $\gamma > 0$ such that $\varrho_{s(\cdot)}(\gamma) < \infty$. Also, all results that are solely based on the theorem hold under this weaker condition on p. The condition $1 \in L^{s(\cdot)}$ has been studied by Nekvinda [314, 316] in the context of the Hardy–Littlewood maximal operator. Note that $1 \in L^{s(\cdot)}$ is equivalent to the existence of $\gamma > 0$ with

$$\bar{\varrho}_{s(\cdot)}(\gamma) = \int_{\mathbb{R}^n} \gamma^{\frac{1}{\left|\frac{1}{p(x)} - \frac{1}{p_\infty}\right|}}\, dx < \infty.$$

4.3   The Boundedness of the Maximal Operator

In order to derive more sophisticated results for the spaces $L^{p(\cdot)}(\mathbb{R}^n)$, we have to investigate the Hardy–Littlewood maximal operator M on $L^{p(\cdot)}(\mathbb{R}^n)$. This operator is a powerful tool and we will see that many properties will follow from the boundedness of M. The most central property of the maximal operator is that it is a bounded operator from L^q to L^q when $q \in (1, \infty]$. In this section we prove the variable exponent generalization of this.

Let us start with some notation. Recall that our cubes are always with sides parallel to the axis.

Definition 4.3.1. For a function $f \in L^0(\mathbb{R}^n)$ and an open, bounded set $U \subset \mathbb{R}^n$ (usually a cube or a ball) we define

$$M_U f := \fint_U |f(y)|\, dy = \frac{1}{|U|} \int_U |f(y)|\, dy. \qquad (4.3.2)$$

The (non-centered) *maximal function* Mf of f is defined by

$$Mf(x) := \sup_{Q \ni x} M_Q f = \sup_{Q \ni x} \fint_Q |f(y)|\, dy$$

for all $x \in \mathbb{R}^n$, where the supremum is taken over all cubes (or balls) $Q \subset \mathbb{R}^n$ which contain x. The operator $M \colon f \mapsto Mf$ is called the *Hardy–Littlewood maximal operator* or just *maximal operator*. Furthermore, for $f \in L^s_{\mathrm{loc}}(\mathbb{R}^n)$, $s \in [1, \infty)$, and an open, bounded set $U \subset \mathbb{R}^n$ we define

$$M_{s,U} f := \left(M_U(|f|^s) \right)^{\frac{1}{s}} = \left(\fint_U |f(y)|^s\, dy \right)^{\frac{1}{s}},$$

$$M_s f := \left(M(|f|^s) \right)^{\frac{1}{s}}.$$

Remark 4.3.3. In Definition 4.3.1 it is possible to use balls instead of cubes. Also, we could take the supremum only over those cubes (or balls) which are centered around x, rather than the ones containing x. Up to constants all of these versions are equivalent, e.g. $c_1 M_{\mathrm{balls}} f \leqslant M_{\mathrm{cubes}} f \leqslant c_2 M_{\mathrm{balls}} f$ with c_1, c_2 only depending on the dimension n.

Let us recall some classical results for the maximal operator M, see for example Stein [360]. For $f \in L^1_{\mathrm{loc}}(\mathbb{R}^n)$ the function $Mf \colon \mathbb{R}^n \to [0, \infty]$ is lower semicontinuous and satisfies $|f| \leqslant Mf$ almost everywhere. For any $1 \leqslant q \leqslant \infty$ and $f \in L^q(\mathbb{R}^n)$ the function Mf is almost everywhere finite. Moreover, for $1 < q \leqslant \infty$ the mapping $f \mapsto Mf$ is bounded from $L^q(\mathbb{R}^n)$ to $L^q(\mathbb{R}^n)$. The constant blows up as $q \searrow 1$. Indeed, M is not bounded from $L^1(\mathbb{R}^n)$ to $L^1(\mathbb{R}^n)$. Actually, $Mf \notin L^1(\mathbb{R}^n)$ for every non-zero $f \in L^1(\mathbb{R}^n)$. In the L^1 case we have the weaker result

$$\left\| \lambda \chi_{\{Mf > \lambda\}} \right\|_{L^1(\mathbb{R}^n)} \leqslant c \|f\|_{L^1(\mathbb{R}^n)}, \tag{4.3.4}$$

for $f \in L^1(\mathbb{R}^n)$ and $\lambda > 0$, where c depends only on n. Here, $\chi_{\{Mf > \lambda\}}$ denotes the characteristic function of the set $\{y \in \mathbb{R}^n \colon Mf(y) > \lambda\}$. This set is open, since Mf is lower semicontinuous. If inequality (4.3.4) holds for every $\lambda > 0$, then we say that M is of *weak type* 1.

The weak Lebesgue space w-L^q with $q \in [1, \infty]$ is defined by the quasinorm

$$\|f\|_{\mathrm{w}\text{-}L^q} := \sup_{\lambda > 0} \left\| \lambda \chi_{\{|f| > \lambda\}} \right\|_q.$$

The quasinorm satisfies the triangle inequality $\|f + g\|_{\mathrm{w}\text{-}L^q} \leqslant 2\left(\|f\|_{\mathrm{w}\text{-}L^q} + \|g\|_{\mathrm{w}\text{-}L^q} \right)$, while the other norm properties remain true. Obviously,

"M is of weak type 1" if and only if M maps $L^1(\mathbb{R}^n)$ to w-$L^1(\mathbb{R}^n)$. We have $L^1(\mathbb{R}^n) \hookrightarrow$ w-$L^1(\mathbb{R}^n)$, since $\lambda \chi_{\{|f|>\lambda\}} \leqslant |f|$ for all $f \in L^1(\mathbb{R}^n)$. Another easy embedding follows:

Lemma 4.3.5. *Let* $p \in \mathcal{P}(\mathbb{R}^n)$ *with* $p^- > 1$. *Then* w-$L^1(\mathbb{R}^n) \cap L^\infty(\mathbb{R}^n) \hookrightarrow L^{p(\cdot)}(\mathbb{R}^n)$.

Proof. We assume that $p^- < \infty$, since the claim is trivial otherwise. Let $f \in$ w-$L^1(\mathbb{R}^n) \cap L^\infty(\mathbb{R}^n)$ with $\max\{\|f\|_{\text{w-}L^1(\mathbb{R}^n)}, \|f\|_\infty\} \leqslant 1$. Then

$$\int_{\mathbb{R}^n} \varphi_{p(x)}(|f|)\, dx \leqslant \int_{\mathbb{R}^n} |f|^{p^-}\, dx = \int_0^1 t^{p^- - 1}|\{|f| > t\}|\, dt$$

$$\leqslant \|f\|_{\text{w-}L^1} \int_0^1 t^{p^- - 2}\, dt < \infty. \qquad \square$$

Next we have a version of Theorem 4.2.4 with maximal functions instead of integral averages.

Lemma 4.3.6. *Let* $p \in \mathcal{P}^{\log}(\mathbb{R}^n)$. *Then for any* $m > 0$ *there exists* $\beta \in (0,1)$ *only depending on* m *and* $c_{\log}(p)$ *such that*

$$\varphi_{p(y)}\big(\beta M f(y)\big) \leqslant M\big(\varphi_{p(\cdot)}(f)\big)(y) + h(y),$$

for all $f \in L^{p(\cdot)}(\mathbb{R}^n) + L^\infty(\mathbb{R}^n)$ *with* $\|f\|_{L^{p(\cdot)}(\mathbb{R}^n)+L^\infty(\mathbb{R}^n)} \leqslant 1$ *and all* $y \in \mathbb{R}^n$, *where* $h(y) := M\big((\mathrm{e} + |\cdot|)^{-m}\big)(y)$.

Proof. Let $m > 0$, then from Theorem 4.2.4 it follows that there exists $\beta > 0$ such that

$$\varphi_{p(x)}\left(\beta \fint_Q |f(y)|\, dy\right)$$

$$\leqslant \fint_Q \varphi_{p(y)}(|f(y)|)\, dy + \frac{1}{2}(\mathrm{e} + |x|)^{-m} + \frac{1}{2}\fint_Q (\mathrm{e} + |y|)^{-m}\, dy$$

for $f \in L^{p(\cdot)}(\mathbb{R}^n) + L^\infty(\mathbb{R}^n)$ with $\|f\|_{L^{p(\cdot)}(\mathbb{R}^n)+L^\infty(\mathbb{R}^n)} \leqslant 1$ and all $x \in Q$. We take the supremum over all cubes (or balls) $Q \subset \mathbb{R}^n$ with $x \in Q$ and use that $\varphi_{p(\cdot)}$ is non-decreasing and left-continuous:

$$\varphi_{p(x)}\big(\beta M f(x)\big) \leqslant M\big(\varphi_{p(\cdot)}(f)\big)(x) + \frac{1}{2}(\mathrm{e} + |x|)^{-m} + \frac{1}{2}M\big((\mathrm{e} + |\cdot|)^{-m}\big)(x)$$

$$\leqslant M\big(\varphi_{p(\cdot)}(f)\big)(x) + M\big((\mathrm{e} + |\cdot|)^{-m}\big)(x). \qquad \square$$

If p is bounded and we are working in a bounded domain, then the following simplified version of the previous lemma is often useful:

$$(Mf(x))^{p(x)} \leqslant cM(|f|^{p(\cdot)}) + c. \qquad (4.3.7)$$

This inequality holds under the same assumptions as in the lemma.

We are now ready to prove the main theorem of this section.

Theorem 4.3.8. *Let $p \in \mathcal{P}^{\log}(\mathbb{R}^n)$ with $p^- > 1$. Then there exists $K > 0$ only depending on the dimension n and $c_{\log}(p)$ such that*

$$\|Mf\|_{p(\cdot)} \leqslant K (p^-)' \|f\|_{p(\cdot)}$$

for all $f \in L^{p(\cdot)}(\mathbb{R}^n)$.

Proof. Let $q := \frac{p}{p^-}$, so that $q \in \mathcal{P}^{\log}(\mathbb{R}^n)$ with $q^- = 1$. Let $f \in L^{p(\cdot)}(\mathbb{R}^n)$ with $\|f\|_{p(\cdot)} \leqslant \frac{1}{4}$, and note that $\|f\|_{L^{q(\cdot)}(\mathbb{R}^n) + L^\infty(\mathbb{R}^n)} \leqslant 1$ by Theorem 3.3.11. This and Lemma 4.3.6 imply that

$$\varphi_{q(x)}\left(\tfrac{\beta}{2}Mf(x)\right) \leqslant \tfrac{1}{2}\varphi_{q(x)}\left(\beta Mf(x)\right) \leqslant \tfrac{1}{2}M\left(\varphi_{q(\cdot)}(f)\right)(x) + \tfrac{1}{2}h(x) \quad (4.3.9)$$

with $h(x) := M\left((\mathrm{e} + |\cdot|)^{-m}\right)(x)$, where we choose $m > n$. Furthermore, from Lemma 3.1.6 it follows that

$$\varphi_{p(x)}(t) \leqslant \bar{\varphi}_{p(x)}(t) = \left(\bar{\varphi}_{q(x)}(t)\right)^{p^-} \leqslant \left(\varphi_{q(x)}(2t)\right)^{p^-}$$

for all $t \geqslant 0$ and all $x \in \mathbb{R}^n$. Combining the results above, we find that

$$\begin{aligned}
\varphi_{p(x)}\left(\tfrac{\beta}{4}Mf(x)\right) &\leqslant \left(\varphi_{q(x)}\left(\tfrac{\beta}{2}Mf(x)\right)\right)^{p^-} \\
&\leqslant \left(\tfrac{1}{2}M\left(\varphi_{q(\cdot)}(f)\right) + \tfrac{1}{2}h(x)\right)^{p^-} \\
&\leqslant \tfrac{1}{2}M\left(\varphi_{q(\cdot)}(f)\right)^{p^-} + \tfrac{1}{2}h(x)^{p^-}.
\end{aligned}$$

Integration over $\mathbb{R}^n$ yields

$$\varrho_{p(\cdot)}(\tfrac{\beta}{4}Mf) \leqslant \tfrac{1}{2}\|M(\varphi_{p(\cdot)}(f))\|_{p^-}^{p^-} + \tfrac{1}{2}\|h\|_{p^-}^{p^-}.$$

Since $(\mathrm{e} + |\cdot|)^{-m} \in L^1(\mathbb{R}^n)$ for $m > n$, and M is of weak type 1, we conclude that $M\left((\mathrm{e} + |\cdot|)^{-m}\right) \in \text{w-}L^1(\mathbb{R}^n)$. Hence h^{p^-} is integrable by Lemma 4.3.5. Moreover, $\|f\|_{p(\cdot)} \leqslant \frac{1}{4}$ implies $\|\varphi_{q(\cdot)}(f)\|_{p^-} \leqslant 1$. So the classical result on the boundedness of M on $L^{p^-}(\mathbb{R}^n)$ implies that $\|M(\varphi_{p(\cdot)}(f))\|_{p^-} \leqslant c\,(p^-)'$, with boundedness constant depending only on n. Thus $\varrho_{p(\cdot)}(\frac{\beta}{4}Mf) \leqslant [c\,(p^-)']^{p^-}$, and by Lemma 3.2.5 $\|Mf\|_{p(\cdot)} \leqslant K\,(p^-)'$ for $\|f\|_{p(\cdot)} \leqslant \frac{1}{4}$. The proof is completed by the scaling argument. □

The proof of Theorem 4.3.8 goes back to many authors. The first version goes back to Diening [91], who proved the result for bounded exponents that are constant outside a large ball. This condition has been later relaxed by Cruz-Uribe, Fiorenza, Martell and Pérez [83, Lemma 4.3] to the log-Hölder decay condition and by Nekvinda [314] to the integral condition $1 \in L^{s(\cdot)}$ as in Remark 4.2.8. The boundedness of the exponent was then removed in [95] and [81]. The proof in this book is closest to the one in [95].

Remark 4.3.10. As in Remark 4.2.8 it is possible to replace the decay condition on $\frac{1}{p}$ in Lemma 4.3.6 and Theorem 4.3.8 by the weaker condition $1 \in L^{s(\cdot)}$ with $\frac{1}{s(x)} := \left|\frac{1}{p(x)} - \frac{1}{p_\infty}\right|$.

Using a standard argument we obtain a local version of the previous result. Note that the decay condition is vacuously true if the domain is bounded, so in this case the local log-Hölder condition is sufficient for boundedness.

Corollary 4.3.11. *Let* $p \in \mathcal{P}^{\log}(\Omega)$ *with* $p^- > 1$. *Then there exists* $K > 0$ *only depending on* $c_{\log}(p)$ *and the dimension* n *such that*

$$\|Mf\|_{L^{p(\cdot)}(\Omega)} \leqslant K\,(p^-)'\|f\|_{L^{p(\cdot)}(\Omega)}$$

for all $f \in L^{p(\cdot)}(\Omega)$.

Proof. By Proposition 4.1.7 we extend the exponent to the whole space $\mathbb{R}^n$ with the same infimum and log-Hölder constant. A function $f \in L^{p(\cdot)}(\Omega)$ can be extended to $\mathbb{R}^n$ by zero outside Ω. Denote these extensions by $\tilde{p}$ and $\tilde{f}$, respectively. Then

$$\|Mf\|_{L^{p(\cdot)}(\Omega)} \leqslant \|M\tilde{f}\|_{L^{\tilde{p}(\cdot)}(\mathbb{R}^n)} \leqslant K\,(p^-)'\|\tilde{f}\|_{L^{\tilde{p}(\cdot)}(\mathbb{R}^n)} = K\,(p^-)'\|f\|_{L^{p(\cdot)}(\Omega)}$$

by Theorem 4.3.8. □

If one assumes a weaker modulus of continuity than log-Hölder, then it is still possible to obtain the boundedness of the maximal operator, but the target space is larger than $L^{p(\cdot)}(\mathbb{R}^n)$. Such results have been studied by Mizuta, Shimomura and their colleagues, see, e.g., [298].

The boundedness of the maximal operator on metric measure spaces has been investigated e.g. in [203, 252] in the variable exponent context. The discrete setting has been studied in [315]. Such results are not considered here.

4.4 Weak-Type Estimates and Averaging Operators

We saw in Sect. 4.3 that the maximal operator is of strong type when p is log-Hölder continuous with $p^- > 1$. In order to get around the latter restriction we consider in this section weak-type estimates and averaging operators. Averaging operators have been studied by Edmunds and Nekvinda in [119], but here we undertake a much broader investigation of their properties.

Recall that a sublinear operator on a real vector space X is an operator T which satisfies

$$T(f + g) \leqslant Tf + Tg \quad \text{and} \quad T(tf) = tTf$$

for all $f, g \in X$ and all scalars $t \geqslant 0$.

Definition 4.4.1. Let $\varphi \in \Phi(\mathbb{R}^n)$ and let T be a sublinear operator which maps $L^\varphi(\mathbb{R}^n)$ into the space of measurable functions on $\mathbb{R}^n$. Then we say that T is of *weak type* φ if there exists $K_1 > 0$ such that

$$\big\| \lambda \chi_{\{|Tf| > \lambda\}} \big\|_{L^\varphi(\mathbb{R}^n)} \leqslant K_1 \big\| f \big\|_{L^\varphi(\mathbb{R}^n)}$$

for all $f \in L^\varphi(\mathbb{R}^n)$ and all $\lambda > 0$. We say that T is of *strong type* φ if there exists $K_2 > 0$ such that

$$\big\| Tf \big\|_{L^\varphi(\mathbb{R}^n)} \leqslant K_2 \big\| f \big\|_{L^\varphi(\mathbb{R}^n)}$$

for any $f \in L^\varphi(\mathbb{R}^n)$. If $p \in \mathcal{P}(\mathbb{R}^n)$, then instead of weak type $\varphi_{p(\cdot)}$ and strong type $\varphi_{p(\cdot)}$ we write weak type $p(\cdot)$ and strong type $p(\cdot)$, respectively.

For a operator T and $\lambda > 0$ we obviously have $\lambda \chi_{\{|Tf| > \lambda\}} \leqslant |Tf|$. Therefore, if T is of strong type $p(\cdot)$, then it is also of weak type $p(\cdot)$. For instance, in classical Lebesgue spaces, M is of strong type q for any $q \in (1, \infty]$ and of weak type q for any $q \in [1, \infty]$.

In this section we study weak and strong type results for M in the context of $L^{p(\cdot)}(\mathbb{R}^n)$ with $p \in \mathcal{P}^{\log}(\mathbb{R}^n)$. We first introduce the notation of *averaging operators* $T_{\mathcal{Q}}$ and then deduce from their properties that M is of weak type $p(\cdot)$.

Definition 4.4.2. A family $\mathcal{Q}$ of measurable sets $U \subset \mathbb{R}^n$ is called *locally N-finite*, where $N \in \mathbb{N}$, if

$$\sum_{U \in \mathcal{Q}} \chi_U \leqslant N$$

almost everywhere in $\mathbb{R}^n$. We simply say that $\mathcal{Q}$ is *locally finite* if it is N-locally finite for some $N \in \mathbb{N}$.

Note that a family $\mathcal{Q}$ of open, bounded sets $Q \subset \mathbb{R}^n$ is locally 1-finite if and only if the sets $Q \in \mathcal{Q}$ are pairwise disjoint.

Definition 4.4.3. For a family $\mathcal{Q}$ of open, bounded sets $U \subset \mathbb{R}^n$ we define $T_{\mathcal{Q}}\colon L^1_{\mathrm{loc}}(\mathbb{R}^n) \to L^0(\mathbb{R}^n)$ and $T_{s,\mathcal{Q}}\colon L^s_{\mathrm{loc}}(\mathbb{R}^n) \to L^0(\mathbb{R}^n)$ with $s \in [1,\infty)$ by

$$T_{\mathcal{Q}}f := \sum_{U \in \mathcal{Q}} \chi_U \, M_U f = \sum_{U \in \mathcal{Q}} \chi_U \fint_U |f(y)| \, dy,$$

$$T_{s,\mathcal{Q}}f := \sum_{U \in \mathcal{Q}} \chi_U \, M_{s,U} f = \sum_{U \in \mathcal{Q}} \chi_U \left(\fint_U |f(y)|^s \, dy \right)^{\frac{1}{s}}.$$

The operators $T_{\mathcal{Q}}$ and $T_{s,\mathcal{Q}}$ are called *averaging operator* and *s-averaging operator*, respectively.

Note that $T_{\mathcal{Q}} = T_{1,\mathcal{Q}}$. The functions $T_{\mathcal{Q}}f$ and $T_{s,\mathcal{Q}}f$ are well defined in $L^0(\mathbb{R}^n)$, since $M_{\mathcal{Q}}f \geqslant 0$, but might be infinite at many points or even everywhere. However, if $\mathcal{Q}$ is locally finite and $f \in L^1_{\mathrm{loc}}(\mathbb{R}^n)$ and $g \in L^s_{\mathrm{loc}}(\mathbb{R}^n)$, then $T_{\mathcal{Q}}f \in L^1_{\mathrm{loc}}(\mathbb{R}^n)$ and $T_{s,\mathcal{Q}}g \in L^s_{\mathrm{loc}}(\mathbb{R}^n)$. By Jensen's inequality we have $T_{\mathcal{Q}}f \leqslant T_{s,\mathcal{Q}}$ for $s \geqslant 1$.

Definition 4.4.4. Let $\varphi \in \Phi(\mathbb{R}^n)$. Then for any $t \geqslant 0$ the mapping $x \mapsto \varphi(x,t)$ is non-negative and measurable. Now, for a cube (or ball) $Q \subset \mathbb{R}^n$ and $t \geqslant 0$ we define

$$M_Q\varphi(t) := \fint_Q \varphi(x,t) \, dx.$$

For a measurable function f on $\mathbb{R}^n$ we define a function $\varphi(f)$ on $\mathbb{R}^n$ by $\varphi(f) := \varphi(\cdot, |f(\cdot)|)$, i.e. for all $x \in \mathbb{R}^n$ we set

$$(\varphi(f))(x) = \varphi(x, |f(x)|).$$

By Lemma 2.3.10, the function $\varphi(f)$ is measurable. If $\chi_Q \in L^\varphi$, then $M_Q\varphi$ is a Φ-function. This is certainly the case if φ is locally integrable. Also note that in the sense of (4.3.2) we could write $M_Q\varphi(t) = M_Q(\varphi(\cdot,t))$. Instead of $\int \varphi(x,|f(x)|) \, dx$ we can now write more compactly $\int \varphi(f) \, dx$. Note that whenever we have a generalized Φ-function φ and a measurable function f, then $\varphi(f)$ depends on x via f and φ. For example, by $\varphi_{p(\cdot)}(f)$ we denote the mapping $x \mapsto \varphi_{p(x)}(|f(x)|)$.

With this notation we can write an analogue of Corollary 4.2.5:

$$(M_Q\varphi_{p(\cdot)})(\beta M_Q f) \leqslant M_Q\big(\varphi_{p(\cdot)}(f)\big) + M_Q\big((\mathrm{e} + |\cdot|)^{-m}\big) \tag{4.4.5}$$

with the notation and assumptions of that lemma.

Definition 4.4.6. By $\mathcal{A}$ we denote the set of all generalized Φ-functions φ on $\mathbb{R}^n$ which have the property that the averaging operators $T_{\mathcal{Q}}$ are bounded from $L^\varphi(\mathbb{R}^n)$ to $L^\varphi(\mathbb{R}^n)$ uniformly for all locally 1-finite families $\mathcal{Q}$ of cubes in $\mathbb{R}^n$. The smallest constant K for which

$$\left\|T_{\mathcal{Q}}f\right\|_\varphi \leqslant K\|f\|_\varphi,$$

for all locally 1-finite families of cubes $\mathcal{Q}$ in $\mathbb{R}^n$ and all $f \in L^\varphi(\mathbb{R}^n)$, will be called the $\mathcal{A}$-*constant of* φ. If $\varphi \in \mathcal{A}$, then we say that φ *is of class* $\mathcal{A}$. In the case $\varphi_{p(\cdot)}$, $p \in \mathcal{P}(\mathbb{R}^n)$, we denote $\varphi_{p(\cdot)} \in \mathcal{A}$ simply by $p \in \mathcal{A}$ and call the $\mathcal{A}$-constant of $\varphi_{p(\cdot)}$ also the $\mathcal{A}$-constant of p.

By $\mathcal{A}_{\mathrm{loc}}$ we denote the set of all generalized Φ-functions φ on $\mathbb{R}^n$ which have the property that the averaging operators $T_{\{Q\}}$ over single cubes Q are uniformly bounded from $L^\varphi(\mathbb{R}^n)$ to $L^\varphi(\mathbb{R}^n)$, i.e. $\sup_Q \left\|T_{\{Q\}}f\right\|_\varphi \leqslant K_2\|f\|_\varphi$ for all $f \in L^\varphi(\mathbb{R}^n)$, where the supremum is taken over all cubes $Q \subset \mathbb{R}^n$. The smallest constant K_2 will be called the $\mathcal{A}_{\mathrm{loc}}$-*constant of* φ. If $\varphi \in \mathcal{A}_{\mathrm{loc}}$, then we say that φ *is of class* $\mathcal{A}_{\mathrm{loc}}$. In the case $\varphi_{p(\cdot)}$, $p \in \mathcal{P}(\mathbb{R}^n)$, we denote $\varphi_{p(\cdot)} \in \mathcal{A}_{\mathrm{loc}}$ simply by $p \in \mathcal{A}_{\mathrm{loc}}$.

In Theorem 4.4.8 we will show that each exponent $p \in \mathcal{P}^{\log}$ satisfies $p \in \mathcal{A}$.

Lemma 4.4.7. *Let* $p \in \mathcal{P}^{\log}(\mathbb{R}^n)$. *If* $p \in \mathcal{A}$ *and* $s \geqslant 1$, *then* $sp \in \mathcal{A}$. *If* M *is bounded on* $L^{p(\cdot)}(\mathbb{R}^n)$, *then it is bounded on* $L^{sp(\cdot)}(\mathbb{R}^n)$.

Proof. Using Lemma 3.2.6, $(T_{\mathcal{Q}}f)^s \leqslant (T_{s,\mathcal{Q}}f)^s = T_{\mathcal{Q}}(|f|^s)$ and $p \in \mathcal{A}$ we estimate

$$\left\|T_{\mathcal{Q}}f\right\|^s_{\bar\varphi_{sp(\cdot)}} = \left\|(T_{\mathcal{Q}}f)^s\right\|_{\bar\varphi_{p(\cdot)}} \leqslant \left\|T_{\mathcal{Q}}(|f|^s)\right\|_{\bar\varphi_{p(\cdot)}} \leqslant c\,\left\||f|^s\right\|_{\bar\varphi_{p(\cdot)}} = c\,\|f\|^s_{\bar\varphi_{sp(\cdot)}}.$$

The claim for M follows similarly from $(Mf)^s \leqslant M(|f|^s)$. $\square$

If $\varphi \in \mathcal{A}_{\mathrm{loc}}$ or $\varphi \in \mathcal{A}$, then necessarily $\chi_Q \in L^\varphi$ for all cubes $Q \subset \mathbb{R}^n$. Obviously, $\mathcal{A} \subset \mathcal{A}_{\mathrm{loc}}$. So naturally the question arises if the reverse holds or not. Is it really necessary to consider locally 1-finite families of cubes rather than just single cubes? At least for classical weighted Lebesgue spaces $L^q(\mathbb{R}^n, \omega\,dx)$ as well as weighted Orlicz spaces there is no difference in using families or single cubes. However, we will see in Theorem 5.3.4 that it is in fact not possible to use only single cubes in the general case.

The properties *class* $\mathcal{A}_{\mathrm{loc}}$ and *class* $\mathcal{A}$ will be studied in great detail in Sects. 4.5 and 5.2. We will see in Sects. 5.2 and 5.7 that there is a strong connection between class $\mathcal{A}$ and the boundedness of the maximal operator M. At this point we only mention that classes $\mathcal{A}$ and $\mathcal{A}_{\mathrm{loc}}$ are natural generalizations of the Muckenhoupt classes of weighted Lebesgue spaces (with constant exponents). Using Corollary 4.2.5 we now derive an analogue of the maximal theorem for the averaging operators $T_{\mathcal{Q}}$. Note that the assumption $p^- > 1$ is not needed in this context.

Theorem 4.4.8. *If* $p \in \mathcal{P}^{\log}(\mathbb{R}^n)$, *then* $p \in \mathcal{A}$ *with* $\mathcal{A}$-*constant depending only on* $c_{\log}(p)$ *and* n. *Moreover,*

$$\|T_{\mathcal{Q}}f\|_{p(\cdot)} \leqslant c\, N\, \|f\|_{p(\cdot)}$$

for every locally N-*finite family* $\mathcal{Q}$ *of cubes (or balls) and all* $f \in L^{p(\cdot)}(\mathbb{R}^n)$. *The constant* c *depends only on* $c_{\log}(p)$.

Proof. Let $f \in L^{p(\cdot)}(\mathbb{R}^n)$ with $\|f\|_{p(\cdot)} \leqslant 1$ and let $\mathcal{Q}$ be a locally N-finite family of cubes (or balls). Then by the unit ball property $\varrho_{p(\cdot)}(f) \leqslant 1$. Let $m > n$ be such that $\int_{\mathbb{R}^n}(e+|y|)^{-m}\, dy \leqslant 1$. Choose $\beta \in (0,1)$ as in Corollary 4.2.5. Then

$$\varrho_{p(\cdot)}\left(\frac{1}{2N}\beta T_{\mathcal{Q}}f\right) \leqslant \frac{1}{2N} \sum_{Q \in \mathcal{Q}} \int_Q \varphi_{p(x)}(\beta\, \chi_Q\, M_Q f)\, dx$$

$$\leqslant \frac{1}{2N} \sum_{Q \in \mathcal{Q}} \left(\int_Q \varphi_{p(y)}(|f(y)|)\, dy + \int_Q (e+|y|)^{-m}\, dy \right)$$

$$\leqslant \frac{1}{2} \left(\int_{\mathbb{R}^n} \varphi_{p(x)}(f)\, dx + \int_{\mathbb{R}^n} (e+|x|)^{-m}\, dx \right)$$

$$\leqslant \frac{1}{2}\left(\varrho_{p(\cdot)}(f) + 1\right) \leqslant 1.$$

where we have used that $\mathcal{Q}$ is locally N-finite. This implies $\|T_{\mathcal{Q}}f\|_{p(\cdot)} \leqslant \frac{2N}{\beta}$. A scaling argument yields the $\|T_{\mathcal{Q}}f\|_{p(\cdot)} \leqslant \frac{2N}{\beta}\|f\|_\varphi$. The case $N = 1$ with cubes implies $p \in \mathcal{A}$. $\square$

Remark 4.4.9. As in Remark 4.2.8 it is possible to replace the decay condition on $\frac{1}{p}$ in Corollary 4.2.5, (4.4.5) and Theorem 4.4.8 by the weaker condition $1 \in L^{s(\cdot)}$ with $\frac{1}{s(x)} := |\frac{1}{p(x)} - \frac{1}{p_\infty}|$.

For any locally 1-finite family of cubes $\mathcal{Q}$ in $\mathbb{R}^n$ and $f \in L^1_{\mathrm{loc}}(\mathbb{R}^n)$ the inequality $|T_{\mathcal{Q}}f| \leqslant Mf$ holds almost everywhere. Therefore, $\|T_{\mathcal{Q}}f\|_\varphi \leqslant \|Mf\|_\varphi$ and we conclude that $\varphi \in \mathcal{A}$ whenever M is of strong type φ. Thus the class $\mathcal{A}$ is weaker than the boundedness of M. This fact was rather simple. More interesting is the following theorem, which shows that a weaker version of the converse statement is true.

Theorem 4.4.10. *If* $\varphi \in \mathcal{A}$, *then* M *is of weak type* φ *and the constant depends only on the* $\mathcal{A}$-*constant of* φ *and the dimension* n.

Proof. In the proof we need the centered maximal operator, $M_{\mathrm{center}}f(x)$, where the supremum is taken over all cubes with center x. This is indicated by the notation Q_x. Note that

$$M_{\mathrm{center}}f \leqslant Mf \leqslant 2^n M_{\mathrm{center}}f \qquad (4.4.11)$$

for every $f \in L^1_{\text{loc}}(\mathbb{R}^n)$. Therefore, it suffices to prove the theorem with M replaced by M_{center}.

Fix $f \in L^{\varphi(\cdot)}(\mathbb{R}^n)$ with $\|f\|_\varphi \leqslant 1$ and $\lambda > 0$. Let $\Omega_\lambda := \{M_{\text{center}}f > \lambda\}$. Then Ω_λ is open, since M_{center} is lower semicontinuous. Let K be a compact subset of Ω_λ. For every $x \in K$ there exists a cube Q_x with center x such that $M_{Q_x}f > \lambda$. From the family $\{Q_x\}_{x \in K}$ we can select by the Besicovitch covering theorem, Theorem 1.4.6, locally 1-finite families $\mathcal{Q}_1, \ldots, \mathcal{Q}_{\xi_n}$, which together cover K. The natural number ξ_n depends only on the dimension n. Then almost everywhere

$$\lambda \chi_K \leqslant \sum_{m=1}^{\xi_n} \sum_{Q \in \mathcal{Q}_m} \lambda \chi_Q \leqslant \sum_{m=1}^{\xi_n} \sum_{Q \in \mathcal{Q}_m} \fint_Q |f(y)| \, dy \, \chi_Q = \sum_{m=1}^{\xi_n} T_{\mathcal{Q}_m} f.$$

This and $\varphi \in \mathcal{A}$ imply

$$\|\lambda \chi_K\|_\varphi \leqslant \left\| \sum_{m=1}^{\xi_n} T_{\mathcal{Q}_m} f \right\|_\varphi \leqslant \sum_{m=1}^{\xi_n} \|T_{\mathcal{Q}_m} f\|_\varphi \leqslant \xi_n \, c_0 \, \|f\|_\varphi,$$

where c_0 is the $\mathcal{A}$-constant of φ. Now, let $K_j \subset\subset \Omega_\lambda$ with $K_j \nearrow \Omega_\lambda$. By the previous inequality and monotone convergence, Theorem 2.3.17, we conclude that

$$\|\lambda \chi_{\Omega_\lambda}\|_\varphi = \lim_{j \to \infty} \|\lambda \chi_{K_j}\|_\varphi \leqslant \xi_n c_0 \|f\|_\varphi. \qquad \square$$

Theorems 4.4.8 and 4.4.10 have the following immediate consequence.

Corollary 4.4.12. *Let $p \in \mathcal{P}^{\log}(\mathbb{R}^n)$. Then M is of weak type $p(\cdot)$ with constant depending on p only via $c_{\log}(p)$.*

We have seen that the uniform continuity of $T_{\mathcal{Q}}$ from L^φ to L^φ for all locally 1-finite families of cubes $\mathcal{Q}$ in $\mathbb{R}^n$ implies that M is of weak-type φ. The following lemma is a weaker version of the converse direction.

Lemma 4.4.13. *Let $\varphi \in \Phi(\mathbb{R}^n)$. If M is of weak type φ, then $\varphi \in \mathcal{A}_{\text{loc}}$.*

Proof. Let $f \in L^\varphi(\mathbb{R}^n)$ and let $Q \subset \mathbb{R}^n$ be a cube. If $\lambda := \frac{1}{2} M_Q f > 0$, then

$$\|T_{\{Q\}} f\|_\varphi = \|\chi_Q \, M_Q f\|_\varphi \leqslant 2 \, \|\chi_{\{Mf > \lambda\}} \lambda\|_\varphi \leqslant c \|f\|_\varphi,$$

where we have used that M is of weak type φ. $\qquad \square$

The condition $\varphi \in \mathcal{A}_{\text{loc}}$ can also be characterized in terms of the norms of characteristic functions, see Theorem 4.5.7.

We have seen in Corollary 4.4.12 that M is weak type $p(\cdot)$ if $p \in \mathcal{P}^{\log}(\mathbb{R}^n)$. In particular, we have the norm estimate $\|\lambda \chi_{\{Mf>\lambda\}}\|_{p(\cdot)} \leqslant K\|f\|_{p(\cdot)}$ for all $f \in L^{p(\cdot)}(\mathbb{R}^n)$. However, often it is better not to work with the norm but rather the modular $\varrho_{p(\cdot)}$, since it behaves more like an integral. For example the Sobolev embedding $W^{1,p(\cdot)} \hookrightarrow L^{p^*(\cdot)}$ in Sect. 8.3 is based on weak type estimates for the modular $\varrho_{p(\cdot)}$.

Proposition 4.4.14. *Let $p \in \mathcal{P}^{\log}(\mathbb{R}^n)$. Then for any $m > 0$ there exists $\beta \in (0,1)$ depending only on the dimension n, m and $c_{\log}(p)$ such that*

$$\varrho_{p(\cdot)}(\beta \lambda \chi_{\{Mf>\lambda\}}) \leqslant \varrho_{p(\cdot)}(f) + \int\limits_{\{Mf>4^{-n}\lambda\}} (e + |x|)^{-m} \, dx$$

for all $f \in L^{p(\cdot)}(\mathbb{R}^n) + L^\infty(\mathbb{R}^n)$ with $\|f\|_{L^{p(\cdot)}(\mathbb{R}^n)+L^\infty(\mathbb{R}^n)} \leqslant 1$ and all $\lambda > 0$.

Proof. We prove the claim for the centered maximal operator. The other case follows from point-wise equivalence (4.4.11) of the centered and the non-centered maximal operator.

Fix $f \in L^{p(\cdot)}(\mathbb{R}^n)$ with $\|f\|_{p(\cdot)} \leqslant 1$ and $\lambda > 0$. Then $\{Mf > \lambda\}$ is open, since M is lower semicontinuous. Let K be an arbitrary compact subset of $\{Mf > \lambda\}$. For every $x \in K$ there exists a cube Q_x with center x such that $M_{Q_x}f > \lambda$. Note that $Q_x \subset \{Mf > 2^{-n}\lambda\}$ by (4.4.11). From the family $\{Q_x \colon x \in K\}$ we can select by the Besicovitch covering theorem, Theorem 1.4.6, a locally ξ_n-finite family $\mathcal{Q}$, which covers K. The natural number ξ_n only depends on the dimension n. Let $m > 0$. By Corollary 4.2.5 there exists $\beta > 0$ such that

$$\int\limits_Q \varphi_{p(y)}(\beta M_Q f) \, dy \leqslant \int\limits_Q \varphi_{p(y)}(f(y)) \, dy + \int\limits_Q (e + |y|)^{-m} \, dy$$

for all $Q \in \mathcal{Q}$. Now, the convexity of the modular (2.1.5) and $M_Q f > \lambda$ for $Q \in \mathcal{Q}$ imply that

$$\varrho_{p(\cdot)}(N^{-1}\beta\lambda\chi_K) \leqslant N^{-1} \sum_{Q \in \mathcal{Q}} \int\limits_Q \varphi_{p(x)}(\beta M_Q f) \, dy.$$

With the previous estimate we get

$$\varrho_{p(\cdot)}(N^{-1}\beta\lambda\chi_K) \leqslant N^{-1} \sum_{Q \in \mathcal{Q}} \left(\int\limits_Q \varphi_{p(y)}(f(y)) \, dy + \int\limits_Q (e + |y|)^{-m} \, dy \right)$$

$$\leqslant \varrho_{p(\cdot)}(f) + \int\limits_{\{Mf>2^{-n}\lambda\}} (e + |y|)^{-m} \, dy,$$

where we used that $\mathcal{Q}$ is locally N-finite. Let $K_j \subset\subset \{Mf > \lambda\}$ with $K_j \nearrow$ $\{Mf > \lambda\}$. Then monotone convergence (Lemma 3.2.8) implies that

$$\varrho_{p(\cdot)}(N^{-1}\beta\lambda\chi_{\{Mf>\lambda\}}) \leqslant \varrho_{p(\cdot)}(f) + \int\limits_{\{Mf>2^{-n}\lambda\}} (\mathrm{e} + |y|)^{-m}\, dy. \qquad \square$$

In Theorem 4.4.8 we have seen that $p \in \mathcal{P}^{\log}(\mathbb{R}^n)$ gives control over averaging operators over locally finite families of cubes in the sense that $\|\sum_{Q\in\mathcal{Q}} \chi_Q M_Q f\|_{p(\cdot)} \leqslant c\|f\|_{p(\cdot)}$. In this situation we distribute the averages of $|f|$ exactly on the same cubes, where the average is calculated. However, sometimes it is useful to take the average on one locally finite family and transfer it another locally finite family that is similar in some sense. This is the purpose of the following theorem. It is a stronger version of Theorem 4.4.8 and will be used for example for the extension of variable exponent Sobolev functions in Sect. 8.5.

Theorem 4.4.15. *Let $p \in \mathcal{P}^{\log}(\mathbb{R}^n)$ and $\lambda \geqslant 1$. Let $\mathcal{Q}$ be a locally N-finite family of cubes (or balls) such that to every $Q \in \mathcal{Q}$, there is associated a cube Q^* with $Q \subset \lambda Q^*$. Further assume that $\sum_{Q\in\mathcal{Q}} \chi_{Q^*} \leqslant N$. Then*

$$\left\| \sum_{Q\in\mathcal{Q}} \chi_Q M_{Q^*} f \right\|_{p(\cdot)} \leqslant c\,\lambda^n N \|f\|_{p(\cdot)}$$

for all $f \in L^{p(\cdot)}(\mathbb{R}^n)$, where c only depends on $c_{\log}(p)$ and n.

Proof. Let $f \in L^{p(\cdot)}(\mathbb{R}^n)$ with $\|f\|_{p(\cdot)} \leqslant 1$. Then by the unit ball property $\varrho_{p(\cdot)}(f) \leqslant 1$. Let $m > n$ be such that $\int_{\mathbb{R}^n}(\mathrm{e}+|y|)^{-m}\, dy \leqslant 1$. Choose $\beta \in (0,1)$ as in Theorem 4.2.4. Let $Q \in \mathcal{Q}$. Then by assumption $Q \subset \lambda Q^*$. We apply Theorem 4.2.4 to the cube λQ^* and the function $f\chi_{Q^*}$ and integrate over $x \in Q$. This gives

$$\int\limits_Q \varphi_{p(x)}\left(\beta \fint_{\lambda Q^*} \chi_{Q^*}|f|\, dy\right) dx \leqslant |Q| \fint_{\lambda Q^*} \varphi_{p(y)}(\chi_{Q^*}|f|)\, dy$$

$$+ \frac{1}{2}\int\limits_Q (\mathrm{e}+|x|)^{-m}\, dx + \frac{1}{2}|Q| \fint_{\lambda Q^*} \chi_{Q^*}(y)(\mathrm{e}+|y|)^{-m}\, dy,$$

where we have used $\{0 < |\chi_{Q^*} f| \leqslant 1\} \subset Q^*$. Since $\lambda \geqslant 1$, it follows that

$$\int\limits_{Q} \varphi_{p(y)}\big(\beta\lambda^{-n}\chi_Q M_{Q^*}f\big)\,dy$$

$$\leqslant \int\limits_{Q^*} \varphi_{p(y)}(|f|)\,dy + \frac{1}{2}\int\limits_{Q}(e+|x|)^{-m}\,dx + \frac{1}{2}\int\limits_{Q^*}(e+|y|)^{-m}\,dy,$$

Then the convexity of the modular implies that

$$\varrho_{p(\cdot)}\bigg(\frac{\beta}{2N\lambda^n}\sum_{Q\in\mathcal{Q}}\chi_Q M_{Q^*}f\bigg) \leqslant \frac{1}{2N}\sum_{Q\in\mathcal{Q}}\int\limits_{Q}\varphi_{p(y)}\big(\beta\lambda^{-n}\chi_Q M_{Q^*}f\big)\,dy$$

$$\leqslant \frac{1}{2}\varrho_{p(\cdot)}(f) + \frac{1}{2}\int\limits_{\mathbb{R}^n}(e+|z|)^{-m}\,dz$$

$$\leqslant 1,$$

where we also used that $\mathcal{Q}$ is locally N-finite and $\sum_{Q\in\mathcal{Q}}\chi_{Q^*}\leqslant N$. This implies that $\big\|\sum_{Q\in\mathcal{Q}}\chi_Q M_{Q^*}f\big\|_{p(\cdot)} \leqslant \frac{2N\lambda^n}{\beta}$. A scaling argument completes the proof. $\qquad\qquad\square$

Remark 4.4.16. If $\mathcal{P}^{\log}(\mathbb{R}^n)$ with $p^- > 1$, then by Theorem 4.3.8 the maximal operator M is bounded from $L^{p(\cdot)}(\mathbb{R}^n)$ to $L^{p(\cdot)}(\mathbb{R}^n)$. In this case Theorem 4.4.15 is a simple consequence of the estimate $\chi_Q M_{Q^*}f \leqslant \lambda^{-n}\chi_Q Mf$ and the boundedness of M. However, we cannot use this argument in the general case $\mathcal{P}^{\log}(\mathbb{R}^n)$ with $p^- \geqslant 1$.

Using the choice $f := \sum_{Q\in\mathcal{Q}}\chi_{Q^*}t_Q$ in Theorem 4.4.15 we immediately get the following result.

Corollary 4.4.17. Let $p \in \mathcal{P}^{\log}(\mathbb{R}^n)$ and $\lambda \geqslant 1$. Let $\mathcal{Q}$ be a locally N-finite family of cubes (or balls) such that to every $Q \in \mathcal{Q}$, there is associated a cube Q^* such that $Q \subset \lambda Q^*$. Further assume that $\sum_{Q\in\mathcal{Q}}\chi_{Q^*}\leqslant N$. Then

$$\bigg\|\sum_{Q\in\mathcal{Q}}\chi_Q t_Q\bigg\|_{p(\cdot)} \leqslant c\,\lambda^n N\bigg\|\sum_{Q\in\mathcal{Q}}\chi_{Q^*}t_Q\bigg\|_{p(\cdot)}$$

for all families $\{t_Q\} \in \mathbb{R}^{\mathcal{Q}}$, where c only depends on $c_{\log}(p)$ and n.

4.5 Norms of Characteristic Functions

In classical Lebesgue spaces $\|\chi_E\|_q = |E|^{\frac{1}{q}}$. We would like to generalize this to the case of a variable exponent $p \in \mathcal{P}(\mathbb{R}^n)$. For general p it is not clear how to replace $\frac{1}{q}$ in the formula by something in terms of p. However, it turns out that if $p \in \mathcal{A}_{\mathrm{loc}}$, then $\|\chi_Q\|_{p(\cdot)} \approx |Q|^{\frac{1}{p_Q}}$ for any cube Q, where p_Q is the harmonic mean from Definition 4.2.6.

Before we come to the proof of $\|\chi_Q\|_{p(\cdot)} \approx |Q|^{\frac{1}{p_Q}}$ for cubes Q we need a few auxiliary results. First we note that the calculation

$$\frac{1}{(p_E)'} = 1 - \frac{1}{p_E} = 1 - \fint_E \frac{1}{p(y)}\,dy = \fint_E \frac{1}{p'(y)}\,dy = \frac{1}{(p')_E}.$$

This implies that $(p_E)' = (p')_E$ for every measurable $E \subset \mathbb{R}^n$ with $|E| \in (0, \infty)$. Thus we do not have to distinguish between $(p_E)'$ and $(p')_E$ and just write p'_E for short.

Lemma 4.5.1. *Let* $p \in \mathcal{P}(\mathbb{R}^n)$ *and let* $E \subset \mathbb{R}^n$ *be measurable with* $|E| \in (0, \infty)$. *Then*

$$\bar{\varphi}_{p_E}\left(\frac{t}{2}\right) \leqslant \widetilde{\varphi}_{p_E}(t) \leqslant \fint_E \widetilde{\varphi}_{p(y)}(t)\,dy \leqslant \fint_E \bar{\varphi}_{p(y)}(t)\,dy,$$

$$\frac{1}{2}\widetilde{\varphi}_{p_E}^{-1}(t) \leqslant \bar{\varphi}_{p_E}^{-1}(t) \leqslant \fint_E \bar{\varphi}_{p(y)}^{-1}(t)\,dy \leqslant \fint_E \widetilde{\varphi}_{p(y)}^{-1}(t)\,dy$$

for all $t \geqslant 0$.

Proof. The mappings $\frac{1}{q} \mapsto \widetilde{\varphi}_q(y, t)$ and $\frac{1}{q} \mapsto \bar{\varphi}_q^{-1}(y, t)$ are convex for all $y \in \mathbb{R}^n$ and $t \geqslant 0$ due to Lemmas 3.1.4 and 3.1.13. Thus, Jensen's inequality implies the middle parts of the inequalities. The remaining inequalities follow with the help of Lemmas 3.1.6 and 3.1.12. $\square$

Lemma 4.5.2. *Let* $p \in \mathcal{P}(\mathbb{R}^n)$ *and let* $E \subset \mathbb{R}^n$ *be measurable with* $|E| \in (0, \infty)$. *Then for all* $t \geqslant 0$

$$t \leqslant \fint_E \varphi_{p(x)}^{-1}(t)\,dx \fint_E \varphi_{p'(x)}^{-1}(t)\,dx.$$

Proof. It suffices to prove the case $t > 0$. By Lemma 3.1.11,

$$\varphi_{p(x)}^{-1}(t) \geqslant \frac{t}{\varphi_{p'(x)}^{-1}(t)}$$

for all $t > 0$ and all $x \in E$. Therefore

$$\fint_E \varphi_{p(x)}^{-1}(t)\,dx \geqslant t \fint_E \frac{1}{\varphi_{p'(x)}^{-1}(t)}\,dx \geqslant \frac{t}{\fint_E \varphi_{p'(x)}^{-1}(t)\,dx},$$

where we have used in the last step Jensen's inequality for $z \mapsto 1/z$. This proves the lemma. $\square$

We can now calculate $\|\chi_Q\|_{p(\cdot)}$ for cubes Q.

Lemma 4.5.3. *Let $p \in \mathcal{A}_{\mathrm{loc}}$ and let A be the $\mathcal{A}_{\mathrm{loc}}$-constant of p. Then*

$$1 \leqslant |Q| \int_Q \varphi_{p(x)'}^{-1}\left(\tfrac{1}{|Q|}\right) dx \int_Q \varphi_{p(x)}^{-1}\left(\tfrac{1}{|Q|}\right) dx \leqslant 2\|\chi_Q\|_{p(\cdot)} \int_Q \varphi_{p(x)}^{-1}\left(\tfrac{1}{|Q|}\right) dx \leqslant 2A$$

$$(4.5.4)$$

and

$$\frac{1}{6}\varphi_{p_Q}^{-1}(|Q|) \leqslant |Q| \int_Q \varphi_{p'(x)}^{-1}\left(\tfrac{1}{|Q|}\right) dx \leqslant 2\|\chi_Q\|_{p(\cdot)} \leqslant 4A\varphi_{p_Q}^{-1}(|Q|) \qquad (4.5.5)$$

for all cubes $Q \subset \mathbb{R}^n$. Moreover,

$$\frac{1}{12}|Q|^{\frac{1}{p_Q}} \leqslant \|\chi_Q\|_{p(\cdot)} \leqslant 4A|Q|^{\frac{1}{p_Q}}, \qquad (4.5.6)$$

where we use the usual convention $\lambda^{\frac{1}{\infty}} = \lambda^0 = 1$ for all $0 < \lambda < \infty$ if $p_Q = \infty$.

Proof. Let $Q \subset \mathbb{R}^n$ be a cube. Define $f := \chi_Q\,\varphi_{p(\cdot)}^{-1}(1/|Q|)$ and $g := \chi_Q\,\varphi_{p'(\cdot)}^{-1}(1/|Q|)$, and note that $\varrho_{p(\cdot)}(f) \leqslant 1$ and $\varrho_{p'(\cdot)}(g) \leqslant 1$ by (3.1.9). Thus, $\|f\|_{p(\cdot)} \leqslant 1$ and $\|g\|_{p'(\cdot)} \leqslant 1$ by the unit ball property.

The first inequality in (4.5.4) follows directly from Lemma 4.5.2. The second one follows from Hölder's inequality (Lemma 3.2.20):

$$|Q| \int_Q \varphi_{p'(x)}^{-1}\left(\tfrac{1}{|Q|}\right) dx = \int_Q g\,dx \leqslant 2\|\chi_Q\|_{p(\cdot)}\|g\|_{p'(\cdot)} \leqslant 2\|\chi_Q\|_{p(\cdot)}.$$

The third inequality can be derived as follows

$$\|\chi_Q\|_{p(\cdot)} \int_Q \varphi_{p(x)}^{-1}\left(\tfrac{1}{|Q|}\right) dx = \|\chi_Q\|_{p(\cdot)} M_Q f = \|T_{\{Q\}}f\|_{p(\cdot)} \leqslant A\|f\|_{p(\cdot)} \leqslant A.$$

Since $\int_Q \varphi_{p(x)}^{-1}\left(\tfrac{1}{|Q|}\right) dx > 0$, we obtain

$$|Q| \int_Q \varphi_{p'(x)}^{-1}\left(\tfrac{1}{|Q|}\right) dx \leqslant 2\|\chi_Q\|_{p(\cdot)} \leqslant \frac{2K}{\int_Q \varphi_{p(x)}^{-1}\left(\tfrac{1}{|Q|}\right) dx}$$

from the second and third inequalities of (4.5.4). By Lemmas 4.5.1, 3.1.11 and 3.1.12 we estimate the left-hand side:

$$|Q| \int_Q \varphi_{p'(x)}^{-1}\left(\tfrac{1}{|Q|}\right) dx \geqslant \tfrac{1}{2}|Q|\varphi_{p'_Q}^{-1}\left(\tfrac{1}{|Q|}\right) \geqslant \tfrac{1}{2}\frac{1}{\varphi_{p_Q}^{-1}\left(\tfrac{1}{|Q|}\right)} \geqslant \tfrac{1}{6}\varphi_{p_Q}^{-1}(|Q|).$$

Combining these two inequalities yields the first inequality in (4.5.5). Similarly we derive

$$\frac{K}{f_Q \varphi_{p(\cdot)}^{-1}\left(\tfrac{1}{|Q|}\right) dx} \leqslant \frac{2A}{\varphi_{p_Q}^{-1}\left(\tfrac{1}{|Q|}\right) dx} \leqslant 2A\varphi_{p_Q}^{-1}(|Q|),$$

which gives the second inequality in (4.5.5). If $\varphi_{p(\cdot)} = \bar{\varphi}_{p(\cdot)}$, then (4.5.5) directly implies

$$\frac{1}{6}|Q|^{\frac{1}{p_Q}} \leqslant \|\chi_Q\|_{\bar{\varphi}_{p(\cdot)}} \leqslant 4A|Q|^{\frac{1}{p_Q}}$$

which is stronger than (4.5.6). This and (3.2.2) imply the result in the case $\varphi_{p(\cdot)} = \tilde{\varphi}_{p(\cdot)}$. $\square$

Note that Lemma 4.5.3 remains valid if we use balls rather than cubes with a possible change of constants. This is due to the fact that every cube is contained in a ball of similar size and vice versa.

Theorem 4.5.7. *Let $p \in \mathcal{P}(\mathbb{R}^n)$. Then the following are equivalent:*

(a) $p \in \mathcal{A}_{\mathrm{loc}}$
(b) $p' \in \mathcal{A}_{\mathrm{loc}}$.
(c) $\|\chi_Q\|_{p(\cdot)} \|\chi_Q\|_{p'(\cdot)} \approx |Q|$ *uniformly for all cubes $Q \subset \mathbb{R}^n$.*
(d) $\|\chi_Q\|_{p(\cdot)} \approx |Q|^{\frac{1}{p_Q}}$ *and* $\|\chi_Q\|_{p'(\cdot)} \approx |Q|^{\frac{1}{p_Q'}}$ *uniformly for all cubes $Q \subset \mathbb{R}^n$.*

The statement remains true if we replace cubes by balls.

Proof. (a) $\Leftrightarrow$ (b): For all:w non-negative $f, g \in L^1_{\mathrm{loc}}(\mathbb{R}^n)$ we have $\int g T_{\{Q\}} f \, dx = \int f T_{\{Q\}} g \, dx$. Thus by the norm conjugate formula (Corollary 3.2.14) $T_{\{Q\}}$ is bounded on $L^{p(\cdot)}$ if and only if it is bounded on $L^{p'(\cdot)}$.

(c) $\Rightarrow$ (a): Using Hölder's inequality (Lemma 3.2.20) we get

$$\|T_{\{Q\}}f\|_{p(\cdot)} = \|\chi_Q M_Q f\|_{p(\cdot)} = \|\chi_Q\|_{p(\cdot)} |Q|^{-1} \int_{\mathbb{R}^n} \chi_Q(y)|f(y)| \, dy$$

$$\leqslant \|\chi_Q\|_{p(\cdot)} 2 |Q|^{-1} \|\chi_Q\|_{p'(\cdot)} \|f\|_{p(\cdot)}$$

for all $f \in L^{p(\cdot)}$. Now, $\|\chi_Q\|_{p(\cdot)} \|\chi_Q\|_{p'(\cdot)} \approx |Q|$ yields the boundedness of $T_{\{Q\}}$.

(a) $\Rightarrow$ (c): We estimate with the norm conjugate formula of $L^{p(\cdot)}$ (Corollary 3.2.14)

$$\|\chi_Q\|_{p(\cdot)}\|\chi_Q\|_{p'(\cdot)} \leqslant 2\|\chi_Q\|_{p(\cdot)} \sup_{\|g\|_{p(\cdot)}\leqslant 1} \int \chi_Q g\, dx$$

$$= 2 \sup_{\|g\|_{p(\cdot)}\leqslant 1} |Q|\, \|\chi_Q\, M_Q g\|_{p(\cdot)}$$

$$\leqslant 2A \sup_{\|g\|_{p(\cdot)}\leqslant 1} |Q|\, \|g\|_{p(\cdot)}$$

$$= 2A|Q|$$

uniformly for all cubes $Q \subset \mathbb{R}^n$, where A is the $\mathcal{A}_{\mathrm{loc}}$-constant of p.

(a) + (b) $\Rightarrow$ (d): This follows from Lemma 4.5.3 and $\frac{1}{p'_Q} = 1 - \frac{1}{p_Q}$.

(d) $\Rightarrow$ (c): This is obvious. $\square$

Remark 4.5.8. If φ is a proper generalized Φ-function on $\mathbb{R}^n$, then the norm conjugate formula is valid for $L^\varphi(\mathbb{R}^n)$ and $L^{\varphi^*}(\mathbb{R}^n)$ (Corollary 2.7.5). In such a situation the equivalence of (a), (b) and (c) in the previous theorem remains valid. More precisely, if $\varphi \in \mathcal{A}$ with $\mathcal{A}_{\mathrm{loc}}$-constant A, then $\|\chi_Q\|_\varphi \|\chi_Q\|_{\varphi^*} \leqslant A|Q|$ for all cubes Q.

In Theorem 4.5.7 and Remark 4.5.8 it is also possible to use balls instead of cubes. By Theorem 4.4.8, $p \in \mathcal{A} \subset \mathcal{A}_{\mathrm{loc}}$ if $p \in \mathcal{P}^{\log}(\mathbb{R}^n)$. Combining this with the previous theorem yields:

Corollary 4.5.9. *Let* $p \in \mathcal{P}^{\log}(\mathbb{R}^n)$. *Then* $\|\chi_Q\|_{p(\cdot)} \approx |Q|^{\frac{1}{p_Q}}$ *for every cube (or ball)* $Q \subset \mathbb{R}^n$. *More concretely,*

$$\|\chi_Q\|_{p(\cdot)} \approx \begin{cases} |Q|^{\frac{1}{p(x)}} & \text{if } |Q| \leqslant 2^n \text{ and } x \in Q, \\ |Q|^{\frac{1}{p_\infty}} & \text{if } |Q| \geqslant 1 \end{cases}$$

for every cube (or ball) $Q \subset \mathbb{R}^n$. *The implicit constants only depend on* $c_{\log}(p)$.

Proof. The first claim follows from the previous theorem. It remains only to prove that $|Q|^{\frac{1}{p_Q}} \approx |Q|^{\frac{1}{p(x)}}$ for small cubes and $|Q|^{\frac{1}{p_Q}} \approx |Q|^{\frac{1}{p_\infty}}$ for large cubes. The former claim follows directly from Lemma 4.1.6 since $p_Q^- \leqslant p_Q \leqslant p_Q^+$. The latter follows if we prove that

$$\left| \frac{1}{p_Q} - \frac{1}{p_\infty} \right| \log |Q| \leqslant c$$

for all cubes with $|Q| \geqslant 1$. By the triangle inequality and the decay condition we have

$$\left|\frac{1}{p_Q} - \frac{1}{p_\infty}\right| \leqslant \fint_Q \left|\frac{1}{p(x)} - \frac{1}{p_\infty}\right| dx \leqslant \fint_Q \frac{c_2}{\log(e + |x|)} dx.$$

Denote $R = \operatorname{diam} Q$. Note that the integrand increases if we move from a arbitrary cube to a cube of the same size centered at the origin. For simplicity we then move to a ball and change to scaled spherical coordinates (with $r = Rs$, $s \in [0,1]$):

$$\left|\frac{1}{p_Q} - \frac{1}{p_\infty}\right| \leqslant \fint_{B(0,R)} \frac{c_2}{\log(e + |x|)} dx = \int_0^1 \frac{c_2 s^{n-1}}{\log(e + Rs)} ds.$$

Since $\frac{s^{n-1}\log(e+R)}{\log(e+Rs)} \leqslant c$ uniformly in $s \in [0,1]$ and $R > 0$, we see that the integral on the right-hand side is bounded by $c\frac{1}{\log(e+R)}$. Since $\frac{\log|Q|}{\log(e+R)} \leqslant c$, the claim follows. $\square$

4.6 Mollification and Convolution

In this section we again start with the key estimate for norms of averages, Theorem 4.2.4. Now we move in the direction of mollifications. Convolutions were first considered in this context by Samko [340], although the sufficiency of the log-Hölder condition was established by Diening [91].

Lemma 4.6.1. *Let $p \in \mathcal{A}$. Then*

$$\left\|f * \frac{\chi_Q}{|Q|}\right\|_{p(\cdot)} \leqslant \left\||f| * \frac{\chi_Q}{|Q|}\right\|_{p(\cdot)} \leqslant 3^n A\|f\|_{p(\cdot)}$$

for all $f \in L^{p(\cdot)}(\mathbb{R}^n)$ and all cubes (or balls) $Q \subset \mathbb{R}^n$ with center at 0. Here A is the $\mathcal{A}$-constant of p.

Proof. Let $Q \subset \mathbb{R}^n$ be a cube with center at 0 and $f \in L^{p(\cdot)}(\mathbb{R}^n)$. For $k \in \mathbb{Z}^n$ let $Q_k := \ell(Q)k + Q$, be the translation of Q by the vector $\ell(Q)k$. Then the cubes $\{Q_k\}_k$ are disjoint and cover $\mathbb{R}^n$ (up to a null set). Moreover, we can split the set $\{3Q_k\}_k$ into 3^n locally 1-finite families $\mathcal{Q}_j$, $j = 1, \ldots, 3^n$. For every $k \in \mathbb{Z}^n$ and every $x \in Q_k$ we have

$$f * \frac{\chi_Q}{|Q|} \leqslant |f| * \frac{\chi_Q}{|Q|} \leqslant \frac{1}{|Q|} \int_{3Q_k} |f(y)| \, dy = 3^n M_{3Q_k} f \leqslant 3^n \sum_{j=1}^{3^n} T_{\mathcal{Q}_j} f,$$

where we have used that Q has center at zero. Therefore,

$$\left\| f * \frac{\chi_Q}{|Q|} \right\|_{p(\cdot)} \leqslant \left\| |f| * \frac{\chi_Q}{|Q|} \right\|_{p(\cdot)} \leqslant 3^n \sum_{j=1}^{3^n} \left\| T_{Q_j} f \right\|_{p(\cdot)} \leqslant 3^n A \| f \|_{p(\cdot)},$$

where A is the $\mathcal{A}$-constant of p.   □

With this lemma we can prove convolution estimates for bell shaped functions.

Definition 4.6.2. A function $\psi \in L^1(\mathbb{R}^n)$ with $\psi \geqslant 0$ is called *bell shaped* if it is radially decreasing and radially symmetric. The function $\Psi(x) := \sup_{y \notin B(0,|x|)} |f(y)|$ is called the *least bell shaped majorant of f*.

Recall that we defined

$$\psi_\varepsilon(x) := \frac{1}{\varepsilon^n} \psi\left(\frac{x}{\varepsilon}\right),$$

for $\varepsilon > 0$.

Lemma 4.6.3 (Mollification). *Let $p \in \mathcal{A}$ or $p \in \mathcal{P}^{\log}(\mathbb{R}^n)$, and let $\psi \in L^1(\mathbb{R}^n)$. Assume that the least bell shaped majorant Ψ of ψ is integrable. Then*

$$\| f * \psi_\varepsilon \|_{p(\cdot)} \leqslant c K \| \Psi \|_1 \| f \|_{p(\cdot)}$$

for all $f \in L^{p(\cdot)}(\mathbb{R}^n)$, where K is the $\mathcal{A}$-constant of p and c depends only on n. Moreover, $|f * \psi_\varepsilon| \leqslant 2 \| \Psi \|_1 Mf$ for all $f \in L^1_{\mathrm{loc}}(\mathbb{R}^n)$.

Proof. By Theorem 4.4.8 we always have $p \in \mathcal{A}$. We may assume without loss of generality that $f, \psi \geqslant 0$. Since $\psi(x) \leqslant \Psi(x)$, we may further assume that ψ is already bell shaped. Any bell shaped function ψ can be approximated from above by functions of type

$$h := \sum_{k=1}^{\infty} a_k \frac{\chi_{B_k}}{|B_k|},$$

where $a_k \in [0, \infty)$, B_k are balls with center at zero and

$$\| h \|_{L^1(\mathbb{R}^n)} = \sum_{k=1}^{\infty} a_k \leqslant 2 \| \psi \|_{L^1(\mathbb{R}^n)}.$$

Let $h_\varepsilon(x) := \varepsilon^{-n} h(x/\varepsilon)$. Then $0 \leqslant \psi_\varepsilon \leqslant h_\varepsilon$ and

$$h_\varepsilon = \sum_{k=1}^{\infty} a_k \frac{\chi_{\varepsilon B_k}}{|\varepsilon B_k|}.$$

Using Lemma 4.6.1 we estimate

$$\left\| f * \psi_\varepsilon \right\|_{p(\cdot)} \leqslant \left\| f * h_\varepsilon \right\|_{p(\cdot)} = \left\| \sum_{k=1}^\infty |f| * \left(a_k \frac{\chi_{\varepsilon B_k}}{|\varepsilon B_k|} \right) \right\|_{p(\cdot)}$$

$$\leqslant \sum_{k=1}^\infty a_k \left\| |f| * \frac{\chi_{\varepsilon B_k}}{|\varepsilon B_k|} \right\|_{p(\cdot)}$$

$$\leqslant \sum_{k=1}^\infty a_k \, 3^n K \left\| f \right\|_{p(\cdot)}$$

$$\leqslant 2 \, 3^n \, K \left\| \psi \right\|_1 \left\| f \right\|_{p(\cdot)}.$$

Analogously, for $f \in L^1_{\mathrm{loc}}(\mathbb{R}^n)$ we estimate pointwise

$$|f * \psi_\varepsilon| \leqslant \sum_{k=1}^\infty |f| * \left(a_k \frac{\chi_{\varepsilon B_k}}{|\varepsilon B_k|} \right) \leqslant \sum_{k=1}^\infty a_k Mf \leqslant 2 \left\| \psi \right\|_1 Mf. \qquad \square$$

Using the previous lemma we can in fact get better control of $f * \psi_\varepsilon$ when ε is small as we show in the following results.

Theorem 4.6.4 (Mollification). *Let $p \in \mathcal{A}$ and $\psi \in L^1(\mathbb{R}^n)$. Assume that the least bell shaped majorant Ψ of ψ is integrable and $\int_{\mathbb{R}^n} \psi(x)\, dx = 1$. Then $f * \psi_\varepsilon \to f$ a.e. as $\varepsilon \to 0$ for $f \in L^{p(\cdot)}(\mathbb{R}^n)$. If additionally $p^+ < \infty$, then $f * \psi_\varepsilon \to f$ in $L^{p(\cdot)}(\mathbb{R}^n)$.*

Proof. Let $f \in L^{p(\cdot)}$ with $\|f\|_{p(\cdot)} \leqslant 1$. By Theorem 3.3.11 we can split f into $f = f_0 + f_1$ with $f_0 \in L^1$ and $f_1 \in L^\infty$. From [359, Theorem 2, p. 62] we deduce $f_j * \psi_\varepsilon \to f_j$ almost everywhere, $j = 0, 1$. This proves $f * \psi_\varepsilon \to f$ almost everywhere.

It remains to prove $\|f * \psi_\varepsilon - f\|_{p(\cdot)} \to 0$ for $\varepsilon \to 0$ if $p^+ < \infty$. Let $\delta > 0$ be arbitrary. Then by density of simple functions in $L^{p(\cdot)}(\mathbb{R}^n)$, Corollary 3.4.10, we can find a simple function g with $\|f - g\|_{p(\cdot)} \leqslant \delta$. This implies that

$$\|f * \psi_\varepsilon - f\|_{p(\cdot)} \leqslant \|g * \psi_\varepsilon - g\|_{p(\cdot)} + \|(f - g) * \psi_\varepsilon - (f - g)\|_{p(\cdot)} =: (I) + (II).$$

Since g is a simple function, we have $g \in L^1(\mathbb{R}^n) \cap L^{p^+}(\mathbb{R}^n)$. Thus the classical theorem on mollification, see [359] again, implies that $g * \psi_\varepsilon \to g$ in $L^1(\mathbb{R}^n) \cap L^{p^+}(\mathbb{R}^n)$. Thus $g * \psi_\varepsilon \to g$ in $L^{p(\cdot)}(\mathbb{R}^n)$ by Theorem 3.3.11. This proves $(I) \to 0$ for $\varepsilon \to 0$. On the other hand, Lemma 4.6.3 implies that

$$(II) = \|(f - g) * \psi_\varepsilon - (f - g)\|_{p(\cdot)} \leqslant c \|f - g\|_{p(\cdot)} \leqslant c\,\delta.$$

This implies

$$\limsup_{\varepsilon \to 0} \|f * \psi_\varepsilon - f\|_{p(\cdot)} \leqslant c\,\delta.$$

Since $\delta > 0$ was arbitrary, this yields $\|f * \psi_\varepsilon - f\|_{p(\cdot)} \to 0$ as $\varepsilon \to 0$.   □

The previous theorem allows us to use the usual proof based on convolution to prove the density of $C_0^\infty(\Omega)$ in $L^{p(\cdot)}(\Omega)$ is the usual fashion. Since this result is weaker than Theorem 3.4.12, the proof, which is an easy modification of the standard proof, is omitted.

Corollary 4.6.5. *If* $p \in \mathcal{A}$ *with* $p^+ < \infty$, *then* $C_0^\infty(\Omega)$ *is dense in* $L^{p(\cdot)}(\Omega)$.

The density of $C_0^\infty(\Omega)$ in $L^{p(\cdot)}(\Omega)$ was used in Corollary 3.4.13 to improve the norm conjugate formula to test functions from $C_0^\infty(\Omega)$. This was done under the restriction $p^- > 1$, which ensured the density of $C_0^\infty(\Omega)$ in $L^{p'(\cdot)}(\Omega)$. The mollification estimates allow us to replace the assumption $p^- > 1$ with $p \in \mathcal{A}$ at the cost of extra constants in the formula.

Corollary 4.6.6 (Norm conjugate formula). *Let* $\Omega \subset \mathbb{R}^n$ *be open and* $p \in \mathcal{A}$ *or* $p \in \mathcal{P}^{\log}(\Omega)$. *Then*

$$c\,\|f\|_{p(\cdot)} \leqslant \sup_{g \in C_0^\infty(\Omega)\,:\,\|g\|_{p'(\cdot)} \leqslant 1} \int_\Omega |f|\,|g|\,d\mu \leqslant 2\,\|f\|_{p(\cdot)}$$

for all $f \in L^0(\Omega)$, *where* c *only depends on the* $\mathcal{A}$*-constant of* p.

Proof. The upper estimate follows by Hölder's inequality. For $f \in L^{p(\cdot)}$ let $\|\|f\|\|$ denote the supremum in the formula. Due to the norm conjugate formula, Corollary 3.2.14, it suffices to prove

$$\int_\Omega |f||g|\,dx \leqslant c\,\|\|f\|\|\,\|g\|_{p'(\cdot)}$$

for all simple functions $g \in S$ with compact support. Let ψ_ε denote a standard mollifier family on $\mathbb{R}^n$. Let $g_j := g * \psi_{1/j}$ for $j \in \mathbb{N}$; then $g_j \to g$ almost everywhere and $\|g_j\|_{p'(\cdot)} \leqslant c\,\|g\|_{p'(\cdot)}$ by Lemma 4.6.3. Since g has compact support and spt $\psi_{1/j} \subset B(0, 1/j)$, there exists $\Omega' \subset\subset \Omega$ and j_0 such that spt $g_j \subset \Omega'$ for all $j \geqslant j_0$. As g is bounded, the function $\chi_{\Omega'}\|g\|_\infty$ is an integrable majorant of $|g_j|$. Since $L^{p(\cdot)}(\Omega') \hookrightarrow L^1(\Omega')$ (by Corollary 3.3.4), $|f|\chi_{\Omega'}\|g\|_\infty$ is an L^1 majorant of $|f||g_j|$. Therefore, by dominated convergence, $\int_\Omega |f||g_j|\,dx \to \int_\Omega |f||g|\,dx$ and hence

$$\int_\Omega |f||g|\,dx = \lim_j \int_\Omega |f||g_j|\,dx \leqslant \lim_j \|\|f\|\|\,\|g_j\|_{p'(\cdot)} \leqslant c\,\|\|f\|\|\,\|g\|_{p'(\cdot)}.\quad □$$

4.7 Necessary Conditions for Boundedness*

Since the maximal operator is a central tool in later chapters, it is obviously of interest to obtain its boundedness under optimal assumptions.

We have seen in Theorem 4.4.10 that the maximal operator M is of weak type $p(\cdot)$ if $p \in \mathcal{P}^{\log}(\mathbb{R}^n)$. So the natural question arises if it is also of strong type $p(\cdot)$. Theorem 4.3.8 gives an affirmative answer to this question if $p \in \mathcal{P}^{\log}(\mathbb{R}^n)$ and $p^- > 1$. Thus the question arises if the latter condition can be weakened. The theory of classical Lebesgue spaces, i.e. with constant exponent, shows that the boundedness of M cannot be expected for all $p \in \mathcal{P}^{\log}(\mathbb{R}^n)$. In particular M is not bounded from $L^1(\mathbb{R}^n)$ to $L^1(\mathbb{R}^n)$. Indeed, M is bounded from $L^q(\mathbb{R}^n)$ to $L^q(\mathbb{R}^n)$ if and only if $1 < q \leqslant \infty$. It is natural to conjecture that the boundedness of M from $L^{p(\cdot)}(\mathbb{R}^n)$ to $L^{p(\cdot)}(\mathbb{R}^n)$ would also require p to be bigger than 1. But in the context of variable exponents this could either mean $p > 1$ almost everywhere or $p^- > 1$, where the second condition is obviously the stronger one. The next theorem shows that indeed the boundedness of M requires $p^- > 1$.

Theorem 4.7.1. *Let $p \in \mathcal{P}(\mathbb{R}^n)$ be such that M is bounded from $L^{p(\cdot)}(\mathbb{R}^n)$ to $L^{p(\cdot)}(\mathbb{R}^n)$. Then $p^- > 1$.*

Proof. It suffices to prove the claim for $\varphi_{p(\cdot)} = \bar{\varphi}_{p(\cdot)}$. The proof will rely on the following fact: although M is bounded from $L^q(\mathbb{R}^n)$ to $L^q(\mathbb{R}^n)$ for all $q \in (1, \infty]$, the constant blows up as $q \to 1$. We show that $p^- = 1$ also implies a blow up of the boundedness constant of M.

Assume for a contradiction that $p^- = 1$ and that $M\colon L^{p(\cdot)}(\Omega) \hookrightarrow L^{p(\cdot)}(\Omega)$ is bounded with constant $K \geqslant 1$.

Fix $\varepsilon \in (0, 1)$. Since $p^- = 1$, the set $\left\{ \frac{1}{p} > \frac{1}{1+\varepsilon/2} \right\}$ has positive measure and therefore some point z_0 has measure density 1, i.e.

$$\lim_{Q \to \{z_0\}} \frac{\left| Q \cap \left\{ \frac{1}{p} > \frac{1}{1+\varepsilon/2} \right\} \right|}{|Q|} = 1,$$

where the limit is taken over all cubes containing z_0. Therefore, there exists a cube $Q_0 \subset \Omega$ with $z_0 \in Q_0$ and $\ell(Q_0) \leqslant 1$ such that $\frac{1}{p_{Q_0}} = \fint_{Q_0} \frac{1}{p(y)} \, dy \geqslant 1/(1+\varepsilon)$, i.e. $p_{Q_0} \leqslant 1 + \varepsilon$. Here $\ell(Q_0)$ denotes the side length of Q_0. Let $m \in \mathbb{N}$ be large and split Q_0 into $N := 2^{mn}$ disjoint cubes $Q_1, \ldots, Q_N$ of side length $\ell(Q_j) = 2^{-m} \ell(Q_0)$. By renumbering we assume without loss of generality that

$$p_{Q_1} = \min_{1 \leqslant j \leqslant N} p_{Q_j}.$$

In particular $p_{Q_1} \leqslant p_{Q_0} \leqslant 1 + \varepsilon < 2$. Define $f \in L^1_{\mathrm{loc}}(Q_0)$ by

$$f := \tfrac{1}{4} K^{-2} |Q_1|^{-1/p_{Q_1}} \chi_{Q_1}.$$

Then Lemma 4.5.3 and the fact that the $\mathcal{A}$-constant of p is smaller than the boundedness constant of M imply $\|f\|_{p(\cdot)} = \tfrac{1}{4} K^{-2} |Q_1|^{-1/p_{Q_1}} \|\chi_{Q_1}\|_{p(\cdot)} \leqslant K^{-1}$. Especially, we have $f \in L^{p(\cdot)}(\mathbb{R}^n)$ and $\|Mf\|_{p(\cdot)} \leqslant K \|f\|_{p(\cdot)} \leqslant 1$. We arrive at a contradiction by showing that $\varrho_{p(\cdot)}(\beta\, Mf)$ is large if $\varepsilon > 0$ is small enough and $m \in \mathbb{N}$ is large enough, for fixed $\beta \in (0,1)$.

Let x_j denote the center of Q_j for $j = 1, \dots, N$. Then for $j = 2, \dots, N$, and for all $y \in Q_j$ one easily checks that

$$\beta\, Mf(y) \geqslant \underbrace{c\,\beta\, K^{-2} |Q_1|^{1-1/p_{Q_1}}}_{=:K_2} |x_j - x_1|^{-n},$$

where c depends only on the dimension n. Therefore, by Lemma 4.5.1,

$$\int\limits_{Q_0 \setminus Q_1} \bar{\varphi}_{p(y)}\big(\beta\, Mf(y)\big)\, dy \geqslant \sum_{j=2}^{\infty} \int\limits_{Q_j} \bar{\varphi}_{p(y)}\Big(K_2 |x_j - x_1|^{-n}\Big)\, dy$$

$$\geqslant \sum_{j=2}^{\infty} |Q_j|\, \bar{\varphi}_{p_{Q_j}}\Big(\tfrac{1}{2} K_2 |x_j - x_1|^{-n}\Big).$$

Since $p_{Q_j} \geqslant p_{Q_1}$, we have $t^{p_{Q_1}} = \bar{\varphi}_{p_{Q_1}}(t) \leqslant \bar{\varphi}_{p_{Q_j}}(t) + 1$. Hence,

$$\int\limits_{Q_0 \setminus Q_1} \bar{\varphi}_{p(y)}\big(\beta\, Mf(y)\big)\, dy \geqslant \sum_{j=2}^{\infty} |Q_j| \left(\big(\tfrac{1}{2} K_2 |x_j - x_1|^{-n}\big)^{p_{Q_1}} - 1 \right)$$

$$\geqslant -|Q_0| + \sum_{j=2}^{\infty} |Q_j| \big(\tfrac{1}{2} K_2 |x_j - x_1|^{-n}\big)^{p_{Q_1}}$$

$$\geqslant -1 + c \int\limits_{Q_0 \setminus Q_1} \big(K_2 |y - x_1|^{-n}\big)^{p_{Q_1}}\, dy.$$

We can essentially calculate the integral in the previous estimate:

$$\int\limits_{Q_0 \setminus Q_1} |y - x_1|^{-np_{Q_1}}\, dy \approx \int\limits_{\ell(Q_1)}^{\ell(Q_0)/2} r^{n-1-np_{Q_1}}\, dr$$

$$\approx \frac{1}{(p_{Q_1} - 1)n} \big(|Q_1|^{1-p_{Q_1}} - (|Q_0|/2^n)^{1-p_{Q_1}}\big).$$

We use this and the expression of K_2 in our previous estimate and conclude that

$$1 + \int_{Q_0} \bar{\varphi}_{p(y)}\big(\beta \, Mf(y)\big) \, dy$$

$$\geqslant c \, \beta \, K^{-2p_{Q_1}} \frac{|Q_1|^{p_{Q_1}-1}}{p_{Q_1}-1} \big(|Q_1|^{1-p_{Q_1}} - (|Q_0|/2^n)^{1-p_{Q_1}}\big)$$

$$= \frac{\beta \, K^{-2p_{Q_1}}}{p_{Q_1}-1} \big(1 - 2^{-n(m-1)(p_{Q_1}-1)}\big),$$

where we used that $|Q_1|/|Q_0| = 2^{-nm}$ in the last step. Now we choose m so large that $2^{-n(m-1)(p_{Q_1}-1)} \leqslant 1/2$ and recall that $1 \leqslant p_{Q_1} \leqslant 1+\varepsilon < 2$. Then

$$1 + \int_{Q_0} \bar{\varphi}_{p(y)}\big(\beta \, Mf(y)\big) \, dy \geqslant c \, \beta \, K^{-4}\varepsilon^{-1}.$$

As $\varepsilon \to 0$ this contradicts $\|Mf\|_{p(\cdot)} \leqslant 1$, which means that the assumption $p^- = 1$ was wrong, as was to be shown. $\qquad\square$

Remark 4.7.2. The proof of Theorem 4.7.1 shows that even a stronger, localized result holds: let $\Omega \subset \mathbb{R}^n$ be an open, non-empty set. For $f \in L^{p(\cdot)}(\Omega)$ and $x \in \Omega$ define

$$M_\Omega f(x) := \sup_{Q:\, x\in Q, 2Q\subset\Omega} M_Q f,$$

where the supremum is taken over all cubes (or balls) with $2Q \subset \Omega$. If M_Ω is bounded from $L^{p(\cdot)}(\Omega)$ to $L^{p(\cdot)}(\Omega)$, then $p_\Omega^- > 1$.

We have seen that log-Hölder continuity implies $p \in \mathcal{A}$ and the boundedness of the maximal operator M for $p^- > 1$. Since these are fundamental tools for many results, it is natural to ask if log-Hölder continuity is optimal or if it can be weakened. The answer to this question is quite subtle, since log-Hölder continuity is in some sense optimal and in another sense not. We will see later in Sect. 5.1 that there exists a variable exponent $p \in \mathcal{P}(\mathbb{R}^n)$ with is neither continuous at zero nor at infinity but still M is bounded from $L^{p(\cdot)}(\mathbb{R}^n)$ to $L^{p(\cdot)}(\mathbb{R}^n)$. Therefore, log-Hölder continuity is only sufficient and not necessary for the boundedness of M.

However, in some sense log-Hölder continuity is optimal. We show that if the local log-Hölder continuity is replaced by a weaker uniform modulus of continuity or if the decay log-Hölder condition is replaced by a weaker decay condition, then the this new condition is not sufficient in the sense that there exists a variable exponent with this modulus of continuity such that M is not bounded from $L^{p(\cdot)}(\mathbb{R}^n)$ to $L^{p(\cdot)}(\mathbb{R}^n)$.

Furthermore, we show in Lemma 4.7.3 that the crucial pointwise estimate in Theorem 4.2.4 and Lemma 4.3.6 can only hold if the variable exponent is local log-Hölder continuous. So the locally log-Hölder continuity is necessary for this kind of pointwise estimate. It follows from Remark 4.3.10 that the log-Hölder decay condition is not necessary for the pointwise estimate in Lemma 4.3.6 but can be replaced by the weaker integral condition of Nekvinda, see Remark 4.2.8.

Lemma 4.7.3. *Let* $p \in \mathcal{P}(\mathbb{R}^n)$ *with* $1 < p^- \leqslant p^+ < \infty$, $\beta > 0$ *and* $h \in L^\infty(\mathbb{R}^n)$ *such that*

$$\varphi_{p(x)}\left(\beta \fint_Q |f(y)|\,dy\right) \leqslant \fint_Q \varphi_{p(y)}(|f(y)|)\,dy + h(x),$$

$$\varphi_{p'(x)}\left(\beta \fint_Q |g(y)|\,dy\right) \leqslant \fint_Q \varphi_{p'(y)}(|g(y)|)\,dy + h(x)$$

for all $f \in L^{p(\cdot)}(\mathbb{R}^n)$ *with* $\|f\|_{p(\cdot)} \leqslant 1$, *all* $g \in L^{p'(\cdot)}(\mathbb{R}^n)$ *with* $\|g\|_{p'(\cdot)} \leqslant 1$, *all cubes* $Q \subset \mathbb{R}^n$, *and all* $x \in Q$. *Then* $\frac{1}{p}$ *is locally* log-*Hölder continuous.*

Proof. It suffices to prove the case $\varphi_{p(\cdot)} = \bar\varphi_{p(\cdot)}$. Let $x_1, x_2 \in \mathbb{R}^n$ with $|x_1 - x_2| \leqslant 1$. Let Q be a cube containing x_1 and x_2 with $|x_1 - x_2|^n \approx |Q|$ and $|Q| \leqslant 1$. Let $f := \frac{1}{4}\chi_Q\,(e + 1/|Q|)^{\frac{1}{p(\cdot)}}$. Then $\varrho_{p(\cdot)}(f) \leqslant \frac{1}{4}|Q|(e + 1/|Q|) \leqslant 1$, so $\|f\|_{p(\cdot)} \leqslant 1$. Note that

$$\fint_Q \varphi_{p(y)}(|f(y)|)\,dy \leqslant \frac{1}{4}\left(e + \frac{1}{|Q|}\right).$$

So the first inequality in the assumptions implies that

$$\left(\beta \fint_Q \left(e + \frac{1}{|Q|}\right)^{\frac{1}{p(y)}}\,dy\right)^{p(x_1)} \leqslant \frac{1}{4}\left(e + \frac{1}{|Q|}\right) + \|h\|_\infty \leqslant c\left(e + \frac{1}{|Q|}\right),$$

where c depends on $\|h\|_\infty$. As a consequence

$$\fint_Q \left(e + \frac{1}{|Q|}\right)^{\frac{1}{p(y)} - \frac{1}{p(x_1)}}\,dy \leqslant c(\beta).$$

Note that the mapping $s \mapsto (e + |Q|)^s$ is convex, so by Jensen's inequality

$$\left(e + \frac{1}{|Q|}\right)^{\fint_Q \frac{1}{p(y)} - \frac{1}{p(x_1)}\,dy} \leqslant c.$$

This implies that

$$\fint_Q \frac{1}{p(y)} - \frac{1}{p(x_1)}\, dy \leqslant \frac{c}{\log(e + \frac{1}{|Q|})}.$$

If we repeat the calculations with p' and x_2, then we obtain

$$\fint_Q \frac{1}{p(x_2)} - \frac{1}{p(y)}\, dy \leqslant \frac{c}{\log(e + \frac{1}{|Q|})}.$$

Adding the last two inequalities leads to

$$\frac{1}{p(x_1)} - \frac{1}{p(x_2)} \leqslant \frac{c}{\log(e + \frac{1}{|Q|})} \leqslant \frac{c}{\log(e + \frac{1}{|x_1-x_2|})}.$$

Switching x_1 and x_2 then gives

$$\left| \frac{1}{p(x_2)} - \frac{1}{p(x_1)} \right| \leqslant \frac{c}{\log(e + \frac{1}{|x_1-x_2|})}. \qquad \square$$

The following lemma shows that $p \in \mathcal{A}_{\mathrm{loc}}$ implies an estimate on the mean oscillation of $\frac{1}{p}$ which is very similar to the log-Hölder continuity condition (both the local and the decay condition). See Theorem 4.5.7 for equivalent conditions for $\varphi \in \mathcal{A}_{\mathrm{loc}}$.

Proposition 4.7.4. *Let $p \in \mathcal{A}_{\mathrm{loc}}$. Then*

$$\fint_Q \fint_Q \left| \frac{1}{p(y)} - \frac{1}{p(z)} \right| dy\, dz \leqslant \frac{c}{\log(e + |Q| + \frac{1}{|Q|})} \qquad (4.7.5)$$

for all cubes $Q \subset \mathbb{R}^n$, where the constant depends on the $\mathcal{A}_{\mathrm{loc}}$-constant of p and n.

Proof. For a cube Q define $f := \chi_Q |Q|^{-1/p}$ and $g := \chi_Q |Q|^{-1/p'}$. Then $\varrho_{p(\cdot)}(f) \leqslant 1$ and $\varrho_{p'(\cdot)}(g) \leqslant 1$, which implies $\|f\|_{p(\cdot)} \leqslant 1$ and $\|g\|_{p'(\cdot)} \leqslant 1$. So with Hölder's inequality and the boundedness of $T_{\{Q\}}$, we get

$$|Q| M_Q(f) M_Q(g) = |Q| \fint_Q M_Q(f) M_Q(g)\, dx \leqslant 2 \|T_{\{Q\}} f\|_{p(\cdot)} \|T_{\{Q\}} g\|_{p'(\cdot)} \leqslant c.$$

By definition of f and g this implies that

$$\fint_Q \fint_Q |Q|^{\frac{1}{p(z)} - \frac{1}{p(y)}} \, dy \, dz \leqslant c.$$

By symmetry in y and z we get

$$\fint_Q \fint_Q |Q|^{\frac{1}{p(z)} - \frac{1}{p(y)}} + |Q|^{\frac{1}{p(y)} - \frac{1}{p(z)}} \, dy \, dz \leqslant c.$$

From this we conclude that

$$\fint_Q \fint_Q \left(e + |Q| + \frac{1}{|Q|} \right)^{\left| \frac{1}{p(y)} - \frac{1}{p(z)} \right|} \, dy \, dz \leqslant c,$$

The mapping $s \mapsto (e + |Q| + 1/|Q|)^s$ is convex, so by Jensen's inequality

$$\left(e + |Q| + \frac{1}{|Q|} \right)^{\fint_Q \fint_Q \left| \frac{1}{p(y)} - \frac{1}{p(z)} \right| \, dy \, dz} \leqslant c.$$

Taking the logarithm, we obtain the claim.   □

Remark 4.7.6. If $p \in \mathcal{P}^{\log}(\mathbb{R}^n)$, then we know by Theorem 4.4.8 that $p \in \mathcal{A}$ and by Lemma 4.4.12 that M is of weak-type $p(\cdot)$. If $p^- > 1$, then by Theorem 4.7.1 also the maximal operator is bounded from $L^{p(\cdot)}(\mathbb{R}^n)$ to $L^{p(\cdot)}(\mathbb{R}^n)$. All of these properties are stronger than the assumption $p \in \mathcal{A}_{\mathrm{loc}}$ of Proposition 4.7.4, see Lemma 4.4.13.

With Proposition 4.7.4 we can show now that neither the local log-Hölder continuity condition nor the log-Hölder decay condition can be replaced by weaker versions in terms of modulus of continuity and a decay condition.

Let us begin with the local log-Hölder condition. Let ω_2 be a modulus of continuity that is weaker than the one for local log-Hölder continuity, denoted by ω, in the sense that $\lim_{t \to 0} \omega_2(t)/\omega(t) = \infty$. Further, let $\eta \in C_0^\infty(-1,1)$ with $\chi_{B(0,1/2)} \leqslant \eta \leqslant \chi_{B(0,1)}$. Define $q \in \mathcal{P}(\mathbb{R})$ by

$$\frac{1}{q(y)} := \begin{cases} \frac{1}{2} & \text{for } y \leqslant 0, \\ \min\left\{ \frac{1}{2} + \eta(y)\omega_2(y), \frac{2}{3} \right\} & \text{for } y \geqslant 0. \end{cases}$$

Then q is continuous with modulus of continuity ω_2, $\frac{3}{2} \leqslant q^- \leqslant q^+ \leqslant 2$, and $q(y) = 2$ for all $|y| \geqslant 1$. But q fails to satisfy (4.7.5) uniformly for all cubes $Q \subset \mathbb{R}^n$. To see this, consider the cubes $Q_t = (-t, t)$ for $t \in (0, 1/2)$. Then

$$\fint_{Q_t} \fint_{Q_t} \left| \frac{1}{q(y)} - \frac{1}{q(z)} \right| dy \, dz \geqslant c t^{-2} \int_{-t}^{0} \int_{t/2}^{t} \omega_2(y) \, dy \, dz \geqslant c \omega_2(t/2).$$

Due to the assumptions on ω_2 the inequality (4.7.5) cannot be uniform with respect to $t \in (0, 1/2)$. This example goes back to Pick and Růžička [323], who used it to prove the unboundedness of the maximal operator for such an exponent.

Let us now turn to the optimality of the decay condition. Let $\psi_2 \colon [0, \infty) \to (0, \infty)$ be a non-increasing function with $\lim_{t \to \infty} (\psi_2(t) \log(e + t)) = \infty$ and $\lim_{t \to \infty} \psi_2(t) = 0$. Since we are only interested in the decay condition, we can assume without loss of generality that ψ_2 is smooth and that $\psi_2(0) \leqslant 1/4$. Define $s \in \mathcal{P}(\mathbb{R})$ by

$$\frac{1}{s(y)} := \begin{cases} \frac{1}{2} & \text{for } y \leqslant 0, \\ \frac{1}{2} + (1 - \eta(y))\psi_2(t) & \text{for } y \geqslant 0. \end{cases}$$

Then s is smooth, $\frac{4}{3} \leqslant s^- \leqslant s^+ \leqslant 2$, and $\frac{1}{s(t)} - \frac{1}{2}$ decays as $\psi_2(t)$ for $t \to \infty$. But s fails to satisfy (4.7.5) uniformly for all cubes $Q \subset \mathbb{R}^n$. To see this, consider the cubes $Q_t = (-t, t)$ for $t \to \infty$. Then

$$\fint_{Q_t} \fint_{Q_t} \left| \frac{1}{s(y)} - \frac{1}{s(z)} \right| dy \, dz \geqslant c t^{-2} \int_{-t}^{0} \int_{1}^{t} \psi_2(y) \, dy \, dz \geqslant c \psi_2(t).$$

Due to the assumptions on ψ_2 the inequality (4.7.5) cannot be uniform with respect to $t \to \infty$. This example goes back to Cruz-Uribe, Fiorenza, Martell, and Pérez [83], who used it to prove the unboundedness of the maximal operator for such an exponent.

Remark 4.7.7. Let $p \in \mathcal{A}_{\mathrm{loc}}$. Then it follows from Proposition 4.7.4 that $\frac{1}{p} \in \mathrm{BMO}_{\log(e + 1/t)}$. Here $\mathrm{BMO}_{\omega(t)}(\mathbb{R}^n)$ consists of all functions $h \in L^1_{\mathrm{loc}}(\mathbb{R}^n)$ such that

$$\fint_{B} \fint_{B} |h(y) - h(z)| \, dy \, dz \leqslant c \omega \big(|B|^{\frac{1}{n}} \big)$$

for all balls $B \subset \mathbb{R}^n$, where ω is some modulus of continuity. For a modulus of continuity ω, let $C^{0,\omega(t)}(\mathbb{R}^n)$ denote the space of continuous functions on $\mathbb{R}^n$ with uniform modulus of continuity ω. It is easy to see that $C^{0,\omega(t)}(\mathbb{R}^n)$ is a subspace of $\mathrm{BMO}_{\omega(t)}(\mathbb{R}^n)$. It is a well known result of Campanato [65] that $C^{0,t^\alpha}(\mathbb{R}^n) = \mathrm{BMO}_{t^\alpha}(\mathbb{R}^n)$ for $\alpha \in (0, 1]$. The relation between $C^{0,\omega(t)}(\mathbb{R}^n)$ and $\mathrm{BMO}_{\omega(t)}(\mathbb{R}^n)$ for arbitrary modulus of continuity has been studied intensively by Spanne [358]. He showed that the spaces $C^{0,\omega(t)}(\mathbb{R}^n)$ and $\mathrm{BMO}_{\omega(t)}(\mathbb{R}^n)$ coincide if and only if

$$\int\limits_0^1 \frac{\omega(t)}{t}\, dt < \infty.$$

Since

$$\int\limits_0^1 \frac{1}{t\log(e + \frac{1}{t})}\, dt = \infty,$$

this shows that the spaces $C^{0,\log}(\mathbb{R}^n)$ and $\mathrm{BMO}_{1/\log(e+1/t)}(\mathbb{R}^n)$ disagree. Thus, Proposition 4.7.4 does not imply that log-Hölder continuity is necessary for $p \in \mathcal{A}_{\mathrm{loc}}$. Indeed, we will see in Sect. 5.1 that there exists a variable exponent p with $\frac{1}{p} \in \mathrm{BMO}_{1/\log(e+1/t)}(\mathbb{R}^n)$ and $\frac{1}{p} \notin C^{\log}(\mathbb{R}^n)$ such that M is bounded from $L^{p(\cdot)}(\mathbb{R}^n)$ to $L^{p(\cdot)}(\mathbb{R}^n)$, which in particular implies $p \in \mathcal{A}_{\mathrm{loc}}$.

4.8 Preimage of the Maximal Operator*

In Theorem 4.3.8 we saw that the maximal operator is bounded on $L^{p(\cdot)}(\mathbb{R}^n)$ when $p \in \mathcal{P}^{\log}(\mathbb{R}^n)$ with $p^- > 1$. In Theorem 4.7.1 we further saw that the condition $p^- > 1$ is necessary for the boundedness. Since $M : L^1 \nrightarrow L^1$, the latter result is not so surprising. One may ask, however, what space M maps onto $L^{p(\cdot)}(\mathbb{R}^n)$? This question was answered in [95] and as extensions we mention [207, 279]. Earlier papers on the maximal operator in the case $p^- = 1$ include [165, 215].

The relevant classical results [359, Sect. 1] are that

$Mf \in L^p$ if and only if $f \in L^p$ $(p > 1)$ and $Mf \in L^1$ if and only if $f \in L\log L$.

The latter result is, of course, restricted to bounded domains: if $Mf \in L^1(\mathbb{R}^n)$, then $f \equiv 0$.

How, then, can we characterize the space $M^{-1}[L^{p(\cdot)}]$? On an intuitive level we need some kind of modified scale of spaces $\widetilde{L}^{p(\cdot)}$ where $p = 1$ corresponds to $L\log L$, not L^1, if we want to characterize functions f for which $Mf \in L^{p(\cdot)}$. It is possible to construct such a space within the framework of Orlicz–Musielak spaces.

We need a function which behaves like a logarithm when $p = 1$ and fades away when $p > 1$. Since the embedding constant of $M : L^p \hookrightarrow L^p$ is p', the function

$$\min\big\{p', \log(e + |t|)\big\},$$

would be a natural choice. Unfortunately, it does not yield a convex modular. The following variant fixes this problem:

$$\psi_p(t) = \begin{cases} \log(e + |t|), & \text{for } |t| < e^{p'} - e \\ 2p' - \frac{e^{p'}}{e+|t|}p', & \text{for } |t| \geqslant e^{p'} - e. \end{cases}$$

Note that $t \mapsto t^p \psi_p(t)$ is convex on $[0, \infty)$ and that

$$\frac{1}{2}\psi_p(t) \leqslant \min\left\{p', \log(e + |t|)\right\} \leqslant \psi_p(t),$$

so ψ_p is equivalent up to a constant to the natural choice of the modular. The norm $\|f\|_{L^{p(\cdot)}\psi_{p(\cdot)}[L]}$ is given by the generalized Φ-function

$$\Phi(x, t) = |t|^{p(x)}\psi_{p(x)}(t).$$

Note that $\|f\|_{L^{p(\cdot)}\psi_{p(\cdot)}[L]} \approx \|f\|_{L^{p(\cdot)}}$ if $p^- > 1$, but the constant of proportionality blows up as $p^- \to 1$.

We are now ready to state the main theorem of this section:

Theorem 4.8.1. *Let* $B \subset \mathbb{R}^n$ *be a ball and let* $p \in \mathcal{P}^{\log}(B)$. *Then*

$$\|f\|_{L^{p(\cdot)}\psi_{p(\cdot)}[L](B)} \approx \|Mf\|_{L^{p(\cdot)}(B)}.$$

Here we will prove the sufficiency of the conditions. The essential new result needed for the proof is the following proposition. The trick in its proof is to use a reverse triangle inequality to recombine terms that were originally split using the triangle inequality. This is possible since our exponent tends to 1 in the critical parts of the domain.

Proposition 4.8.2. *Let* $\Omega \subset \mathbb{R}^n$ *be bounded and let* $p \in \mathcal{P}^{\log}(\Omega)$ *with* $1 \leqslant p^- \leqslant p^+ \leqslant 2$. *Then there exists a constant* c *depending only on* p, Ω *and the dimension* n *such that*

$$\|Mf\|_{L^{p(\cdot)}(\Omega)} \leqslant c\|f\|_{L^{p(\cdot)}\psi_{p(\cdot)}[L](\Omega)}$$

for every $f \in L^{p(\cdot)}\psi_{p(\cdot)}[L](\Omega)$.

Proof. By a scaling argument, it suffices to consider such non-negative functions f that

$$\int_\Omega f(x)^{p(x)}\psi_{p(x)}(f(x))\,dx \leqslant 1.$$

Then we must show that $\|Mf\|_{L^{p(\cdot)}} \leqslant c$, which is equivalent to $\varrho_{L^{p(\cdot)}}(Mf) \leqslant c$.

We split f into small and large parts as follows:

$$f_s := f\chi_{\{f \leqslant e^{p'(\cdot)} - e\}} \quad \text{and} \quad f_l := f\chi_{\{f > e^{p'(\cdot)} - e\}}.$$

Note that $\varrho_{p(\cdot)}(Mf) \leqslant c\,\varrho_{p(\cdot)}(Mf_s) + c\,\varrho_{p(\cdot)}(Mf_l)$. By (4.3.7) we have

$$Mf_s(x)^{p(x)} \leqslant c\,M\big[f_s^{p(\cdot)}\big](x) + c$$

for $x \in \Omega$. Then the embedding $L \log L \hookrightarrow L^1$ implies that

$$\int_\Omega Mf_s(x)^{p(x)}\,dx \leqslant c \int_\Omega f_s(y)^{p(y)} \log\left(e + f_s(y)^{p(y)}\right) dy + |\Omega|$$

$$\leqslant c\,p^+ \int_\Omega f_s(y)^{p(y)} \log\left(e + f_s(y)\right) dy + |\Omega|$$

$$\leqslant c \int_\Omega f(y)^{p(y)} \psi_{p(y)}(f(y))\,dy + |\Omega| \leqslant c.$$

This takes care of f_s

Next we treat f_l. Let us define $r_i := 1 + 1/i$ for $i \geqslant 1$ and $\Omega_i := \{r_i < p \leqslant r_{i-1}\}$ so that $\bigcup_{i=1}^\infty \Omega_i = \{p > 1\}$. The sequences $(1)_{i=0}^\infty$ and $(r_i)_{i=0}^\infty$ satisfy the criterion of Corollary 4.1.9 so we conclude that $l^1 \cong l^{(r_i)}$. We fix $K > 0$ so that

$$\sum_{i=1}^\infty x_i \leqslant K \quad \text{whenever} \quad \sum_{i=1}^\infty x_i^{r_i} \leqslant L \tag{4.8.3}$$

where L will be specified later.

Define $f_i := f_l\,\chi_{\Omega_i}$ and $p_i := \max\{r_i, p\}$ for $i \geqslant 2$. Since $f_l = 0$ in $\{p = 1\}$, we see that $\sum_{i=1}^\infty f_i = f_l$. By the subadditivity of the maximal operator, the triangle inequality and the embedding $L^{p_i(\cdot)}(\Omega) \hookrightarrow L^{p(\cdot)}(\Omega)$ (Corollary 3.3.4), we conclude that

$$\|Mf_l\|_{p(\cdot)} \leqslant \sum_{i=2}^\infty \|Mf_i\|_{p(\cdot)} \leqslant 2(1 + |\Omega|) \sum_{i=2}^\infty \|Mf_i\|_{p_i(\cdot)}.$$

Next, Theorem 4.3.8 and the norm-modular inequality, Lemma 3.2.5, imply that

$$\|Mf_l\|_{p(\cdot)} \leqslant c \sum_{i=2}^\infty (r_i - 1)^{-1} \|f_i\|_{L^{p_i(\cdot)}(\Omega_i)} \leqslant c \sum_{i=2}^\infty (r_i - 1)^{-1} \varrho_{L^{p_i(\cdot)}(\Omega_i)}(f_i)^{\frac{1}{r_i - 1}}.$$

By (4.8.3) the right-hand side is bounded by K provided we show that

$$\sum_{i=2}^{\infty}(r_i - 1)^{-r_i-1}\varrho_{L^{p_i(\cdot)}(\Omega_i)}(f_i) \leqslant L. \tag{4.8.4}$$

But for this we just need to estimate as follows:

$$\sum_{i=2}^{\infty}(r_i - 1)^{-r_i-1}\varrho_{L^{p_i(\cdot)}(\Omega_i)}(f_i) \leqslant c\sum_{i=2}^{\infty}\int_{\Omega_i}(r_i - 1)^{-1}f_i(x)^{p_i(x)}dx$$

$$\leqslant c\sum_{i=2}^{\infty}\int_{\Omega_i}(p(x) - 1)^{-1}f(x)^{p(x)}dx$$

$$\leqslant c\varrho_{L^{p(\cdot)}\psi_{p(\cdot)}[L](\Omega)}(f) \leqslant c,$$

where we have used $n^{1/n} \leqslant c$ in the first inequality. We collect the implicit absolute constants from these estimates and define L by them. Thus (4.8.4) holds, and so (4.8.3) concludes the proof.   □

To prove the main theorem of the section we use the following lemma, whose proof can be found in [95].

Lemma 4.8.5. *Let $p \in \mathcal{P}^{\log}(\Omega)$ with $1 \leqslant p^- \leqslant p^+ \leqslant \infty$ in a ball $\Omega \subset \mathbb{R}^n$. If $Mf \in L^{p(\cdot)}(\Omega)$, then $f \in L^{p(\cdot)}\psi_{p(\cdot)}[L](\Omega)$.*

Proof of Theorem 4.8.1. Lemma 4.8.5 states that if $Mf \in L^{p(\cdot)}(B)$ for a ball B, then $f \in L^{p(\cdot)}\psi_{p(\cdot)}[L](B)$. We proceed to show how the reverse implication can be pieced together from the previous results.

By a scaling argument, it suffices to consider such non-negative functions f that

$$\int_B f(x)^{p(x)}\psi_{p(x)}(f(x))\,dx \leqslant 1.$$

Then we must show that $\|Mf\|_{L^{p(\cdot)}} \leqslant c$, which is equivalent to $\varrho_{L^{p(\cdot)}(B)}(\beta\,Mf) \leqslant c$, $\beta \in (0,1)$. Denote $B_a^- = \{p < a\}$, $B_a^+ = \{p \geqslant a\}$ and

$$d_0 = \min\{\mathrm{dist}(B_{4/3}^-, B_{5/3}^+), \mathrm{dist}(B_{5/3}^-, B_2^+)\}.$$

The uniform continuity of the exponent implies that $d_0 > 0$. Denote by D the set of points $x \in B$ for which $Mf(x) = \fint_{B(x,r)} f(y)\,dy$ with some $r \geqslant d_0$. We note that

$$\varrho_{L^{p(\cdot)}(B)}(\beta\,Mf) \leqslant \varrho_{L^{p(\cdot)}(B_{5/3}^-\backslash D)}(Mf) + \varrho_{L^{p(\cdot)}(B_{5/3}^+\backslash D)}(\beta\,Mf) + \varrho_{L^{p(\cdot)}(D)}(\beta\,Mf).$$

For $x \in D$ we have $Mf(x) \leqslant cd_0^{-n}\|f\|_1$ and so $\varrho_{L^{p(\cdot)}(D)}(\beta\,Mf) \leqslant c$. For the other two terms only points where $p < 2$ or $p \geqslant 4/3$ affect the maximal function. Thus Theorem 4.3.8 implies that $\varrho_{L^{p(\cdot)}(B_{5/3}^+\backslash D)}(\beta\,Mf) \leqslant c$ and Proposition 4.8.2 implies that $\varrho_{L^{p(\cdot)}(B_{5/3}^-\backslash D)}(Mf) \leqslant c$, which completes the proof.   □

Chapter 5
The Generalized Muckenhoupt Condition*

The boundedness of the maximal operator M is closely linked to very important properties of the spaces $L^{p(\cdot)}$. Indeed, we will see in Chaps. 6 and 8 that the boundedness of M is needed for the Sobolev embeddings $W^{1,p(\cdot)} \hookrightarrow L^{p^*(\cdot)}$, boundedness singular integrals on $L^{p(\cdot)}$ and Korn's inequality. Moreover, the extrapolation result in Sect. 7.2 shows that most of the results for weighted Lebesgue spaces can be generalized to the setting of Lebesgue spaces with variable exponents as long as the maximal operator M is bounded.

It is the aim of this chapter to provide a full characterization of the boundedness of M which is closely related to the concept of Muckenhoupt classes. It is clear from the estimate $T_{\mathcal{Q}} f \leqslant M f$ for locally 1-finite families of cubes that the boundedness of M on $L^{p(\cdot)}$ implies $p \in \mathcal{A}$. In Theorem 5.7.2 we prove the reverse statement: if $p \in \mathcal{P}(\mathbb{R}^n)$ with $1 < p^- \leqslant p^+ < \infty$ and $p \in \mathcal{A}$, then M is bounded on $L^{p(\cdot)}$. The condition $p^- > 1$ is necessary for the boundedness of M by Theorem 4.7.1. However, we need the condition $p^+ < \infty$, since we use duality arguments and $p^+ < \infty$ is equivalent to $(p')^- > 1$. We also present partial results for the case of generalized Φ-functions. These results are extensions of [93].

5.1 Non Sufficiency of log-Hölder Continuity*

We saw in Theorem 4.3.8 that $p \in \mathcal{P}^{\log}(\mathbb{R}^n)$ with $p^- > 1$ is sufficient for the boundedness of M on $L^{p(\cdot)}(\mathbb{R}^n)$. We have seen in Sect. 4.7 that the condition $p^- > 1$ is necessary for the boundedness of M. As a consequence of Proposition 4.7.4 it was shown in Sect. 4.7 that neither the local log-Hölder continuity condition nor the log-Hölder decay condition can be replaced by weaker moduli of continuity. However, the log-Hölder continuity is not necessary for the boundedness of M, although the stronger pointwise estimate does imply the local log-Hölder continuity (Lemma 4.7.3).

One of the problems is summarized in Remark 4.7.7, namely that

$$C^{0,1/(\log(e+1/t))}(\mathbb{R}^n) \subsetneq BMO^{1/(\log(e+1/t))}(\mathbb{R}^n).$$

L. Diening et al., *Lebesgue and Sobolev Spaces with Variable Exponents*, Lecture Notes in Mathematics 2017, DOI 10.1007/978-3-642-18363-8_5, © Springer-Verlag Berlin Heidelberg 2011

In this context is has been observed by Nekvinda that the counterexample of Pick and Růžička showing "almost optimality" (Sect. 4.7) relies strongly on the non-symmetry of the exponent: p is constant in $(-\infty)$-direction and decays slightly slower than $\log(e + |x|)$ at $+\infty$. Nekvinda (private communication) found symmetric, continuous exponents $p(x) = p(|x|)$, which decay slower at ∞ than $\log(e + |x|)$ but still guarantee the boundedness of M on $L^{p(\cdot)}(\mathbb{R}^n)$. Based on this observation and his own study on sufficient conditions it has been conjectured by Diening [93] that there exist even exponents p, which are discontinuous and have no limit at infinity, but for which M is bounded on $L^{p(\cdot)}(\mathbb{R}^n)$. Such an example was constructed by Lerner [267] in a very clever way. He showed that the boundedness of M in $L^{p(\cdot)}(\mathbb{R}^n)$ is closely related to the class of pointwise multipliers for BMO. In this section we present this result. We follow [267] rather closely, although some proofs have been simplified at the expense of worse constants.

The result is related to the well-known function space BMO consisting of functions of bounded mean oscillation. More precisely, we set

$$M_Q^\sharp f := M_Q(f - \langle f \rangle_Q) = \fint_Q |f(x) - \langle f \rangle_Q|\, dx$$

for a cube Q and define a norm on a subset of $L^1_{\mathrm{loc}}(\mathbb{R}^n)$ by

$$\|f\|_{\mathrm{BMO}} := \sup_{Q \subset \mathbb{R}^n} M_Q^\sharp f.$$

The space BMO consists of those functions with $\|f\|_{\mathrm{BMO}} < \infty$.

Let $Q_1 \subset Q_2 \subset \mathbb{R}^n$ be two cubes. Let j be the smallest integer for which $e^{j+1} Q_1 \supset Q_2$. Then $n(j-1) \leqslant \log(|Q_2|/|Q_1|) \leqslant n(j+1)$. Thus we conclude that

$$|\langle f \rangle_{Q_1} - \langle f \rangle_{Q_2}| \leqslant |f_{e^j Q_1} - \langle f \rangle_{Q_2}| + \sum_{k=0}^{j-1} |\langle f \rangle_{e^{k+1} Q_1} - \langle f \rangle_{e^k Q}|$$

$$\leqslant M_{e^j Q_1}\big(|f - \langle f \rangle_{Q_2}|\big) + \sum_{k=0}^{j-1} M_{e^k Q_1}\big(|f - \langle f \rangle_{e^{k+1} Q}|\big)$$

$$\leqslant e^n M_{Q_2}\big(|f - \langle f \rangle_{Q_2}|\big) + e^n \sum_{k=0}^{j-1} M_{e^{k+1} Q_1}\big(|f - \langle f \rangle_{e^{k+1} Q}|\big)$$

$$\leqslant e^n j\, \|f\|_{\mathrm{BMO}}$$

$$\leqslant \frac{e^n}{n}\left(1 + \log \frac{|Q_2|}{|Q_1|}\right) \|f\|_{\mathrm{BMO}}.$$

Let us denote by Q_0 the unit cube centered at zero. Let $Q \subset \mathbb{R}^n$ be some cube, and let $\widetilde{Q}$ be the smallest cube containing Q and Q_0. Applying the previous inequality, we find that

$$|\langle f\rangle_Q - \langle f\rangle_{Q_0}| \leqslant |\langle f\rangle_Q - \langle f\rangle_{\widetilde{Q}}| + |\langle f\rangle_{Q_0} - \langle f\rangle_{\widetilde{Q}}| \leqslant c\Big(1 + \log\frac{|\widetilde{Q}|}{|Q|} + \log|\widetilde{Q}|\Big)\|f\|_{\mathrm{BMO}}.$$

In addition, we can estimate the size of $\widetilde{Q}$ by $\operatorname{diam}\widetilde{Q} \leqslant |\operatorname{cen}(Q)| + \frac{1}{2}$ $\operatorname{diam} Q + \frac{1}{2}$, where $\operatorname{cen}(Q)$ denotes the center of Q. Hence we conclude that

$$|\langle f\rangle_Q - \langle f\rangle_{Q_0}| \leqslant c\underbrace{\log\big(\mathrm{e} + |\operatorname{cen}(Q)| + |Q| + |Q|^{-1}\big)}_{=:L(Q)}\|f\|_{\mathrm{BMO}}. \qquad (5.1.1)$$

We need the following result of Coifman and Rochberg [77].

Proposition 5.1.2. *Let* $Mf < \infty$ *almost everywhere on* $\mathbb{R}^n$. *Then* $\log(Mf) \in BMO(\mathbb{R}^n)$, *and* $\|\log(Mf)\|_{BMO} \leqslant \gamma_n$, *where* γ_n *only depends on* n.

We need one more observation before the main lemma of the section. For $f \in L^{p(\cdot)}(\mathbb{R}^n)$ with $\|f\|_{p(\cdot)} \leqslant 1$ we have $f \in L^{p_{2Q_0}^-}(2Q_0)$ and hence

$$\int_{Q_0} \log(1 + Mf(x))\,dx \leqslant c\int_{Q_0} 1 + M_< f(x)^{p_{2Q_0}^-}\,dx + c\,|Q_0|\sup M_\geqslant f(x) \leqslant c,$$

$$(5.1.3)$$

where $M_< f(x) = \sup_{|Q|<1} M_Q f$ and $M_\geqslant f(x) = \sup_{|Q|\geqslant 1} M_Q f$. Now we get to the lemma:

Lemma 5.1.4. *Let* p *and* r *be measurable functions, with* $1 < p^- \leqslant p^+ < \infty$. *Let* $f \in L^{p(\cdot)}(\mathbb{R}^n)$ *with* $\|f\|_{p(\cdot)} \leqslant 1$ *and set* $\tilde{f} := f + \chi_{Q_0}$, *where* Q_0 *is the unit cube centered at* 0. *Then*

$$\|r \log M\tilde{f}\|_{BMO} \leqslant c_{n,p^-}\big(\|r\|_\infty + \sup_Q L(Q)\, M_Q^\sharp r\big).$$

Proof. We start by observing that $f\chi_{\{|f|>1\}} \in L^1(\mathbb{R}^n)$ and hence

$$M\tilde{f} \leqslant M\chi_{Q_0} + 1 + M(f\chi_{\{|f|>1\}}) < \infty \quad \text{a.e.}$$

Set $\psi := \log(M\tilde{f})$. It follows by Proposition 5.1.2 that $\|\psi\|_{\mathrm{BMO}} \leqslant \gamma_n$. Since $\psi \leqslant \log(1 + Mf)$, (5.1.3) implies that $\psi_{Q_0} \leqslant c$. Thus (5.1.1) gives

$$|\psi_Q| \leqslant |\psi_Q - \psi_{Q_0}| + |\psi_{Q_0}| \leqslant c\,L(Q)\,\gamma_n + c \leqslant c\,L(Q).$$

We now estimate the oscillation of $r\psi$ and use this inequality:

$$
\begin{aligned}
M_Q^\sharp(r\psi) &\leqslant 2 \inf_{c \in \mathbb{R}} \fint_Q |r\psi - c|\, dx \\
&\leqslant 2 \fint_Q |r\psi - r_Q \psi_Q|\, dx \\
&\leqslant 2 \fint_Q |r(\psi - \psi_Q)|\, dx + 2\,|\psi_Q|\, M_Q^\sharp r \\
&\leqslant 2\|r\|_\infty \|\psi\|_{\mathrm{BMO}} + c\, L(Q)\, M_Q^\sharp r.
\end{aligned}
$$

The claim follows from this when we take the supremum over Q and use $\|\psi\|_{\mathrm{BMO}} \leqslant \gamma_n$.   □

The proof of Lerner's theorem is based on Muckenhoupt weights. Recall the definition of the A_2 norm:

$$
\|\omega\|_{A_2} := \sup_Q \fint_Q \omega\, dx \fint_Q \frac{1}{\omega}\, dx,
$$

where the supremum is taken over all cubes Q in $\mathbb{R}^n$. The next lemma combined with our knowledge from the previous result allows us to conclude that in some cases $M\tilde{f}^{r(x)}$ is an A_2 weight.

Lemma 5.1.5. *There exists a constant c_n for which $\|\psi\|_{BMO} < c_n$ implies that $\|e^\psi\|_{A_2} \leqslant 4$.*

Proof. The John–Nirenberg inequality [223] implies that there exists a constant ξ_n such that

$$
\left| \{|\psi - \langle\psi\rangle_Q| > \lambda\} \right| \leqslant 2\,|Q|\, e^{-\frac{\lambda}{\xi_n \|\psi\|_{\mathrm{BMO}}}}.
$$

For $\|\psi\|_{\mathrm{BMO}} \leqslant \frac{1}{3\xi_n}$ we then obtain

$$
\int_Q e^{|\psi - \langle\psi\rangle_Q|}\, dx = |Q| + \int_1^\infty \left| \{ e^{|\psi(\cdot) - \langle\psi\rangle_Q|} > \lambda \} \right| d\lambda \leqslant |Q| + 2\,|Q| \int_1^\infty \lambda^{-3} d\lambda = 2\,|Q|.
$$

Therefore the claim follows for $c_n = \frac{1}{3\xi_n}$ by the inequality

$$
\begin{aligned}
\|e^\psi\|_{A_2} &= \sup_{Q \subset \mathbb{R}^n} M_Q(e^\psi) M_Q(e^{-\psi}) = \sup_{Q \subset \mathbb{R}^n} M_Q(e^{\psi - \langle\psi\rangle_Q}) M_Q(e^{\langle\psi\rangle_Q - \psi}) \\
&\leqslant \sup_{Q \subset \mathbb{R}^n} \left(M_Q(e^{|\psi - \langle\psi\rangle_Q|}) \right)^2 \leqslant 4.
\end{aligned}
$$

                                                                                                □

Theorem 5.1.6. *There exists a small constant $\mu_n > 0$ depending only on the dimension n such that $M \colon L^{p(\cdot)}(\mathbb{R}^n) \to L^{p(\cdot)}(\mathbb{R}^n)$ is bounded when $p = 2 - r$, $r \geqslant 0$ is measurable and $\|r\|_\infty + \sup_Q L(Q) M_Q^\sharp r \leqslant \mu_n$.*

Proof. Let $\mu_n := \min\{c_n/c_{n,3/2}, 1/2\}$, where c_n and $c_{n,3/2}$ are the constants from Lemmas 5.1.5 and 5.1.4, respectively. Suppose r and p are as in the statement of the theorem and $f \in L^{p(\cdot)}(\mathbb{R}^n)$ has norm at most one. Then Lemma 5.1.4 implies that $\|r \log M\tilde{f}\|_{\mathrm{BMO}} \leqslant c_n$. Hence it follows from Lemma 5.1.5 that $\exp(r \log M\tilde{f}) = (M\tilde{f})^{r(\cdot)}$ is an A_2 weight; the definition of A_2 directly implies that also $M\tilde{f}^{-r(\cdot)}$ is an A_2 weight. Therefore, the weighted L^2 boundedness of the maximal operator [305] yields

$$
\int\limits_{\mathbb{R}^n} |Mf(x)|^{p(x)}\, dx \leqslant \int\limits_{\mathbb{R}^n} |M\tilde{f}(x)|^{2-r(x)}\, dx = \int\limits_{\mathbb{R}^n} |M\tilde{f}(x)|^2\, |M\tilde{f}(x)|^{-r(x)}\, dx
$$

$$
\leqslant c \int\limits_{\mathbb{R}^n} |\tilde{f}(x)|^2\, |M\tilde{f}(x)|^{-r(x)}\, dx.
$$

Since $M\tilde{f} \geqslant f$ a.e. and $r \geqslant 0$, we obtain $|M\tilde{f}(x)|^{-r(x)} \leqslant |\tilde{f}(x)|^{-r(x)}$ and hence

$$
\int\limits_{\mathbb{R}^n} |Mf(x)|^{p(x)}\, dx \leqslant c \int\limits_{\mathbb{R}^n} |\tilde{f}(x)|^{p(x)}\, dx \leqslant c|Q_0| + \int\limits_{\mathbb{R}^n} |f(x)|^{p(x)}\, dx \leqslant c.
$$

Thus $Mf \in L^{p(\cdot)}(\mathbb{R}^n)$, as claimed. $\square$

Remark 5.1.7. The condition $r \geqslant 0$ in Theorem 5.1.6 is not needed. If $p = 2 + s$ with $\|s\|_\infty + \sup_Q L(Q) M_Q^\sharp s \leqslant \mu_n$, then $q := \frac{2(2+s)}{2+s^+}$ satisfies the assumptions of Theorem 5.1.6, which implies the boundedness of M on $L^{q(\cdot)}(\mathbb{R}^n)$. So Lemma 4.4.7 proves the boundedness of M on $L^{p(\cdot)}(\mathbb{R}^n)$.

Kapanadze and Kopaliani showed that it is possible in bounded domains to require the smallness condition in terms of μ_n only for small cubes [228].

Example 5.1.8. Lerner [267] showed that

$$
p(x) := 2 - a(1 + \sin(\log\log(e + |x| + 1/|x|)))
$$

satisfies the requirement of Theorem 5.1.6 if $a > 0$ is small enough. This is an example of an exponent $p \in \mathcal{A}$ which is discontinuous at zero and infinity. In particular, the log-Hölder continuity of $\frac{1}{p}$ is not needed for $p \in \mathcal{A}$.

Remark 5.1.9. Based on these results Lerner raised the question of whether the smallness condition on μ_n in Theorem 5.1.6 is really needed: is $\mu_n < \infty$

already sufficient for the boundedness of M. He showed that the smallness of r in Theorem 5.1.6 expressed by μ_n should be understood relatively to the 2 used in the construction $p = 2 - r$. Instead of the smallness of r it is also possible to take $p = p_0 - r$ with p_0 sufficiently large. Alternatively, we could take $p = \lambda(1 + r)$ with $\lambda \in (1, \infty)$ large. This gives rise to several questions:

(a) Is it possible to remove the smallness assumption on μ_n in Theorem 5.1.6?
(b) Let $p \in \mathcal{P}(\mathbb{R}^n)$ with $1 < p^- \leqslant p^+ < \infty$ and $\alpha > 0$ such that M is bounded on $L^{\alpha + p(\cdot)}(\mathbb{R}^n)$. Does this implies the boundedness of M on $L^{p(\cdot)}(\mathbb{R}^n)$?
(c) Let $p \in \mathcal{P}(\mathbb{R}^n)$ with $1 < p^- \leqslant p^+ < \infty$ and $\lambda > 1$ such that M is bounded on $L^{\lambda p(\cdot)}(\mathbb{R}^n)$. Does this implies the boundedness of M on $L^{p(\cdot)}(\mathbb{R}^n)$?

However, it has been shown by Lerner [268] and Kopaliani [256] that all of these questions have a negative answer. We present slightly modified examples:

Let $\theta(t) := \min\{\max\{0, t + \frac{1}{2}\}, 1\}$, so that θ is Lipschitz with constant 1, $\theta(t) = 0$ for $t \leqslant -\frac{1}{2}$ and $\theta = 1$ for $t \geqslant \frac{1}{2}$. Let $p(x) := 2 + 6\,\theta(\sin(\pi \log \log(1/|x|)))$ on $(-1, 1)$. Note that $p = 2$ when the sine is less than $-\frac{1}{2}$ and $p = 8$ when the sine is greater than $\frac{1}{2}$. It follows as in [267] that $\|r\|_\infty + \sup_Q L(Q)M_Q^\sharp r < \infty$, where $r := 2 - p$.

For $k \in \mathbb{N}$ define $a_k := \exp(-\exp(2k + \frac{5}{4}))$, $b_k := \exp(-\exp(2k + \frac{3}{4}))$, $c_k := \exp(-\exp(2k + \frac{1}{4}))$, $d_k := \exp(-\exp(2k - \frac{1}{4}))$, and note that $p(x) = 8$ on (a_k, b_k) and $p(x) = 2$ on (c_k, d_k). Also, a simple calculation shows that $8a_k < 4b_k < 2c_k < d_k$ for all $k \in \mathbb{N}$. Let $Q_k := (0, d_k)$; then $1/p_{Q_k} \geqslant \frac{1}{2}\frac{d_k - c_k}{d_k} \geqslant \frac{1}{4}$ and therefore $p_{Q_k} \leqslant 4$. Thus, for every $\lambda > 1$,

$$\int\limits_{Q_k} \left(\frac{1}{\lambda |Q_k|^{-1/p_{Q_k}}}\right)^{p(x)} dx \geqslant \lambda^{-8} \int\limits_{a_k}^{b_k} d_k^{-\frac{8}{4}}\, dx \geqslant \lambda^{-8}(b_k - a_k)d_k^{-2} \geqslant \frac{1}{2}\lambda^{-8} d_k^{-1},$$

which tends to ∞ as $k \to \infty$. As a consequence there exists no constant $\lambda > 1$ such that $\|\chi_{Q_k}\|_{p(\cdot)} \leqslant \lambda |Q|^{1/p_{Q_k}}$. However, by Theorem 4.5.7 this would be a consequence of $p \in A_{\text{loc}}$, so it follows that $p \notin A_{\text{loc}}$. This means that the answer to question (a) is in the negative.

Moreover, there exists $p_0 \in (1, \infty)$ and $\lambda_0 > 1$ such that $(p_0 + p)/\lambda_0$ satisfies the condition of Theorem 5.1.6 so that $(p_0 + p)/\lambda_0 \in \mathcal{A}$. By Lemma 4.4.7 we obtain $p_0 + p \in \mathcal{A}$. Since $p \notin \mathcal{A}$, this shows that (b) does not hold.

We show that (c) does not hold by contradiction, so assume that (c) holds. Define $q \in \mathcal{P}(\mathbb{R}^n)$ by $q := \frac{2\lambda_1 - 1}{\lambda_1}(\lambda_1 p)'$ for large $\lambda_1 > 1$. Then $q = 2 + \frac{2 - p}{\lambda_1 p - 1}$ satisfies the assumptions of Remark 5.1.7 and therefore $q \in \mathcal{A}$. Now, (c) implies $(\lambda_1 p)' \in \mathcal{A}$, Lemma 5.2.2 below implies $\lambda_1 p \in \mathcal{A}$, and (c) implies $p \in \mathcal{A}$. This is the desired contradiction.

5.2   Class $\mathcal{A}^*$

In this section we study class $\mathcal{A}$ in greater detail. Recall, that $\varphi \in \mathcal{A}$ if the averaging operators $T_{\mathcal{Q}}$ are bounded from $L^\varphi(\mathbb{R}^n)$ to $L^\varphi(\mathbb{R}^n)$ uniformly for all locally 1-finite families $\mathcal{Q}$ of cubes in $\mathbb{R}^n$, where

$$T_{\mathcal{Q}}f = \sum_{Q \in \mathcal{Q}} \chi_Q\, M_Q f = \sum_{Q \in \mathcal{Q}} \chi_Q \fint_Q |f(y)|\, dy$$

for $f \in L^1_{\mathrm{loc}}(\mathbb{R}^n)$. See Definitions 4.4.3 and 4.4.6 for more details.

The averaging operators have the interesting property that they are selfdual with respect to non-negative functions, i.e.

$$\int_{\mathbb{R}^n} T_{\mathcal{Q}}f\, g\, dx = \sum_{Q \in \mathcal{Q}} |Q| M_Q f M_Q g = \int_{\mathbb{R}^n} f\, T_{\mathcal{Q}} g\, dx \qquad (5.2.1)$$

for all $f \in L^\varphi(\mathbb{R}^n)$, $g \in L^{\varphi^*}(\mathbb{R}^n)$ with $f, g \geqslant 0$ and every locally 1-finite family $\mathcal{Q}$ of cubes. As a consequence class $\mathcal{A}$ behaves well with respect to duality.

Lemma 5.2.2. *Let $\varphi \in \Phi(\mathbb{R}^n)$ be proper. Then $\varphi \in \mathcal{A}$ if and only if $\varphi^* \in \mathcal{A}$. Moreover, the $\mathcal{A}$-constants of φ and φ^* are comparable up to a factor of 4.*

Proof. Let $\varphi \in \mathcal{A}$. Then by (5.2.1) and the norm conjugate formula and Hölder's inequality,

$$\|T_{\mathcal{Q}}\|_{L^{\varphi^*} \to L^{\varphi^*}} \leqslant 2 \sup_{\|g\|_{\varphi^*} \leqslant 1} \sup_{\|f\|_\varphi \leqslant 1} \int_{\mathbb{R}^n} T_{\mathcal{Q}} g\, |f|\, dx$$

$$= 2 \sup_{\|f\|_\varphi \leqslant 1} \sup_{\|g\|_{\varphi^*} \leqslant 1} \int_{\mathbb{R}^n} |g|\, T_{\mathcal{Q}} f\, dx \leqslant 4 \|T_{\mathcal{Q}}\|_{L^\varphi \to L^\varphi}.$$

The reverse estimate follows by replacing φ with φ^* since $(\varphi^*)^* = \varphi$. $\square$

Let us point out that the concept of *class $\mathcal{A}$* is a generalization of the Muckenhoupt classes. Recall that for $1 < q < \infty$ a weight ω is in the Muckenhoupt class A_q if and only if

$$\fint_Q \omega(x)\, dx \left(\fint_Q \omega(x)^{\frac{1}{1-q}}\, dx \right)^{q-1} \leqslant C \qquad (5.2.3)$$

uniformly for all cubes $Q \subset \mathbb{R}^n$. This condition is equivalent to the following:

$$\|T_{\{Q\}}f\|_{L^q(\omega\, dx)} = \|\chi_Q M_Q f\|_{L^q(\omega\, dx)} \leqslant C \|f\|_{L^q(\omega\, dx)} \qquad (5.2.4)$$

uniformly for all cubes $Q \subset \mathbb{R}^n$ and $f \in L^q(\omega\,dx)$. Indeed, if (5.2.4) holds, then the choice $f := \omega^{\frac{1}{1-q}}$ implies (5.2.3). On the other hand if (5.2.3) holds, then by Hölder's inequality

$$
\begin{aligned}
\|\chi_Q\, M_Q f\|_{L^q(\omega\,dx)} &= \frac{1}{|Q|} \int\limits_Q f\,dx \left(\int\limits_Q \omega\,dx \right)^{\frac{1}{q}} \\
&\leqslant \frac{1}{|Q|} \left(\int\limits_Q |f|^q\,\omega\,dx \right)^{\frac{1}{q}} \left(\int\limits_Q \omega^{\frac{1}{1-q}}\,dx \right)^{\frac{q-1}{q}} \left(\int\limits_Q \omega\,dx \right)^{\frac{1}{q}} \\
&\leqslant C\,\|f\|_{L^q(\omega\,dx)}.
\end{aligned}
$$

Hence, (5.2.4) is equivalent to $\omega \in A_q$. Let us translate this into the language of Musielak–Orlicz spaces. Define $\varphi(x,t) := t^q \omega(x)$. Then $\omega \in A_q$ if and only if $\varphi \in \mathcal{A}_{\mathrm{loc}}$. For more information on A_p weights we refer to [174].

The difference between $\varphi \in \mathcal{A}_{\mathrm{loc}}$ and $\varphi \in \mathcal{A}$ is that the uniform boundedness of the averaging operators is only required for single cubes rather than locally 1-finite families of cubes. But in the context of $L^q(\omega\,dx)$ there is no difference. Indeed, let $\mathcal{Q}$ be a locally 1-finite family of cubes and let $\omega \in A_q$ with $1 < q < \infty$. Then the boundedness of the operators $T_{\{Q\}}$ imply

$$
\begin{aligned}
\left\| \sum_{Q \in \mathcal{Q}} \chi_Q M_Q f \right\|^q_{L^q(\omega\,dx)} &= \sum_{Q \in \mathcal{Q}} \|T_{\{Q\}} f\|^q_{L^q(\omega\,dx)} \\
&\leqslant \sum_{Q \in \mathcal{Q}} c\,\|\chi_Q\, f\|^q_{L^q(\omega\,dx)} \\
&\leqslant c\,\|f\|^q_{L^q(\omega\,dx)}.
\end{aligned}
$$

Hence, for $\varphi(x,t) = t^q \omega(x)$ the three conditions $\omega \in A_q$, $\varphi \in \mathcal{A}_{\mathrm{loc}}$ and $\varphi \in \mathcal{A}$ are equivalent. It is therefore reasonable to say that *class $\mathcal{A}_{\mathrm{loc}}$ and $\mathcal{A}$ are generalizations of the Muckenhoupt classes.* In particular, if M is bounded on $L^q_\omega(\mathbb{R}^n)$, then $\varphi \in \mathcal{A}$ and therefore $w \in A_q$. Surprisingly, the reverse holds: if $w \in A_q$ with $1 < q < \infty$, then M is bounded on $L^q(\omega\,dx)$. This is the fundamental result of Muckenhoupt, see for example [305]. The aim of this chapter is, to provide a similar result for the spaces $L^\varphi(\mathbb{R}^n)$ and $L^{p(\cdot)}(\mathbb{R}^n)$. So far, we have shown (cf. Theorem 4.4.10 and Lemma 4.4.13) that

$$
M \text{ is strong type } \varphi \;\;\Rightarrow\;\; \varphi \in \mathcal{A} \;\;\Rightarrow\;\; M \text{ is weak type } \varphi \;\;\Rightarrow\;\; \varphi \in \mathcal{A}_{\mathrm{loc}}.
$$

Note that the first and the last implication is strict, as can be seen by the example $p = 1$ and Theorem 5.3.4 below.

As an extension of Definition 4.4.4 we introduce the following notation.

Definition 5.2.5. Let $\varphi \in \Phi(\mathbb{R}^n)$ and $s \in [1,\infty)$. Then we define $M_{s,Q}\varphi : \mathbb{R} \to [0,\infty]$ by

$$M_{s,Q}\varphi(t) := \left(\fint_Q \left(\varphi(x,t) \right)^s dx \right)^{\frac{1}{s}} \tag{5.2.6}$$

for all $t \geqslant 0$ and every cube (open set) $Q \subset \mathbb{R}^n$. We also write $M_Q\varphi$ instead of $M_{1,Q}\varphi$ (see Definition 4.4.4).

Lemma 5.2.7. *If $\varphi \in \Phi(\mathbb{R}^n)$ be proper, then $t \mapsto M_Q\varphi(t)$ is a Φ-function for every cube (open set) Q.*

Proof. It follows from $M_Q\varphi(t) = |Q|^{-1} \int_Q \varphi(x,t)\,dx = |Q|^{-1}\varrho_\varphi(t\chi_Q)$ and Lemma 2.3.10 that $M_Q\varphi$ is a semimodular on $\mathbb{R}$. Since φ is proper and $|Q| < \infty$, we have $\chi_Q \in L^\varphi$. This implies that $\mathbb{R}_{M_Q\varphi} = \mathbb{R}$, so by Lemma 2.3.2 follows that $M_Q\varphi$ is a Φ-function. $\square$

In the following we denote by $\mathcal{Y}_1^n$ the set of all locally 1-finite families of cubes in $\mathbb{R}^n$. If $\mathcal{Q} \in \mathcal{Y}_1^n$, then $\mu(Q) := |Q|$ defines a natural, atomic measure on $\mathcal{Q}$. Analogously to Definition 2.3.9 we define $\Phi(\mathcal{Q})$ to be the set of generalized Φ-functions on the measure space $(\mathcal{Q}, \mu)$.

Let e_Q denote the function on $\mathcal{Q}$ that is one at Q and zero elsewhere. Since $\mathcal{Q}$ is at most countable, every function on $\mathcal{Q}$ is μ-measurable, i.e.

$$L^0(\mathcal{Q}) := L^0(\mathcal{Q},\mu) = \left\{ \sum_{Q \in \mathcal{Q}} t_Q e_Q : t_Q \in \mathbb{R} \right\} = \mathbb{R}^{\mathcal{Q}}$$

If $\psi \in \Phi(\mathcal{Q})$, then the semimodular induced by ψ on $L^0(\mathcal{Q})$ is given by

$$\psi\left(\sum_{Q \in \mathcal{Q}} t_Q e_Q \right) := \sum_{Q \in \mathcal{Q}} |Q| \psi(Q, t_Q).$$

We denote the corresponding Musielak–Orlicz space by $l^\psi(\mathcal{Q})$. Recall that the norm

$$\left\| \sum_{Q \in \mathcal{Q}} t_Q e_Q \right\|_{l^\psi(\mathcal{Q})} := \inf \left\{ \lambda > 0 : \sum_{Q \in \mathcal{Q}} |Q|\, \psi(Q, |t_Q/\lambda|) \leqslant 1 \right\}.$$

makes $l^\psi(\mathcal{Q})$ a Banach space (Theorem 2.3.13).

Let $\varphi \in \Phi(\mathbb{R}^n)$. Then it follows from Lemma 5.2.7 that $M_Q\varphi \in \Phi(\mathcal{Q})$. Since $\varphi^* \in \Phi(\mathbb{R}^n)$, also $M_Q\varphi^* := M_Q(\varphi^*) \in \Phi(\mathcal{Q})$. Therefore, the conjugate functions $(M_Q\varphi)^*$ and $(M_Q\varphi^*)^*$ are also from $\Phi(\mathcal{Q})$. Of particular interest for us is the function $(M_Q\varphi^*)^*$. In contrast to $M_Q\varphi$ the function $(M_Q\varphi^*)^*$ involves two conjugations: Pass to the conjugate function φ^* (defined on $\mathbb{R}^n$), take the mean average M_Q (now defined on Q), and pass again to the conjugate function (defined on Q). If $\varphi(x,t)$ is independent of x (translation invariant case), then $(M_Q\varphi^*)^* = (\varphi^*)^* = \varphi = M_Q\varphi$. This is not the case if $\varphi(x,t)$ is x-dependent. So in a sense, we can use the difference of $M_Q\varphi$ and $(M_Q\varphi^*)^*$ to measure the failure of translation invariance of the space. The function $(M_Q\varphi^*)^*$ might look awkward at first glance, but the following lemma provides a concrete characterization of the "abstract" function $(M_Q\varphi^*)^*$.

Lemma 5.2.8. *Let $\varphi \in \Phi(\mathbb{R}^n)$ be proper. Then*

$$(M_Q\varphi^*)^*(t) = \inf_{f \in L^0 : M_Q f \geqslant t} M_Q(\varphi(\cdot, f))$$

$$= \inf_{f \in L^0 : M_Q f = t} M_Q(\varphi(\cdot, f)) \leqslant (M_Q\varphi)(t)$$

for every $t \geqslant 0$ and all cubes (or open sets) Q.

Proof. Let us denote the four terms in the claim by (I), (II), (III) and (IV). Obviously, $(II) \leqslant (III)$. The choice $f = \chi_Q t$ implies $(III) \leqslant (IV)$. We estimate using Young's inequality

$$(M_Q\varphi^*)^*(M_Q f) = \sup_{u \geqslant 0} \left(u\, M_Q f - (M_Q\varphi^*)(u) \right)$$

$$= \sup_{u \geqslant 0} \left(M_Q(u\, f) - (M_Q\varphi^*)(u) \right)$$

$$\leqslant \sup_{u \geqslant 0} \left(M_Q\big(\varphi(\cdot, f)\big) + (M_Q\varphi^*)(u) - (M_Q\varphi^*)(u) \right)$$

$$= M_Q\big(\varphi(\cdot, f)\big)$$

proving $(I) \leqslant (II)$. We define $\psi(t) := (III)$. It follows easily from the convexity of φ that ψ is convex. This and the already proven estimate $(M_Q\varphi^*)^*(t) \leqslant \psi(t) \leqslant (M_Q\varphi)(t)$ for all $t \geqslant 0$ ensures that ψ is a Φ-function. For the conjugate of ψ we calculate

$$\psi^*(u) = \sup_{t \geqslant 0} \left(ut - \inf_{f \in L^0 : M_Q f = t} M_Q\big(\varphi(\cdot, f)\big) \right)$$

$$= \sup_{t \geqslant 0} \sup_{f \in L^0 : M_Q f = t} \left(ut - M_Q\big(\varphi(\cdot, f)\big) \right)$$

$$= \sup_{f \in L^0} \left(uM_Q f - M_Q\big(\varphi(\cdot, f)\big) \right)$$

$$= \frac{1}{|Q|} \sup_{f \in L^0} \left(\int_Q uf\, dx - \int_Q \varphi(\cdot, f)\, dx \right)$$

$$= \frac{1}{|Q|} (\varrho_\varphi)^* (J_{u\chi_Q})$$

for $u \geqslant 0$. By Theorem 2.7.4 we have $(\varrho_\varphi)^*(J_{u\chi_Q}) = \varrho_{\varphi^*}(u\chi_Q)$, so

$$\psi^*(u) = (M_Q \varphi^*)(u)$$

for all $u \geqslant 0$. This proves $\psi(t) = (M_Q\varphi^*)^*(t)$ since $(\psi^*)^* = \psi$ and so
$(III) = (I)$.   □

Remark 5.2.9. If φ is a generalized N-function, then the infima in Lemma
5.2.8 are attained. Indeed, for $t \geqslant 0$ let $u_t := \big((M_Q\varphi^*)^*\big)'(t)$ and $f_t(x) :=$
$(\varphi^*)'(x, u_t)$ for $x \in Q$. Then by Remark 2.6.9 $(M_Q\varphi^*)^*(t) = t\,u_t - (M_Q\varphi^*)(u_t)$
and $u_t\,f_t(x) = \varphi(x, f_t(x)) + \varphi^*(x, u_t)$ for $x \in Q$. As a consequence $M_Q f_t =$
$M_Q\big((\varphi^*)'(u_t)\big) = (M_Q\varphi^*)'(u_t) = t$, where we have used that $(M_Q\varphi^*)'$ is the
right-continuous inverse of $((M_Q\varphi^*)^*)'$. So f is admissible in the infima in
Lemma 5.2.8 and

$$
\begin{aligned}
(M_Q\varphi^*)^*(t) &= t\,u_t - (M_Q\varphi^*)(u_t) \\
&= M_Q(f_t\,u_t) - (M_Q\varphi^*)(u_t) \\
&= M_Q\big(\varphi(\cdot, f_t)\big) + (M_Q\varphi^*)(u_t) - (M_Q\varphi^*)(u_t) \\
&= M_Q\big(\varphi(\cdot, f_t)\big).
\end{aligned}
$$

Thus the infima are attained by the function f_t.

We have the following relation between $M_Q\varphi$ and $(M_Q\varphi^*)^*$.

Lemma 5.2.10. *Let $\varphi \in \Phi(\mathbb{R}^n)$ be proper. Then*

$$
\begin{aligned}
(M_Q\varphi^*)^*(t) &\leqslant (M_Q\varphi)(t), \\
(M_Q\varphi)^*(t) &\leqslant (M_Q\varphi^*)(t).
\end{aligned}
$$

for all $t \geqslant 0$ and all cubes (or open sets) Q.

Proof. The first estimate follows from Lemma 5.2.8 with $f := \chi_Q\,t$. The
second one follows from the first, if we replace φ by φ^* and use $\varphi^{**} = \varphi$ and
$(M_Q\varphi^*)^{**} = M_Q\varphi^*$.   □

We have seen in Theorem 3.2.13 that $L^\varphi(\mathbb{R}^n)$ is a Banach function space
if $\varphi \in \Phi(\mathbb{R}^n)$ is proper. We will see in the following that φ inherits this
property to $M_Q\varphi$.

Lemma 5.2.11. *Let $\varphi \in \Phi(\mathbb{R}^n)$ be proper and $\mathcal{Q} \in \mathcal{Y}_1^n$. Then $M_Q\varphi \in \Phi(\mathcal{Q})$
is proper. In particular, $l^{M_Q\varphi}(\mathcal{Q})$ is a Banach function space.*

Moreover, if φ is a generalized N-function, then so is $M_Q\varphi$.

Proof. Due to Corollary 2.7.9 we have to show that the simple functions S
are contained in $l^{M_Q\varphi}(\mathcal{Q})$ and $l^{(M_Q\varphi)^*}(\mathcal{Q})$. We begin with $S \subset l^{M_Q\varphi}(\mathcal{Q})$.
It suffices to show that characteristic functions of measurable sets with

finite measures are contained in $l^{M_Q\varphi}$. Let $\mathcal{U} \subset \mathcal{Q}$ with $\mu(\mathcal{U}) < \infty$ and let $t > 0$. We have to show $\sum_{Q \in \mathcal{U}} e_Q \in l^{M_Q\varphi}(\mathcal{Q})$. Let $A := \bigcup_{Q \in \mathcal{U}} Q \subset \mathbb{R}^n$. Then $|A| = \mu(\mathcal{Q}) < \infty$. Since φ is proper we have $\chi_A \in L^\varphi(\mathbb{R}^n)$. Thus $\varrho_{M_Q\varphi}(t \sum_{Q \in \mathcal{U}} e_Q) = \sum_{Q \in \mathcal{Q}} |Q| M_Q\varphi(t) = \varrho_\varphi(t \chi_A) < \infty$ for some $t > 0$. It follows that $\sum_{Q \in \mathcal{U}} e_Q \in l^{M_Q\varphi}(\mathcal{Q})$. This proves $S \subset l^{M_Q\varphi}(\mathcal{Q})$. Since φ^* is also proper it similarly follows that $S \subset l^{M_Q\varphi^*}(\mathcal{Q})$. Now, $(M_Q\varphi)^* \leqslant M_Q\varphi^*$ (Lemma 5.2.10) implies $S \subset l^{(M_Q\varphi)^*}(\mathcal{Q})$.

Now assume that φ is a generalized N-function. It remains to show that $\lim_{t \to 0} \frac{M_Q\varphi(t)}{t} = 0$ and $\lim_{t \to \infty} \frac{M_Q\varphi(t)}{t} = \infty$. Fix $Q \in \mathcal{X}^n$. Since $M_Q\varphi$ is proper, there exists $\lambda > 0$ with $(M_Q\varphi)(\lambda) < \infty$. Now $\lim_{t \to 0} \frac{\varphi(x,t)}{t} = 0$ for all $x \in Q$ and the theorem of dominated convergence with majorant $\frac{\varphi(\cdot,\lambda)}{\lambda} \in L^1(Q)$ implies that $\lim_{t \to 0} \frac{M_Q\varphi(t)}{t} = 0$. On the other hand $\lim_{t \to \infty} \frac{\varphi(x,t)}{t} = \infty$ for all $x \in Q$ and the theorem of monotone convergence implies that $\lim_{t \to \infty} \frac{M_Q\varphi(t)}{t} = \infty$. $\square$

As a consequence of Lemma 5.2.11 also $M_Q\varphi^*$, $(M_Q\varphi)^*$ and $(M_Q\varphi^*)^*$ are proper if $\varphi \in \Phi(\mathbb{R}^n)$ is proper.

Corollary 5.2.12. *Let $\varphi \in \Phi(\mathbb{R}^n)$, $\mathcal{Q} \in \mathcal{Y}_1^n$ and $s \geqslant 1$. If φ^s is proper, then $M_{s,Q}\varphi \in \Phi(\mathcal{Q})$.*

Proof. If follows from Lemma 5.2.11 that $M_Q\varphi, M_Q(\varphi^s) \in \Phi(\mathcal{Q})$. Since $\varphi(x,\cdot)$ is convex and $f \mapsto M_{s,Q}f$ is convex, so is $M_{s,Q}\varphi$. The left-continuity of $M_Q(\varphi^s)$ implies the left-continuity of $M_{s,Q}\varphi$. Moreover, $\lim_{t \to 0}(M_{s,Q}\varphi)(t) = 0$ and $\lim_{t \to \infty}(M_{s,Q}\varphi))(t) = \infty$ follow from the corresponding limits of $M_Q(\varphi^s)$. Thus, $M_{s,Q}\varphi \in \Phi(\mathcal{X}^n)$. $\square$

The following lemma is a generalization of Lemma 5.2.8.

Lemma 5.2.13. *Let $\varphi \in \Phi(\mathbb{R}^n)$ be proper and $\mathcal{Q} \in \mathcal{Y}_1^n$. Then*

$$\varrho_{(M_Q\varphi^*)^*}\left(\sum_{Q \in \mathcal{Q}} t_Q e_Q\right) = \inf_{f \in L^0 : M_Q f = t_Q} \varrho_\varphi(f)$$

and

$$\left\| \sum_{Q \in \mathcal{Q}} t_Q e_Q \right\|_{l^{(M_Q\varphi^*)^*}} = \inf_{f \in L^0 : M_Q f = t_Q} \|f\|_\varphi. \tag{5.2.14}$$

The infimum is taken over all $f \in L^0$ which satisfy $M_Q f = t_Q$ for all $Q \in \mathcal{Q}$.

Proof. The first estimate follows when we apply Lemma 5.2.8 in each cube Q. In particular, for all $f \in L^\varphi$,

$$\varrho_{(M_Q\varphi^*)^*}\left(\sum_{Q \in \mathcal{Q}} M_Q f\, e_Q\right) \leqslant \varrho_\varphi(f).$$

This and the unit ball property proves the case "$\leqslant$" of (5.2.14). Let us prove equality. If the left-hand side of (5.2.14) is zero, then the right-hand side is zero since $f = 0$. Therefore, we can assume by a scaling argument that the left-hand side of (5.2.14) is equal to 1 and we have the show that the right-hand side is smaller or equal to 1. The unit ball property implies $\varrho_{(M_Q\varphi^*)^*}(\sum_{Q\in\mathcal{Q}} \lambda t_Q e_Q) \leqslant 1$. Thus for every $\lambda > 1$ we conclude using (2.1.5) that $\varrho_{(M_Q\varphi^*)^*}(\sum_{Q\in\mathcal{Q}} \frac{t_Q}{\lambda} e_Q) \leqslant \frac{1}{\lambda} < 1$. So by the first estimate of the lemma there exists $g_\lambda \in L^0$ with $M_Q g_\lambda = t_Q/\lambda$ such that $\varrho_\varphi(g_\lambda) < 1$. This proves $\|g_\lambda\|_\varphi \leqslant 1$ by the unit ball property. The function $f_\lambda := \lambda g_\lambda$ satisfies $M_Q f_\lambda = t_Q$ and $\|f_\lambda\|_\varphi \leqslant \lambda$. Since $\lambda > 1$ was arbitrary, it follows that the right-hand side of (5.2.14) is bounded by 1. $\square$

We are now able to characterize class $\mathcal{A}$ in terms of the generalized Φ-functions $M_Q\varphi$ and $(M_Q\varphi^*)^*$.

Theorem 5.2.15. *Let $\varphi \in \Phi(\mathbb{R}^n)$ be proper. Then φ is of class $\mathcal{A}$ if and only if the embeddings*

$$l^{(M_Q\varphi^*)^*}(\mathcal{Q}) \hookrightarrow l^{M_Q\varphi}(\mathcal{Q})$$

are uniformly bounded for all $\mathcal{Q} \in \mathcal{Y}_1^n$. Moreover, for every $\mathcal{Q} \in \mathcal{Y}_1^n$

$$\|T_\mathcal{Q}\|_{L^\varphi \to L^\varphi} = \|\mathrm{Id}\|_{l^{(M_Q\varphi^*)^*}(\mathcal{Q}) \hookrightarrow l^{M_Q\varphi}(\mathcal{Q})}.$$

In particular, the $\mathcal{A}$-constant of φ equals

$$\sup_{\mathcal{Q}\in\mathcal{Y}_1^n} \|\mathrm{Id}\|_{l^{(M_Q\varphi^*)^*}(\mathcal{Q}) \hookrightarrow l^{M_Q\varphi}(\mathcal{Q})}.$$

Proof. For all $f \in L^\varphi$,

$$\varrho_{M_Q\varphi}\left(\sum_{Q\in\mathcal{Q}} M_Q f \, e_Q\right) = \sum_{Q\in\mathcal{Q}} |Q|(M_Q\varphi)(M_Q f) = \varrho_\varphi(T_\mathcal{Q} f)$$

and as a consequence of the unit ball property

$$\left\|\sum_{Q\in\mathcal{Q}} M_Q f \, e_Q\right\|_{l^{M_Q\varphi}} = \|T_\mathcal{Q} f\|_\varphi. \tag{5.2.16}$$

Assume first that $l^{(M_Q\varphi^*)^*}(\mathcal{Q}) \hookrightarrow l^{M_Q\varphi}(\mathcal{Q})$ and with embedding constant A. Then by Lemma 5.2.13

$$\|T_\mathcal{Q} f\|_\varphi = \left\|\sum_{Q\in\mathcal{Q}} M_Q f e_Q\right\|_{l^{M_Q\varphi}} \leqslant A \left\|\sum_{Q\in\mathcal{Q}} M_Q f e_Q\right\|_{l^{(M_Q\varphi^*)^*}} \leqslant A \|f\|_\varphi$$

for all $f \in L^\varphi$.

Assume now, that $\varphi \in \mathcal{A}$ with $\mathcal{A}$-constant A_2. Then for all $f \in L^\varphi$, which satisfy $M_Q f = t_Q$ for all $Q \in \mathcal{Q}$, it holds that

$$\left\| \sum_{Q \in \mathcal{Q}} t_Q e_Q \right\|_{l^{M_Q \varphi}} = \left\| \sum_{Q \in \mathcal{Q}} M_Q f e_Q \right\|_{l^{M_Q \varphi}} = \| T_Q f \|_\varphi \leqslant A_2 \| f \|_\varphi,$$

where we used (5.2.16) in the second step. Taking the infimum over all such f proves that

$$\left\| \sum_{Q \in \mathcal{Q}} t_Q e_Q \right\|_{l^{M_Q \varphi}} \leqslant A_2 \left\| \sum_{Q \in \mathcal{Q}} t_Q e_Q \right\|_{l^{(M_Q \varphi^*)^*}},$$

where we used Lemma 5.2.13.   □

Let us introduce the following useful notation about embeddings.

Definition 5.2.17. By $\mathcal{X}^n$ we denote the set of all open cubes in $\mathbb{R}^n$. Let $\varphi, \psi : \mathcal{X}^n \times \mathbb{R}_\geqslant \to \mathbb{R}_\geqslant$ be generalized Φ-functions on $\mathcal{Y}_1^n$. Then we say that ψ is *dominated by* φ and write $\psi \preceq \varphi$ if the embeddings

$$l^{\varphi(Q)}(\mathcal{Q}) \hookrightarrow l^{\psi(Q)}(\mathcal{Q})$$

are uniformly bounded with respect to $\mathcal{Q} \in \mathcal{Y}_1^n$. We write $\psi \cong \varphi$ if $\psi \preceq \varphi$ and $\varphi \preceq \psi$.

If $\psi \preceq \kappa$ and $\kappa \preceq \varphi$, then $\psi \preceq \varphi$. With the new notation we can rewrite Theorem 5.2.15 as follows:

Theorem 5.2.18. *Let $\varphi \in \Phi(\mathbb{R}^n)$ be proper. Then φ is of class $\mathcal{A}$ if and only if $M_Q \varphi \preceq (M_Q \varphi^*)^*$.*

Remark 5.2.19. Let φ and ψ be generalized Φ-functions on $\mathcal{Y}_1^n$ such that for every $\mathcal{Q} \in \mathcal{Y}_1^n$ the functions $(Q, t) \mapsto \varphi(Q, t)$ and $(Q, t) \mapsto \psi(Q, t)$ as elements of $\Phi(\mathcal{Q})$ are proper. Then $\varphi \preceq \psi$ is equivalent to $\psi^* \preceq \varphi^*$. Indeed, if $l^{\varphi(Q)}(\mathcal{Q}) \hookrightarrow l^{\psi(Q)}(\mathcal{Q})$, then $(l^{\varphi(Q)}(\mathcal{Q}))' \hookrightarrow (l^{\psi(Q)}(\mathcal{Q}))'$ and therefore $(l^{\varphi^*(Q)}(\mathcal{Q})) \hookrightarrow (l^{\psi^*(Q)}(\mathcal{Q}))$ using Theorem 2.7.4. This implies $\psi^* \preceq \varphi^*$.

Remark 5.2.20. Let $\varphi \in \Phi(\mathbb{R}^n)$ be proper. Then $(M_Q \varphi^*)^* \leqslant M_Q \varphi$ holds by Lemma 5.2.10 for every cube Q. Thus, $M_Q \varphi \preceq (M_Q \varphi^*)^*$ from Theorem 5.2.18 is equivalent to $M_Q \varphi \cong (M_Q \varphi^*)^*$.

Remark 5.2.21. Let $\varphi \in \Phi(\mathbb{R}^n)$ be proper. We have seen in Lemma 5.2.2 that $\varphi \in \mathcal{A}$ if and only if $\varphi^* \in \mathcal{A}$. This "stability under conjugation" also follows from Theorem 5.2.18, since $M_Q \varphi \preceq (M_Q \varphi^*)^*$ is by conjugation (see Remark 5.2.19) equivalent to $M_Q \varphi^* \preceq (M_Q \varphi)^* = (M_Q (\varphi^*)^*)^*$.

5.3   Class $\mathcal{A}$ for Variable Exponent Lebesgue Spaces*

In the case of variable exponent Lebesgue spaces $L^{p(\cdot)}(\mathbb{R}^n)$ we can provide another characterization of class $\mathcal{A}$, which avoids the use of $(M_Q\varphi^*_{p(\cdot)})^*$. This characterization is based on the following refined version of Lemma 5.2.10:

Lemma 5.3.1. *Let* $p \in \mathcal{P}(\mathbb{R}^n)$. *Then*

$$(M_Q\widetilde{\varphi}^*_{p(\cdot)})^*(t) \leqslant \widetilde{\varphi}_{p_Q}(t) \leqslant (M_Q\widetilde{\varphi}_{p(\cdot)})(t) \tag{5.3.2}$$

for all $t \geqslant 0$ *and all cubes (or open sets)* Q. *Recall that* $\frac{1}{p_Q} = \fint_Q \frac{1}{p(x)}\,dx$ *for every cube* Q.

Proof. The second inequality is just Lemma 4.5.1. If we apply this inequality to $(\widetilde{\varphi}_{p(\cdot)})^* = \widetilde{\varphi}_{p'(\cdot)}$, then $\widetilde{\varphi}_{p'_Q}(t) \leqslant (M_Q\widetilde{\varphi}^*_{p(\cdot)})(t)$ all $t \geqslant 0$. So by conjugation $(M_Q\widetilde{\varphi}^*_{p(\cdot)})^*(t) \leqslant (\widetilde{\varphi}_{p'_Q})^*(t) = \widetilde{\varphi}_{p_Q}(t)$.   □

The use of φ_{p_Q} enables us to avoid $(M_Q\varphi^*)^*$, as can be seen in (c) of the following theorem.

Theorem 5.3.3. *Let* $p \in \mathcal{P}(\mathbb{R}^n)$. *Then the following statements are equivalent:*

(a) $p \in \mathcal{A}$.
(b) $M_Q\varphi_{p(\cdot)} \preceq (M_Q\varphi^*_{p(\cdot)})^*$.
(c) $M_Q\varphi_{p(\cdot)} \preceq \varphi_{p_Q}$ *and* $M_Q\varphi_{p'(\cdot)} \preceq \varphi_{p'_Q}$.

Proof. It suffices to prove the theorem for $\varphi_{p(\cdot)} = \widetilde{\varphi}_{p(\cdot)}$. The equivalence of (a) and (b) is just the statement of Theorem 5.2.18. If (b) holds, then $M_Q\widetilde{\varphi}_{p(\cdot)} \cong \widetilde{\varphi}_{p_Q} \cong (M_Q\widetilde{\varphi}^*_{p(\cdot)})^*$ by Lemma 5.3.1. If (c) holds, then also $M_Q\widetilde{\varphi}_{p(\cdot)} \cong \widetilde{\varphi}_{p_Q}$ and $M_Q\widetilde{\varphi}_{p'(\cdot)} \cong \widetilde{\varphi}_{p'_Q}$ by Lemma 5.3.1. Conjugation of the second equivalence and $\widetilde{\varphi}_{p'(\cdot)} = \widetilde{\varphi}^*_{p(\cdot)}$ gives $(M_Q\widetilde{\varphi}^*_{p(\cdot)})^* \cong \widetilde{\varphi}_{p_Q}$. Combining the results we get $(M_Q\widetilde{\varphi}^*_{p(\cdot)})^* \cong \widetilde{\varphi}_{p_Q} \cong M_Q\widetilde{\varphi}_{p(\cdot)}$.   □

We now see that the uniform boundedness of the averaging $T_{\{Q\}}$ over single cubes Q is not enough to ensure the boundedness of the averaging operators $T_{\mathcal{Q}}$ uniformly with respect to locally 1-finite families of cubes. Especially, the uniform boundedness of $T_{\{Q\}}$ on $L^{p(\cdot)}(\mathbb{R})$ with respect to single cubes cannot imply boundedness of M on $L^{p(\cdot)}(\mathbb{R})$. This is reflected in the following counterexample. We refer to Definition 7.3.2 below for the definition of $\mathcal{G}$.

Theorem 5.3.4. *There exists an exponent* $p \in \mathcal{P}(\mathbb{R})$ *which is uniformly Lipschitz continuous and satisfies* $\frac{3}{2} \leqslant p^- \leqslant p^+ \leqslant 3$ *such that* $p \in \mathcal{A}_{\mathrm{loc}} \setminus \mathcal{A}$.

Proof. Let $\eta \in C_0^\infty(B(0,1/2))$ with $\|\eta\|_\infty \leqslant \frac{1}{6}$, $\eta \neq 0$, $\int \eta(x)\,dx = 0$, and $\eta(x) = -\frac{1}{6}$ for all $x \in B(0,1/4)$. Let $x_j := \exp(10\,j^2)$ for $j \in \mathbb{N}$. Define $p \in \mathcal{P}(\mathbb{R})$ by $\frac{1}{p(x)} := \frac{1}{2} + \sum_{j=1}^\infty \eta(x - x_j)$. Then p is uniformly Lipschitz continuous and satisfies $\frac{3}{2} \leqslant p^- \leqslant p^+ \leqslant 3$.

First, we show that $p \notin \mathcal{A}$. Let $Q_j := B(x_j, 2)$ and $\mathcal{Q} := \{Q_j : j \in \mathbb{N}\}$. For $(a_k) \in l^3(\mathbb{N})$ define a function $G(\{a_k\})$ by $G(\{a_k\})(x) := \sum_{j=1}^{\infty} a_j \chi_{B(0,1/4)}$ $(x - x_j)$ for $x \in \mathbb{R}$. If $p \in \mathcal{A}$, then $\|T_{\mathcal{Q}}(G(\{a_k\}))\|_{p(\cdot)} \leqslant K \|G(\{a_k\})\|_{p(\cdot)}$ for some $K > 0$. Now, $\|G(\{a_k\})\|_{p(\cdot)} = \|\{a_k\}\|_{l^3(\mathbb{N})}$. On the other hand $T_{\mathcal{Q}}(G(\{a_k\})) \geqslant \frac{1}{8} \sum_{j=1}^{\infty} |a_j| \chi_{B(0,1/4)}(x - x_j - 1)$, so

$$\left\| T_{\mathcal{Q}}(G(\{a_k\})) \right\|_{p(\cdot)} \geqslant \left\| \frac{1}{8} \sum_{j=1}^{\infty} |a_j| \chi_{B(0,1/4)}(x - x_j - 1) \right\|_{p(\cdot)} = \frac{1}{8\sqrt{2}} \|\{a_k\}\|_{l^2(\mathbb{N})}.$$

Therefore, if $p \in \mathcal{A}$, we have shown that $\|\{a_k\}\|_{l^2(\mathbb{N})} \leqslant 8\sqrt{2} K \|\{a_k\}\|_{l^3(\mathbb{N})}$ for all sequences a_k. This is impossible and thus $p \notin \mathcal{A}$.

Second, we show that M is not of weak type $p(\cdot)$. We proceed again by contradiction. So let us assume that $\|\lambda \chi_{\{Mf > \lambda\}}\|_{p(\cdot)} \leqslant K \|f\|_{p(\cdot)}$ for all $f \in L^{p(\cdot)}(\mathbb{R})$. Let $(a_k) \in l^3(\mathbb{N})$ and define $G(\{a_k\})$ as above, so $\|G(\{a_k\})\|_{p(\cdot)} = \|\{a_k\}\|_{l^3(\mathbb{N})}$. Then for $\lambda > 0$ define $J := \{j \in \mathbb{N} : |a_j| > \lambda\}$. Then as above $M(G(\{a_k\})) \geqslant \sum_{j \in J} \frac{1}{8} \lambda \chi_{B(0,1/4)}(x - x_j)$. This proves that

$$\left\| \frac{1}{8} \lambda \chi_{\{Mf > \frac{1}{8}\lambda\}} \right\|_{p(\cdot)} \geqslant \left\| \frac{1}{8} \lambda \sum_{j \in J} \chi_{B(0,1/4)}(x - x_j - 1) \right\|_{p(\cdot)} = \frac{1}{8\sqrt{2}} \left(\sum_{j \in J} \lambda^2 \right)^{\frac{1}{2}}$$

$$= \frac{1}{8\sqrt{2}} \left(\sum_{j \in \mathbb{N} : |a_j| > \lambda} \lambda^2 \right)^{\frac{1}{2}}.$$

Since $\lambda > 0$ was arbitrary, this proves $\|\{a_k\}\|_{\mathrm{w}\text{-}l^2(\mathbb{N})} \leqslant 8\sqrt{2} K \|\{a_k\}\|_{l^3(\mathbb{N})}$ where $\|\{a_k\}\|_{\mathrm{w}\text{-}l^2} := \sup_{\lambda > 0} \|\lambda \chi_{\{|a_k| > \lambda\}}\|_{\ell^2}$. This is the desired contradiction.

Third, we show that $p \in \mathcal{A}_{\mathrm{loc}}$. We have to show that for some constant $K_2 > 0$ and for all cubes Q

$$\|T_Q(f \chi_Q)\|_{p(\cdot)} \leqslant K_2 \|f \chi_Q\|_{p(\cdot)}.$$

We start with small cubes. So let Q be a cube with $\mathrm{diam}\, Q \leqslant 2$. If Q does not intersect any of the sets $B(x_j, 1)$, then $p(x) = 2$ for all $x \in Q$, then there is nothing to show, since T_Q is bounded on $L^2(\mathbb{R})$. Otherwise, Q intersects exactly one of the sets $B(x_j, 1)$, whose index we call j_0. Define $q \in \mathcal{P}(\mathbb{R})$ by $\frac{1}{q(x)} := \frac{1}{2} + \eta(x - x_{j_0})$, then $q(x) = p(x)$ for all $x \in Q$. By definition of q we have $q \in \mathcal{P}^{\log}(\mathbb{R})$ with $\frac{3}{2} \leqslant q^- \leqslant q^+ \leqslant 3$. So by Theorem 4.4.8, $T_{\{Q\}}$ is bounded on $L^{q(\cdot)}(\mathbb{R})$ with continuity constant independent of j_0. Since $p = q$ on Q, the operator $T_{\{Q\}}$ is also bounded on $L^{p(\cdot)}(\mathbb{R})$. This proves the claim for small cubes, i.e. $\mathrm{diam}\, Q \leqslant 2$.

Now, let Q be a large cube, i.e. $\mathrm{diam}(Q) \geqslant 2$. Then there exists a cube W with $Q \subset W \subset 2Q$ such that every $B(x_j, 1)$ that intersects W is completely contained in W. By definition of p and $\int \eta(x) \, dx = 0$, it follows that

$\frac{1}{p_W} = \fint_W \frac{1}{p(x)}\,dx = \frac{1}{2}$. Since $T_Q f \leqslant 2T_W f$, it suffices to show the bounded-ness of T_W. Due to Theorem 4.5.7 is suffices to show $\|\chi_W\|_{p(\cdot)} \leqslant c\,|W|^{\frac{1}{p_W}}$ and $\|\chi_W\|_{p'(\cdot)} \leqslant c\,|W|^{\frac{1}{p'_W}}$ with constants independent of W. Fix $k \in \mathbb{N}$ such that $\exp(k-1) \leqslant \operatorname{diam}(W) \leqslant \exp(k)$. Then by definition of x_k fewer than $\sqrt{k}+1$ of the cubes $Q_j = B(x_j, 2)$ intersect W. So $U := W \cap \bigcup_{j=1}^\infty Q_j$ satisfies $|U| \leqslant 2\,(\sqrt{k}+1)$. We estimate

$$\int_W |W|^{-\frac{p(x)}{p_W}}\,dx = \int_U |W|^{-\frac{p(x)}{2}}\,dx + \int_{W \setminus U} |W|^{-1}\,dx$$

$$\leqslant |U|\,|W|^{-\frac{p^-}{2}} + 1$$

$$\leqslant 2\,(\sqrt{k}+1)\,|\exp(k)|^{-\frac{3}{4}} + 1$$

$$\leqslant c.$$

This gives $\|\chi_W\|_{p(\cdot)} \leqslant c\,|W|^{\frac{1}{p_W}}$. The proof of $\|\chi_W\|_{p'(\cdot)} \leqslant c\,|W|^{\frac{1}{p'_W}}$ is analo-gous. This finishes the argument for large cubes. $\qquad\square$

Remark 5.3.5. Kopaliani gave in [256] a two-dimensional example of an exponent $p \in \mathcal{A} \setminus \mathcal{A}_{\mathrm{loc}}$ of the form $p(x, y) = q(x)$ with $q \in \mathcal{P}^{\log}(\mathbb{R})$.

Kopaliani [255] and Lerner [268] showed that if $p \in \mathcal{A}_{\mathrm{loc}}$ is such that the support of $p - p_\infty$ has finite Lebesgue measure, then $p \in \mathcal{A}$.

5.4 Class $\mathcal{A}_\infty{}^*$

In this section we define an analogy of the classical Muckenhoupt class $\mathcal{A}_\infty$. We show that, as in the case of classical Muckenhoupt class $\mathcal{A}_\infty$, our new condition $\mathcal{A}_\infty$ implies an improvement of integrability, i.e. we prove that $\varphi \in \mathcal{A}_\infty$ implies $\psi \in \mathcal{A}_\infty$ with $\varphi(\cdot, t) \approx \psi(\cdot, t^s)$ for some $s > 1$. This is the analogue of the reverse Hölder's inequality for (classical) Muckenhoupt weights.

Definition 5.4.1. By $\mathcal{A}_\infty$ we denote the set of all generalized Φ-functions on $\mathbb{R}^n$ which have the following property: For every $0 < \alpha < 1$ there exists $0 < \beta < 1$ such that if $N \subset \mathbb{R}^n$ (measurable) and $\mathcal{Q} \in \mathcal{Y}_1^n$ satisfy

$$|N \cap Q| \leqslant \alpha|Q| \qquad \text{for all } Q \in \mathcal{Q}, \tag{5.4.2}$$

then

$$\left\| \sum_{Q \in \mathcal{Q}} t_Q\,\chi_{N \cap Q} \right\|_\varphi \leqslant \beta \left\| \sum_{Q \in \mathcal{Q}} t_Q\,\chi_Q \right\|_\varphi \tag{5.4.3}$$

for any sequence $\{t_Q\}_{Q\in\mathcal{Q}} \in \mathbb{R}^\mathcal{Q}$. The smallest constant β for $\alpha = \frac{1}{2}$ is called the $\mathcal{A}_\infty$-constant of φ.

We show below in Lemma 5.4.4 that the class $\mathcal{A}_\infty$ coincides with the Muckenhoupt class A_∞ of the classical weighted Lebesgue spaces. Let us recall one of the equivalent characterizations of the Muckenhoupt class A_∞: $\omega \in A_\infty$ if and only if for every $\varepsilon > 0$ there exists $\delta > 0$ such that $\omega(N) < \delta\,\omega(Q)$ for all cubes Q and all $N \subset Q$ with $|N| < \varepsilon\,|Q|$. (Here we used the notation $\omega(N) = \int_N \omega(x)\,dx$.)

Lemma 5.4.4. *Let ω be a weight on $\mathbb{R}^n$, $1 \leqslant q < \infty$, and $\varphi(x,t) = t^q\,\omega(x)$ for every $x \in \mathbb{R}^n$ and $t \geqslant 0$. Then $\varphi \in \mathcal{A}_\infty$ if and only if $\omega \in A_\infty$.*

Proof. Assume that $\varphi \in \mathcal{A}_\infty$ and let $\varepsilon > 0$. Further, let the cube Q and $N \subset Q$ be such that $|N| < \varepsilon|Q|$. Since $\varphi \in \mathcal{A}_\infty$ there exists $\beta > 0$ (only depending on ε) such that $\|\chi_{N\cap Q}\|_\varphi \leqslant \beta\,\|\chi_Q\|_\varphi$. Since $N \cap Q = N$, we obtain that

$$\omega(N) = \|\chi_N\|_\varphi^q \leqslant \beta^q\,\|\chi_Q\|_\varphi^q = \beta^q\,\omega(Q).$$

The choice $\delta := \beta^q$ in the definition of A_∞ shows that $\omega \in A_\infty$.

Assume now that $\omega \in A_\infty$ and $\alpha > 0$. Let $\mathcal{Q} \in \mathcal{Y}_1^n$ and $N \subset \mathbb{R}^n$ be such that (5.4.2) is satisfied. Further, let $\{t_Q\}_{Q\in\mathcal{Q}} \in \mathbb{R}^\mathcal{Q}$. Since $\omega \in A_\infty$, there exists $\delta > 0$ (only depending on α) such that for every cube Q the inequality $|N \cap Q| < \alpha\,|Q|$ implies $\omega(N \cap Q) < \delta\,\omega(Q)$. Hence,

$$\left\|\sum_{Q\in\mathcal{Q}} t_Q\,\chi_{N\cap Q}\right\|_\varphi^q = \sum_{Q\in\mathcal{Q}} |t_Q|^q\,\omega(N \cap Q) \leqslant \delta \sum_{Q\in\mathcal{Q}} |t_Q|^q\,\omega(Q) = \delta\left\|\sum_{Q\in\mathcal{Q}} t_Q\,\chi_Q\right\|_\varphi^q.$$

So the choice $\beta = \delta^{\frac{1}{q}}$ gives $\varphi \in \mathcal{A}_\infty$. $\square$

We would like to mention that the original definition of $\mathcal{A}_\infty$ from [93], denoted below by $\mathcal{A}'_\infty$, slightly differs from Definition 5.4.1 above: (5.4.2) is reversed to (5.4.6) and the condition for the Φ-function is modified accordingly. However, if φ satisfies the Δ_2-condition, then both definitions agree.

Definition 5.4.5. By $\mathcal{A}'_\infty$ we denote the set of all generalized Φ-functions on $\mathbb{R}^n$, which have the following property: For every $0 < \alpha_2 < 1$ there exists $0 < \beta_2 < 1$ such that if $P \subset \mathbb{R}^n$ (measurable) and $\mathcal{Q} \in \mathcal{Y}_1^n$ satisfy

$$|P \cap Q| \geqslant \alpha_2|Q| \qquad \text{for all } Q \in \mathcal{Q}, \tag{5.4.6}$$

then

$$\left\|\sum_{Q \in \mathcal{Q}} t_Q \chi_{P \cap Q}\right\|_\varphi \geqslant \beta_2 \left\|\sum_{Q \in \mathcal{Q}} t_Q \chi_Q\right\|_\varphi \qquad (5.4.7)$$

for any sequence $\{t_Q\}_{Q \in \mathcal{Q}} \in \mathbb{R}^{\mathcal{Q}}$.

The relation between $\mathcal{A}_\infty$ and $\mathcal{A}'_\infty$ is the following:

Lemma 5.4.8. *Let* $\varphi \in \Phi(\mathbb{R}^n)$.

(a) *If* $\varphi \in \mathcal{A}_\infty$, *then* $\varphi \in \mathcal{A}'_\infty$. *Moreover,* $\beta_2(\alpha_2)$ *in Definition 5.4.5 only depends on* $\beta(\alpha)$.

(b) *If* $\varphi \in \mathcal{A}'_\infty$ *and* φ *satisfies the* Δ_2*-condition, then* $\varphi \in \mathcal{A}_\infty$. *Moreover,* $\beta(\alpha)$ *in Definition 5.4.1 only depends on* $\beta_2(\alpha_2)$ *and the* Δ_2*-constant of* φ.

Proof. (a): Let $\alpha' \in (0,1)$ and let P, $\mathcal{Q}$ satisfy (5.4.6). Define $N := \bigcup_{Q \in \mathcal{Q}} (Q \setminus P)$. Then N and $\mathcal{Q}$ satisfy (5.4.2). Let $\alpha := \alpha'$ and let β be the constant from the definition of $\mathcal{A}_\infty$. Then

$$\left\|\sum_{Q \in \mathcal{Q}} t_Q \chi_Q\right\|_\varphi \leqslant \left\|\sum_{Q \in \mathcal{Q}} t_Q \chi_{P \cap Q}\right\|_\varphi + \left\|\sum_{Q \in \mathcal{Q}} t_Q \chi_{N \cap Q}\right\|_\varphi$$

$$\leqslant \left\|\sum_{Q \in \mathcal{Q}} t_Q \chi_{P \cap Q}\right\|_\varphi + \beta \left\|\sum_{Q \in \mathcal{Q}} t_Q \chi_Q\right\|_\varphi.$$

This proves (5.4.7) with $\beta_2 := (1 - \beta)$.

(b): Let K be the Δ_2-constant of φ. Let $\alpha \in (0,1)$ and let N, $\mathcal{Q}$ satisfy (5.4.2). Define $P := \bigcup_{Q \in \mathcal{Q}}(Q \setminus N)$. Then P, $\mathcal{Q}$ satisfy (5.4.6). Let $\alpha' := \alpha$. Then $\varphi \in \mathcal{A}'_\infty$ implies the existence of $\beta' \in (0,1)$ (only depending on α') such that (5.4.7) holds. Without loss of generality we can assume $\left\|\sum_{Q \in \mathcal{Q}} t_Q \chi_Q\right\|_\varphi = 1$. So $\varrho_\varphi(\sum_{Q \in \mathcal{Q}} t_Q \chi_Q) = 1$ by the unit ball property. Then (5.4.7) implies that

$$\left\|\sum_{Q \in \mathcal{Q}} t_Q \chi_{P \cap Q}\right\|_\varphi \geqslant \beta_2.$$

It follows by Lemma 2.4.2 that there exists $\beta_3 \in (0,1)$ (only depending on β_2 and K) such that

$$\varrho_\varphi\left(\sum_{Q \in \mathcal{Q}} t_Q \chi_{P \cap Q}\right) \geqslant \beta_3.$$

This and $\varrho_\varphi(\sum_{Q \in \mathcal{Q}} t_Q \chi_Q) = 1$ imply that

$$\varrho_\varphi\left(\sum_{Q\in\mathcal{Q}} t_Q\,\chi_{N\cap Q}\right) = \varrho_\varphi\left(\sum_{Q\in\mathcal{Q}} t_Q\,\chi_Q\right) - \varrho_\varphi\left(\sum_{Q\in\mathcal{Q}} t_Q\,\chi_{P\cap Q}\right) \leqslant 1 - \beta_3.$$

If follows by Lemma 2.4.3 that there exists $\beta_4 \in (0,1)$ which only depends on β_3 and K such that

$$\left\|\sum_{Q\in\mathcal{Q}} t_Q\,\chi_{N\cap Q}\right\|_\varphi \leqslant \beta_4.$$

This proves (5.4.3) with $\beta := \beta_4$.   □

It is important for us that class $\mathcal{A}_\infty$ and $\mathcal{A}'_\infty$ are weaker than $\mathcal{A}$.

Lemma 5.4.9. *Let* $\varphi \in \Phi(\mathbb{R}^n)$. *If* $\varphi \in \mathcal{A}$, *then* $\varphi \in \mathcal{A}'_\infty$. *Moreover,* $\beta_2(\alpha_2)$ *in Definition 5.4.5 only depends on the $\mathcal{A}$-constant of* φ.

Proof. Let K be the $\mathcal{A}$-constant of φ. Let α_2, $\mathcal{Q}$ and P as in (5.4.6). Let $f := \sum_{Q\in\mathcal{Q}} s_Q\,\chi_{P\cap Q}$ with $\|f\|_\varphi < \infty$. Then

$$\alpha_2 \sum_{Q\in\mathcal{Q}} s_Q\,\chi_Q \leqslant \sum_{Q\in\mathcal{Q}} s_Q\,\frac{|P\cap Q|}{|Q|}\,\chi_Q = \sum_{Q\in\mathcal{Q}} (M_Q f)\,\chi_Q = T_Q f.$$

Since φ is of class $\mathcal{A}$,

$$\alpha_2 \left\|\sum_{Q\in\mathcal{Q}} s_Q\,\chi_Q\right\|_\varphi \leqslant \|T_Q f\|_\varphi \leqslant K\,\|f\|_\varphi = K \left\|\sum_{Q\in\mathcal{Q}} s_Q\,\chi_{P\cap Q}\right\|_\varphi.$$

This is (5.4.7) with $\beta_2 := \alpha_2/K$. Thus φ is of class $\mathcal{A}'_\infty$.   □

In view of Lemma 5.4.8 this has the following direct consequence:

Corollary 5.4.10. *Let* $\varphi \in \Phi(\mathbb{R}^n)$ *satisfy the Δ_2-condition. If* $\varphi \in \mathcal{A}$, *then* $\varphi \in \mathcal{A}_\infty$. *Moreover,* $\beta(\alpha)$ *in Definition 5.4.5 and the $\mathcal{A}_\infty$-constant only depend on the $\mathcal{A}$-constant and the Δ_2-constant of* φ.

For the proof of the next lemma it is convenient to work with dyadic cubes.

Definition 5.4.11. We say that the cube Q is *dyadic* if there exists $\overline{k} = (k_1,\ldots,k_d) \in \mathbb{Z}^n$ and $z \in \mathbb{Z}$ such that $Q = 2^z\big((0,1)^n + \overline{k}\big)$. Let Q_0 be a cube and let $\tau : \mathbb{R}^n \to \mathbb{R}^n$ be the affine mapping $\tau(x) = r\,x + x_0$, $r > 0$, $x_0 \in \mathbb{R}^n$ that maps Q_0 onto the unit cube $(0,1)^n$. We say that the cube Q is *Q_0-dyadic*, if $\tau(Q)$ is dyadic. For $q \geqslant 1$ we define the *Q-dyadic maximal function $M_q^{\Delta,Q}$* by

$$M_q^{\Delta,Q}f(x) := \sup_{\substack{Q'\ni x \\ \text{and } Q' \text{ is } Q\text{-dyadic}}} M_{Q',q}f.$$

In the special case $q = 1$ we define $M^{\Delta,Q}f := M_1^{\Delta,Q}f$. Moreover, the $(0,1)^n$-dyadic maximal functions will simply be denoted M_q^Δ and M^Δ.

Note that $M^{\Delta,Q}$ has the same properties as the usual dyadic maximal function. Let $\Omega \subset \mathbb{R}^n$ be an open set. Then $Q_1 \subset \Omega$ is called a maximal Q-dyadic cube of Ω if and only if Q_1 is Q-dyadic and there exists no Q-dyadic cube Q_2 with $Q_1 \subsetneqq Q_2 \subset \Omega$. If $Q = (0,1)^n$ we just speak of a maximal dyadic cube of Ω. Note that every maximal Q-dyadic cube Q_1 of the set $\{M^{\Delta,Q}f > \lambda\}$, with $f \in L^1_{\text{loc}}(\mathbb{R}^n)$ and $\lambda > 0$, satisfies $\lambda < M_{Q_1}f \leqslant 2^n\lambda$.

Lemma 5.4.12. *Let $\varphi \in \Phi(\mathbb{R}^n)$ with $\varphi \in \mathcal{A}_\infty$. Then there exists $\delta > 0$ and $K \geqslant 1$ which only depend on the $\mathcal{A}_\infty$-constant of φ such that*

$$\left\| \sum_{Q\in\mathcal{Q}} t_Q \left| \frac{f}{M_Q f} \right|^\delta \chi_Q \right\|_\varphi \leqslant K \left\| \sum_{Q\in\mathcal{Q}} t_Q \chi_Q \right\|_\varphi$$

for all $\mathcal{Q} \in \mathcal{Y}_1^n$, all $\{t_Q\}_{Q\in\mathcal{Q}}$, $t_Q \geqslant 0$, and all $f \in L^1_{\text{loc}}(\mathbb{R}^n)$ with $M_Q f \neq 0$, $Q \in \mathcal{Q}$.

Proof. Let $\mathcal{Q} \in \mathcal{Y}_1^n$, $\{t_Q\}_{Q\in\mathcal{Q}}$ with $t_Q \geqslant 0$, and $f \in L^1_{\text{loc}}$ with $M_Q f \neq 0$ for all $Q \in \mathcal{Q}$. We will fix $\delta > 0$ and $K \geqslant 1$ later. For all $Q \in \mathcal{Q}$ we define $f_Q \in L^1_{\text{loc}}(\mathbb{R}^n)$ by $f_Q := f\chi_Q$. Since Q is Q-dyadic, f_Q is zero outside of Q and $M_Q f > 0$, we obtain

$$\{M^{\Delta,Q}f_Q > \tfrac{2}{3}M_Q f\} = Q.$$

Let

$$E_Q^k := \left\{ x \in \mathbb{R}^n \ : \ M^{\Delta,Q}f_Q(x) > \tfrac{2}{3}2^{(n+1)k}M_Q f \right\},$$

where $k \in \mathbb{N}_0$. Then

$$E_Q^{k+1} \subset E_Q^k \subset \cdots \subset E_Q^0 = Q.$$

Auxiliary claim: For every maximal Q-dyadic cube V of E_Q^{k-1}

$$|E_Q^k \cap V| \leqslant \tfrac{1}{2}|V|.$$

Proof of auxiliary claim. Let V be a maximal Q-dyadic cube of E_Q^{k-1} and let W be a maximal Q-dyadic cube of E_Q^k that intersects V. Since $E_Q^k \subset E_Q^{k-1}$, $W \subset V$ so that $W \subset E_Q^k \cap V$. Since W is maximal Q-dyadic in E_Q^k there holds (special property of the *dyadic* maximal function) $M_W f_Q > \frac{2}{3} 2^{(n+1)k} M_Q f$. This implies that

$$|W| M_Q f \leqslant \frac{3}{2} 2^{-(n+1)k} \int\limits_W |f_Q| \, dx.$$

Summing over all maximal Q-dyadic cubes W of E_Q^k that intersect V yields that

$$|E_Q^k \cap V| M_Q f \leqslant \frac{3}{2} 2^{-(n+1)k} \int\limits_V |f_Q| \, dx.$$

Since V is maximal Q-dyadic in E^{k-1}, $M_{2V} f_Q \leqslant \frac{2}{3} 2^{(n+1)(k-1)} M_Q f$, where $2V$ exceptionally denotes the dyadic cube containing V with twice the side-length of V. Thus

$$\int\limits_V |f_Q| \, dx \leqslant 2^n \, |V| \, M_{2V} f_Q \leqslant \frac{2}{3} 2^n \, 2^{(n+1)(k-1)} |V| \, M_Q f.$$

Using $M_Q f \neq 0$ we derive from our estimates

$$|E_Q^k \cap V| \leqslant \frac{1}{2} |V|.$$

This proves the auxiliary claim and we continue the original proof.   □

Let $\{V_{Q,l}^{k-1}\}_l$ be the collection of maximal Q-dyadic cubes of E_Q^{k-1}. Then

$$|E_Q^k \cap V_{Q,l}^{k-1}| \leqslant \frac{1}{2} |V_{Q,l}^{k-1}|.$$

Since $E_Q^{k-1} \subset Q$ and the family $\mathcal{Q}$ is pairwise disjoint, it follows that the collection $\{V_{Q,l}^{k-1}\}_{Q,l}$ is pairwise disjoint with respect to Q, l. Let

$$G^k := \bigcup_{Q \in \mathcal{Q}} E_Q^k,$$
$$\Omega^k := \bigcup_{Q,l} V_{Q,l}^{k-1}.$$

Then

$$|G^k \cap V_{Q,l}^{k-1}| = |E_Q^k \cap V_{Q,l}^{k-1}| \leqslant \frac{1}{2} |V_{Q,l}^{k-1}|.$$

Thus it follows from the definition of $\mathcal{A}_\infty$ that Ω^k to get

$$\left\|\sum_{Q\in\mathcal{Q}}\sum_l t_Q\,\chi_{G^k\cap V^{k-1}_{Q,l}}\right\|_\varphi \leqslant \beta\left\|\sum_{Q\in\mathcal{Q}}\sum_l t_Q\,\chi_{V^{k-1}_{Q,l}}\right\|_\varphi,$$

where $\beta\in(0,1)$ is the $\mathcal{A}_\infty$-constant of φ.

Since $\bigcup_l V^{k-1}_{Q,l}=E^{k-1}_Q$,

$$\left\|\sum_{Q\in\mathcal{Q}} t_Q\,\chi_{G^k\cap E^{k-1}_Q}\right\|_\varphi \leqslant \beta\left\|\sum_{Q\in\mathcal{Q}} t_Q\,\chi_{E^{k-1}_Q}\right\|_\varphi.$$

The definition of G^k and the monotonicity of E^k_Q imply $G^k\cap E^{k-1}_Q = E^k_Q\cap E^{k-1}_Q = E^k_Q$. Thus,

$$\left\|\sum_{Q\in\mathcal{Q}} t_Q\,\chi_{E^k_Q}\right\|_\varphi \leqslant \beta\left\|\sum_{Q\in\mathcal{Q}} t_Q\,\chi_{E^{k-1}_Q}\right\|_\varphi.$$

By induction

$$\left\|\sum_{Q\in\mathcal{Q}} t_Q\,\chi_{E^k_Q}\right\|_\varphi \leqslant \beta^k\left\|\sum_{Q\in\mathcal{Q}} t_Q\,\chi_{E^0_Q}\right\|_\varphi = \beta^k\left\|\sum_{Q\in\mathcal{Q}} t_Q\,\chi_Q\right\|_\varphi.$$

From this and the definition of E^k_Q we conclude that

$$\left\|\sum_{Q\in\mathcal{Q}} t_Q\left|\frac{f_Q}{M_Q f}\right|^\delta \chi_{E^k_Q\setminus E^{k+1}_Q}\right\|_\varphi \leqslant \left\|\sum_{Q\in\mathcal{Q}} t_Q\left(\frac{M^{\Delta,Q}f_Q}{M_Q f}\right)^\delta \chi_{E^k_Q\setminus E^{k+1}_Q}\right\|_\varphi$$

$$\leqslant \left\|\sum_{Q\in\mathcal{Q}} t_Q\left(\tfrac{2}{3}\,2^{(n+1)(k+1)}\right)^\delta \chi_{E^k_Q\setminus E^{k+1}_Q}\right\|_\varphi$$

$$\leqslant 2^{(n+1)(k+1)\delta}\left\|\sum_{Q\in\mathcal{Q}} t_Q\,\chi_{E^k_Q}\right\|_\varphi$$

$$\leqslant 2^{(n+1)(k+1)\delta}\beta^k\left\|\sum_{Q\in\mathcal{Q}} t_Q\,\chi_Q\right\|_\varphi.$$

We fix $\delta>0$ such that $\varepsilon:=2^{(n+1)\delta}\beta<1$ and $(n+1)\delta\leqslant 1$. In particular, $\delta>0$ only depends on β and the dimension n. Then

$$\left\|\sum_{Q\in\mathcal{Q}} t_Q\left|\frac{f_Q}{M_Q f}\right|^\delta \chi_{E^k_Q\setminus E^{k+1}_Q}\right\|_\varphi \leqslant 2\,\varepsilon^k\left\|\sum_{Q\in\mathcal{Q}} t_Q\,\chi_Q\right\|_\varphi.$$

This implies with the monotonicity of E_Q^k that

$$
\left\| \sum_{Q \in \mathcal{Q}} t_Q \left| \frac{f_Q}{M_Q f} \right|^\delta \chi_{E_Q^0} \right\|_\varphi = \left\| \sum_{k=0}^\infty \sum_{Q \in \mathcal{Q}} t_Q \left| \frac{f_Q}{M_Q f} \right|^\delta \chi_{E_Q^k \setminus E_Q^{k+1}} \right\|_\varphi
$$

$$
\leqslant \sum_{k=0}^\infty \left\| \sum_{Q \in \mathcal{Q}} t_Q \left| \frac{f_Q}{M_Q f} \right|^\delta \chi_{E_Q^k \setminus E_Q^{k+1}} \right\|_\varphi
$$

$$
\leqslant \sum_{k=0}^\infty 2 \, \varepsilon^k \left\| \sum_{Q \in \mathcal{Q}} t_Q \, \chi_Q \right\|_\varphi
$$

$$
= \frac{2}{1 - \varepsilon} \left\| \sum_{Q \in \mathcal{Q}} t_Q \, \chi_Q \right\|_\varphi .
$$

Since $E_Q^0 = Q$, this is the claim with $K = \frac{2}{1-\varepsilon}$.   □

With the help of this lemma we now derive improved properties for $\varphi \in \mathcal{A}$, which correspond to the left-openness of the classical Muckenhoupt classes. We need the s-averaging operators $T_{s,\mathcal{Q}}$ from Definition 4.4.3.

Theorem 5.4.13. *Let $\varphi \in \Phi(\mathbb{R}^n)$ be proper such that φ^* satisfy the Δ_2-condition. Then there exists $s > 1$ which only depends on the Δ_2-constant of φ^* and the $\mathcal{A}$-constant of φ, such that $T_{s,\mathcal{Q}}$ is uniformly bounded on L^φ with respect to $\mathcal{Q} \in \mathcal{Y}_1^n$. Moreover, the boundedness constant of $T_{s,\mathcal{Q}}$ only depends on the Δ_2-constant and the $\mathcal{A}$-constant of φ and the $\mathcal{A}_\infty$-constant of φ^*.*

Proof. Due to Lemma 5.2.2 and Corollary 5.4.10 it follows from $\varphi \in \mathcal{A}$ that $\varphi^* \in \mathcal{A}$ and $\varphi^* \in \mathcal{A}_\infty$, where the $\mathcal{A}$-constant and the $\mathcal{A}_\infty$-constant of φ^* are bounded in terms of the $\mathcal{A}$-constant and the Δ_2-constant of φ^*. Let $\delta > 0$ and $K > 0$ be as in Lemma 5.4.12 and define $s := 1 + \delta$.

Let $\mathcal{Q} \in \mathcal{Y}_1^n$ and $g \in L^\varphi(\mathbb{R}^n)$. We want to show $\|T_{s,\mathcal{Q}} g\|_\varphi \leqslant c \|g\|_\varphi$. Without loss of generality we can assume that $M_{s,Q} g \neq 0$ for all $Q \in \mathcal{Q}$. In particular, $M_Q g \neq 0$ for all $Q \in \mathcal{Q}$. We use the norm conjugate formula to estimate $\|T_{s,\mathcal{Q}} g\|_\varphi$, so let $h \in L^{\varphi^*}(\mathbb{R}^n)$. Then

$$
\langle T_{s,\mathcal{Q}} \, g, h \rangle \leqslant \left\langle \sum_{Q \in \mathcal{Q}} \chi_Q M_{s,Q} g, |h| \right\rangle \leqslant \left\langle \sum_{Q \in \mathcal{Q}} \chi_Q \frac{(M_{s,Q} g)^s}{(M_Q g)^{s-1}}, |h| \right\rangle
$$

$$
= \left\langle \sum_{Q \in \mathcal{Q}} \chi_Q \frac{M_Q(|g|^s)}{(M_Q g)^{s-1}}, |h| \right\rangle = \left\langle T_{\mathcal{Q}} \left(\sum_{Q \in \mathcal{Q}} \chi_Q \frac{|g|^s}{(M_Q g)^{s-1}} \right), |h| \right\rangle
$$

$$
= \left\langle \sum_{Q \in \mathcal{Q}} \chi_Q \frac{|g|^s}{(M_Q g)^{s-1}}, T_{\mathcal{Q}} |h| \right\rangle
$$

$$
= \left\langle \sum_{Q \in \mathcal{Q}} \chi_Q |g|, \sum_Q \chi_Q M_Q h \left(\frac{|g|}{M_Q |g|} \right)^{s-1} \right\rangle .
$$

By Hölder's inequality

$$\langle T_{s,Q}\, g, h\rangle \leqslant 2\, \|g\|_\varphi \left\| \sum_Q \chi_Q M_Q h \left(\frac{|g|}{M_Q g}\right)^{s-1} \right\|_{\varphi^*}.$$

With the help of Lemma 5.4.12 and $\varphi, \varphi^* \in \mathcal{A}$ it follows that

$$\langle T_{s,Q}\, g, h\rangle \leqslant 2K\, \|g\|_\varphi\, \|T_Q h\|_{\varphi^*} \leqslant 2K\, \|g\|_\varphi\, \|h\|_{\varphi^*}.$$

The norm conjugate formula concludes the proof.   □

Remark 5.4.14. Let ψ, φ be proper, generalized Φ-functions on $\mathbb{R}^n$ with $\psi(\cdot, t) \approx \varphi(\cdot, t^s)$ for some $s \geqslant 1$. Then it follows as in Lemma 3.2.6 that $\|f\|_\psi^s \approx \||f|^s\|_\varphi$.

The previous theorem is our first analogue to the left-openness of classical Muckenhoupt classes. Indeed, if we apply it to the function $\varphi(x, t) = t^q\, \omega(x)$ with $1 < q < \infty$ and $\omega \in A_q$, it would imply that $\omega \in A_{q/s}$. The following theorem is another version of the left-openness.

Theorem 5.4.15. *Let $\varphi \in \Phi(\mathbb{R}^n)$ be proper such that φ^* satisfies the Δ_2-condition. Suppose that $\psi \in \Phi(\mathbb{R}^n)$ is proper and $s \in (0, 1]$ such that $\psi(x, t) \approx \varphi(x, t^s)$ for all $x \in \mathbb{R}^n$ and $t > 0$. There exists $s_0 \in (0, 1)$ which only depends on the $\mathcal{A}$-constant of φ and the Δ_2-constant of φ^*, such that $\psi \in \mathcal{A}$ if $s \geqslant s_0$.*

Proof. Due to Theorem 5.4.13 there exists $s_0 \in (0, 1)$ such that $T_{s_0,Q}$ is uniformly bounded on L^φ with respect to $Q \in \mathcal{Y}_1^n$. Let $s \in [s_0, 1]$ and $\psi(\cdot, t) \approx \varphi(\cdot, t^s)$. Then $T_{s,Q}$ is also uniformly bounded on L^φ. By assumption on ψ and Remark 5.4.14 we have $\|f\|_\varphi^s \approx \||f|^s\|_\psi$ for every $f \in L^\varphi$. Hence,

$$\|T_Q g\|_\psi = \left\| \left(T_{s,Q}(|g|^{1/s})\right)^s \right\|_\psi \leqslant c \left\| T_{s,Q}(|g|^{1/s}) \right\|_\varphi^s \leqslant c \left\| |g|^{1/s} \right\|_\varphi^s \leqslant c\, \|g\|_\psi.$$

Hence, $\psi \in \mathcal{A}$.   □

Remark 5.4.16. Note that Theorem 5.4.15 is the counterpart to Lemmas 5.5.9 and 4.4.7. Here, the latter tell us about the easy situation, i.e. $s \geqslant 1$, while Theorem 5.4.15 considers the difficult part of the left-openness, i.e. $0 < s < 1$.

In the case of Lebesgue spaces with variable exponent with exponent $p^- > 1$ it is clear that for every $\varphi(x, t) = t^{p(x)}$ there exists a function ψ with $\psi(x, t) \approx \varphi(x, t^s)$ and $s \in (0, 1)$. Just take $\psi(x, t) = t^{q(x)}$ with $s \in (1/p^-, 1)$ and $q(x) := sp(x)$ for all x. In the general case the existence of such a ψ with $s \in (0, 1)$ is not obvious. However, if φ^* satisfy the Δ_2-condition, then the existence of such a function ψ follows from the results from [239, Lemmas 1.2.2 and 1.2.3].

If we apply Theorems 5.4.13 and 5.4.15 to the variable exponent Lebesgue spaces, we immediately get the following result.

Theorem 5.4.17. *Let $p \in \mathcal{P}(\mathbb{R}^n)$ with $1 < p^- \leqslant p^+ \leqslant \infty$ and $p \in \mathcal{A}$. Then there exists $s \in (1, p^-)$ such that $\varphi_{p(\cdot)/s} \in \mathcal{A}$ and the operators $T_{s,\mathcal{Q}}$ are uniformly bounded on $L^{p(\cdot)}(\mathbb{R}^n)$ with respect to $\mathcal{Q} \in \mathcal{Y}_1^n$.*

5.5   A Sufficient Condition for the Boundedness of M *

Let φ be a proper generalized Φ-function on $\mathbb{R}^n$ such that M is bounded from $L^\varphi(\mathbb{R}^n)$ to $L^\varphi(\mathbb{R}^n)$. Then $\varphi \in \mathcal{A}$ and it follows from Theorem 5.2.18 that $M_\mathcal{Q}\varphi \preceq (M_\mathcal{Q}\varphi^*)^*$. Therefore, $M_\mathcal{Q}\varphi \preceq (M_\mathcal{Q}\varphi^*)^*$ is necessary for the boundedness of M. In the following we define a new relation $\ll$ which is stronger than $\preceq$ and use it to state a sufficient condition for the boundedness of M.

Definition 5.5.1. Let $\varphi, \psi : \mathcal{X}^n \times \mathbb{R}_\geqslant \to \mathbb{R}_\geqslant$ be generalized Φ-functions on $\mathcal{Y}_1^n$. We say that ψ is *strongly dominated* by φ, $\psi \ll \varphi$, if for every $A_1 > 0$ there exist $A_2 > 0$ such that the following holds:

For all families $\mathcal{Q}_j \in \mathcal{Y}_1^n$, $j \in \mathbb{Z}$, with

$$\sum_{k=-\infty}^{\infty} \sum_{Q \in \mathcal{Q}_k} |Q| \, \varphi(Q, 2^k) \leqslant A_1 \tag{5.5.2}$$

there holds

$$\sum_{k=-\infty}^{\infty} \sum_{Q \in \mathcal{Q}_k} |Q| \, \psi(Q, 2^k) \leqslant A_2. \tag{5.5.3}$$

Remark 5.5.4. Let φ be a proper, generalized Φ-function on $\mathbb{R}^n$ that satisfies the Δ_2-condition. Then due to the Δ_2-condition it is possible to replace 2^k in Definition 5.5.1 by α^k for any $\alpha > 1$. Also due to the Δ_2-condition it suffices to verify Definition (5.5.1) for a single choice of $A_1 > 0$.

The next lemma shows that *strong domination* $\ll$ is stronger than *domination* $\preceq$, as the name indicates.

Lemma 5.5.5. *Let $\varphi, \psi : \mathcal{X}^n \times \mathbb{R}_\geqslant \to \mathbb{R}_\geqslant$ be generalized Φ-functions on $\mathcal{Y}_1^n$. If $\varphi \ll \psi$, then $\varphi \preceq \psi$.*

Proof. Let $A_1 := 1$ and choose $A_2 > 0$ such that (5.5.2) implies (5.5.3). Let $\mathcal{Q} \in \mathcal{Y}_1^n$ and $\bar{t} \in l^\varphi(\mathcal{Q})$ with $\|\bar{t}\|_{l^\varphi(\mathcal{Q})} \leqslant 1$, so that

$$\sum_{Q \in \mathcal{Q}} \varphi(Q, t_Q) \leqslant 1.$$

For every $Q \in \mathcal{Q}$ there exists $k_Q \in \mathbb{N}$ such that $2^{k_Q} \leqslant t_Q < 2^{k_Q+1}$. Define $\mathcal{Q}_k \in \mathcal{Y}_1^n$ for $k \in \mathbb{Z}$ by $\mathcal{Q}_k := \{Q \in \mathcal{Q} : k_Q = k\}$. Then

$$\sum_{k=-\infty}^{\infty} \sum_{Q \in \mathcal{Q}_k} \varphi(Q, 2^k) \leqslant \sum_{Q \in \mathcal{Q}} \varphi(Q, t_Q) \leqslant 1.$$

Let $m \in \mathbb{Z}$ such that $2^{m-1} < A_2 \leqslant 2^m$. Then

$$\sum_{k=-\infty}^{\infty} \sum_{Q \in \mathcal{Q}_k} \psi(Q, 2^{k-m}) \leqslant 2^{-m} \sum_{k=-\infty}^{\infty} \sum_{Q \in \mathcal{Q}_k} \psi(Q, 2^k) \leqslant 2^{-m} A_2 \leqslant 1.$$

This gives

$$\sum_{Q \in \mathcal{Q}} \psi(Q, 2^{-1-m} t_Q) \leqslant \sum_{k=-\infty}^{\infty} \sum_{Q \in \mathcal{Q}_k} \psi(Q, 2^{k-m}) \leqslant 1.$$

This proves $\|\bar{t}\|_{l^\varphi(\mathcal{Q})} \leqslant 2^{m+1} \leqslant 4A_2$. Therefore, $l^\varphi(\mathcal{Q}) \hookrightarrow l^\psi(\mathcal{Q})$ uniformly in $\mathcal{Q} \in \mathcal{Y}_1^n$. In other words $\varphi \preceq \psi$. $\qquad \square$

From Theorem 5.2.18 we know that $\varphi \in \mathcal{A}$ is equivalent to $M_Q \varphi \preceq (M_Q \varphi^*)^*$. Since by the previous Lemma $M_Q \varphi \ll (M_Q \varphi^*)^*$ is a stronger assumption, we can define a more restrictive class than $\mathcal{A}$, namely $\mathcal{A}_{\text{strong}}$.

Definition 5.5.6. By $\mathcal{A}_{\text{strong}}$ we denote the set of all generalized Φ-functions φ on $\mathbb{R}^n$ such that $M_Q \varphi \ll (M_Q \varphi^*)^*$.

In the context of Lebesgue spaces with variable exponents there is a close connection between $\mathcal{A}$ and $\mathcal{A}_{\text{strong}}$. Indeed, we show in Theorem 5.7.1 that $p \in \mathcal{A}$ is equivalent to $p \in \mathcal{A}_{\text{strong}}$ as long as $p \in \mathcal{P}(\mathbb{R}^n)$ with $1 < p^- \leqslant p^+ < \infty$.

Lemma 5.5.7. Let $\varphi \in \mathcal{A}$ be proper. For every $\lambda > 0$ and $f \in L^\varphi(\mathbb{R}^n)$ there exists $\mathcal{Q}_1, \ldots, \mathcal{Q}_{\xi_n} \in \mathcal{Y}_1^n$, where ξ_n only depends on the dimension n, such that

$$M_Q f > 2^{-n} \lambda \quad \text{for all } Q \in \mathcal{Q}_j, \ j = 1, \ldots, \xi_n,$$

and

$$\int \varphi(x, \lambda) \chi_{\{Mf > \lambda\}} \, dx \leqslant 2 \sum_{j=1}^{\xi_n} \sum_{Q \in \mathcal{Q}_j} |Q| \, (M_Q \varphi)(\lambda),$$

where c only depends on the Δ_2-constant of φ and the dimension n.

Proof. Let $f \in L^\varphi(\mathbb{R}^n)$ and $\lambda > 0$. According to Theorem 4.4.10, M is of weak type φ, since $\varphi \in \mathcal{A}$. In particular $\int_{\mathbb{R}^n} \varphi(x, \lambda) \chi_{\{Mf>\lambda\}} \, dx < \infty$. Due to the absolute continuity of the integral there exists a compact set $K \subset\subset \{Mf > \lambda\}$ such that

$$\int_{\mathbb{R}^n} \varphi(x, \lambda) \chi_{\{Mf>\lambda\}} \, dx \leqslant 2 \int_{\mathbb{R}^n} \varphi(x, \lambda) \chi_K \, dx. \tag{5.5.8}$$

Since K is compact, it is bounded. For every $x \in K$ we have $Mf(x) > \lambda$, so by (4.4.11), $M_{\text{center}} f(x) > 2^{-n}\lambda$. Thus, there exists a cube Q_x with center x such that $M_{Q_x} f > 2^{-n}\lambda$. The collection $\{Q_x\}_{x \in K}$ covers the compact set K. From the family $\{Q_x\}_{x \in K}$ we can select by the Besicovitch covering theorem, Theorem 1.4.6, locally 1-finite families $\mathcal{Q}_1, \ldots, \mathcal{Q}_{\xi_n}$, which together cover K. The natural number ξ_n only depends on the dimension n. We estimate

$$\int_{\mathbb{R}^n} \varphi(x, \lambda) \chi_K \, dx \leqslant \int_{\mathbb{R}^n} \varphi(x, \lambda) \sum_{j=1}^{\xi_n} \sum_{Q \in \mathcal{Q}_j} \chi_Q \, dx = \sum_{j=1}^{\xi_n} \sum_{Q \in \mathcal{Q}_j} |Q| \, (M_Q \varphi)(\lambda).$$

This and (5.5.8) prove the assertion. $\square$

Lemma 5.5.9. *Let* ψ, φ *be proper, generalized Φ-functions on* $\mathbb{R}^n$ *with* $\psi(\cdot, t) \approx \varphi(\cdot, t^s)$ *for some* $s \geqslant 1$. *Then*

$$(M_Q \psi)(t) \approx (M_Q \varphi)(t^s),$$
$$(M_Q \psi^*)^*(t) \geqslant c \, (M_Q \varphi^*)^*(t^s). \tag{5.5.10}$$

If $\psi \in \mathcal{A}$, *then* $\varphi \in \mathcal{A}$. *If* $\psi \in \mathcal{A}_{\text{strong}}$, *then* $\varphi \in \mathcal{A}_{\text{strong}}$.

Proof. The estimate $(M_Q \psi)(t) \approx (M_Q \varphi)(t^s)$ is an immediate consequence of $\psi(t) \approx \varphi(t^s)$. We estimate with Lemma 5.2.8.

$$(M_Q \varphi^*)^*(t^s) = \inf_{f : M_Q f \geqslant t^s} M_Q(\varphi(\cdot, f))$$
$$\geqslant c \inf_{f : M_Q f \geqslant t^s} M_Q(\psi(\cdot, |f|^{1/s}))$$
$$= c \inf_{g : M_{s,Q} g \geqslant t} M_Q(\psi(\cdot, g))$$
$$\geqslant c \inf_{f : M_Q g \geqslant t} M_Q(\psi(\cdot, g))$$
$$= c \, (M_Q \psi^*)^*(t).$$

The claims for $\varphi \in \mathcal{A}$ and $\varphi \in \mathcal{A}_{\text{strong}}$ follow from (5.5.10), Theorem 5.2.18, and Definition 5.5.1. $\square$

Corollary 5.5.11. *If $p \in \mathcal{A}_{\text{strong}}$ and $s \geqslant 1$, then $sp \in \mathcal{A}_{\text{strong}}$.*

We are now prepared to prove the main result of this section. The technique is similar to the real interpolation result of Marcinkiewicz (see for example [360]). However, due to the additional difficulties in the context of Musielak–Orlicz spaces we have to rely on the class $\mathcal{A}_{\text{strong}}$ rather than $\mathcal{A}$. Note that it would be sufficient to use $\mathcal{A}$ if we would stay in the context of weighted Lebesgue spaces.

Theorem 5.5.12. *Let ψ_0 and φ be proper, generalized N-functions on $\mathbb{R}^n$ such that ψ_0, φ, ψ_0^* and φ^* satisfy the Δ_2-condition and $\psi_0(x,t) \approx \varphi(x,t^{s_0})$ for all $x \in \mathbb{R}^n$ and $t \geqslant 0$ and for some $s_0 \in (0,1)$. If $\psi_0 \in \mathcal{A}_{\text{strong}}$, then M is bounded from $L^\varphi(\mathbb{R}^n)$ to $L^\varphi(\mathbb{R}^n)$.*

Proof. For $s_1 > 1$ define $\psi_1(\cdot,t) := \varphi(\cdot,t^{s_1})$. Then ψ_1 is an N-function and $\psi_1(\cdot,t) = \varphi(\cdot,t^{s_1}) \approx \psi_0(\cdot,t^{s_1/s_0})$. Moreover $\psi_j(\cdot,t) \approx \varphi(\cdot,t^{s_j})$ for $j = 0,1$. From $(M_Q\psi_0) \ll (M_Q\psi_0^*)^*$ and $\psi_1(\cdot,t) = \psi_0(\cdot,t^{s_1/s_0})$ we deduce by Lemma 5.5.9 that $(M_Q\psi_1) \ll (M_Q\psi_1^*)^*$. It suffices to show that there exists $A > 0$ such that for all $f \in L^\varphi(\mathbb{R}^n)$

$$\int_{\mathbb{R}^n} \varphi(x,f) \, dx \leqslant 1 \quad \Rightarrow \quad \int_{\mathbb{R}^n} \varphi(x,Mf) \, dx \leqslant A.$$

Let $f \in L^\varphi(\mathbb{R}^n)$ with $\int \varphi(x,f) \, dx \leqslant 1$. We estimate

$$\int_{\mathbb{R}^n} \varphi(x,Mf) \, dx = \int_0^\infty \int_{\mathbb{R}^n} \varphi'(x,\lambda) \, \chi_{\{Mf>\lambda\}} \, dx \, d\lambda$$

$$\leqslant \int_0^\infty \sum_{k=-\infty}^\infty \varphi(x,2^{k+2}) \, \chi_{\{Mf>2^{k+1}\}} \, dx.$$

For $k \in \mathbb{Z} > 0$ define $f_{0,k}, f_{1,k} : \mathbb{R}^n \to \mathbb{R}$ by

$$f_{0,k} := f \, \chi_{\{|f|>2^{k+1}\}},$$

$$f_{1,k} := f \, \chi_{\{|f|\leqslant 2^{k+1}\}}.$$

Then

$$\{Mf > 2^{k+1}\} \subset \{Mf_{0,k} > 2^k\} \cup \{Mf_{1,k} > 2^k\}.$$

Hence

$$\int_{\mathbb{R}^n} \varphi(x, Mf)\, dx \leqslant c \sum_{j=0}^{1} \sum_{k=-\infty}^{\infty} \int_{\mathbb{R}^n} \varphi(x, 2^{k+2})\, \chi_{\{Mf_{j,k} > 2^k\}}\, dx$$

$$\leqslant c \sum_{j=0}^{1} \sum_{k=-\infty}^{\infty} \int_{\mathbb{R}^n} \varphi(x, 2^k)\, \chi_{\{Mf_{j,k} > 2^k\}}\, dx,$$

where we used that φ satisfies the Δ_2-condition. Due to Lemma 5.5.7 there exist for each $k \in \mathbb{Z}$ and $j = 0, 1$ families $\mathcal{Q}_{j,k,1}, \ldots, \mathcal{Q}_{j,k,\xi_n}$, where ξ_n only depends on the dimension d, such that

$$M_Q f > 2^k \qquad \text{for all } Q \in \bigcup_{l=1}^{\xi_n} \mathcal{Q}_{j,k,l}$$

$$\int_{\mathbb{R}^n} \varphi(x, 2^k)\, \chi_{\{Mf_{j,k} > 2^k\}}\, dx \leqslant 2 \sum_{l=1}^{\xi_n} \sum_{Q \in \mathcal{Q}_{j,k,l}} |Q|(M_Q\varphi)(2^k).$$

Hence,

$$\int_{\mathbb{R}^n} \varphi(x, Mf)\, dx \leqslant c \sum_{j=0}^{1} \sum_{l=1}^{\xi_n} \sum_{k=-\infty}^{\infty} \sum_{Q \in \mathcal{Q}_{j,k,l}} |Q|\,(M_Q\varphi)(2^k)$$

$$\leqslant c \sum_{j=0}^{1} \sum_{l=1}^{\xi_n} \sum_{k=-\infty}^{\infty} \sum_{Q \in \mathcal{Q}_{j,k,l}} |Q|\,(M_Q\psi_j)(2^{k/s_j}). \tag{5.5.13}$$

We will show that for any $j = 0, 1$ and $l = 1, \ldots, \xi_n$

$$(I_{j,l}) := \sum_{k=-\infty}^{\infty} \sum_{Q \in \mathcal{Q}_{j,k,l}} |Q|\,(M_Q\psi_j^*)^*(2^{k/s_j}) \leqslant c, \tag{5.5.14}$$

where c does not depend on f, j, and l. Once we have proven this, $M_Q\psi_j \ll (M\psi_j^*)^*$ for $j = 0, 1$ and Remark 5.5.4 imply the boundedness of the right-hand side of (5.5.13). This concludes the proof of the theorem. It remains to prove (5.5.14).

The definition of $f_{j,k}$ and $0 < s_0 < 1 < s_1$ imply, for $j = 0, 1$ and $k \in \mathbb{Z}$, that

$$0 \leqslant |f_{j,k}|^{1-1/s_j}\, 2^{k(1/s_j-1)}\, \chi_{\{f_{j,k} \neq 0\}} \leqslant 1, \tag{5.5.15}$$

where we use the convention that the term in the middle is zero outside the set $\{f_{j,k} \neq 0\}$ regardless of whether the factor in front of it is undefined. This, the convexity of ψ_j and $\psi_j(\cdot, t) \approx \varphi(\cdot, t^{s_j})$ imply

$$\psi_j\left(|f_{j,k}|\,2^{k(1/s_j-1)}\right) \leqslant \psi_j\left(|f_{j,k}|^{1/s_j}\right)|f_{j,k}|^{1-1/s_j}\,2^{k(1/s_j-1)}\,\chi_{\{f_{j,k}\neq 0\}}$$

$$\leqslant c\,\varphi\left(\cdot,|f_{j,k}|\right)|f_{j,k}|^{1-1/s_j}\,2^{k(1/s_j-1)}\,\chi_{\{f_{j,k}\neq 0\}} \quad (5.5.16)$$

We use $M_Q f_{j,k} > 2^k$, Lemma 5.2.8, (5.5.16), and the definition of $f_{j,k}$ to estimate

$$(I_{j,l}) \leqslant c\sum_{k=-\infty}^{\infty}\sum_{Q\in\mathcal{Q}_{j,k,l}}|Q|\,(M_Q\psi_j^*)^*\left(2^{k(1/s_j-1)}\,M_Q f_{j,k}\right)$$

$$\leqslant c\sum_{k=-\infty}^{\infty}\sum_{Q\in\mathcal{Q}_{j,k,l}}|Q|\,M_Q\left(\psi_j\left(f_{j,k}\,2^{k(1/s_j-1)}\right)\right)$$

$$\leqslant c\sum_{k=-\infty}^{\infty}\int_{\mathbb{R}^n}\left(\psi_j\left(f_{j,k}\,2^{k(1/s_j-1)}\right)\right)dx$$

$$\leqslant c\sum_{k=-\infty}^{\infty}\int_{\mathbb{R}^n}\varphi(|f_{j,k}|)\,|f_{j,k}|^{1-1/s_j}\,2^{k(1/s_j-1)}\,\chi_{\{f_{j,k}\neq 0\}}\,dx$$

$$= c\int_{\mathbb{R}^n}\varphi(x,|f|)\,|f|^{1-1/s_j}\sum_{k=-\infty}^{\infty}2^{k(1/s_j-1)}\,\chi_{\{f_{j,k}=f,f\neq 0\}}\,dx.$$

By definition of $f_{j,k}$ and $0 < s_0 < 1 < s_1$,

$$\sum_{k=-\infty}^{\infty}2^{k(1/s_0-1)}\chi_{\{f_{0,k}=f,f\neq 0\}} = \sum_{k=-\infty}^{\infty}2^{k(1/s_0-1)}\chi_{\{|f|>2^{k+1}\}} \leqslant c\,|f|^{1/s_0-1},$$

$$\sum_{k=-\infty}^{\infty}2^{k(1/s_1-1)}\chi_{\{f_{1,k}=f,f\neq 0\}} = \sum_{k=-\infty}^{\infty}2^{k(1/s_1-1)}\chi_{\{0<|f|\leqslant 2^{k+1}\}} \leqslant c\,|f|^{1/s_1-1},$$

where c depends on s_0 and s_1. In combination with the previous estimate this gives

$$(I_{j,l}) \leqslant c\int_{\mathbb{R}^n}\psi_j\left(|f|^{1/s_j}\right)dx \leqslant c\int_{\mathbb{R}^n}\varphi(x,|f|)\,dx.$$

This proves the theorem. $\square$

If we apply Theorem 5.5.12 to the variable exponent Lebesgue spaces, we immediately get the following result.

Corollary 5.5.17. *Let $p \in \mathcal{P}(\mathbb{R}^n)$ with $1 < p^- \leqslant p^+ < \infty$. If there exists $s \in (1/p^-,1)$ such that $sp \in \mathcal{A}_{\text{strong}}$, then M is bounded from $L^{p(\cdot)}(\mathbb{R}^n)$ to $L^{p(\cdot)}(\mathbb{R}^n)$.*

If $\varphi \in \Phi(\mathbb{R}^n)$ is proper such that M is bounded on $L^\varphi(\mathbb{R}^n)$, then obviously $\varphi \in \mathcal{A}$. The following theorem extends this result to the stronger condition $\mathcal{A}_{\text{strong}}$.

Theorem 5.5.18. *Let φ be a proper, generalized N-function on $\mathbb{R}^n$ such that φ and φ^* satisfy the Δ_2-condition. If M is bounded from $L^\varphi(\mathbb{R}^n)$ to $L^\varphi(\mathbb{R}^n)$, then $\varphi \in \mathcal{A}_{\text{strong}}$.*

Proof. Let $\mathcal{Q}_j \in \mathcal{Y}_1^n$, $j \in \mathbb{Z}$, be such that

$$\sum_{k=-\infty}^{\infty} \sum_{Q \in \mathcal{Q}_j} |Q| \left(M_Q \varphi^*\right)^* (2^j) \leqslant 1.$$

For every $j \in \mathbb{Z}$ and $Q \in \mathcal{Q}_j$ we can choose due to Remark 5.2.9 a function $f_Q \in L^1(Q)$ such that $M_Q f_Q = 2^j$ and $(M_Q \varphi^*)^*(M_Q f_Q) = M_Q(\varphi(f_Q))$. Set

$$f := \varphi^{-1}\left(\cdot, \sum_{j=-\infty}^{\infty} \sum_{Q \in \mathcal{Q}_j} \chi_Q \, \varphi(\cdot, f_Q)\right).$$

Then

$$\int_{\mathbb{R}^n} \varphi(x, f) \, dx = \int_{\mathbb{R}^n} \sum_{j=-\infty}^{\infty} \sum_{Q \in \mathcal{Q}_j} \chi_Q \, \varphi(x, f_Q) \, dx$$

$$= \sum_{j=-\infty}^{\infty} \sum_{Q \in \mathcal{Q}_j} |Q| \, M_Q\big(\varphi(\cdot, f_Q)\big)$$

$$= \sum_{j=-\infty}^{\infty} \sum_{Q \in \mathcal{Q}_j} |Q| \, (M_Q \varphi^*)^*(2^j)$$

$$\leqslant 1.$$

The boundedness of M on L^φ implies $\int_{\mathbb{R}^n} \varphi(Mf) \, dx \leqslant c$. Then

$$M_Q f \geqslant M_Q\left(\varphi^{-1}\left(\cdot, \chi_Q \varphi(\cdot, f_Q)\right)\right) = M_Q f_Q = 2^j$$

every $j \in \mathbb{Z}$ and $Q \in \mathcal{Q}_j$. In particular $Q \subset \{Mf > 2^j\}$ and consequently $\bigcup_{Q \in \mathcal{Q}_j} Q \subset \{Mf > 2^j\}$. This implies that

$$\sum_{j=-\infty}^{\infty} \sum_{Q \in \mathcal{Q}_j} |Q| \, (M_Q \varphi)(2^j) \leqslant \sum_{j=-\infty}^{\infty} \int \varphi(2^{j+1}) \, \chi_{\{Mf > 2^j\}} \, dx$$

$$\leqslant c \int_{\mathbb{R}^n} \varphi(x, Mf) \, dx$$

$$\leqslant c.$$

This proves the assertion. $\square$

Remark 5.5.19. Let us consider the case of classical weighted Lebesgue spaces, i.e. $\varphi(x,t) = t^p \omega(x)$ with $1 \leqslant p < \infty$. Then $\varphi \in \mathcal{A}$ immediately implies $\varphi \in \mathcal{A}_{\text{strong}}$: Indeed, by Theorem 5.2.18 $M_Q\varphi \preceq (M_Q\varphi^*)^*$. For a cube Q let $t_{0,Q} := 1/\|\chi_Q\|_\varphi$; then it follows from Lemma 5.7.14 that

$$
(M_Q\varphi)(t) \approx \left(\frac{t}{t_{0,Q}}\right)^p (M_Q\varphi)(t_{0,Q}) \approx \left(\frac{t}{t_{0,Q}}\right)^p
$$
$$
\approx \left(\frac{t}{t_{0,Q}}\right)^p (M_Q\varphi^*)^*(t_{0,Q}) \approx (M_Q\varphi^*)^*(t)
$$

uniformly in $t > 0$ and Q. Thus (5.6.14) holds (for the choice $\varphi(Q, 2^k) := (M_Q\varphi)(2^k)$ and $\psi(Q, 2^k) := (M_Q\varphi^*)^*(2^k)$) with $b := 0$. Therefore, $M_Q\varphi \ll (M_Q\varphi^*)^*$ and $\varphi \in \mathcal{A}_{\text{strong}}$.

As a consequence we can deduce from our theorems the well known left-openness results for the Muckenhoupt classes A_p: Let $\omega \in \mathcal{A}_p$ with $1 < p < \infty$ and let $\varphi(x,t) := t^p \omega(x)$. Due to the remarks after equation (5.2.4) we know that $\varphi \in \mathcal{A}$ if and only if $\omega \in A_p$, so $\varphi \in \mathcal{A}$. Due to Theorem 5.4.15 and $p > 1$ we find $q \in (1, p)$ such that $\psi(x,t) := t^q \omega(x)$ satisfies $\psi \in \mathcal{A}$. This proves that $\omega \in A_q$ and proves the left-openness of the Muckenhoupt class A_p. By the considerations in the beginning of this remark we know that $\psi \in \mathcal{A}$ implies $\psi \in \mathcal{A}_{\text{strong}}$. So we can use Theorem 5.5.12 to conclude that M is bounded on $L^\varphi = L^p(\omega\,dx)$. Overall, we have shown that $\omega \in A_p$ with $p > 1$ implies that M is bounded on $L^p(\omega\,dx)$.

Remark 5.5.20. Let φ be a proper, generalized N-function on $\mathbb{R}^n$ such that φ and φ^* satisfy the Δ_2-condition. We know that $\varphi \in \mathcal{A}$ and $\varphi \in \mathcal{A}_{\text{strong}}$ (see Theorem 5.5.18) are both necessary for the boundedness of the Hardy–Littlewood maximal operator M from $L^\varphi(\mathbb{R}^n)$ to $L^\varphi(\mathbb{R}^n)$. However, it is an open problem if $\varphi \in \mathcal{A}$ or $\varphi \in \mathcal{A}_{\text{strong}}$ is in general sufficient for the boundedness of M. Theorems 5.4.13 and 5.4.13 provide the necessary left-openness result for $\mathcal{A}$. But in Theorem 5.5.12 we need the corresponding left-openness for $\mathcal{A}_{\text{strong}}$. In the case of weighted Lebesgue spaces, i.e. $\varphi(x,t) = \omega(x)\,t^q$, it is easily seen (Remark 5.5.19) that $\varphi \in \mathcal{A}$ if and only if $\varphi \in \mathcal{A}_{\text{strong}}$. In the case of Lebesgue spaces with variable exponents the situation is much more difficult. However, in Sect. 5.7 we will see that if $p \in \mathcal{A}$ and $p^- > 1$, then $\varphi \in \mathcal{A}_{\text{strong}}$. The proof for this result is elaborate. Due to these two fundamental examples we conjecture that $\varphi \in \mathcal{A}$ is sufficient for the boundedness of M. This of course is topic to further research.

5.6   Characterization of (Strong-)Domination*

In this section we characterize the property of domination and strong domination in a "pointwise" sense, i.e. for proper generalized Φ-functions φ, ψ on $\mathcal{X}^n \times \mathbb{R}_\geqslant$ with $\psi \preceq \varphi$ or $\psi \ll \varphi$ we estimate $\psi(Q,t)$ in terms of $\varphi(Q,t)$.

This is similar to the characterization of the embeddings in Theorem 2.8.1. We need this in Sect. 5.7 in order to show that domination is equivalent to strong domination in the context of Lebesgue spaces with variable exponents $L^{p(\cdot)}(\mathbb{R}^n)$ with $p^- > 1$. We begin with a general lemma.

Lemma 5.6.1. *Let X be an arbitrary set. Let Y be a subset of the power set of X such that $M_1 \subset M_2 \in Y$ implies $M_1 \in Y$. Let $\varphi, \psi : X \to \mathbb{R}_{\geqslant}$ and $A_1 > 0$ and $A_2, A_3 \geqslant 0$ be such that*

$$\sum_{\omega \in M} \varphi(\omega) \leqslant A_1 \qquad \Rightarrow \qquad \sum_{\omega \in M} \psi(\omega) \leqslant A_2 \sum_{\omega \in M} \varphi(\omega) + A_3 \qquad (5.6.2)$$

for all $M \in Y$. Then there exists $b : X \to \mathbb{R}_{\geqslant}$ such that

$$\varphi(\omega) \leqslant \frac{A_1}{4} \qquad \Rightarrow \qquad \psi(\omega) \leqslant \max\left\{\frac{4 A_3}{A_1}, 2 A_2\right\} \varphi(\omega) + b(\omega) \qquad (5.6.3)$$

for all $\omega \in X$, and

$$\sup_{M \in Y} \sum_{\omega \in M} b(\omega) \leqslant A_3. \qquad (5.6.4)$$

Proof. For $\omega \in X$, $\gamma, \delta > 0$ define

$$G(\omega, \gamma, \delta) := \begin{cases} \psi(\omega) - \dfrac{\gamma}{2} \varphi(\omega) & \text{if } \varphi(\omega) < \min\left\{\delta, \gamma^{-1} \psi(\omega)\right\}, \\ 0 & \text{otherwise.} \end{cases}$$

Then $G(\omega, \gamma, \delta) \geqslant 0$.
Claim 1: For all $\omega \in X$

$$\varphi(\omega) < \delta \qquad \Rightarrow \qquad \psi(\omega) \leqslant \gamma \varphi(\omega) + G(\omega, \gamma, \delta). \qquad (5.6.5)$$

Proof of Claim 1. We prove the claim by contradiction. Assume there exists $\omega \in X$ with $\varphi(\omega) \leqslant \delta$ and $\psi(\omega) > \gamma \varphi(\omega) + G(\omega, \gamma, \delta)$. Especially, $\psi(\omega) - \frac{\gamma}{2} \varphi(\omega) > G(\omega, \gamma, \delta)$. From this and the definition of $G(\omega, \gamma, \delta)$ we deduce $\varphi(\omega) \geqslant \min\{\delta, \gamma^{-1}\psi(\omega)\}$. Since $\varphi(\omega) < \delta$, this implies $\varphi(\omega) \geqslant \gamma^{-1}\psi(\omega)$. Thus $\psi(\omega) \leqslant \gamma \varphi(\omega) \leqslant \gamma \varphi(\omega) + G(\omega, \gamma, \delta)$ which contradicts the assumptions. ☐

Claim 2: Let $\delta_0 := A_1/4$, $\gamma_0 := \max\{4 A_3/A_1, 2 A_2\}$. Then

$$\sup_{M \in Y} \sum_{\omega \in M} G(\omega, \gamma_0, \delta_0) \leqslant A_3.$$

Proof of Claim 2. We prove the claim by contradiction, so assume that the claim does not hold. Then there exists $M_0 \in Y$ such that

$$\sum_{\omega \in M_0} G(\omega, \gamma_0, \delta_0) > A_3.$$

Therefore there exists $M_1 \subset M_0$ and $\omega_0 \in M_1$ such that

$$G(\omega, \gamma_0, \delta_0) > 0 \qquad \text{for all } \omega \in M_1,$$

$$\sum_{\omega \in M_1 \setminus \{\omega_0\}} G(\omega, \gamma_0, \delta_0) \leqslant A_3,$$

$$\sum_{\omega \in M_1} G(\omega, \gamma_0, \delta_0) > A_3. \tag{5.6.6}$$

Since $M_1 \subset M_0 \in Y$, $M_1 \in Y$. From the positivity of G we deduce that M_1 is at most countable and that

$$G(\omega, \gamma_0, \delta_0) = \psi(\omega) - \frac{\gamma_0}{2} \varphi(\omega) \qquad \text{for all } \omega \in M_1, \tag{5.6.7}$$

$$\varphi(\omega) < \min\{\delta_0, \gamma_0^{-1} \psi(\omega)\} \qquad \text{for all } \omega \in M_1.$$

This implies that

$$\sum_{\omega \in M_1} \varphi(\omega) \leqslant \delta_0 + \sum_{\omega \in M_1 \setminus \{\omega_0\}} \gamma_0^{-1} \psi(\omega)$$

$$= \delta_0 + \sum_{\omega \in M_1 \setminus \{\omega_0\}} \gamma_0^{-1} \left(G(\omega, \gamma_0, \delta_0) + \frac{\gamma_0}{2} \varphi(\omega) \right)$$

$$\leqslant \delta_0 + \gamma_0^{-1} A_3 + \frac{1}{2} \sum_{\omega \in M_1 \setminus \{\omega_0\}} \varphi(\omega).$$

Note that this inequality remains true if we replace M_1 by an arbitrary finite subset $M \subset M_1$. For all such sets the last term is finite and can be absorbed by the left-hand side. By exhausting M_1 by finite subsets we can pass back to M_1. We get

$$\sum_{\omega \in M_1} \varphi(\omega) \leqslant 2\delta_0 + 2\gamma_0^{-1} A_3 \leqslant A_1. \tag{5.6.8}$$

On the other hand (5.6.7), (5.6.6), and $\gamma_0 \geqslant 2A_2$ imply that

$$\sum_{\omega \in M_1} \psi(\omega) = \sum_{\omega \in M_1} \left(G(\omega, \gamma_0, \delta_0) + \frac{\gamma_0}{2} \varphi(\omega) \right) > A_3 + A_2 \sum_{\omega \in M_1} \varphi(\omega).$$

Now this and (5.6.8) contradict (5.6.2). This proves Claim 2. $\square$

We complete the proof by choosing $b(Q) := G(Q, \gamma_0, \delta_0)$. Then the claim follows from Claims 1 and 2 in view of the definition of G. $\square$

Suppose there exist $b : X \to \mathbb{R}_{\geqslant}$, $A_1 > 0$, and $A_2, A_3 \geqslant 0$ such that (5.6.3) and (5.6.4) hold. Then

$$\sum_{\omega \in M} \varphi(\omega) \leqslant \frac{A_1}{4} \quad \Rightarrow \quad \sum_{\omega \in M} \psi(\omega) \leqslant \max\left\{\frac{4\,A_3}{A_1}, 2\,A_2\right\} \sum_{\omega \in M} \varphi(\omega) + A_3$$

for all $M \in Y$.

Definition 5.6.9. For $b : \mathcal{X}^n \to \mathbb{R}_{\geqslant}$ we define

$$\|b\|_{\mathcal{Y}_1^n, 1} := \sup_{\mathcal{Q} \in \mathcal{Y}_1^n} \sum_{Q \in \mathcal{Q}} |Q|\, b(Q) \quad \text{and} \quad \|b\|_{\mathcal{Y}_1^n, \infty} := \sup_{Q \in \mathcal{Q}} b(Q).$$

Theorem 5.6.10. *Let φ, ψ be proper, generalized Φ-functions on $\mathcal{Y}_1^n$ such that $\varphi, \varphi^*, \psi, \psi^*$ satisfy the Δ_2-condition. Then $\varphi \preceq \psi$ if and only if there exists $b : \mathcal{X}^n \to \mathbb{R}_{\geqslant}$ with $\|b\|_{\mathcal{Y}_1^n, 1} \leqslant A_2$ such that*

$$|Q|\, \varphi(Q, t) \leqslant \frac{A_1}{4} \quad \Rightarrow \quad \psi(Q, t) \leqslant \frac{4\,A_2}{A_1} \varphi(Q, t) + b(Q) \qquad (5.6.11)$$

for all $Q \in \mathcal{X}^n$ and all $t \geqslant 0$.

Proof. Assume first that $\varphi \preceq \psi$. Let $X := \mathcal{X}^n$ and $Y := \mathcal{Y}_1^n$. Then X and Y are admissible for Lemma 5.6.1. For $u : \mathcal{X}^n \to \mathbb{R}_{\geqslant}$ define $\varphi_{[u]}, \psi_{[u]} : \mathcal{X}^n \to \mathbb{R}_{\geqslant}$ by

$$\varphi_{[u]}(Q) := |Q|\, \varphi\big(Q, u(Q)\big), \qquad \psi_{[u]}(Q) := |Q|\, \psi\big(Q, u(Q)\big).$$

Since $\varphi \preceq \psi$,

$$\sum_{Q \in \mathcal{Q}} \varphi_{[u]}(Q) \leqslant A_1 \quad \Rightarrow \quad \sum_{Q \in \mathcal{Q}} \psi_{[u]}(Q) \leqslant A_2 \qquad (5.6.12)$$

for all $\mathcal{Q} \in \mathcal{Y}_1^n$. Thus we can apply Lemma 5.6.1 to X, Y, and $\varphi_{[u]}, \psi_{[u]}$. Hence there exists $a_{[u]} : \mathcal{X}^n \to \mathbb{R}_{\geqslant}$ with $\|a_{[u]}\|_{\mathcal{Y}_1^n, 1} \leqslant A_2$ such that

$$\varphi_{[u]}(Q) \leqslant \frac{A_1}{4} \quad \Rightarrow \quad \psi_{[u]}(Q) \leqslant \frac{4\,A_2}{A_1} \varphi_{[u]}(Q) + a_{[u]}(Q)$$

for all $Q \in \mathcal{X}^n$.

Thus for all $Q \in \mathcal{X}^n$

$$|Q|\varphi\big(Q, u(Q)\big) \leqslant \frac{A_1}{4} \Rightarrow |Q|\psi\big(Q, u(Q)\big) \leqslant \frac{4\,A_2}{A_1} |Q|\varphi\big(Q, u(Q)\big) + a_{[u]}(Q).$$

Define $b : \mathcal{X}^n \times \mathbb{R}_{\geqslant} \to \mathbb{R}$ by

$$
b(Q,t) := \begin{cases} |Q|^{-1} \displaystyle\inf_{\substack{u : \mathcal{X}^n \to \mathbb{R}_{\geqslant} \\ \text{with } u(Q)=t}} a_{[u]}(Q) & \text{if } |Q|\,\varphi(Q,t) \leqslant \dfrac{A_1}{4}, \\ 0 & \text{otherwise.} \end{cases}
$$

Then for all $Q \in \mathcal{X}^n$ and all $t \geqslant 0$

$$
|Q|\,\varphi(Q,t) \leqslant \frac{A_1}{4} \qquad \Rightarrow \qquad \psi(Q,t) \leqslant \frac{4\,A_2}{A_1}\,\varphi(Q,t) + b(Q,t)
$$

and for all $\mathcal{Q} \in \mathcal{Y}_1^n$ and all sequences $\{t_Q\}_{Q \in \mathcal{Q}}$, $t_Q \geqslant 0$, holds

$$
\sup_{Q \in \mathcal{Q}} \sum_{Q \in \mathcal{Q}} |Q|\,b(Q,t_Q) \leqslant A_2.
$$

Finally, we define $b : \mathcal{X}^n \to \mathbb{R}_{\geqslant}$ by

$$
b(Q) := \sup_{t \geqslant 0} b(Q,t).
$$

Then the claim follows directly from the previous two estimates.

If on the other hand there exists $b : \mathcal{X}^n \to \mathbb{R}_{\geqslant}$ and $A_1, A_2 > 0$ such that $\|a_{[u]}\|_{\mathcal{Y}_1^n,1} \leqslant A_2$ and (5.6.11) hold, then

$$
\sum_{Q \in \mathcal{Q}} |Q|\,\varphi(Q,t_Q) \leqslant \frac{A_1}{4} \qquad \Rightarrow \qquad \sum_{Q \in \mathcal{Q}} |Q|\,\psi(Q,t_Q) \leqslant 2\,A_2,
$$

The strong Δ_2-condition for φ and ψ implies that $\varphi \preceq \psi$. $\qquad\square$

Theorem 5.6.13. *Let* $\varphi, \psi : \mathcal{X}^n \times \mathbb{R}_{\geqslant} \to \mathbb{R}_{\geqslant}$. *Then* $\psi \ll \varphi$ *if and only if there exists* $b : \mathcal{X}^n \times \{2^k : k \in \mathbb{Z}\} \to \mathbb{R}_{\geqslant}$ *such that*

$$
|Q|\,\varphi(Q,2^k) \leqslant \frac{A_1}{4} \quad \Rightarrow \quad \psi(Q,2^k) \leqslant \frac{4\,A_2}{A_1}\,\varphi(Q,2^k) + b(Q,2^k) \qquad (5.6.14)
$$

for all $Q \in \mathcal{X}^n$ *and* $k \in \mathbb{Z}$, *and*

$$
\sum_{k=-\infty}^{\infty} \sum_{Q \in \mathcal{Q}_k} |Q|\,b(Q,2^k) \leqslant A_2 \qquad (5.6.15)
$$

for all $\mathcal{Q}_k \in \mathcal{Y}_1^n$ *and all sequences* $\{t_Q\}_{Q \in \mathcal{Q}_k}$ *with* $t_Q \geqslant 0$.

Proof. Assume that $\psi \ll \varphi$, and let A_1 and A_2 be the constants from (5.5.2) and (5.5.3). We want to prove (5.6.14) and (5.6.15). Let $X := \mathcal{X}^n \times \{2^k : k \in \mathbb{Z}\}$ and define $\pi_k : X \to \mathcal{X}^n$ by

$$\pi_k(M) := \{Q \in \mathcal{X}^n : (k, Q) \in M\}.$$

Further let

$$Y := \{M \subset X : \pi_k(M) \in \mathcal{Y}_1^n \text{ for all } k \in \mathbb{Z}\}.$$

Then X, Y are admissible for Lemma 5.6.1 and so there exists $b : X \to \mathbb{R}_\geqslant$ which satisfies (5.6.14).

If on the other hand (5.6.14) holds, then summation of (5.6.14) over $k \in \mathbb{Z}$ and $Q \in \mathcal{Q}_k$ yield that

$$\sum_{k=-\infty}^{\infty} \sum_{Q \in \mathcal{Q}_k} |Q| \varphi(Q, 2^k) \leqslant \frac{A_1}{4} \quad \Rightarrow \quad \sum_{k=-\infty}^{\infty} \sum_{Q \in \mathcal{Q}_k} |Q| \psi(Q, 2^k) \leqslant 2A_2,$$

i.e. $\psi \ll \varphi$. $\square$

5.7   The Case of Lebesgue Spaces with Variable Exponents*

It is clear that for the boundedness of M on $L^{p(\cdot)}(\mathbb{R}^n)$ it is necessary that p is of class $\mathcal{A}$. We want to show in this section that $p \in \mathcal{A}$ is also sufficient for the boundedness of M on $L^{p(\cdot)}(\mathbb{R}^n)$ as long as $1 < p^- \leqslant p^+ < \infty$. With the results so far this is not directly possible and there is still a small gap to close. Let us point out what remains to show: Assume that $p \in \mathcal{A}$. Then due to Theorem 5.4.15 it is possible to choose $s \in (1/p^-, 1)$ such that $q \in \mathcal{A}$ with $q(x) := sp(x)$ for all $x \in \mathbb{R}^n$. If additionally $q \in \mathcal{A}_{\text{strong}}$, then we can use Theorem 5.5.12 to conclude that M is bounded on $L^{p(\cdot)}(\mathbb{R}^n)$. So we have $q \in \mathcal{A}$ but need $q \in \mathcal{A}_{\text{strong}}$. We close this gap by showing that $q \in \mathcal{A}$ implies $q \in \mathcal{A}_{\text{strong}}$. The proof of this result relies on the special structure of $L^{p(\cdot)}$.

Theorem 5.7.1. *Let $p \in \mathcal{P}(\mathbb{R}^n)$ with $1 < p^- \leqslant p^+ < \infty$. Then $p \in \mathcal{A}$ if and only if $p \in \mathcal{A}_{\text{strong}}$.*

Before we get to the proof of this theorem let us present the following important implication.

Theorem 5.7.2. *Let $p \in \mathcal{P}(\mathbb{R}^n)$ with $1 < p^- \leqslant p^+ < \infty$. Then the following are equivalent*

(a) $p \in \mathcal{A}$.
(b) $p' \in \mathcal{A}$.
(c) M is bounded on $L^{p(\cdot)}(\mathbb{R}^n)$.
(d) M_{s_1} is bounded on $L^{p(\cdot)}(\mathbb{R}^n)$ for some $s_1 > 1$ ("left-openness").
(e) M is bounded on $L^{sp(\cdot)}(\mathbb{R}^n)$ for some $s \in (1/p^-, 1)$ ("left-openness").
(f) M is bounded on $L^{p'(\cdot)}(\mathbb{R}^n)$.

Proof. (a) $\Leftrightarrow$ (b): It suffices to show the equivalence for $\widetilde{\varphi}_{p(\cdot)}$. Due to Lemma 3.1.3 we know that $(\widetilde{\varphi}_{p(\cdot)})^* = \widetilde{\varphi}_{p'(\cdot)}$, so the equivalence follows directly from Lemma 5.2.2.
(c) $\Rightarrow$ (a): This is obvious, since $T_Q f \leqslant Mf$.
(a) $\Rightarrow$ (c): Let $p \in \mathcal{A}$. Then by Theorem 5.4.17 there exists $s \in (1/p^-, 1)$ such that $q \in \mathcal{A}$ with $q(x) := sp(x)$ for all $x \in \mathbb{R}^n$. So Theorem 5.7.1 implies that $q \in \mathcal{A}_{\text{strong}}$, which, by Corollary 5.5.17, implies that M is bounded on $L^{p(\cdot)}(\mathbb{R}^n)$.
(a) $\Rightarrow$ (e): By the argument of the previous case, $q \in \mathcal{A}$. So the implication "(a) $\Leftrightarrow$ (c)" implies that M is bounded on $L^{q(\cdot)}(\mathbb{R}^n)$.
(d) $\Leftrightarrow$ (e): It suffices to prove this for $\varphi_{p(\cdot)} = \bar{\varphi}_{p(\cdot)}$. Let $s = 1/s_1$; then the claim follows from the identity

$$\|M_{s_1}f\|_{p(\cdot)} = \left\|(M(|f|^{s_1}))^{\frac{1}{s_1}}\right\|_{p(\cdot)} = \left\|M(|f|^{s_1})\right\|_{\frac{p(\cdot)}{s_1}}^{\frac{1}{s_1}} = \left\|M(|f|^{s_1})\right\|_{sp(\cdot)}^{\frac{1}{s_1}}.$$

(d) $\Rightarrow$ (c): This follows from $Mf \leqslant M_{s_1}f$, since $s_1 > 1$.
(b) $\Leftrightarrow$ (f): This follows from (a) $\Leftrightarrow$ (c) with p replaced by p'. $\square$

Remark 5.7.3. A careful tracking of the constants reveals that the operator norm of M in $L^{p(\cdot)}$ and $L^{p'(\cdot)}$ in the previous theorem only depends on n, p^-, p^+ and the $\mathcal{A}$-constant of p. Moreover, the operator norm of M on $L^{sp(\cdot)}$ and M_{s_1} on $L^{p(\cdot)}$ depend additionally on s_1 and s.

Before we get to the proof of Theorem 5.7.1 we need some auxiliary results.

Lemma 5.7.4. *Let $\varphi \in \mathcal{A}$ be proper and satisfy the Δ_2-condition. Then there exists $s > 1$ such that $(M_{s,Q}\varphi)(t) \preceq (M_Q\varphi)(t)$.*

Proof. Due to Corollary 5.4.10 we have $\varphi \in \mathcal{A}_\infty$. So by Lemma 5.4.12 there exists $\delta > 0$ and $A \geqslant 1$ such that

$$\left\|\sum_{Q \in \mathcal{Q}} t_Q \left|\frac{f}{M_Q f}\right|^\delta \chi_Q\right\|_\varphi \leqslant A \left\|\sum_{Q \in \mathcal{Q}} t_Q \chi_Q\right\|_\varphi \tag{5.7.5}$$

for all $\mathcal{Q} \in \mathcal{Y}_1^n$, all $\{t_Q\}_{Q \in \mathcal{Q}}$ with $t_Q \geqslant 0$, and all $f \in L^1_{\text{loc}}$ with $M_Q f \neq 0$, $Q \in \mathcal{Q}$. Define $s := 1 + \delta$. Let $\mathcal{Q} \in \mathcal{Y}_1^n$ and $\{u_Q\}_{Q \in \mathcal{Q}}$ with $u_Q > 0$ be such that

$$\sum_{Q \in \mathcal{Q}} |Q| \, (M_Q \varphi)(u_Q) \leqslant 1$$

so that

$$\varrho_\varphi \Big(\sum_{Q \in \mathcal{Q}} \chi_Q \, u_Q \Big) \leqslant 1 \quad \text{and} \quad \Big\| \sum_{Q \in \mathcal{Q}} \chi_Q \, u_Q \Big\|_\varphi \leqslant 1. \tag{5.7.6}$$

We have to show that

$$\sum_{Q \in \mathcal{Q}} |Q| \, (M_{s,Q} \varphi) \Big(\frac{u_Q}{A_2} \Big) \leqslant 1, \tag{5.7.7}$$

where $A_2 \geqslant 1$ does not depend on $\mathcal{Q}$ or $\{u_Q\}_{Q \in \mathcal{Q}}$. Define $f \in L^1_{\mathrm{loc}}(\mathbb{R}^n)$ by

$$f := \sum_{Q \in \mathcal{Q}} \chi_Q \, \varphi \Big(\cdot, \frac{u_Q}{2A} \Big),$$

so that $M_Q f \neq 0$ for all $Q \in \mathcal{Q}$. Now (5.7.5) implies that

$$\Big\| \sum_{Q \in \mathcal{Q}} u_Q \Big| \frac{f}{M_Q f} \Big|^\delta \chi_Q \Big\|_\varphi \leqslant A.$$

The unit ball property and the convexity of φ imply that

$$1 \geqslant \sum_{Q \in \mathcal{Q}} \int_Q \varphi \Big(\frac{u_Q}{A} \Big| \frac{f}{M_Q f} \Big|^\delta \Big) \, dx \geqslant 2 \sum_{Q \in \mathcal{Q}} \int_Q \varphi \Big(\frac{u_Q}{2A} \Big) \Big| \frac{f}{M_Q f} \Big|^\delta \chi_{\{|f| \geqslant M_Q f\}} \, dx.$$

On the other hand (5.7.6) implies that

$$1 \geqslant \sum_{Q \in \mathcal{Q}} \int_Q \varphi \Big(\frac{u_Q}{A} \Big) \, dx \geqslant 2 \sum_{Q \in \mathcal{Q}} \int_Q \varphi \Big(\frac{u_Q}{2A} \Big) \Big| \frac{f}{M_Q f} \Big|^\delta \chi_{\{|f| < M_Q f\}} \, dx.$$

Combining the two previous estimates we obtain, with $s = 1 + \delta$,

$$1 \geqslant \sum_{Q \in \mathcal{Q}} \int_Q \varphi \Big(\frac{u_Q}{2A} \Big) \Big| \frac{f}{M_Q f} \Big|^\delta \, dx$$

$$= \sum_{Q \in \mathcal{Q}} \int_Q \varphi \Big(\frac{u_Q}{2A} \Big)^{1+\delta} dx \, \Big((M_Q \varphi) \Big(\frac{u_Q}{2A} \Big) \Big)^{-\delta}$$

$$= \sum_{Q \in \mathcal{Q}} |Q| \left((M_{s,Q}\varphi)\left(\frac{u_Q}{2A}\right) \right)^{1+\delta} \left((M_Q\varphi)\left(\frac{u_Q}{2A}\right) \right)^{-\delta}$$

$$\geqslant \sum_{Q \in \mathcal{Q}} |Q| (M_{s,Q}\varphi)\left(\frac{u_Q}{2A}\right).$$

This proves (5.7.7), and concludes the proof. $\square$

Remark 5.7.8. Lemma 5.7.4 in particular shows that $M_{s,Q}\varphi$ is a generalized Φ-function on $\mathcal{Q}$ for every $\mathcal{Q} \in \mathcal{Y}_1^n$. Indeed, the convexity and the left-continuity of $M_{s,Q}\varphi$ follow as in Corollary 5.2.12. Now, $\lim_{t\to\infty}(M_{s,Q}\varphi))(t) = \infty$ follows from $M_{s,Q}\varphi(t) \geqslant M_Q\varphi(t)$ and the corresponding limit for $M_Q\varphi(t)$. The limit $\lim_{t\to 0}(M_{s,Q}\varphi)(t) = 0$, however, follows from the estimate $M_{s,Q}\varphi \preceq M_Q\varphi$ of Lemma 5.7.4.

Note that Lemma 5.7.4 is not restricted to the case of Lebesgue spaces with variable exponents. The lemma provides a kind of reverse Hölder estimate. In the case of weighted (classical) Lebesgue spaces, i.e. $\varphi(x,t) = t^q \omega(x)$, it matches exactly the reverse Hölder estimate for Muckenhoupt weights $\omega \in A_q$. Let us summarize our results so far.

Lemma 5.7.9. *Let φ be a proper, generalized N-function on $\mathbb{R}^n$ such that φ and φ^* satisfy the Δ_2-condition. Then the following conditions are equivalent:*

(a) *$\varphi \in \mathcal{A}$.*
(b) *$M_Q\varphi \preceq (M_Q\varphi^*)^*$.*
(c) *There exists $s > 1$ such that $M_{s,Q}\varphi \preceq M_Q\varphi \preceq (M_Q\varphi^*)^* \preceq (M_{s,Q}\varphi^*)^*$.*

Proof. The implication (a) $\Leftrightarrow$ (b) follows from Theorem 5.2.18 while (c) $\Rightarrow$ (b) is obvious. To show (a), (b) $\Rightarrow$ (c), let $\varphi \in \mathcal{A}$ or equivalently $M_Q\varphi \preceq (M_Q\varphi^*)^*$. Then by Lemma 5.2.2 also $\varphi^* \in \mathcal{A}$. Thus Lemma 5.4.9 implies that φ and φ^* are of class $\mathcal{A}_\infty$. Hence, by Lemma 5.7.4 there exists $s > 1$ such that $M_{s,Q}\varphi \preceq M_Q\varphi$ and $M_{s,Q}\varphi^* \preceq M_Q\varphi^*$. From Remark 5.2.19 it follows that $(M_Q\varphi^*)^* \preceq (M_{s,Q}\varphi^*)^*$, which completes the proof. $\square$

For the proof of Theorem 5.7.1 we need more auxiliary results.

Lemma 5.7.10. *Let $\varphi \in \Phi(\mathbb{R}^n)$ be proper and satisfy the Δ_2-condition. Then $M_{s,Q}\varphi$ and $(M_{s,Q}\varphi^*)^*$ satisfy the Δ_2-condition with the same Δ_2-constant as φ.*

Proof. Let K denote the Δ_2-constant of φ. The estimate $\varphi(2t) \leqslant K\varphi(t)$ immediately implies $(M_{s,Q}\varphi)(2t) \leqslant K(M_{s,Q}\varphi)(t)$.

Due to Lemma 2.6.4 the estimate $\varphi(x, 2t) \leqslant K\varphi(x,t)$ for all $x \in \mathbb{R}^n$ and $t \geqslant 0$ is equivalent to $\varphi^*(x, t/2) \geqslant K\varphi^*(x, t/K)$. This implies $(M_{s,Q}\varphi^*)(t/2) \geqslant K(M_{s,Q}\varphi^*)(t/K)$, which is by Lemma 2.6.4 equivalent to $(M_{s,Q}\varphi^*)^*(2t) \leqslant K (M_{s,Q}\varphi^*)^*(t)$. $\square$

Lemma 5.7.11. *Let $\psi_j : \mathcal{X}^n \times \mathbb{R}_{\geqslant} \to \mathbb{R}_{\geqslant}$, $j = 1, 2$ be proper, generalized N-functions with $\psi_1 \cong \psi_2$. Furthermore, let ψ_1, ψ_2, ψ_1^* and ψ_2^* satisfy the Δ_2-condition. Then for all $d_1, D_1 > 0$ there exist $d_2, D_2 > 0$ such that the following holds: If $Q \in \mathcal{Y}_1^n$ and $\{t_Q\}_{Q \in \mathcal{Q}}$ with $t_Q \geqslant 0$ satisfy*

$$d_1 \leqslant \sum_{Q \in \mathcal{Q}} |Q| \, \psi_1(t_Q) \leqslant D_1 \qquad (5.7.12)$$

then

$$d_2 \leqslant \sum_{Q \in \mathcal{Q}} |Q| \, \psi_2(t_Q) \leqslant D_2. \qquad (5.7.13)$$

Proof. Since ψ_1 and ψ_2 satisfy the Δ_2-condition, it suffices to prove the case $d_1 = D_1 = 1$. In particular we have $\sum_{Q \in \mathcal{Q}} |Q| \, \psi_1(t_Q) = 1$. Let $A_2 > 0$ be the constant from the definition of $\psi_2 \preceq \psi_1$ and $\psi_1 \preceq \psi_2$ with $A_1 := 1$. Further let $C_0 > 0$ be such that $\psi_2(Q, 2t) \leqslant C_0 \psi_2(Q, t)$ for all $Q \in \mathcal{X}^n$ and $t \geqslant 0$. Let $\mathcal{Q} \in \mathcal{Y}_1^n$ and $\{t_Q\}_{Q \in \mathcal{Q}}$ with $t_Q \geqslant 0$ be such that (5.7.12) holds. Then the second inequality of (5.7.13) holds with $D_2 := A_2$. Let $m \in \mathbb{N}$ such that $2^m \geqslant A_2$ and let $d_2 := C_0^{-m}$. We proceed by contradiction. Assume that $\sum_{Q \in \mathcal{Q}} |Q| \, \psi_2(t_Q) < d_2$. Then, by the Δ_2-condition and convexity,

$$\sum_{Q \in \mathcal{Q}} |Q| \, \psi_2(2^m t_Q) \leqslant C_0^m \sum_{Q \in \mathcal{Q}} |Q| \, \psi_2(t_Q) < C_0^m d_2 = 1,$$

$$\sum_{Q \in \mathcal{Q}} |Q| \, \psi_1(2^m t_Q) \geqslant 2^m \sum_{Q \in \mathcal{Q}} |Q| \, \psi_1(t_Q) = 2^m \geqslant A_2.$$

This contradicts the choice of A_1, A_2 for $\psi_1 \preceq \psi_2$. This proves the lemma. $\square$

Lemma 5.7.14. *Let φ be a proper, generalized N-function on $\mathbb{R}^n$ and $s \geqslant 1$. Furthermore, let φ^s and $(\varphi^*)^s$ be proper and let φ and φ^* satisfy the Δ_2-condition and $(M_{s,Q}\varphi) \preceq (M_{s,Q}\varphi^*)^*$. Then, uniformly in $Q \in \mathcal{X}^n$,*

$$|Q| \, (M_{s,Q}\varphi) \left(\frac{1}{\|\chi_Q\|_\varphi} \right) \approx 1, \qquad |Q| \, (M_{s,Q}\varphi^*)^* \left(\frac{1}{\|\chi_Q\|_\varphi} \right) \approx 1.$$

Proof. For $Q \in \mathcal{X}^n$ define $t_{0,Q} := 1/\|\chi_Q\|_\varphi$. Then

$$|Q| \, (M_Q\varphi)(t_{0,Q}) = \int_Q \varphi \left(\frac{1}{\|\chi_Q\|_\varphi} \right) dx = 1. \qquad (5.7.15)$$

By Jensen's inequality we have $M_Q\varphi \leqslant M_{s,Q}\varphi$ and $M_Q\varphi^* \leqslant M_{s,Q}\varphi^*$. Thus, $(M_{s,Q}\varphi^*)^* \leqslant (M_Q\varphi^*)^*$ by conjugation. From $(M_Q\varphi^*)^* \preceq M_Q\varphi$ (Lemma 5.2.10) we deduce

$$M_Q\varphi \cong \psi$$

where $\psi : \mathcal{X}^n \times \mathbb{R}_{\geqslant} \to \mathbb{R}_{\geqslant}$ is either $M_{s,Q}\varphi$ or $(M_{s,Q}\varphi^*)^*$. Thus (5.7.15) and Lemma 5.7.11 prove the lemma. $\qquad\qquad\qquad\qquad\qquad\qquad\qquad\qquad\qquad\quad\square$

Lemma 5.7.16. *Let $p \in \mathcal{P}(\mathbb{R}^n)$ with $1 < p^- \leqslant p^+ < \infty$. Further assume $M_{s,Q}\varphi_{p(\cdot)} \preceq (M_{s,Q}\varphi^*_{p(\cdot)})^*$ for some $s \geqslant 1$. Define $\alpha_s : \mathcal{X}^n \times \mathbb{R}_{>0} \to \mathbb{R}_{>0}$ by*

$$\alpha_s(Q,t) := \frac{(M_{s,Q}\varphi_{p(\cdot)})(t)}{(M_{s,Q}\varphi^*_{p(\cdot)})^*(t)}. \qquad (5.7.17)$$

Then, uniformly in $Q \in \mathcal{X}^n$ and $t > 0$,

$$\alpha_s\left(Q, \frac{1}{\|\chi_Q\|_{p(\cdot)}}\right) \approx 1, \qquad\qquad \alpha_s(Q,1) \approx 1. \qquad (5.7.18)$$

Moreover, there exists $C_5 \geqslant 1$ such that for all $Q \in \mathcal{X}^n$

$$\begin{aligned}
\alpha_s(Q,t_2) &\leqslant C_5\,(\alpha_s(Q,t_1)+1) \quad \text{for } 0 < t_1 \leqslant t_2 \leqslant 1, \\
\alpha_s(Q,t_3) &\leqslant C_5\,(\alpha_s(Q,t_4)+1) \quad \text{for } 1 \leqslant t_3 \leqslant t_4.
\end{aligned} \qquad (5.7.19)$$

Furthermore, for all $C_6, C_7 > 0$ there exists $C_8 \geqslant 1$ such that for all $Q \in \mathcal{X}^n$

$$t \in \left[C_6 \min\left\{1, \frac{1}{\|\chi_Q\|_{p(\cdot)}}\right\}, C_7 \max\left\{1, \frac{1}{\|\chi_Q\|_{p(\cdot)}}\right\}\right] \;\Rightarrow\; \alpha_s(Q,t) \leqslant C_8.$$

Proof. It suffices to prove the claim for $\varphi_{p(\cdot)} = \bar{\varphi}_{p(\cdot)}$. The first part of (5.7.18) follows from Lemma 5.7.14. Recall that $(\bar{\varphi}^*_{p(\cdot)})' = (\bar{\varphi}'_{p'(\cdot)})^{-1}$. Due to Lemma 2.6.11 applied to $(M_{s,Q}\bar{\varphi}^*_{p(\cdot)})$ and the Δ_2-condition,

$$\left(M_{s,Q}\bar{\varphi}^*_{p(\cdot)}\right)^*\left((M_{s,Q}(\bar{\varphi}^*_{p(\cdot)})')(t)\right) \approx \left(M_{s,Q}\bar{\varphi}^*_{p(\cdot)}\right)^*\left(\frac{M_{s,Q}\bar{\varphi}^*_{p(\cdot)}(t)}{t}\right)$$

$$\approx \left(M_{s,Q}\bar{\varphi}^*_{p(\cdot)}\right)(t) \approx t\left(M_{s,Q}(\bar{\varphi}^*_{p(\cdot)})'\right)(t).$$

Thus

$$\begin{aligned}
\alpha_s\left(Q, (M_{s,Q}(\bar{\varphi}^*_{p(\cdot)})')(t)\right) &= \frac{(M_{s,Q}\bar{\varphi}_{p(\cdot)})\left((M_{s,Q}(\bar{\varphi}^*_{p(\cdot)})')(t)\right)}{\left(M_{s,Q}\bar{\varphi}^*_{p(\cdot)}\right)^*\left((M_{s,Q}(\bar{\varphi}^*_{p(\cdot)})')(t)\right)} \\
&\approx \frac{(M_{s,Q}\bar{\varphi}_{p(\cdot)})\left((M_{s,Q}(\bar{\varphi}^*_{p(\cdot)})')(t)\right)}{t\left(M_{s,Q}(\bar{\varphi}^*_{p(\cdot)})'\right)(t)} \\
&\approx \frac{(M_{s,Q}\bar{\varphi}'_{p(\cdot)})\left((M_{s,Q}(\bar{\varphi}^*_{p(\cdot)})')(t)\right)}{t} \\
&\approx \left(\fint_Q\left(\fint_Q t^{\frac{s(p(y)-p(z))}{(p(z)-1)(p(y)-1)}}\,dz\right)^{\frac{1}{p(y)-1}}\,dy\right)^{\frac{1}{s}}.
\end{aligned}$$

Since $(M_{s,Q}(\bar{\varphi}^*_{p(\cdot)})')(1) \approx 1$ this proves (5.7.18). Define

$$\beta_s^>(Q,t) := \left(\fint_Q \left(\fint_Q t^{\frac{s(p(y)-p(z))}{(p(z)-1)(p(y)-1)}} \chi_{\{p(y)>p(z)\}} \, dz \right)^{\frac{1}{p(y)-1}} dy \right)^{\frac{1}{s}},$$

$$\beta_s^\leqslant(Q,t) := \left(\fint_Q \left(\fint_Q t^{\frac{s(p(y)-p(z))}{(p(z)-1)(p(y)-1)}} \chi_{\{p(y)\leqslant p(z)\}} \, dz \right)^{\frac{1}{p(y)-1}} dy \right)^{\frac{1}{s}},$$

then

$$\alpha_s\left(Q, (M_{s,Q}(\bar{\varphi}^*_{p(\cdot)})')(t)\right) \approx \beta_s^>(Q,t) + \beta_s^\leqslant(Q,t) \quad \text{uniformly in } Q,t,$$

$0 \leqslant \beta_s^>(Q,t) \leqslant 1$ for $0 < t \leqslant 1$,

$0 \leqslant \beta_s^\leqslant(Q,t) \leqslant 1$ for $t \geqslant 1$,

$\beta_s^>(Q,t)$ is increasing on $[1,\infty)$,

$\beta_s^\leqslant(Q,t)$ is decreasing on $(0,1]$,

where we have used that $1 < p^- \leqslant p^+ < \infty$. Thus there exists $C_5 \geqslant 1$ such that

$$\alpha_s\left(Q, (M_{s,Q}(\bar{\varphi}^*_{p(\cdot)})')(t_2)\right) \leqslant C_5\left(\alpha_s\left(Q, (M_{s,Q}(\bar{\varphi}^*_{p(\cdot)})')(t_1)\right)+1\right) \text{for } 0<t_1\leqslant t_2\leqslant 1,$$

$$\alpha_s\left(Q, (M_{s,Q}(\bar{\varphi}^*_{p(\cdot)})')(t_3)\right) \leqslant C_5\left(\alpha_s\left(Q, (M_{s,Q}(\bar{\varphi}^*_{p(\cdot)})')(t_4)\right)+1\right) \text{for } 1\leqslant t_3\leqslant t_4.$$

This, $(M_{s,Q}(\bar{\varphi}^*_{p(\cdot)})')(1) \approx 1$, and the strong Δ_2-condition prove (5.7.19). The last claim follows immediately from (5.7.18), (5.7.19), and the Δ_2-condition. This completes the proof. □

Lemma 5.7.20. *Let $p \in \mathcal{P}(\mathbb{R}^n)$ with $1 < p^- \leqslant p^+ < \infty$. Further assume $M_{s,Q}\varphi_{p(\cdot)} \preceq (M_{s,Q}\varphi^*_{p(\cdot)})^*$ for some $s > 1$. Then there exists $b : \mathcal{X}^n \to \mathbb{R}_\geqslant$ and $K > 0$ such that $\|b\|_{\mathcal{Y}_1^n,1} + \|b\|_{\mathcal{Y}_1^n,\infty} < \infty$ and*

$$|Q|\, (M_{s,Q}\varphi^*_{p(\cdot)})^*(t) \leqslant 1$$
$$\Rightarrow \quad (M_{s,Q}\varphi_{p(\cdot)})(t) \leqslant K\,(M_{s,Q}\varphi^*_{p(\cdot)})^*(Q,t) + b(Q)\chi_{\{t<1\}}$$

for all $Q \in \mathcal{X}^n$ and all $t \geqslant 0$.

Proof. Due to Theorem 5.6.10 there exist $b_2 : \mathcal{X}^n \to \mathbb{R}_\geqslant$ with $\|b_2\|_{\mathcal{Y}_1^n,1} < \infty$ and $K_2 > 0$ such that, for all $Q \in \mathcal{X}^n$ and all $t \geqslant 0$,

$$|Q|\, (M_{s,Q}\varphi^*_{p(\cdot)})^*(t) \leqslant 1$$
$$\Rightarrow (m_{s,Q}\varphi_{p(\cdot)})(t) \leqslant K_2\,(M_{s,Q}\varphi^*_{p(\cdot)})^*(Q,t) + b_2(Q). \tag{5.7.21}$$

Assume that $|Q|\,(M_{s,Q}\varphi_{p(\cdot)}^*)^*(t) \leqslant 1$. Then due to Lemma 5.7.14 and the Δ_2-condition of $M_{s,Q}\varphi_{p(\cdot)}$ there exists $A \geqslant 0$ (independent of Q and t) such that $t \leqslant A/\|\chi_Q\|_{p(\cdot)}$. Now Lemma 5.7.16 holds for some $C_8 \geqslant 1$ with the choice $C_6 := 1$, $C_7 := A$. Let $K := \max\{C_8, K_2\}$ and define $b : \mathcal{X}^n \to \mathbb{R}_{\geqslant 0}$ by

$$b(Q) := \min\{(M_{s,Q}\varphi_{p(\cdot)})(1), b_2(Q)\}. \tag{5.7.22}$$

Since $(M_{s,Q}\varphi_{p(\cdot)})(1) \approx 1$, $\|b\|_{\mathcal{Y}_1^n,1} + \|b\|_{\mathcal{Y}_1^n,\infty} < \infty$. If $0 \leqslant t \leqslant 1$ then by (5.7.21) and (5.7.22) we obtain the claim.

If on the other hand $1 < t \leqslant A/\|\chi_Q\|_{p(\cdot)}$, then by Lemma 5.7.16 we deduce $\alpha_s(Q,t) \leqslant C_8$. The definition of α_s and $C_8 \leqslant K$ immediately imply the claim without the term $b(Q)$. $\square$

Lemma 5.7.23. *Let Ω either be $\mathbb{R}^n$ or $\mathcal{X}^n$ and let $r, s > 0$. Let φ be a generalized N-function on Ω such that φ and φ^* satisfy the Δ_2-condition. Let $\gamma : \Omega \times \mathbb{R}_{\geqslant} \to \mathbb{R}_{\geqslant}$ be defined by*

$$\gamma(t) := \int\limits_0^t \left(\varphi'(u^{\frac{1}{r}})\right)^s du.$$

Then γ is a generalized N-function on Ω with

$$\frac{\gamma(\omega, t^r)}{t^r} \approx \left(\frac{\varphi(\omega, t)}{t}\right)^s,$$

$$\frac{\gamma^*(\omega, t^s)}{t^s} \approx \left(\frac{\varphi^*(\omega, t)}{t}\right)^r$$

uniformly in $\omega \in \Omega$ and $t > 0$. Furthermore, γ and γ^* satisfy the Δ_2-condition. If ψ is another N-function on Ω such that

$$\frac{\psi(\omega, t^r)}{t^r} \approx \left(\frac{\varphi(\omega, t)}{t}\right)^s \tag{5.7.24}$$

uniformly in $\omega \in \Omega$ and $t > 0$, then ψ^* is an N-function on Ω and

$$\frac{\psi^*(\omega, t^s)}{t^s} \approx \left(\frac{\varphi^*(\omega, t)}{t}\right)^r \tag{5.7.25}$$

uniformly in $\omega \in \Omega$ and $t > 0$. Moreover, ψ and ψ^* satisfy the Δ_2-condition.

Proof. Since all following calculations are uniform with respect to ω, we omit the dependence on ω. From the definition of γ it follows immediately that γ is an N-function on Ω. This implies

$$\frac{\gamma(t^r)}{t^r} \approx \gamma'(t^r) = \left(\varphi'(t)\right)^s \approx \left(\frac{\varphi(t)}{t}\right)^s. \tag{5.7.26}$$

Here we used Remark 2.6.7, i.e. $\gamma(t) \approx \gamma'(t)\,t$ uniformly in $t \geqslant 0$. From $\gamma'(t^r) = \varphi'(t)^s$ we deduce $\varphi'^{-1}(t)^r = (\gamma')^{-1}(t^s)$. Thus

$$(\varphi^*)'(t)^r = (\varphi')^{-1}(t)^r = (\gamma')^{-1}(t^s) = (\gamma^*)'(t^s).$$

Hence

$$\left(\frac{\varphi^*(t)}{t}\right)^r \approx \left((\varphi^*)'(t)\right)^r = (\gamma^*)'(t^s) \approx \frac{\gamma^*(t^s)}{t^s}. \tag{5.7.27}$$

Since φ and φ^* satisfy the strong Δ_2-condition, we immediately deduce from (5.7.26) and (5.7.27) that γ and γ^* satisfy the Δ_2-condition. From (5.7.24) and (5.7.26) we deduce that $\psi \approx \gamma$. Thus there exist $c_0, c_1 > 0$ with

$$c_0\,\gamma(t) \leqslant \psi(t) \leqslant c_1\,\gamma(t).$$

Thus by Lemma 2.6.4

$$c_1\,\gamma^*\left(\frac{t}{c_1}\right) \leqslant \psi^*(t) \leqslant c_0\,\gamma^*\left(\frac{t}{c_0}\right).$$

Since γ^* satisfies the Δ_2-condition, this implies $\gamma^* \approx \psi^*$. Overall, we have shown $\gamma \approx \psi$ and $\gamma^* \approx \psi^*$. So (5.7.25) and the Δ_2-condition follow from properties of γ. This proves the lemma. $\qquad\square$

Lemma 5.7.28. *Let $p \in \mathcal{P}(\mathbb{R}^n)$ with $1 < p^- \leqslant p^+ < \infty$. Further assume $M_{s_2,Q}\varphi_{p(\cdot)} \preceq (M_{s_2,Q}\varphi^*_{p(\cdot)})^*$ for some $1 \leqslant s_2$. Let α_1, α_{s_2} be defined as in (5.7.17). Then uniformly in $Q \in \mathcal{X}^n$ and $t > 0$*

$$\left(\alpha_{s_2}\left(Q, t^{\frac{1}{s_2}}\right)\right)^{s_2} \approx \alpha_1(Q, t).$$

Proof. Note that for any $r \geqslant 1$, $\varphi_{p(x)}(t^r) \approx \varphi_{p(x)}(t)^r$ and $\varphi^*_{p(x)}(t^r) \approx (\varphi^*_{p(x)}(t))^r$ uniformly in $x \in \mathbb{R}^n$ and $t \geqslant 0$, since $1 < p^- \leqslant p^+ < \infty$. Thus

$$(M_Q\varphi_{p(\cdot)})(t) \approx (M_Q\varphi_{p(\cdot)})(t) \approx \left((M_{s_2,Q}\varphi_{p(\cdot)})\left(t^{\frac{1}{s_2}}\right)\right)^{s_2}, \tag{5.7.29}$$

$$(M_Q\varphi^*_{p(\cdot)})(t) \approx M_Q\varphi^*_{p(\cdot)})(t) \approx \left((M_{s_2,Q}\varphi^*_{p(\cdot)})\left(t^{\frac{1}{s_2}}\right)\right)^{s_2}, \tag{5.7.30}$$

uniformly in $Q \in \mathcal{X}^n$ and $t \geqslant 0$. It is easy to see that $(Q, t) \mapsto ((M_{s_2, Q} \varphi_{p(\cdot)}^*)$ $(t^{\frac{1}{s_2}}))^{s_2}$ is a generalized N-function. Lemma 5.7.10 implies that this N-function and its conjugate satisfy the Δ_2-condition. Thus it follows from (5.7.30) and Lemma 5.7.23 that

$$(M_Q \varphi_{p(\cdot)}^*)^*(t) \approx \left((M_{s_2, Q} \varphi_{p(\cdot)}^*)^* \left(t^{\frac{1}{s_2}} \right) \right)^{s_2}.$$

This and (5.7.29) imply

$$\left(\alpha_{s_2} \left(Q, t^{\frac{1}{s_2}} \right) \right)^{s_2} \approx \alpha_1(Q, t). \square$$

We are now ready to prove the equivalence of $\mathcal{A}$ and $\mathcal{A}_{\text{strong}}$

Proof of Theorem 5.7.1. Due to Lemma 5.7.9 there exists $s_2 > 1$ with

$$M_{s_2, Q} \varphi_{p(\cdot)} \preceq M_Q \varphi_{p(\cdot)} \preceq (M_Q \varphi_{p(\cdot)}^*)^* \preceq (M_{s_2, Q} \varphi_{p(\cdot)}^*)^*.$$

Due to Lemma 5.7.20 there exist $b_2 : \mathcal{X}^n \to \mathbb{R}_{\geqslant}$ with $\|b_2\|_{\mathcal{Y}_1^n, 1} + \|b_2\|_{\mathcal{Y}_1^n, \infty} < \infty$ and $K_2 \geqslant 1$ such that the inequality of the lemma holds (for the choice $s = s_2$ and $b = b_2$). Due to Lemma 5.7.14 and the strong Δ_2-condition of $(M_{s_2, Q} \varphi_{p(\cdot)}^*)^*$ there exists $0 < D_2 \leqslant 1$ (independent of Q and t) such that $t \leqslant D_2 / \|\chi_Q\|_{p(\cdot)}$ implies $|Q| (M_{s_2, Q} \varphi_{p(\cdot)}^*)^*(t) \leqslant 1$. Due to Lemma 5.7.16 there exists $C_8 \geqslant 1$ such that

$$t \in \left[D_2 \min \left\{ 1, \frac{1}{\|\chi_Q\|_{p(\cdot)}} \right\}, \max \left\{ 1, \frac{1}{\|\chi_Q\|_{p(\cdot)}} \right\} \right] \Rightarrow \alpha_{s_2}(Q, t) \leqslant C_8.$$

$$(5.7.31)$$

Moreover, by Lemma 5.7.28 and the strong Δ_2-condition there exists $A_0 \geqslant 1$ such that for all $Q \in \mathcal{X}^n$ and all $t > 0$

$$\alpha_1(Q, t) \leqslant A_0 \left(\alpha_{s_2} \left(Q, t^{\frac{1}{s_2}} \right) \right)^{s_2}. \qquad (5.7.32)$$

Define $K_1 := A_0 \left(\max \left\{ 2 K_2, C_8 \right\} \right)^{s_2}$.

Auxiliary claim: For all $Q \in \mathcal{X}^n$ and $t > 0$ with

$$|Q| (M_Q \varphi_{p(\cdot)}^*)^*(t) \leqslant 1, \qquad (5.7.33)$$

we have

$$(M_Q \varphi_{p(\cdot)})(t) \leqslant \begin{cases} \max \left\{ K_1 \left(M_Q \varphi_{p(\cdot)}^* \right)^* (Q,t), \, 2\, b_2(Q)\, t^{1-\frac{1}{s_2}} \right\} & \text{for } 0 < t < 1, \\ K_1 \left(M_Q \varphi_{p(\cdot)}^* \right)^* (Q,t) & \text{for } t \geqslant 1. \end{cases}$$

Proof of auxiliary claim. Assume that (5.7.33) is satisfied, then by Jensen's inequality

$$|Q| \left(M_{s_2, Q} \varphi_{p(\cdot)}^* \right)^* (t) \leqslant 1.$$

If $t \geqslant 1$, then by Lemma 5.7.20 and Jensen's inequality

$$(M_Q \varphi_{p(\cdot)})(t) \leqslant (M_{s_2, Q} \varphi_{p(\cdot)})(t) \leqslant K_2 \left(M_{s_2, Q} \varphi_{p(\cdot)} \right)(t) \leqslant K_1 \left(M_Q \varphi_{p(\cdot)}^* \right)^* (Q,t),$$

so the claim holds in this case. If $0 < t < 1$ and $\alpha_1(Q,t) \leqslant K_1$, then

$$(M_Q \varphi_{p(\cdot)})(t) = \alpha_1(Q,t) \left(M_Q \varphi_{p(\cdot)}^* \right)^* (t) \leqslant K_1 \left(M_Q \varphi_{p(\cdot)}^* \right)^* (t),$$

so the claim holds also in this case. It remains to consider the case

$$0 < t < 1 \quad \text{and} \quad \alpha_1(Q,t) > K_1.$$

From (5.7.32) we deduce that

$$A_0 \left(\max \left\{ 2\, K_2, C_8 \right\} \right)^{s_2} = K_1 < \alpha_1(Q,t) \leqslant A_0 \left(\alpha_{s_2} \left(Q, t^{\frac{1}{s_2}} \right) \right)^{s_2}.$$

Especially,

$$0 < t^{\frac{1}{s_2}} < 1 \quad \text{and} \quad \alpha_{s_2}(Q, t^{\frac{1}{s_2}}) > \max \left\{ 2\, K_2, C_8 \right\}. \tag{5.7.34}$$

From (5.7.31) we deduce that

$$0 < t^{\frac{1}{s_2}} < \frac{D_2}{\|\chi_Q\|_{p(\cdot)}}.$$

Now the choice of D_2 implies that

$$|Q| \left(M_{s_2, Q} \varphi_{p(\cdot)}^* \right)^* \left(t^{\frac{1}{s_2}} \right) \leqslant 1.$$

From Lemma 5.7.20 we deduce that

$$\left(M_{s_2,Q}\varphi_{p(\cdot)}\right)\!\left(t^{\frac{1}{s_2}}\right) \leqslant K_2\left(M_{s_2,Q}\varphi^*_{p(\cdot)}\right)^*\!\left(t^{\frac{1}{s_2}}\right) + b_2(Q)$$

$$= \frac{K_2}{\alpha_{s_2}\!\left(Q, t^{\frac{1}{s_2}}\right)}\left(M_{s_2,Q}\varphi_{p(\cdot)}\right)\!\left(t^{\frac{1}{s_2}}\right) + b_2(Q).$$

Since $\alpha_{s_2}\!\left(Q, t^{\frac{1}{s_2}}\right) \geqslant 2\,K_2$ by (5.7.34), we can absorb the first term of the right-hand side on the left-hand side, whence

$$\left(M_{s_2,Q}\varphi_{p(\cdot)}\right)\!\left(t^{\frac{1}{s_2}}\right) \leqslant 2\,b_2(Q). \qquad (5.7.35)$$

It follows from $0 < t < 1$, $s_2 > 1$, Jensen's inequality, the convexity of $\varphi_{p(\cdot)}$, and (5.7.35) that

$$\left(M_Q\varphi_{p(\cdot)}\right)(t) \leqslant \left(M_{s_2,Q}\varphi_{p(\cdot)}\right)(t)$$

$$\leqslant \left(M_{s_2,Q}\varphi_{p(\cdot)}\right)\!\left(t^{\frac{1}{s_2}}\right) t^{1-\frac{1}{s_2}}$$

$$\leqslant 2\,b_2(Q)\,t^{1-\frac{1}{s_2}}.$$

So the claim holds also in this case, which completes the proof of the auxiliary claim. □

We now deduce from the auxiliary claim that $\left(M_Q\varphi_{p(\cdot)}\right) \ll \left(M_Q\varphi^*_{p(\cdot)}\right)^*$. Let $\mathcal{Q}_k \in \mathcal{Y}_1^n$ for $k \in \mathbb{Z}$ be such that

$$\sum_{k=-\infty}^{\infty} \sum_{Q \in \mathcal{Q}_k} |Q|\left(M_Q\varphi^*_{p(\cdot)}\right)^*(2^k) \leqslant 1.$$

Then by the auxiliary claim

$$\sum_{k=-\infty}^{\infty} \sum_{Q \in \mathcal{Q}_k} |Q|\left(M_Q\varphi_{p(\cdot)}\right)(2^k)$$

$$\leqslant \sum_{k=-\infty}^{-1} \sum_{Q \in \mathcal{Q}_k} |Q|\,\max\left\{K_1\left(M_Q\varphi^*_{p(\cdot)}\right)^*(2^k), 2\,b_2(Q)\,2^{k\left(1-\frac{1}{s_2}\right)}\right\}$$

$$+ \sum_{k=0}^{\infty} \sum_{Q \in \mathcal{Q}_k} |Q|\,K_1\left(M_Q\varphi^*_{p(\cdot)}\right)^*(2^k)$$

$$\leqslant K_1 \sum_{k=-\infty}^{\infty} \sum_{Q \in \mathcal{Q}_k} |Q|\left(M_Q\varphi^*_{p(\cdot)}\right)^*(2^k) + \sum_{k=-\infty}^{-1} 2\,\|b_2\|_{\mathcal{Y}_1^n, 1}\,2^{k\left(1-\frac{1}{s_2}\right)}$$

$$\leqslant K_1 + c(s_2)\,\|b_2\|_{\mathcal{Y}_1^n, 1},$$

where we have used $s_2 > 1$. This and Remark 5.5.4 prove $M_Q \varphi_{p(\cdot)} \ll (M_Q \varphi^*_{p(\cdot)})^*$. Thus, $p \in \mathcal{A}_{\text{strong}}$. □

5.8   Weighted Variable Exponent Lebesgue Spaces*

A measurable function $\omega : \mathbb{R}^n \to (0, \infty)$ is called a *weight*. For $p \in \mathcal{P}(\mathbb{R}^n)$ we define

$$\varphi_{p(\cdot),\omega}(x,t) := \varphi_{p(\cdot)}\big(x, t\omega(x)\big) = \varphi_{p(x)}\big(t\omega(x)\big).$$

We define the corresponding Musielak–Orlicz by

$$L_\omega^{p(\cdot)}(\Omega) := L^{\varphi_{p(\cdot),\omega}}(\Omega).$$

The norm $\|\cdot\|_{p(\cdot),\omega}$ of $L_\omega^{p(\cdot)}(\Omega)$ satisfies

$$\|f\|_{p(\cdot),\omega} = \|f\omega\|_{p(\cdot)}. \tag{5.8.1}$$

We want to examine for which weights the maximal operator is bounded from $L_\omega^{p(\cdot)}(\mathbb{R}^n)$ to $L_\omega^{p(\cdot)}(\mathbb{R}^n)$. Kokilashvili, Samko and their collaborators have proved several boundedness results with particular classes of weights: initially in the case of power-type weights [248,346–348,365] and more recently in the case of weights which are controlled by power-type functions [25,240,244,245, 247,249,250,327,339]. Other investigations with such weights include [57,66, 123,230,273,286]; more general metric measure spaces have been studied for instance in [167,199,204,251,301]. The discrete weighted case was studied in [315].

A more comprehensive framework was recently introduced in [97], along the line of Muckenhoupt in the constant exponent case. This allowed us to characterize weights for which the maximal operator is bounded when $p \in \mathcal{P}^{\log}$ and $1 < p^- \leqslant p^+ < \infty$. In this section we give a new proof for this characterization based on our results on classes $\mathcal{A}$ and $\mathcal{G}$.

Note that the proofs in this section are partly based on results of Sect. 7.3 below.

The conjugate function of $\varphi_{p(\cdot),\omega}$ is given by

$$\varphi^*_{p(\cdot),\omega}(x,t) = \varphi^*_{p(\cdot)}\big(x, t/\omega(x)\big) = \varphi^*_{p(x)}\big(t/\omega(x)\big).$$

Moreover, we have $\widetilde{\varphi}^*_{p(\cdot),\omega} = \widetilde{\varphi}_{p'(\cdot),1/\omega}$. Unfortunately, for general weights the spaces $L_\omega^{p(\cdot)}$ is not a Banach function space, since simple functions need not be contained in $L_\omega^{p(\cdot)}$. However, if $\omega \in L_{\text{loc}}^{p(\cdot)}$, then all characteristic functions of cubes are contained in $L_\omega^{p(\cdot)}$. As a consequence the restriction $L_\omega^{p(\cdot)}(Q)$

is a Banach function space for every cube Q. Hence, the norm conjugate formula holds for $L^{p(\cdot)}_\omega(Q)$ in the form of Corollary 2.7.5. Using monotone convergence we conclude that the norm conjugate formula also holds for $L^{p(\cdot)}_\omega(\mathbb{R}^n)$ if $\omega \in L^{p(\cdot)}_{\mathrm{loc}}$ even if $L^{p(\cdot)}_\omega(\mathbb{R}^n)$ might not be a Banach function space.

We already know from Sect. 4.4 that we need $\varphi_{p(\cdot),\omega} \in \mathcal{A}$ for the boundedness of the maximal operator. On the other hand, if $\varphi_{p(\cdot),\omega} \in \mathcal{A}$, then M is of weak type $\varphi_{p(\cdot),\omega}$. However, to verify the condition $\varphi_{p(\cdot),\omega} \in \mathcal{A}$, we need to check the uniform boundedness of the averaging operators T_Q with respect to all locally 1-finite families of cubes in $\mathbb{R}^n$. The following observation allows us to reduce this condition to single cubes.

Lemma 5.8.2. *Let $p \in \mathcal{G}$ and let ω be a weight. Then $\varphi_{p(\cdot),\omega} \in \mathcal{G}$. If $\varphi_{p(\cdot),\omega} \in \mathcal{A}_{\mathrm{loc}}$, then $\varphi_{p(\cdot),\omega} \in \mathcal{A}$.*

Proof. The property $\varphi_{p(\cdot),\omega} \in \mathcal{G}$ follows immediately from the definition of $\mathcal{G}$ and (5.8.1). If $\varphi_{p(\cdot),\omega} \in \mathcal{A}_{\mathrm{loc}}$, then Corollary 7.3.7 implies $\varphi_{p(\cdot),\omega} \in \mathcal{A}$. $\square$

By Theorem 4.5.7 we know that $\varphi_{p(\cdot),\omega} \in \mathcal{A}_{\mathrm{loc}}$ if and only if

$$\|\chi_Q\|_{\varphi_{p(\cdot),\omega}} \|\chi_Q\|_{\varphi^*_{p(\cdot),\omega}} \approx |Q|.$$

In other words $\varphi_{p(\cdot),\omega} \in \mathcal{A}_{\mathrm{loc}}$ if and only if

$$\|\chi_Q \omega\|_{p(\cdot)} \|\chi_Q/\omega\|_{p'(\cdot)} \approx |Q|. \tag{5.8.3}$$

Lemma 5.8.4. *Let $p \in \mathcal{A}_{\mathrm{loc}}$ and let ω be a weight. Then $\varphi_{p(\cdot),\omega} \in \mathcal{A}_{\mathrm{loc}}$ if and only if*

$$M_{p(\cdot),Q}(\omega)M_{p'(\cdot),Q}(\omega^{-1}) \approx 1$$

uniformly for all cubes (or balls) $Q \subset \mathbb{R}^n$.

Proof. Due to Theorem 4.5.7, $\|\chi_Q\|_{p(\cdot)} \|\chi_Q\|_{p'(\cdot)} \approx |Q|$. On the other hand $\varphi_{p(\cdot),\omega} \in \mathcal{A}_{\mathrm{loc}}$ is by (5.8.3) equivalent to $\|\chi_Q\omega\|_{p(\cdot)} \|\chi_Q\omega^{-1}\|_{p'(\cdot)} \approx |Q|$. Taking the quotient of these equivalences proves the claim. $\square$

Due to Theorem 4.5.7, $\|\chi_Q\|_{p(\cdot)} \|\chi_Q\|_{p'(\cdot)} \approx |Q|$. This and Hölder's inequality applied to $\fint_Q fg\,dx$ yields that

$$M_Q(fg) = \fint_Q |fg|\,dx \leqslant c\,M_{p(\cdot),Q}f\,M_{p'(\cdot),Q}g \tag{5.8.5}$$

for $p \in \mathcal{A}_{\mathrm{loc}}$ and all $f \in L^{p(\cdot)}_{\mathrm{loc}}(\mathbb{R}^n)$ and $g \in L^{p'(\cdot)}_{\mathrm{loc}}(\mathbb{R}^n)$, where the constant only depends on the $\mathcal{A}_{\mathrm{loc}}$-constant of p.

The following result is the main result of this section and goes back to Cruz-Uribe, Diening and Hästo [82]. It has been shown first using a different technique by Diening and Hästo [97] with the notation as in Remark 5.8.10

Theorem 5.8.6. *Let* $p \in \mathcal{P}^{\log}(\mathbb{R}^n)$ *with* $1 < p^- \leqslant p^+ < \infty$. *Then* M *is bounded from* $L_\omega^{p(\cdot)}(\mathbb{R}^n)$ *to* $L_\omega^{p(\cdot)}(\mathbb{R}^n)$ *if and only if* $\varphi_{p(\cdot),\omega} \in \mathcal{A}_{\mathrm{loc}}$.

Before we get to the proof of Theorem 5.8.6 we need the following Calderón–Zygmund decomposition which goes back to Aimar and Macías [16, Lemma 2].

Lemma 5.8.7. *Let* $p \in \mathcal{P}^{\log}(\mathbb{R}^n)$ *with* $p^+ < \infty$, $b \geqslant 2 \cdot 15^{3n}$, *and* $f \in L^{p(\cdot)}(\mathbb{R}^n)$. *Let* $D_k := \{b^{k+1} \geqslant Mf > b^k\}$ *for* $k \in \mathbb{Z}$, *where* M *denotes the non-centered maximal function with respect to balls.*

Then $\mathbb{R}^n = \bigcup_{k \in \mathbb{Z}} D_k$ *up to measure zero and there exists a family* $\{B_i^k\}_{k \in \mathbb{Z}, i \in \mathbb{N}}$ *of balls such that the following holds.*

(a) $D_k \subset \bigcup_{i \in \mathbb{N}} 5B_i^k$ *for all* $k \in \mathbb{Z}$.
(b) $B_i^k \cap B_j^k = \emptyset$ *for* $i \neq j$.
(c) *For* $\gamma \geqslant 5$,

$$b^{k+1} \geqslant \fint_{B_i^k} |f| \, dx > b^k \geqslant \fint_{\gamma B_i^k} |f| \, dx.$$

(d) *Let* $I_i^k := \{(l,j) \in \mathbb{Z} \times \mathbb{N} : l \geqslant k+1, 5B_j^l \cap 5B_i^k \neq \emptyset\}$ *and define* $A_i^k := \bigcup_{(l,j) \in I_i^k} 5B_j^l$. *Then* $|A_i^k| \leqslant \frac{1}{2}|B_i^k|$.
(e) *Let* $F_i^k := B_i^k \setminus A_i^k$. *Then the family* $\{F_i^k\}_{i,k}$ *is disjoint.*

Proof. Define $\Omega_k := \{Mf > b^k\}$. Since $p \in \mathcal{P}^{\log}$, we know that M is of weak type $p(\cdot)$, by Corollary 4.4.12. Therefore $\|\chi_{D_k}\|_{p(\cdot)} \leqslant \|\chi_{\Omega_k}\|_{p(\cdot)} \leqslant c \|f\|_{p(\cdot)}/b^k$ and it follows by Lemma 3.2.12 that $|D_k| \leqslant |\Omega_k| < \infty$ and that $|\Omega_k| \to 0$ for $k \to \infty$. Especially, $\mathbb{R}^n = \bigcup_k D_k$ up to measure zero. For every $x \in D_k$ we can choose a ball B_x such that $b^{k+1} \geqslant \fint_{B_x} |f| \, dx > b^k \geqslant \fint_{\gamma B_x} |f| \, dx$ for every $\gamma \geqslant 5$. Denote by $\mathcal{W}_k$ the set of all such balls. Since M is the non-centered maximal function, we have $B \subset \Omega_k$ for every $B \in \mathcal{W}_k$. So $|\Omega_k| < \infty$ implies that $\sup \{\mathrm{diam}(B) : B \in \mathcal{W}_k\} < \infty$. Thus we can apply the basic covering theorem, Theorem 1.4.5, to find a countable, pair-wise disjoint family subfamily $\{B_i^k\}_i$ of $\mathcal{W}_k$ such that $D_k \subset \bigcup_{i \in \mathbb{N}} 5B_i^k$. This subfamily satisfies (a)–(c).

Let $(l,j) \in I_i^k$. Suppose for a contradiction that $5B_j^l \not\subset 15B_i^k$. Since $5B_j^l \cap 5B_i^k \neq \emptyset$, this implies that $\mathrm{diam}\, B_i^k \leqslant \mathrm{diam}\, B_j^l$. Let B be the ball concentric with B_i^k and diameter equal to $15 \,\mathrm{diam}\, B_j^l$. Then $B_j^l \subset B$ and $B = \gamma B_i^k$ for some $\gamma \geqslant 5$. Hence, by (c),

$$b^l \leqslant \fint_{B_j^l} |f| \, dx \leqslant 15^n \fint_{\gamma B_i^k} |f| \, dx \leqslant 15^n b^k.$$

Since $l \geqslant j + 1$ and $b > 15$, this is a contradiction and so we conclude that $5B_j^l \subset 15B_i^k$. This, (b) and (c) imply that

$$
\begin{aligned}
|A_j^k| &\leqslant \sum_{(l,j) \in I_i^k} |5B_j^l| < \sum_{(l,j) \in I_i^k} 5^n b^{-l} \int_{B_j^l} |f| \, dx \leqslant \sum_{l \geqslant k+1} 5^n b^{-l} \int_{15B_i^k} |f| \, dx \\
&\leqslant 5^n \frac{b^{-k-1}}{1 - 1/b} |15B_i^k| \fint_{15B_i^k} |f| \, dx \leqslant 5^n 15^n \frac{b^{-1}}{1 - 1/b} |B_i^k| \leqslant \frac{1}{2} |B_i^k|,
\end{aligned}
$$

i.e. (d) holds.

Since the family $\{B_i^k\}_i$ is disjoint, it follows that $\{F_i^k\}_i$ is disjoint. Suppose that $F_j^l \cap F_i^k \neq \emptyset$, and assume by symmetry that $l \geqslant j + 1$. Since $F_j^l \subset 5B_j^l$, this means that $(j, l) \in I_i^k$; but, by the definition of I_i^k, this implies that $5B_j^l \subset A_i^k$, so that $5B_j^l \cap F_i^k = \emptyset$ contradicting the assumption. Therefore, also the family $\{F_i^k\}_{i,k}$ is disjoint with respect to both indices. $\square$

Proof of Theorem 5.8.6. Since $p \in \mathcal{P}^{\log}(\mathbb{R}^n)$, we have $p \in \mathcal{G} \cap \mathcal{A}$ by Theorems 4.4.8 and 7.3.22. Let $\varphi_{p(\cdot),\omega} \in \mathcal{A}_{\mathrm{loc}}$, so that $\varphi_{p(\cdot),\omega} \in \mathcal{A}$ by Lemma 5.8.2 and by conjugation $\varphi_{p'(\cdot),1/\omega} \in \mathcal{A}$. Then the left-openness of $\mathcal{A}$ (Theorem 5.4.15) ensures the existence of $s \in (\min\{1/p^-, 1/(p')^-\}, 1)$ such that $\varphi_{sp(\cdot),\omega^{1/s}}, \varphi_{sp'(\cdot),\omega^{-1/s}} \in \mathcal{A}$, where we have used that $\bar\varphi_{sp(\cdot),\omega^{1/s}}(t) = \bar\varphi_{p(\cdot),\omega}(t^s)$ and $\bar\varphi_{sp'(\cdot),\omega^{-1/s}}(t) = \bar\varphi_{p'(\cdot),\omega}(t^s)$. Define $u, v \in \mathcal{P}^{\log}(\mathbb{R}^n)$ by

$$
\frac{1}{u'(x)} = s - \frac{1}{p(x)} \qquad \text{and} \qquad \frac{1}{v(x)} = s - \frac{1}{p'(x)}
$$

for all $x \in \mathbb{R}^n$. Since $s \in (\min\{1/p^-, 1/(p')^-\}, 1)$ the exponents u and v are well defined. Moreover, $u' = \frac{1}{s}(sp)'$ and $v := \frac{1}{s}(sp')'$ and

$$
\frac{p'(x)}{v'(x)} = p'(x)(1-s) + 1 \geqslant (p')^-(1-s) + 1,
$$

$$
\frac{p(x)}{u(x)} = p(x)(1-s) + 1 \geqslant p^-(1-s) + 1.
$$

Thus, by Theorem 7.3.27, $M_{u(\cdot)}$ is bounded on $L^{p(\cdot)}(\mathbb{R}^n)$ and $M_{v'(\cdot)}$ is bounded on $L^{p'(\cdot)}(\mathbb{R}^n)$.

Let $f \in L^{p(\cdot)}(\mathbb{R}^n)$ and b, D_k, B_j^k and F_j^k be as in Lemma 5.8.7. Further let $g \in L^{p'(\cdot)}(\mathbb{R}^n)$ and abbreviate $\hat{B}_j^k := 5B_j^k$. Then by (a) and (c) of the lemma it follows that

$$\int_{\mathbb{R}^n} |Mf||g|\, dx \leqslant \sum_k \int_{D_k} b^{k+1} |g|\, dx$$

$$\leqslant \sum_{j,k} b^{k+1} \int_{\hat{B}_j^k} |g|\, dx$$

$$\leqslant b \sum_{j,k} \fint_{B_j^k} |f|\, dx \int_{\hat{B}_j^k} |g|\, dx$$

$$= b\, 5^n \sum_{j,k} |\hat{B}_j^k|\, M_{\hat{B}_j^k} f\, M_{\hat{B}_j^k} g$$

We use (5.8.5) with exponents u and v to get

$$\int_{\mathbb{R}^n} |Mf||g|\, dx$$

$$\leqslant c \sum_{j,k} |\hat{B}_j^k|\, M_{u(\cdot),\hat{B}_j^k}(f\omega) M_{u'(\cdot),\hat{B}_j^k}(\omega^{-1}) M_{v'(\cdot),\hat{B}_j^k}(g\omega^{-1}) M_{v(\cdot),\hat{B}_j^k}(\omega).$$

We claim that

$$M_{u'(\cdot),\hat{B}_j^k}(\omega^{-1}) M_{v(\cdot),\hat{B}_j^k}(\omega) \approx 1. \qquad (5.8.8)$$

This estimate corresponds to the reverse Hölder estimate of (classical) Muckenhoupt weights. Lemma 5.8.4 applied to the exponents $sp(\cdot)$ and $sp'(\cdot)$ and weights $\omega^{1/s}$ and $\omega^{-1/s}$, respectively, implies that

$$M_{sp(\cdot),\hat{B}_j^k}(\omega^{1/s}) M_{(sp(\cdot))',\hat{B}_j^k}(\omega^{-1/s}) \approx 1,$$

$$M_{sp'(\cdot),\hat{B}_j^k}(\omega^{-1/s}) M_{(sp'(\cdot))',\hat{B}_j^k}(\omega^{1/s}) \approx 1.$$

Using $\|h\|_{sq(\cdot)}^s \approx \||h|^s\|_{q(\cdot)}$ for any $h \in L^{sq(\cdot)}(\mathbb{R}^n)$ we rewrite this as

$$M_{p(\cdot),\hat{B}_j^k}(\omega) M_{u'(\cdot),\hat{B}_j^k}(\omega^{-1}) \approx 1,$$

$$M_{p'(\cdot),\hat{B}_j^k}(\omega^{-1}) M_{v(\cdot),\hat{B}_j^k}(\omega) \approx 1.$$

This combined with $M_{p(\cdot),\hat{B}_j^k}(\omega) M_{p'(\cdot),\hat{B}_j^k}(\omega^{-1}) \approx 1$ (Lemma 5.8.4) implies (5.8.8). Therefore,

$$\int_{\mathbb{R}^n} |Mf||g|\, dx \leqslant c \sum_{j,k} |\hat{B}_j^k|\, M_{u(\cdot),\hat{B}_j^k}(f\omega)\, M_{v'(\cdot),\hat{B}_j^k}(g\omega^{-1}).$$

Using that $|\hat{B}_j^k| = 5^n |B_j^k| \leqslant 2 \cdot 5^n |F_j^k|$ and that $\{F_j^k\}_{j,k}$ is locally 3-finite we get

$$\int\limits_{\mathbb{R}^n} |Mf||g| \, dx \leqslant c \int\limits_{\mathbb{R}^n} \sum_{j,k} \chi_{F_j^k} M_{u(\cdot)}(f\omega) \, M_{v'(\cdot)}(g\omega^{-1}) \, dx$$

$$\leqslant c \int\limits_{\mathbb{R}^n} M_{u(\cdot)}(f\omega) \, M_{v'(\cdot)}(g\omega^{-1}) \, dx.$$

Now, Hölder's inequality with p and p' and the boundedness of $M_{u(\cdot)}$ on $L^{p(\cdot)}(\mathbb{R}^n)$ and of $M_{v'(\cdot)}$ on $L^{p'(\cdot)}(\mathbb{R}^n)$ imply that

$$\int\limits_{\mathbb{R}^n} |Mf||g| \, dx \leqslant c \, \|f\omega\|_{p(\cdot)} \|g\omega^{-1}\|_{p'(\cdot)} = c \, \|f\|_{p(\cdot),\omega} \|g\|_{p'(\cdot),\omega^{-1}}.$$

The result follows by the norm conjugate formula of $L_\omega^{p(\cdot)}$. $\qquad\qquad\square$

Remark 5.8.9. Note that in Theorem 5.8.6 we do not require explicitly $\varphi_{p(\cdot),\omega} \in \mathcal{A}$ but only $\varphi_{p(\cdot),\omega} \in \mathcal{A}_{\mathrm{loc}}$ (although the later follows automatically by the theorem). This appears to be in contrast to the fact that $p \in \mathcal{A}_{\mathrm{loc}}$ and $1 < p^- \leqslant p^+ < \infty$ does not imply $p \in \mathcal{A}$ (see Theorem 5.3.4).

However, it has been shown by Kopaliani [255] that $p \in \mathcal{A}_{\mathrm{loc}}$ implies $p \in \mathcal{A}$ if one additionally requires that p in constant outside a large set (i.e. around ∞). Based on arguments as in Sect. 7.3 this extra requirement can be relaxed to the log-Hölder decay condition. This decay condition is also responsible in Theorem 5.8.6 for the implication $\varphi_{p(\cdot),\omega} \in \mathcal{A}_{\mathrm{loc}} \Rightarrow \varphi_{p(\cdot),\omega} \in \mathcal{A}$.

Remark 5.8.10. The definition above of $L_\omega^{p(\cdot)}$ considers the weights as multipliers. This approach fits nicely to the theory of Banach function spaces. However, it is also possible to treat the weights as change of measure, which leads to the spaces $L^{p(\cdot)}(\Omega, \sigma \, dx)$, where σ is a weight. This approach leads naturally to the theory of Muckenhoupt weights, see (5.2.3). Analogously, we define as in [97] the variable exponent Muckenhoupt classes $A_{p(\cdot)}$ and $A_{p(\cdot),\mathrm{loc}}$ to consist of those weights σ, which satisfy $\sigma \, dx \in \mathcal{A}$ and $\sigma \, dx \in \mathcal{A}_{\mathrm{loc}}$, respectively. Note that for bounded exponents $\varphi_{p(\cdot),\omega} \in \mathcal{A}$ and $\varphi_{p(\cdot),\omega} \in \mathcal{A}_{\mathrm{loc}}$ if and only if $\omega^{\frac{1}{p}} \in A_{p(\cdot)}$ and $\omega^{\frac{1}{p}} \in A_{p(\cdot),\mathrm{loc}}$, respectively.

Chapter 6
Classical Operators

In this section we treat some of the most important operators of harmonic analysis in a variable exponent context. The results build on the boundedness of the maximal operator. We treat the Riesz potential operator, the sharp maximal function and singular integral operators in the three sections of the chapter. Several further operators are considered in Sect. 7.2. These results are applied in the second part of the book for instance to prove Sobolev embeddings and in the third part to prove existence and regularity of solutions to certain PDEs.

6.1 Riesz Potentials

In this section we derive natural generalizations of boundedness results for the Riesz potential operator in the context of variable exponents. Riesz potential operators have been studied in the variable exponent context e.g. in [22, 92, 115, 118, 163, 341]. Our proof is based on Hedberg's trick.

Definition 6.1.1. Let $0 < \alpha < n$. For measurable f we define $I_\alpha f : \mathbb{R}^n \to [0, \infty]$ by

$$I_\alpha f(x) := \int\limits_{\mathbb{R}^n} \frac{|f(y)|}{|x - y|^{n-\alpha}} \, dy.$$

The operator I_α is called the *Riesz potential operator* and the kernel $|x|^{\alpha-n}$ is called the *Riesz kernel*.

If the function f is defined on Ω only, then the integral should be taken over Ω, i.e. $I_\alpha f = I_\alpha(\chi_\Omega f)$ using the zero extension.

L. Diening et al., *Lebesgue and Sobolev Spaces with Variable Exponents,*
Lecture Notes in Mathematics 2017, DOI 10.1007/978-3-642-18363-8_6,
© Springer-Verlag Berlin Heidelberg 2011

Definition 6.1.2. For $k \in \mathbb{Z}$ we define the *averaging operator* T_k by

$$T_k f := \sum_{\substack{Q \text{ dyadic} \\ \text{diam}(Q)=2^{-k}}} \chi_Q M_{2Q} f$$

for all $f \in L^1_{\text{loc}}(\mathbb{R}^n)$.

Remark 6.1.3. Note that T_k can be written as the sum of 2^n-averaging operators $T_{\mathcal{Q}_j}$ with locally 1-finite families of shifted dyadic cubes, namely, each $2Q$ is split into 2^n disjoint cubes of the same size as Q. This property ensures the boundedness of T_k on $L^\varphi(\mathbb{R}^n)$ if $\varphi \in \mathcal{A}$.

The following result relates part of the Riesz potential operator I_α to the Hardy–Littlewood maximal operator M.

Lemma 6.1.4. *Let $x \in \mathbb{R}^n$, $\delta > 0$, $0 < \alpha < n$, and $f \in L^1_{\text{loc}}(\mathbb{R}^n)$. Then*

$$\int\limits_{B(x,\delta)} \frac{|f(y)|}{|x-y|^{n-\alpha}} \, dy \leqslant c(\alpha) \, \delta^\alpha \sum_{k=0}^\infty 2^{-\alpha k} T_{k+k_0} f(x) \leqslant c(\alpha) \, \delta^\alpha M f(x),$$

where $k_0 \in \mathbb{Z}$ is chosen such that $2^{-k_0-1} \leqslant \delta \leqslant 2^{-k_0}$.

Proof. We split the integration domain into annuli and use the definition of T_k:

$$\int\limits_{B(x,\delta)} \frac{|f(y)|}{|x-y|^{n-\alpha}} \, dy \leqslant \sum_{k=1}^\infty (\delta 2^{-k})^{\alpha-n} \int\limits_{2^{-k}\delta \leqslant |x-y| < 2^{-k+1}\delta} |f(y)| \, dy$$

$$\leqslant c \sum_{k=1}^\infty (\delta 2^{-k})^\alpha \fint\limits_{|x-y| < 2^{-k+1}\delta} |f(y)| \, dy$$

$$\leqslant c \, \delta^\alpha \sum_{k=0}^\infty 2^{-\alpha k} T_{k+k_0} f(x)$$

This is the first inequality. Since $T_{k+k_0} f \leqslant M f$, the second inequality follows by convergence of the geometric series. $\square$

For $0 \leqslant \alpha < \frac{n}{p^+}$ and $p \in \mathcal{P}(\mathbb{R}^n)$ we define $p^\sharp \in \mathcal{P}(\mathbb{R}^n)$ point-wise by

$$\frac{1}{p^\sharp(y)} := \frac{1}{p(y)} - \frac{\alpha}{n} \quad \text{for } y \in \mathbb{R}^n.$$

Then $p^\sharp \in \mathcal{P}(\mathbb{R}^n)$ with

$$1 < \frac{np^-}{n - \alpha p^-} = (p^\sharp)^- \leqslant (p^\sharp)^+ = \frac{np^+}{n - \alpha p^+} < \infty.$$

Also, it is clear that $p^\sharp \in \mathcal{P}^{\log}(\mathbb{R}^n)$ with $c_{\log}(p) = c_{\log}(p^\sharp)$ if $p \in \mathcal{P}^{\log}(\mathbb{R}^n)$.

Lemma 6.1.5. *Let* $p \in \mathcal{P}^{\log}(\mathbb{R}^n)$ *with* $1 < p^- \leqslant p^+ < \frac{n}{\alpha}$ *for* $\alpha \in (0, n)$. *Let* $x \in \mathbb{R}^n$, $\delta > 0$, *and* $f \in L^{p(\cdot)}(\mathbb{R}^n)$ *with* $\|f\|_{p(\cdot)} \leqslant 1$. *Then*

$$\int\limits_{\mathbb{R}^n \setminus B(x,\delta)} \frac{|f(y)|}{|x - y|^{n-\alpha}}\, dy \leqslant c \left(\frac{n - \alpha}{n - \alpha p^+} \right)^{\frac{n-\alpha}{n}} |B(x,\delta)|^{-\frac{1}{p^\sharp_{B(x,\delta)}}},$$

where the constant depends on $c_{\log}(p)$.

Proof. Set $B := B(x,\delta)$. We start with Hölder's inequality and take into account that $\|f\|_{p(\cdot)} \leqslant 1$:

$$\int\limits_{\mathbb{R}^n \setminus B} \frac{|f(y)|}{|x - y|^{n-\alpha}}\, dy \leqslant 2 \|f\|_{p(\cdot)} \big\| \chi_{\mathbb{R}^n \setminus B} |x - \cdot|^{\alpha - n} \big\|_{p'(\cdot)}$$

$$\leqslant 2 \big\| \chi_{\mathbb{R}^n \setminus B} |x - \cdot|^{-n} \big\|_{q(\cdot)}^{\frac{n-\alpha}{n}},$$

where $q := \frac{n-\alpha}{n} p'$. Next we note that

$$M(\chi_B |B|^{-1})(y) \geqslant \fint\limits_{B(y,2|x-y|)} \chi_B(z)|B|^{-1}\, dz$$

$$= \big| B(y, 2|x - y|) \big|^{-1}$$

$$= c\,|x - y|^{-n}$$

for all $y \in \mathbb{R}^n \setminus B$. Therefore

$$c\,\chi_{\mathbb{R}^n \setminus B}(y)|x - y|^{-n} \leqslant M(\chi_B |B|^{-1})(y)$$

for all $y \in \mathbb{R}^n$. Combining the previous estimates, we find that

$$\int\limits_{\mathbb{R}^n \setminus B} \frac{|f(y)|}{|x - y|^{n-\alpha}}\, dy \leqslant c \big\| M(\chi_B |B|^{-1}) \big\|_{q(\cdot)}^{\frac{n-\alpha}{n}}$$

$$= c\,|B|^{\frac{\alpha - n}{n}} \big\| M(\chi_B) \big\|_{q(\cdot)}^{\frac{n-\alpha}{n}}$$

$$\leqslant c\,\big((q^-)' \big)^{\frac{n-\alpha}{n}} |B|^{\frac{\alpha - n}{n}} \|\chi_B\|_{q(\cdot)}^{\frac{n-\alpha}{n}},$$

where we used Theorem 4.3.8 for the boundedness of M on $L^{q(\cdot)}(\mathbb{R}^n)$, which holds since $q \in \mathcal{P}^{\log}(\mathbb{R}^n)$ satisfies $1 < q^- \leqslant q^+ < \infty$.

Given that χ_B takes only values 0 and 1, we conclude that

$$\|\chi_B\|_{q(\cdot)}^{\frac{n-\alpha}{n}} = \|\chi_B\|_{p'(\cdot)} \leqslant c\,|B|^{\frac{1}{p'_B}},$$

where the second estimate follows from Corollary 4.5.9. Combining these estimates yields

$$\int_{\mathbb{R}^n \setminus B} \frac{|f(y)|}{|x-y|^{n-\alpha}}\, dy \leqslant c\,((q^-)')^{\frac{n-\alpha}{n}}\,|B|^{\frac{\alpha-n}{n}+\frac{1}{p'_B}}$$

$$\leqslant c\left(\frac{n-\alpha}{n-\alpha p^+}\right)^{\frac{n-\alpha}{n}}\,|B|^{-\frac{1}{p^\sharp_B}}. \qquad \square$$

For future reference we record the following point-wise estimate:

Proposition 6.1.6. *Let $\Omega \subset \mathbb{R}^n$ be a bounded, open set and $0 < \alpha < n$. Let $p \in \mathcal{P}^{\log}(\Omega)$ with $1 < p^- \leqslant p^+ < \frac{n}{\alpha}$. If $k \geqslant \max\left\{\frac{p^+}{n-\alpha p^+}, 1\right\}$, then*

$$I_\alpha f(x) \leqslant c\,k^{\frac{1}{(p^+)'}} Mf(x)^{1-\frac{\alpha}{n}p(x)}.$$

for every $f \in L^{p(\cdot)}(\Omega)$ with $\|f\|_{p(\cdot)} \leqslant 1$ and every $x \in \Omega$. The constant depends only on α, n, $c_{\log}(p)$, and $\mathrm{diam}(\Omega)$.

Proof. Let $x \in \Omega$ and let $0 < \delta \leqslant 2\,\mathrm{diam}\,\Omega$ be a number to be specified later. We extend p to $\mathbb{R}^n$ by Proposition 4.1.7. Then it follows from Lemmas 6.1.4 and 6.1.5 that

$$I_\alpha f(x) \leqslant c\,\delta^\alpha Mf(x) + c\,k^{\frac{1}{(p^+)'}} \delta^{-\frac{n}{p^\sharp_{B(x,\delta)}}}. \tag{6.1.7}$$

Since $\delta \leqslant 2\,\mathrm{diam}\,\Omega$, $\delta^{-n/p^\sharp_{B(x,\delta)}} \approx \delta^{-n/p^\sharp(x)}$.

If $[Mf(x)]^{-p(x)/n} < 2\,\mathrm{diam}(\Omega)$, we choose $\delta = [Mf(x)]^{-p(x)/n}$. Then inequality (6.1.7) gives the claim. On the other hand, if $[Mf(x)]^{-p(x)/n} \geqslant 2\,\mathrm{diam}(\Omega)$, we choose $\delta = 2\,\mathrm{diam}(\Omega)$. Now we have $\delta^\alpha \leqslant [Mf(x)]^{-(\alpha p(x))/n}$, so the claim follows directly from Lemma 6.1.4. $\qquad \square$

Lemma 6.1.8. *Let $p \in \mathcal{P}^{\log}(\mathbb{R}^n)$, $0 < \alpha < n$, and $1 < p^- \leqslant p^+ < \frac{n}{\alpha}$. For any $m > n$ there exists $c > 0$ only depending on $c_{\log}(p)$, p^+, α, and n such that*

$$I_\alpha f(x)^{p^\sharp(x)} \leqslant c\,Mf(x)^{p(x)} + h(x),$$

for all $f \in L^{p(\cdot)}$ with $\|f\|_{p(\cdot)} \leqslant 1$ and all $x \in \mathbb{R}^n$, where $h \in L^1(\mathbb{R}^n) \cap L^\infty(\mathbb{R}^n)$.

Proof. Define $q \in \mathcal{P}^{\log}(\mathbb{R}^n)$ by $q := \frac{n-\alpha}{n} p'$ and note that $1 < q^- \leqslant q^+ < \infty$. Therefore, by Theorem 4.3.8, M is bounded on $L^{q(\cdot)}(\mathbb{R}^n)$. Recall that $(p_B)' = (p')_B$ and $(p_B)^\sharp = (p^\sharp)_B$. Let $\|f\|_{p(\cdot)} \leqslant 1$ and $x \in \mathbb{R}^n$. By Lemma 6.1.4,

$$\int\limits_{B(x,\delta)} \frac{|f(y)|}{|x-y|^{n-\alpha}}\, dy \leqslant c\,\delta^\alpha Mf(x)$$

for every $\delta > 0$. We can assume that $f \neq 0$, since the claim is obvious for $f = 0$. Fix $\delta := \big(Mf(x)\big)^{-\frac{p(x)}{n}}$, so that $\big(\delta^\alpha Mf(x)\big)^{p^\sharp(x)} = Mf(x)^{p(x)}$. Then

$$\left(\int\limits_{B(x,\delta)} \frac{|f(y)|}{|x-y|^{n-\alpha}}\, dy \right)^{p^\sharp(x)} \leqslant c\, Mf(x)^{p(x)}.$$

We then estimate the remaining part of the integral, outside the ball $B(x,\delta)$. By Lemma 6.1.5,

$$\int\limits_{\mathbb{R}^n \setminus B(x,\delta)} \frac{|f(y)|}{|x-y|^{n-\alpha}}\, dy \leqslant c\,|B(x,\delta)|^{-\frac{1}{p^\sharp_{B(x,\delta)}}}.$$

Continuing with Lemma 4.2.7 for the exponent $p^\sharp$, we see that there exist $c > 0$ and $m > n$ such that

$$\left(\int\limits_{\mathbb{R}^n \setminus B(x,\delta)} \frac{|f(y)|}{|x-y|^{n-\alpha}}\, dy \right)^{p^\sharp(x)}$$

$$\leqslant c\,|B(x,\delta)|^{-1} + c\,(e+|x|)^{-m} + c\left(\fint\limits_{B(x,\delta)} (e+|y|)^{-m}\, dy \right)^{p^-}.$$

By definition of δ, the first term on the right-hand side equals $c\,Mf(x)^{p(x)}$. Let us set $h'(x) := c\,M((e + |\cdot|)^{-m})(x)^{p^-}$. Since $M((e+|\cdot|)^{-m}) \in$ w-$L^1(\mathbb{R}^n) \cap L^\infty(\mathbb{R}^n)$, it belongs to $L^{p^-}(\mathbb{R}^n)$ (Lemma 4.3.5), hence $h' \in L^1(\mathbb{R}^n)$. Then we have seen that both parts of the Riesz potential can been estimated by $c\,Mf(x)^{p(x)} + h(x)$, where $h := h' + (e + |\cdot|)^{-m} \in L^1(\mathbb{R}^n) \cap L^\infty(\mathbb{R}^n)$. □

Theorem 6.1.9. *Let* $p \in \mathcal{P}^{\log}(\mathbb{R}^n)$, $0 < \alpha < n$, *and* $1 < p^- \leqslant p^+ < \frac{n}{\alpha}$. *Then*

$$\|I_\alpha f\|_{p^\sharp(\cdot)} \leqslant c(n,\alpha,p)\,\|f\|_{p(\cdot)},$$

where the constant depends on p *only via* $c_{\log}(p)$, p^- *and* p^+.

Proof. Let $h \in \text{w-}L^1(\mathbb{R}^n) \cap L^\infty(\mathbb{R}^n)$ be as in Lemma 6.1.8. Let $\|f\|_{p(\cdot)} \leqslant 1$ and thus $\varrho_{p(\cdot)}(f) \leqslant 1$ by the unit ball property

Integrating the inequality in Lemma 6.1.8 over $x \in \mathbb{R}^n$ yields

$$\varrho_{p^\sharp(\cdot)}(I_\alpha f) \leqslant \varrho_{p(\cdot)}(Mf) + \varrho_1(h) \leqslant \varrho_{p(\cdot)}(Mf) + c.$$

By Theorem 4.3.8, M is bounded on $L^{p(\cdot)}(\mathbb{R}^n)$ and so $\varrho_{p(\cdot)}(f) \leqslant 1$ implies $\varrho_{p(\cdot)}(Mf) \leqslant c$, where we used $p^+ < \infty$. Hence $\varrho_{p^\sharp(\cdot)}(I_\alpha f) \leqslant c$ and therefore $\|I_\alpha f\|_{p^\sharp(\cdot)} \leqslant c$, where we used $(p^\sharp)^+ < \infty$. Since I_α is sublinear, a scaling argument completes the proof. $\square$

Remark 6.1.10. As in Remark 4.2.8, it is possible to replace the log-Hölder decay condition of $\frac{1}{p}$ in Lemma 6.1.8 and Theorem 6.1.9 by the weaker condition $1 \in L^{s(\cdot)}$ with $\frac{1}{s(x)} := \left| \frac{1}{p(x)} - \frac{1}{p_\infty} \right| = \left| \frac{1}{p^\sharp(x)} - \frac{1}{(p^\sharp)_\infty} \right|$.

Next we prove a weak type estimate for the Riesz potential. The proof shares the idea with Proposition 4.4.14.

Theorem 6.1.11. *Let* $\Omega \subset \mathbb{R}^n$ *be a bounded, open set. Suppose that* $p \in \mathcal{P}^{\log}(\Omega)$, $0 < \alpha < n$, *and* $1 < p^- \leqslant p^+ < \frac{n}{\alpha}$. *Let* $f \in L^{p(\cdot)}(\Omega)$ *be such that* $2(1 + |\Omega|)\|f\|_{p(\cdot)} \leqslant 1$. *Then for every* $t > 0$ *we have*

$$\int\limits_{\{I_\alpha f > t\}} t^{p^\sharp(x)} dx \leqslant c \int\limits_\Omega |f(y)|^{p(y)} dy + \left| \{x \in \Omega : 0 < |f| \leqslant 1\} \right|.$$

Proof. We obtain by Proposition 6.1.6 that

$$\left\{ I_\alpha f(x) > t \right\} \subset \left\{ c\,[Mf(x)]^{\frac{p(x)}{p^\sharp(x)}} > t \right\} =: E.$$

For every $z \in E$ we choose $B_z := B(z, r)$ such that $c\,(M_{B_z} f)^{\frac{p(z)}{p^\sharp(z)}} > t$. Let $x \in B_z$ and raise this inequality to the power $p^\sharp(x)$. Let us write $q(x) := p(z)p^\sharp(x)/p^\sharp(z)$. Assume first that $q(x) \geqslant p(x)$, i.e. $p(x) \geqslant p(z)$. Since $2(1 + |\Omega|)\|f\|_{p(\cdot)} \leqslant 1$ we get $M_{B_z} f \leqslant |B_z|^{-1}$, and thus we obtain

$$t^{p^\sharp(x)} \leqslant c\,(M_{B_z} f)^{p(x)} (M_{B_z} f)^{q(x)-p(x)} \leqslant c\,(M_{B_z} f)^{p(x)} |B_z|^{p(x)-q(x)}$$
$$= c\,(M_{B_z} f)^{p(x)} |B_z|^{\frac{\alpha p(x)(p(z)-p(x))}{n-\alpha p(x)}}.$$

The term $|B_z|^{\frac{\alpha p(x)(p(z)-p(x))}{n-\alpha p(x)}}$ is uniformly bounded since p is log-Hölder continuous. By Theorem 4.2.4 this yields

$$t^{p^\sharp(x)} \leqslant c\,M_{B_z}\left(|f|^{p(\cdot)} + \chi_{\{0<|f|\leqslant 1\}} \right)$$

for every $x \in B_z$ such that $q(x) \geqslant p(x)$. Assume now that $q(x) < p(x)$. By Theorem 4.2.4 we obtain

$$t^{p^\sharp(x)} \leqslant c \left(M_{B_z}\left(|f|^{p(\cdot)} + \chi_{\{0<|f|\leqslant 1\}}\right) \right)^{\frac{q(x)}{p(x)}} \leqslant c M_{B_z}\left(|f|^{p(\cdot)} + \chi_{\{0<|f|\leqslant 1\}}\right),$$

where the last inequality follows since $M_{B_z}\left(|f|^{p(\cdot)} + \chi_{\{0<|f|\leqslant 1\}}\right) \geqslant 1$ and $q(x)/p(x) < 1$.

By the Besicovitch covering theorem (Theorem 1.4.6) there is a countable covering subfamily (B_i), with the bounded overlap-property. Thus

$$\int_E t^{p^\sharp(x)} dx \leqslant \sum_{i=1}^\infty \int_{B_i} t^{p^\sharp(x)} dx \leqslant c \sum_{i=1}^\infty \int_{B_i} \fint_{B_i} |f(y)|^{p(y)} + \chi_{\{0<|f|\leqslant 1\}}(y)\, dy\, dx$$

$$= c \sum_{i=1}^\infty \int_{B_i} |f(y)|^{p(y)} + \chi_{\{0<|f|\leqslant 1\}}(y)\, dy$$

$$\leqslant c \int_\Omega |f(y)|^{p(y)} dy + |\{x \in \Omega : 0 < |f| \leqslant 1\}|. \qquad \square$$

The following Jensen type inequality will be needed later in Sect. 8.2 to prove a special type Poincaré inequality, Proposition 8.2.11. The result is from [350].

Lemma 6.1.12 (Jensen inequality with singular measure). *Let $p \in \mathcal{P}^{\log}(\mathbb{R}^n)$. For every $m > 0$ there exists $\beta \in (0,1)$ only depending on m and $c_{\log}(p)$ such that*

$$\varphi_{p(x)} \left(\beta \int_B \frac{|f(y)|}{r\,|x-y|^{n-1}}\, dy \right) \leqslant \int_B \frac{\varphi_{p(y)}(|f(y)|)}{r\,|x-y|^{n-1}}\, dy + M\left((e + |\cdot|)^{-m}\right)(x)$$

for every $R > 0$, $B := B(z,r)$, $x \in B$ and every $f \in L^{p(\cdot)}(B) \cap L^\infty(B)$ with $\|f\|_{L^{p(\cdot)}(B) + L^\infty(B)} \leqslant 1$.

Proof. Fix $r > 0$ and let B be a ball with radius r. Define annuli $A_k := \{y \in B : 2^{-k} \leqslant |x-y| R^{-1} \leqslant 2^{1-k}\}$ for $k \geqslant 0$. As in the proof of Lemma 6.1.4 we split the B into the annuli A_k and obtain

$$\int_B \frac{|f(y)|}{r\,|x-y|^{n-1}}\, dy \leqslant c_1 \sum_{k=0}^\infty 2^{-k} \fint_{B(x,2^{1-k})} \chi_{A_k} |f|\, dy.$$

Let $\beta_2 > 0$ be the β from the key estimate Lemma 4.2.4. Since $\sum_{k=0}^{\infty} 2^{-k} \leqslant 1$, it follows by convexity that

$$(I) := \varphi_{p(x)}\left(\frac{\beta_2}{c_1} \int_B \frac{|f(y)|}{r\,|x-y|^{n-1}}\,dy\right) \leqslant \sum_{k=1}^{\infty} 2^{-k} \varphi_{p(x)}\left(\beta_2 \fint_{B(x,2^{1-k})} \chi_{A_k} |f|\,dy\right).$$

So the key estimate, Lemma 4.2.4, yields

$$(I) \leqslant \sum_{k=1}^{\infty} 2^{-k} \Bigg(\fint_{B(x,2^{1-k})} \chi_{A_k} \varphi_{p(y)}(|f|)\,dy$$

$$+ \frac{1}{2}(e+|x|)^{-m} + \frac{1}{2} \fint_{B(x,2^{1-k})} (e+|y|)^{-m}\,dy \Bigg)$$

$$\leqslant \fint_B \frac{\varphi_{p(y)}(|f(y)|)}{r\,|x-y|^{n-1}}\,dy + M\big((e+|\cdot|)^{-m}\big)(x). \qquad \square$$

6.2 The Sharp Operator $M^\sharp f$

In this section we prove the fundamental estimate that the norm $\|f\|_{p(\cdot)}$ of an $L^{p(\cdot)}(\mathbb{R}^n)$ function f (scalar or vectorial) can be estimated by $\|M^\sharp f\|_{p(\cdot)}$ as long as $p \in \mathcal{A}$ satisfies $1 < p^- \leqslant p^+ < \infty$. We introduce the following maximal operators: let $0 < s < \infty$ and $f \in L^s_{\mathrm{loc}}(\mathbb{R}^n)$. Then for all balls (or cubes) B we define

$$M_{s,B}f = \left(\fint_B |f(y)|^s\,dy\right)^{\frac{1}{s}}, \qquad\qquad M_s f(x) = \sup_{B \ni x} M_{s,B}f,$$

$$M^\sharp_{s,B}f := \left(\fint_B |f(y)-f_B|^s\,dy\right)^{\frac{1}{s}}, \qquad\qquad M^\sharp_s f(x) := \sup_{B \ni x} M^\sharp_{s,B}f,$$

for all $x \in \mathbb{R}^n$, where the supremum is taken over all balls (or cubes) $B \subset \mathbb{R}^n$ which contain x. Then the *sharp operator* $M^\sharp f$ is defined by $M^\sharp f := M^\sharp_1 f$. Note that $f \in \mathrm{BMO}(\mathbb{R}^n)$ if and only if $M^\sharp f \in L^\infty(\mathbb{R}^n)$.

Further note that $M_1 f = Mf$ and that $M_{s_1} f \leqslant M_{s_2} f$ and $M^\sharp_{s_1} f \leqslant M^\sharp_{s_2} f$ for $s_1 \leqslant s_2$ by Jensen's or Hölder's inequality. Since $|M^\sharp f| \leqslant 2\,Mf$, it is easy to see that $f \mapsto M^\sharp f$ is bounded on $L^s(\mathbb{R}^n)$ for all $1 < s < \infty$, and that this generalizes to the variable exponent context due to Theorem 5.7.2:

Lemma 6.2.1. *For $p \in \mathcal{A}$ with $1 < p^- \leqslant p^+ < \infty$, $M^\sharp f$ is bounded on $L^{p(\cdot)}(\mathbb{R}^n)$.*

A more interesting fact about $M^\sharp f$ is that

$$\|f\|_s \leqslant c\,\|M^\sharp f\|_s, \tag{6.2.2}$$

for all $f \in L^s(\mathbb{R}^n)$ and constant $s \in (1, \infty)$ (cf. [360]). Hence $f \in L^s(\mathbb{R}^n)$ if and only if $M^\sharp f \in L^s(\mathbb{R}^n)$ when $1 < s < \infty$. We now generalize this to the variable exponent context.

The original proof of (6.2.2) uses the grand maximal function, however, in our case we can circumvent this by the following lemma, which combines [360, Sect. II.2.1] and [360, Sect. IV.2].

Lemma 6.2.3. *For all $f \in L^\infty(\mathbb{R}^n)$ and $g \in C_{0,0}^\infty(\mathbb{R}^n)$,*

$$|\langle f, g \rangle| \leqslant c\,\langle M^\sharp f, Mg \rangle.$$

Using approximation by smooth functions, this is generalized to Lebesgue functions:

Lemma 6.2.4. *Let $p \in \mathcal{A}$ with $1 < p^- \leqslant p^+ < \infty$. Then*

$$|\langle f, g \rangle| \leqslant c\,\langle M^\sharp f, Mg \rangle$$

for all $f \in L^{p(\cdot)}(\mathbb{R}^n)$ and $g \in L^{p'(\cdot)}(\mathbb{R}^n)$

Proof. Let $f \in L^{p(\cdot)}(\mathbb{R}^n)$ and $g \in L^{p'(\cdot)}(\mathbb{R}^n)$. Then by the Theorem 3.4.12 and Proposition 3.4.14 there exist sequences $(f_i) \subset C_0^\infty(\mathbb{R}^n)$ and $(g_i) \subset C_{0,0}^\infty(\mathbb{R}^n)$ such that $f_i \to f$ in $L^{p(\cdot)}(\mathbb{R}^n)$ and $g_i \to g$ in $L^{p'(\cdot)}(\mathbb{R}^n)$. Since $M^\sharp$ is a bounded operator, $M^\sharp f_i \to M^\sharp f$ in $L^{p(\cdot)}(\mathbb{R}^n)$. As $p' \in \mathcal{A}$ by Theorem 5.7.2, the maximal operator is bounded on $L^{p'(\cdot)}(\mathbb{R}^n)$ and so $Mg_i \to Mg$ in $L^{p'(\cdot)}(\mathbb{R}^n)$. Thus it follows from Lemma 6.2.3 that

$$|\langle f, g \rangle| = \lim_{n \to \infty} |\langle f_i, g_i \rangle| \leqslant \lim_{n \to \infty} c\,\langle M^\sharp f_i, Mg_i \rangle = c\,|\langle M^\sharp f, Mg \rangle|. \qquad \square$$

Theorem 6.2.5. *If $p \in \mathcal{A}$ with $1 < p^- \leqslant p^+ < \infty$, then*

$$\|f\|_{p(\cdot)} \leqslant c\,\|M^\sharp f\|_{p(\cdot)}$$

for all $f \in L^{p(\cdot)}(\mathbb{R}^n)$.

Proof. From Lemma 6.2.4 it follows that

$$|\langle f, g \rangle| \leqslant c \, \|M^{\sharp} f\|_{p(\cdot)} \|M g\|_{p'(\cdot)} \leqslant c \, \|M^{\sharp} f\|_{p(\cdot)} \|g\|_{p'(\cdot)}$$

for all $f \in L^{p(\cdot)}(\mathbb{R}^n)$ and $g \in L^{p'(\cdot)}(\mathbb{R}^n)$. The assertion now follows from the norm conjugate formula (Corollary 3.2.14). □

6.3 Calderón–Zygmund Operators

We now turn to the examination of singular integral operators on $L^{p(\cdot)}(\mathbb{R}^n)$. These operators are very useful in the study of partial differential equations. In a classical application they are used to prove that the Poisson problem $-\Delta u = f$ with $f \in L^q(\mathbb{R}^n)$ has a solution $u \in W^{2,q}(\mathbb{R}^n)$. We will generalize this statement to the variable exponent setting in Sect. 14.1. Singular integral operators have been studied in the variable exponent setting for instance in [231, 241–243, 247].

Definition 6.3.1. A kernel k on $\mathbb{R}^n \times \mathbb{R}^n$ is a locally integrable complex-valued function k, defined off the diagonal. We say that k satisfies *standard estimates* if there exists $\delta > 0$ and $c > 0$, such that the following inequalities hold for all distinct $x, y \in \mathbb{R}^n$ and all $z \in \mathbb{R}^n$ with $|x - z| < \frac{1}{2}|x - y|$:

$$|k(x, y)| \leqslant c \, |x - y|^{-n},$$
$$|k(x, y) - k(z, y)| \leqslant c \, |x - z|^{\delta} |x - y|^{-n-\delta}, \qquad (6.3.2)$$
$$|k(y, x) - k(y, z)| \leqslant c \, |x - z|^{\delta} |x - y|^{-n-\delta}.$$

In this case we call k a *standard kernel*.

We say that a linear and bounded operator $T : C_0^{\infty}(\mathbb{R}^n) \to \mathcal{D}'(\mathbb{R}^n)$, where $\mathcal{D}'$ is the space of distributions, is *associated to a kernel* k, if

$$\langle Tf, g \rangle = \int_{\mathbb{R}^n} \int_{\mathbb{R}^n} k(x, y) f(y) g(x) \, dx \, dy,$$

whenever $f, g \in C_0^{\infty}(\mathbb{R}^n)$ with $\mathrm{spt}(f) \cap \mathrm{spt}(g) = \emptyset$. We say that T is a *singular integral operator* if T is associated to a standard kernel. If in addition T extends to a bounded, linear operator on $L^2(\mathbb{R}^n)$, then it is a *Calderón–Zygmund operator*. The following result is proved in [74, 76].

Proposition 6.3.3. *Every Calderón–Zygmund operator is of strong type s, $s \in (1, \infty)$, and of weak type 1.*

Since most of the classical convolution operators are defined by principal value integrals, we define the *truncated kernels* k_{ε} for $\varepsilon > 0$ by

$$k_\varepsilon(x, y) := \begin{cases} k(x, y) & \text{for } |x - y| > \varepsilon, \\ 0 & \text{for } |x - y| \leqslant \varepsilon. \end{cases}$$

Furthermore we define for $\varepsilon > 0$

$$T_\varepsilon f(x) := \int_{\mathbb{R}^n} k_\varepsilon(x, y) f(y) \, dy,$$

and we say that T_ε is associated to the kernel k_ε. In [63] it is shown that

Proposition 6.3.4. *Let k be a kernel on $\mathbb{R}^n \times \mathbb{R}^n$. Assume that $N(x, z) := k(x, x - z)$ is homogeneous of degree $-n$ in z and that*

(a) *for every x, $N(x, z)$ is integrable over the sphere $|z| = 1$ and its integral equals zero; and*
(b) *for some $\sigma > 1$ and every x, $|N(x, z)|^\sigma$ is integrable over the sphere $|z| = 1$ and its integral is bounded uniformly with respect to x.*

Then for every $s \in [\sigma', \infty)$ the operators T_ε are uniformly bounded on $L^s(\mathbb{R}^n)$ with respect to $\varepsilon > 0$. Moreover,

$$Tf(x) := \lim_{\varepsilon \to 0^+} T_\varepsilon f(x) \tag{6.3.5}$$

exists almost everywhere and $T_\varepsilon f \to Tf$ in $L^s(\mathbb{R}^n)$. In particular T is bounded on $L^s(\mathbb{R}^n)$.

Remark 6.3.6. If k is a standard kernel which satisfies (a) and (b) in Proposition 6.3.4, and T is defined by the principal value integral of (6.3.5), then T is a Calderón–Zygmund operator: Indeed, due to Proposition 6.3.4 the operator T is bounded on $L^s(\mathbb{R}^n)$ for all $s \in [\sigma', \infty)$ and $T_\varepsilon f \to Tf$ almost everywhere and in measure for all $f \in L^s(\mathbb{R}^n)$. Moreover, (6.3.2) implies for all $x, y, z \in \mathbb{R}^n$ with $|x - z| < r$ that

$$\int_{|y-x| \geqslant 2r} |k(y, x) - k(y, z)| \, dy \leqslant c \int_{|y-x| \geqslant 2r} |x - z|^\delta |x - y|^{-n-\delta} \, dy$$

$$\leqslant c \, r^\delta \int_{|w| \geqslant 2r} |w|^{-n-\delta} \, dw$$

$$\leqslant c.$$

Thus, [360, Corollary I.7.1, p. 33] implies that the operators T_ε are of weak type 1 uniformly with respect to $\varepsilon > 0$, i.e.

$$\left| \{ |T_\varepsilon f| > \alpha \} \right| \leqslant c \frac{\|f\|_1}{\alpha}$$

for all $f \in L^1(\mathbb{R}^n) \cap L^{\sigma'}(\mathbb{R}^n)$ and all $\alpha > 0$. Thus

$$\left|\{|Tf| > \alpha\}\right| \leqslant \left|\{|(T_\varepsilon - T)f| > \tfrac{\alpha}{2}\}\right| + \left|\{|T_\varepsilon f| > \tfrac{\alpha}{2}\}\right|$$

$$\leqslant \left|\{|(T_\varepsilon - T)f| > \tfrac{\alpha}{2}\}\right| + c\frac{\|f\|_1}{\alpha}$$

for all $\alpha > 0$. Since $T_\varepsilon f \to Tf$ in measure, this implies that T is of weak type 1. Since T is also bounded on $L^s(\mathbb{R}^n)$ for all s with $s \in [\sigma', \infty)$, this implies that T is bounded on $L^2(\mathbb{R}^n)$. Thus T is a Calderón–Zygmund operator.

Although the class of Calderón–Zygmund operators will be our model case we use the following definition of Alvarez and Pérez [29].

Definition 6.3.7 (Condition (D)). For a kernel k on $\mathbb{R}^n \times \mathbb{R}^n$ we define

$$D_{B(x_0,r)}k(y) := \fint_{B(x_0,r)} \fint_{B(x_0,r)} |k(z,y) - k(x,y)|\, dx\, dz.$$

The kernel k is said to satisfy *condition (D)* if there are constants $c, N > 0$ such that

$$\sup_{r>0} \int_{|y-x_0|>Nr} |f(y)| D_{B(x_0,r)}k(y)\, dy \leqslant c\, Mf(x_0)$$

for all $f \in C_0^\infty(\mathbb{R}^n)$ and $x_0 \in \mathbb{R}^n$.

Note that a standard kernel satisfies condition (D). This is easily seen by the following argument: Let $r > 0$ and $B := B(x_0, r)$. If $|x_0 - y| > 5r$, then $|x - z| < 2r < \tfrac{1}{2}|x - y|$ and $|x - y| > \tfrac{4}{5}|x_0 - y|$ for $x, z \in B$. Thus, by (6.3.2),

$$D_B k(y) \leqslant \fint_B \fint_B c\, |x - z|^\delta |x - y|^{-n-\delta}\, dx\, dz \leqslant c\, r^\delta |x_0 - y|^{-n-\delta}.$$

Let $A := 10B \setminus 5B$ denote an annulus. Then

$$\int_{2^j A} |f|\, D_B k\, dy \leqslant c \int_{2^j A} |f|\, \left(r/|y - x_0|\right)^\delta |y - x_0|^{-n}\, dy \leqslant c\, 2^{-\delta j} Mf(x_0)$$

for $j \geqslant 0$, since $\left(r/|y - x_0|\right)^\delta \approx 2^{-\delta j}$ and $|y - x_0|^{-n} \approx |2^j A|^{-1}$ for $y \in 2^j A$. Therefore

$$\int\limits_{\mathbb{R}^n \setminus 5B} |f| \, D_B k \, dy = \sum_{j=0}^{\infty} \int\limits_{2^j A} |f| \, D_B k \, dy \leqslant c \sum_{j=0}^{\infty} 2^{-\delta j} Mf(x_0) = c(\delta) \, Mf(x_0).$$

Thus condition (D) is verified.

In [29] it is shown that

Proposition 6.3.8. *Let T be an operator associated to a kernel k on $\mathbb{R}^n \times \mathbb{R}^n$ satisfying condition (D). Let us suppose that T extends to a bounded operator from $L^1(\mathbb{R}^n)$ to $w\text{-}L^1(\mathbb{R}^n)$. Then, for each $0 < s < 1$, there exists $c = c(s) > 0$ such that*

$$\left(M^\sharp(|Tf|^s)\right)^{\frac{1}{s}}(x) \leqslant c \, Mf(x)$$

for all $f \in C_0^\infty(\mathbb{R}^n)$ and $x \in \mathbb{R}^n$.

We now show the boundedness of such operators on $L^{p(\cdot)}(\mathbb{R}^n)$ under the usual assumptions.

Theorem 6.3.9. *Let $p \in \mathcal{A}$ with $1 < p^- \leqslant p^+ < \infty$. Let T be an operator associated to a kernel k on $\mathbb{R}^n \times \mathbb{R}^n$ satisfying condition (D). Let us suppose that T extends to a bounded operator from $L^1(\mathbb{R}^n)$ to $w\text{-}L^1(\mathbb{R}^n)$. Then T is of strong type $p(\cdot)$.*

Proof. By Theorem 5.7.2 there exists $0 < s < 1$ such that $\left(\frac{p}{s}\right)' \in \mathcal{A}$. Then it follows from Theorem 6.2.5 that

$$\|Tf\|_{p(\cdot)} = \left\| |Tf|^s \right\|_{\frac{p(\cdot)}{s}}^{\frac{1}{s}} \leqslant c \left\| M^\sharp(|Tf|^s) \right\|_{\frac{p(\cdot)}{s}}^{\frac{1}{s}} = c \left\| (M^\sharp(|Tf|^s))^{\frac{1}{s}} \right\|_{p(\cdot)}.$$

From Proposition 6.3.8 it follows that

$$\left(M^\sharp(|Tf|^s)\right)^{\frac{1}{s}}(x) \leqslant c \, Mf(x)$$

for all $f \in C_0^\infty(\mathbb{R}^n)$ and all $x \in \mathbb{R}^n$. Thus $\|Tf\|_{p(\cdot)} \leqslant c \|Mf\|_{p(\cdot)} \leqslant c \|f\|_{p(\cdot)}$ for all $f \in C_0^\infty(\mathbb{R}^n)$. Since $C_0^\infty(\mathbb{R}^n)$ is dense in $L^{p(\cdot)}(\mathbb{R}^n)$ (Theorem 3.4.12), this proves the theorem.   □

Since every Calderón–Zygmund kernel is a standard kernel, it satisfies condition (D), and hence we immediately obtain the following

Corollary 6.3.10. *Let T be a Calderón–Zygmund operator with kernel k on $\mathbb{R}^n \times \mathbb{R}^n$ and let $p \in \mathcal{A}$ with $1 < p^- \leqslant p^+ < \infty$. Then T is bounded on $L^{p(\cdot)}(\mathbb{R}^n)$.*

To transfer the statements about the principal value integral $\lim_{\varepsilon \to 0^+} T_\varepsilon f$ to the spaces $L^{p(\cdot)}(\mathbb{R}^n)$ we make use of the maximal truncated operator T^*, which is defined by

$$T^* f(x) := \sup_{\varepsilon > 0} |T_\varepsilon f(x)|.$$

It is shown e.g. in [74] that

Proposition 6.3.11. *Let T be a Calderón–Zygmund operator with kernel k on $\mathbb{R}^n \times \mathbb{R}^n$. Then there exists $c = c(s) > 0$, such that*

$$T^* f(x) \leqslant c \left(M(Tf)(x) + Mf(x) \right)$$

for all $f \in L^s(\mathbb{R}^n)$ and all $x \in \mathbb{R}^n$, where $1 \leqslant s < \infty$.

Corollary 6.3.12. *Let $p \in \mathcal{A}$ with $1 < p^- \leqslant p^+ < \infty$, and let T be a Calderón–Zygmund operator with kernel k on $\mathbb{R}^n \times \mathbb{R}^n$. Then T^* is bounded on $L^{p(\cdot)}(\mathbb{R}^n)$.*

Proof. Let $f \in C_0^\infty(\mathbb{R}^n)$ with $\|f\|_{p(\cdot)} \leqslant 1$. By Theorem 5.7.2 the operator M is bounded on $L^{p(\cdot)}(\mathbb{R}^n)$ and by Corollary 6.3.10 $\|Tf\|_{p(\cdot)} \leqslant c$. Therefore $\|M(Tf)\|_{p(\cdot)} + \|Mf\|_{p(\cdot)} \leqslant c$. Now, Proposition 6.3.11 yields $\|T^* f\|_{p(\cdot)} \leqslant c$. Since $C_0^\infty(\mathbb{R}^n)$ is dense in $L^{p(\cdot)}(\mathbb{R}^n)$ (Theorem 3.4.12), the same holds also for arbitrary $f \in L^{p(\cdot)}(\mathbb{R}^n)$ with $\|f\|_{p(\cdot)} \leqslant 1$. The last restriction is removed by a scaling argument, which proves the corollary. $\square$

Corollary 6.3.13. *Let k be a standard kernel on $\mathbb{R}^n \times \mathbb{R}^n$ which satisfies condition (a) and (b) of Proposition 6.3.4. Let $p \in \mathcal{A}$ with $1 < p^- \leqslant p^+ < \infty$. Then the operators T_ε are uniformly bounded on $L^{p(\cdot)}(\mathbb{R}^n)$ with respect to $\varepsilon > 0$. Moreover,*

$$Tf(x) := \lim_{\varepsilon \to 0^+} T_\varepsilon f(x) \tag{6.3.14}$$

exists almost everywhere and $T_\varepsilon f \to Tf$ in $L^{p(\cdot)}(\mathbb{R}^n)$. In particular, T is bounded on $L^{p(\cdot)}(\mathbb{R}^n)$.

Proof. By Corollary 6.3.10 and Corollary 6.3.12 the operators T and T^* are bounded on $L^{p(\cdot)}(\mathbb{R}^n)$. Since $|T_\varepsilon f(x)| \leqslant T^* f(x)$ for all $f \in L^{p(\cdot)}(\mathbb{R}^n)$ and all $x \in \mathbb{R}^n$ by definition of T^*, it follows that the operators T_ε are uniformly bounded on $L^{p(\cdot)}(\mathbb{R}^n)$ with respect to $\varepsilon > 0$. Now fix $g \in C_0^\infty(\mathbb{R}^n)$. By Proposition 6.3.4, $\lim_{\varepsilon \to 0^+} T_\varepsilon g = Tg$ in $L^{p^-}(\mathbb{R}^n)$ and $L^{p^+}(\mathbb{R}^n)$, hence also in $L^{p(\cdot)}(\mathbb{R}^n)$ by Theorem 3.3.11. By the density of smooth functions in $L^{p(\cdot)}(\mathbb{R}^n)$ (Theorem 3.4.12) and the boundedness of T and T_ε, the claim follows also for $f \in L^{p(\cdot)}(\mathbb{R}^n)$. Due to Theorem 3.3.11 each $f \in L^{p(\cdot)}(\mathbb{R}^n)$ belongs to $L^{p^-}(\mathbb{R}^n) + L^{p^+}(\mathbb{R}^n)$, which together with Proposition 6.3.4 shows that the convergence in (6.3.14) holds also almost everywhere. $\square$

Chapter 7
Transfer Techniques

This chapter is a collection of various techniques with the common theme "transfer". In other words we study methods which allow us to take results from one setting and obtain corresponding results in another setting "for free". The best known example of such a technique is interpolation, which has played an important unifying role in the development of the theory of constant exponent spaces [362–364]. Complex interpolation is presented in Sect. 7.1. Unfortunately, interpolation is not so useful in the variable exponent setting, since it is not possible to interpolate from constant exponents to variable exponents. Therefore other techniques are also included, namely, extrapolation (Sect. 7.2), a tool for going from bounded to unbounded domains (Sect. 7.3), and a tool for going from results in balls to results in John domains (Sect. 7.4).

7.1 Complex Interpolation

Interpolation is a useful tool for linear, bounded operators on Banach spaces. The idea is the following: if S is a linear operator which is bounded as an operator from X_0 to Y_0 and as an operator from X_1 to Y_1, then automatically S is linear and bounded as an operator from $X_{[\theta]}$ to $Y_{[\theta]}$ for all $0 < \theta < 1$, where $X_{[\theta]}$ and $Y_{[\theta]}$ are the complex interpolation spaces of the Banach couples (X_0, X_1) and (Y_0, Y_1). Here *Banach couple* means that both spaces are continuously embedded into the same Hausdorff topological vector space. To use this result, we need to know what Lebesgue space $[L^{p(\cdot)}, L^{q(\cdot)}]_{[\theta]}$ is. In this section we characterize this interpolation space.

We refer to Bergh and Löfström [45] for an exposition on interpolation theory. Given $p_0, p_1 \in \mathcal{P}(A, \mu)$, we define $p_\theta \in \mathcal{P}(A, \mu)$ point-wise by the expression

$$\frac{1}{p_\theta} := \frac{1 - \theta}{p_0} + \frac{\theta}{p_1}$$

for $\theta \in [0, 1]$.

L. Diening et al., *Lebesgue and Sobolev Spaces with Variable Exponents*,
Lecture Notes in Mathematics 2017, DOI 10.1007/978-3-642-18363-8_7,
© Springer-Verlag Berlin Heidelberg 2011

Lemma 7.1.1. *If $q_0, q_1 \in [1, \infty]$ and $\theta \in (0, 1)$, then*

$$\bar{\varphi}_{q_\theta}^{-1}(t) = \left(\bar{\varphi}_{q_0}^{-1}(t)\right)^{1-\theta} \left(\bar{\varphi}_{q_1}^{-1}(t)\right)^{\theta}$$

for all $t \geqslant 0$.

Proof. If $q_0, q_1 < \infty$, then the claim follows directly from $\bar{\varphi}_q^{-1}(t) = t^{\frac{1}{q}}$ for $q \in [1, \infty)$ and $t \geqslant 0$. The assertion is also obvious if $q_0 = q_1 = \infty$. Now, let $q_0 < \infty$ and $q_1 = \infty$. Since $\varphi_\infty^{-1}(t) = \chi_{(0,\infty)}(t)$ we have

$$\left(\bar{\varphi}_{q_0}^{-1}(t)\right)^{1-\theta} \left(\bar{\varphi}_\infty^{-1}(t)\right)^{\theta} = \left(\bar{\varphi}_{q_0}^{-1}(t)\right)^{1-\theta} = t^{\frac{1-\theta}{q_0}} = \bar{\varphi}_{q_\theta}^{-1}(t)$$

for all $t \geqslant 0$, as claimed. $\square$

We recall the definition of the norm in the interpolation space $[L^{p(\cdot)}, L^{q(\cdot)}]_{[\theta]}$. Let $S := \{z \in \mathbb{C} : 0 < \operatorname{Re} z < 1\}$, so that $\overline{S} = \{z \in \mathbb{C} : 0 \leqslant \operatorname{Re} z \leqslant 1\}$, where $\operatorname{Re} z$ is the real part of z. Let $\mathcal{F}$ be the space of functions on $\overline{S}$ with values in $L^{p_0(\cdot)} + L^{p_1(\cdot)}$ which are analytic on S and bounded and continuous on $\overline{S}$ such that $F(\mathrm{i}t)$ and $F(1 + \mathrm{i}t)$ tend to zero for $|t| \to \infty$. (Recall that i denotes the imaginary unit. Also, F is analytic with values in a Banach space means that $\frac{d}{d\bar{z}}F = 0$ in the Banach space.) For $F \in \mathcal{F}$ we set

$$\|F\|_{\mathcal{F}} := \sup_{t \in \mathbb{R}} \max \left\{ \left\|F(\mathrm{i}t)\right\|_{p_0(\cdot)}, \left\|F(1 + \mathrm{i}t)\right\|_{p_1(\cdot)} \right\}.$$

Then we define

$$\|f\|_{[\theta]} := \inf \left\{ \|F\|_{\mathcal{F}} : F \in \mathcal{F} \text{ and } f = F(\theta) \right\}.$$

We are now prepared to study the space $[L^{p(\cdot)}, L^{q(\cdot)}]_{[\theta]}$.

Theorem 7.1.2 (Complex interpolation). *Let $p_0, p_1 \in \mathcal{P}(A, \mu)$ and $\theta \in (0, 1)$. Then*

$$\left[L^{p_0(\cdot)}(A, \mu), L^{p_1(\cdot)}(A, \mu)\right]_{[\theta]} \cong L^{p_\theta(\cdot)}(A, \mu)$$

and $\|g\|_{[\theta]} \leqslant \|g\|_{p_\theta(\cdot)} \leqslant 4\|g\|_{[\theta]}$ if $\varphi_{p(\cdot)} = \bar{\varphi}_{p(\cdot)}$.

Proof. We proceed along the lines of [45]. For $z \in \overline{S}$ define $\bar{\varphi}_{[z]} \in \Phi(A, \mu)$ by

$$\bar{\varphi}_{[z]}^{-1}(y, t) := \left(\bar{\varphi}_{p_0(y)}^{-1}(t)\right)^{1-z} \left(\bar{\varphi}_{p_1(y)}^{-1}(t)\right)^{z}.$$

Then $z \mapsto \bar{\varphi}_{[z]}^{-1}$ is analytic on S and continuous on $\overline{S}$. Note that $\bar{\varphi}_{[\theta]}^{-1} = \bar{\varphi}_{p_\theta(\cdot)}^{-1}$ if $\theta \in [0, 1]$ by Lemma 7.1.1.

Let $\theta \in (0,1)$ and $g \in L^{p_\theta(\cdot)}$ with $\|g\|_{p_\theta(\cdot)} \leqslant 1$. Then $\overline{\varrho}_{p_\theta(\cdot)}(g) \leqslant 1$ by the unit ball property and therefore $\varphi_{p_\theta(y)}(|g(y)|)$ is a.e. finite. For $\varepsilon > 0$, $z \in \overline{S}$, and $y \in A$, define

$$F_\varepsilon(z; y) = \exp(-\varepsilon + \varepsilon z^2 - \varepsilon \theta^2)\, \overline{\varphi}_{[z]}^{-1}\Big(y, \overline{\varphi}_{p_\theta(y)}\big(|g(y)|\big)\Big)\operatorname{sgn} g(y).$$

Then $F_\varepsilon(\theta; y) = \exp(-\varepsilon)\, g(y)$. Clearly, $F_\varepsilon(z; y)$ is analytic in z for a.e. y. Hence, $\frac{d}{d\bar{z}}\mathcal{F}_\varepsilon(z) = 0$ in $L^{p_0(\cdot)} + L^{p_1(\cdot)}$. Note that

$$\begin{aligned}
\big|\exp\big(-\varepsilon + \varepsilon(\mathrm{i}t)^2 - \varepsilon\,\theta^2\big)\big| &= \big|\exp\big(\varepsilon(-1 - t^2 - \theta^2)\big)\big| \leqslant 1, \\
\big|\exp\big(-\varepsilon + \varepsilon(1 + \mathrm{i}t)^2 - \varepsilon\,\theta^2\big)\big| &= \big|\exp\big(\varepsilon(-t^2 - \theta^2)\big)\big| \leqslant 1
\end{aligned} \tag{7.1.3}$$

for all $t \in \mathbb{R}$. Thus

$$\begin{aligned}
\big|F_\varepsilon(\mathrm{i}t; y)\big| &\leqslant \overline{\varphi}_{p_0(y)}^{-1}\Big(\overline{\varphi}_{p_\theta(y)}\big(|g(y)|\big)\Big), \\
\big|F_\varepsilon(1 + \mathrm{i}t; y)\big| &\leqslant \overline{\varphi}_{p_1(y)}^{-1}\Big(\overline{\varphi}_{p_\theta(y)}\big(|g(y)|\big)\Big)
\end{aligned}$$

for all $y \in A$ and all $t \in \mathbb{R}$. If we apply $\overline{\varphi}_{p_0}$ and $\overline{\varphi}_{p_1}$ to the penultimate line and the last line, respectively, use $\varphi(\varphi^{-1}(t)) \leqslant t$ from (3.1.9), and integrate over $y \in A$, then we get

$$\begin{aligned}
\overline{\varrho}_{p_0(\cdot)}\big(F_\varepsilon(\mathrm{i}t)\big) &\leqslant \overline{\varrho}_{p_\theta(\cdot)}(g) \leqslant 1, \\
\overline{\varrho}_{p_1(\cdot)}\big(F_\varepsilon(1 + \mathrm{i}t)\big) &\leqslant \overline{\varrho}_{p_\theta(\cdot)}(g) \leqslant 1
\end{aligned}$$

for all $t \in \mathbb{R}$. Hence by the unit ball property

$$\begin{aligned}
\|F_\varepsilon(\mathrm{i}t)\|_{p_0(\cdot)} &\leqslant 1, \\
\|F_\varepsilon(1 + \mathrm{i}t)\|_{p_1(\cdot)} &\leqslant 1
\end{aligned}$$

for all $t \in \mathbb{R}$. Thus

$$\|F_\varepsilon\|_{\mathcal{F}} = \sup_{t \in \mathbb{R}} \max\big\{\|F_\varepsilon(\mathrm{i}t)\|_{p_0(\cdot)}, \|F_\varepsilon(1 + \mathrm{i}t)\|_{p_1(\cdot)}\big\} \leqslant 1.$$

Since $F_\varepsilon(\theta) = \exp(-\varepsilon)\, g$, this implies that $\|\exp(-\varepsilon)\, g\|_{[\theta]} \leqslant 1$. As $\varepsilon > 0$ was arbitrary, we deduce $\|g\|_{[\theta]} \leqslant 1$. A scaling argument yields $\|g\|_{[\theta]} \leqslant \|g\|_{p_\theta(\cdot)}$.

Assume now that $\theta \in (0,1)$ and $\|g\|_{[\theta]} < 1$. By definition of $\|\cdot\|_{[\theta]}$ there exists $F : \overline{S} \to (L^{p_0(\cdot)} + L^{p(\cdot)})$ with $\|F\|_{\mathcal{F}} \leqslant 1$ such that F is analytic on S and continuous on $\overline{S}$ with $\|F\|_{\mathcal{F}} < 1$, $F(\mathrm{i}t)$ and $F(1 + \mathrm{i}t)$ tend to zero for $|t| \to \infty$, and $F(\theta) = g$. So the unit ball property implies $\sup_{t \in \mathbb{R}} \max\big\{\overline{\varrho}_{p_0(\cdot)}(F(\mathrm{i}t)), \overline{\varrho}_{p_1(\cdot)}(F(1 + \mathrm{i}t))\big\} \leqslant 1$. Let b be a simple function on $\mathbb{R}^n$ with $\|b\|_{p_\theta'(\cdot)} \leqslant 1$. Then $\varrho_{p_\theta'(\cdot)}(b) \leqslant 1$ by the unit ball property and using $\bar{\varphi}^*_{p_\theta}(t) \leqslant \bar{\varphi}_{p'_\theta}(t)$ (Lemma 3.1.3) we see that $\bar{\varphi}^*_{p_\theta(y)}(|b(y)|)$ is finite almost everywhere. For $\varepsilon > 0$ define $h_{b,\varepsilon} : \overline{S} \to (L^{p'_0(\cdot)} \cap L^{p'_1(\cdot)})$ and $H_{b,\varepsilon} : \overline{S} \to \mathbb{C}$ by

$$h_{b,\varepsilon}(z,y) := \exp(-\varepsilon + \varepsilon z^2 - \varepsilon \theta^2)\, (\bar{\varphi}^*_{[z]})^{-1}\Big(y, \bar{\varphi}^*_{p_\theta(y)}(|b(y)|)\Big)\, \mathrm{sgn}\, g(y),$$

$$H_{b,\varepsilon}(z) := \int_A h_{b,\varepsilon}(z,y)F(z;y)\, d\mu(y),$$

where we define $(\bar{\varphi}^*)_{[z]}$ analogously to $\bar{\varphi}_{[z]}$ by

$$(\bar{\varphi}^*)^{-1}_{[z]}(y,t) = \left(\left(\bar{\varphi}^*_{p_0(y)}\right)^{-1}(t)\right)^{1-z}\left(\left(\bar{\varphi}^*_{p_1(y)}\right)^{-1}(t)\right)^{z}.$$

Then $H_{b,\varepsilon}$ is analytic on S and continuous on $\overline{S}$, $H_{b,\varepsilon}(it)$ and $H_{b,\varepsilon}(1+it)$ tend to zero for $|t| \to \infty$, and

$$H_{b,\varepsilon}(\theta) \leqslant \int_A h_{b,\varepsilon}(\theta,y)F(\theta;y)\, d\mu(y) = \int_A \exp(-\varepsilon)\, b(y)g(y)\, d\mu(y).$$

We estimate using (7.1.3), Young's inequality, $\varphi(\varphi^{-1}(t)) \leqslant t$, and $\bar{\varphi}^*_{p_\theta}(t) \leqslant \bar{\varphi}_{p'_\theta}(t)$ from Lemma 3.1.3:

$$\big|H_{b,\varepsilon}(it)\big| = \int_A \big|h_{b,\varepsilon}(it,y)F(it;y)\big|\, d\mu(y)$$

$$\leqslant \int_A (\bar{\varphi}^*_{p_0(y)})^{-1}\Big(y, \bar{\varphi}^*_{p_\theta(y)}(|b(y)|)\Big)\, |F(it;y)|\, d\mu(y)$$

$$\leqslant \int_A \bar{\varphi}^*_{p_\theta(y)}(|b(y)|) + \bar{\varphi}^*_{p_0(y)}(|F(it;y)|)\, d\mu(y)$$

$$\leqslant \overline{\varrho}_{p'_\theta(\cdot)}(b) + \overline{\varrho}_{p_0(\cdot)}(F(it))$$

$$\leqslant 1 + 1 = 2.$$

Analogously,

$$\big|H_{b,\varepsilon}(1+it)\big| \leqslant \overline{\varrho}_{p'_\theta(\cdot)}(b) + \overline{\varrho}_{p_1(\cdot)}(F(1+it)) \leqslant 2.$$

Overall, we know that $H_{b,\varepsilon}$ is analytic on S, continuous on $\overline{S}$ and bounded on the boundary of S by 2. The three line theorem [45, Lemma 1.1.2] implies that $H_{b,\varepsilon}$ is globally bounded by 2. In particular,

$$|H_{b,\varepsilon}(\theta)| = \int_A \exp(-\varepsilon)\, (\bar{\varphi}^*_{p_\theta(y)})^{-1}\Big(\bar{\varphi}^*_{p_\theta(y)}(|b(y)|)\Big)\, |g(y)|\, dy \leqslant 2.$$

Since $\bar{\varphi}^*_{p_\theta(y)}\big(|b(y)|\big)$ is finite almost everywhere, we have the lower estimate $(\bar{\varphi}^*_{p_\theta(y)})^{-1}(\bar{\varphi}^*_{p_\theta(y)}|b(y)|)) \geqslant |b(y)|$ almost everywhere using (3.1.10). Letting $\varepsilon \to 0$, we thus find that

$$\int_A |b(y)||g(y)|\, dy \leqslant 2,$$

for all simple functions b with $\|b\|_{p'_\theta(\cdot)} \leqslant 1$. So the norm conjugate formula (Corollary 3.2.14) yields $\|g\|_{p_\theta(\cdot)} \leqslant 4$. □

Theorem 7.1.2 has the following consequence.

Corollary 7.1.4. *Let $p_0, p_1 \in \mathcal{P}(A, \mu)$ and $\theta \in (0,1)$. Let S be a linear, bounded mapping $S : L^{p_j(\cdot)}(A, \mu) \to L^{p_j(\cdot)}(A, \mu)$ for $j = 0, 1$. Then $S : L^{p_\theta(\cdot)}(A, \mu) \to L^{p_\theta(\cdot)}(A, \mu)$ is bounded. Moreover,*

$$\|S\|_{L^{p_\theta(\cdot)} \to L^{p_\theta(\cdot)}} \leqslant 4\, \|S\|^{1-\theta}_{L^{p_0(\cdot)} \to L^{p_0(\cdot)}} \|S\|^{\theta}_{L^{p_1(\cdot)} \to L^{p_1(\cdot)}}$$

for $\varphi_{p(\cdot)} = \bar{\varphi}_{p(\cdot)}$.

Proof. Let $X := \big[L^{p_0(\cdot)}(A, \mu), L^{p_1(\cdot)}(A, \mu)\big]_{[\theta]}$ with $\|\cdot\|_{[\theta]}$. Then by the complex interpolation theorem of Riesz–Thorin (e.g., [45, Theorem 4.1.2]), we obtain

$$\|S\|_{X \to X} \leqslant \|S\|^{1-\theta}_{L^{p_0(\cdot)} \to L^{p_0(\cdot)}} \|S\|^{\theta}_{L^{p_1(\cdot)} \to L^{p_1(\cdot)}}.$$

By Theorem 7.1.2 we conclude that

$$\|Sf\|_{p_\theta(\cdot)} \leqslant 4\, \|Sf\|_{[\theta]} \leqslant 4\, \|S\|^{1-\theta}_{L^{p_0(\cdot)} \to L^{p_0(\cdot)}} \|S\|^{\theta}_{L^{p_1(\cdot)} \to L^{p_1(\cdot)}} \|f\|_{[\theta]}$$
$$\leqslant 4\, \|S\|^{1-\theta}_{L^{p_0(\cdot)} \to L^{p_0(\cdot)}} \|S\|^{\theta}_{L^{p_1(\cdot)} \to L^{p_1(\cdot)}} \|f\|_{p_\theta(\cdot)},$$

for all $f \in L^{p_\theta(\cdot)}$. □

Using Corollary 7.1.4 we obtain that $\mathcal{A}$ is convex with respect to the reciprocal of the exponents:

Corollary 7.1.5. *If $p_0, p_1 \in \mathcal{A}$, then $p_\theta \in \mathcal{A}$ for every $\theta \in (0,1)$.*

Proof. If $p_0, p_1 \in \mathcal{A}$, then by definition the averaging operators T_Q are bounded from $L^{p_j(\cdot)}(\mathbb{R}^n)$ to $L^{p_j(\cdot)}(\mathbb{R}^n)$ with $j = 0, 1$ uniformly for all locally 1-finite families of cubes Q in $\mathbb{R}^n$. So by Corollary 7.1.4 these operators are also bounded from $L^{p_\theta(\cdot)}(\mathbb{R}^n)$ to $L^{p_\theta(\cdot)}(\mathbb{R}^n)$, which implies $p_\theta \in \mathcal{A}$. □

Under suitable circumstances, complex interpolation also generates compact mappings if one of the two initial mappings is compact. In what follows we need the following version:

Corollary 7.1.6. *Let* $q_0, q_1 \in \mathcal{P}(A, \mu)$, $\theta \in (0, 1)$ *and let* X *be a Banach space. Let* S *be a linear mapping with*

$$S : X \hookrightarrow L^{q_0(\cdot)}(A, \mu) \quad and \quad S : X \hookrightarrow\hookrightarrow L^{q_1(\cdot)}(A, \mu)$$

Then $S : X \hookrightarrow\hookrightarrow L^{q_\theta(\cdot)}(A, \mu)$.

Proof. The result follows as in Corollary 7.1.4, but now we use the abstract theorem of Calderón [62] for interpolation of Banach spaces instead of the Riesz–Thorin theorem; Calderón's result says that if

$$S : X \hookrightarrow Y_0 \quad and \quad S : X \hookrightarrow\hookrightarrow Y_1$$

then $S : X \hookrightarrow\hookrightarrow [Y_0, Y_1]_\theta$ (here (Y_0, Y_1) is a Banach couple).   □

7.2 Extrapolation Results

For extrapolation we need to consider Muckenhoupt weights, which were already introduced in Chap. 5. Recall that a locally integrable function $w \colon \mathbb{R}^n \to (0, \infty)$ is an A_p-weight, $1 < p < \infty$, if

$$\fint_Q w(x) \, dx \left(\fint_Q w(x)^{1-p'} \, dx \right)^{p-1} \leqslant c$$

for every cube Q. Also, w is an A_1-weight if $Mw \leqslant cw$ almost everywhere. The smallest constant c for which the inequality holds is known as the A_p-constant of the weight. Recall also that A_p is increasing in p, i.e. $A_p \subset A_q$ if $p \leqslant q$. Moreover, the class has a self-improving property: if $w \in A_p$, then there exists $q < p$ so that $w \in A_q$. For more information on A_p weights we refer to [174].

The following is the main result of [83, Theorem 1.3] by Cruz-Uribe, Fiorenza, Martell and Pérez:

Theorem 7.2.1 (Extrapolation theorem). *Given a family* $\mathcal{F}$ *of pairs of measurable functions and an open set* $\Omega \subset \mathbb{R}^n$, *suppose for some* $p_0 \in (0, \infty)$ *that*

$$\int_\Omega |f(x)|^{p_0} w(x) \, dx \leqslant c_1 \int_\Omega |g(x)|^{p_0} w(x) \, dx \quad for \ every \ (f, g) \in \mathcal{F},$$

and for every weight $w \in A_1$, *where* $c_1 > 0$ *depends only on the* A_1*-constant of* w. *Let* $p \in \mathcal{P}(\Omega)$ *be a bounded exponent with* $p^- \geqslant p_0$ *such that the maximal operator is bounded on* $L^{(p(\cdot)/p_0)'}(\Omega)$. *Then*

$$\|f\|_{L^{p(\cdot)}(\Omega)} \leqslant c_2 \|g\|_{L^{p(\cdot)}(\Omega)},$$

for all $(f, g) \in \mathcal{F}$, *where* c_2 *only depends on the operator norm of* M *on* $L^{(p(\cdot)/p_0)'}(\Omega)$ *and* c_1.

Proof. It suffices to prove the claim for $\varphi_{p(\cdot)} = \bar{\varphi}_{p(\cdot)}$. The proof depends only on the following two specific properties of the norm:

$$\|f\|_{p(\cdot)}^r = \big\| |f|^r \big\|_{\frac{p(\cdot)}{r}},$$

for any $r \in (0, p^-]$ and the norm conjugate formula, Corollary 3.2.14.

Define $s \in \mathcal{P}(\Omega)$ by $s := p/p_0$ and let $h \in L^{s'(\cdot)}(\Omega)$. By assumption $\|Mh\|_{s'(\cdot)} \leqslant A\|h\|_{s'(\cdot)}$ for some constant independent of h. Thus we may define the *Rubio de Francia operator* $\mathcal{R}$ by

$$\mathcal{R}h(x) := \sum_{k=0}^{\infty} \frac{M^k h(x)}{(2A)^k},$$

where M^k denotes the iterated maximal operator, i.e. $M^0 = \text{id}$ and $M^i = M \circ M^{i-1}$ for $i \geqslant 1$. We note the following properties of the Rubio de Francia operator which are easily proved:

(a) $|h| \leqslant \mathcal{R}h$.
(b) $\|\mathcal{R}h\|_{s'(\cdot)} \leqslant 2\|h\|_{s'(\cdot)}$.
(c) $M(\mathcal{R}h) \leqslant 2A\mathcal{R}h$, so that $\mathcal{R}h$ is an A_1-weight, with A_1-constant independent of $h \in L^{s'(\cdot)}(\Omega)$.

From these properties and Hölder's inequality it follows that

$$\int_{\Omega} |f|^{p_0} |h| \, dx \leqslant \int_{\Omega} |f|^{p_0} \mathcal{R}h \, dx$$

$$\leqslant c_1 \int_{\Omega} |g|^{p_0} \mathcal{R}h \, dx$$

$$\leqslant 2c_1 \big\| |g|^{p_0} \big\|_{s(\cdot)} \|\mathcal{R}h\|_{s'(\cdot)}$$

$$\leqslant 4Ac_1 \|g\|_{p(\cdot)}^{p_0} \|h\|_{s'(\cdot)}.$$

Now the norm conjugate formula (Corollary 3.2.14) yields that $|f|^{p_0}$ in $L^{s(\cdot)}(\Omega)$ with

$$\big\| |f|^{p_0} \big\|_{s(\cdot)} \leqslant 4Ac_1 \|g\|_{p(\cdot)}^{p_0}.$$

The claim follows since $\big\| |f|^{p_0} \big\|_{s(\cdot)} = \|f\|_{p(\cdot)}^{p_0}$. $\qquad\square$

Remark 7.2.2. In contrast to [83, Theorem 1.3] the estimate $\||f|^{p_0}\|_{s(\cdot)} \leqslant c_2\|g\|_{p(\cdot)}^{p_0}$ is valid for all $(f, g) \in \mathcal{F}$. In the original version the estimate only holds for those $(f, g) \in \mathcal{F}$ with $f \in L^{p(\cdot)}(\Omega)$. We overcome this technicality by the use of the norm conjugate formula Corollary 3.2.14, which is valid for all measurable functions. Moreover, the use of the associate space instead of the dual space allows us to include the case $p^- = p_0$. This for example allows us in Corollary 7.2.5 below to handle the case $p^- = 1$.

A similar argument, for which we refer the reader to [83], yields a version of the extrapolation theorem with different exponents on f and g [83, Theorem 1.8], as well as a result valid for vector valued functions [83, Corollary 1.10]:

Theorem 7.2.3. *Given a family $\mathcal{F}$ of pairs of measurable functions and an open set $\Omega \subset \mathbb{R}^n$, suppose for some fixed $0 < p_0 < q_0 < \infty$ that*

$$\left(\int_\Omega |f(x)|^{q_0} w(x)\, dx\right)^{\frac{1}{q_0}} \leqslant c_1 \left(\int_\Omega |g(x)|^{p_0} w(x)^{\frac{p_0}{q_0}}\, dx\right)^{\frac{1}{p_0}} \quad \text{for every } (f, g) \in \mathcal{F},$$

and every weight $w \in A_1$, where c_1 depends only on the A_1-constant of w.

Let $p \in \mathcal{P}(\Omega)$ be a variable exponent with $p_0 \leqslant p^-$ and $\frac{1}{p_0} - \frac{1}{q_0} < \frac{1}{p^+}$, and define the exponent $q \in \mathcal{P}(\Omega)$ through

$$\frac{1}{q(x)} - \frac{1}{p(x)} = \frac{1}{q_0} - \frac{1}{p_0}.$$

Assume also that the maximal operator M is bounded on $L^{(q(\cdot)/q_0)'}(\Omega)$. Then

$$\|f\|_{q(\cdot)} \leqslant c_2\|g\|_{p(\cdot)},$$

for all $(f, g) \in \mathcal{F}$, where c_2 only depends on the operator norm of M in $L^{(q(\cdot)/q_0)'}(\Omega)$ and c_1.

Theorem 7.2.4. *Under the assumptions of Theorem 7.2.1 we also have*

$$\left\|\, \||f_j\|_{l^q}\, \right\|_{p(\cdot)} \leqslant C \left\|\, \||g_j\|_{l^q}\, \right\|_{p(\cdot)}$$

for every sequence $\{(f_j, g_j)\} \subset \mathcal{F}$.

As an example of an application of the extrapolation results we give here another proof of the boundedness of the Riesz potential operator and of the fractional maximal operator,

$$M_\alpha f(x) := \sup_{r>0} r^{\alpha-n} \int_{B(x,r)} |f(y)|\, dy.$$

To apply the extrapolation results, we need the corresponding weighted estimates in the constant exponent case. For these operators it was shown by Muckenhoupt and Wheeden [306] that

$$\left(\int_{\mathbb{R}^n} |I_\alpha f(x)|^q w(x)\, dx\right)^{1/q} \leqslant c \left(\int_{\mathbb{R}^n} |f(x)|^p w(x)^{p/q}\, dx\right)^{1/p}$$

and

$$\left(\int_{\mathbb{R}^n} M_\alpha f(x)^q w(x)\, dx\right)^{1/q} \leqslant c \left(\int_{\mathbb{R}^n} |f(x)|^p w(x)^{p/q}\, dx\right)^{1/p}$$

where $w \in A_{1+q/p'}$ and c depends on the $A_{1+q/p'}$-constant of w, thus in particular for every $w \in A_1$. As in Corollary 4.3.11 this result extends trivially to the case $\Omega \subset \mathbb{R}^n$.

Suppose now that p is a variable exponent with $p^+ < n/\alpha$ and define q by $1/q(x) = 1/p(x) - \alpha/n$. If there is a constant $q_0 \in (\frac{n}{n-\alpha}, \infty)$ for which the maximal operator is of strong type $(q/q_0)'$, then all the assumptions of Theorem 7.2.3 are satisfied. We showed in Theorem 5.7.2 that the maximal operator is bounded on $L^{p(\cdot)}(\mathbb{R}^n)$ if and only if it is bounded on $L^{p'(\cdot)}(\mathbb{R}^n)$. Taking also this into account, we obtain:

Corollary 7.2.5. Let $p \in \mathcal{A}$ with $p^+ < \frac{n}{\alpha}$. Then

$$\|I_\alpha f\|_{p^\sharp(\cdot)} \leqslant c \|f\|_{p(\cdot)} \quad and \quad \|M_\alpha f\|_{p^\sharp(\cdot)} \leqslant c \|f\|_{p(\cdot)}.$$

Based on the same results and appropriate constant exponent, weighted estimates, boundedness results for several other operators were given in [83]. They read as follows:

Corollary 7.2.6. Let $p \in \mathcal{A}$ with $1 < p^- \leqslant p^+ < \infty$. Then

$$\|Mf\|_{p(\cdot)} \leqslant c \|M^\sharp f\|_{p(\cdot)}.$$

For the definition of $M^\sharp f$ and a direct proof of this result see Theorem 6.2.5.

Corollary 7.2.7. Let $p \in \mathcal{A}$ with $1 < p^- \leqslant p^+ < \infty$. Then

$$\|K * f\|_{p(\cdot)} \leqslant c \|f\|_{p(\cdot)},$$

where K is a standard singular integral kernel (cf. Definition 6.3.1).

Results for Fourier multipliers, square functions and commutators were also given in [83].

As a final example we present a result for vector valued maximal inequalities [83, Corollary 2.1], which was proved for the first time in the variable exponent setting by this method.

Corollary 7.2.8. *Let $p \in \mathcal{A}$ with $1 < p^- \leqslant p^+ < \infty$. Then*

$$\left\| \left(\sum_{j=0}^{\infty} Mf_j^{\,q} \right)^{1/q} \right\|_{p(\cdot)} \leqslant c \left\| \left(\sum_{j=0}^{\infty} |f_j|^q \right)^{1/q} \right\|_{p(\cdot)}$$

for every $q \in (1, \infty)$.

Corollary 7.2.9. *Let k be a kernel on $\mathbb{R}^n$ of the form*

$$k(x) := \frac{P(x/|x|)}{|x|^n}, \tag{7.2.10}$$

where $P \in L_0^r(\partial B(0,1))$ for some $r \in (1, \infty]$ (with the $(n-1)$-dimensional Hausdorff measure). Moreover, let $p \in \mathcal{P}^{\log}(\mathbb{R}^n)$ with $p^- > r'$. Then the operator T, defined for almost everywhere by

$$Tf(x) := \lim_{\varepsilon \to 0} \int_{(B(x,\varepsilon))^c} k(x-y)f(y)\,dy,$$

is bounded on $L^{p(\cdot)}(\mathbb{R}^n)$.

7.3 Local-to-Global Results

In this section we present a simple and convenient method to pass from local to global results. The idea is simply to generalize the following two properties of the Lebesgue norm:

$$\|f\|_q = \left(\sum_i \|\chi_{\Omega_i} f\|_q^q \right)^{\frac{1}{q}}$$

$$\sum_i \|\chi_{\Omega_i} f\|_q \|\chi_{\Omega_i} g\|_{q'} \leqslant \|f\|_q \|g\|_{q'}$$

for any partition of $\mathbb{R}^n$ into measurable sets Ω_i and $1 \leqslant q < \infty$ with obvious modification for $q = \infty$.

First we note that all terms with $|\Omega_i| = 0$ can be dropped from the sums, since their contribution is zero. If each Ω_i has positive measure, then we can restate the first equation as

$$\|f\|_q = \left\| \sum_i \chi_{\Omega_i} \frac{\|\chi_{\Omega_i} f\|_q}{\|\chi_{\Omega_i}\|_q} \right\|_q.$$

This raises the question if for some variable exponents

$$\|f\|_{p(\cdot)} \approx \left\| \sum_i \chi_{\Omega_i} \frac{\|\chi_{\Omega_i} f\|_{p(\cdot)}}{\|\chi_{\Omega_i}\|_{p(\cdot)}} \right\|_{p(\cdot)} \tag{7.3.1}$$

$$\sum_i \|\chi_{\Omega_i} f\|_{p(\cdot)} \|\chi_{\Omega_i} g\|_{p'(\cdot)} \leqslant c \|f\|_{p(\cdot)} \|g\|_{p'(\cdot)}.$$

Unfortunately, these estimate cannot hold for arbitrary choices of the sets Ω_i. However, we show in Theorem 7.3.22 that they hold when $p \in \mathcal{P}^{\log}(\mathbb{R}^n)$ and (Ω_i) is a locally N-finite family of cubes (or balls).

The first generalizations of these estimates to the variable exponent context were due to Kopaliani and Hästö. Kopaliani showed in [254] that if $p \in \mathcal{P}^{\log}([0,1])$, then (7.3.1) holds for all $f \in L^{p(\cdot)}([0,1])$ and $g \in L^{p'(\cdot)}([0,1])$. He later generalized this to the case $\mathcal{P}^{\log}(\mathbb{R}^n)$, $f \in L^{p(\cdot)}(\mathbb{R}^n)$ and $g \in L^{p'(\cdot)}(\mathbb{R}^n)$. Hästö showed in [216] that

$$\|f\|_{p(\cdot)} \approx \left(\sum_i \|\chi_{\Omega_i} f\|_{p(\cdot)}^{p_\infty} \right)^{\frac{1}{p_\infty}}$$

if $p \in \mathcal{P}^{\log}(\mathbb{R}^n)$ and Ω_i are cubes of the same size.

The equivalence and the estimate in (7.3.1) are very powerful tools. They allow us to extend many estimates and results known for cubes or balls to more complicated domains. This transfer technique is explained in detail in Sect. 7.4.

Definition 7.3.2. By $\mathcal{G}$ we denote the set of all generalized Φ-functions φ on $\mathbb{R}^n$ which have the property that

$$\sum_{Q \in \mathcal{Q}} \|\chi_Q f\|_\varphi \|\chi_Q g\|_{\varphi^*} \leqslant K \|f\|_\varphi \|g\|_{\varphi^*}$$

for all $f \in L^\varphi(\mathbb{R}^n)$, $g \in L^{\varphi^*}(\mathbb{R}^n)$, and all locally 1-finite families $\mathcal{Q}$ of cubes. The smallest constant K, is called the $\mathcal{G}$-constant of φ. If $\varphi \in \mathcal{G}$, then we say that φ is of class $\mathcal{G}$. In the case $\varphi_{p(\cdot)}$, we denote $\varphi_{p(\cdot)} \in \mathcal{G}$ simply by $p \in \mathcal{G}$.

The name "class $\mathcal{G}$" is derived from the works of Berezhnoi for ideal Banach spaces [44]. In the notation of Berezhnoi $\varphi \in \mathcal{G}$ is just $(L^\varphi, L^{\varphi^*}) \in \mathbf{G}(\mathcal{X}^n)$, where $\mathcal{X}^n$ is the set of all cubes in $\mathbb{R}^n$.

The following result shows that the second part of (7.3.1) implies the first one.

Theorem 7.3.3. *Assume that $\varphi \in \mathcal{G}$ and $\mathcal{Q}$ is a locally 1-finite family of cubes. Then*

$$\frac{1}{4K}\left\|\sum_{Q\in\mathcal{Q}}\chi_Q f\right\|_\varphi \leqslant \left\|\sum_{Q\in\mathcal{Q}}\chi_Q \frac{\|\chi_Q f\|_\varphi}{\|\chi_Q\|_\varphi}\right\|_\varphi \leqslant 4K\left\|\sum_{Q\in\mathcal{Q}}\chi_Q f\right\|_\varphi \qquad (7.3.4)$$

for all $f \in L^\varphi(\mathbb{R}^n)$.

Proof. Using the norm conjugate formula and Hölder's inequality, we estimate

$$\left\|\sum_{Q\in\mathcal{Q}}\chi_Q f\right\|_\varphi \leqslant 2\sup_{\|g\|_{\varphi^*}\leqslant 1}\sum_{Q\in\mathcal{Q}}\int_Q \chi_Q|f|\,|g|\,dx$$

$$\leqslant 4\sup_{\|g\|_{\varphi^*}\leqslant 1}\sum_{Q\in\mathcal{Q}}\|\chi_Q f\|_\varphi\|\chi_Q g\|_{\varphi^*}$$

$$= 4\sup_{\|g\|_{\varphi^*}\leqslant 1}\sum_{Q\in\mathcal{Q}}\left\|\chi_Q\frac{\|\chi_Q f\|_\varphi}{\|\chi_Q\|_\varphi}\right\|_\varphi\|\chi_Q g\|_{\varphi^*}.$$

Then the upper bound follows from $\varphi \in \mathcal{G}$.

Using the same tools, we also estimate

$$\left\|\sum_{Q\in\mathcal{Q}}\chi_Q \frac{\|\chi_Q f\|_\varphi}{\|\chi_Q\|_\varphi}\right\|_\varphi \leqslant 2\sup_{\|g\|_{\varphi^*}\leqslant 1}\sum_{Q\in\mathcal{Q}}\int_Q \frac{\|\chi_Q f\|_\varphi}{\|\chi_Q\|_\varphi}g\,dx$$

$$\leqslant 4\sup_{\|g\|_{\varphi^*}\leqslant 1}\sum_{Q\in\mathcal{Q}}\|\chi_Q f\|_\varphi\|\chi_Q g\|_{\varphi^*}$$

$$\leqslant 4K\left\|\sum_{Q\in\mathcal{Q}}\chi_Q f\right\|_\varphi. \qquad \square$$

We show now that the estimate of Theorem 7.3.3 is stable under duality if we additionally assume $\varphi \in \mathcal{A}_{\mathrm{loc}}$.

Proposition 7.3.5. *Let $\varphi \in \mathcal{A}_{\mathrm{loc}}$ with $\mathcal{A}_{\mathrm{loc}}$-constant A and let $\mathcal{Q}$ be a locally 1-finite family of cubes (or balls). If (7.3.4) holds for all $f \in L^\varphi(\mathbb{R}^n)$, then*

$$\frac{1}{16AK}\left\|\sum_{Q\in\mathcal{Q}}\chi_Q g\right\|_{\varphi^*} \leqslant \left\|\sum_{Q\in\mathcal{Q}}\chi_Q \frac{\|\chi_Q g\|_{\varphi^*}}{\|\chi_Q\|_\varphi}\right\|_{\varphi^*} \leqslant 256K^2\left\|\sum_{Q\in\mathcal{Q}}\chi_Q g\right\|_{\varphi^*}.$$

holds for all $g \in L^{\varphi^*}(\mathbb{R}^n)$.

Proof. We begin with the first estimate of the claim. Let $\|f\|_\varphi \leqslant 1$. Then

$$\int_{\mathbb{R}^n}\left|\sum_{Q\in\mathcal{Q}}\chi_Q g\right||f|\,dx \leqslant \sum_{Q\in\mathcal{Q}}\int_Q |g|\,|f|\,dx \leqslant 2\sum_{Q\in\mathcal{Q}}\|\chi_Q g\|_{\varphi^*}\|\chi_Q f\|_\varphi.$$

Since $\varphi \in \mathcal{A}_{\mathrm{loc}}$, it follows by Remark 4.5.8 that $\|\chi_Q\|_\varphi \|\chi_Q\|_{\varphi^*} \leqslant A|Q|$ for all cubes $Q \subset \mathbb{R}^n$, where A is the $\mathcal{A}_{\mathrm{loc}}$-constant of φ. Using this, Hölder's inequality and (7.3.4) we estimate

$$
\int_{\mathbb{R}^n} |\sum_{Q \in \mathcal{Q}} \chi_Q g| |f| \, dx \leqslant 2A \sum_{Q \in \mathcal{Q}} \int_Q \frac{\|\chi_Q g\|_{\varphi^*}}{\|\chi_Q\|_{\varphi^*}} \frac{\|\chi_Q f\|_\varphi}{\|\chi_Q\|_\varphi} \, dx
$$

$$
\leqslant 4A \left\| \sum_{Q \in \mathcal{Q}} \chi_Q \frac{\|\chi_Q g\|_{\varphi^*}}{\|\chi_Q\|_{\varphi^*}} \right\|_{\varphi^*} \left\| \sum_{Q \in \mathcal{Q}} \chi_Q \frac{\|\chi_Q f\|_\varphi}{\|\chi_Q\|_\varphi} \right\|_\varphi
$$

$$
\leqslant 16AK \left\| \sum_{Q \in \mathcal{Q}} \chi_Q \frac{\|\chi_Q g\|_{\varphi^*}}{\|\chi_Q\|_{\varphi^*}} \right\|_{\varphi^*} \|f\|_\varphi.
$$

If we take the supremum over all f with $\|f\|_\varphi \leqslant 1$, then the norm conjugate formula gives the first estimate of the claim.

Let us consider the second estimate of the claim. Let $h \in L^\varphi$ with $\|h\|_\varphi \leqslant 1$ and $z_Q \in L^\varphi(Q)$ with $\|z_Q\|_\varphi \leqslant 1$. Then

$$
(I) := \sum_{Q \in \mathcal{Q}} \int_{\mathbb{R}^n} \frac{\chi_Q(x)}{\|\chi_Q\|_{\varphi^*}} |h(x)| \int_{\mathbb{R}^n} \chi_Q(y) |g(y)| |z_Q(y)| \, dy \, dx
$$

$$
= \int_{\mathbb{R}^n} |g(y)| \sum_{Q \in \mathcal{Q}} \frac{\chi_Q(y)}{\|\chi_Q\|_{\varphi^*}} |z_Q(y)| \int_Q |h(x)| \, dx \, dy
$$

$$
\leqslant 2 \|g\|_{\varphi^*} \left\| \sum_{Q \in \mathcal{Q}} \chi_Q \frac{|z_Q(y)|}{\|\chi_Q\|_{\varphi^*}} \int_Q |h(x)| \, dx \right\|_\varphi,
$$

where we used Fubini's theorem and Hölder's inequality. Now, the first inequality in (7.3.4) implies

$$
(I) \leqslant 8K \|g\|_{\varphi^*} \left\| \sum_{Q \in \mathcal{Q}} \chi_Q \frac{\|z_Q\|_\varphi}{\|\chi_Q\|_\varphi \|\chi_Q\|_{\varphi^*}} \int_Q |h(x)| \, dx \right\|_\varphi.
$$

So with $\|z_Q\|_\varphi \leqslant 1$, Hölder's inequality and the second estimate in (7.3.4) we get

$$
(I) \leqslant 16K \|g\|_{\varphi^*} \left\| \sum_{Q \in \mathcal{Q}} \chi_Q \frac{\|\chi_Q h\|_\varphi}{\|\chi_Q\|_\varphi} \right\|_\varphi
$$

$$
\leqslant 64K^2 \|g\|_{\varphi^*} \|h\|_\varphi.
$$

If we first take in (I) the supremum over all $z_Q \in L^\varphi(Q)$ with $\|z_Q\|_\varphi \leqslant 1$ and then take supremum over all $h \in L^\varphi$ with $\|h\|_\varphi \leqslant 1$, then by the norm conjugate formula

$$(I) \geqslant \frac{1}{2} \sum_{Q \in \mathcal{Q}} \int_{\mathbb{R}^n} \frac{\chi_Q(x)}{\|\chi_Q\|_{\varphi^*}} |h(x)| \, \|\chi_Q g\|_{\varphi^*} \, dx$$

$$\geqslant \frac{1}{4} \left\| \sum_{Q \in \mathcal{Q}} \chi_Q \frac{\|\chi_Q g\|_{\varphi^*}}{\|\chi_Q\|_{\varphi^*}} \right\|_{\varphi^*}.$$

Combining the upper and lower estimate for (I) proves the second estimate of the claim.   □

We show now that the estimate of Theorem 7.3.3 implies that φ belongs to the class $\mathcal{G}$ if we additionally assume $\varphi \in \mathcal{A}_{\mathrm{loc}}$.

Proposition 7.3.6. *Let* $\varphi \in \mathcal{A}_{\mathrm{loc}}$. *If* (7.3.4) *holds for all locally* 1-*finite families* $\mathcal{Q}$ *of cubes (or balls) and all* $f \in L^\varphi(\mathbb{R}^n)$, *then* $\varphi \in \mathcal{G}$.

Proof. Let $\mathcal{Q}$ be a locally 1-finite family of cubes (or balls), $f \in L^\varphi$ and $g \in L^{\varphi^*}$. From $\varphi \in \mathcal{A}_{\mathrm{loc}}$ it follows by Proposition 7.3.5 that (7.3.4) holds for φ^*. The requirement $\varphi \in \mathcal{A}_{\mathrm{loc}}$ also yields, as in the proof of Proposition 7.3.5,

$$\sum_{Q \in \mathcal{Q}} \|\chi_Q f\|_\varphi \|\chi_Q g\|_{\varphi^*} \leqslant A \sum_{Q \in \mathcal{Q}} \int_Q \frac{\|\chi_Q f\|_\varphi}{\|\chi_Q\|_\varphi} \frac{\|\chi_Q g\|_{\varphi^*}}{\|\chi_Q\|_{\varphi^*}} \, dx.$$

We now use Hölder's inequality and (7.3.4) for φ and φ^* to conclude

$$\sum_{Q \in \mathcal{Q}} \|\chi_Q f\|_\varphi \|\chi_Q g\|_{\varphi^*} \leqslant 2A \left\| \sum_{Q \in \mathcal{Q}} \chi_Q \frac{\|\chi_Q f\|_\varphi}{\|\chi_Q\|_\varphi} \right\|_\varphi \left\| \sum_{Q \in \mathcal{Q}} \chi_Q \frac{\|\chi_Q g\|_{\varphi^*}}{\|\chi_Q\|_{\varphi^*}} \right\|_{\varphi^*}$$

$$\leqslant c \|f\|_\varphi \|g\|_{\varphi^*}.$$

In particular, $\varphi \in \mathcal{G}$.   □

We know from Theorem 5.3.4 that $p \in \mathcal{A}_{\mathrm{loc}}$ is not enough to ensure $p \in \mathcal{A}$. However, the following Corollary shows that if $p \in \mathcal{G}$, then $\mathcal{A}_{\mathrm{loc}}$ and $\mathcal{A}$ are equivalent.

Corollary 7.3.7. $\mathcal{G} \cap \mathcal{A}_{\mathrm{loc}} \subset \mathcal{A}$.

Proof. If $\varphi \in \mathcal{G} \cap \mathcal{A}_{\mathrm{loc}}$, then Theorem 7.3.3, $\varphi \in \mathcal{A}_{\mathrm{loc}}$ and again Theorem 7.3.3 imply

$$\|T_{\mathcal{Q}}f\|_\varphi = \left\|\sum_{Q\in\mathcal{Q}}\chi_Q M_Q f\right\|_\varphi \approx \left\|\sum_{Q\in\mathcal{Q}}\chi_Q \frac{\|\chi_Q M_Q f\|_\varphi}{\|\chi_Q\|_\varphi}\right\|_\varphi$$

$$\leqslant c\left\|\sum_{Q\in\mathcal{Q}}\chi_Q \frac{\|\chi_Q f\|_\varphi}{\|\chi_Q\|_\varphi}\right\|_\varphi \leqslant c\|f\|_\varphi$$

for all $f \in L^\varphi(\mathbb{R}^n)$ and all locally 1-finite families $\mathcal{Q}$ of cubes.   □

Remark 7.3.8. The exponent $p = 2 + \chi_{(0,\infty)}$ satisfies $p \in \mathcal{G}\setminus\mathcal{A}_{\mathrm{loc}} \subset \mathcal{G}\setminus\mathcal{A}$.

Recall that the exponent p in Theorem 5.3.4 satisfies $p \in \mathcal{A}_{\mathrm{loc}}\setminus\mathcal{A}$; since $\mathcal{G}\cap\mathcal{A}_{\mathrm{loc}}\subset\mathcal{A}$ by the previous corollary, we conclude that $p \in \mathcal{A}_{\mathrm{loc}}\setminus\mathcal{G}$. Thus neither of the classes $\mathcal{A}_{\mathrm{loc}}$ and $\mathcal{G}$ is a subset of the other.

Even more, the example $p(x) := 2 - a(1 + \sin(\log\log(e + |x| + 1/|x|)))$ of Lerner, see Example 5.1.8, satisfies $p \in \mathcal{A}\setminus\mathcal{G}$ for sufficiently small a: We already know from the example that $p \in \mathcal{A}$. Now, choose intervals Q_j, A_j and B_j with $p_{A_j}^- = 2 - \frac{1}{4}a$ and $p_{B_j}^+ = 2 - \frac{3}{4}a$ and $|A_j| = |B_j| = 1$ such that $A_j, B_j \subset Q_j$ and the Q_j are pairwise disjoint. For any sequence (t_j), define $f := \sum_{j=1}^\infty t_j\chi_{A_j}$ and $g := \sum_{j=1}^\infty t_j\chi_{B_j}$. Then $\|f\|_{p(\cdot)} \geqslant c\|t_j\|_{l^{2-\frac{1}{4}a}}$ and $\|g\|_{p(\cdot)} \leqslant c\|t_j\|_{l^{2-\frac{3}{4}a}}$. Moreover,

$$\left\|\sum_{j=1}^\infty \chi_{Q_j}\frac{\|\chi_{Q_j}f\|_{p(\cdot)}}{\|\chi_{Q_j}\|_{p(\cdot)}}\right\|_{p(\cdot)} \approx \left\|\sum_{j=1}^\infty \chi_{Q_j}\frac{\|\chi_{Q_j}g\|_{p(\cdot)}}{\|\chi_{Q_j}\|_{p(\cdot)}}\right\|_{p(\cdot)},$$

since $\|\chi_{Q_j}f\|_{p(\cdot)} \approx \|\chi_{Q_j}g\|_{p(\cdot)}$. So if $p \in \mathcal{G}$, then Theorem 7.3.3 would imply $\|t_j\|_{l^{2-\frac{1}{4}a}} \leqslant c\|t_j\|_{l^{2+\frac{3}{4}a}}$ for all sequences t_j, but this is not possible. Hence, $p \notin \mathcal{G}$.

For $f \in L^\varphi_{\mathrm{loc}}(\mathbb{R}^n)$ and an open, bounded set $U \subset \mathbb{R}^n$ we define

$$M_{\varphi,U}f := \frac{\|f\chi_U\|_\varphi}{\|\chi_U\|_\varphi}. \tag{7.3.9}$$

For a family $\mathcal{Q}$ of open, bounded sets $U \subset \mathbb{R}^n$ we define the *averaging operator* $T_{\varphi,\mathcal{Q}}\colon L^\varphi_{\mathrm{loc}}(\mathbb{R}^n) \to L^0(\mathbb{R}^n)$ by

$$T_{\varphi,\mathcal{Q}}f := \sum_{U\in\mathcal{Q}}\chi_U M_{\varphi,U}f = \sum_{Q\in\mathcal{Q}}\chi_U \frac{\|f\chi_U\|_\varphi}{\|\chi_U\|_\varphi}.$$

The function $T_{\varphi,\mathcal{Q}}f$ is well defined in $L^0(\mathbb{R}^n)$, since $M_{\varphi,Q}f \geqslant 0$, but $T_{\varphi,\mathcal{Q}}f$ might be infinite at many points of even everywhere. However, if $\mathcal{Q}$ is locally finite and $f \in L^\varphi_{\mathrm{loc}}(\mathbb{R}^n)$, then $T_{\varphi,\mathcal{Q}}f \in L^1_{\mathrm{loc}}(\mathbb{R}^n)$.

Furthermore, we define the maximal operator M_φ by

$$M_\varphi f(x) := \sup_{Q \ni x} M_{\varphi,Q} f,$$

for $f \in L^\varphi_{\mathrm{loc}}(\mathbb{R}^n)$ and $x \in \mathbb{R}^n$, where the supremum is taken over all cubes (or balls) containing x. With this terminology the estimate in Theorem 7.3.3 can be rewritten as $\left\| \sum_{Q \in \mathcal{Q}} \chi_Q f \right\|_\varphi \approx \left\| T_{\varphi,\mathcal{Q}} f \right\|_\varphi$ for $\varphi \in \mathcal{G}$ and all locally 1-finite families $\mathcal{Q}$ of cubes (or balls).

For $p \in \mathcal{P}(\mathbb{R}^n)$ and $f \in L^{p(\cdot)}_{\mathrm{loc}}(\mathbb{R}^n)$ we define

$$M_{p(\cdot),\mathcal{Q}} f := M_{\varphi_{p(\cdot)},\mathcal{Q}} f, \qquad T_{p(\cdot),\mathcal{Q}} f := T_{\varphi_{p(\cdot)},\mathcal{Q}} f, \qquad M_{p(\cdot)} f := M_{\varphi_{p(\cdot)}} f.$$

We can either use $\bar{\varphi}_{p(\cdot)}$ or $\widetilde{\varphi}_{p(\cdot)}$, since

$$\frac{1}{2} M_{\widetilde{\varphi}_{p(\cdot)},\mathcal{Q}} f \leqslant M_{\bar{\varphi}_{p(\cdot)},\mathcal{Q}} f \leqslant 2 M_{\widetilde{\varphi}_{p(\cdot)},\mathcal{Q}} f$$

by (3.2.2). We observe that

$$M_{\bar{\varphi}_q,\mathcal{Q}} f = \left(\fint_Q |f(y)|^q \, dy \right)^{\frac{1}{q}}$$

for $q \in [1,\infty)$. Therefore the definition of $M_{q,Q}$ by (7.3.9) agrees (up to a constant if $\varphi_q = \widetilde{\varphi}_q$) with the one in Definition 4.3.1.

Lemma 7.3.10. *Let $q \in [1,\infty]$, let $Q \subset \mathbb{R}^n$ be a cube (or ball), and let $f \in L^q(Q)$. Then*

$$\varphi_q(M_{q,Q} f) = \fint_Q \varphi_q(|f(y)|) \, dy.$$

Proof. It suffices to prove the equation for $\bar{\varphi}$, since $\bar{\varphi}_q = c_q \widetilde{\varphi}_q$ for some constant. The case $1 \leqslant q < \infty$ is simple. Let $q = \infty$. If $\|\chi_Q f\|_\infty \leqslant 1$, then both sides of the claim are zero. If $\|\chi_Q f\|_\infty > 1$, then both sides of the claim are infinity. $\qquad \square$

Lemma 7.3.11. *Let $p \in \mathcal{P}^{\log}(\mathbb{R}^n)$. Then for every $m > 0$ there exists $\beta \in (0,1)$ which only depends on $c_{\log}(p)$, n and m such that*

$$\varphi_{p(x)}(\beta s) \leqslant \varphi_{p(y)}(s) + \frac{1}{2}(e + |x|)^{-m} + \frac{1}{2}(e + |y|)^{-m},$$

$$\varphi_{p(x)}(\beta t) \leqslant \varphi_{p_\infty}(t) + (e + |x|)^{-m},$$

$$\varphi_{p_\infty}(\beta t) \leqslant \varphi_{p(x)}(t) + (e + |x|)^{-m},$$

for all $s \in [0, \max\{1, 1/\|\chi_Q\|_{p(\cdot)}\}]$, $t \in [0,1]$, and every $x, y \in Q$, where $Q \subset \mathbb{R}^n$ is a cube or a ball.

Proof. Assume without loss of generality that $\varphi_{p(\cdot)} = \tilde{\varphi}_{p(\cdot)}$. Let us begin with the case $t = s \in [0,1]$. If $p(x) \leqslant p(y)$, then by Lemmas 3.2.15 and 4.2.3

$$\tilde{\varphi}_{p(x)}(\beta t) \leqslant \tilde{\varphi}_{p(y)}(t) + \tilde{\varphi}_{q(x,y)}(\beta) \leqslant \tilde{\varphi}_{p(y)}(t) + \tilde{\varphi}_{s(x)}(\beta^{\frac{1}{2}}) + \tilde{\varphi}_{s(y)}(\beta^{\frac{1}{2}})$$

for $\beta > 0$. Now, the first claim follows from Proposition 4.1.8 for suitable $\beta \in (0,1)$. The two other assertions follow analogously.

If on the other hand $p(x) \geqslant p(y)$, then the claim follows for any $\beta \leqslant \frac{1}{2}$ by Lemma 3.1.6.

It remains to consider the case $1 < t \leqslant 1/\|\chi_Q\|_{p(\cdot)}$. In particular, $|Q| < 1$ and we can apply Corollary 4.5.9 to find $\gamma_1 \in (0,1)$ which only depends on $c_{\log}(p)$ such that

$$\varphi_{p(x)}\left(\frac{\gamma_1}{\|\chi_Q\|_{p(\cdot)}}\right) \leqslant \varphi_{p(y)}\left(\frac{1}{\|\chi_Q\|_{p(\cdot)}}\right)$$

for all $x, y \in Q$. Choose $\alpha \in (0,1]$ such that $t = \|\chi_Q\|_{p(\cdot)}^{-\alpha}$. Then

$$\varphi_{p(x)}(\gamma_1 t) \leqslant \left(\varphi_{p(x)}\left(\frac{\gamma_1}{\|\chi_Q\|_{p(\cdot)}}\right)\right)^\alpha \leqslant \left(\varphi_{p(y)}\left(\frac{1}{\|\chi_Q\|_{p(\cdot)}}\right)\right)^\alpha \leqslant \varphi_{p(y)}(t)$$

for all $x, y \in \mathbb{R}^n$. $\qquad\square$

Lemma 7.3.12. *Let $p \in \mathcal{P}^{\log}(\mathbb{R}^n)$. Then for every $m > 0$ there exists $\beta \in (0,1)$ which only depends on $c_{\log}(p)$, n and m such that*

$$\varphi_{p(x)}\left(\beta M_{p(\cdot),Q} f\right) \leqslant \fint_Q \varphi_{p(y)}(|f(y)|) \, dy + \fint_Q (e + |y|)^{-m} \, dy, \qquad (7.3.13)$$

$$\fint_Q \varphi_{p(y)}(\beta |f(y)|) \, dy \leqslant \varphi_{p(x)}\left(M_{p(\cdot),Q} f\right) + \fint_Q (e + |y|)^{-m} \, dy. \qquad (7.3.14)$$

for every cube (or ball) $Q \subset \mathbb{R}^n$ with $|Q| \leqslant 2^n$, all $x \in Q$, and all $f \in L^{p(\cdot)}(\mathbb{R}^n) + L^\infty(\mathbb{R}^n)$ with $\|\chi_Q f\|_{L^{p(\cdot)}(\mathbb{R}^n) + L^\infty(\mathbb{R}^n)} \leqslant 1$.

Proof. Assume without loss of generality that $\varphi_{p(\cdot)} = \tilde{\varphi}_{p(\cdot)}$. Since $M_{p(\cdot),Q}$ is subadditive it suffices to prove the claim independently for $\|\chi_Q f\|_{p(\cdot)} \leqslant 1$ and $\|\chi_Q f\|_\infty \leqslant 1$. Since $L^\infty(Q) \hookrightarrow L^{p(\cdot)}(\Omega)$, we always have $\|\chi_Q f\|_{p(\cdot)} < \infty$. If $\|\chi_Q f\|_{p(\cdot)} = 0$, then $\chi_Q f = 0$ and there is nothing to show. So we assume in the following that $\|\chi_Q f\|_{p(\cdot)} > 0$.

For the sake of simplicity we start with the case $p^+ < \infty$ and thus have $\bar{\varphi}_{p(y)}(t) = t^{p(y)}$. Later we show how to deduce the case $p^+ = \infty$ from this by a limiting argument. For that it is certainly important that the constants in the estimates do not depend on p^+.

Since $|Q| \leqslant 2^n$ we conclude from Lemma 3.2.11 that $\|\chi_Q\|_{p(\cdot)} \leqslant 2^{n+1}$. If $\|\chi_Q f\|_{p(\cdot)} \leqslant 1$, then $M_{p(\cdot),Q} f \leqslant \|\chi_Q\|_{p(\cdot)}^{-1}$. If $\|\chi_Q f\|_\infty \leqslant 1$, then $M_{p(\cdot),Q} f \leqslant 1$. Therefore, in both case $M_{p(\cdot),Q} f \leqslant \max\{1, \|\chi_Q\|_{p(\cdot)}^{-1}\}$ and we can apply Lemma 7.3.11 with $s = M_{p(\cdot),Q} f$. Thus there exists $\alpha \in (0,1)$ such that

$$(\alpha M_{p(\cdot),Q} f)^{p(z_1)} \leqslant (M_{p(\cdot),Q} f)^{p(z_2)} + \frac{1}{2}(e + |z_1|)^{-m} + \frac{1}{2}(e + |z_2|)^{-m}$$

for every $z_1, z_2 \in Q$. Since $|Q| \leqslant 2^n$, we have $e + |z_1| \approx e + |z_2| \approx e + |y|$ for all $y \in Q$. Therefore, the previous estimate implies

$$(\alpha M_{p(\cdot),Q} f)^{p(z_1)} \leqslant (M_{p(\cdot),Q} f)^{p(z_2)} + c_1 \fint_Q (e + |y|)^{-m} \, dy \qquad (7.3.15)$$

for every $z_1, z_2 \in Q$ and some $c_1 \geqslant 1$.

By Corollary 4.5.9 with $|Q| \leqslant 2^n$

$$\frac{1}{\gamma} |Q|^{\frac{1}{p(y)}} \leqslant \|\chi_Q\|_{p(\cdot)} \leqslant \gamma |Q|^{\frac{1}{p(y)}}$$

for all $y \in Q$, where $\gamma \geqslant 1$ only depends on the $c_{\log}$-constant of $\frac{1}{p}$. Since we assumed that $p^+ < \infty$, the continuity of the modular gives $1 = \varrho_{p(\cdot)}(\chi_Q f / \|\chi_Q\|_{p(\cdot)})$. This together with the previous equivalence implies that

$$\fint_Q \left(\frac{|f(y)|}{\gamma M_{p(\cdot),Q} f} \right)^{p(y)} dy \leqslant 1 \leqslant \fint_Q \left(\frac{\gamma |f(y)|}{M_{p(\cdot),Q} f} \right)^{p(y)} dy. \qquad (7.3.16)$$

From now on let $x \in Q$ be fixed. We distinguish in the following two cases:

Case 1: $(\alpha M_{p(\cdot),Q} f)^{p(x)} \leqslant 2c_1 \fint_Q (e + |y|)^{-m} \, dy$.

In this case (7.3.13) holds with $\beta = \alpha/(2c_1)$, and it remains to show (7.3.14). Using that (7.3.15) also holds for f replaced by αf we estimate

$$(\alpha^2 M_{p(\cdot),Q} f)^{p(z)} \leqslant 3c_1 \fint_Q (e + |y|)^{-m} \, dy$$

for all $z \in Q$. Using this in the first inequality of (7.3.16), we obtain

$$\fint_Q \left(\frac{\alpha^2 |f(y)|}{\gamma}\right)^{p(y)} dy \leqslant 3\,c_1 \fint_Q (e + |y|)^{-m}\, dy.$$

This proves (7.3.14) with $\beta = \alpha^2/(3c_1\gamma)$.

Case 2: $2c_1 \fint_Q (e + |y|)^{-m}\, dy \leqslant (\alpha M_{p(\cdot),Q} f)^{p(x)}$.
In this case we deduce from (7.3.15) that

$$\frac{2}{3}(\alpha^2 M_{p(\cdot),Q} f)^{p(z)} \leqslant (\alpha M_{p(\cdot),Q} f)^{p(x)} \leqslant 2\,(M_{p(\cdot),Q} f)^{p(z)} \qquad (7.3.17)$$

for all $z \in Q$. Using this in the second inequality (7.3.16), we obtain

$$\left(\frac{\alpha}{2} M_{p(\cdot),Q} f\right)^{p(x)} \leqslant \fint_Q (\gamma |f(y)|)^{p(y)}\, dy.$$

This applied to $\gamma^{-1} f$ proves (7.3.13) with $\beta = \alpha/(2\gamma)$. Using the other inequality from (7.3.17) in (7.3.16), we find that (7.3.14) holds with $\beta = \frac{2}{3}\frac{\alpha}{\gamma}$.

We have proved that the lemma holds in the case $p^+ < \infty$ with β independent of p^+. We explain now how the case $p^+ = \infty$ can be recovered from this. For $N > 1$ define $p_N \in \mathcal{P}^{\log}(\mathbb{R}^n)$ by $p_N := \min\{p, N\}$ so that $c_{\log}(p_N) \leqslant c_{\log}(p)$. Let $f_N := f\chi_{\{p \leqslant N\}}$. Then

$$\|\chi_Q f_N\|_{L^{p_N(\cdot)}(\mathbb{R}^n) + L^\infty(\mathbb{R}^n)} = \|\chi_Q f_N\|_{L^{p(\cdot)}(\mathbb{R}^n) + L^\infty(\mathbb{R}^n)}$$

Therefore, there exists $\beta \in (0,1)$ independent of N such that (7.3.13) and (7.3.14) hold with f and p replaced by f_N and p_N, respectively. Now, $\|\chi_Q f_N\|_{p_N(\cdot)} = \|\chi_Q f_N\|_{p(\cdot)} \to \|\chi_Q f\|_{p(\cdot)}$ for $N \to \infty$ by the Fatou property of $L^{p(\cdot)}(\mathbb{R}^n)$. This and $\|\chi_Q\|_{p_N(\cdot)} \to \|\chi_Q\|_{p(\cdot)}$ by Theorem 3.5.7 give $M_{p_N(\cdot),Q} f_N \to M_{p(\cdot),Q} f$. Using this, the continuity of $q \mapsto \widetilde{\varphi}_q(t)$, and $|f_N| \nearrow |f|$ for $N \to \infty$, we can easily pass to the limit $N \to \infty$ in (7.3.13) and (7.3.14) by monotone convergence. $\qquad \square$

Corollary 7.3.18. *Let $p \in \mathcal{P}^{\log}(\mathbb{R}^n)$. If $\mathcal{Q}$ is a locally 1-finite family of cubes with $|Q| \leqslant 2^n$ for all $Q \in \mathcal{Q}$, then*

$$\|T_{p(\cdot),\mathcal{Q}} f\|_{p(\cdot)} \approx \left\|\sum_{Q \in \mathcal{Q}} \chi_Q f\right\|_{p(\cdot)}$$

for all $f \in L^{p(\cdot)}_{\mathrm{loc}}(\mathbb{R}^n)$, where the implicit constant depends only on $c_{\log}(p)$ and n.

Proof. Choose $m > n$ and let $\beta > 0$ be as in Lemma 7.3.12. It suffices to prove the claim under the condition that one of the two terms is finite. We begin with the case that the term on the right-hand side is finite. By scaling we can assume that $\left\|\sum_{Q\in\mathcal{Q}}\chi_Q f\right\|_{p(\cdot)} \leqslant 1$. Thus $\|\chi_Q f\|_{p(\cdot)} \leqslant 1$ for each $Q \in \mathcal{Q}$. Integrating (7.3.13) over Q with respect to x and summing over $Q \in \mathcal{Q}$ we get

$$
\varrho_{p(\cdot)}\big(\beta T_{p(\cdot),\mathcal{Q}} f\big) \leqslant \varrho_{p(\cdot)}\bigg(\sum_{Q\in\mathcal{Q}}\chi_Q f\bigg) + \int\limits_{\mathbb{R}^n}(e+|x|)^{-m}\,dx \leqslant c.
$$

A scaling argument therefore yields $\|T_{p(\cdot),\mathcal{Q}} f\|_{p(\cdot)} \leqslant c/\beta \|f\|_{p(\cdot)}$. Assume now $\|T_{p(\cdot),\mathcal{Q}} f\|_{p(\cdot)} \leqslant 1$. Again we obtain $\|\chi_Q f\|_{p(\cdot)} \leqslant 1$ for each $Q \in \mathcal{Q}$. Now we proceed as before with (7.3.14) instead of (7.3.13) and obtain the other inequality. $\square$

Lemma 7.3.19. *Let $p \in \mathcal{P}^{\log}(\mathbb{R}^n)$. If $\mathcal{Q}$ is a locally 1-finite family of cubes with $|Q| \geqslant 1$ for every $Q \in \mathcal{Q}$, then*

$$
\bigg\|\sum_{Q\in\mathcal{Q}}\chi_Q t_Q\bigg\|_{p_\infty} \approx \bigg\|\sum_{Q\in\mathcal{Q}}\chi_Q t_Q\bigg\|_{p(\cdot)}
$$

for any family $t_Q \geqslant 0$ with $Q \in \mathcal{Q}$, where the implicit constant depends only on $c_{\log}(p)$ and n.

Proof. We assume without loss of generality that $\varphi_{p(\cdot)} = \bar{\varphi}_{p(\cdot)}$. By Lemma 3.2.11 we have $\|\chi_Q\|_{p(\cdot)} \geqslant 1$ and $\|\chi_Q\|_{p_\infty} \geqslant 1$ for every $Q \in \mathcal{Q}$.

Let $m > n$ be such that $\int_{\mathbb{R}^n}(e+|y|)^{-m}\,dy \leqslant 1$ and let $\beta > 0$ be as in Lemma 7.3.11. We begin with the case that the right-hand side is finite. By scaling we can assume that $\left\|\sum_{Q\in\mathcal{Q}}\chi_Q t_Q\right\|_{p(\cdot)} \leqslant 1$. In particular, we have $0 \leqslant t_Q\|\chi_Q\|_{p(\cdot)} \leqslant 1$, so $t_Q \in [0,1]$ since $\|\chi_Q\|_{p(\cdot)} \geqslant 1$. Thus Lemma 7.3.11 implies that

$$
\varrho_{p_\infty}\bigg(\beta\sum_{Q\in\mathcal{Q}}\chi_Q t_Q\bigg) \leqslant \varrho_{p(\cdot)}\bigg(\sum_{Q\in\mathcal{Q}}\chi_Q t_Q\bigg) + \int\limits_{\mathbb{R}^n}(e+|y|)^{-m}\,dy \leqslant 2.
$$

This and a scaling argument yield $\left\|\sum_{Q\in\mathcal{Q}}\chi_Q t_Q\right\|_{p_\infty} \leqslant \frac{1}{2\beta}\left\|\sum_{Q\in\mathcal{Q}}\chi_Q t_Q\right\|_{p(\cdot)}$. The estimate $\left\|\sum_{Q\in\mathcal{Q}}\chi_Q t_Q\right\|_{p(\cdot)} \leqslant \frac{1}{2\beta}\left\|\sum_{Q\in\mathcal{Q}}\chi_Q t_Q\right\|_{p_\infty}$ follows analogously. $\square$

We are now in a position to prove a version of Lemma 7.3.11 without the size restriction on the cubes.

Theorem 7.3.20. *Let* $p \in \mathcal{P}^{\log}(\mathbb{R}^n)$. *Then for every* $m > 0$ *there exists* $\beta \in (0,1)$ *which only depends on* $c_{\log}(p)$, n *and* m *such that*

$$\varphi_{p(x)}\big(\beta M_{p(\cdot),Q}f\big) \leqslant \fint_Q \varphi_{p(y)}(|f(y)|)\,dy + \fint_Q (e+|y|)^{-m}\,dy + (e+|x|)^{-m},$$

$$\fint_Q \varphi_{p(y)}(\beta|f(y)|)\,dy \leqslant \varphi_{p(x)}\big(M_{p(\cdot),Q}f\big) + \fint_Q (e+|y|)^{-m}\,dy + (e+|x|)^{-m}$$

for every cube $Q \subset \mathbb{R}^n$, *all* $x \in Q$, *and all* $f \in L^{p(\cdot)}(\mathbb{R}^n) + L^\infty(\mathbb{R}^n)$ *with* $\|\chi_Q f\|_{L^{p(\cdot)}(\mathbb{R}^n)+L^\infty(\mathbb{R}^n)} \leqslant 1$.

Proof. Exactly as in the proof of Lemma 7.3.12 it suffices to consider the case $\varphi_{p(\cdot)} = \bar\varphi_{p(\cdot)}$ and $p^+ < \infty$ with β independent of p^+, since we can recover the case $p^+ = \infty$ by a limiting argument. By Lemma 7.3.12 it suffices to consider the case $|Q| \geqslant 2^n$.

Let $m > n$. As in Lemma 7.3.12 we obtain $M_{p(\cdot),Q}f \leqslant \max\{1, 1/\|\chi_Q\|_{p(\cdot)}\}$. Since $\|\chi_Q\|_{p(\cdot)} \leqslant 1$ by Lemma 3.2.11, we find that $M_{p(\cdot),Q}f \leqslant 1$.

As $M_{p(\cdot),Q}$ is subadditive it suffices to prove the claim independently for $\|\chi_Q f\|_{p(\cdot)} \leqslant 1$ and $\|\chi_Q f\|_\infty \leqslant 1$. Choose a locally 1-finite family $\mathcal{W}$ of cubes such that $1 \leqslant |W| \leqslant 2^n$ for every $W \in \mathcal{W}$ and $\bigcup_{W \in \mathcal{W}} W = Q$ up to a nullset. Define $g := T_{p(\cdot),\mathcal{W}}f$. If $\|\chi_Q f\|_\infty \leqslant 1$, then $M_{p(\cdot),W}f \leqslant 1$ and consequently $\|g\|_\infty \leqslant 1$. If $\|\chi_Q f\|_{p(\cdot)} \leqslant 1$, then $\|g\|_\infty \leqslant 1$, since $\|\chi_W\|_{p(\cdot)} \geqslant 1$ for every $W \in \mathcal{W}$ by Lemma 3.2.11. So in both cases $\|g\|_\infty \leqslant 1$. This immediately implies that $M_{p(\cdot),Q}g \leqslant 1$.

By Corollary 7.3.18, $\|\chi_Q f\|_{p(\cdot)} \approx \|T_{p(\cdot),\mathcal{W}}f\|_{p(\cdot)}$ so that $M_{p(\cdot),Q}f \approx M_{p(\cdot),Q}g$. By Lemma 7.3.19 for g and Corollary 4.5.9, $\|g\|_{p_\infty} \approx \|g\|_{p(\cdot)}$ and $\|\chi_Q\|_{p_\infty} \approx \|\chi_Q\|_{p(\cdot)}$. Hence $M_{p_\infty,Q}g \approx M_{p(\cdot),Q}g$. Let then $\beta_1 \in (0,1)$ be such that $\beta_1 M_{p(\cdot),Q}f \leqslant M_{p_\infty,Q}g$.

Let $\beta_2, \beta_3 \in (0,1)$ be the β of Lemmas 7.3.11 and 7.3.12, respectively. Note that the former lemma is applicable to g since $\|g\|_\infty \leqslant 1$. Then Lemmas 7.3.10, 7.3.11, and 7.3.12 (the last one applied to $M_{p(\cdot),W}f$ on each $W \in \mathcal{W}$) imply that

$$\bar\varphi_{p_\infty}\big(\beta_1\beta_2\beta_3 M_{p(\cdot),Q}f\big) \leqslant \bar\varphi_{p_\infty}\big(\beta_2\beta_3 M_{p_\infty,Q}g\big) = \fint_Q \bar\varphi_{p_\infty}(\beta_2\beta_3|g(y)|)\,dy$$

$$\leqslant \fint_Q \bar\varphi_{p(y)}(|\beta_3 g(y)|)\,dy + \fint_Q (e+|y|)^{-m}\,dy$$

$$\leqslant \fint_Q \bar\varphi_{p(y)}(|f(y)|)\,dy + 2\fint_Q (e+|y|)^{-m}\,dy.$$

An application of Lemma 7.3.11 gives us the first inequality of the theorem. The same steps in reversed order we yield the other inequality. $\square$

Corollary 7.3.21. *Let* $p \in \mathcal{P}^{\log}(\mathbb{R}^n)$ *and* $\mathcal{Q}$ *be a locally N-finite family of cubes or balls* $Q \subset \mathbb{R}^n$. *Then*

$$\|T_{p(\cdot),\mathcal{Q}}f\|_{p(\cdot)} \approx \left\| \sum_{Q \in \mathcal{Q}} \chi_Q f \right\|_{p(\cdot)}$$

for all $f \in L^{p(\cdot)}_{\mathrm{loc}}(\mathbb{R}^n)$. *The implicit constants only depend on* $c_{\log}(p)$, n *and* N. *The dependence on N is linear.*

Proof. Choose $m > n$ and let $\beta > 0$ be as in Theorem 7.3.20. It suffices to prove the claim under the condition that one of the two terms is finite. We begin with the case that the right-hand side is finite. By scaling we can assume that $\|\sum_{Q \in \mathcal{Q}} \chi_Q f\|_{p(\cdot)} \leqslant 1$. Thus we get that $\varrho_{p(\cdot)}(\chi_\Omega f) \leqslant 1$, where $\Omega := \cup_{Q \in \mathcal{Q}} Q$, and that $\|\chi_Q f\|_{p(\cdot)} \leqslant 1$ for each $Q \in \mathcal{Q}$. Using that the family $\mathcal{Q}$ is locally N-finite, the convexity of $\varphi_{p(\cdot)}$, and the first inequality in Theorem 7.3.20, we get

$$\varrho_{p(\cdot)}\left(\frac{\beta}{N}T_{p(\cdot),\mathcal{Q}}f\right) = \int_{\mathbb{R}^n} \varphi_{p(x)}\left(\frac{\beta}{N}\sum_{Q \in \mathcal{Q}}\chi_Q(x)M_{p(\cdot),Q}f\right) dx$$

$$\leqslant \frac{1}{N}\sum_{Q \in \mathcal{Q}}\int_Q \varphi_{p(x)}(\beta M_{p(\cdot),Q}f)\, dx$$

$$\leqslant \frac{1}{N}\sum_{Q \in \mathcal{Q}}\left(\int_Q \varphi_{p(x)}(f)\, dx + 2\int_Q (e + |x|)^{-m}\, dx\right)$$

$$\leqslant \varrho_{p(\cdot)}(\chi_\Omega f) + 2\int_{\mathbb{R}^n}(e + |x|)^{-m}\, dx \leqslant c.$$

A scaling argument gives $\|T_{p(\cdot),\mathcal{Q}}f\|_{p(\cdot)} \leqslant cN/\beta\|f\|_{p(\cdot)}$. The other inequality is proved similarly, by the other inequality of Theorem 7.3.20. $\square$

Theorem 7.3.22. *If* $p \in \mathcal{P}^{\log}(\mathbb{R}^n)$, *then* $p \in \mathcal{G}$ *and the $\mathcal{G}$-constant only depends on* $c_{\log}(p)$ *and* n.

Proof. Remark 4.1.5 implies that $p' \in \mathcal{P}^{\log}(\mathbb{R}^n)$. Thus Corollary 7.3.21 yields $\|T_{p(\cdot),\mathcal{Q}}f\|_{p(\cdot)} \approx \|\sum_{Q \in \mathcal{Q}}\chi_Q f\|_{p(\cdot)}$ and $\|T_{p'(\cdot),\mathcal{Q}}g\|_{p'(\cdot)} \approx \|\sum_{Q \in \mathcal{Q}}\chi_Q g\|_{p'(\cdot)}$ for all $f \in L^{p(\cdot)}_{\mathrm{loc}}(\mathbb{R}^n)$ and $g \in L^{p'(\cdot)}_{\mathrm{loc}}(\mathbb{R}^n)$. Moreover, by Theorem 4.4.8, $p \in \mathcal{A}$ and therefore Theorem 4.5.7 implies that $|Q| \approx \|\chi_Q\|_{p(\cdot)}\|\chi_Q\|_{p'(\cdot)}$. Thus

$$\sum_{Q \in \mathcal{Q}} \|\chi_Q f\|_{p(\cdot)}\|\chi_Q g\|_{p'(\cdot)} \approx \int_{\mathbb{R}^n}\sum_{Q \in \mathcal{Q}}\chi_Q \frac{\|\chi_Q f\|_{p(\cdot)}}{\|\chi_Q\|_{p(\cdot)}}\frac{\|\chi_Q g\|_{p'(\cdot)}}{\|\chi_Q\|_{p'(\cdot)}}\, dx$$

$$\leqslant \int_{\mathbb{R}^n} T_{p(\cdot),\mathcal{Q}}f\, T_{p'(\cdot),\mathcal{Q}}g\, dx$$

$$\leqslant c \, \|T_{p(\cdot),\mathcal{Q}}f\|_{p(\cdot)} \|T_{p(\cdot),\mathcal{Q}}g\|_{p'(\cdot)}$$
$$\approx \|f\|_{p(\cdot)} \|g\|_{p'(\cdot)}. \qquad\qquad \square$$

Using this result, we easily obtain the following version of the result from [216]:

Corollary 7.3.23. *Let $p \in \mathcal{P}^{\log}(\mathbb{R}^n)$. Then*

$$\|f\|_{p(\cdot)} \approx \big\| \|\chi_Q f\|_{p(\cdot)} \big\|_{l^{p_\infty}}$$

for every $f \in L_{\mathrm{loc}}^{p(\cdot)}(\mathbb{R}^n)$ and every locally 1-finite partition of $\mathbb{R}^n$ into a family $\mathcal{Q}$ of cubes (up to measure zero) satisfying $1 \leqslant |Q|$ for every $Q \in \mathcal{Q}$. The implicit constants only depend on $c_{\log}(p)$ and n.

Proof. Using Theorem 7.3.22, Lemma 7.3.19, and we obtain

$$\|f\|_{p(\cdot)} \approx \|T_{p(\cdot),\mathcal{Q}}f\|_{p(\cdot)} \approx \|T_{p(\cdot),\mathcal{Q}}f\|_{p_\infty} = \left(\sum_{Q \in \mathcal{Q}} |Q| \frac{\|\chi_Q f\|_{p(\cdot)}^{p_\infty}}{\|\chi_Q\|_{p(\cdot)}^{p_\infty}} \right)^{\frac{1}{p_\infty}}.$$

Since $\|\chi_Q\|_{p(\cdot)}^{p_\infty} \approx |Q|$ by Corollary 4.5.9, the right-hand side is equivalent to $\big\| \|\chi_Q f\|_{p(\cdot)} \big\|_{l^{p_\infty}}$, so the claim follows. $\qquad \square$

Corollary 7.3.24. *Let $p \in \mathcal{P}^{\log}(\mathbb{R}^n)$, let $\mathcal{Q}$ be a locally N-finite family $\mathcal{Q}$ of cubes (or balls) in $\mathbb{R}^n$, and let $f_Q \in L^{p(\cdot)}(Q)$. Then*

$$\left\| \sum_{Q \in \mathcal{Q}} \chi_Q f_Q \right\|_{p(\cdot)} \leqslant c \left\| \sum_{Q \in \mathcal{Q}} \chi_Q \frac{\|\chi_Q f_Q\|_{p(\cdot)}}{\|\chi_Q\|_{p(\cdot)}} \right\|_{p(\cdot)}, \qquad (7.3.25)$$

with a constant only depending on $c_{\log}(p)$, N and n.

Proof. We denote $f := \sum_{Q \in \mathcal{Q}} \chi_Q f_Q$. From the assumptions $f \in L_{\mathrm{loc}}^{p(\cdot)}(\mathbb{R}^n)$. Thus we obtain from the norm conjugate formula (Corollary 3.2.14), Hölder's inequality, and Theorem 7.3.22 in the last step, that

$$\|f\|_{p(\cdot)} \leqslant 2 \sup_{\|g\|_{p'(\cdot)} \leqslant 1} \int |f| \, |g| \, dx$$

$$\leqslant 4 \sup_{\|g\|_{p'(\cdot)} \leqslant 1} \sum_{Q \in \mathcal{Q}} \|\chi_Q f_Q\|_{p(\cdot)} \|\chi_Q g\|_{p'(\cdot)}$$

$$= 4 \sup_{\|g\|_{p'(\cdot)} \leqslant 1} \sum_{Q \in \mathcal{Q}} \left\| \chi_Q \frac{\|\chi_Q f_Q\|_{p(\cdot)}}{\|\chi_Q\|_{p(\cdot)}} \right\|_{p(\cdot)} \|\chi_Q g\|_{p'(\cdot)}$$

$$\leqslant 4 \sup_{\|g\|_{p'(\cdot)} \leqslant 1} \sum_{Q \in \mathcal{Q}} \left\| \chi_Q \sum_{B \in \mathcal{Q}} \chi_B \frac{\|\chi_B f_B\|_{p(\cdot)}}{\|\chi_B\|_{p(\cdot)}} \right\|_{p(\cdot)} \|\chi_Q g\|_{p'(\cdot)}$$

$$\leqslant c \sup_{\|g\|_{p'(\cdot)} \leqslant 1} \left\| \sum_{B \in Q} \chi_B \frac{\|\chi_B f_B\|_{p(\cdot)}}{\|\chi_B\|_{p(\cdot)}} \right\|_{p(\cdot)} \|g\|_{p'(\cdot)}$$

$$\leqslant c \left\| \sum_{B \in Q} \chi_B \frac{\|\chi_B f_B\|_{p(\cdot)}}{\|\chi_B\|_{p(\cdot)}} \right\|_{p(\cdot)}. \qquad \square$$

If $q \in [1, \infty)$, then by Jensen's inequality $M_Q f \leqslant M_{q,Q} f$ since $M_{q,Q} f = (f_Q |f(y)|^q dx)^{1/q}$. This is in general not true if we replace $M_{q,Q}$ by $M_{p(\cdot),Q}$, however, we have a generalization for the case $\mathcal{A}_{\text{loc}}$:

Lemma 7.3.26. *If $p \in \mathcal{A}_{\text{loc}}$, then $M_Q f \leqslant c M_{p(\cdot),Q} f$ and $Mf \leqslant c M_{p(\cdot)} f$ for all $f \in L^{p(\cdot)}_{\text{loc}}(\mathbb{R}^n)$, where c only depends on the $\mathcal{A}_{\text{loc}}$-constant of p.*

Proof. Let $f \in L^{p(\cdot)}_{\text{loc}}(\mathbb{R}^n)$. Hölder's inequality and Theorem 4.5.7 imply

$$M_Q f = \frac{1}{|Q|} \int_Q |f(x)| \, dx \leqslant 2 \frac{\|\chi_Q f\|_{p(\cdot)} \|\chi_Q\|_{p'(\cdot)}}{|Q|} \leqslant c \frac{\|\chi_Q f\|_{p(\cdot)}}{\|\chi_Q\|_{p'(\cdot)}} = c M_{p(\cdot),Q} f,$$

which proves the assertion for M_Q. Taking the supremum over all cubes Q proves the assertion for M. $\qquad \square$

If $r \in (1, \infty]$ and $q \in [1, r)$, then it follows from the boundedness of M from $L^{r/q}(\mathbb{R}^n)$ to $L^{r/q}(\mathbb{R}^n)$ that M_q is bounded from $L^r(\mathbb{R}^n)$ to $L^r(\mathbb{R}^n)$. The following theorem generalizes this feature to variable exponents.

Theorem 7.3.27. *Let $p, q, s \in \mathcal{P}^{\log}(\mathbb{R}^n)$ such that $p = qs$ and $s^- > 1$. Then $M_{q(\cdot)}$ is bounded from $L^{p(\cdot)}(\mathbb{R}^n)$ to $L^{p(\cdot)}(\mathbb{R}^n)$. The operator norm of $M_{q(\cdot)}$ depends only on $c_{\log}(p)$, $c_{\log}(q)$, $c_{\log}(s)$, and s^-.*

Proof. We prove the claim for $\varphi_{p(\cdot)} = \bar{\varphi}_{p(\cdot)}$. The case $\varphi_{p(\cdot)} = \tilde{\varphi}_{p(\cdot)}$ then follows by Lemma 3.1.6. Fix $m > n$. Let $f \in L^{p(\cdot)}(\mathbb{R}^n)$ with $\|f\|_{p(\cdot)} \leqslant \frac{1}{2}$; then by Theorem 3.3.11, $\|f\|_{L^{q(\cdot)}(\mathbb{R}^n) + L^\infty(\mathbb{R}^n)} \leqslant 1$.

Let β_1 be the β of Theorem 7.3.20 for $q(\cdot)$. Taking the supremum over all cubes Q containing x in Theorem 7.3.20 implies that

$$\bar{\varphi}_{q(x)}(\beta_1 M_{p(\cdot)} f) \leqslant M\big(\bar{\varphi}_{q(\cdot)}(f)\big)(x) + 2 M\big((e + |\cdot|)^{-m}\big)(x)$$

for all $x \in \mathbb{R}^n$. Let $\beta_2 \in (0, 1)$ be such that Lemma 4.3.6 holds with β_2 for the exponent s/s^-. We multiply the previous inequality by $\frac{1}{4}\beta_2$, use the convexity of $\bar{\varphi}_{q(x)}$, apply the convex function $\bar{\varphi}_{s(x)}$ to it, and use Lemma 4.3.6 for s/s^-:

$$\bar{\varphi}_{p(x)}(\tfrac{1}{4}\beta_1\beta_2 M_{p(\cdot)} f) \leqslant \bar{\varphi}_{s(x)}\Big(\tfrac{1}{4}\beta_2 \varphi_{q(x)}(\beta_1 M_{p(\cdot)} f)\Big)$$

$$\leqslant \bar{\varphi}_{s(x)}\Big(\tfrac{1}{4}\beta_2\Big(M\big(\bar{\varphi}_{q(\cdot)}(f)\big)(x) + 2 M\big((e + |\cdot|)^{-m}\big)(x)\Big)\Big)$$

$$\leqslant \bar{\varphi}_{s(x)}\Big(\tfrac{1}{2}\beta_2 M\big(\bar{\varphi}_{q(\cdot)}(f)\big)(x)\Big) + \bar{\varphi}_{s(x)}\big(\beta_2 M\big((e + |\cdot|)^{-m}\big)(x)\big)$$

$$\leqslant \big(M(\bar{\varphi}_{p(\cdot)/s^-}(f))(x)\big)^{s^-} + c\big(M(\bar{\varphi}_{s(\cdot)/s^-}((e + |\cdot|)^{-m}))(x)\big)^{s^-}.$$

We integrate this over $x \in \mathbb{R}^n$, use the boundedness of M from $L^{s^-}(\mathbb{R}^n)$ to $L^{s^-}(\mathbb{R}^n)$, and obtain

$$\overline{\varrho}_{p(\cdot)}\left(\tfrac{1}{4}\beta_1\beta_2 M_{p(\cdot)}f\right) \leqslant c\overline{\varrho}_{p(\cdot)}(f) + c\overline{\varrho}_{s(\cdot)}\left((e + |\cdot|)^{-m}\right) \leqslant c.$$

This implies $\|M_{p(\cdot)}f\|_{q(\cdot)} \leqslant c/(\beta_1\beta_2)$. The claim follows by a scaling argument. □

7.4 Ball/Cubes-to-John

In the previous section we presented results for going from a regular bounded domain (e.g., cube) to the whole space $\mathbb{R}^n$. In this section we consider to what extent we can deduce from results on cubes results on more general domains. In contrast to the previous section, we have to worry about the boundary of our domain. Hence we need to restrict our attention to fairly nice domains. We use John domains in the sense of Martio and Sarvas [283]. Several equivalent characterizations for John domains can be found in [311].

Definition 7.4.1. A bounded domain $\Omega \subset \mathbb{R}^n$ is called an α-*John domain*, $\alpha > 0$, if there exists $x_0 \in \Omega$ (the *John center*) such that each point in Ω can be joined to x_0 by a rectifiable path γ (the *John path*) parametrized by arc-length such that

$$B\left(\gamma(t), \frac{1}{\alpha}t\right) \subset \Omega$$

for all $t \in [0, \ell(\gamma)]$, where $\ell(\gamma)$ is the length of γ. The ball $B(x_0, \frac{1}{2\alpha}\operatorname{diam}(\Omega))$ is called the *John ball*.

Example 7.4.2. There are many examples of John domains. Clearly every ball is a 1-John domain and every cube is a $\sqrt{n}$-John domain with its center as John center. Also every bounded domain that satisfies the uniform interior cone condition is a John domain. John domains may possess fractal boundaries or internal cusps while external cusps are excluded. The interior of Koch's snowflake is a John domain.

In a bounded α-John domain any point can be selected as the John center, possibly with different α. By definition, the John ball B satisfies $\Omega \subset 2\alpha B$ and therefore $|B| \leqslant |\Omega| \leqslant (2\alpha)^n|B|$.

For future reference, we also define here domains with $C^{k,\lambda}$-boundary, where $C^{k,\lambda}$ is the space of k times continuously differentiable functions with λ-Hölder continuous k-th derivative. A domain with $C^{0,1}$-boundary is called a *Lipschitz domain*. Any bounded Lipschitz domain is a John domain, so that the results derived in this section apply in particular to Lipschitz domains.

Definition 7.4.3. We say that a domain $\Omega \subset \mathbb{R}^n$ has $C^{k,\lambda}$-*boundary*, $k \in \mathbb{N}_0$, $\lambda \in (0, 1]$, if there exists.

Cartesian coordinate systems X_j, $j = 1, \ldots, m$,

$$X_j = (x_{j,1}, \ldots, x_{j,n-1}, x_{j,n}) =: (x_j', x_{j,n}),$$

positive real numbers $\alpha, \beta > 0$, and m functions $a_j \in C^{k,\lambda}([-\alpha, \alpha]^{n-1})$, $j = 1, \ldots, m$, such that the sets

$$\Lambda^j := \left\{ (x_j', x_{j,n}) \in \mathbb{R}^n \ : \ |x_j'| < \alpha \,, x_{j,n} = a_j(x_j') \right\},$$
$$V_+^j := \left\{ (x_j', x_{j,n}) \in \mathbb{R}^n \ : \ |x_j'| < \alpha \,, a_j(x_j') < x_{j,n} < a_j(x_j') + \beta \right\},$$
$$V_-^j := \left\{ (x_j', x_{j,n}) \in \mathbb{R}^n \ : \ |x_j'| < \alpha \,, a_j(x_j') - \beta < x_{j,n} < a_j(x_j') \right\},$$

satisfy:

$$\Lambda^j \subset \partial\Omega,, \quad V_+^j \subset \Omega, \quad V_-^j \subset \mathbb{R}^n \setminus \Omega, \quad j = 1, \ldots, m, \quad \bigcup_{j=1}^m \Lambda^j = \partial\Omega.$$

Example 7.4.4. Every ball and cube is a Lipschitz domain. There exist John domains which are not Lipschitz, for instance the unit ball with a segment removed, $B(0, 1) \setminus [0, e_1)$, where e_1 is a unit vector.

We now return to the more general class of John domains.

Lemma 7.4.5. *Let Ω be a bounded α-John domain and let $p \in \mathcal{P}^{\log}(\Omega)$. Then*

$$\|1\|_{L^{p(\cdot)}(\Omega)} \|1\|_{L^{p'(\cdot)}(\Omega)} \approx |\Omega|,$$

where the constant only depends on $c_{\log}(p)$, α, and n.

Proof. Extend p by Proposition 4.1.7 to $\mathbb{R}^n$ preserving $c_{\log}(p)$. Let B be the John ball of Ω, then $B \subset \Omega \subset 2\alpha B$. By Theorem 4.5.7 we find that

$$\|1\|_{L^{p(\cdot)}(\Omega)} \|1\|_{L^{p'(\cdot)}(\Omega)} \leqslant \|1\|_{L^{p(\cdot)}(2\alpha B)} \|1\|_{L^{p'(\cdot)}(2\alpha B)} \leqslant c\,|2\alpha B| \leqslant c\,(2\alpha)^n |\Omega|.$$

On the other hand by Hölder's inequality $|\Omega| \leqslant 2\,\|1\|_{L^{p(\cdot)}(\Omega)} \|1\|_{L^{p'(\cdot)}(\Omega)}.$ $\quad\square$

Closely related to the definition of a John domain is the following definition using a chain condition.

Definition 7.4.6. Let $\Omega \subset \mathbb{R}^n$ and let $\sigma_2 > \sigma_1 > 1$. Then we say that Ω satisfies the *emanating chain condition* with constants σ_1 and σ_2 if there exists a countable covering $\mathcal{Q}$ of Ω consisting of open cubes (or balls) such that:

(B1) We have $\sigma_1 Q \subset \Omega$ for all $Q \in \mathcal{Q}$ and $\sum_{Q \in \mathcal{Q}} \chi_{\sigma_1 Q} \leqslant \sigma_2 \chi_\Omega$ on $\mathbb{R}^n$.

(B2) If Ω is bounded:

There exists a central cube $Q_\infty \in \mathcal{Q}$ with $\Omega \subset \sigma_2 Q_\infty$. For every $Q \in \mathcal{Q}$ there exists a *chain* $Q_1, \ldots, Q_m$ *emanating from* Q *ending in* Q_∞: $Q_1, \ldots, Q_m$ are pairwise distinct such that $Q_1 = Q$, $Q_m = Q_\infty$ and $Q_i \subset \sigma_2 Q_j$ for $1 \leqslant i \leqslant j \leqslant m$. Moreover, $Q_i \cap Q_{i+1}$ contains a ball B_i such that $Q_i \cup Q_{i+1} \subset \sigma_2 B_i$ for $i = 1, \ldots, m - 1$.

If Ω is unbounded:

For every $Q \in \mathcal{Q}$ there exists an unending *chain* $Q_1, Q_2, \ldots$ *emanating from* Q: $Q_1, Q_2, \ldots$ are pairwise distinct such that $Q_1 = Q$ and $Q_i \subset \sigma_2 Q_j$ for $1 \leqslant i \leqslant j$. Moreover, $Q_i \cap Q_{i+1}$ contains a ball B_i such that $Q_i \cup Q_{i+1} \subset \sigma_2 B_i$ for $i \geqslant 1$.

(B3) The set $\{Q \in \mathcal{Q} : Q \cap K \neq \emptyset\}$ if finite for every compact subset $K \subset \Omega$.

The family $\mathcal{Q}$ is called the *chain-covering* of Ω.

If Ω is bounded, then the emanating chain condition is equivalent to the *Boman chain condition*, which appears in the preprint of Boman [54]. In his version only the conditions (B1) and (B2) were included. The technical but useful property (B3) was added by Diening, Růžička and Schumacher [108].

Remark 7.4.7. Let Ω be a domain satisfying the emanating chain condition with chain-covering $\mathcal{W}$. Then it is possible, see [108, Remark 3.15], to choose the balls B_k in (B2) from a family $\mathcal{B}$ of balls with $\sum_{B \in \mathcal{B}} \chi_B \leqslant \sigma_2 \chi_\Omega$.

It has been shown by Buckley, Koskela and Lu [60] that a bounded domain satisfies the emanating chain condition if and only if it is a John domain. The constants σ_1 and σ_2 only depend on α and n, or vice versa. In [108] it is shown that the equivalence holds also when condition (B3) is included.

Remark 7.4.8. An unbounded domain $\Omega \subset \mathbb{R}^n$ is an α-*John domain*, if there exists an increasing sequences of α-John domains Ω_k with $\bigcup_{k=1}^\infty \Omega_k = \Omega$. The whole space and the half space are 1-John domains.

It has been shown in [108, Theorem 3.12, Remark 3.14] that also unbounded John domains satisfy the emanating chain condition. However, the converse is not true. Indeed, the aperture domain

$$\{(x_1, x_2) \in \mathbb{R}^2 : x_2 \neq 0 \text{ or } |x_1| < 1\}$$

and domains with (at least two) conical outlets satisfy the emanating chain condition but are not John domains.

It is often convenient to prove a result in a simple domain such as a cube. The next result allows us to transfer it to domains satisfying the John or emanating chain conditions.

Theorem 7.4.9 (Decomposition theorem). *Let $\Omega \subset \mathbb{R}^n$ be a domain satisfying the emanating chain condition with constants σ_1 and σ_2 and with chain covering $\mathcal{Q}$. Then there exists a family of linear operators S_Q : $L_0^\infty(\Omega) \to L_0^\infty(Q)$ with $Q \in \mathcal{Q}$ which also maps $C_{0,0}^\infty(\Omega) \to C_{0,0}^\infty(Q)$, such that for every $p \in \mathcal{P}^{\log}(\Omega)$ with $p^- > 1$ the following holds:*

(a) *For each $Q \in \mathcal{Q}$, the operator norm of $S_Q : L_0^{p(\cdot)}(\Omega) \hookrightarrow L_0^{p(\cdot)}(Q)$ depends only on α, p^-, $c_{\log}(p)$ and n.*
(b) *For every $f \in L_0^{p(\cdot)}(\Omega)$, $|S_Q f| \leqslant c \sigma_2 \chi_Q Mf$ a.e.*
(c) *The family $S_Q f$ is a decomposition of f, i.e.*

$$f = \sum_{Q \in \mathcal{Q}} S_Q f \qquad a.e. \ and \ in \ L_{\mathrm{loc}}^1(\Omega)$$

for all $f \in L_0^{p(\cdot)}(\Omega)$. If $p^+ < \infty$, then the sum converges unconditionally *in $L_0^{p(\cdot)}(\Omega)$, i.e. every permutation of the series converges in $L_0^{p(\cdot)}(\Omega)$.*
(d) *For all $f \in L_0^{p(\cdot)}(\Omega)$*

$$\left\| \sum_{Q \in \mathcal{Q}} \chi_Q \frac{\|S_Q f\|_{L_0^{p(\cdot)}(Q)}}{\|\chi_Q\|_{p(\cdot)}} \right\|_{p(\cdot)} \approx \|f\|_{L_0^{p(\cdot)}(\Omega)}$$

with constant only depending on α, p^-, $c_{\log}(p)$ and n.
(e) *For all $f \in L_0^{p(\cdot)}(\Omega)$ and $g \in L^{p'(\cdot)}(\Omega)$*

$$\int_\Omega fg \, dx = \sum_{Q \in \mathcal{Q}} \int_Q S_Q f \, g \, dx = \sum_{Q \in \mathcal{Q}} \int_Q S_Q f \, (g - \langle g \rangle_Q) \, dx.$$

(f) *If Ω is bounded and $f \in C_{0,0}^\infty(\Omega)$, then $\{Q \in \mathcal{Q} : S_Q f \neq 0\}$ is finite.*

Proof. Most of the proof is the same as the one in [108]. We only sketch the ideas of those parts. The case of bounded and unbounded domains are treated simultaneously. Let $p \in \mathcal{P}^{\log}(\Omega)$ with $p^- > 1$. We use the notation $\mathcal{Q} = \{Q_i : i \in \mathbb{N}_0\}$, $Q_{i,k}$, $k = 1, \ldots, m_i$, for the chain emanating from Q_i and $B_{i,k}$ for the corresponding balls as in Definition 7.4.6. Within the proof the operators are denoted by S_i instead of S_{Q_i}.

We begin with the construction of our operators S_i. Let $f \in L_0^\infty(\Omega)$. Due to (B1) and $\Omega = \bigcup_{i=0}^\infty Q_i$, there exists a smooth partition of unity $\{\xi_i\}_{i \geqslant 0}$ subordinate to the covering $\{Q_i\}_{i \geqslant 0}$, cf. [12, Theorem 3.15]. Due to Remark 7.4.7 we can assume that the $B_{i,k}$ are from a family $\mathcal{B}$ of balls which satisfies $\sum_{B \in \mathcal{B}} \chi_B \leqslant \sigma_2 \chi_\Omega$. For every $B \in \mathcal{B}$ let $\eta_B \in C_0^\infty(B)$ with $\eta_B \geqslant 0$, $\int \eta_B \, dx = 1$, and $\|\eta_B\|_\infty \leqslant c/|B|$, where $c = c(n)$. For $B_{i,k} \in \mathcal{B}$ we define $\eta_{i,k} := \eta_{B_{i,k}}$. In the case of a bounded domain Ω we pick a function $\eta_0 \in C^\infty(Q_\infty)$ with $\eta_0 \geqslant 0$, $\int \eta_0 \, dx = 1$ and $\|\eta_0\|_\infty \leqslant c/|Q_\infty|$, where $c = c(n)$. Then we define $\eta_{i,m_i} := \eta_0$ for every $i \geqslant 0$. We define $S_i f$ for $f \in L_0^{p(\cdot)}(\Omega)$ by

$$S_i f := \xi_i f - \int_{Q_i} \xi_i f \, dx \, \eta_{i,0} + \sum_{\substack{j \geq 0 \\ j \neq i}} \left(\int_{Q_j} \xi_j f \, dx \sum_{\substack{k:0<k\leq m_j \\ Q_{j,k}=Q_i}} (\eta_{j,k-1} - \eta_{j,k}) \right).$$

The sum over j could be restricted to all j such that Q_i is contained in the chain emanating from Q_j, since for all other j the sum over k is empty. Note that the sum over k consists of at most one summand, since all cubes in a chain are pairwise different. Since the sum over j may still be countable, it is a priori not clear that $S_i f$ is well defined. However, using (B1) and (B2) it has been shown in [108, Theorem 4.2] that the $S_i f$ are well defined, satisfy

$$|S_i f| \leq c \sigma_2 \chi_{Q_i} M f, \quad \text{and} \quad \sum_{i=0}^{\infty} S_i f = f \tag{7.4.10}$$

almost everywhere and $\int S_i f \, dx = 0$. Corollary 4.3.11 gives $Mf \in L^{p(\cdot)}(\Omega)$. Thus the operators S_i are bounded from $L_0^{p(\cdot)}(\Omega)$ to $L_0^{p(\cdot)}(Q_i)$ yielding (a). Since $Mf \in L^{p(\cdot)}(\Omega)$ and $L^{p(\cdot)}(\Omega) \hookrightarrow L^1_{\mathrm{loc}}(\Omega)$, we deduce further from (7.4.10) that $\sum_{i=0}^{\infty} S_i f = f$ in $L^1_{\mathrm{loc}}(\Omega)$. This proves (b). Furthermore, it has been shown in [108, Theorem 4.2] that S_i maps $C_{0,0}^{\infty}(\Omega)$ to $C_{0,0}^{\infty}(Q_i)$. The special choice $p = \infty$ proves that S_i also maps $L_0^{\infty}(\Omega)$ to $L_0^{\infty}(Q_i)$.

Let $f \in L_0^{p(\cdot)}(\Omega)$ with $p^+ < \infty$ and let σ be a permutation of $\mathbb{N}_0$. Then as in [108] it follows that

$$\left\| f - \sum_{i=0}^{k} S_{\sigma(i)} f \right\|_{L_0^{p(\cdot)}(\Omega)} \leq c \left\| \sum_{i=k+1}^{\infty} \chi_{W_{\sigma(i)}} M f \right\|_{L_0^{p(\cdot)}(\Omega)}.$$

Using (B1), $p^+ < \infty$ and Corollary 4.3.11 the right-hand side tends to zero for $k \to \infty$. This proves the unconditional convergence of $\sum_{i=0}^{\infty} S_i f$ in $L_0^{p(\cdot)}(\Omega)$ for $f \in L_0^{p(\cdot)}(\Omega)$ with $p^+ < \infty$. We have proved (c).

For every $f \in L_0^{p(\cdot)}(\Omega)$ and $g \in L^{p'(\cdot)}(\Omega)$ we obtain by (c):

$$\int_{\Omega} fg \, dx = \int_{\Omega} \sum_{i=0}^{\infty} S_i f \, g \, dx.$$

Since $\sum_{i=0}^{\infty} |S_i f| \leq c \, M f \in L^{p(\cdot)}(\Omega)$ by Corollary 4.3.11 and $g \in L^{p'(\cdot)}(\Omega)$, the sum $\sum_{i=0}^{\infty} S_i f \, g$ converges not only almost everywhere but also in $L^1(\Omega)$. Thus,

$$\int_{\Omega} fg \, dx = \sum_{i=0}^{\infty} \int_{Q_i} S_i f \, g \, dx = \sum_{i=0}^{\infty} \int_{Q_i} S_i f \, (g - \langle g \rangle_{Q_i}) \, dx,$$

which proves (e).

For $f \in L_0^{p(\cdot)}(\Omega)$ we estimate using (b), Theorem 7.3.22, (B1) and Corollary 4.3.11

$$\left\| \sum_{i=0}^{\infty} \chi_{Q_i} \frac{\|S_i f\|_{L_0^{p(\cdot)}(Q_i)}}{\|\chi_{Q_i}\|_{p(\cdot)}} \right\|_{p(\cdot)} \leqslant c \left\| \sum_{i=0}^{\infty} \chi_{Q_i} \frac{\|\chi_{Q_i} M f\|_{p(\cdot)}}{\|\chi_{Q_i}\|_{p(\cdot)}} \right\|_{p(\cdot)}$$

$$\leqslant c \left\| M f \right\|_{L_0^{p(\cdot)}(\Omega)}$$

$$\leqslant c \left\| f \right\|_{L_0^{p(\cdot)}(\Omega)}.$$

This proves one part of (d). The other part follows immediately from Corollary 7.3.24 and (c).

The proof of (f) is given in [108, Theorem 4.2]. It is based on the fact that for $f \in C_{0,0}^{\infty}$ only finitely many chains start in the support of f due to (B3) and that by the boundedness of Ω those chains have finite length.   □

The *decomposition theorem* has the following useful consequences.

Proposition 7.4.11. *Let* $\Omega \subset \mathbb{R}^n$ *be a domain satisfying the emanating chain condition with constants* σ_1 *and* σ_2 *and with chain covering* $\mathcal{Q}$. *Let* $p \in \mathcal{P}^{\log}(\mathbb{R}^n)$ *with* $p^+ < \infty$. *Then*

$$\|f - \langle f \rangle_\Omega\|_{L_0^{p(\cdot)}(\Omega)} \approx \left\| \sum_{Q \in \mathcal{Q}} \chi_Q \frac{\|f - \langle f \rangle_Q\|_{L^{p(\cdot)}(Q)}}{\|\chi_Q\|_{p(\cdot)}} \right\|_{L^{p(\cdot)}(\Omega)}$$

for all $f \in L^{p(\cdot)}(\Omega)$. *The constant only depends on* p^+, $c_{\log}(p)$ *and the chain condition constants.*

Proof. Let $f \in L^{p(\cdot)}(\Omega)$ and $g \in L_0^{p'(\cdot)}(\Omega)$. We use (e) of the decomposition theorem for the space $L_0^{p'(\cdot)}(\Omega)$, Hölder's inequality, and Theorem 7.3.22 to conclude that

$$\int_\Omega (f - \langle f \rangle_\Omega) \, g \, dx = \sum_{Q \in \mathcal{Q}} \int_Q (f - \langle f \rangle_Q) S_Q g \, dx$$

$$\leqslant 2 \sum_{Q \in \mathcal{Q}} \|f - \langle f \rangle_Q\|_{L^{p(\cdot)}(Q)} \|\chi_Q S_Q g\|_{L_0^{p'(\cdot)}(\Omega)}$$

$$\leqslant c \left\| \sum_{Q \in \mathcal{Q}} \chi_Q \frac{\|f - \langle f \rangle_Q\|_{L^{p(\cdot)}(Q)}}{\|\chi_Q\|_{p(\cdot)}} \right\|_{p(\cdot)} \left\| \sum_{Q \in \mathcal{Q}} \chi_Q S_Q g \right\|_{L_0^{p'(\cdot)}(\Omega)}$$

Note that $\sum_{Q \in \mathcal{Q}} \chi_Q S_Q g = g$, so the right-hand side simplifies. Then we take the supremum over g with $\|g\|_{L_0^{p'(\cdot)}(\Omega)} \leqslant 1$ and use the norm conjugate formula (Corollary 3.2.14):

$$\|f - \langle f \rangle_\Omega\|_{L_0^{p(\cdot)}(\Omega)} \approx \sup_{\|g\|_{L_0^{p'(\cdot)}(\Omega)} \leqslant 1} \int_\Omega (f - \langle f \rangle_\Omega) g \, dx$$

$$\leqslant c \left\| \sum_{Q \in \mathcal{Q}} \chi_Q \frac{\|f - \langle f \rangle_Q\|_{L^{p(\cdot)}(Q)}}{\|\chi_Q\|_{p(\cdot)}} \right\|_{p(\cdot)}.$$

This proves the first direction of the claim. Since $p \in \mathcal{P}^{\log}(\Omega)$ we have $p \in \mathcal{A}$ and therefore

$$\|f - \langle f \rangle_Q\|_{L^{p(\cdot)}(Q)} \leqslant \|f - \langle f \rangle_\Omega\|_{L^{p(\cdot)}(Q)} + \|\chi_Q M_Q (f - \langle f \rangle_\Omega)\|_{L^{p(\cdot)}(Q)}$$

$$\leqslant c \|f - \langle f \rangle_\Omega\|_{L^{p(\cdot)}(Q)}$$

This and Theorem 7.3.22 prove the other direction of the claim.   □

In Sect. 8.2 after Theorem 8.2.4 we will show that the decomposition theorem and its consequences easily yield the Poincaré inequality. In Sect. 14.3 we will use it to prove estimates for the divergence equation, the negative norm theorem and Korn's inequality.

Let us finish the section by showing how the operators S_Q, defined on $L_0^{p(\cdot)}(\Omega)$, can be modified in such a way that they are defined on $L^{p(\cdot)}(\Omega)$.

Lemma 7.4.12. *Let $\Omega \subset \mathbb{R}^n$ be a bounded α-John domain, and let $p \in \mathcal{P}^{\log}(\Omega)$. Let $\eta \in L^\infty(\Omega)$ with $\int_\Omega \eta \, dx = 1$. Then $V_\eta : f \mapsto \eta \int_\Omega f \, dx$ and $U_\eta := I - V_\eta$ are bounded, linear mappings from $L^{p(\cdot)}(\Omega)$ to $L^{p(\cdot)}(\Omega)$ and to $L_0^{p(\cdot)}(\Omega)$, respectively. The operator norms depend only on $c_{\log}(p)$, α, and $|\Omega| \|\eta\|_\infty$. Moreover, the mapping U_η is onto and if $\eta \in C_0^\infty(\Omega)$, then $U_\eta : C_0^\infty(\Omega) \to C_{0,0}^\infty(\Omega)$.*

Proof. To prove the boundedness of V_η, we note by Hölder's inequality that $V_\eta f = \eta \int_\Omega f \, dy \leqslant 2\|\eta\|_\infty \|f\|_{L^{p(\cdot)}(\Omega)} \|1\|_{L^{p'(\cdot)}(\Omega)}$ almost everywhere. Thus by Lemma 7.4.5 we get

$$\|V_\eta f\|_{L^{p(\cdot)}(\Omega)} \leqslant 2\|\eta\|_\infty \|f\|_{L^{p(\cdot)}(\Omega)} \|1\|_{L^{p'(\cdot)}(\Omega)} \|1\|_{L^{p(\cdot)}(\Omega)} \approx |\Omega| \, \|\eta\|_\infty \|f\|_{L^{p(\cdot)}(\Omega)}.$$

This proves the boundedness of the operators U_η and V_η. The mapping U_η is onto, since it is the identity on $L_0^{p(\cdot)}(\Omega)$. Since $\int_\Omega \eta \, dx = 1$, we have $\int_\Omega U_\eta f \, dx = 0$. That U_η maps $C_0^\infty(\Omega)$ into $C_{0,0}^\infty(\Omega)$ for $\eta \in C_0^\infty(\Omega)$ is obvious.   □

Remark 7.4.13. If $\Omega \subset \mathbb{R}^n$ is a bounded John domain and $p \in \mathcal{P}^{\log}(\Omega)$ satisfies $p^- > 1$, then we can combine Lemma 7.4.12 with Theorem 7.4.9 to extend our operators $S_Q : L_0^{p(\cdot)}(\Omega) \to L_0^{p(\cdot)}(Q)$ to $\hat{S}_Q : L^{p(\cdot)}(\Omega) \to L_0^{p(\cdot)}(Q)$

for $Q \in \mathcal{Q}$. Indeed, let η be as in Lemma 7.4.12, then the operators $\hat{S}_Q :=$ $S_Q \circ U_\eta$ have the desired property. Moreover,

$$\sum_{Q \in \mathcal{Q}} \hat{S}_Q f = \sum_{Q \in \mathcal{Q}} S_Q(U_\eta f) = U_\eta f = f - \eta \int_\Omega f \, dx$$

with unconditional convergence in $L_0^{p(\cdot)}(\Omega)$ for $p^+ < \infty$. It is easily seen that the operators $\hat{S}_Q$ satisfy (a), (b), and (d) of Theorem 7.4.9.

If additionally $\eta \in C_0^\infty(\Omega)$, then $\hat{S}_Q : C_0^\infty(\Omega) \to C_{0,0}^\infty(\Omega)$ and the $\hat{S}_Q$ also satisfy (f) of Theorem 7.4.9.

The following result will also be useful for applications.

Lemma 7.4.14. *Let $\Omega \subset \mathbb{R}^n$ be a bounded John domain, let $p \in \mathcal{P}^{\log}(\Omega)$ and let $\eta \in L^\infty(\Omega)$ satisfy $\int_\Omega \eta \, dx = 1$. Then*

$$\|f\|_{L^{p(\cdot)}(\Omega)} \leqslant c \|f - \langle f \rangle_\Omega\|_{L_0^{p(\cdot)}(\Omega)} + 4 \|1\|_{L^{p(\cdot)}(\Omega)} \left| \int_\Omega f \eta \, dx \right|$$

for all $f \in L^{p(\cdot)}(\Omega)$. The constant depends only on $c_{\log}(p)$, α and $|\Omega| \|\eta\|_\infty$.

Proof. For $f \in L^{p(\cdot)}(\Omega)$ we estimate using the norm conjugate formula (Corollary 3.2.14)

$$\|f\|_{L^{p(\cdot)}(\Omega)} \leqslant 2 \sup_{\|g\|_{L^{p'(\cdot)}(\Omega)} \leqslant 1} \left| \int_\Omega f g \, dx \right|$$

$$= 2 \sup_{\|g\|_{L^{p'(\cdot)}(\Omega)} \leqslant 1} \left| \int_\Omega f U_\eta g \, dx + \int_\Omega f \eta \, dx \int_\Omega g \, dy \right|,$$

where $U_\eta g = g - \eta \int_\Omega g \, dy \in L_0^{p'(\cdot)}(\Omega)$. With Lemma 7.4.12 and Hölder's inequality we get

$$\|f\|_{L^{p(\cdot)}(\Omega)} \leqslant c \sup_{\|h\|_{L_0^{p'(\cdot)}(\Omega)} \leqslant 1} \left| \int_\Omega f h \, dx \right| + 2 \sup_{\|g\|_{L^{p'(\cdot)}(\Omega)} \leqslant 1} \left| \int_\Omega f \eta \, dx \right| \int_\Omega |g| \, dy$$

$$\leqslant c \|f - \langle f \rangle_\Omega\|_{L_0^{p(\cdot)}(\Omega)} + 4 \|1\|_{L^{p(\cdot)}(\Omega)} \left| \int_\Omega f \eta \, dx \right|. \qquad \square$$

Part II
Sobolev Spaces

Chapter 8
Introduction to Sobolev Spaces

In this chapter we begin our study of Sobolev functions. The Sobolev space is
a vector space of functions with weak derivatives. One motivation of studying
these spaces is that solutions of partial differential equations belong naturally
to Sobolev spaces (cf. Part III). In Sect. 8.1 we study functional analysis-type
properties of Sobolev spaces, in particular we show that the Sobolev space
is a Banach space and study its basic properties as reflexivity, separability
and uniform convexity. In Sect. 8.2 we prove several versions of the Poincaré
inequality under various assumption on the regularity of the exponent. In
Sect. 8.3 we study Sobolev embeddings of the type $W^{1,p(\cdot)} \hookrightarrow L^{p^*(\cdot)}$ in the
case that p is log-Hölder continuous and $1 \leqslant p^- \leqslant p^+ < n$ or $n < p^- \leqslant
p^+ < \infty$. In Sect. 8.4 compact embeddings $W^{1,p(\cdot)} \hookrightarrow\hookrightarrow L^{p^*(\cdot)-\varepsilon}$ are proved.
In Sect. 8.5 we show that Sobolev functions defined in a sufficiently smooth
domain can be extended to the whole space. In the last section, Sect. 8.6, we
study Sobolev embeddings in the limit cases when either $p^+ = n$ or $n = p^-$.

There are some topics which could not be included. This includes the
theory of variable exponent Sobolev spaces on metric measure spaces studied
for example in [28, 162, 168, 198, 199, 204, 301]; a mean continuity type result
by Fiorenza [155]; and Hardy's inequality in Sobolev space [192, 286, 287]
(Hardy's inequality in Lebesgue spaces has been considered, e.g., in [109, 114,
117, 344]).

8.1 Basic Properties

In this section we define Sobolev spaces and prove functional analysis-type
properties for them. The results are from [91] by Diening, [149] by Fan and
Zhao, and [258] by Kováčik and Rákosník.

Let $\Omega \subset \mathbb{R}^n$ be an open set. We start by recalling the definition of weak
derivatives.

Definition 8.1.1. Assume that $u \in L^1_{\text{loc}}(\Omega)$. Let $\alpha := (\alpha_1, \ldots, \alpha_n) \in \mathbb{N}_0^n$ be a multi-index. If there exists $g \in L^1_{\text{loc}}(\Omega)$ such that

$$\int_\Omega u \frac{\partial^{\alpha_1 + \ldots + \alpha_n} \psi}{\partial^{\alpha_1} x_1 \cdots \partial^{\alpha_n} x_n} \, dx = (-1)^{\alpha_1 + \ldots + \alpha_n} \int_\Omega \psi g \, dx$$

for all $\psi \in C_0^\infty(\Omega)$, then g is called a *weak partial derivative* of u with respect to α. The function g is denoted by $\partial_\alpha u$ or by $\frac{\partial^{\alpha_1 + \ldots + \alpha_n} u}{\partial^{\alpha_1} x_1 \cdots \partial^{\alpha_n} x_n}$. Moreover, we write ∇u to denote the *weak gradient* $\left(\frac{\partial u}{\partial x_1}, \ldots, \frac{\partial u}{\partial x_n} \right)$ of u and we write short $\partial_j u$ for $\frac{\partial u}{\partial x_j}$ with $j = 1, \ldots, n$. More generally we write $\nabla^k u$ to denote the tensor with entries $\partial_\alpha u$, $|\alpha| = k$.

If a function u has classical derivatives then they are also weak derivatives of u. Also by definition $\nabla u = 0$ almost everywhere in an open set where $u = 0$.

Definition 8.1.2. The function $u \in L^{p(\cdot)}(\Omega)$ belongs to the space $W^{k,p(\cdot)}(\Omega)$, where $k \in \mathbb{N}_0$ and $p \in \mathcal{P}(\Omega)$, if its weak partial derivatives $\partial_\alpha u$ with $|\alpha| \leqslant k$ exist and belong to $L^{p(\cdot)}(\Omega)$. We define a semimodular on $W^{k,p(\cdot)}(\Omega)$ by

$$\varrho_{W^{k,p(\cdot)}(\Omega)}(u) := \sum_{0 \leqslant |\alpha| \leqslant k} \varrho_{L^{p(\cdot)}(\Omega)}(\partial_\alpha u)$$

which induces a norm by

$$\|u\|_{W^{k,p(\cdot)}(\Omega)} := \inf \left\{ \lambda > 0 : \varrho_{W^{k,p(\cdot)}(\Omega)} \left(\frac{u}{\lambda} \right) \leqslant 1 \right\}.$$

For $k \in \mathbb{N}$ the space $W^{k,p(\cdot)}(\Omega)$ is called *Sobolev space* and its elements are called *Sobolev functions*. Clearly $W^{0,p(\cdot)}(\Omega) = L^{p(\cdot)}(\Omega)$.

Remark 8.1.3. It is also possible to define the semimodular $\varrho_{W^{k,p(\cdot)}(\Omega)}$ on the larger set $W^{k,1}_{\text{loc}}(\Omega)$ or even $L^1_{\text{loc}}(\Omega)$. Then $W^{k,p(\cdot)}(\Omega)$ is just the corresponding semimodular space.

We define local Sobolev spaces as usual:

Definition 8.1.4. A function u belongs to $W^{k,p(\cdot)}_{\text{loc}}(\Omega)$ if $u \in W^{k,p(\cdot)}(U)$ for every open $U \subset\subset \Omega$. We equip $W^{k,p(\cdot)}_{\text{loc}}(\Omega)$ with the initial topology induced by the embeddings $W^{k,p(\cdot)}_{\text{loc}}(\Omega) \hookrightarrow W^{k,p(\cdot)}(U)$ for all open $U \subset\subset \Omega$.

Sobolev functions, as Lebesgue functions, are defined only up to measure zero and thus we identify functions that are equal almost everywhere. If the set Ω is clear from the content, we abbreviate $\|u\|_{W^{k,p(\cdot)}(\Omega)}$ to $\|u\|_{k,p(\cdot)}$ and $\varrho_{W^{k,p(\cdot)}(\Omega)}$ to $\varrho_{k,p(\cdot)}$.

Remark 8.1.5. (i) Note that in $W^{k,p(\cdot)}(\Omega)$

$$\sum_{m=0}^{k} \varrho_{L^{p(\cdot)}(\Omega)}(|\nabla^m u|) \quad \text{and} \quad \sum_{m=0}^{k} \||\nabla^m u|\|_{L^{p(\cdot)}(\Omega)}$$

define a semimodular and a norm equivalent to the Sobolev semimodular and the Sobolev norm, respectively. We abbreviate $\||\nabla^m u|\|_{L^{p(\cdot)}(\Omega)}$ as $\|\nabla^m u\|_{L^{p(\cdot)}(\Omega)}$, $m \in \mathbb{N}$.

(ii) One easily proves, using Lemma 3.2.4, that for each 1-finite partition $(\Omega_i)_{i\in\mathbb{N}}$ of Ω we have

$$\|u\|_{W^{k,p(\cdot)}(\Omega)} \leqslant \sum_{i=1}^{\infty} \|u\|_{W^{k,p(\cdot)}(\Omega_i)}$$

for all $u \in W^{k,p(\cdot)}(\Omega)$.

Theorem 8.1.6. *Let $p \in \mathcal{P}(\Omega)$. The space $W^{k,p(\cdot)}(\Omega)$ is a Banach space, which is separable if p is bounded, and reflexive and uniformly convex if $1 < p^- \leqslant p^+ < \infty$.*

Proof. We proof only the case $k = 1$, the proof for the general case is similar. We first show that the Sobolev spaces is a Banach space; for that let (u_i) be a Cauchy sequence in $W^{1,p(\cdot)}(\Omega)$. We have to show that there exists $u \in W^{1,p(\cdot)}(\Omega)$ such that $u_i \to u$ in $W^{1,p(\cdot)}(\Omega)$ as $i \to \infty$. Since the Lebesgue space $L^{p(\cdot)}(\Omega)$ is a Banach space (Theorem 3.2.7), there exist $u, g_1, \ldots, g_n \in L^{p(\cdot)}(\Omega)$ such that $u_i \to u$ and $\partial_j u_i \to g_j$ in $L^{p(\cdot)}(\Omega)$ for every $j = 1, \ldots, n$. Let $\psi \in C_0^\infty(\Omega)$. Since u_i is in $W^{1,p(\cdot)}(\Omega)$ we have

$$\int_\Omega u_i \, \partial_j \psi \, dx = - \int_\Omega \psi \, \partial_j u_i \, dx.$$

Strong convergence in $L^{p(\cdot)}(\Omega)$ implies weak convergences and hence we have

$$\int_\Omega u_i \, \partial_j \psi \, dx \to \int_\Omega u \, \partial_j \psi \, dx \quad \text{and} \quad \int_\Omega \psi \, \partial_j u_i \, dx \to \int_\Omega \psi \, g_j \, dx$$

as $i \to \infty$. Thus the right-hand sides on the previous line yield that $(g_1, \ldots, g_n)$ is the weak gradient of u. It follows that $u \in W^{1,p(\cdot)}$ and $u_j \to u$ in $W^{1,p(\cdot)}$.

By Theorem 3.4.4, $L^{p(\cdot)}(\Omega)$ is separable if $p^+ < \infty$ and by Theorem 3.4.7, $L^{p(\cdot)}(\Omega)$ is reflexive if $1 < p^- \leqslant p^+ < \infty$. By the mapping $u \mapsto (u, \nabla u)$, the space $W^{1,p(\cdot)}(\Omega)$ is a closed subspace of $L^{p(\cdot)}(\Omega) \times \left(L^{p(\cdot)}(\Omega)\right)^n$. Thus $W^{1,p(\cdot)}(\Omega)$ is separable if $p^+ < \infty$, and reflexive if $1 < p^- \leqslant p^+ < \infty$ by Proposition 1.4.4.

For the uniform convexity we note that $W^{1,p(\cdot)}(\Omega)$ satisfies the Δ_2-condition provided that $p^+ < \infty$. The $L^{p(\cdot)}$-modular is uniformly convex for $p^- > 1$ by Theorem 2.4.11 and the proof of Theorem 3.4.9. Thus $\varrho_{W^{k,p(\cdot)}}$ is uniform convex as a sum of uniform convex modulars (Lemma 2.4.16). Thus $W^{1,p(\cdot)}(\Omega)$ is uniform convex with its own norm by Theorem 2.4.14.   □

A normed space X has the *Banach–Saks property* if $\frac{1}{m}\sum_{i=1}^{m} u_i \to u$ whenever $u_i \rightharpoonup u$. By [226] every uniformly convex space has the Banach–Saks property. This together with the previous theorem implies the following corollary.

Corollary 8.1.7. *Let $p \in \mathcal{P}(\Omega)$ with $1 < p^- \leqslant p^+ < \infty$. Then the Sobolev space $W^{k,p(\cdot)}(\Omega)$ has the Banach–Saks property.*

Lemma 8.1.8. *Let $p \in \mathcal{P}(\Omega)$. Then $W^{k,p(\cdot)}(\Omega) \hookrightarrow W_{\mathrm{loc}}^{k,p^-}(\Omega)$. If $|\Omega| < \infty$, then $W^{k,p(\cdot)}(\Omega) \hookrightarrow W^{k,p^-}(\Omega)$.*

Proof. This follows immediately from the embedding $L^{p(\cdot)}(\Omega) \hookrightarrow L^{p^-}(\Omega)$, see Corollary 3.3.4.   □

A (real valued) function space is a *lattice* if the point-wise minimum and maximum of any two of its elements belong to the space. Next we show that the variable exponent Sobolev space of first order has this property.

Proposition 8.1.9. *Let $p \in \mathcal{P}(\Omega)$. If $u, v \in W^{1,p(\cdot)}(\Omega)$, then $\max\{u,v\}$ and $\min\{u,v\}$ are in $W^{1,p(\cdot)}(\Omega)$ with*

$$\nabla \max(u,v)(x) = \begin{cases} \nabla u(x), & \text{for almost every } x \in \{u \geqslant v\}; \\ \nabla v(x), & \text{for almost every } x \in \{v \geqslant u\}; \end{cases}$$

and

$$\nabla \min(u,v)(x) = \begin{cases} \nabla u(x), & \text{for almost every } x \in \{u \leqslant v\}; \\ \nabla v(x), & \text{for almost every } x \in \{v \leqslant u\}. \end{cases}$$

In particular, $|u|$ belongs to $W^{1,p(\cdot)}(\Omega)$ and $|\nabla|u|| = |\nabla u|$ almost everywhere in Ω.

Proof. It suffices to prove the assertions for $\max\{u,v\}$ since $\min\{u,v\} = -\max\{-u,-v\}$. By Lemma 8.1.8 we know that $W^{1,p(\cdot)}(\Omega) \hookrightarrow W_{\mathrm{loc}}^{1,1}(\Omega)$ and so the formulas for $\nabla \max(u,v)$ and $\nabla \min(u,v)$ follow from [219, Theorem 1.20]. We next note that $\|\max\{u,v\}\|_{p(\cdot)} \leqslant \|u\|_{p(\cdot)} + \|v\|_{p(\cdot)}$ and $\|\nabla \max\{u,v\}\|_{p(\cdot)} \leqslant \|\nabla u\|_{p(\cdot)} + \|\nabla v\|_{p(\cdot)}$. Thus it follows that $\max\{u,v\} \in W^{1,p(\cdot)}(\Omega)$. Analogously, we get $\min\{u,v\} \in W^{1,p(\cdot)}(\Omega)$. The claims for $|u|$ follow by noting that $|u| = \max\{u,0\} - \min\{u,0\}$.   □

Note that the previous proposition yields that $\nabla u = 0$ almost everywhere in a set where u is constant.

We close this section by defining Sobolev spaces with zero boundary values and proving basic properties for them.

Definition 8.1.10. Let $p \in \mathcal{P}(\Omega)$ and $k \in \mathbb{N}$. The *Sobolev space* $W_0^{k,p(\cdot)}(\Omega)$ *with zero boundary values* is the closure of the set of $W^{k,p(\cdot)}(\Omega)$-functions with compact support, i.e.

$$\left\{ u \in W^{k,p(\cdot)}(\Omega) \colon u = u\chi_K \text{ for a compact } K \subset \Omega \right\}$$

in $W^{k,p(\cdot)}(\Omega)$.

Warning 8.1.11. *The closure of $C_0^\infty(\Omega)$ in the space $W^{k,p(\cdot)}(\Omega)$ is denoted by $H_0^{k,p(\cdot)}(\Omega)$.*

Clearly $C_0^\infty(\Omega) \subset W_0^{k,p(\cdot)}(\Omega)$. Later in Sect. 11.2 we will study in more detail Sobolev functions with zero boundary values. We will show in Proposition 11.2.3 that if p is bounded and smooth functions are dense in the Sobolev space then $W_0^{k,p(\cdot)}(\Omega) = H_0^{k,p(\cdot)}(\Omega)$. In particular we will obtain that if $p \in \mathcal{P}^{\log}(\Omega)$ is bounded, then $W_0^{k,p(\cdot)}(\Omega) = H_0^{k,p(\cdot)}(\Omega)$ (Corollary 11.2.4).

Remark 8.1.12. In contrast to $H_0^{1,p(\cdot)}(\Omega)$, the space $W_0^{1,p(\cdot)}(\Omega)$ has the following fundamental property: if $u \in W^{1,p(\cdot)}(\Omega)$ and v is a Lipschitz continuous function with compact support in Ω, then $uv \in W_0^{1,p(\cdot)}(\Omega)$. In Sect. 11.5 we will see that for certain exponents p the product uv need not to be in $H_0^{1,p(\cdot)}(\Omega)$ and thus it may hold that $H_0^{1,p(\cdot)}(\Omega) \subsetneq W_0^{1,p(\cdot)}(\Omega)$.

Theorem 8.1.13. *Let $p \in \mathcal{P}(\mathbb{R}^n)$. The space $W_0^{k,p(\cdot)}(\Omega)$ is a Banach space, which is separable if p is bounded, and reflexive and uniformly convex if $1 < p^- \leqslant p^+ < \infty$.*

Proof. Since $W_0^{k,p(\cdot)}(\Omega)$ is a closed subspace of $W^{k,p(\cdot)}(\Omega)$, the properties follow by Proposition 1.4.4 and Theorem 8.1.6. $\square$

Lemma 8.1.14. *Let $p \in \mathcal{P}(\mathbb{R}^n)$ and $u \in W_0^{k,p(\cdot)}(\Omega)$. Then u extended by zero to $\mathbb{R}^n \setminus \Omega$ belongs to $W^{k,p(\cdot)}(\mathbb{R}^n)$.*

Proof. Let $u \in W^{k,p(\cdot)}(\Omega)$ with compact support, i.e. there exists a compact set $K \subset \Omega$ such that $u = \chi_K u$ almost everywhere. We define $\mathcal{E}u$ to be the extension of u (as a measurable function) by zero outside of Ω. We claim that $\mathcal{E}u \in W^{k,p(\cdot)}(\mathbb{R}^n)$ and $\partial_\alpha \mathcal{E}(u) = \mathcal{E}(\partial_\alpha u)$ almost everywhere for $|\alpha| \leqslant k$.

Choose $\eta \in C_0^\infty(\Omega)$ such that $\chi_K \leqslant \eta \leqslant \chi_\Omega$. Then for all $\psi \in C_0^\infty(\mathbb{R}^n)$ and $|\alpha| \leqslant k$ we have

$$\int_{\mathbb{R}^n} \mathcal{E}u \, \partial_\alpha \psi \, dx = \int_\Omega u \, \partial_\alpha(\psi\eta) \, dx = (-1)^{|\alpha|} \int_\Omega (\partial_\alpha u) \, \psi\eta \, dx$$

$$= (-1)^{|\alpha|} \int_{\mathbb{R}^n} \left(\mathcal{E}(\partial_\alpha u)\right) \psi \, dx,$$

where we used that $u = 0$ and $\partial_\alpha u = 0$ outside of K and $\eta = 1$ on K. This proves $\partial_\alpha \mathcal{E}(u) = \mathcal{E}(\partial_\alpha u)$. Since $\mathcal{E}(\partial_\alpha u) \in L^{p(\cdot)}(\mathbb{R}^n)$, it follows that $\mathcal{E}u \in W^{k,p(\cdot)}(\mathbb{R}^n)$. Moreover, $\|u\|_{W^{k,p(\cdot)}(\Omega)} = \|\mathcal{E}u\|_{W^{k,p(\cdot)}(\mathbb{R}^n)}$, so $\mathcal{E}$ is a isometry on the set of compactly supported $W^{k,p(\cdot)}(\Omega)$ functions. Since those functions are by definition dense in $W_0^{k,p(\cdot)}(\Omega)$, the extension operator is also an isometry from $W_0^{k,p(\cdot)}(\Omega)$ to $W^{k,p(\cdot)}(\mathbb{R}^n)$. In particular, $u \in W_0^{k,p(\cdot)}(\Omega)$ implies $\mathcal{E}u \in W^{k,p(\cdot)}(\mathbb{R}^n)$. $\square$

8.2   Poincaré Inequalities

We start this section by showing that for log-Hölder continuous exponents we get the Poincaré inequality with a constant proportional to diam(Ω). After that we give a relatively mild condition on the exponent for the Poincaré inequality to hold. We also show that this condition is, in a certain sense, the best possible.

Concerning the regularity of the domain we consider in particular bounded John domains (cf. Definition 7.4.1). The constant exponent Poincaré inequality is known to hold for more irregular domains but the inequality is mostly used in John domains.

We recall the following well-known lemma that estimates u in terms of the Riesz potential, due to Bojarski [52, Chap. 6]. For completeness we provide a proof. Recall that I_1 denotes the Riesz potential operator (cf. Definition 6.1.1); note also the convention that $I_1 f$ denotes $I_1(\chi_\Omega f)$ if the function f is defined only in Ω.

Lemma 8.2.1. (a) *For every* $u \in W_0^{1,1}(\Omega)$, *the inequality*

$$|u| \leqslant c \, I_1|\nabla u|$$

holds a.e. in Ω *with the constant c depending only on the dimension n.*
(b) *If* $\Omega \subset \mathbb{R}^n$ *is a bounded α-John domain, then there exists a ball $B \subset \Omega$ and a constant c such that*

$$|u(x) - \langle u \rangle_B| \leqslant c \, I_1|\nabla u|(x)$$

holds a.e. in Ω for every $u \in W^{1,1}(\Omega)$. The ball B satisfies $|B| \leqslant |\Omega| \leqslant c'|B|$ and the constants c and c' depend only on the dimension n and α.

Proof. We prove only (b). The proof of (a) is similar, see for example [176, Lemma 7.14, p. 154].

We consider first claim (b) when Ω is a ball. Assume that $u \in C^\infty(\Omega) \cap W^{1,1}(\Omega)$. We have

$$u(x) - u(y) = -\int_0^{|x-y|} \nabla u\left(x + r\frac{y-x}{|y-x|}\right) \cdot \frac{y-x}{|y-x|}\, dr.$$

Integrating with respect to y over Ω and dividing the result by $|\Omega|$, we obtain

$$u(x) - \langle u\rangle_\Omega = -\frac{1}{|\Omega|}\int_\Omega \int_0^{|x-y|} \nabla u\left(x + r\frac{y-x}{|y-x|}\right) \cdot \frac{y-x}{|y-x|}\, dr\, dy.$$

Using the notation

$$D(z) = \begin{cases} \nabla u(z) & \text{if } z \in \Omega; \\ 0 & \text{if } z \notin \Omega; \end{cases}$$

we obtain

$$|u(x) - \langle u\rangle_\Omega| \leqslant \frac{1}{|\Omega|} \int_{\{|x-y|\leqslant \operatorname{diam}(\Omega)\}} \int_0^\infty \left|D\left(x + r\frac{y-x}{|y-x|}\right)\right| dr\, dy$$

$$= \frac{1}{|\Omega|} \int_0^\infty \int_{\partial B(0,1)} \int_0^{\operatorname{diam}(\Omega)} |D(x+rw)|\varrho^{n-1}\, d\varrho\, dw\, dr$$

$$\leqslant \int_0^\infty \int_{\partial B(0,1)} |D(x+rw)|\, dw\, dr$$

$$= \int_\Omega \frac{|D(y)|}{|x-y|^{n-1}}\, dy = I_1|\nabla u|(x).$$

This concludes the proof in the ball when u is smooth. For $u \in W^{1,1}(\Omega)$, we take smooth approximations ψ_i such that $\psi_i \to u$ in $W^{1,1}(\Omega)$ and almost everywhere. Then $\langle\psi_i\rangle_\Omega \to \langle u\rangle_\Omega$ and $I_1|\nabla\psi_i|(x) \to I_1|\nabla u|(x)$, where we also used the continuity of the Riesz potential in $L^1(\Omega)$ (cf. [280, Theorem 1.1.31]). This yields the claim for u.

Suppose now that Ω is a bounded α-John domain. Then Ω satisfies the emanating chain condition with constants depending only on α. Let Q_0 be the central emanating ball. If $x \in Q_0$, then the claim follows by what was just proved. Otherwise, let $(Q_j)_{j=0}^m$ be the emanating chain connecting x and Q_0. Let B_j be the balls in the intersection $Q_j \cap Q_{j+1}$ as in the Definition 7.4.1. Then

$$|u(x) - \langle u \rangle_{Q_0}| \leqslant |u(x) - \langle u \rangle_{Q_m}| + \sum_{j=0}^{m-1} |\langle u \rangle_{Q_{j+1}} - \langle u \rangle_{B_j}| + |\langle u \rangle_{B_j} - \langle u \rangle_{Q_j}|$$

$$\leqslant I|\nabla u|(x) + 2 \sum_{j=0}^{m-1} |\langle u \rangle_{B_j} - \langle u \rangle_{Q_j}|$$

Let us estimate the second term:

$$|\langle u \rangle_{B_j} - \langle u \rangle_{Q_j}| \leqslant \fint_{B_j} |u - \langle u \rangle_{Q_j}| \, dy$$

$$\leqslant \sigma_2^n \fint_{Q_j} |u - \langle u \rangle_{Q_j}| \, dy$$

$$\leqslant c \, \sigma_2^n \operatorname{diam}(Q_j) \fint_{Q_j} |\nabla u| \, dy.$$

Since $|x - y| \leqslant \sigma_2 \operatorname{diam}(Q_j)$ and (Q_j) has overlap less than or equal to σ_1, we obtain

$$\sum_{j=0}^{m-1} \operatorname{diam}(Q_j) \fint_{Q_j} |\nabla u| \, dy \leqslant c \sigma_2^{n-1} \sum_{j=0}^{m-1} \int_{Q_j} \frac{|\nabla u|}{|x-y|^{n-1}} \, dy \leqslant c \, \sigma_1 \sigma_2^{n-1} I|\nabla u|(x).$$

The assertion follows when we combine the previous three estimates.   □

Remark 8.2.2. One easily checks that the assertions in Lemma 8.2.1 (b) also holds for $u \in L^1_{\mathrm{loc}}(\Omega)$ with $|\nabla u| \in L^{p(\cdot)}(\Omega)$.

Lemma 8.2.3. *Let Ω be a bounded α-John domain and let $p \in \mathcal{P}^{\log}(\Omega)$. If $A \subset \Omega$ has positive finite measure, then*

$$c \frac{|A|}{|\Omega|} \|u - \langle u \rangle_A\|_{L^{p(\cdot)}(\Omega)} \leqslant \|u - \langle u \rangle_\Omega\|_{L^{p(\cdot)}(\Omega)} \leqslant c \|u - \langle u \rangle_A\|_{L^{p(\cdot)}(\Omega)}$$

for $u \in L^{p(\cdot)}(\Omega)$, where c depends on the dimension n, $c_{\log}(p)$ and α.

Proof. By the triangle inequality, $\|u - \langle u \rangle_\Omega\|_{L^{p(\cdot)}(\Omega)} \leqslant \|u - \langle u \rangle_A\|_{L^{p(\cdot)}(\Omega)} + \|\langle u \rangle_A - \langle u \rangle_\Omega\|_{L^{p(\cdot)}(\Omega)}$. We estimate the second term by Hölder's inequality:

$$\|\langle u\rangle_A - \langle u\rangle_\Omega\|_{L^{p(\cdot)}(\Omega)} = |\langle u\rangle_A - \langle u\rangle_\Omega|\, \|1\|_{L^{p(\cdot)}(\Omega)}$$

$$= |\Omega|^{-1}\|u - \langle u\rangle_A\|_{L^1(\Omega)}\|1\|_{L^{p(\cdot)}(\Omega)}$$

$$\leqslant c\,\frac{\|1\|_{L^{p'(\cdot)}(\Omega)}\|1\|_{L^{p(\cdot)}(\Omega)}}{|\Omega|}\|u - \langle u\rangle_A\|_{L^{p(\cdot)}(\Omega)}.$$

Since $p \in \mathcal{P}^{\log}(\Omega)$, the fraction in the last estimate is bounded by a constant according to Lemma 7.4.5. The lower bound is proved similarly.   □

Theorem 8.2.4 (Poincaré inequality). *Let $p \in \mathcal{P}^{\log}(\Omega)$ or $p \in \mathcal{A}$.*

(a) *For every $u \in W_0^{1,p(\cdot)}(\Omega)$, the inequality*

$$\|u\|_{L^{p(\cdot)}(\Omega)} \leqslant c\,\mathrm{diam}(\Omega)\|\nabla u\|_{L^{p(\cdot)}(\Omega)}$$

holds with a constant c depending only on the dimension n and $c_{\log}(p)$.
(b) *If Ω is a bounded α-John domain, then*

$$\|u - \langle u\rangle_\Omega\|_{L^{p(\cdot)}(\Omega)} \leqslant c\,\mathrm{diam}(\Omega)\|\nabla u\|_{L^{p(\cdot)}(\Omega)}$$

for $u \in W^{1,p(\cdot)}(\Omega)$. The constant c depends only on the dimension n, α and $c_{\log}(p)$.

Proof. We prove only the latter case. The proof for the first case is similar; the only difference is to use in Lemma 8.2.1 case (a) instead of case (b). We note that $p \in \mathcal{P}^{\log}(\Omega)$ can be extended to $\mathbb{R}^n$ so that $p \in \mathcal{A}$ (Proposition 4.1.7 and Theorem 4.4.8).

By Lemma 8.2.1 (b) and Lemma 6.1.4 we obtain

$$|u(x) - \langle u\rangle_B| \leqslant I_1|\nabla u|(x) \leqslant c\,\mathrm{diam}(\Omega)\sum_{k=0}^{\infty}2^{-k}T_{k+k_0}|\nabla u|(x)$$

for every $u \in W^{1,p(\cdot)}(\Omega)$ and almost every $x \in \Omega$, where $k_0 \in \mathbb{Z}$ is chosen such that $2^{-k_0-1} \leqslant \mathrm{diam}(\Omega) \leqslant 2^{-k_0}$. Since $p \in \mathcal{A}$, the averaging operator T_{k+k_0} is bounded on $L^{p(\cdot)}(\Omega)$ (cf. Remark 6.1.3). Using also the triangle inequality, we obtain

$$\|u - \langle u\rangle_B\|_{L^{p(\cdot)}(\Omega)} \leqslant c\,\mathrm{diam}(\Omega)\sum_{k=0}^{\infty}2^{-k}\|T_{k+k_0}|\nabla u|\|_{L^{p(\cdot)}(\Omega)}$$

$$\leqslant c\,\mathrm{diam}(\Omega)\|\nabla u\|_{L^{p(\cdot)}(\Omega)}.$$

The estimate for $\|u - \langle u\rangle_\Omega\|_{L^{p(\cdot)}(\Omega)}$ follows from this and Lemma 8.2.3.   □

In the case that $p \in \mathcal{P}^{\log}(\mathbb{R}^n)$ with $p^+ < \infty$ we give an alternative proof for Theorem 8.2.4 (b) based on the decomposition Theorem 7.4.9. The proof is not self-contained since it uses the variable exponent Poincaré inequality in cubes.

Proof of Theorem 8.2.4 (b) for $p \in \mathcal{P}^{\log}(\mathbb{R}^n)$ with $p^+ < \infty$. Let $p \in \mathcal{P}^{\log}(\mathbb{R}^n)$ with $p^+ < \infty$. Let $\mathcal{Q}$ be the chain covering of Ω. Let $u \in W^{1,p(\cdot)}(\Omega)$ with $\int_\Omega u \, dx = 0$. Then Proposition 7.4.11, the Poincaré inequality in cubes, $\operatorname{diam}(Q) \leqslant \operatorname{diam}(\Omega)$ for $Q \in \mathcal{Q}$, and Theorem 7.3.22 yield

$$
\begin{aligned}
\|u\|_{L_0^{p(\cdot)}(\Omega)} &\leqslant c \left\| \sum_{Q \in \mathcal{Q}} \chi_Q \frac{\|u - \langle u \rangle_Q\|_{L^{p(\cdot)}(Q)}}{\|\chi_Q\|_{p(\cdot)}} \right\|_{L^{p(\cdot)}(\Omega)} \\
&\leqslant c \left\| \sum_{Q \in \mathcal{Q}} \chi_Q \frac{\operatorname{diam}(Q) \|\nabla u\|_{L^{p(\cdot)}(Q)}}{\|\chi_Q\|_{p(\cdot)}} \right\|_{L^{p(\cdot)}(\Omega)} \\
&\leqslant c \operatorname{diam}(\Omega) \left\| \sum_{Q \in \mathcal{Q}} \chi_Q \frac{\|\nabla u\|_{L^{p(\cdot)}(Q)}}{\|\chi_Q\|_{p(\cdot)}} \right\|_{L^{p(\cdot)}(\Omega)} \\
&\leqslant c \operatorname{diam}(\Omega) \|\nabla u\|_{L^{p(\cdot)}(\Omega)}. \qquad\qquad \square
\end{aligned}
$$

Theorem 8.2.4 (a) immediately yields.

Corollary 8.2.5. *Let Ω be bounded and $p \in \mathcal{P}^{\log}(\Omega)$ or $p \in \mathcal{A}$. For every $u \in W_0^{1,p(\cdot)}(\Omega)$ the inequality*

$$
\|\nabla u\|_{L^{p(\cdot)}(\Omega)} \leqslant \|u\|_{W^{1,p(\cdot)}(\Omega)} \leqslant \left(1 + c \operatorname{diam}(\Omega)\right) \|\nabla u\|_{L^{p(\cdot)}(\Omega)}
$$

holds with constant c depending only on the dimension n, and $c_{\log}(p)$.

Theorem 8.2.4 (b) and Remark 8.2.2 yields the following corollary, which will be important in Sect. 12.2.

Corollary 8.2.6. *Let Ω be a bounded α-John domain and let $p \in \mathcal{P}^{\log}(\Omega)$ or $p \in \mathcal{A}$. Furthermore, let $A \subset \Omega$ be such that $|A| \approx |\Omega|$. Then*

$$
\|u - \langle u \rangle_A\|_{L^{p(\cdot)}(\Omega)} \leqslant c \operatorname{diam}(\Omega) \|\nabla u\|_{L^{p(\cdot)}(\Omega)}
$$

for $u \in L^1_{\mathrm{loc}}(\Omega)$ with $|\nabla u| \in L^{p(\cdot)}(\Omega)$. The constant c depends only on the dimension n, α and $c_{\log}(p)$.

Let us next consider modular versions of the Poincaré inequality. In the constant exponent case there is an obvious connection between modular and norm versions of the inequality, which does not hold in the variable exponent context. Indeed, the following one-dimensional example shows that the Poincaré inequality can not, in general, hold in a modular form.

Example 8.2.7. Let $p : (-2, 2) \to [2, 3]$ be a Lipschitz continuous exponent that equals 3 in $(-2, -1) \cup (1, 2)$, 2 in $(-\frac{1}{2}, \frac{1}{2})$ and is linear elsewhere. Let u_λ be a Lipschitz function such that $u_\lambda(\pm 2) = 0$, $u_\lambda = \lambda$ in $(-1, 1)$ and $|u'_\lambda| = \lambda$ in $(-2, -1) \cup (1, 2)$. Then

$$\frac{\overline{\varrho}_{p(\cdot)}(u_\lambda)}{\overline{\varrho}_{p(\cdot)}(u'_\lambda)} = \frac{\int_{-2}^{2} |u_\lambda|^{p(x)} \, dx}{\int_{-2}^{2} |u'_\lambda|^{p(x)} \, dx} \geqslant \frac{\int_{-1/2}^{1/2} \lambda^2 \, dx}{2 \int_{-2}^{-1} \lambda^3 \, dx} = \frac{1}{2\lambda} \to \infty$$

as $\lambda \to 0^+$.

In fact, Fan, Zhao and Zhang [145] have shown that the modular Poincaré inequality $\varrho_{p(\cdot)}(u) \leqslant c\varrho_{p(\cdot)}(\nabla u)$ does not hold if p is continuous and has a minimum or maximum. Allegretto [21] has shown that the inequality holds if there exists a function $\xi \in W^{1,1}(\Omega)$ such that $\nabla p \cdot \nabla \xi \geqslant 0$ and $\nabla \xi \neq 0$. This holds if p is suitably monotone. Next we prove another version of the modular Poincaré inequality; our inequality applies for log-Hölder continuous exponents, and it includes the radius of the domain; the price to pay is an additional, additive term on the right-hand side. This kind of Poincaré inequality has been used in [211].

Proposition 8.2.8. Let $p \in \mathcal{P}^{\log}(\Omega)$ be a bounded exponent.

(a) Let Ω be bounded. For $m > 0$ there exist a constant c depending on the dimension n, $c_{\log}(p)$, m, and p^+ such that

$$\int_\Omega \left(\frac{|u|}{\operatorname{diam}(\Omega)} \right)^{p(x)} dx \leqslant c \int_\Omega |\nabla u|^{p(x)} \, dx + c \int_{B(z, \operatorname{diam}(\Omega))} (\mathrm{e} + |x|)^{-m} \, dx$$

for all $u \in W_0^{1,p(\cdot)}(\Omega)$ with $\|\nabla u\|_{L^{p(\cdot)}(\Omega)} \leqslant 1$ and all $z \in \Omega$.

(b) Let Ω be a bounded α-John domain. For $m > 0$ there exist a constant c depending on the dimension n, $c_{\log}(p)$, m, p^+, α, $|\Omega|$ and $\operatorname{diam}(\Omega)$ such that

$$\int_\Omega \left(\frac{|v - \langle v \rangle_B|}{\operatorname{diam}(\Omega)} \right)^{p(x)} dx \leqslant c \int_\Omega |\nabla v|^{p(x)} \, dx + c \int_{B(z, \operatorname{diam}(\Omega))} (\mathrm{e} + |x|)^{-m} \, dx$$

for all $v \in W^{1,p(\cdot)}(\Omega)$ with $\|\nabla v\|_{L^{p(\cdot)}(\Omega)} \leqslant 1$ and all $z \in \Omega$. The ball B is from Lemma 8.2.1.

Proof. Assume that $u \in W_0^{1,p(\cdot)}(\Omega)$ is extended by zero outside Ω (Lemma 8.1.14). By Lemmas 8.2.1 (a) and 6.1.4 we obtain

$$|u(x)| \leqslant c \int_\Omega \frac{|\nabla u(y)|}{|x - y|^{n-1}} dy \leqslant c \operatorname{diam}(\Omega) \sum_{k=0}^{\infty} 2^{-k} T_{k+k_0} |\nabla u|(x),$$

where $k_0 \in \mathbb{Z}$ is chosen such that $2^{-k_0-1} \leqslant \operatorname{diam}(\Omega) \leqslant 2^{-k_0}$. Exactly the same estimate holds with u replaced by $v - \langle v \rangle_B$ using Lemma 8.2.1 (b), so it suffices to derive the estimate of the first claim involving u.

We divide by $\operatorname{diam}(\Omega)$ and raise both sides of this inequality to the power $p(x)$, integrate over Ω, and use $p^+ < \infty$ to obtain

$$\int_\Omega \left(\frac{|u|}{\operatorname{diam}(\Omega)} \right)^{p(x)} dx \leqslant c \int_\Omega \left(\sum_{k=0}^\infty 2^{-k} T_{k+k_0} |\nabla u| \right)^{p(x)} dx$$

$$\leqslant c \sum_{k=0}^\infty 2^{-k} \int_\Omega \left(T_{k+k_0} |\nabla u| \right)^{p(x)} dx,$$

where we used convexity in the second step. Since $\varrho_{L^{p(\cdot)}(\Omega)}(\nabla u) \leqslant 1$ by the unit ball property, we may use Corollary 4.2.5 for $|\nabla u| \chi_\Omega$ on the ball $B(z, \operatorname{diam}(\Omega))$ and get

$$\int_\Omega \left(T_{k+k_0} |\nabla u| \right)^{p(x)} dx \leqslant c \int_\Omega |\nabla u|^{p(x)} dx + c \int_{B(z,\operatorname{diam}(\Omega))} (e + |x|)^{-m} dx,$$

where we also used that the dyadic cubes $2Q$ are locally N-finite. Combining the last two inequality proves the claim.   □

Remark 8.2.9. If $p \in \mathcal{P}^{\log}(\Omega)$ with no restriction on p^+, then the first estimate in Proposition 8.2.8 reads

$$\int_\Omega \varphi_{p(x)} \left(\beta \frac{|u|}{\operatorname{diam}(\Omega)} \right) dx \leqslant \int_\Omega \varphi_{p(x)} (|\nabla u|) dx + \int_{B(z,\operatorname{diam}(\Omega))} (e + |x|)^{-m} dx$$

for some $\beta \in (0,1)$ and all $u \in W_0^{1,p(\cdot)}(\Omega)$ with $\|\nabla u\|_{L^{p(\cdot)}(\Omega)} \leqslant 1$. The constant β depends only on $c_{\log}(p)$, m and n. The second estimate in Proposition 8.2.8 has to be changed accordingly. The proof is the same.

Remark 8.2.10. It is possible to replace $\langle u \rangle_B$ in Lemma 8.2.1 (b) and Proposition 8.2.8 by $\langle u \rangle_\Omega$ or even $\langle u \rangle_A$, where $A \subset \Omega$ is a set with $|A| \approx |\Omega|$. Indeed, it follows from Jensen's inequality and Corollary 8.2.6 for $p = 1$ that

$$|\langle u \rangle_A - \langle u \rangle_B| \leqslant \fint_A |u - \langle u \rangle_B| \, dy \leqslant c \fint_\Omega |u - \langle u \rangle_B| \, dy \leqslant c \operatorname{diam}(\Omega) \fint_\Omega |\nabla u| \, dy$$

for all $u \in W^{1,1}(\Omega)$. In particular, this implies

$$|\langle u \rangle_A - \langle u \rangle_B| \leqslant c \int_\Omega \frac{|\nabla u(y)|}{(\operatorname{diam}(\Omega))^{n-1}} \, dy \leqslant c \, I_1(\nabla u)(x).$$

for any $x \in \Omega$. This proves the modified version of Lemma 8.2.1 (b). With this new estimate we get the modified version of Proposition 8.2.8 with no change in the proof.

The following improvement of Proposition 8.2.8 is useful in the study of $p(\cdot)$-minimizers and is the starting point for reverse Hölder estimates. The result is from Schwarzacher [350].

Proposition 8.2.11. *Let* $p \in \mathcal{P}^{\log}(\mathbb{R}^n)$ *satisfy* $1 < p^- \leqslant p^+ < \infty$ *and let* $s \leqslant p^-$ *satisfy* $s \in [1, \frac{n}{n-1})$. *Then for every* $m > 0$ *there exist a constant* c *depending on* n, $c_{\log}(p)$, m, *and* s *such that*

$$\fint_{B_R} \left(\frac{|u|}{R} \right)^{p(x)} dx \leqslant c \left(\fint_{B_R} |\nabla u|^{\frac{p(\cdot)}{s}} dx \right)^s + c \fint_{B_R} (e + |x|)^{-ms} dx$$

$$\fint_{B_R} \left(\frac{|v - \langle v \rangle_{B_R}|}{R} \right)^{p(x)} dx \leqslant c \left(\fint_{B_R} |\nabla v|^{\frac{p(\cdot)}{s}} dx \right)^s + c \fint_{B_R} (e + |x|)^{-ms} dx$$

for every ball B_R *with radius* R, *and every* $u \in W_0^{1, \frac{p(\cdot)}{s}}(B_R)$, $v \in W^{1, \frac{p(\cdot)}{s}}(B_R)$ *with* $\|\nabla u\|_{L^{p(\cdot)/s} + L^\infty}, \|\nabla v\|_{L^{q(\cdot)/s} + L^\infty} \leqslant 1$.

Proof. By Jensen's inequality the case $s > 1$ implies the case $s = 1$ and thus we may assume that $s > 1$. By Lemma 8.2.1 we have, for $x \in B_R$,

$$|u(x)| \leqslant c \int_{B_R} \frac{|\nabla u(y)|}{|x - y|^{n-1}} dy, \qquad |v(x) - \langle v \rangle_{B_R}| \leqslant c \int_{B_R} \frac{|\nabla v(y)|}{|x - y|^{n-1}} dy.$$

Starting from here the proofs for u and v are the same, so we just present the estimate for u. With the previous estimate, the help of Lemma 6.1.12 applied to $p(\cdot)/s$ and $p^+ < \infty$ we get

$$(I) := \fint_{B_R} \left(\frac{|u(x)|}{R} \right)^{p(x)} dx \leqslant c(p^+) \fint_{B_R} \left(\int_{B_R} \frac{|\nabla u(y)|}{R|x - y|^{n-1}} dy \right)^{p(x)} dx$$

$$\leqslant c \fint_{B_R} \left(\int_{B_R} \frac{|\nabla u(y)|^{\frac{p(y)}{s}}}{R|x - y|^{n-1}} dy \right)^s dx + c \fint_{B_R} \left(M((e + |\cdot|)^{-m})(x) \right)^s dx$$

$$=: (II) + (III).$$

In order to estimate (II) we set $J := \int_{B_R} |\nabla u|^{p(\cdot)/s} dx$. We can assume $J > 0$ in the following, since otherwise $\nabla u = 0$ and there is nothing to estimate. We apply Jensen's inequality for the probability measure $\mu := |\nabla u|^{p(\cdot)/s}/J$ and the convex function $t \mapsto t^s$, then we use Fubini's theorem to change the integration order and $\int_{B_R} \frac{dx}{|x-y|^{s(n-1)}} \approx R^{-s(n-1)+n}$ for a.e. $y \in B_R$ using $s < \frac{n}{n-1}$:

$$(II) \leqslant c \fint_{B_R} J^{s-1} \int_{B_R} \frac{|\nabla u(y)|^{\frac{p(y)}{s}}}{R^s |x-y|^{s(n-1)}} \, dy \, dx$$

$$\leqslant c \, J^{s-1} R^{n-sn} \fint_{B_R} |\nabla u(y)|^{\frac{p(y)}{s}} \, dy \leqslant c \left(\fint_{B_R} |\nabla u(y)|^{\frac{p(y)}{s}} \, dy \right)^s.$$

Since $s > 1$, we can use the boundedness of M on $L^s(B_R)$ to conclude

$$(III) \leqslant c \fint_{B_R} (\mathrm{e} + |x|)^{-ms} \, dx$$

Combining the estimates for (I)–(III) we obtain the result. □

Remark 8.2.12. Exactly as in the Remarks 8.2.9 and 8.2.10 it is possible to modify Proposition 8.2.11 to include the case $p^+ = \infty$: we have to replace the constant c on the right-hand side by a constant β on the left-hand side. As in Proposition 8.2.8 and Remark 8.2.10 we can replace the integration domain B_R by a bounded John domain Ω and the mean value $\langle v \rangle_{B_R}$ by $\langle v \rangle_A$ for any $A \subset \Omega$ with $|A| \approx |\Omega|$.

In the remainder of the section, we generalize the norm-type Poincaré inequalities to more general exponents. Again, there is a price to pay, namely, we do not get the factor "diam(Ω)" on the right-hand side. These results are from Harjulehto and Hästö [188].

Let us recall the following constant exponent result, which in the case $r < n$ and $q = r^*$ is due to B. Bojarski [52, (6.6)] and in the case $q < r^*$ follows from it by Hölder's inequality. By r^* we denote the Sobolev conjugate exponent of $r < n$, $r^* := nr/(n-r)$.

Lemma 8.2.13. Let $\Omega \subset \mathbb{R}^n$ be a bounded α-John domain. If $1 \leqslant r < n$ and $r \leqslant q \leqslant r^*$ or if $r \geqslant n$ and $q < \infty$, then

$$\|u - \langle u \rangle_\Omega\|_{L^q(\Omega)} \leqslant c \, |\Omega|^{\frac{1}{n} + \frac{1}{q} - \frac{1}{r}} \|\nabla u\|_{L^r(\Omega)}$$

for all $u \in W^{1,r}(\Omega)$. In the first case the constant c depends only on n, r and α, while in the second case it depends also on q.

Using the previous constant exponent Sobolev–Poincaré inequality we are able to prove the Poincaré inequality in bounded John domains for variable exponents.

Lemma 8.2.14. Let $\Omega \subset \mathbb{R}^n$ be a bounded α-John domain. If $p \in \mathcal{P}(\Omega)$ is bounded with $p^+ \leqslant (p^-)^*$ or $p^- \geqslant n$, then there exists a constant c depending on the dimension n, p^-, p^+, and α such that for every $u \in W^{1,p(\cdot)}(\Omega)$ we have

$$\|u - \langle u \rangle_\Omega\|_{L^{p(\cdot)}(\Omega)} \leqslant c\,(1 + |\Omega|)^2 |\Omega|^{\frac{1}{n} + \frac{1}{p^+} - \frac{1}{p^-}} \|\nabla u\|_{L^{p(\cdot)}(\Omega)}.$$

Proof. Assume first that $p^+ \leqslant (p^-)^*$. Since $p(x) \leqslant p^+ \leqslant (p^-)^*$ we deduce from Corollary 3.3.4 and Lemma 8.2.13 that

$$\|u - \langle u \rangle_\Omega\|_{p(\cdot)} \leqslant 2\big(1 + |\Omega|\big)\|u - \langle u \rangle_\Omega\|_{p^+}$$
$$\leqslant c(n, p^-, \alpha)\big(1 + |\Omega|\big)|\Omega|^{\frac{1}{n} + \frac{1}{p^+} - \frac{1}{p^-}} \|\nabla u\|_{p^-}$$
$$\leqslant c(n, p^-, \alpha)\big(1 + |\Omega|\big)^2 |\Omega|^{\frac{1}{n} + \frac{1}{p^+} - \frac{1}{p^-}} \|\nabla u\|_{p(\cdot)}.$$

Let $p^- \geqslant n$. We choose a constant $q \in [1, n)$ such that $p^+ = q^*$. We obtain the result by using similar arguments than in the previous case. The only difference is that the constant in the second inequality in the above chain of inequalities is $c(n, p^+, \alpha)$. □

The following lemma will be generalized (with proof) in Proposition 10.2.10 to the variable exponent case; this version can be found e.g. in [219, Example 2.12, p. 35].

Lemma 8.2.15. *For a constant $q \in (1, n)$, arbitrary $x \in \mathbb{R}^n$ and $R > r > 0$ we have*

$$\inf \int\limits_{B(x,R)} |\nabla u|^q \, dx = c \left|\frac{q - n}{q - 1}\right|^{q-1} \left|R^{\frac{q-n}{q-1}} - r^{\frac{q-n}{q-1}}\right|^{1-q},$$

where the infimum is taken over all $u \in C_0^\infty(B(x, R))$ with $u|_{B(x,r)} = 1$. Here the constant c depends only on the dimension n.

The following proposition shows that for general non-constant exponents the Poincaré inequality does not hold.

Proposition 8.2.16. *Let B be a unit ball in the plane. For every $q_1 \in [1, 2)$ and $q_2 \in (2, \infty)$ there exists $p \in \mathcal{P}(B)$ with $p^- = q_1$ and $p^+ = q_2$ such that the norm-version of the Poincaré inequality,*

$$\|u - \langle u \rangle_\Omega\|_{L^{p(\cdot)}(B)} \leqslant c\,\|\nabla u\|_{L^{p(\cdot)}(B)},$$

does not hold.

Proof. Our aim is to construct a sequence of functions in $B \subset \mathbb{R}^2$ for which the constant in the Poincaré inequality goes to infinity. Let $e_1 = (1, 0)$, $B_i := B(2^{-i}e_1, \frac{1}{4}2^{-i}) \subset B$ and $B_i' := B(2^{-i}e_1, \frac{1}{8}2^{-i^2}) \subset B$ for every $i \in \mathbb{N}$. Let $u_i \in C_0^\infty(B_i)$ with $u_i|_{\overline{B}_i'} = 1$. Define $p := q_2$ in every B_i' and $p := q_1$ otherwise in B with positive first coordinate. Since $\nabla u_i = 0$ in B_i' we obtain

$$\|\nabla u_i\|_{L^{p(\cdot)}(B_i)} = \|\nabla u_i\|_{L^{q_1}(B_i)}.$$

Let $\tilde{B}_i := B(-2^{-i}e_1, \frac{1}{4}2^{-i})$. We extend u_i to B as an odd function of the first coordinate in $\tilde{B}_i$ and by zero elsewhere. We extend p to B as an even function of the first coordinate. We denote these extensions by $\tilde{u}_i$ and $\frac{p^*}{n'}$.

By Lemma 8.2.15 we may choose the functions u_i such that

$$\|\nabla \tilde{u}_i\|_{L^{\frac{p^*(\cdot)}{n'}}(B)} \leqslant c(q_1) \left| \left(\tfrac{1}{4}2^{-i} \right)^{\frac{q_1-2}{q_1-1}} - \left(\tfrac{1}{8}2^{-i^2} \right)^{\frac{q_1-2}{q_1-1}} \right|^{\frac{1-q_1}{q_1}}.$$

For large i, the right-hand side is approximately equal to $c(q_1)2^{-i^2\frac{2-q_1}{q_1}}$. Since $\langle \tilde{u}_i \rangle_B = 0$, we obtain

$$\|\tilde{u}_i - \langle \tilde{u}_i \rangle_B\|_{L^{\frac{p^*(\cdot)}{n'}}(B)} = \|\tilde{u}_i\|_{L^{\frac{p^*(\cdot)}{n'}}(B)} \geqslant |B_i'|^{\frac{1}{q_2}} \geqslant c\, 2^{-i^2\frac{2}{q_2}}.$$

Combining the previous two inequalities, we find that

$$\frac{\|\tilde{u}_i - \langle \tilde{u}_i \rangle_B\|_{L^{\frac{p^*(\cdot)}{n'}}(B)}}{\|\nabla \tilde{u}_i\|_{L^{\frac{p^*(\cdot)}{n'}}(B)}} \geqslant c(q_1)\, 2^{i^2(\frac{2}{q_1}-1-\frac{2}{q_2})} \to \infty$$

as $i \to \infty$ if $\frac{2}{q_1} - 1 - \frac{2}{q_2} > 0$, that is, if $q_2 > \frac{2q_1}{2-q_1} > 2$. $\square$

The following theorem shows that the condition $p^+ \leqslant (p^-)^*$ in Lemma 8.2.14 can be replaced by a set of local conditions.

Theorem 8.2.17. *Let $\Omega \subset \mathbb{R}^n$ be a bounded John domain and $p \in \mathcal{P}(\Omega)$ be bounded. Assume that there exist John domains D_i, $i = 1, \ldots, j$, such that $\Omega = \cup_{i=1}^j D_i$ and either $p_{D_i}^+ \leqslant (p_{D_i}^-)^*$ or $p_{D_i}^- \geqslant n$ for every i. Then there exists a constant c such that*

$$\|u - \langle u \rangle_\Omega\|_{L^{p(\cdot)}(\Omega)} \leqslant c\, \|\nabla u\|_{L^{p(\cdot)}(\Omega)}$$

for every $u \in W^{1,p(\cdot)}(\Omega)$. The constant c depends on n, $\mathrm{diam}(\Omega)$, $|D_i|$, p and the John constants of Ω and D_i, $i = 1, \ldots, j$.

Proof. Using the triangle inequality we obtain

$$\|u - \langle u \rangle_\Omega\|_{L^{p(\cdot)}(\Omega)} \leqslant \sum_{i=1}^j \|u - \langle u \rangle_\Omega\|_{L^{p(\cdot)}(D_i)}$$

$$\leqslant \sum_{i=1}^j \|u - \langle u \rangle_{D_i}\|_{L^{p(\cdot)}(D_i)} + \sum_{i=1}^j \|\langle u \rangle_\Omega - \langle u \rangle_{D_i}\|_{L^{p(\cdot)}(D_i)}.$$

We estimate the first part of the sum using Lemma 8.2.14. This yields for every $i = 1, \ldots, j$

$$\|u - \langle u \rangle_{D_i}\|_{L^{p(\cdot)}(D_i)} \leqslant c\,\|\nabla u\|_{L^{p(\cdot)}(D_i)} \leqslant c\,\|\nabla u\|_{L^{p(\cdot)}(\Omega)}$$

with constants depending on $n, p_{D_i}^+, p_{D_i}^-, |D_i|, \alpha_i$, where α_i is the John constant of D_i. Therefore it remains only to estimate the sum of the terms $\|\langle u \rangle_\Omega - \langle u \rangle_{D_i}\|_{L^{p(\cdot)}(D_i)}$. We use the constant exponent Poincaré inequality (in the third inequality):

$$
\begin{aligned}
\|\langle u \rangle_\Omega - \langle u \rangle_{D_i}\|_{L^{p(\cdot)}(D_i)} &\leqslant \|1\|_{L^{p(\cdot)}(D_i)} \fint_{D_i} |u(x) - \langle u \rangle_\Omega|\, dx \\
&\leqslant \|1\|_{L^{p(\cdot)}(D_i)} |D_i|^{-1} \int_\Omega |u(x) - \langle u \rangle_\Omega|\, dx \\
&\leqslant c(n, \operatorname{diam}(\Omega), \alpha) |D_i|^{-1} \|1\|_{L^{p(\cdot)}(D_i)} \|\nabla u\|_{L^1(\Omega)} \\
&\leqslant c(n, \operatorname{diam}(\Omega), \alpha) |D_i|^{-1} \|1\|_{L^{p(\cdot)}(D_i)} \|\nabla u\|_{L^{p(\cdot)}(\Omega)}
\end{aligned}
$$

for every $i = 1, \ldots, j$. Here α is the John constant of Ω. By Corollary 3.3.4 $\|1\|_{L^{p(\cdot)}(D_i)}$ depends only on p and $|D_i|$, which completes the proof. $\qquad \square$

Next we prove the Poincaré inequality for Sobolev functions with zero boundary values using Lemma 8.2.14.

Theorem 8.2.18. *Let Ω be bounded. Assume that $p \in \mathcal{P}(\Omega)$ and there exists $\delta > 0$ such that for every $x \in \Omega$ either*

$$
p_{B(x,\delta) \cap \Omega}^+ \leqslant \frac{n\, p_{B(x,\delta) \cap \Omega}^-}{n - p_{B(x,\delta) \cap \Omega}^-} \qquad or \qquad p_{B(x,\delta) \cap \Omega}^- \geqslant n. \tag{8.2.19}
$$

Alternatively, assume that p is uniformly continuous in Ω. Then the inequality

$$\|u\|_{L^{p(\cdot)}(\Omega)} \leqslant c\,\|\nabla u\|_{L^{p(\cdot)}(\Omega)},$$

holds for every $u \in W_0^{1,p(\cdot)}(\Omega)$. Here the constant c depends on p, Ω, δ and the dimension n.

Proof. Note that if p is continuous in $\overline{\Omega}$ or uniformly continuous in Ω, then p satisfies the first set of conditions of the theorem for some $\delta > 0$.

By the assumptions there exist $x_1, \ldots, x_j$ and $\delta > 0$ such that

$$\Omega \subset \bigcup_{i=1}^j B(x_i, \delta).$$

and each ball $B(x_i, \delta)$ satisfies either of the two inequalities in (8.2.19). We write $B_i := B(x_i, \delta)$ and denote by χ_i the characteristic function of $B_i \cap \Omega$. In each B_i we define $p_i(x) := p(x)\chi_i + p^-_{B_i \cap \Omega}(1 - \chi_i)$. Then in each B_i either $p_i^+ \leqslant (p_i^-)^*$ or $p_i^- \geqslant n$. Let $\tilde{u}$ be the zero extension of $u \in W_0^{1,p(\cdot)}(\Omega)$ to $\mathbb{R}^n \setminus \Omega$ (Lemma 8.1.14). By the triangle inequality we obtain

$$\|u\|_{L^{p(\cdot)}(\Omega)} \leqslant \left\| \tilde{u} \sum_{i=1}^j \chi_i \right\|_{L^{p(\cdot)}(\Omega)} \leqslant \sum_{i=1}^j \|\tilde{u}\|_{L^{p_i(\cdot)}(B_i)}$$

$$\leqslant \sum_{i=1}^j \|\tilde{u} - \langle \tilde{u} \rangle_{B_i}\|_{L^{p_i(\cdot)}(B_i)} + \sum_{i=1}^j |\langle \tilde{u} \rangle_{B_i}| \, \|1\|_{L^{p_i(\cdot)}(B_i)}.$$

We estimate the first sum on the right-hand side of the previous inequality. By Lemma 8.2.14 we obtain

$$\|\tilde{u} - \langle \tilde{u} \rangle_{B_i}\|_{L^{p_i(\cdot)}(B_i)} \leqslant c\,(1 + |B_i|)^2 |B_i|^{\frac{1}{n} + \frac{1}{p_i^+} - \frac{1}{p_i^-}} \|\nabla \tilde{u}\|_{L^{p_i(\cdot)}(B_i)}$$

$$\leqslant c\,(1 + |B_i|)^2 |B_i|^{\frac{1}{n} + \frac{1}{p_i^+} - \frac{1}{p_i^-}} \|\nabla u\|_{L^{p(\cdot)}(\Omega)}.$$

for every $i = 1, \ldots j$. To estimate the second sum, $\sum_{i=1}^j |\langle \tilde{u} \rangle_{B_i}| \, \|1\|_{L^{p_i(\cdot)}(B_i)}$, in the above inequality we use Lemma 3.2.12 to estimate $\|1\|_{L^{p_i(B_i)}}$ by a constant depending only on p and δ. Further, the constant exponent Poincaré inequality implies that

$$|\langle \tilde{u} \rangle_{B_i}| \leqslant \frac{c(n)}{\delta^n} \int_\Omega |u| \, dx \leqslant \frac{c}{\delta^n} \operatorname{diam}(\Omega) \int_\Omega |\nabla u| \, dx$$

$$\leqslant \frac{c(n)}{\delta^n} \operatorname{diam}(\Omega)(1 + |\Omega|)\|\nabla u\|_{L^{p(\cdot)}(\Omega)},$$

again for every $i = 1, \ldots j$. Combining the last three estimates yields the assertion.   □

Remark 8.2.20. Assume that Ω is convex and $p \in \mathcal{P}(\Omega)$ is uniform continuous (or $p \in C(\overline{\Omega})$). As in the proof of Theorem 8.2.18 we may cover Ω by finitely many balls $B(x_i, \delta)$ so that (8.2.19) holds. Since Ω is convex so is $B(x_i, \delta) \cap \Omega$ and thus it is a John domain. Hence Theorem 8.2.17 yields that the Poincaré inequality

$$\|u - \langle u \rangle_\Omega\|_{L^{p(\cdot)}(\Omega)} \leqslant c\,\|\nabla u\|_{L^{p(\cdot)}(\Omega)}$$

holds for every $u \in W^{1,p(\cdot)}(\Omega)$.

8.3 Sobolev-Poincaré Inequalities and Embeddings

In this section we assume that the exponent p is log-Hölder continuous with $1 \leqslant p^- \leqslant p^+ < n$. We prove that the Sobolev-Poincaré inequality holds for general Sobolev functions in bounded John domains and for Sobolev functions with zero boundary values in open sets. Bounded John domains are almost the right class of irregular domains for the constant exponent Sobolev–Poincaré inequality, see [52,59]. We give an example which shows that Sobolev embeddings do not hold for every continuous p. We close this section by studying the Sobolev embedding in the case $p^- > n$. Sobolev embeddings in the case $p^+ = n$ need a target space that is not a variable exponent Lebesgue space, and are studied in Sect. 8.6 together with the other limit case $p^- = n$.

We define the *Sobolev conjugate exponent* point-wise, i.e.

$$p^*(x) := \frac{np(x)}{n - p(x)}$$

when $p(x) < n$ and $p^*(x) = \infty$ otherwise. This section is based on [92, 190, 258], see also [121, 122, 141, 166, 246, 285, 299, 302].

Theorem 8.3.1. *Let $p \in \mathcal{P}^{\log}(\Omega)$ satisfy $1 \leqslant p^- \leqslant p^+ < n$.*

(a) *For every $u \in W_0^{1,p(\cdot)}(\Omega)$, the inequality*

$$\|u\|_{L^{p^*(\cdot)}(\Omega)} \leqslant c \, \|\nabla u\|_{L^{p(\cdot)}(\Omega)}$$

holds with a constant c depending only on the dimension n, $c_{\log}(p)$, and p^+.

(b) *If Ω is a bounded α-John domain, then*

$$\|u - \langle u \rangle_\Omega\|_{L^{p^*(\cdot)}(\Omega)} \leqslant c \, \|\nabla u\|_{L^{p(\cdot)}(\Omega)}$$

for $u \in W^{1,p(\cdot)}(\Omega)$. The constant c depends only on the dimension n, α, $c_{\log}(p)$ and p^+.

If we add the extraneous assumption $p^- > 1$, then we immediately obtain a proof using results on operators that we proved earlier: the inequalities follow from Lemma 8.2.1, Lemma 8.2.3 and Theorem 6.1.9; the constant in this case also depends on p^-.

We obtain the Sobolev embedding as a corollary.

Corollary 8.3.2. *Let Ω be a bounded α-John domain and let $p \in \mathcal{P}^{\log}(\Omega)$. Let $q \in \mathcal{P}(\Omega)$ be bounded and assume that $q \leqslant p^*$. Then*

$$W^{1,p(\cdot)}(\Omega) \hookrightarrow L^{q(\cdot)}(\Omega),$$

where the embedding constant depends only on α, $|\Omega|$, n, $c_{\log}(p)$ and q^+.

Proof. Let $r \in (1, n)$ be such that $r^* \geqslant q^+$. Corollary 3.3.4 and Lemma 3.2.12 yield

$$
\begin{aligned}
\|u\|_{q(\cdot)} &\leqslant \|u - \langle u \rangle_\Omega\|_{q(\cdot)} + \|\langle u \rangle_\Omega\|_{q(\cdot)} \\
&\leqslant 2(1 + |\Omega|)\|u - \langle u \rangle_\Omega\|_{\min\{p^*(\cdot), r^*\}} + \max\{|\Omega|^{\frac{1}{q^+} - 1}, 1\}\|u\|_1 \\
&\leqslant 2(1 + |\Omega|) \max\{|\Omega|^{\frac{1}{q^+} - 1}, 1\} \left(\|u - \langle u \rangle_\Omega\|_{\min\{p^*(\cdot), r^*\}} + \|u\|_{p(\cdot)} \right).
\end{aligned}
$$

Since $\min\{p^*(\cdot), r^*\} \in \mathcal{P}^{\log}(\Omega)$, Theorem 8.3.1 (b) and Corollary 3.3.4 yield

$$
\|u - \langle u \rangle_\Omega\|_{\min\{p^*(\cdot), r^*\}} \leqslant c \|\nabla u\|_{\min\{p(\cdot), r\}} \leqslant c (1 + |\Omega|)\|\nabla u\|_{p(\cdot)}.
$$

The claim follows by combining these two inequalities.   □

Now we move on to the complete proof, covering also the case $p^- = 1$. In this case the Riesz potential is not strong type $p(\cdot)$, i.e. Theorem 6.1.9 is not available. Our proof is based on the weak type estimate for the Riesz potential. We first give a proof of Theorem 8.3.1 (b) in which the constant additionally depends on $\operatorname{diam}(\Omega)$.

Lemma 8.3.3. *Let Ω be a bounded α-John domain and let $p \in \mathcal{P}^{\log}(\Omega)$ satisfy $1 \leqslant p^- \leqslant p^+ < n$. Then*

$$
\|u - \langle u \rangle_\Omega\|_{L^{p^*(\cdot)}(\Omega)} \leqslant c \|\nabla u\|_{L^{p(\cdot)}(\Omega)}
$$

for every $u \in W^{1,p(\cdot)}(\Omega)$. The constant c depends only on the dimension n, p^+, $c_{\log}(p)$, α and $\operatorname{diam}(\Omega)$.

Proof. By a scaling argument we may assume without loss of generality that $(1 + |\Omega|)\|\nabla u\|_{p(\cdot)} \leqslant 1$. We need to show that $\varrho_{L^{p^*(\cdot)}(\Omega)}(|u - \langle u \rangle_\Omega|)$ is uniformly bounded. For every $j \in \mathbb{Z}$ we set $\Omega_j := \{x \in \Omega : 2^j < |u(x) - \langle u \rangle_\Omega| \leqslant 2^{j+1}\}$ and $v_j := \max\{0, \min\{|u - \langle u \rangle_\Omega| - 2^j, 2^j\}\}$. From Proposition 8.1.9 follows $v_j \in W^{1,p(\cdot)}(\Omega)$. By Lemma 8.2.1 (b) we have

$$
|v_j(x) - \langle v_j \rangle_B| \leqslant c I_1 |\nabla v_j|(x)
$$

for almost every $x \in \Omega$. Here the radius of $B \subset \Omega$ depends on α. We obtain by the pointwise inequality $v_j \leqslant |u - \langle u \rangle_\Omega|$ and by the constant exponent Poincaré inequality Lemma 8.2.13 that

$$
\begin{aligned}
v_j(x) &\leqslant |v_j(x) - \langle v_j \rangle_B| + \langle v_j \rangle_B \leqslant c I_1 |\nabla v_j|(x) + \fint_B |u - \langle u \rangle_\Omega| \, dx \\
&\leqslant c I_1 |\nabla v_j|(x) + c \int_\Omega |u - \langle u \rangle_\Omega| \, dx \leqslant c I_1 |\nabla v_j|(x) + c \int_\Omega |\nabla u| \, dx \qquad (8.3.4) \\
&\leqslant c I_1 |\nabla v_j|(x) + c (1 + |\Omega|)\|\nabla u\|_{p(\cdot)} \leqslant c_1 (I_1 |\nabla v_j|(x) + 1).
\end{aligned}
$$

For the rest of this proof we fix the constant c_1 to denote the constant on the last line. It depends only on n, α and $\mathrm{diam}(\Omega)$.

Using the definition of Ω_j we get

$$\int\limits_{\Omega} |u(x) - \langle u \rangle_\Omega|^{p^*(x)}\,dx = \sum_{j=-\infty}^{\infty} \int\limits_{\Omega_j} |u(x) - \langle u \rangle_\Omega|^{p^*(x)}\,dx$$

$$\leqslant \sum_{j=-\infty}^{\infty} \int\limits_{\Omega_j} 2^{(j+1)p^*(x)}\,dx\,.$$

For every $x \in \Omega_{j+1}$ we have $v_j(x) = 2^j$ and thus obtain by (8.3.4) the pointwise inequality $c_1 I_1 |\nabla v_j|(x) + c_1 > 2^j$ for almost every $x \in \Omega_{j+1}$. Note that if $a + b > c$, then $a > \frac{1}{2}c$ or $b > \frac{1}{2}c$. Thus

$$\sum_{j=-\infty}^{\infty} \int\limits_{\Omega_j} 2^{(j+1)p^*(x)}\,dx$$

$$\leqslant \sum_{j=-\infty}^{\infty} \int\limits_{\{x \in \Omega_j\,:\,c_1 I_1|\nabla v_j|(x)+c_1 > 2^{j-1}\}} 2^{(j+1)p^*(x)}\,dx$$

$$\leqslant \sum_{j=-\infty}^{\infty} \int\limits_{\{x \in \Omega\,:\,c_1 I_1|\nabla v_j|(x) > 2^{j-2}\}} 2^{(j+1)p^*(x)}\,dx$$

$$+ \sum_{j=-\infty}^{\infty} \int\limits_{\{x \in \Omega_j\,:\,c_1 > 2^{j-2}\}} 2^{(j+1)p^*(x)}\,dx.$$

Since $(1 + |\Omega|)\|\nabla u\|_{p(\cdot)} \leqslant 1$, we obtain by Theorem 6.1.11 for the first term on the right-hand side that

$$\sum_{j=-\infty}^{\infty} \int\limits_{\{x \in \Omega\,:\,c_1 I_1|\nabla v_j(y)|(x) > 2^{j-2}\}} 2^{(j+1)p^*(x)}\,dx$$

$$\leqslant c \sum_{j=-\infty}^{\infty} \left(\int\limits_{\Omega} |\nabla v_j(y)|^{p(y)}\,dy + |\{0 < |\nabla v_j| \leqslant 1\}| \right)$$

$$\leqslant c \sum_{j=-\infty}^{\infty} \left(\int\limits_{\Omega_j} |\nabla u|^{p(y)}\,dy + |\Omega_j| \right)$$

$$= c \int\limits_{\Omega} |\nabla u|^{p(y)}\,dy + c\,|\Omega|.$$

Let j_0 be the largest integer satisfying $c_1 > 2^{j_0-2}$. Hence

$$\sum_{j=-\infty}^{\infty} \int_{\{x \in \Omega_j : c_1 > 2^{j-2}\}} 2^{(j+1)p^*(x)} \, dx \leqslant \int_\Omega \sum_{j=-\infty}^{j_0} 2^{(j+1)p^*(x)} \, dx \leqslant c\,|\Omega|,$$

which concludes the proof. $\qquad\qquad\qquad\qquad\qquad\qquad\qquad\qquad\qquad\square$

Next we use the local-to-global trick to generalize the previous lemma, removing the dependence of the constant on $\mathrm{diam}(\Omega)$.

Proof of Theorem 8.3.1. We prove only part (a) of the theorem; the second part follows by essentially identical arguments.

Let $u \in W_0^{1,p(\cdot)}(\Omega)$ and extend it by 0 to $\mathbb{R}^n \setminus \Omega$ (Lemma 8.1.14). By a scaling argument we may assume that $\|u\|_{W^{1,p(\cdot)}(\mathbb{R}^n)} = 1$. Let (Q_j) be a partition of $\mathbb{R}^n$ into unit cubes. As was noted in Example 7.4.2, every Q_j is a John domain with the same constant. Thus Lemma 8.3.3 implies that

$$\|u\|_{L^{p^*(\cdot)}(Q_j)} \leqslant \|u - \langle u\rangle_{Q_j}\|_{L^{p^*(\cdot)}(Q_j)} + |\langle u\rangle_{Q_j}|\,\|1\|_{L^{p^*(\cdot)}(Q_j)}$$
$$\leqslant c\,\|\nabla u\|_{L^{p(\cdot)}(Q_j)} + c\,|\langle u\rangle_{Q_j}|.$$

Next we apply Corollary 7.3.23, the previous inequality, and the triangle inequality in $\ell^{p_\infty^*}$:

$$\|u\|_{L^{p^*(\cdot)}(\mathbb{R}^n)} \leqslant c\left(\sum_j \|u\|_{L^{p^*(\cdot)}(Q_j)}^{p_\infty^*}\right)^{1/p_\infty^*}$$
$$\leqslant c\left(\sum_j \left(\|\nabla u\|_{L^{p(\cdot)}(Q_j)} + |\langle u\rangle_{Q_j}|\right)^{p_\infty^*}\right)^{1/p_\infty^*}$$
$$\leqslant c\left(\sum_j \|\nabla u\|_{L^{p(\cdot)}(Q_j)}^{p_\infty^*}\right)^{1/p_\infty^*} + c\left(\sum_j |\langle u\rangle_{Q_j}|^{p_\infty^*}\right)^{1/p_\infty^*}.$$

Note that we end up with the wrong power after the inequality for using Corollary 7.3.23 a second time: we would want the norm to be raised to the power of p_∞ instead of p_∞^*. However, since $\|\nabla u\|_{L^{p(\cdot)}(Q_j)} \leqslant \|u\|_{W^{1,p(\cdot)}(\mathbb{R}^n)} = 1$ and $p_\infty \leqslant p_\infty^*$, we conclude that $\|\nabla u\|_{L^{p(\cdot)}(Q_j)}^{p_\infty^*} \leqslant \|\nabla u\|_{L^{p(\cdot)}(Q_j)}^{p_\infty}$. Then we can use Corollary 7.3.23 again:

$$\sum_j \|\nabla u\|_{L^{p(\cdot)}(Q_j)}^{p_\infty^*} \leqslant \sum_j \|\nabla u\|_{L^{p(\cdot)}(Q_j)}^{p_\infty} \approx \|\nabla u\|_{L^{p(\cdot)}(\mathbb{R}^n)}^{p_\infty} = 1.$$

It remains to control $\sum_j |\langle u\rangle_{Q_j}|^{p_\infty^*}$. For this we define an auxiliary function $v := |u| * \chi_{Q(0,1/2)}$. Then $|\langle u\rangle_{Q_j}| \leqslant c\,|\langle v\rangle_{Q_j}|$, so it suffices to consider the sum over $|\langle v\rangle_{Q_j}|$. Using also Hölder's inequality, we conclude that

$$\sum_j |\langle u \rangle_{Q_j}|^{p_\infty^*} \leqslant c \sum_j |\langle v \rangle_{Q_j}|^{p_\infty^*} \leqslant c \sum_j \int_{Q_j} |v(x)|^{p_\infty^*}\, dx = c \int_{\mathbb{R}^n} |v(x)|^{p_\infty^*}\, dx.$$

Then it follows from the constant exponent Sobolev inequality (see for example [176, Chap. 7]) that

$$\left(\sum_j |\langle u \rangle_{Q_j}|^{p_\infty^*} \right)^{1/p_\infty^*} \leqslant c\, \|v\|_{L^{p_\infty^*}(\mathbb{R}^n)} \leqslant c\, \|\nabla v\|_{L^{p_\infty}(\mathbb{R}^n)}.$$

Next we notice that

$$|\nabla v(x)| \leqslant \int_{B(x,1)} |\nabla| u(y)|\, | \, dy \leqslant 2 \|\nabla u\|_{p(\cdot)} \|1\|_{L^{p'(\cdot)}(B(x,1))} \leqslant c < \infty,$$

so $|\nabla v| \in L^\infty(\mathbb{R}^n)$. Since $L^{p_\infty} \cap L^\infty \cong L^{p(\cdot)} \cap L^\infty$ (Lemma 3.3.12), it follows that $\|\nabla v\|_{L^{p_\infty}(\mathbb{R}^n)} \leqslant c\, \|\nabla v\|_{L^{p(\cdot)}(\mathbb{R}^n)} + c \leqslant c\, \|\nabla u\|_{L^{p(\cdot)}(\mathbb{R}^n)} + c \leqslant c$, where we used the boundedness of convolution (Lemma 4.6.1) in the second step.   □

Remark 8.3.5. Using Hölder's inequality we see that the Sobolev-Poincaré inequality implies the Poincaré inequality also in the variable exponent context. By a suitable choice of intermediate exponent, we can relax the condition $1 \leqslant p^- \leqslant p^+ < n$ to arbitrary bounded exponents in this case.

The following proposition is due to Kováčik and Rákosník [258]. The exponent p^* is the best possible for constant p, see for example [11, Example 5.25]; using this we show that it is also the best possible for a variable continuous exponent p.

Proposition 8.3.6. Let $p, q \in \mathcal{P}(\Omega) \cap C(\Omega)$ satisfy $p^+ < n$. If $W^{1,p(\cdot)}(\Omega) \hookrightarrow L^{q(\cdot)}(\Omega)$, then $q \leqslant p^*$.

Proof. Suppose that $q(x) > p^*(x)$ for some $x \in \Omega$. By the continuity of p and q, there exist $s \in (1,n)$, $t \in (1,\infty)$ and $r > 0$ such that

$$p^*(y) < s^* < t < q(y)$$

for every $y \in B(x,r)$. By Corollary 3.3.4, $W^{1,s}(B(x,r)) \hookrightarrow W^{1,p(\cdot)}(B(x,r))$ and $L^{q(\cdot)}(B(x,r)) \hookrightarrow L^t(B(x,r))$. Since $s^* < t$ we have $W^{1,s}(B(x,r)) \not\hookrightarrow L^t(B(x,r))$. Thus $W^{1,p(\cdot)}(B(x,r)) \not\hookrightarrow L^{q(\cdot)}(B(x,r))$, which is a contradiction, and so $q(x) \leqslant p^*(x)$.   □

Next we construct a continuous exponent for which the Sobolev embedding $W^{1,p(\cdot)}(\Omega) \hookrightarrow L^{p^*(\cdot)}(\Omega)$ does not hold.

Proposition 8.3.7. *Let $\Omega \subset \mathbb{R}^2$ be the intersection of the upper half-plane and the unit disk. There exists a continuous exponent $p \in \mathcal{P}(\Omega)$ with $1 < p^- \leqslant p^+ < 2$ such that*

$$W^{1,p(\cdot)}(\Omega) \not\hookrightarrow L^{p^*(\cdot)}(\Omega).$$

Proof. Fix t and s such that $1 < t < s < 2$ and define $f(\tau) := 2(\frac{\tau}{t} - 1)$ for $\tau \in [t, 2]$. Denoting by (r, ψ) spherical coordinates in Ω (with $\psi \in (0, \pi)$) we define the variable exponent p as follows:

$$p(r, \psi) := \begin{cases} t, & \text{if } \psi \geqslant r^{f(t)} = 1; \\ \tau, & \text{for } \tau \in (t, s) \text{ satisfying } \psi = r^{f(\tau)}; \\ s, & \text{if } \psi \leqslant r^{f(s)}. \end{cases}$$

We consider the function $u(x) = |x|^\mu$, where $\mu := \frac{s-2}{t}$. Note that u does not belong to $L^{p^*(\cdot)}(\Omega)$, because

$$\overline{\varrho}_{p^*(\cdot)}(u) = \int_\Omega |x|^{\mu p^*(x)} \, dx \geqslant \int_0^1 \int_0^{r^{f(s)}} r^{\frac{2s\mu}{2-s}} r \, d\psi dr$$

$$= \int_0^1 r^{\frac{2s\mu}{2-s} + f(s) + 1} \, dr = \infty.$$

The last equality follows since $\frac{2s\mu}{2-s} + f(s) + 1 = -1$. However, u belongs to $W^{1,p(\cdot)}(\Omega)$. We easily calculate that $|\nabla u(x)| = |\mu||x|^{\mu-1}$. Since $|\mu| < 1$, we find that

$$\int_\Omega |\nabla u(x)|^{p(x)} \, dx < \int_0^1 \int_0^\pi r^{(\mu-1)p(r,\psi)} \, d\psi \, r \, dr.$$

We first estimate the parts of the domain where $p(x) = t$ or $p(x) = s$:

$$\int_0^1 \int_1^\pi r^{(\mu-1)t} d\psi \, r \, dr = (\pi - 1) \int_0^1 r^{(\mu-1)t+1} dr < \infty,$$

since $(\mu - 1)t + 1 > -1$, and

$$\int_0^1 \int_0^{r^{f(s)}} r^{(\mu-1)s} d\psi \, r \, dr = \int_0^1 r^{(\mu-1)s+f(s)+1} dr < \infty$$

since $(\mu - 1)s + f(s) + 1 > -1$. Let us denote the integral over these parts by $K < \infty$.

In the remaining part we have $p(r, \psi) = \tau$ and $\psi = r^{f(\tau)}$. Solving this equation for τ we find that $p(r, \psi) = (\frac{1}{2}\frac{\log \psi}{\log r} + 1)\, t$. Thus we have

$$\int_0^1 \int_0^\pi r^{(\mu-1)p(r,\psi)}\, d\psi\, r\, dr \leqslant K + \int_0^1 \int_0^1 e^{(\mu-1)(\frac{1}{2}\log \psi + \log r)t}\, d\psi\, r\, dr$$

$$= K + \int_0^1 \int_0^1 \psi^{(\mu-1)t/2}\, d\psi\, r^{(\mu-1)t+1}\, dr$$

$$= K + \frac{2}{s-t}\frac{1}{s-t} < \infty.$$

So we have shown that $|\nabla u| \in L^{p(\cdot)}(\Omega)$. For our function u we find that $|u(x)| = \frac{1}{|\mu|}|\nabla u(x)||x| \leqslant \frac{1}{|\mu|}|\nabla u(x)|$ and so it also follows that $u \in L^{p(\cdot)}(\Omega)$, and we are done. $\qquad \square$

If a variable exponent is locally greater than the dimension n, then we have locally an embedding to a constant exponent Sobolev space with exponent greater than n (Lemma 8.1.8) and thus $W^{1,p(\cdot)}(\Omega) \subset C(\Omega)$. It follows immediately that if $p \in \mathcal{P}(\Omega)$ is continuous and $p(x) > n$ for every $x \in \Omega$, then $W^{1,p(\cdot)}(\Omega) \subset C(\Omega)$. In fact, we can obtain a better result in terms of a Hölder-type continuity modulus with varying exponent. The following result was originally proved by Edmunds and Rákosník [121]; we present here a new, much simpler proof. This question has been more systematically studied by Almeida and Samko in [27, 28].

Theorem 8.3.8. *We write* $\delta(x) := \min\{1, \mathrm{dist}\{x, \partial\Omega\}\}$. *Let* $p \in \mathcal{P}^{\log}(\Omega)$ *satisfy* $p^- > n$. *Then there exists a constant* c *such that*

$$\sup_{y \in B(x,\delta(x))} \frac{|u(x) - u(y)|}{|x-y|^{1-\frac{n}{p(x)}}} \leqslant c\, \|\nabla u\|_{L^{p(\cdot)}(\Omega)}$$

for every $u \in W^{1,p(\cdot)}(\Omega)$ *and for every* $x \in \Omega$. *The constant depends only on the dimension* n, p^- *and* $c_{\log}(p)$.

Proof. Let $x \in \Omega$. If $|x - y| < \delta(x)$, then there exists $r < \delta(x)$ such that $\frac{1}{2}r < |x-y| < r$. Denote $B := B(x, r)$. Since

$$W^{1,p(\cdot)}(B) \hookrightarrow W^{1,p_B^-}(B),$$

we obtain by the constant exponent result [129, Theorem 3, p. 143] that

$$|u(x) - u(y)| \leqslant c\, r^{1-\frac{n}{p_B^-}}\, \|\nabla u\|_{L^{p_B^-}(B)}$$

$$\leqslant c\, (1 + |B(0,1)|)\, r^{1-\frac{n}{p_B^-}}\, \|\nabla u\|_{L^{p(\cdot)}(\Omega)}.$$

Since $r < 1$ the log-Hölder continuity of p implies that $r^{1-\frac{n}{p_B}} \leqslant c\, r^{1-\frac{n}{p(x)}}$. Using $r < 2|x - y|$ we obtain the claim. $\qquad\qquad\qquad\qquad\qquad\qquad\square$

8.4 Compact Embeddings

We start this section by examining compact embeddings of $W^{1,p(\cdot)}(\Omega)$ into $L^{p(\cdot)}(\Omega)$. This section is based on Diening [92], see also Fan, Zhao and Zhao [152] for further results. We first prove a quantitative version of Theorem 4.6.4 (b) for Sobolev functions.

Lemma 8.4.1. *Let $p \in \mathcal{A}$ be bounded, and let ψ be a standard mollifier. There exists $A > 0$, depending only on the $\mathcal{A}$-constant of p, such that*

$$\|u * \psi_\varepsilon - u\|_{p(\cdot)} \leqslant \varepsilon\, A\, \|\nabla u\|_{p(\cdot)}$$

for all $u \in W^{1,p(\cdot)}(\mathbb{R}^n)$ and all $\varepsilon > 0$.

Proof. Let $u \in C_0^\infty(\mathbb{R}^n)$. Using the properties of the mollifier, Fubini's theorem and a change of variables we deduce

$$u * \psi_\varepsilon(x) - u(x) = \int_{\mathbb{R}^n} \int_0^1 \psi_\varepsilon(y) \nabla u(x - ty) \cdot y\, dt\, dy$$

$$= \int_0^1 \int_{\mathbb{R}^n} \psi_{\varepsilon t}(y) \nabla u(x - y) \cdot \frac{y}{t}\, dy\, dt.$$

This yields the pointwise estimate

$$\left| u * \psi_\varepsilon(x) - u(x) \right| \leqslant \varepsilon \int_0^1 \int_{\mathbb{R}^n} |\psi_{\varepsilon t}(y)|\, |\nabla u(x - y)|\, dy\, dt$$

$$= \varepsilon \int_0^1 |\nabla u| * |\psi_{\varepsilon t}|(x)\, dt$$

for all $u \in C_0^\infty(\mathbb{R}^n)$. Since $\operatorname{spt} \psi_{t\varepsilon} \subset B(0, \varepsilon)$ for all $t \in [0,1]$, the estimate is of a local character. Due to the density of $C_0^\infty(\mathbb{R}^n)$ in $W_{\mathrm{loc}}^{1,1}(\mathbb{R}^n)$ the same estimate holds almost everywhere for all $u \in W_{\mathrm{loc}}^{1,1}(\mathbb{R}^n)$. Hence, it holds in particular for all $u \in W^{1,p(\cdot)}(\mathbb{R}^n)$. The pointwise estimate thus yields a norm inequality, which due to the properties of the Bochner integral implies

$$\|u * \psi_\varepsilon - u\|_{p(\cdot)} \leqslant \varepsilon \left\| \int_0^1 |\nabla u| * |\psi_{\varepsilon t}| \, dt \right\|_{p(\cdot)} \leqslant \varepsilon \int_0^1 \left\| |\nabla u| * |\psi_{\varepsilon t}| \right\|_{p(\cdot)} dt.$$

By Lemma 4.6.3 we obtain $\left\| |\nabla u| * |\psi_{\varepsilon t}| \right\|_{p(\cdot)} \leqslant K \|\psi\|_1 \|\nabla u\|_{p(\cdot)}$. Now, the claim follows due to $\|\psi\|_1 = 1$. $\square$

Theorem 8.4.2. *Let $\Omega \subset \mathbb{R}^n$ be a bounded domain and let $p \in \mathcal{P}^{\log}(\Omega)$ or $p \in \mathcal{A}$ be bounded. Then*

$$W_0^{1,p(\cdot)}(\Omega) \hookrightarrow\hookrightarrow L^{p(\cdot)}(\Omega).$$

Proof. By Theorem 4.4.8 we always have $p \in \mathcal{A}$. Let $u_k \rightharpoonup u$ in $W_0^{1,p(\cdot)}(\Omega)$. We write $v_k := u_k - u$ and hence $v_k \rightharpoonup 0$ in $W_0^{1,p(\cdot)}(\Omega)$. Thus $\|v_k\|_{1,p(\cdot)}$ is uniformly bounded. Furthermore, we extend the functions v_k by zero outside of Ω (Theorem 8.1.14). We have to show that $v_k \to 0$ in $L^{p(\cdot)}(\Omega)$. Let ψ_ε be the standard mollifier. Then we have $v_k(x) = (v_k - \psi_\varepsilon * v_k)(x) + \psi_\varepsilon * v_k(x)$ and Lemma 8.4.1 implies

$$\begin{aligned}
\|v_k\|_{p(\cdot)} &\leqslant \|v_k - v_k * \psi_\varepsilon\|_{p(\cdot)} + \|v_k * \psi_\varepsilon\|_{p(\cdot)} \\
&\leqslant c\varepsilon \|\nabla v_k\|_{p(\cdot)} + \|v_k * \psi_\varepsilon\|_{p(\cdot)}.
\end{aligned} \tag{8.4.3}$$

Since $v_k \rightharpoonup 0$ and $\varepsilon > 0$ is fixed we obtain

$$v_k * \psi_\varepsilon(x) = \int_{\mathbb{R}^n} \psi_\varepsilon(x - y) v_k(y) \, dy \to 0$$

as $k \to \infty$. Let $\Omega_\varepsilon := \{x \in \mathbb{R}^n : \operatorname{dist}(x, \Omega) \leqslant \varepsilon\}$. Thus $v_k * \psi_\varepsilon(x) = 0$ for all $x \in \mathbb{R}^n \setminus \Omega_\varepsilon$. By Hölder's inequality we obtain for all $x \in \Omega_\varepsilon$ that

$$\left| v_k * \psi_\varepsilon(x) \right| = \left| \int_{\mathbb{R}^n} \psi_\varepsilon(x - y) v_k(y) \, dy \right| \leqslant c \|v_k\|_{p(\cdot)} \|\psi_\varepsilon(x - \cdot)\|_{p'(\cdot)}.$$

Since $\psi \in C_0^\infty(\mathbb{R}^n)$ we have $|\psi| \leqslant c$ and thus $|\psi_\varepsilon| \leqslant c\varepsilon^{-n}$. This yields $\|\psi_\varepsilon(x - \cdot)\|_{p'(\cdot)} \leqslant c\varepsilon^{-n} \|\chi_{\Omega_\varepsilon}\|_{p'(\cdot)} \leqslant c(\varepsilon, p)$ independently of $x \in \mathbb{R}^n$ and k. Using the uniform boundedness of v_k in $L^{p(\cdot)}$ we all together proved

$$\left| v_k * \psi_\varepsilon(x) \right| \leqslant c(\varepsilon, p) \chi_{\Omega_\varepsilon}(x)$$

for all $x \in \mathbb{R}^n$. Since $\chi_{\Omega_\varepsilon} \in L^{p(\cdot)}(\mathbb{R}^n)$ and $v_k * \psi_\varepsilon(x) \to 0$ almost everywhere, we obtain by the theorem of dominated convergence that $v_k * \psi_\varepsilon \to 0$ in $L^{p(\cdot)}(\mathbb{R}^n)$ as $k \to \infty$. Hence it follows from (8.4.3) that

$$\limsup_{k\to\infty} \|v_k\|_{p(\cdot)} \leqslant c\,\varepsilon \limsup_{k\to\infty} \|\nabla v_k\|_{p(\cdot)}.$$

Since $\varepsilon > 0$ was arbitrary and $\|\nabla v_k\|_{p(\cdot)}$ was uniformly bounded this yields that $v_k \to 0$ in $L^{p(\cdot)}(\mathbb{R}^n)$ and thus $u_k \to u$ in $L^{p(\cdot)}(\Omega)$. □

In fact, it is easy to improve the previous result to deal with higher exponents in the Lebesgue space:

Corollary 8.4.4. *Let Ω be a bounded domain and let $p \in \mathcal{P}^{\log}(\Omega)$ satisfy $p^+ < n$. Then*

$$W_0^{1,p(\cdot)}(\Omega) \hookrightarrow\hookrightarrow L^{p^*(\cdot)-\varepsilon}(\Omega)$$

for every $\varepsilon \in (0, n')$.

We give two alternative proofs. The first one is based on interpolation. The later one is a straight forward, but it uses Corollary 3.3.4 for the exponent $\frac{\alpha p^*(\cdot)(p^*(\cdot)-\varepsilon)}{\varepsilon}$ that is less than 1. However, it is easy to check that Corollary 3.3.4 holds also in this case.

Proof 1. We extend the exponent p to the whole space by Proposition 4.1.7. Due to Theorems 8.3.1 and 8.4.2 we have $W_0^{1,p(\cdot)}(\Omega) \hookrightarrow\hookrightarrow L^{p(\cdot)}(\Omega)$ and $W_0^{1,p(\cdot)}(\Omega) \hookrightarrow L^{p^*(\cdot)}(\Omega)$, respectively. Therefore, by complex interpolation, see Theorem 7.1.2, $W_0^{1,p(\cdot)}(\Omega) \hookrightarrow\hookrightarrow L^{q_\theta(\cdot)}(\Omega)$, where $q_\theta \in \mathcal{P}(\Omega)$ is defined for $\theta \in (0,1)$ by $\frac{1}{q_\theta} = \frac{1-\theta}{p^*} + \frac{\theta}{p} = \frac{1}{p} - \frac{1-\theta}{n}$. Instead of interpolation, one can also use Hölder's inequality in this argument. For θ small enough, we have $p^* - \varepsilon \leqslant q_\theta \leqslant p^*$ and therefore by Corollary 3.3.4 $L^{q_\theta(\cdot)}(\Omega) \hookrightarrow L^{p^*(\cdot)-\varepsilon}(\Omega)$. The claim follows as the composition of a compact and a bounded embedding is compact. □

Proof 2. Fix $\varepsilon \in (0, n')$. As in the previous proof, it suffices to show that $u_k \to 0$ in $L^{p^*(\cdot)-\varepsilon}(\Omega)$ whenever $u_k \rightharpoonup 0$ in $W_0^{1,p(\cdot)}(\Omega)$. Define $\alpha := \varepsilon((p^+)^*)^{-2}$. Then an application of Hölder's inequality (3.2.22) yields

$$\|u_k\|_{p^*(\cdot)-\varepsilon} \leqslant 2\| |u_k|^\alpha \|_{\frac{p^*(\cdot)(p^*(\cdot)-\varepsilon)}{\varepsilon}} \| |u_k|^{1-\alpha} \|_{p^*(\cdot)}.$$

By Corollary 3.3.4 and 8.3.1, $\| |u_k|^{1-\alpha} \|_{p^*(\cdot)} < \infty$. On the other hand,

$$\| |u_k|^\alpha \|_{\frac{p^*(\cdot)(p^*(\cdot)-\varepsilon)}{\varepsilon}} = \|u_k\|^\alpha_{\frac{\alpha p^*(\cdot)(p^*(\cdot)-\varepsilon)}{\varepsilon}} \leqslant 2(1 + |\Omega|)\|u_k\|_1 \to 0$$

since $W_0^{1,p(\cdot)}(\Omega) \hookrightarrow\hookrightarrow L^1(\Omega)$. □

When p is (almost) continuous, it is possible to prove Theorem 8.4.2 by a different argument. We present the method here for general Sobolev functions.

Theorem 8.4.5. *Let $\Omega \subset \mathbb{R}^n$ be a bounded convex open set. Let $p \in \mathcal{P}(\Omega)$ and assume that there exists $\delta > 0$ such that*

$$p^+_{B(x,\delta) \cap \Omega} < \left(p^-_{B(x,\delta) \cap \Omega}\right)^* \quad or \quad p^-_{B(x,\delta) \cap \Omega} \geqslant n$$

for every $x \in \Omega$. Then

$$W^{1,p(\cdot)}(\Omega) \hookrightarrow\hookrightarrow L^{p(\cdot)}(\Omega).$$

Proof. We may cover Ω by finitely many balls B_i, with radius δ, such that either $p^+_{B_i \cap \Omega} < (p^-_{B_i \cap \Omega})^*$ or $p^-_{B_i \cap \Omega} \geqslant n$. Let $\varepsilon > 0$ be so small that $p \leqslant (p^-_{B_i \cap \Omega})^* - \varepsilon$ in each $B_i \cap \Omega$. We obtain

$$W^{1,p(\cdot)}(B_i \cap \Omega) \hookrightarrow W^{1,p^-_{B_i \cap \Omega}}(B_i \cap \Omega)$$
$$\hookrightarrow\hookrightarrow L^{(p^-_{B_i \cap \Omega})^* - \varepsilon}(B_i \cap \Omega) \hookrightarrow L^{p(\cdot)}(B_i \cap \Omega),$$

where the convexity of $B_i \cap \Omega$ is used in the second embedding. Since there was only finitely many balls, we obtain the claim. $\square$

It is well-known that the embedding $W^{1,p} \hookrightarrow\hookrightarrow L^q$ holds in the constant exponent case if and only if $q < p^*$. Surprisingly, we sometimes have a compact embedding in the variable exponent case even if $q(x) = p^*(x)$ at some point. This was first shown by Kurata and Shioji [259] and generalized to more complicated sets by Mizuta, Ohno, Shimomura and Shioji in [300]. We present here, without proof, only a simplified version of [300, Theorem 3.4].

Theorem 8.4.6. *Let $p \in \mathcal{P}^{\log}(\mathbb{R}^n)$ be bounded, let $q \in \mathcal{P}(\mathbb{R}^n)$ be bounded and suppose that*

$$q(x) \leqslant p^*(x) - \frac{\omega(|x|)}{\log(e + 1/|x|)}$$

where $\omega \colon [0, \infty) \to [0, \infty)$ is increasing and continuous with $\omega(0) = 0$. Then $W^{1,p(\cdot)}(\mathbb{R}^n) \hookrightarrow\hookrightarrow L^{q(\cdot)}(\mathbb{R}^n)$.

Let us point out that the exponent q in the previous theorem cannot be locally log-Hölder continuous, i.e. $q \notin \mathcal{P}^{\log}(\mathbb{R}^n)$.

8.5   Extension Operator

In this section we study extension operators for variable exponent Sobolev functions. We show that for certain domains Ω there exists a bounded extension operator $\mathcal{E}$ from $W^{m,p(\cdot)}(\Omega)$ to $W^{m,p(\cdot)}(\mathbb{R}^n)$ for every $m \in \mathbb{N}_0$ and all $p \in \mathcal{A}$.

Definition 8.5.1. A domain $\Omega \subset \mathbb{R}^n$ is called an (ε, ∞)-*domain*, $0 < \varepsilon \leqslant 1$, if every pair of points $x, y \in \Omega$ can be joined by a rectifiable path γ parametrized by arc-length such that

$$\ell(\gamma) \leqslant \frac{1}{\varepsilon}|x - y|,$$

$$B\left(\gamma(t), \frac{\varepsilon|x - \gamma(t)||y - \gamma(t)|}{|x - y|}\right) \subset \Omega \quad \text{for all } t \in [0, \ell(\gamma)],$$

where $\ell(\gamma)$ is the length of γ.

These domains are also known as *uniform domains* [283] or *Jones domains*. Every bounded Lipschitz domain and the half-space are (ε, ∞)-domain for some value of ε. The boundary of an (ε, ∞)-domain can be non-rectifiable as in the case of the Koch snowflake domain. For us it is important to know that $|\partial\Omega| = 0$ for every (ε, ∞)-domain Ω [224, Lemma 2.3].

We start with an easy extension result which follows from extrapolation. The theorem works also in the case $\mathcal{P}^{\log}(\Omega)$, since we can then extend p to all of $\mathbb{R}^n$ by Proposition 4.1.7.

Theorem 8.5.2. *Let Ω be an (ε, ∞)-domain and suppose that $p \in \mathcal{A}$ with $1 < p^- \leqslant p^+ < \infty$. If $u \in W^{k,p(\cdot)}(\Omega)$, then it can be extended to a function in $W^{k,p(\cdot)}(\mathbb{R}^n)$, with*

$$\|u\|_{W^{k,p(\cdot)}(\mathbb{R}^n)} \leqslant c\,\|u\|_{W^{k,p(\cdot)}(\Omega)},$$

where the constant c depends on n, ε, p^+, p^- and the $\mathcal{A}$-constant of p.

Proof. Let $\Lambda : W^{k,1}(\Omega) \hookrightarrow W^{k,1}(\mathbb{R}^n)$ be the extension operator of Jones [224]. Chua generalized Jones' extension result to the weighted case:

$$\int\limits_{\mathbb{R}^n} |\partial_\alpha(\Lambda u)|\, w\, dx \leqslant c \int\limits_{\Omega} |g|\, w\, dx$$

for all $w \in A_1$, where $|\alpha| \leqslant k$, $g := \sum_{|\beta| \leqslant k} |\partial_\beta u|$ and the constant depends on w only through $\|w\|_{A_1}$ [75, Theorem 1.1]. Note also that M is bounded by Theorem 5.7.2 since $p \in \mathcal{A}$. Hence the extrapolation theorem, Theorem 7.2.1, implies that

$$\|\partial_\alpha(\Lambda u)\|_{L^{p(\cdot)}(\mathbb{R}^n)} \leqslant c\,\|g\|_{L^{p(\cdot)}(\Omega)}.$$

This applies to every $|\alpha| \leqslant k$, so by the triangle inequality we obtain

$$\sum_{|\alpha| \leqslant k} \|\partial_\alpha(\Lambda u)\|_{L^{p(\cdot)}(\mathbb{R}^n)} \leqslant c\,\|g\|_{L^{p(\cdot)}(\Omega)} \leqslant c \sum_{|\beta| \leqslant k} \|\partial_\beta u\|_{L^{p(\cdot)}(\Omega)},$$

which implies the claim. $\qquad\qquad\qquad\qquad\qquad\qquad\qquad\qquad\qquad\qquad\square$

The disadvantage of the previous theorem is that there is a condition on p^- and p^+. Next we show that these restrictions are not necessary. The technique is based on the works of Jones [224] for classical Sobolev spaces and of Chua [75] for A_p-weighted Sobolev spaces. The results presented below where first developed by Fröschl [159].

The following proposition shows that every open set can be decomposed (up to measure zero) into a union of cubes whose lengths are proportional to the distance from the boundary of the open set. For a proof see for example [359, Theorem 1, p.167] or [179, Appendix J]. We say that two open sets A and B *touch* if $A \cap B = \emptyset$ and $\overline{A} \cap \overline{B} \neq \emptyset$.

Proposition 8.5.3 (Dyadic Whitney decomposition). *Let $\Omega \subsetneq \mathbb{R}^n$ be an open non-empty set. Then there exists a countable family $\mathcal{F}$ of dyadic cubes such that*

(a) $\bigcup_{Q \in \mathcal{F}} \overline{Q} = \Omega$ *and the cubes from $\mathcal{F}$ are pairwise disjoint.*
(b) $\sqrt{n}\ell(Q) < \operatorname{dist}(Q, \Omega^{\complement}) \leqslant 4\sqrt{n}\ell(Q)$ *for all $Q \in \mathcal{F}$.*
(c) *If $\overline{Q}, \overline{Q'} \in \mathcal{F}$ intersect, then*

$$\frac{1}{4} \leqslant \frac{\ell(Q)}{\ell(Q')} \leqslant 4.$$

(d) *For given $Q \in \mathcal{F}$, there exists at most 12^n cubes $Q' \in \mathcal{F}$ touching Q.*

Using this Whitney decomposition the following property for (ε, ∞)-domains has been shown in [224, Lemmas 2.4 and 2.8].

Lemma 8.5.4. *Let $\Omega \subsetneq \mathbb{R}^n$ be a non-empty (ε, ∞)-uniform domain, and let $\mathcal{W}_1$ and $\mathcal{W}_2$ denote the dyadic Whitney decomposition of Ω and $\mathbb{R}^n \setminus \overline{\Omega}$. Further, let*

$$\mathcal{W}_3 := \left\{ Q \in \mathcal{W}_2 : \ell(Q) \leqslant \frac{\varepsilon \operatorname{diam}(\Omega)}{16n} \right\}.$$

Then for every $Q \in \mathcal{W}_3$ there exists a reflected cube $Q^* \in \mathcal{W}_1$ such that

$$1 \leqslant \frac{\ell(Q^*)}{\ell(Q)} \leqslant 4,$$
$$\operatorname{dist}(Q, Q^*) \leqslant c\,\ell(Q),$$

where $c = c(n, \varepsilon)$. Moreover, if $Q, Q_2 \in \mathcal{W}_3$ touch, then there exists a chain $F_{Q,Q_2} = \{Q^* = S_1, S_2, \ldots, S_{j_Q} = Q_2^*\}$ of touching cubes in $\mathcal{W}_2$ with $j_Q \leqslant j_{\max}(\varepsilon, n)$ connecting Q^* and Q_2^* with

$$\frac{1}{4} \leqslant \frac{\ell(S_j)}{\ell(S_{j+1})} \leqslant 4 \qquad \text{for } j = 1, \ldots, j_Q - 1.$$

Observe that all cubes in a chain F_{Q_1,Q_2} are of comparable size.

We need some preparation before we can define our extension operator. Let $\omega \in C_0^\infty([0,1]^n)$ with $\omega \geqslant 0$ and $\int_{\mathbb{R}^n} \omega(x)\,dx = 1$. For $Q \in \mathcal{W}_1$ let $\omega_Q \in C_0^\infty(Q)$ be defined by $\omega_Q := \ell(Q)^{-n}\omega\circ L_Q^{-1}$, where L_Q is the affine linear mapping from $[0,1]^n$ onto Q. In particular, $\omega_Q \geqslant 0$ and $\int_{\mathbb{R}^n} \omega_Q\,dx = 1$. For $m \in \mathbb{N}_0$ and $v \in W^{m,1}(Q)$ let $\Pi_Q^m v$ denote the Q-averaged Taylor polynomial of degree m, i.e.

$$(\Pi_Q^m v)(x) = \sum_{|\alpha|\leqslant m} \int_Q \omega_Q(y)\nabla^\alpha v(y)\frac{(x-y)^\alpha}{\alpha!}\,dy. \tag{8.5.5}$$

In the definition of Π_Q^m it suffices to assume $v \in L_{\mathrm{loc}}^1(Q)$, if all derivatives are moved by partial integration to $\omega(y)$ and $(x-y)^\alpha$ using also that $\omega \in C_0^\infty(Q)$. If $\alpha \in \mathbb{N}_0^n$ with $0 \leqslant |\alpha| \leqslant m$, then $\partial_\alpha \Pi_Q^m v = \Pi_Q^{m-|\alpha|}(\partial_\alpha v)$. The averaged Taylor polynomial has the following nice properties [75, Theorem 4.7]. If $Q \in \mathcal{W}_1$, $1 \leqslant q \leqslant \infty$ and $0 \leqslant |\beta| \leqslant k \leqslant m$, then

$$\begin{aligned} \left\|\Pi_Q^m v\right\|_{L^q(Q)} &\leqslant c\left\|v\right\|_{L^q(Q)} && \text{for all } v \in L^q(Q),\\ \left\|\partial_\beta(v - \Pi_Q^m v)\right\|_{L^q(Q)} &\leqslant c\,\ell(Q)^{k-|\beta|}\left\|\nabla^k v\right\|_{L^q(Q)} && \text{for all } v \in W^{k,q}(Q). \end{aligned} \tag{8.5.6}$$

We only need the case $q = 1$.

We need a partition of unity for $\mathcal{W}_3$, see Theorem 1.4.7. For each $Q \in \mathcal{W}_3$ choose $\varphi_Q \in C_0^\infty(\frac{17}{16}Q)$ with $0 \leqslant \varphi_Q \leqslant 1$ such that

$$\begin{aligned} \sum_{Q\in\mathcal{W}_3} \varphi_Q &= 1 && \text{for all } x \in \bigcup_{Q\in\mathcal{W}_3} Q,\\ 0 \leqslant \sum_{Q\in\mathcal{W}_3} \varphi_Q &\leqslant 1,\\ |\nabla^k\varphi_Q| &\leqslant c\,\ell(Q)^{-k} && \text{for all } 0 \leqslant k \leqslant m. \end{aligned}$$

Then $\varphi_Q\varphi_{Q'} \neq 0$ if and only if $\overline{Q} \cap \overline{Q'} \neq \emptyset$.

We can now define our extension operator $\mathcal{E}^m : L^1(\Omega) \to L_{\mathrm{loc}}^1(\mathbb{R}^n)$ by

$$\mathcal{E}^m v := \begin{cases} v & \text{on } \Omega,\\ \sum_{Q\in\mathcal{W}_3} \varphi_Q \Pi_Q^m v & \text{on } \mathbb{R}^n \setminus \overline{\Omega}. \end{cases} \tag{8.5.7}$$

Chua showed in [75, Theorem 1.2], using the method presented by Jones [224], that the operator $\mathcal{E}^m$ defines a bounded extension operator from $W^{m,q}(\Omega)$ to $W^{m,q}(\mathbb{R}^n)$ for every $1 \leqslant q < \infty$. It is important for our considerations that the extension operator $\mathcal{E}^m$ is of local type in the sense that for every ball

$B_1 \subset \mathbb{R}^n$, there exists a ball $B_2 \subset \mathbb{R}^n$ such that $\mathcal{E}^m u = \mathcal{E}^m v$ on B_1 if $u = v$ on $B_2 \cap \Omega$. Since $W^{m,p(\cdot)}(\Omega \cap B_2) \hookrightarrow W^{m,1}(\Omega \cap B_2)$ for every ball B_2, we can apply Chua's result for $q = 1$ to conclude that $\mathcal{E}^m v \in W^{m,1}_{\mathrm{loc}}(\mathbb{R}^n)$ for every $v \in W^{m,p(\cdot)}(\Omega)$. So we already know that $\mathcal{E}^m v$ has weak derivatives up to order m in L^1_{loc} and it remains to prove the estimates of these derivatives in $L^{p(\cdot)}(\mathbb{R}^n)$. This saves us the trouble of approximation by smooth functions.

We begin with some local estimates.

Lemma 8.5.8. *If $Q \in \mathcal{W}_3$, then*

$$\left\| \nabla^k \mathcal{E}^m v \right\|_{L^\infty(Q)} \leqslant c |Q|^{-1} \left\| \nabla^k v \right\|_{L^1(F(Q))}$$

for all $0 \leqslant k \leqslant m$ with $c = c(\varepsilon, n, m)$, where

$$F(Q) := \bigcup_{\substack{Q_2 \in \mathcal{W}_3 \\ Q \text{ and } Q_2 \text{ touch}}} \bigcup_{S \in F_{Q,Q_2}} S.$$

Proof. Let $Q \in \mathcal{W}_3$ and $k \in \mathbb{N}_0^n$ with $0 \leqslant |k| \leqslant m$. Using $\sum_{Q_2 \in \mathcal{W}_3} \varphi_{Q_2} = 1$ on Q, the product rule and the estimates for $\partial_\beta \varphi_{Q_2}$, we get

$$\left\| \nabla^k \mathcal{E}^m v \right\|_{L^\infty(Q)} = \left\| \nabla^k \sum_{Q_2 \in \mathcal{W}_3} \varphi_{Q_2} \Pi_{Q_2}^m v \right\|_{L^\infty(Q)}$$

$$\leqslant \left\| \nabla^k \sum_{Q_2 \in \mathcal{W}_3 \,:\, \overline{Q_2} \cap \overline{Q} \neq \emptyset} \varphi_{Q_2} \left(\Pi_{Q_2}^m v - \Pi_Q^m v \right) \right\|_{L^\infty(Q)} + \left\| \nabla^k \Pi_Q^m v \right\|_{L^\infty(Q)}$$

$$\leqslant c \sum_{j=0}^k \ell(Q)^{j-k} \sum_{Q_2 \in \mathcal{W}_3 \,:\, \overline{Q_2} \cap \overline{Q} \neq \emptyset} \left\| \nabla^j \left(\Pi_{Q_2}^m v - \Pi_Q^m v \right) \right\|_{L^\infty(Q)} + \left\| \nabla^k \Pi_Q^m v \right\|_{L^\infty(Q)}.$$

The terms in the norms are polynomials of order at most m. For any polynomial z of order m, $\|z\|_{L^\infty(Q)} \leqslant c |Q|^{-1} \|z\|_{L^1(Q_2)}$ if Q and Q_2 are cubes of similar size and with a maximal distance of Q and Q_2 comparable to the size of Q. This is a consequence of the fact that all norms on a finite dimensional space are equivalent. The independence of c of Q and Q_2 follows by a scaling argument. We use this fact and $\partial_\alpha \Pi_Q^m v = \Pi_Q^{m-|\alpha|}(\partial_\alpha v)$ for $0 \leqslant |\alpha| \leqslant m$, to estimate

$$\left\| \nabla^j \left(\Pi_{Q_2}^m v - \Pi_Q^m v \right) \right\|_{L^\infty(Q)} \leqslant c |Q|^{-1} \left\| \Pi_{Q_2}^{m-j}(\nabla^j v) - \Pi_Q^{m-j}(\nabla^j v) \right\|_{L^1(Q)}$$

considering also that Q and Q_2 are of comparable size. Let $F_{Q,Q_2} = \{S_1, \dots, S_{j_Q}\}$ denote the chain connecting Q^* and Q_2^*. Then

$$\left\|\Pi_{Q_2}^{m-j}(\nabla^j v) - \Pi_Q^{m-j}(\nabla^j v)\right\|_{L^1(Q)} \leqslant \sum_{i=1}^{j_Q-1}\left\|\Pi_{S_{i+1}}^{m-j}(\nabla^j v) - \Pi_{S_i}^{m-j}(\nabla^j v)\right\|_{L^1(Q)}$$

$$\leqslant 2\sum_{i=1}^{j_Q}\left\|\Pi_{S_i}^{m-j}(\nabla^j v) - P_i\right\|_{L^1(Q)}$$

for any function P_i. We then set $P_i := \Pi_{S_i \cup S_{i+1}}^{m-j}(\nabla^j v)$.

Again we use the fact that we are working with polynomials. Namely let P be a polynomial of degree m and let E and F be measurable subsets of a cube Q with $|E|, |F| \geqslant \gamma|Q|$, for some $\gamma > 0$, then $\|P\|_{L^1(E)} \leqslant c(\gamma, m)\|P\|_{L^1(F)}$. For the proof of this fact see [224, Lemma 2.1]. Using this in the first step, and (8.5.6) in the last step, we get

$$\left\|\Pi_{S_i}^{m-j}(\nabla^j v) - P_i\right\|_{L^1(Q)}$$
$$\leqslant c\left\|\Pi_{S_{i+1}}^{m-j}(\nabla^j v) - \Pi_{S_i \cup S_{i+1}}^{m-j}(\nabla^j v)\right\|_{L^1(S_i)}$$
$$\leqslant c\left\|\nabla^j v - \Pi_{S_{i+1}}^{m-j}(\nabla^j v)\right\|_{L^1(S_i)} + \left\|\nabla^j v - \Pi_{S_i \cup S_{i+1}}^{m-j}(\nabla^j v)\right\|_{L^1(S_i \cup S_{i+1})}$$
$$\leqslant c\,\ell(Q)^{k-j}\left\|\nabla^k v\right\|_{L^1(\bigcup_i S_i)}$$

where we have also used that all cubes in the chain F_{Q,Q_2} are of comparable size. On the other hand it follows from the L^∞-L^1 estimate for polynomials and (8.5.6) that

$$\left\|\nabla^k \Pi_Q^m v\right\|_{L^\infty(Q)} \leqslant c\,|Q|^{-1}\left\|\Pi_Q^{m-k}\nabla^k v\right\|_{L^1(Q)} \leqslant c\,|Q|^{-1}\left\|\nabla^k v\right\|_{L^1(Q)}.$$

Combining the above estimates we get

$$\left\|\nabla^k \mathcal{E}^m v\right\|_{L^\infty(Q)} \leqslant c\,|Q|^{-1} \sum_{Q_2 \in \mathcal{W}_3\,:\,\overline{Q_2} \cap \overline{Q} \neq \emptyset}\left\|\nabla^k v\right\|_{L^1(\bigcup F_{Q,Q_2})}$$
$$+ c\,|Q|^{-1}\left\|\nabla^k v\right\|_{L^1(Q)},$$

which yields the claim by the definition of $F(Q)$. $\qquad\square$

In order to use that $p \in \mathcal{A}$, we have to reformulate the previous lemma in terms of averaging operators.

Corollary 8.5.9. *There exists $c = c(\varepsilon, n, m) > 0$ such that*

$$\sum_{Q \in \mathcal{W}_3} \chi_Q |\nabla^k \mathcal{E}^m v| \leqslant c \sum_{Q \in \mathcal{W}_3} \chi_Q M_{Q_*}\big(\underbrace{T_{\frac{17}{16}\mathcal{W}_1} \circ \cdots \circ T_{\frac{17}{16}\mathcal{W}_1}}_{(j_{\max}+1)\text{-times}} \circ T_{\mathcal{W}_1}(\nabla^k v)\big),$$

where $j_{\max}$ is the maximal chain length in Lemma 8.5.4.

Proof. We show that it follows from the definition of $F(Q)$ that

$$\frac{1}{|Q|}\int_{F(Q)}|\nabla^k v|\,dx \leqslant c\Big(\underbrace{T_{\frac{17}{16}W_1}\circ\cdots\circ T_{\frac{17}{16}W_1}}_{(j_{\max}+1)\text{-times}}\circ T_{W_1}(\nabla^k v)\Big)(z)$$

for all $z \in Q^*$. First, the application of T_{W_1} ensures that the L^1-averages over all cubes participating in a chain starting from Q are calculated. Then these averages are accumulated at Q^* by passing them along the chains by the $(j_{\max}+1)$-fold application of $T_{\frac{17}{16}W_1}$, where we use that the neighboring cubes in $\frac{17}{16}W_1$ have sufficiently big overlap. Finally, this value is transported by the operator $\sum_{Q\in W_3}\chi_Q M_{Q^*}$ from Q^* to Q. Together with Lemma 8.5.8 the claim follows.   □

We turn to the case $Q \in W_2 \setminus W_3$.

Lemma 8.5.10. *Let $Q \in W_2 \setminus W_3$, then*

$$\big\|\nabla^k \mathcal{E}^m v\big\|_{L^\infty(Q)} \leqslant c\,|Q|^{-1}\sum_{j=0}^{k}\ell(Q)^{j-k}\sum_{Q_2\in W_3\,:\,\overline{Q_2}\cap\overline{Q}\neq\emptyset}\big\|\nabla^j v\big\|_{L^1(Q_2)}$$

for all $0 \leqslant k \leqslant m$ with $c = c(\varepsilon, n, m)$.

Proof. Let $Q \in W_2 \setminus W_3$ and $k \in \mathbb{N}_0^n$ with $0 \leqslant k \leqslant m$. Using the product rule and the estimates for $\partial_\beta \varphi_{Q_2}$, we get

$$\big\|\nabla^k \mathcal{E}^m v\big\|_{L^\infty(Q)} = \Big\|\nabla^k \sum_{Q_2\in W_3}\varphi_{Q_2}\Pi_{Q_2}^m v\Big\|_{L^\infty(Q)}$$
$$\leqslant c\sum_{j=0}^{k}\ell(Q)^{j-k}\sum_{Q_2\in W_3\,:\,\overline{Q_2}\cap\overline{Q}\neq\emptyset}\big\|\nabla^j \Pi_{Q_2}^m v\big\|_{L^\infty(Q)}$$

using also that Q and Q_2 are of comparable size. Now, the claim follows as in the proof of Lemma 8.5.8.   □

As before we reformulate this in terms of averaging operators.

Corollary 8.5.11. *Let $Q \in W_2 \setminus W_3$, then*

$$\sum_{Q\in W_2\setminus W_3}\chi_Q|\nabla^k \mathcal{E}^m v| \leqslant c\sum_{j=0}^{k}\ell(Q)^{j-k}\,T_{\frac{17}{16}W_1}\circ T_{\frac{17}{16}W_1}\Big(\sum_{Q\in W_3}\chi_Q M_{Q^*}(\nabla^j v)\Big)$$

for all $0 \leqslant k \leqslant m$ with $c = c(\varepsilon, n, m)$.

Proof. The proof is similar to the one of Corollary 8.5.9. The application of $\sum_{Q \in \mathcal{W}_3} \chi_Q M_{Q^*}$ ensures that the L^1-averages over all cubes in $\mathcal{W}_1$ are calculated and transported from Q^* to Q. Then the values of the neighboring cubes Q_2 are transported by two applications of $T_{\frac{17}{16}\mathcal{W}_1}$ to Q. Together with Lemma 8.5.10 the claim follows. □

We are now prepared to prove our extension result.

Theorem 8.5.12. *Let Ω be an (ε, ∞)-domain and $m \in \mathbb{N}_0$. Then for every $p \in \mathcal{P}^{\log}(\mathbb{R}^n)$ the operator $\mathcal{E}^m$ defined in (8.5.7) is a bounded extension operator from $W^{m,p(\cdot)}(\Omega)$ to $W^{m,p(\cdot)}(\mathbb{R}^n)$. In particular,*

$$\|\mathcal{E}^m v\|_{W^{m,p(\cdot)}(\mathbb{R}^n)} \leqslant c \, \|v\|_{W^{m,p(\cdot)}(\Omega)} \tag{8.5.13}$$

for all $v \in W^{m,p(\cdot)}(\Omega)$, where c only depends on ε, $\operatorname{diam}(\Omega)$, n, m, and $c_{\log}(p)$. Moreover, if Ω is unbounded, then

$$\|\nabla^k \mathcal{E}^m v\|_{L^{p(\cdot)}(\mathbb{R}^n)} \leqslant c \, \|\nabla^k v\|_{L^{p(\cdot)}(\Omega)} \tag{8.5.14}$$

for all $v \in W^{m,p(\cdot)}(\Omega)$ and $0 \leqslant k \leqslant m$, where c only depends on ε, n, m, and $c_{\log}(p)$.

Proof. Let $v \in W^{m,p(\cdot)}(\Omega)$. Then $\mathcal{E}^m v \in W^{m,1}_{\mathrm{loc}}(\mathbb{R}^n)$ by the discussions before Lemma 8.5.8. It follows from Proposition 8.5.3 that the families $\mathcal{W}_1$, $\frac{17}{16}\mathcal{W}_1$ and $\frac{17}{16}\mathcal{W}_2$ are locally $(12^n + 1)$-finite. By Theorem 4.4.8 the corresponding averaging operators $T_{\mathcal{W}_1}$, $T_{\frac{17}{16}\mathcal{W}_1}$, $T_{\frac{17}{16}\mathcal{W}_2}$ are bounded on $L^{p(\cdot)}(\mathbb{R}^n)$. Moreover, it follows from $p \in \mathcal{P}^{\log}(\mathbb{R}^n)$ and Theorem 4.4.15 that the operator $f \mapsto \sum_{Q \in \mathcal{Q}} \chi_Q M_{Q^*} f$ is bounded on $L^{p(\cdot)}(\mathbb{R}^n)$. Now, the estimates for $\nabla^k \mathcal{E}^m v$ on $\mathbb{R}^n \setminus \Omega$ follow from Corollaries 8.5.9 and 8.5.11 using also that $\ell(Q) > \frac{\varepsilon}{16n} \operatorname{diam}(\Omega)$ for all $Q \in \mathcal{W}_2 \setminus \mathcal{W}_3$. The estimates on Ω follow from $\mathcal{E}^m v = v$ on Ω. If Ω is unbounded, then $\bigcup_{Q \in \mathcal{W}_3} Q = \mathbb{R}^n \setminus \Omega$ up to measure zero. So in this case it suffices to rely on Corollary 8.5.9, which results in the sharper estimates and the independence of the constants of $\operatorname{diam}(\Omega)$. □

Remark 8.5.15. The dependence of the constant in (8.5.13) on $\operatorname{diam}(\Omega)$ is similar as in Chua's paper [75]: the constant blows up as $\operatorname{diam}(\Omega) \to 0$.

Remark 8.5.16. The previous theorem can be directly generalized to the case of (ε, δ)-domains (see [75,224] for the definition). These sets are similar to (ε, ∞)-domains, except that (ε, δ)-domains do not have to be connected and the conditions on γ in Definition 8.5.1 have to be checked only for $x, y \in \Omega$ with $|x - y| \leqslant \delta$. Theorem 8.5.12 remains valid for such (ε, δ)-domains whose connected components have a diameter bounded away from zero.

8.6 Limiting Cases of Sobolev Embeddings*

The results so far have dealt mostly with the case $p^+ < n$, and to some extent with $p^- > n$. Obviously, it would be interesting to have one result which deals with all cases, irrespective of how p relates to n. Unfortunately, it has not yet been possible to archive such a result. In this section we consider results which are somewhat closer to the general case, in that they allow the critical value n to be reached, if not crossed; i.e. we consider the cases $p^- = n$ and $p^+ = n$.

Assume first that $n \leqslant p^- \leqslant p^+ < \infty$. Harjulehto and Hästö studied a simple sufficient condition for the embedding $W^{1,p(\cdot)}(\Omega) \hookrightarrow L^\infty(\Omega)$ to hold in a regular domain [188, Theorem 4.6]. We state here their result in a ball.

Theorem 8.6.1. *Suppose that B is a ball. If $p \in \mathcal{P}(B)$ is bounded such that*

$$p(x) \geqslant n + (n - 1 + \varepsilon) \frac{\log \log(c/\operatorname{dist}(x, \partial B))}{\log(c/\operatorname{dist}(x, \partial B))}$$

for some fixed $\varepsilon > 0$ and sufficiently large constant $c > 0$, then $W^{1,p(\cdot)}(B) \hookrightarrow L^\infty(B)$.

Futamura, Mizuta and Shimomura studied the Riesz potential in a similar situation. They showed in [166] that if

$$p(x) := \frac{n}{\alpha} + a \frac{\log(\log(e + 1/|x|))}{\log(1/|x|)},$$

where $a > \frac{n-\alpha}{\alpha^2}$, then the Riesz potential I_α is continuous at the origin. Note that this gives the same condition as in the previous theorem when $\alpha = 1$.

The next example and remark show that the growth condition in Theorem 8.6.1 is essentially sharp; we do not know what happens when $\varepsilon = 0$.

Example 8.6.2. Let $B := B\left(0, \frac{1}{16}\right) \subset \mathbb{R}^n$, $\varepsilon \in (0, n - 1)$ and suppose that

$$p(x) := n + (n - 1 - \varepsilon) \frac{\log \log(1/|x|)}{\log(1/|x|)}$$

for $x \in B \setminus \{0\}$ and $p(0) > n$. We show that $W^{1,p(\cdot)}(B) \not\subset C(B)$.

Define $u(x) = \cos\left(\log |\log |x||\right)$ for $x \in B \setminus \{0\}$ and $u(0) = 0$. Clearly u is not continuous at the origin. So we have to show that $u \in W^{1,p(\cdot)}(B)$. It is clear that u has partial derivatives, except at the origin. Since u is bounded it follows that $u \in L^{p(\cdot)}(B)$. We next estimate the gradient:

$$|\nabla u(x)| \leqslant \left| \sin\left(\log |\log |x||\right) \frac{1}{|x| \log |x|} \right| \leqslant \left| \frac{1}{|x| \log |x|} \right|.$$

We therefore find that

$$\int_B |\nabla u(x)|^{p(x)}\, dx \leqslant \int_B \frac{dx}{(|x||\log|x||)^{p(x)}}$$

$$= c \int_0^{\frac{1}{16}} \frac{r^{n-1}}{(r|\log r|)^{p(r)}}\, dr$$

$$\leqslant c \sum_{j=2}^{\infty} \int_{e^{-j-1}}^{e^{-j}} \frac{r^{n-1}}{(r|\log r|)^{p(r)}}\, dr.$$

Since $1/(r|\log r|) > 1$ we may increase the exponent p for an upper bound. In the annulus $B(0, e^{-j}) \setminus B(0, e^{-j-1})$ we have $j \leqslant \log(1/|x|) \leqslant j+1$. Since $y \mapsto \log(y)/y$ is decreasing we find that

$$p(x) \leqslant n + (n - 1 - \varepsilon)\frac{\log j}{j}$$

in the same annulus. We can therefore continue our previous estimate by

$$\int_B |\nabla u(x)|^{p(x)}\, dx \leqslant c \sum_{j=2}^{\infty} \int_{e^{-j-1}}^{e^{-j}} \frac{r^{n-1}\, dr}{(r|\log r|)^{n+(n-1-\varepsilon)\log(j)/j}}$$

$$\leqslant c \sum_{j=2}^{\infty} \int_{e^{-j-1}}^{e^{-j}} \frac{e^{-j(n-1)}\, dr}{((j-1)e^{-j-1})^{n+(n-1-\varepsilon)\log(j)/j}}$$

$$\leqslant c \sum_{j=2}^{\infty} e^{(n-1-\varepsilon)\log(j)} (j-1)^{-n-(n-1-\varepsilon)\log(j)/j}$$

$$= c \sum_{j=2}^{\infty} j^{-1-\varepsilon}(j-1)^{-(n-1-\varepsilon)\log(j)/j} \leqslant c \sum_{j=2}^{\infty} j^{-1-\varepsilon} < \infty.$$

Remark 8.6.3. Define $D := B\left(0, \frac{1}{16}\right) \setminus \{0\} \subset \mathbb{R}^n$ and let p be as in Example 8.6.2. Then the standard example $u(x) := \log|\log(x)|$ shows that $W^{1,p(\cdot)}(D) \not\hookrightarrow L^\infty(D)$, the calculations being as in the example.

If $p^- > n$, then Sobolev functions are known to be continuous. The case $p^- \leqslant n$ is more precarious. In the paper [164], Futamura and Mizuta showed that points where $p(x) = n$ are points of continuity of a weakly monotone Sobolev function provided p approaches n sufficiently fast (from below) in some neighborhood of x.

Next we want to generalize the Sobolev embedding Theorem 8.3.1 to the case $1 \leqslant p^- \leqslant p^+ = n$. Recall that the relevant classical result says that $W_0^{1,n}(\Omega) \hookrightarrow \exp L^{n'}(\Omega)$.

We follow the approach of Harjulehto and Hästö [190]. We define

$$F_p^*(t) := \sum_{j=1}^{\lfloor \frac{p^*}{n'} \rfloor - 1} \frac{1}{j!} |t|^{jn'} + \frac{1}{\lfloor \frac{p^*}{n'} \rfloor !} |t|^{p^*}$$

for $1 \leqslant p \leqslant n$, with the understanding that the last term disappears if $p = n$ and hence $F_n^*(t) = \exp(|t|^{n'}) - 1$. Here $\lfloor x \rfloor$ is the largest natural number that is less than or equal to x. Note that the function $F_{p(\cdot)}^*$ does not satisfy the Δ_2-condition if $p^+ = n$. Using this function we define a new Orlicz–Musielak space.

Definition 8.6.4. We define a convex modular by setting

$$\varrho_{L_*^{p(\cdot)}(\Omega)}(f) := \int_\Omega F_{p(x)}^*(f(x))\, dx,$$

where p is a variable exponent satisfying $p^+ \leqslant n$. The norm in $L_*^{p(\cdot)}(\Omega)$ is defined as usual,

$$\|f\|_{L_*^{p(\cdot)}(\Omega)} := \inf_{\lambda > 0} \left\{ \lambda > 0 : \varrho_{L_*^{p(\cdot)}(\Omega)}\left(\frac{f}{\lambda}\right) \leqslant 1 \right\}.$$

Let $\Omega \subset \mathbb{R}^n$ be bounded. This new variable exponent Lebesgue space of exponential type has the following obvious properties:

(a) If $p \in [1, n)$ is a constant, then $L_*^p(\Omega) \cong L^{p^*}(\Omega)$.
(b) If $p = n$, then $L_*^n(\Omega) \cong \exp L^{n'}$.

Thus we always have $W^{1,p}(\Omega) \hookrightarrow L_*^p(\Omega)$ for a constant exponent $1 \leqslant p \leqslant n$. We further note that $L_*^{p(\cdot)}(\Omega) \cong L^{p^*(\cdot)}(\Omega)$ if p is an exponent with $p^+ < n$. Thus also $W_0^{1,p(\cdot)}(\Omega) \hookrightarrow L_*^{p(\cdot)}(\Omega)$ for p in the same range (Theorem 8.3.1). Next we show that this embedding holds also when the restriction $p^+ < n$ is relaxed to $p^+ \leqslant n$.

Theorem 8.6.5. Let $\Omega \subset \mathbb{R}^n$ be bounded. Suppose that $p \in \mathcal{P}^{\log}(\Omega)$ with $1 \leqslant p^- \leqslant p^+ \leqslant n$. Then

$$\|u\|_{L_*^{p(\cdot)}(\Omega)} \leqslant c \|\nabla u\|_{p(\cdot)}$$

for every $u \in W_0^{1,p(\cdot)}(\Omega)$. The constant depends only on the dimension n, p and $\mathrm{diam}(\Omega)$.

One can also derive an analogous estimate for $\|u - \langle u \rangle_\Omega\|_{L_*^{p(\cdot)}(\Omega)}$ which holds for all $u \in W^{1,p(\cdot)}(\Omega)$. See [190] for details.

The proof of Theorem 8.6.5 is based on Theorem 8.3.1 ($1 \leqslant p^- \leqslant p^+ < n$) and the following proposition ($1 < p^- \leqslant p^+ \leqslant n$).

Proposition 8.6.6. *Suppose that $p \in \mathcal{P}^{\log}(\Omega)$ satisfies $1 < p^- \leqslant p^+ \leqslant n$. Then $\|u\|_{L_*^{p(\cdot)}(\Omega)} \leqslant c \|\nabla u\|_{p(\cdot)}$ for every $u \in W_0^{1,p(\cdot)}(\Omega)$. The constant depends only on n, p^-, $c_{\log}(p)$ and $|\Omega|$.*

Proof. In this proof it is necessary to keep close track on the dependence of constants on various exponents. We will therefore make the dependence on the upper bounds of the exponents explicit in our constants.

Let $u \in W_0^{1,p(\cdot)}(\Omega)$ be a function with $2(1 + |\Omega|)\|\nabla u\|_{p(\cdot)} \leqslant 1$. Then the claim follows if we can prove that $\varrho_{L_*^{p(\cdot)}(\Omega)}(\lambda u) \leqslant 1$ for some constant, which is $\lambda > 0$ independent of u. Recall that, $|u(x)| \leqslant c(n)I_1|\nabla u(x)|$ for almost every $x \in \Omega$ by Lemma 8.2.1 (a). Thus

$$
\begin{aligned}
\varrho_{L_*^{p(\cdot)}(\Omega)}(\lambda u) = \int_\Omega \sum_{j=1}^{\lfloor \frac{p^*(x)}{n'} \rfloor - 1} \frac{1}{j!} |\lambda u|^{jn'} \, dx \\
+ \int_{\{p<n\}} \frac{1}{\lfloor \frac{p^*(x)}{n'} \rfloor !} |\lambda u|^{p^*(x)} dx \\
\leqslant c \sum_{j=1}^{\infty} \frac{1}{j!} \int_{\{j \leqslant \frac{p^*}{n'}\}} (\lambda I_1 |\nabla u|)^{jn'} dx \\
+ \int_{\{p<n\}} \frac{1}{\lfloor \frac{p^*(x)}{n'} \rfloor !} (\lambda I_1 |\nabla u|)^{p^*(x)} dx.
\end{aligned}
$$

Fix the variable exponent q_j in such a way that $q_j^*(x) := \min\{jn', p^*(x)\}$ in Ω. Since $q_j \leqslant p$ we have $\|\nabla u\|_{q_j(\cdot)} \leqslant 2(1 + |\Omega|)\|\nabla u\|_{p(\cdot)} \leqslant 1$, and since $q_j^* = jn'$ in $\{j \leqslant \frac{p^*}{n'}\}$ we have

$$
\int_{\{j \leqslant \frac{p^*}{n'}\}} (\lambda I_1 |\nabla u|)^{jn'} dx \leqslant \lambda^{jn'} \int_\Omega (I_1 |\nabla u|)^{q_j^*(x)} \, dx.
$$

Now Lemma 6.1.6 applied with the exponent q_j and $k = \max\{\frac{j}{n-1}, 1\} \leqslant j$ yields

$$
\left(I_1 |\nabla u|(x)\right)^{q^*(x)} \leqslant c^{jn'} i^{\frac{q_j^*(x)}{(q+)'}} \left(M|\nabla u|(x)\right)^{q_j(x)}.
$$

Since $q_j^* \leqslant jn'$, we easily derive that $q_j^*(x)/(q_j^+)' \leqslant j - 1$. Hence we obtain

$$
\int\limits_{\{j \leqslant \frac{p^*}{n'}\}} (\lambda I_1 |\nabla u|)^{jn'} dx \leqslant (c_0 \lambda)^{jn'} j^{j-1} \int\limits_\Omega (M|\nabla u|)^{q_j(x)} dx
$$

$$
\leqslant c\, c_0^j \lambda^{in'} j^j \left(\int\limits_\Omega (M|\nabla u|)^{p(x)} dx + |\Omega| \right).
$$

By Corollary 4.3.11 the maximal operator is bounded, and hence we conclude that $\varrho_{L^{p(\cdot)}(\Omega)}(M|\nabla u|) \leqslant c$. It follows that

$$
\frac{1}{j!} \int\limits_{\{j \leqslant \frac{p^*}{n'}\}} (\lambda I_1 |\nabla u|)^{jn'} dx \leqslant c \frac{1}{j!} c_0^j \lambda^{jn'} j^j \leqslant c j^{-j-1/2} e^j c_0^j \lambda^{jn'} j^j \leqslant c c_1^j \lambda^{jn'}
$$

where we used Stirling's formula in the second step. The right-hand-side is bounded by 4^{-j} if we choose $\lambda \leqslant (c^{n'} 4 c_1)^{-1/n'}$. Therefore, we have control of the sum in the previous estimate:

$$
\sum_{j=1}^\infty \frac{1}{j!} \int\limits_{\{j \leqslant \frac{p^*}{n'}\}} (\lambda I_1 |\nabla u|)^{jn'} dx \leqslant \sum_{j=1}^\infty 4^{-j} \leqslant \frac{1}{2}.
$$

It remains to estimate the term

$$
\int\limits_{\{p<n\}} \frac{1}{\left\lfloor \frac{p^*(x)}{n'} \right\rfloor !} (\lambda I_1 |\nabla u|)^{p^*(x)} dx = \sum_{j=1}^\infty \frac{1}{j!} \int\limits_{\{j \leqslant \frac{p^*}{n'} < j+1\}} (\lambda I_1 |\nabla u|)^{p^*(x)} dx
$$

$$
\leqslant \sum_{j=1}^\infty \frac{1}{j!} \int\limits_\Omega \lambda^{jn'} (I_1 |\nabla u|)^{p_j^*(x)} dx, \qquad (8.6.7)
$$

where $p_j(x) := \min\left\{ p(x), \frac{nj+n}{n+j} \right\}$. Since $p_j \leqslant p$, we note that $\|\nabla u\|_{p_j(\cdot)} \leqslant 2(1+|\Omega|) \|\nabla u\|_{p(\cdot)} \leqslant 1$. By Lemma 6.1.6 we have

$$
\left(I_1 |\nabla u|(x) \right)^{p_j^*(x)} \leqslant c\, c_2^{p_j^*(x)} k^{p_j^*(x)/(p_j^+)'} \left(M|\nabla u|(x) \right)^{p_j(x)},
$$

where $k = \max\{ p_j^+/(n-p_j^+), 1 \}$. Since $p_j \leqslant \frac{nj+n}{n+j}$, we conclude that $k \leqslant j+1$ and $\frac{p_j^*(x)}{(p_j^+)'} \leqslant \frac{n(p_j^+ - 1)}{n - p_j^+} \leqslant j$. Therefore

$$
\int\limits_\Omega \left(I_1 |\nabla u|(x) \right)^{p_j^*(x)} dx \leqslant c\, (c_2 k)^j \int\limits_\Omega \left(M|\nabla u|(x) \right)^{p_j(x)} dx \leqslant c\, (c_2 j)^j,
$$

where we used the same arguments for M as in the previous paragraph. Using this in (8.6.7), with Stirling's formula as before, gives

$$\int_{\{p<n\}} \frac{1}{\lfloor \frac{p^*(x)}{n'} \rfloor !} (\lambda I_1 |\nabla u|)^{p^*(x)} dx \leqslant c \sum_{j=1}^{\infty} \frac{1}{j!} \lambda^{jn'} (c_2 j)^j \leqslant \frac{1}{2},$$

provided λ is chosen small enough. This completes the proof. $\qquad\square$

Proof of Theorem 8.6.5. We choose a Lipschitz function $\varphi : \Omega \to \mathbb{R}$ with $0 \leqslant \varphi \leqslant 1$, $\varphi = 1$ in $p^{-1}([1, \frac{4}{3}])$ and $\varphi = 0$ in $\Omega \setminus p^{-1}([1, \frac{5}{3}])$. This can be done since $p^{-1}([1, \frac{4}{3}])$ and $p^{-1}([\frac{5}{3}, n])$ are closed disjoint sets in Ω. Let $\psi := 1 - \varphi$. We write $A := \{\varphi > 0\}$ and $B := \{\psi > 0\}$, and define $p_1 := \min\{p, \frac{5}{3}\}$ and $p_2 := \max\{\frac{4}{3}, p\}$. Then $p_1 = p$ in A and $p_2 = p$ in B. To prove the claim, we calculate:

$$\begin{aligned}
\|u\|_{L_*^{p(\cdot)}(\Omega)} &\leqslant \|\varphi u\|_{L_1^{p*(\cdot)}(\Omega)} + \|\psi u\|_{L^{p_2(\cdot),*}(\Omega)} \\
&\leqslant c \|\nabla(\varphi u)\|_{L^{p_1(\cdot)}(\Omega)} + c \|\nabla(\psi u)\|_{L^{p_2(\cdot)}(\Omega)} \\
&\leqslant c \|u\|_{W^{1,p(\cdot)}(\Omega)},
\end{aligned}$$

where the second step follows from Theorem 8.3.1 and Proposition 8.6.6. Finally, we see that $\|u\|_{W^{1,p(\cdot)}(\Omega)} \leqslant c \|\nabla u\|_{L^{p(\cdot)}(\Omega)}$ by the Poincaré inequality, Theorem 8.2.4 (a). $\qquad\square$

Chapter 9
Density of Regular Functions

This chapter deals with the delicate question of when every function in a Sobolev space can be approximated by a more regular function, such as a smooth or Lipschitz continuous function. For the Lebesgue space, this question was solved in Theorem 3.4.12. An important fact is that log-Hölder continuity is sufficient for density of smooth functions. This is shown in Sect. 9.1. However, for the density question log-Hölder continuity is by no means necessary. Despite the contributions of many researchers, there remain substantial gaps in our understanding of this question. Indeed, it is fair to say that the results are in a transitory state and will hopefully be improved and unified in the future. We nevertheless endeavor to present in Sects. 9.2–9.4 the current state of the art, theorems by Fan, Wang and Zhao [142], Hästö [214] and Zhikov [397]. We also include an overview of the most important techniques and ideas used in the proofs.

The density of smooth functions was one of the questions that was considered early on (from 1986) in the context of minimizers of variational integrals. If such a minimizer turns out not to be smooth, then we say that the Lavrentiev phenomenon occurs. This was precisely the object of Zhikov's [392–394] studies. It turns out that the Lavrentiev phenomenon is related to the density of smooth functions in the corresponding function space [393].

Density of smooth functions was also one of the first questions investigated in the function space setting, as early as 1992, by Edmunds and Rákosník [120]. Samko [342, 343] and Diening [91] have shown, independently, that log-Hölder continuity of the exponent is sufficient for the density of smooth functions.

Zhikov [392] was also first to present a counter example to density, for a piece-wise constant exponent. Hästö [212] later gave a counter example with a uniformly continuous exponent. Other features of the Sobolev space determined by his exponent are that not quasievery point is a Lebesgue point (see Sect. 11.5), and that not every minimizer of the Dirichlet energy integral is continuous.

L. Diening et al., *Lebesgue and Sobolev Spaces with Variable Exponents*,
Lecture Notes in Mathematics 2017, DOI 10.1007/978-3-642-18363-8_9,
© Springer-Verlag Berlin Heidelberg 2011

9.1 Basic Results on Density

We start with some density results which hold under minimal restrictions on the exponent. Correspondingly, the spaces which are dense do not consist of very regular functions.

Lemma 9.1.1. *If $p \in \mathcal{P}(\Omega)$ is bounded, then bounded Sobolev functions are dense in $W^{1,p(\cdot)}(\Omega)$.*

Proof. Let $u \in W^{1,p(\cdot)}(\Omega)$. We define the truncation $u_m \in W^{1,p(\cdot)}(\Omega)$ (cf. Proposition 8.1.9) by

$$u_m(x) := \max\{\min\{u(x), m\}, -m\}$$

for $m > 0$. For

$$\int_\Omega |g|^{p(x)}\, dx < \infty,$$

we obtain $|\{x \in \Omega : |g(x)|^{p(x)} \geqslant m\}| \to 0$ as $m \to \infty$. If $|u(x)| \geqslant 1$ then $|u(x)| \leqslant |u(x)|^{p(x)}$ and thus $\{x \in \Omega : |u(x)| \geqslant m\} \subset \{x \in \Omega : |u(x)|^{p(x)} \geqslant m\}$ for all $m \geqslant 1$. Hence

$$\varrho_{1,p(\cdot)}\big(u - u_m\big) \leqslant \int_{\{x\in\Omega:|u(x)|\geqslant m\}} \big(|u|^{p(x)} + |\nabla u|^{p(x)}\big)\, dx \to 0$$

as $m \to \infty$. Since p is bounded this yields that $\|u - u_m\|_{W^{1,p(\cdot)}(\Omega)} \to 0$. □

Theorem 9.1.2. *If $p \in \mathcal{P}(\mathbb{R}^n)$ is bounded, then Sobolev functions with compact support in $\mathbb{R}^n$ are dense in $W^{k,p(\cdot)}(\mathbb{R}^n)$, $k \in \mathbb{N}$.*

Proof. Let us denote $B_t := B(0,t)$, $t \geqslant 1$. Let $\psi_r \in C_0^\infty(\mathbb{R}^n)$ be a cut-off function with $\psi_r = 1$ on B_r, $\psi_r = 0$ on $\mathbb{R}^n \setminus B_{r+1}$, $0 \leqslant \psi_r(x) \leqslant 1$ and $|\nabla^m \psi_r| \leqslant c$, $m = 0, \ldots, k$. We show that $u\psi_r \to u$ in $W^{k,p(\cdot)}(\mathbb{R}^n)$ as $r \to \infty$. Note first that

$$\|u - u\psi_r\|_{W^{k,p(\cdot)}(\mathbb{R}^n)} \leqslant \|u\|_{W^{k,p(\cdot)}(\mathbb{R}^n\setminus B_{r+1})} + \|u - u\psi_r\|_{W^{k,p(\cdot)}(B_{r+1}\setminus B_r)}.$$

The absolute continuity of the integral implies that

$$\varrho_{W^{k,p(\cdot)}(\mathbb{R}^n\setminus B_{r+1})}(u) \to 0$$

as $r \to \infty$. Since $p^+ < \infty$ we get that $\|u\|_{W^{k,p(\cdot)}(\mathbb{R}^n\setminus B_{r+1})} \to 0$ as $r \to \infty$. To handle the second term in the above inequality we observe that

$$\big|\nabla^m u - \nabla^m(u\psi_r)\big| \leqslant (1 - \psi_r)|u| + \sum_{j=1}^m |\nabla^j \psi_r|\,|\nabla^{m-j} u| \leqslant c \sum_{j=0}^m |\nabla^j u|$$

for $m = 0, \ldots, k$. This, the absolute continuity of the integral, and $p^+ < \infty$ imply

$$\varrho_{W^{k,p(\cdot)}(B_{r+1} \setminus B_r)}(u) \leqslant c \, \varrho_{W^{k,p(\cdot)}(\mathbb{R}^n \setminus B_r)}(u) \to 0$$

as $r \to \infty$. Since $p^+ < \infty$ also the second term in the above inequality converges to 0 as $r \to \infty$. □

Corollary 9.1.3. *If $p \in \mathcal{P}(\mathbb{R}^n)$ is bounded, then $W_0^{k,p(\cdot)}(\mathbb{R}^n) = W^{k,p(\cdot)}(\mathbb{R}^n)$, $k \in \mathbb{N}$.*

Combining the proofs of Lemma 9.1.1 and of Theorem 9.1.2 for $k = 1$ we immediately deduce:

Corollary 9.1.4. *If $p \in \mathcal{P}(\mathbb{R}^n)$ is bounded, then bounded Sobolev functions with compact support in $\mathbb{R}^n$ are dense in $W^{1,p(\cdot)}(\mathbb{R}^n)$.*

We now give an example where p is unbounded and functions with compact support are not dense in $W^{1,p(\cdot)}(\mathbb{R})$.

Example 9.1.5. Let $p(x) := \max\{|x|, 1\}$ and $u(x) := 1/2$ for every $x \in \mathbb{R}$. Then

$$\int_{-\infty}^{\infty} |u(x)|^{p(x)} \, dx = 2 \int_{1}^{\infty} \left(\frac{1}{2}\right)^x dx + 1 < \infty$$

and hence $u \in W^{1,p(\cdot)}(\mathbb{R})$. Let g be a function with compact support in $(-a, a)$. We find that

$$\varrho_{p(\cdot)}\left(\frac{u-g}{\lambda}\right) \geqslant \int_{a}^{\infty} \left(\frac{1}{2\lambda}\right)^x dx = \infty$$

for $0 < \lambda \leqslant \frac{1}{2}$ and hence $\|u - g\|_{L^{p(\cdot)}(\mathbb{R})} \geqslant \frac{1}{2}$.

Next we give short proofs for the density of smooth functions based on the boundedness of convolution. For Lebesgue spaces this has been shown in Corollary 4.6.5.

Theorem 9.1.6. *Let $p \in \mathcal{P}$ be a bounded exponent. If $p \in \mathcal{A}$ or $p \in \mathcal{P}^{\log}$, then $C_0^\infty(\mathbb{R}^n)$ is dense in $W^{k,p(\cdot)}(\mathbb{R}^n)$, $k \in \mathbb{N}$.*

Proof. Let $u \in W^{k,p(\cdot)}(\mathbb{R}^n)$ and let $\varepsilon > 0$ be arbitrary. By Theorem 9.1.2 we may assume that u is has compact support in $\mathbb{R}^n$. Let ψ_ε be the standard mollifier. Thus $u * \psi_\varepsilon$ belongs to $C_0^\infty(\mathbb{R}^n)$ and

$$\nabla^m(u * \psi_\varepsilon) - \nabla^m u = (\nabla^m u) * \psi_\varepsilon - \nabla^m u$$

for $m = 1, \ldots, k$. By Theorem 4.6.4 we obtain that $\|u * \psi_\varepsilon - u\|_{L^{p(\cdot)}(\mathbb{R}^n)} \to 0$
$\varepsilon \to 0$, and

$$\|\nabla^m(u * \psi_\varepsilon) - \nabla^m u\|_{L^{p(\cdot)}(\mathbb{R}^n)} = \|\nabla^m u * \psi_\varepsilon - \nabla^m u\|_{L^{p(\cdot)}(\mathbb{R}^n)} \to 0$$

for $m = 1, \ldots, k$, as $\varepsilon \to 0$.   □

From this we easily derive the density of smooth functions also in "nice" domains. Recall that extension domains include in particular domains with Lipschitz boundary (see Sect. 8.5).

Theorem 9.1.7. *Let Ω be an extension domain. Assume that $p \in \mathcal{P}^{\log}(\Omega)$ satisfies $1 \leqslant p^- \leqslant p^+ < \infty$. Then $C^\infty(\overline{\Omega})$ is dense in $W^{k,p(\cdot)}(\Omega)$, $k \in \mathbb{N}$.*

Proof. We extend first p to all of $\mathbb{R}^n$ by Proposition 4.1.7. Let $u \in W^{k,p(\cdot)}(\Omega)$. Since Ω is an extension domain, we find $\tilde{u} \in W^{k,p(\cdot)}(\mathbb{R}^n)$ with $\|\tilde{u}\|_{W^{k,p(\cdot)}(\mathbb{R}^n)} \leqslant c\|u\|_{W^{k,p(\cdot)}(\Omega)}$. Due to Theorem 9.1.6 we can choose $\tilde{u}_\varepsilon \in C_0^\infty(\mathbb{R}^n)$ with $\tilde{u}_\varepsilon \to \tilde{u}$ in $W^{k,p(\cdot)}(\mathbb{R}^n)$. We set $u_\varepsilon := \tilde{u}_\varepsilon|_\Omega$. Then

$$\|u - u_\varepsilon\|_{W^{1,p(\cdot)}(\Omega)} \leqslant \|\tilde{u} - \tilde{u}_\varepsilon\|_{W^{1,p(\cdot)}(\mathbb{R}^n)} \to 0,$$

so $u_\varepsilon \in C^\infty(\overline{\Omega})$ are the required approximating functions.   □

For arbitrary open sets we still can prove that smooth Sobolev functions are dense. More precisely we have:

Theorem 9.1.8. *Let $p \in \mathcal{P}(\Omega)$ be a bounded exponent. If $p \in \mathcal{P}^{\log}(\Omega)$ or $p \in \mathcal{A}$, then $C^\infty(\Omega) \cap W^{k,p(\cdot)}(\Omega)$ is dense in $W^{k,p(\cdot)}(\Omega)$, $k \in \mathbb{N}$.*

Proof. Let $u \in W^{k,p(\cdot)}(\Omega)$. Fix $\varepsilon > 0$ and define $\Omega_0 := \emptyset$ and

$$\Omega_m = \left\{ x \in \Omega : \operatorname{dist}(x, \partial\Omega) > \tfrac{1}{m} \right\} \cap B(x_0, m)$$

for $m = 1, 2, \ldots$ and a fixed $x_0 \in \Omega$. We write

$$U_m := \Omega_{m+1} \setminus \overline{\Omega}_{m-1} \text{ for } m = 1, 2, \ldots.$$

Let (ξ_m) be a partition of unity corresponding to the covering (U_m), i.e. $\xi_m \in C_0^\infty(U_m)$ and $\sum_{m=1}^\infty \xi_m(x) = 1$ for every $x \in \Omega$ (Theorem 1.4.7). Let ψ_δ be a standard mollifier. For every m there exists δ_m such that

$$\operatorname{spt}\left((\xi_m u) * \psi_{\delta_m}\right) \subset U_m$$

and Theorem 4.6.4 yields, by choosing a smaller δ_m if necessary, that

$$\|(\xi_m u) - (\xi_m u) * \psi_{\delta_m}\|_{W^{k,p(\cdot)}(\Omega)} \leqslant \varepsilon 2^{-m}.$$

We define

$$u_\varepsilon := \sum_{m=1}^{\infty} (\xi_m u) * \psi_{\delta_m}.$$

Every point $x \in \Omega$ has a neighborhood such that the above sum has only finitely many nonzero terms and thus $u_\varepsilon \in C^\infty(\Omega)$. Furthermore, this is an approximating sequence, since

$$\|u - u_\varepsilon\|_{W^{k,p(\cdot)}(\Omega)} \leqslant \sum_{m=1}^{\infty} \|\xi_m u - (\xi_m u) * \psi_{\delta_m}\|_{W^{k,p(\cdot)}(\Omega)} \leqslant \varepsilon. \qquad \square$$

Another simple situation was considered by Harjulehto and Hästö [187], who showed that smooth functions are dense on open intervals of the real line irrespective of the variation of the bounded exponent. This result is based on first approximating the derivative, which is an $L^{p(\cdot)}$ function in this case.

9.2 Density with Continuous Exponents

Zhikov [397] proved that log-Hölder continuity is not a necessary continuity modulus for density of smooth functions, but that in fact a slightly weaker modulus will suffice. We present his results in this section. As far as we know, the optimal modulus of continuity for density is unknown, although Corollary 9.2.7 shows that Zhikov's modulus of continuity is at least close to optimal.

The next simple lemma shows that it suffices to prove the density of Lipschitz functions, since the density of smooth functions always follows from this.

Lemma 9.2.1. *Let $p \in \mathcal{P}(\Omega)$ be bounded. If $W^{1,\infty}(\Omega)$ is dense in $W^{1,p(\cdot)}(\Omega)$, then so are smooth Sobolev functions.*

Proof. By the assumption, it suffices to show that every Lipschitz function can be approximated by smooth functions. Let $u \in W^{1,p(\cdot)}(\Omega)$ be a K-Lipschitz function. Arguing as in Theorem 9.1.1 and Corollary 9.1.4, we may assume that u is bounded and has bounded support in $\mathbb{R}^n$. By McShane extension [289] we may extend u as a bounded K-Lipschitz function to $\mathbb{R}^n$ and by using a suitable cut off function, that is one in spt u, we may assume that the extension has a compact support in $\mathbb{R}^n$. Next, we let u_ε be a standard mollification of u. Clearly $u_\varepsilon \to u$ and $\nabla u_\varepsilon \to \nabla u$ almost everywhere. Since $u_\varepsilon \in L^\infty(\Omega)$, and $|\nabla u_\varepsilon|$ is bounded by K, and all functions vanish outside a bounded set determined by the support of u, we see that the claim follows by the theorem of dominated convergence. $\qquad \square$

Theorem 9.2.2 (Theorem 3.1, [397]). *Suppose that $p \in \mathcal{P}(\mathbb{R}^n)$ has a modulus of continuity ω which satisfies*

$$\int_0^{1/2} t^{-1+\frac{n}{p^-}\omega(t)} \, dt = \infty$$

and that $1 < p^- \leqslant p^+ < \infty$. Then smooth compactly supported functions are dense in the Sobolev space $W^{1,p(\cdot)}(\mathbb{R}^n)$.

Let us briefly consider the condition in the theorem before moving on to the proof. Suppose that p is log-Hölder continuous. Then we may take $\omega(t) = c/\log(e + 1/t)$. The integral in the previous theorem becomes

$$\int_0^{1/2} t^{-1+\frac{n}{p^-}\omega(t)} \, dt = \int_0^{1/2} t^{-1+c/\log(e+1/t)} \, dt \geqslant \int_0^{1/2} t^{-1} e^{-c} \, dt = \infty,$$

so the condition is satisfied. Moreover, some weaker moduli of continuity satisfy the condition as well; for instance ω with

$$\lim_{t \to 0} \omega(t) \frac{\log(1/t)}{\log\log(1/t)} =: c$$

satisfies the condition if and only if $0 < c < \frac{p^-}{n}$.

We now proceed with the proof of the theorem. The proof largely follows the arguments given in [397], although minor changes have been made in the interest of clarity.

Proof of Theorem 9.2.2. By Theorem 9.1.2 we know that Sobolev functions with compact support are dense. Hence it suffices to consider the case of a function $u \in W^{1,p(\cdot)}(\mathbb{R}^n)$ with support in the ball $B(0, R)$. Thus we have that $u \in W^{1,p^-}(\mathbb{R}^n)$ and the following calculations are justified. Define $f := |u| + |\nabla u|$. By dividing the function be a suitable constant, we may assume without loss of generality that $\|f\|_{p(\cdot)} \leqslant 20^{-n}/(2 + 2|B(0, R)|)$.

As usual, we introduce the level-sets of the maximal function:

$$E(\lambda, Mf) := \{x \in \mathbb{R}^n : Mf(x) \geqslant \lambda\}$$

for $\lambda \geqslant 0$. We denote $F_\lambda = \mathbb{R}^n \setminus E(\lambda, Mf)$. By [129, p. 255] we know that u is $c\lambda$-Lipschitz in F_λ, where the constant c depends only on the dimension. By McShane extension [289] we get a $c\lambda$-Lipschitz function u_λ on $\mathbb{R}^n$ which agrees with u on the set F_λ. It follows that $\nabla u = \nabla u_\lambda$ a.e. in F_λ, and hence

$$\int\limits_{\mathbb{R}^n} |\nabla u - \nabla u_\lambda|^{p(x)}\, dx = \int\limits_{E(\lambda, Mf)} |\nabla u - \nabla u_\lambda|^{p(x)}\, dx$$

$$\leqslant c \int\limits_{E(\lambda, Mf)} |\nabla u|^{p(x)}\, dx + c \int\limits_{E(\lambda, Mf)} \lambda^{p(x)}\, dx.$$

Since $|\nabla u| \in L^{p(\cdot)}(\mathbb{R}^n)$ and $|E(\lambda, Mf)| \to 0$, it is clear that the first integral tends to zero as $\lambda \to \infty$. So it remains to control the second term. We do this by showing that there exists a sequence (λ_i) tending to infinity such that

$$\Theta(\lambda_i) := \int\limits_{E(\lambda_i, Mf)} \lambda_i^{p(x)}\, dx \to 0. \qquad (9.2.3)$$

Then the previous inequality implies that $\nabla u_{\lambda_i} \to \nabla u$ in $L^{p(\cdot)}(\mathbb{R}^n)$. Since p is uniformly continuous and the functions vanish outside $B(0, 2R)$, the Poincaré inequality in $B(0, 2R)$ (Theorem 8.2.18) implies that $u_{\lambda_i} \to u$ in $L^{p(\cdot)}(\mathbb{R}^n)$. Hence u_{λ_i} is an approximating Lipschitz sequence, and claim then follows by Lemma 9.2.1. We will show that there exists a continuous non-negative function θ such that

$$\int\limits_2^\infty \theta(\lambda)\, d\lambda = \infty \quad \text{and} \quad \int\limits_2^\infty \theta(\lambda)\Theta(\lambda)\, d\lambda < \infty,$$

where Θ is as in (9.2.3). If $\liminf_{\lambda \to \infty} \Theta(\lambda) > 0$, then $\Theta(\lambda) > c > 0$, so this is not possible. Thus $\liminf_{\lambda \to \infty} \Theta(\lambda) = 0$, which implies the existence of a suitable sequence (λ_i) satisfying (9.2.3).

By the basic covering theorem (cf. Theorem 1.4.5) there exist disjoint balls $B_i := B(x_i, r_i)$, $r_i < 1/10$, such that $\{5B_i\}$ covers $E(\lambda, Mf)$ and

$$\int\limits_{B_i} f\, dx > \lambda |B_i|$$

for each i. We may split B_i into a part where $f < \frac{\lambda}{2}$ and another part which is contained in $E(\frac{\lambda}{2}, f)$. Hence

$$\lambda |B_i| < \int\limits_{B_i \cap E(\frac{\lambda}{2}, f)} f\, dx + \frac{\lambda}{2} |B_i|.$$

Note that we are now using the level sets of f, not of Mf. From this we conclude that

$$\frac{\lambda}{2} \leqslant \frac{1}{|B_i|} \int\limits_{B_i \cap E(\frac{\lambda}{2}, f)} f \, dx$$

We combine this estimate with Hölder's inequality for p^- to deduce that

$$\frac{\lambda}{2} < \left(\frac{1}{|B_i|} \int\limits_{B_i} f^{p^-} \, dx \right)^{\frac{1}{p^-}} \leqslant |B_i|^{-\frac{1}{p^-}} 20^{-n}, \qquad (9.2.4)$$

where we used that $\|f\|_{p^-} \leqslant 2(1 + |B(0,R)|)\|f\|_{p(\cdot)} \leqslant 20^{-n}$ in the second inequality. Define next $p_i := p^+_{5B_i}$. For $\lambda \geqslant 1$ we have

$$\Theta(\lambda) \leqslant \sum_{i=1}^{\infty} \int\limits_{5B_i} \lambda^{p(x)} \, dx \leqslant 5^n \sum_{i=1}^{\infty} \lambda^{p_i} |B_i| \leqslant c \sum_{i=1}^{\infty} \int\limits_{B_i \cap E(\frac{\lambda}{2}, f)} \lambda^{p_i - 1} f(x) \, dx.$$

Since p has continuity modulus ω, we have $\lambda^{p_i} \leqslant c \lambda^{p(x) + \omega(5r_i)}$ for $x \in 5B_i$. Using this in the previous estimate gives

$$\Theta(\lambda) \leqslant c \sum_{i=1}^{\infty} \lambda^{\omega(5r_i)} \int\limits_{B_i \cap E(\frac{\lambda}{2}, f)} \lambda^{p(x)-1} f(x) \, dx.$$

From (9.2.4) we calculate $5r_i \leqslant \lambda^{-p^-/n}$; then we see that $\omega(5r_i) \leqslant \omega(\lambda^{-p^-/n})$. Now our previous estimate becomes

$$\Theta(\lambda) \leqslant c \lambda^{\omega(\lambda^{-p^-/n})} \sum_{i=1}^{\infty} \int\limits_{B_i \cap E(\frac{\lambda}{2}, f)} \lambda^{p(x)-1} f(x) \, dx$$

$$\leqslant c \frac{1}{\theta(\lambda)} \int\limits_{E(\frac{\lambda}{2}, f)} \lambda^{p(x)-2} f(x) \, dx,$$

where we defined $\theta(\lambda) := \lambda^{-1-\omega(\lambda^{-p^-/n})}$. Multiplying this inequality by $\theta(\lambda)$ and integrating with respect to λ over $[2, \infty)$ gives

$$\int\limits_{2}^{\infty} \theta(\lambda) \Theta(\lambda) \, d\lambda \leqslant c \int\limits_{2}^{\infty} \int\limits_{E(\frac{\lambda}{2}, f)} \lambda^{p(x)-2} f(x) \, dx \, d\lambda.$$

From this we continue by

$$\int\limits_{2}^{\infty} \theta(\lambda)\Theta(\lambda)\,d\lambda \leqslant c \int\limits_{2}^{\infty}\int\limits_{\mathbb{R}^n} \lambda^{p(x)-2} f(x)\chi_{E(\frac{\lambda}{2},f)}(x)\,dx\,d\lambda$$

$$= \int\limits_{\mathbb{R}^n}\int\limits_{2}^{\infty} \lambda^{p(x)-2} f(x)\chi_{E(\frac{\lambda}{2},f)}(x)\,d\lambda\,dx$$

$$\leqslant \int\limits_{\mathbb{R}^n} \frac{2^{p(x)-1}}{p(x)-1}\big(f(x)^{p(x)} - f(x)\big)\chi_{\{f\geqslant 1\}}(x)\,dx.$$

Since $1 < p^- \leqslant p^+ < \infty$ and f has bounded support, we conclude that the right-hand side of the previous inequality is finite.

The change of variables $t = \lambda^{-p^-/n}$ shows that the condition $\int \theta\,d\lambda = \infty$ is equivalent with the assumption

$$\int\limits_{0}^{1/2} t^{-1+\frac{n}{p^-}\omega(t)}\,dt = \infty$$

in the theorem, so we have found a suitable function θ and are thus done with the proof. □

The next result, which is [397, Theorem 5.1], gives us an example of a variable exponent Sobolev spaces in which not every function can be approximated by smooth functions. Earlier examples are due to Zhikov [392] and Hästö [212]. By a *sector* we mean a set of the type $\{x \in B(0,1) : \langle x, z\rangle > |x|\}$ for some vector $z \in \mathbb{R}^2 \setminus B(0,1)$.

Theorem 9.2.5. *Consider $\Omega := B(0,1) \subset \mathbb{R}^2$ and let A_1, A_2, A_3 and A_4 be four disjoint sectors numbered in clockwise order. If $p \in \mathcal{P}(\Omega)$ with*

$$\int\limits_{A_1\cup A_3} |x|^{-p'(x)}\,dx < \infty \quad\text{and}\quad \int\limits_{A_2\cup A_4} |x|^{-p(x)}\,dx < \infty,$$

then smooth functions are not dense in $W^{1,p(\cdot)}(\Omega)$.

Proof. We use polar coordinates, (r,θ), and denote $A_i(R) = A_i \cap B(0,R)$. We denote by Θ_i an interval of θ-values such that $\{(r,\theta): r \in (0,1), \theta \in \Theta_i\} = A_i$. Let $u \in C^\infty(\Omega)$. Then

$$u(R,\theta) = \int\limits_{0}^{R} \frac{\partial u}{\partial r}(r,\theta)\,dr + u(0).$$

Integration of the previous equation with respect to θ over Θ_1 gives

$$\int_{\Theta_1} |u(R,\theta) - u(0)| \, d\theta \leqslant \int_{A_1(R)} |\nabla u| \, r^{-1} \, dr$$

$$\leqslant \int_{A_1(R)} |\nabla u|^{p(x)} \, dr + \int_{A_1(R)} r^{-p'(x)} \, dr,$$

where we applied Young's inequality in the second step. By assumption, $|\nabla u|^{p(x)}, r^{-p'(x)} \in L^1$, so we see that

$$\lim_{R \to 0} \fint_{\Theta_1} |u(R,\theta)| \, d\theta = u(0).$$

Since the average over arcs tends to $u(0)$, the same holds for the average over sectors, hence

$$\lim_{R \to 0} \fint_{A_1(R)} |u(x)| \, dx = u(0).$$

The same argument works for $A_3(R)$, hence

$$\lim_{R \to 0} \fint_{A_1(R)} |u(x)| \, dx = \lim_{R \to 0} \fint_{A_3(R)} |u(x)| \, dx.$$

The last claim holds also for Sobolev functions that can be approximated by smooth functions.

However, consider the function $u(r,\theta) = \Psi(\theta)$, where $\Psi \in C^\infty(\mathbb{R})$ is a 2π-periodic function which equals 0 in A_1, 1 in A_3 and whose derivative is supported in $A_2 \cup A_4$. Then $|\nabla u| \leqslant cr^{-1}\chi_{A_2 \cup A_4}$ where c does not depend on r, and the assumption of the theorem implies that $|\nabla u| \in L^{p(\cdot)}(\Omega)$. The function u is bounded, hence in $W^{1,p(\cdot)}(\Omega)$. On the other hand,

$$0 = \lim_{R \to 0} \fint_{\Theta_1} |u(R,\theta)| \, d\theta \neq \lim_{R \to 0} \fint_{\Theta_3} |u(R,\theta)| \, d\theta = 1$$

for this function. Thus, by what was said above, u cannot be approximated by smooth functions.   □

Example 9.2.6. Let $B(0,1) \subset \mathbb{R}^2$ and let A_i, $i = 1,2,3,4$, denote the quadrants in $B(0,1)$. The exponent is defined as $p_1 > 2$ in A_1 and A_3, and $p_2 < 2$ in A_2 and A_4. Then clearly the assumptions of the previous theorem are satisfied. This was the first example of non-density, presented by Zhikov in [392]. A function u from the theorem which cannot be approximated by smooth functions in this setting is shown in Fig. 9.1.

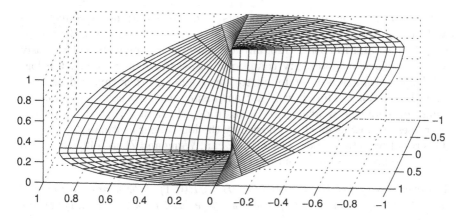

Fig. 9.1 A function which cannot be approximated by a continuous function

Corollary 9.2.7. *Consider $B(0,1) \subset \mathbb{R}^2$ and let A_1, A_2, A_3 and A_4 be four disjoint sectors numbered in clockwise order. Let ω be a modulus of continuity and suppose that $p \in \mathcal{P}(B(0,1))$ satisfies $p(x) := 2 + \omega(|x|)$ on $A_1 \cup A_3$ and $p(x) := 2 - \omega(|x|)$ on $A_2 \cup A_4$. If*

$$\int_0^{1/2} t^{-1+\omega(t)-\omega(t)^2} \, dt < \infty,$$

then smooth functions are not dense in $W^{1,p(\cdot)}(B(0,1))$.

Proof. We check that the exponent p satisfies the conditions of the previous theorem. We have

$$\int_{A_1 \cup A_3} |x|^{-p'(x)} \, dx = c \int_0^1 r^{-(2+\omega(r))'} r \, dr = c \int_0^1 r^{-1+\omega(r)-\frac{\omega(r)^2}{1+\omega(r)}} \, dr < \infty.$$

The other integral is handled similarly.   □

9.3   Density with Discontinuous Exponents*

We have seen in the previous section that we have a quite good understanding of the modulus of continuity necessary for density results. However, it turns out that we have density also in many other cases, including many discontinuous exponents.

One way to approach such exponents is to use smoothing by convolution with an appropriately chosen, anisotropic kernel. This idea was first used by Edmunds and Rákosník [120] already in 1992 for exponents which satisfy a suitable cone monotonicity condition. As described in Sect. 9.1, the log-Hölder condition was later found to imply among other things density of smooth functions. Thus it was not surprising that the cone monotonicity condition could be weakened to a condition which says that the exponent decreases at most as much as allowed by the log-Hölder continuity condition. This result was stated independently by Fan, Wang and Zhao [142] and Hästö [214]. The proof below is from the latter article.

Lemma 9.3.1. *Let $B' := B(0, 7/6)$, $B := B(0, 1)$, $\hat{e}$ be a unit vector, and let $p \in \mathcal{P}(B')$ be bounded. Suppose that there exist $K \geqslant 0$, $r \in (0, 1/12)$ and $h \in (0, 1)$ such that*

$$p(y) - p(x) \geqslant -\frac{K}{\log(e + 1/|x - y|)}$$

for every $x \in B$ when y lies in the cone

$$\bigcup_{0 < t \leqslant r} B(x + t\hat{e}, ht).$$

Then for every $u \in W^{1,p(\cdot)}(B')$ there exists a sequence (u_i) in $C^\infty(B) \cap W^{1,p(\cdot)}(B)$ such that $u_i \to u$ in $W^{1,p(\cdot)}(B)$.

Proof. Let $\psi \in C_0^\infty(B)$ be a non-negative function with $\int_B \psi \, dx = 1$. For an integrable function $u \colon B' \to \mathbb{R}$ and $\delta \in (0, r)$ we define

$$R_\delta u(x) = \int_B u(x + \delta(hz + \hat{e}))\psi(z) \, dz.$$

The usual integration-by-parts argument shows that $R_\delta u$ is smooth in B.

For $u \in L^{p(\cdot)}(B')$ we show that $\|R_\delta u - u\|_{p(\cdot)} \to 0$. In fact it suffices to show that $\varrho_{p(\cdot)}(R_\delta u - u) \to 0$, since p is bounded. Using that ψ is bounded, we estimate

$$\varrho_{p(\cdot)}(R_\delta u - u) = \int_B \left| \int_B \left(u(x + \delta(hz + \hat{e})) - u(x) \right) \psi(z) \, dz \right|^{p(x)} dx$$

$$\leqslant c \int_B \left(\int_B |u(x + \delta(hz + \hat{e})) - u(x)| \, dz \right)^{p(x)} dx =: c \, I$$

Let us denote $B^\delta := B(x + \delta\hat{e}, h\delta)$ and $q := \min\{p_{B^\delta}^-, p_B^-\}$. Using Hölder's inequality for the fixed exponent q and the fact that $\int_{B^\delta} |u(y) - u(x)|^q \, dy \leqslant 1$

for small enough δ (uniformly), we find that

$$
\begin{aligned}
I &= \int_B \left(c \fint_{B^\delta} |u(y) - u(x)|\, dy \right)^{p(x)} dx \\
&\leqslant c \int_B \left(\fint_{B^\delta} |u(y) - u(x)|^q\, dy \right)^{p(x)/q} dx \\
&\leqslant c \int_B (h\delta)^{-np(x)/q} \left(\int_{B^\delta} |u(y) - u(x)|^q\, dy \right)^{p(x)/q} dx \\
&\leqslant c \int_B (h\delta)^{-n(p(x)-q)/q} \fint_{B^\delta} |u(y) - u(x)|^q\, dy\, dx.
\end{aligned}
$$

By assumption we have

$$
q \geqslant p_{B^\delta}^- \geqslant p(x) - \frac{K}{\log(e + 1/|x - y|)} \geqslant p(x) - \frac{K}{\log(e + 1/\delta)}.
$$

Thus $\delta^{q-p(x)}$ is bounded by a constant independent of δ and x; of course, $h^{q-p(x)} \leqslant h^{1-p^+} \leqslant c(h)$ even without the assumption on p in the lemma. Using this, we continue our previous estimate:

$$
\begin{aligned}
&\int_B \left(\int_B |u(x + \delta\,(hz + \hat{e})) - u(x)|\, dz \right)^{p(x)} dx \\
&\qquad \leqslant c \int_B \fint_{B^\delta} |u(y) - u(x)|^q\, dy\, dx \\
&\qquad = c \int_B \int_B |u(x + \delta\,(hz + \hat{e})) - u(x)|^q\, dx\, dz.
\end{aligned}
$$

Now we continue as in the proof by Edmunds–Rákosník. Fix $\varepsilon > 0$. Since $1 + |u(x)|^{p(x)}$ is an integrable function, we can choose $\tau > 0$ such that

$$
\int_V 1 + |u(x)|^{p(x)}\, dx < \varepsilon/2
$$

for every $V \subset B$ with $|V| < \tau$. For a fixed $z \in B$ this implies that

$$
\int_V 2 + |u(x + \delta\,(hz + \hat{e}))|^{p(x+\delta\,(hz+\hat{e}))} + |u(x)|^{p(x)}\, dx < \varepsilon,
$$

for $|V| < \tau$, since the translation of V also satisfies $|V + \delta(hz + \hat{e})| < \tau$. Since u is measurable there exists, by Luzin's theorem, an open set $U \subset B(7/6)$ such that u is continuous in $B(7/6) \setminus U$ and $|U| < \tau/2$. By choosing δ small enough we get for all $x, y \in \overline{B} \setminus U$ with $|x - y| < \delta$ that $|u(y) - u(x)| < \varepsilon$. For $z \in B$ we denote by U_z the set of those points $x \in B$ for which $x \in U$ or $x + \delta(hz + \hat{e}) \in U$. Note that $|U_z| < \tau$ for every z. We find that

$$\int_B \int_B |u(x + \delta(hz + \hat{e})) - u(x)|^q \, dx \, dz$$

$$\leqslant \int_B \int_B \varepsilon^{p_{\overline{B}}} \, dx \, dz + \int_B \int_{U_z} |u(x + \delta(hz + \hat{e})) - u(x)|^q \, dx \, dz$$

$$\leqslant c \int_B \int_B \max\{\varepsilon, \varepsilon^{p^+}\} \, dx \, dz + \int_B \int_{U_z} |u(x + \delta(hz + \hat{e}))|^{p_{\overline{B}\delta}} + |u(x)|^{p_{\overline{B}\delta}} \, dx \, dz$$

$$\leqslant c \max\{\varepsilon, \varepsilon^{p^+}\} + \int_B \int_{U_z} 2 + |u(x + \delta(hz + \hat{e}))|^{p(x + \delta(hz + \hat{e}))} + |u(x)|^{p(x)} \, dx \, dz$$

$$\leqslant \max\{\varepsilon, \varepsilon^{p^+}\} + \int_B \varepsilon \, dz,$$

where, for the third inequality, we used that

$$|u(x)|^{p_{\overline{B}\delta}} \leqslant 1 + |u(x)|^{p(x)},$$

and similarly for $|u(x + \delta(hz + \hat{e}))|^{p_{\overline{B}}}$. Thus we have an upper bound which tends to zero with ε. To complete the proof of the lemma we still have to show that $\|R_\delta u - u\|_{1,p(\cdot)} \to 0$ for a function u in the Sobolev space. This follows easily by applying the $L^{p(\cdot)}$-result to u and $|\nabla u|$ separately, since $\nabla[R_\delta u] = R_\delta[\nabla u]$.   □

The idea of the next proof is to patch up the balls from the previous lemma following the proof of [120, Theorem 1].

Theorem 9.3.2. *Let $\Omega \subset \mathbb{R}^n$. Suppose that for every point $x \in \Omega$ there are four quantities*

$$r_x \in \left(0, \tfrac{1}{2}\min\{1, d(x, \partial\Omega)\}\right),$$

$h_x \in (0, 1)$, $\xi_x \in S(0, 1)$ *and* $K_x \in [0, \infty)$ *such that*

$$p(z) - p(y) \geqslant -\frac{K_x}{\log(e + 1/|y - z|)}$$

for every point $y \in B(x, r_x)$, *where* z *lies in the cone*

$$C(y) = \bigcup_{0 < t \leqslant r_x} B(y + t\xi_x, h_x t).$$

Then $C^\infty(\Omega) \cap W^{1,p(\cdot)}(\Omega)$ *is dense in* $W^{1,p(\cdot)}(\Omega)$.

Proof. Let $B_x := B(x, r_x/10)$ for $x \in \Omega$. By the basic covering theorem (Theorem 1.4.5) we find a countable subfamily consisting of disjoint $B'_i = B_{x_i}$ such that $\bigcup_{i=1}^\infty 5B'_i = \Omega$. We define $B_i = 6B'_i$ and $B^*_i = 7B'_i$. We note that we still have

$$\bigcup_{i=1}^\infty B_i = \bigcup_{i=1}^\infty \overline{B^*_i} = \Omega$$

and we see (by the disjointness of the balls B'_i) that any point $x \in \Omega$ is contained in at most N balls B^*_i. Thus there exists a partition of unity by smooth functions ψ_i such that ψ_i are supported in B_i and $|\nabla \psi_i|$ is bounded by $L_i \geqslant 1$.

Fix $u \in W^{1,p(\cdot)}(\Omega)$ and $\varepsilon > 0$. By Lemma 9.3.1 we can choose $v_i \in W^{1,p(\cdot)}(B_i) \cap C^\infty(B_i)$ such that

$$\|u - v_i\|_{W^{1,p(\cdot)}(B_i)} < 2^{-i}\varepsilon/L_i.$$

Define $v := \sum_{i=1}^\infty \psi_i v_i$. Since at most finitely many of the ψ_i are non-zero in a neighborhood of any point, we see that v is smooth. We easily calculate that

$$\|u - v\|_{W^{1,p(\cdot)}(B_i)} \leqslant \sum_{i=1}^\infty \|\psi_i (u - v_i)\|_{W^{1,p(\cdot)}(B_i)}$$

$$\leqslant \sum_{i=1}^\infty (1 + L_i)\|u - v_i\|_{W^{1,p(\cdot)}(B_i)} \leqslant 2\varepsilon. \qquad \square$$

Notice that if we set $K_x \equiv 0$ in the preceding theorem, then we regain the result of Edmunds and Rákosník.

A special case of the previous theorem is the following:

Corollary 9.3.3. *Let* $r \colon \Omega \to [0, \infty)$ *be monotone in the sense of Edmunds–Rákosník (i.e. satisfy the condition of the previous theorem with* $K_x \equiv 0$*), and let* $q \in \mathcal{P}^{\log}(\Omega)$ *be such that* $p := q + r \geqslant 1$ *a.e. Then* $C^\infty(\Omega) \cap W^{1,p(\cdot)}(Q)$ *is dense in* $W^{1,p(\cdot)}(\Omega)$.

Example 9.3.4. Consider the exponent shown in Fig. 9.2: here we have added the log-Hölder continuous function $1 + x_1 + |x_2|$ (x_i refers to the i-th coordinate of x) to the monotone characteristic function $\chi_{\{x \in Q \colon x_1 < 0\}}$ in the unit square Q. This exponent satisfies the assumptions of the previous corollary.

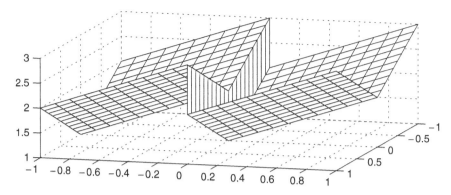

Fig. 9.2 The exponent p which is neither monotone nor continuous at the origin

As is shown by this corollary and example, the previous theorem is stronger than just saying that the exponent is log-Hölder continuous in one part of the domain and monotone in the rest – the exponent in the example satisfies the assumptions of the theorem only because the exponent is allowed to be either monotone or continuous within a single cone. This flexibility is needed at the origin.

Another approach to the density question for discontinuous exponents is to divide the domain into disjoint pieces Ω_i. Suppose we know that smooth functions are dense in each $W^{1,p(\cdot)}(\Omega_i)$. Does it follow that it is also dense in the whole domain? This question was studied by Fan, Wang and Zhao [142].

In general the answer is negative, as shown by Example 9.2.6. In some special cases, however, the answer is affirmative.

Theorem 9.3.5 (Theorems 2.4 and 2.7, [142]). *Let $p \in \mathcal{P}(\Omega)$ be bounded and suppose that there exist sets Ω_i, $i = 1, \dots, k$, such that:*

(a) *Every Ω_i is open and bounded, and the sets are pairwise disjoint.*
(b) $\Omega = \operatorname{int}\left(\bigcup_{i=1}^{k} \overline{\Omega_i}\right)$.
(c) $\Omega \setminus \bigcup_{i=1}^{k} \Omega_i$ *has measure zero.*
(d) $\Omega_i = \Omega \setminus \bigcup_{j \neq i} \overline{\Omega_j}$.

We assume that $p|_{\Omega_i} \in \mathcal{P}^{\log}(\Omega_i)$ and $p_{\Omega_i}^+ \leqslant p_{\Omega_j}^-$ for all $i < j$.

Define, for $i = 1, \dots, k-1$, $Q_i = \Omega \setminus \bigcup_{j=i+1}^{k} \overline{\Omega_j}$ and $Q_i' = \Omega \setminus Q_i$. Also, assume that each Q_i and Q_i' has Lipschitz boundary, and $Q_i \cap Q_i'$ is an $(n-1)$-Lipschitz manifold. Then $C^\infty(\overline{\Omega})$ is dense in $W^{1,p(\cdot)}(\Omega)$.

Let us next consider an example where Theorem 9.2.2 is not applicable.

Example 9.3.6. Let $B(0,1) \subset \mathbb{R}^2$ and let A_i, $i = 1, 2, 3, 4$, denote the quadrants in $B(0,1)$. The exponent is defined by $p := \sum_{i=1}^{4} i\chi_{A_i}$. This exponent is discontinuous on the coordinate axes, and does not satisfy a cone monotonicity property at the origin. However, the conditions of Theorem 9.3.5 are clearly fulfilled, and so smooth functions are indeed dense.

Let us conclude this section by noting that the density of smooth functions is invariant under a bilipschitz change of coordinates:

Recall that the mapping $f \colon A \to \mathbb{R}^n$ is said to be L-*bilipschitz* if

$$\tfrac{1}{L}|x - y| \leqslant |f(x) - f(y)| \leqslant L|x - y|$$

for all $x, y \in A$.

Proposition 9.3.7. *Let $\Omega \subset \mathbb{R}^n$ and let $f \colon \Omega \to \mathbb{R}^n$ be L-bilipschitz. Let $p \in \mathcal{P}(\Omega)$ be a bounded variable exponent such that $C^\infty(\Omega) \cap W^{1,p(\cdot)}(\Omega)$ is dense in $W^{1,p(\cdot)}(\Omega)$. Define $q := p \circ f^{-1}$. Then $C^\infty(f(\Omega)) \cap W^{1,q(\cdot)}(f(\Omega))$ is dense in $W^{1,q(\cdot)}(f(\Omega))$.*

Proof. This is just a change of variables. Let $u \in W^{1,p(\cdot)}(\Omega)$ and define $v := u \circ f^{-1}$. We have

$$\varrho_{1,q(\cdot)}(v) = \int_Q |v(y)|^{q(y)} + |\nabla v(y)|^{q(y)} \, dy$$

$$\leqslant \int_\Omega \big(|u(x)|^{p(x)} + |L\nabla u(x)|^{p(x)} \big) |J_f(x)| \, dx \leqslant L^{n+p^+} \varrho_{1,p(\cdot)}(u).$$

It follows that if $\psi_i \to u$ is an approximating sequence of smooth functions in $W^{1,p(\cdot)}(\Omega)$, then $\psi_i \circ f^{-1}$ is an approximating sequence of locally Lipschitz functions of the function v in $W^{1,p(\cdot)}(f(\Omega))$. Now the claim follows by Lemma 9.2.1 □

9.4 Density of Continuous Functions*

In this section we introduce a method which is based on convolution only in the level-sets of the exponent; the results are from Hästö [214]. Once we restrict our attention to the level-set of the exponent, convolution again becomes a very natural operation which does not impose any additional restrictions on the exponent. However, to patch up our approximations on level-sets we have to assume that the level-sets are bilipschitz images of parallel planes or concentric circles. In contrast to previous results, Theorem 9.4.7 allows us to conclude only that continuous, but not necessarily smooth, functions are dense.

Lemma 9.4.1. *Let $Q := (-1,1)^n$ and let $p \in \mathcal{P}(Q)$ be bounded. Suppose that the exponent p depends only on the n-th coordinate. Then $C(Q) \cap W^{1,p(\cdot)}(Q)$ is dense in $W^{1,p(\cdot)}(Q)$.*

Proof. In this proof dx, $dm_{n-1}(x)$ and $dm_1(x)$ stand for integration with respect to the n-, $(n-1)$- and 1-dimensional Lebesgue measures. We denote the n-th coordinate of $x \in \mathbb{R}^n$ by x_n and use $B := B(0,1) \cap \{x \in \mathbb{R}^n : x_n = 0\}$ to denote the $(n-1)$-dimensional unit ball which lives in the $x_n = 0$ plane.

Let $u \in W^{1,p(\cdot)}(Q)$ and assume first that u has compact support in Q. We consider only ε smaller than the distance between the support of u and ∂Q. Let $\psi \colon B \to [0, \infty)$ be a standard mollifier on $\mathbb{R}^{n-1}$. We define an $(n-1)$-dimensional convolution by

$$u_\varepsilon(x) = \int_B u(x + \varepsilon y) \psi(y) \, dm_{n-1}(y).$$

Then clearly u_ε is continuous (even smooth) in the plane orthogonal to the x_n-axis. Consider two points differing in the x_n coordinate. Using that u is absolutely continuous on almost every line parallel to the coordinate axes (see, e.g., Theorem 11.1.12 below), we find that

$$|u_\varepsilon(x) - u_\varepsilon(x + \delta e_n)| = \left| \int_B [u(x + \varepsilon y) - u(x + \delta e_n + \varepsilon y)] \psi(y) \, dm_{n-1}(y) \right|$$

$$\leqslant c \int_B |u(x + \varepsilon y) - u(x + \delta e_n + \varepsilon y)| \, dm_{n-1}(y)$$

$$\leqslant c \int_B \int_0^\delta |\nabla u(x + \varepsilon y + t e_n)| \, dm_1(t) \, dm_{n-1}(y)$$

$$= c \int_{B \times [0, \delta/\varepsilon]} |\nabla u(x + \varepsilon z)| \, dz.$$

Since $|\nabla u| \in L^1(Q)$, the last integral tends to zero as $\delta \to 0$. Therefore u_ε is uniformly continuous in the e_n direction as well, hence in all of Q.

It remains to show that $u_\varepsilon \to u$ in $W^{1,p(\cdot)}(Q)$. Denote by Q_t the intersection of Q with the hyperplane $\{x \in \mathbb{R}^n : x_n = t\}$. Since p is constant in Q_t (let's call it p_t), we find by the continuity of the shift operator in the fixed exponent Lebesgue space that

$$\int_{Q_t} |u(x + z) - u(x)|^{p_t} \, dm_{n-1}(x) =: F_z(t) \to 0$$

as $z \to 0$ for those values of t with $u \in L^{p_t}(Q_t, m_{n-1})$. Clearly

$$F_{\varepsilon y}(t) \leqslant \int\limits_{Q_t} |u(x + \varepsilon y)|^{p_t} + |u(x)|^{p_t} \, dm_{n-1}(x) = 2 \underbrace{\int\limits_{Q_t} |u(x)|^{p_t} \, dm_{n-1}(x),}_{=:G(t)}$$

so that $\int_{\mathbb{R}} F_{\varepsilon y}(t) \, dt \leqslant 2\varrho_{p(\cdot)}(u) < \infty$. Therefore $u \in L^{p_t}(Q_t, m_{n-1})$ for almost every $t \in \mathbb{R}$ and $F_{\varepsilon y}$ has an integrable upper bound G. Thus we find by Fubini's theorem that

$$\int\limits_{Q} |u_\varepsilon(x) - u(x)|^{p(x)} \, dx \leqslant c \int\limits_{Q} \fint\limits_{\varepsilon B} |u(x + z) - u(x)|^{p(x)} \, dm_{n-1}(z) \, dx$$

$$= \fint\limits_{\varepsilon B} \int\limits_{\mathbb{R}} F_z(x) \, dt \, dm_{n-1}(z).$$

Since $F_z \to 0$ and $F_z \leqslant G \in L^1(\mathbb{R})$ the inner integral tends to 0 as $|z| \to 0$ by the theorem of dominated convergence. Hence it follows that $u_\varepsilon \to u$ in $L^{p(\cdot)}(Q)$.

The approximation result for the gradient is analogous, using the identity

$$\nabla u_\varepsilon(x) = \int\limits_{B} [\nabla u(x + \varepsilon y)] \psi(y) \, dm_{n-1}(y).$$

Therefore the previous argument applies to ∇u, and so $\|\nabla(u - u_\varepsilon)\|_{p(\cdot)} \to 0$. Thus $u_\varepsilon \in W^{1,p(\cdot)}(Q)$ approximate the compactly supported function u.

Let then $u \in W^{1,p(\cdot)}(Q)$ be a general, not compactly supported, function. Let Q_i be the cube centered at the origin with side-length $2 - 2^{1-i}$. Define $A_2 := Q_2$ and $A_i := Q_i \setminus Q_{i-2}$ for larger i. Then we can find a partition of unity by Lipschitz functions ψ_i such that ψ_i is compactly supported in A_i. Let $L_i \geqslant 1$ be the Lipschitz constant of ψ_i. The function $\psi_i u$ has compact support in Q, so the previous argument implies that there exists $v_i \in Q$ supported in A_i such that $\|\psi_i u - v_i\|_{1,p(\cdot)} \leqslant 2^{-i}\varepsilon/L_i$. Then $v = \sum_{i=2}^\infty v_i$ is continuous and

$$\|u - v\|_{1,p(\cdot)} \leqslant \sum_{i=2}^\infty \|\psi_i u - v_i\|_{1,p(\cdot)} \leqslant \sum_{i=2}^\infty (1 + L_i) 2^{-i}\varepsilon/L_i \leqslant 2\varepsilon. \qquad \square$$

Remark 9.4.2. The original proof of this result which appears in [214] was erroneous. The claim $u_\varepsilon \to u$ was not correctly proved. This passage of the proof has been replaced above with a new and correct one.

Remark 9.4.3. If $p^- > 1$, then we can say a bit more about the continuity of u_ε: in the previous proof we derived the estimate

$$|u_\varepsilon(x) - u_\varepsilon(x + \delta e_n)| \leqslant c \int_{B \times [0, \delta/\varepsilon]} |\nabla u(x + \varepsilon y)| dy.$$

Since $|\nabla u| \in L^{p^-}(Q)$, this implies, by Hölder's inequality, that

$$|u_\varepsilon(x) - u_\varepsilon(x + \delta e_n)|$$

$$\leqslant c \left| B \times [0, \tfrac{\delta}{\varepsilon}] \right|^{1 - \frac{1}{p^-}} \left(\int_{B \times [0, \delta/\varepsilon]} |\nabla u(x + \varepsilon y)|^{p^-} dy \right)^{\frac{1}{p^-}} \leqslant c \left(\tfrac{\delta}{\varepsilon} \right)^{1 - \frac{1}{p^-}},$$

so that u_ε is $(1 - \frac{1}{p^-})$-Hölder continuous on compact subsets of Q.

By Proposition 9.3.7 we have

Corollary 9.4.4. *Let* $Q := (0, 1)^n$ *and let* $f : Q \to \mathbb{R}^n$ *be* L*-bilipschitz. Let* $q \in \mathcal{P}(Q)$ *be a bounded variable exponent which depends only on the* n^{th} *coordinate. Define* $p := q \circ f^{-1}$. *Then* $C(f(Q)) \cap W^{1,p(\cdot)}(f(Q))$ *is dense in* $W^{1,p(\cdot)}(f(Q))$.

If an exponent has a strict local extremum, then it will never satisfy the assumptions of the previous corollary. For that we need a different model.

Lemma 9.4.5. *Suppose that the bounded exponent* $p \in \mathcal{P}(B(0, R))$ *depends only on* $|x|$. *Then* $C(B(0, R)) \cap W^{1,p(\cdot)}(B(0, R))$ *is dense in* $W^{1,p(\cdot)}(B(0, R))$.

The main idea of the proof is the same as of the proof of the previous lemma. It is consequently omitted. As before, Proposition 9.3.7 gives

Corollary 9.4.6. *Let* $f : B \to \mathbb{R}^n$ *be* L*-bilipschitz in* $B := B(0, R)$. *Let* $q \in \mathcal{P}(B)$ *be a bounded exponent which depends only on* $|x|$. *Define* $p := q \circ f^{-1}$. *Then* $C(f(B)) \cap W^{1,p(\cdot)}(f(B))$ *is dense in* $W^{1,p(\cdot)}(f(B))$.

We can combine the results from the corollaries in this section. The statement of the following theorem is quite complicated, but the intuition behind it is simple. We must be able to split the domain into regular pieces with sufficient overlap, such that every piece comes from one of the previous corollaries. Notice that the partition of unity need not be uniformly bilipschitz. This means that we can also handle cases where the regularity of the exponent decreases toward the boundary of the domain.

Theorem 9.4.7. *Let* $\Omega \subset \mathbb{R}^n$ *be open and let* (Ω_i) *be an open covering of* Ω *with a subordinate partition of unity by bilipschitz functions* ψ_i *such that the number of indices* i *for which* $\psi_i(x) > 0$ *is locally bounded. Suppose further that for every* i *the set* Ω_i *satisfies the conditions of Corollary 9.4.4 or 9.4.6. Then* $C(\Omega) \cap W^{1,p(\cdot)}(\Omega)$ *is dense in* $W^{1,p(\cdot)}(\Omega)$.

Proof. Denote the bilipschitz constant of ψ_i by L_i. Fix $u \in W^{1,p(\cdot)}(\Omega)$ and $\varepsilon > 0$. By Corollary 9.4.4 or 9.4.6 we conclude that $C(\Omega_i) \cap W^{1,p(\cdot)}(\Omega_i)$ is dense in $W^{1,p(\cdot)}(\Omega_i)$. Therefore we can choose $v_i \in C(\Omega_i) \cap W^{1,p(\cdot)}(\Omega_i)$ so that $\|u - v_i\|_{W^{1,p(\cdot)}(\Omega_i)} \leqslant \varepsilon \, 2^{-i}/L_i$. Then $v = \sum_{i=1}^{\infty} \psi_i v_i$ is continuous and satisfies

$$\|u - v\|_{p(\cdot)} \leqslant \sum_{i=1}^{\infty} \|\psi_i(u - v_i)\|_{W^{1,p(\cdot)}(\Omega_i)} \leqslant \sum_{i=1}^{\infty}(1 + L_i)\|u - v_i\|_{W^{1,p(\cdot)}(\Omega_i)} \leqslant \varepsilon,$$

which proves the assertion.   □

Remark 9.4.8. If the bilipschitz mappings in Theorem 9.4.7 could be replaced by homeomorphisms, then the result would be essentially sharp, since it would cover all cases except when the exponent has a saddle-point, which seem to be the only cases of non-density.

We consider next some examples which show that the results in this section apply in some cases when the previous ones do not.

Example 9.4.9. Let $Q := (-1, 1)^2$ and define $p(x) := 2 - \big(\log(100/|x_2|)\big)^{-a}$ for some $a > 1$. This exponent is shown in Fig. 9.3. A bilipschitz deformation of the exponent $p(x) := 2 + \big(\log(100/|x_2|)\big)^{-a}$ is shown in Fig. 9.4. Corollary 9.4.4 allows us to conclude that continuous functions are dense in $W^{1,p(\cdot)}(Q)$ in both cases.

Example 9.4.10. Define $p(x) := 2 - \big(\log(1/|x|)\big)^{-a}$ for some $a > 1$. Then Lemma 9.4.5 allows us to conclude that continuous functions are dense in $W^{1,p(\cdot)}(B)$, whereas the other results are not applicable. Another example is given by $p(x) := 2 + \sin(1/|x|)$.

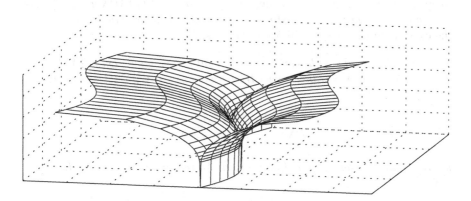

Fig. 9.3 The exponent p with a ridge

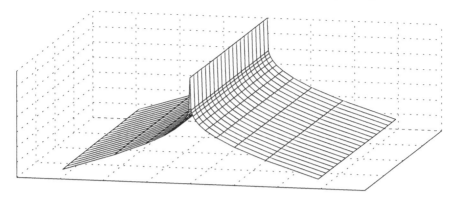

Fig. 9.4 The exponent p with a through

9.5 The Lipschitz Truncation Method*

The purpose of this section is to show that a weakly convergent sequence of Sobolev functions can be approximated by a sequence of Lipschitz functions such that certain additional convergence properties hold. This approximation property holds for a large class of domains Ω. We say that a bounded domain $\Omega \subset \mathbb{R}^n$ has a A-fat complement, $A \geqslant 1$, if

$$|B(z,r)| \leqslant A\,|B(z,r) \setminus \Omega|$$

for all $z \in \partial\Omega$ and all $r \in (0, \operatorname{diam}\Omega)$. Note that a bounded domain $\Omega \subset \mathbb{R}^n$ with Lipschitz continuous boundary has this property.

The proof of the following theorem can be found in Diening, Málek and Steinhauer [102, Theorem 2.3]. It is based on the ideas from Acerbi and Fusco [2] and Landes [262], or the monograph Maý and Ziemer [280]. We roughly speaking restrict v to the set $\{-\theta \leqslant v \leqslant \theta\} \cap \{M(\nabla u) \leqslant \lambda\}$ and denote by $v_{\theta,\lambda}$ the McShane extension (cf. [289]) of this restriction to $\mathbb{R}^n$, for the exact definition see the above references.

Theorem 9.5.1. *Let Ω be bounded with A-fat complement. Let $v \in W_0^{1,1}(\Omega)$. Then for every $\theta, \lambda > 0$ there exist truncations $v_{\theta,\lambda} \in W_0^{1,\infty}(\Omega)$ such that*

$$\|v_{\theta,\lambda}\|_{L^\infty(\Omega)} \leqslant \theta,$$
$$\|\nabla v_{\theta,\lambda}\|_{L^\infty(\Omega)} \leqslant c\,A\,\lambda,$$

where c only depends on the dimension n. Moreover, up to a set of Lebesgue measure zero

$$\{v_{\theta,\lambda} \neq v\} \subset \{Mv > \theta\} \cup \{M(\nabla v) > \lambda\}.$$

Our version of fatness is slightly different than that used in [102], but it implies the desired condition. We then move on to the truncation result itself.

Theorem 9.5.2. *Let Ω be a bounded domain with A-fat complement and let $p \in \mathcal{A}$ or $p \in \mathcal{P}^{\log}(\Omega)$ with $1 < p^- \leqslant p^+ < \infty$. Let $v \in W_0^{1,p(\cdot)}(\Omega)$, fix $\theta > 0$ and let $K \geqslant \|v\|_{W^{1,p(\cdot)}(\Omega)}$. Then there exists a sequence $(\lambda_j)_{j=0}^\infty$ with*

$$2^{2^j} \leqslant \lambda_j \leqslant 2^{2^{j+1}}$$

and functions $v_j \in W_0^{1,\infty}(\Omega)$ such that the following hold for all $j \in \mathbb{N}$:

$$\|v_j\|_{L^\infty(\Omega)} \leqslant \theta,$$

$$\|\nabla v_j\|_{L^\infty(\Omega)} \leqslant c_1 \, A \, c_0 \, K \, \lambda_j,$$

$$\left\| \lambda_j \, \chi_{\{v_j \neq v\}} \right\|_{L^{p(\cdot)}(\Omega)} \leqslant c_0 \frac{\lambda_j}{\theta} \|v\|_{L^{p(\cdot)}(\Omega)} + 2^{1-\frac{j}{p^+}},$$

$$\left\| \nabla v_j \, \chi_{\{v_j \neq v\}} \right\|_{L^{p(\cdot)}(\Omega)} \leqslant c_1 \, A \, c_0 \, K \left(\frac{c_0}{\theta} 2^{2^j} \|v\|_{L^{p(\cdot)}(\Omega)} + 2^{1-\frac{j}{p^+}} \right)$$

and, up to a set of measure zero,

$$\{v_j \neq v\} \subset \{Mv > \theta\} \cup \{M(\nabla v) > c_0 \, K \, \lambda_j\}.$$

Here c_0 denotes the operator norm of the maximal operator in $L^{p(\cdot)}(\mathbb{R}^n)$ and c_1 is the constant from Theorem 9.5.1.

Proof. If $p \in \mathcal{P}^{\log}(\Omega)$ then we extend it by Proposition 4.1.7 to the whole space and denote it again by p. Moreover, by Theorem 4.4.8 we obtain $p \in \mathcal{A}$. We extend v by zero outside of Ω and obtain $v \in W_0^{1,p(\cdot)}(\mathbb{R}^n)$ (Theorem 8.1.14). The assumptions on p imply that the maximal operator is bounded (Theorem 5.7.2) and thus

$$\|Mv\|_{L^{p(\cdot)}(\mathbb{R}^n)} + \|M(\nabla v)\|_{L^{p(\cdot)}(\mathbb{R}^n)} \leqslant c_0 \|v\|_{W^{1,p(\cdot)}(\mathbb{R}^n)} \leqslant c_0 \, K,$$

so that $\|\frac{Mv}{c_0 K}\|_{L^{p(\cdot)}(\Omega)} + \|\frac{M(\nabla v)}{c_0 K}\|_{L^{p(\cdot)}(\Omega)} \leqslant 1$. Next, we observe that for $g \in L^{p(\cdot)}(\mathbb{R}^n)$ with $\|g\|_{L^{p(\cdot)}(\mathbb{R}^n)} \leqslant 1$ we have

$$\sum_{j=1}^\infty \sum_{k=2^j}^{2^{j+1}-1} \int_{\mathbb{R}^n} 2^{kp(x)} \, \chi_{\{2^k < |g|\}} \, dx$$

$$\leqslant \sum_{k=1}^\infty \int_{\mathbb{R}^n} 2^{kp(x)} \, \chi_{\{2^k < |g|\}} \, dx = \sum_{k=1}^\infty \int_{\mathbb{R}^n} \sum_{i=k}^\infty 2^{kp(x)} \, \chi_{\{2^i < |g| \leqslant 2^{i+1}\}} \, dx$$

$$= \sum_{i=1}^{\infty} \sum_{k=1}^{i} \int_{\mathbb{R}^n} 2^{kp(x)} \chi_{\{2^i < |g| \leqslant 2^{i+1}\}} \, dx \leqslant 2^{p^+} \sum_{i=1}^{\infty} \int_{\mathbb{R}^n} 2^{ip(x)} \chi_{\{2^i < |g| \leqslant 2^{i+1}\}} \, dx$$

$$\leqslant 2^{p^+} \int_{\mathbb{R}^n} |g(x)|^{p(x)} \, dx \leqslant 2^{p^+}.$$

This inequality implies, in particular, that each term in the outer sum on the left-hand side is at most 2^{p^+}. Since the inner sum has 2^j terms, we can find for every j an index $k_j \in \{2^j, \ldots, 2^{j+1} - 1\}$ such that

$$\int_{\mathbb{R}^n} 2^{k_j p(x)} \chi_{\{|g| > 2^{k_j}\}} \, dx \leqslant 2^{p^+ - j}.$$

Let us choose the indices k_j for the function $g = \dfrac{M(\nabla v)}{c_0 \, K}$, and denote $\lambda_j := 2^{k_j}$. By construction it is clear that $2^{2^j} \leqslant \lambda_j < 2^{2^{j+1}}$. The previous inequality can be written as

$$\int_{\mathbb{R}^n} \lambda_j^{p(x)} \chi_{\{|M(\nabla v)/(c_0 K)| > \lambda_j\}} \, dx \leqslant 2^{p^+ - j} =: \varepsilon_j^{p^+}.$$

By Lemma 3.2.5, this yields

$$\left\| \lambda_j \, \chi_{\{|M(\nabla v)| > c_0 K \lambda_j\}} \right\|_{L^{p(\cdot)}(\mathbb{R}^n)} \leqslant \varepsilon_j. \tag{9.5.3}$$

For each $j \in \mathbb{N}$ we apply Theorem 9.5.1 with θ and $c_0 K \lambda_j$ and denote $v_j := v_{\theta, c_0 K \lambda_j}$. The theorem directly implies the following conclusions:

$$\|v_j\|_{L^\infty(\Omega)} \leqslant \theta \quad \text{and} \quad \|\nabla v_j\|_{L^\infty(\Omega)} \leqslant c_1 \, A \, c_0 K \lambda_j,$$

and, up to a set of measure zero,

$$\{v_j \neq v\} \subset \{Mv > \theta\} \cup \{M(\nabla v) > c_0 K \, \lambda_j\}.$$

Obviously the second inequality implies that

$$\left\| \nabla v_j \, \chi_{\{v_j \neq v\}} \right\|_{L^{p(\cdot)}(\mathbb{R}^n)} \leqslant c_1 \, A \, c_0 K \left\| \lambda_j \, \chi_{\{v_j \neq v\}} \right\|_{L^{p(\cdot)}(\mathbb{R}^n)}.$$

Thus it remains only to estimate $\left\| \lambda_j \, \chi_{\{v_j \neq v\}} \right\|_{L^{p(\cdot)}(\mathbb{R}^n)}$. The set inclusion above implies that

$$\left\| \lambda_j \, \chi_{\{v_j \neq v\}} \right\|_{L^{p(\cdot)}(\mathbb{R}^n)} \leqslant \left\| \lambda_j \, \chi_{\{Mv > \theta\} \cup \{M(\nabla v) > c_0 K \lambda_j\}} \right\|_{L^{p(\cdot)}(\Omega)}.$$

Then we estimate, using also (9.5.3),

$$\left\| \lambda_j \, \chi_{\{Mv>\theta\} \cup \{M(\nabla v)>c_0 K \lambda_j\}} \right\|_{L^{p(\cdot)}(\mathbb{R}^n)}$$
$$\leqslant \frac{\lambda_j}{\theta} \left\| \theta \, \chi_{\{Mv>\theta\}} \right\|_{L^{p(\cdot)}(\mathbb{R}^n)} + \left\| \lambda_j \, \chi_{\{M(\nabla v)>c_0 K \lambda_j\}} \right\|_{L^{p(\cdot)}(\mathbb{R}^n)}$$
$$\leqslant \frac{\lambda_j}{\theta} \left\| Mv \right\|_{L^{p(\cdot)}(\mathbb{R}^n)} + \varepsilon_j$$

which combined with the continuity of the maximal operator completes the proof. $\qquad\square$

We next show how the theorem can be applied to a sequence to yield nice, uniform conclusions.

Corollary 9.5.4. *Let Ω be a bounded domain with A-fat complement and let $p \in \mathcal{A}$ or $p \in \mathcal{P}^{\log}(\Omega)$ with $1 < p^- \leqslant p^+ < \infty$. Let $v^k \in W_0^{1,p(\cdot)}(\Omega)$ be such that $v^k \rightharpoonup 0$ weakly in $W_0^{1,p(\cdot)}(\Omega)$. Set*

$$K := \sup_k \|v^k\|_{W_0^{1,p(\cdot)}(\Omega)} < \infty,$$
$$\gamma_k := \|v^k\|_{L^{p(\cdot)}(\Omega)} \to 0.$$

Then there exists a null-sequence (ε_j) and for every $j, k \in \mathbb{N}$ there exists a function $v^{k,j} \in W_0^{1,\infty}(\Omega)$ and a number $\lambda_{k,j} \in \left[2^{2^j}, 2^{2^{j+1}} \right]$ such that

$$\lim_{k \to \infty} \left(\sup_{j \in \mathbb{N}} \|v^{k,j}\|_{L^\infty(\Omega)} \right) = 0,$$
$$\|\nabla v^{k,j}\|_{L^\infty(\Omega)} \leqslant c \, A \, K \, \lambda_{k,j} \leqslant c \, A \, K \, 2^{2^{j+1}},$$
$$\limsup_{k \to \infty} \left\| \lambda_{k,j} \, \chi_{\{v^{k,j} \neq v^k\}} \right\|_{L^{p(\cdot)}(\Omega)} \lesssim \varepsilon_j,$$
$$\limsup_{k \to \infty} \left\| \nabla v^{k,j} \, \chi_{\{v^{k,j} \neq v^k\}} \right\|_{L^{p(\cdot)}(\Omega)} \lesssim \varepsilon_j,$$

The constant c depends only on n, p^-, p^+ and the $\mathcal{A}$-constant of p. Moreover, for fixed $j \in \mathbb{N}$, $\nabla v^{k,j} \rightharpoonup 0$ in $L^s(\Omega)$ when $s < \infty$ and $\nabla v^{k,j} \overset{*}{\rightharpoonup} 0$ in $L^\infty(\Omega)$ as $k \to \infty$.

Proof. The compact embedding $W_0^{1,p(\cdot)}(\Omega) \hookrightarrow\hookrightarrow L^{p(\cdot)}(\Omega)$ (Theorem 8.4.2) and $v \rightharpoonup 0$ imply that $v^k \to 0$ in $L^{p(\cdot)}(\Omega)$ and $K < \infty$. Setting $\theta_k := \sqrt{\gamma_k}$, we can apply Theorem 9.5.2 to each function v_k with θ_k. The statements are an immediate consequence of the assertions in that theorem since $\gamma_k \to 0$, $\theta_k \to 0$ and $\gamma_k/\theta_k \to 0$. It remains only to prove the claims regarding the weak convergence of the gradient.

If $\psi \in C_0^\infty(\Omega)$, then

$$\left| \int_\Omega \nabla v^{k,j}\, \psi\, dx \right| = \left| \int_\Omega v^{k,j}\, \nabla \psi\, dx \right| \leqslant \theta_k \left\| \nabla \psi \right\|_{L^1(\Omega)}.$$

Clearly this tends to zero as $k \to \infty$. Since $C_0^\infty(\Omega)$ is dense in $L^{s'}(\Omega)$ for $s' \in [1, \infty)$ it follows that $\langle \nabla v^{k,j}, \psi \rangle \to 0$ as $k \to \infty$, for all $\psi \in L^{s'}(\Omega)$. By definition of weak (star) convergence, this means that $\nabla v^{k,j} \rightharpoonup 0$ in $L^s(\Omega)$ when $s \in (1, \infty)$ and $\nabla v^{k,j} \overset{*}{\rightharpoonup} 0$ in $L^\infty(\Omega)$ when $s = \infty$. The case $s = 1$ follows from the case $s = 2$ and the embedding $L^2(\Omega) \hookrightarrow L^1(\Omega)$. $\qquad \square$

Chapter 10
Capacities

Capacities are needed to understand point-wise behavior of Sobolev functions. They also play an important role in studies of solutions of partial differential equations. In this chapter we study two kinds of capacities: Sobolev capacity in Sect. 10.1 and relative capacity in Sect. 10.2. Both capacities have their advantages. The Sobolev capacity is independent of the underlying set, but extremal functions are difficult to find. The situation for relative capacity is the opposite and it is a Choquet capacity for every measurable exponent. For a constant exponent, our definitions of the Sobolev and relative capacities coincide with the classical ones. In Sect. 10.3 we compare the capacities with each other and in Sect. 10.4 with the variable dimension Hausdorff measure. Later in Sect. 11.1 we define a capacity based on quasicontinuous representatives.

10.1 Sobolev Capacity

In this section we define the Sobolev capacity. The results are based on [194] by Harjulehto, Hästö, Koskenoja and Varonen. Further studies of variable exponent capacities are contained in Alkhutov and Krasheninnikova [20], Harjulehto and Latvala [211], and Mashiyev [284].

Definition 10.1.1. Let $p \in \mathcal{P}(\mathbb{R}^n)$ satisfy $p(x) \in [1, \infty)$ for almost every x. For $E \subset \mathbb{R}^n$ we denote

$$S_{p(\cdot)}(E) := \{u \in W^{1,p(\cdot)}(\mathbb{R}^n) : u \geqslant 1 \text{ in an open set containing } E \ \& \ u \geqslant 0\}.$$

Functions $u \in S_{p(\cdot)}(E)$ are said to be $p(\cdot)$-*admissible* for the capacity of the set E. The *Sobolev* $p(\cdot)$-*capacity* of E is defined by

$$C_{p(\cdot)}(E) := \inf_{u \in S_{p(\cdot)}(E)} \overline{\varrho}_{1,p(\cdot)}(u) = \inf_{u \in S_{p(\cdot)}(E)} \int_{\mathbb{R}^n} |u|^{p(x)} + |\nabla u|^{p(x)} \, dx.$$

In case $S_{p(\cdot)}(E) = \emptyset$, we set $C_{p(\cdot)}(E) = \infty$.

L. Diening et al., *Lebesgue and Sobolev Spaces with Variable Exponents*,
Lecture Notes in Mathematics 2017, DOI 10.1007/978-3-642-18363-8_10,
© Springer-Verlag Berlin Heidelberg 2011

If $u \in S_{p(\cdot)}(E)$, then $\min\{1, u\} \in S_{p(\cdot)}(E)$ and $\overline{\varrho}_{1,p(\cdot)}(\min\{1, u\}) \leqslant \overline{\varrho}_{1,p(\cdot)}(u)$. Thus it is enough to test the Sobolev capacity by $u \in S_{p(\cdot)}(E)$ with $0 \leqslant u \leqslant 1$.

The Sobolev $p(\cdot)$-capacity enjoys all relevant properties of general capacities; specifically, it will be seen that $C_{p(\cdot)}(E)$ defines a Choquet capacity if $1 < p^- \leqslant p^+ < \infty$. We start with some properties that hold for arbitrary measurable exponents $p : \mathbb{R}^n \to [1, \infty)$.

Theorem 10.1.2. *Let $p \in \mathcal{P}(\mathbb{R}^n)$ with $p(x) \in [1, \infty)$ for every x. The set function $E \mapsto C_{p(\cdot)}(E)$ has the following properties:*

(C1) $C_{p(\cdot)}(\emptyset) = 0$.
(C2) *If $E_1 \subset E_2$, then $C_{p(\cdot)}(E_1) \leqslant C_{p(\cdot)}(E_2)$.*
(C3) *If E is a subset of $\mathbb{R}^n$, then*

$$C_{p(\cdot)}(E) = \inf_{\substack{E \subset U \\ U \text{ open}}} C_{p(\cdot)}(U).$$

(C4) *If E_1 and E_2 are subsets of $\mathbb{R}^n$, then*

$$C_{p(\cdot)}(E_1 \cup E_2) + C_{p(\cdot)}(E_1 \cap E_2) \leqslant C_{p(\cdot)}(E_1) + C_{p(\cdot)}(E_2).$$

(C5) *If $K_1 \supset K_2 \supset \cdots$ are compact sets, then*

$$\lim_{i \to \infty} C_{p(\cdot)}(K_i) = C_{p(\cdot)}\left(\bigcap_{i=1}^{\infty} K_i \right).$$

Proof. Since the constant function 0 belongs to $S_{p(\cdot)}(\emptyset)$, assertion (C1) follows.

To prove (C2), let $E_1 \subset E_2$. Then $S_{p(\cdot)}(E_1) \supset S_{p(\cdot)}(E_2)$, and hence by definition

$$C_{p(\cdot)}(E_1) = \inf_{u \in S_{p(\cdot)}(E_1)} \overline{\varrho}_{p(\cdot)}(u) \leqslant \inf_{u \in S_{p(\cdot)}(E_2)} \overline{\varrho}_{p(\cdot)}(u) = C_{p(\cdot)}(E_2).$$

To prove (C3), fix $E \subset \mathbb{R}^n$. By property (C2),

$$C_{p(\cdot)}(E) \leqslant \inf_{\substack{E \subset U \\ U \text{ open}}} C_{p(\cdot)}(U).$$

Fix $\varepsilon > 0$ and a function $u \in S_{p(\cdot)}(E)$ so that $\overline{\varrho}_{1,p(\cdot)}(u) \leqslant C_{p(\cdot)}(E) + \varepsilon$. Denote $U := \mathrm{int}\{u \geqslant 1\}$. Then $E \subset U$, and $C_{p(\cdot)}(U) \leqslant \overline{\varrho}_{1,p(\cdot)}(u) \leqslant C_{p(\cdot)}(E) + \varepsilon$, and thus the claim follows as $\varepsilon \to 0$.

To prove (C4), let $\varepsilon > 0$. Choose $u_1 \in S_{p(\cdot)}(E_1)$ such that $\overline{\varrho}_{1,p(\cdot)}(u_1) \leqslant C_{p(\cdot)}(E_1) + \varepsilon$, and $u_2 \in S_{p(\cdot)}(E_2)$ such that $\overline{\varrho}_{1,p(\cdot)}(u_2) \leqslant C_{p(\cdot)}(E_2) + \varepsilon$. We have $\max\{u_1, u_2\} \in S_{p(\cdot)}(E_1 \cup E_2)$ and $\min\{u_1, u_2\} \in S_{p(\cdot)}(E_1 \cap E_2)$, and, by Theorem 8.1.9,

$$
\int_{\mathbb{R}^n} |\nabla \max\{u_1, u_2\}|^{p(x)} \, dx + \int_{\mathbb{R}^n} |\nabla \min\{u_1, u_2\}|^{p(x)} \, dx
$$

$$
= \int_{\mathbb{R}^n} |\nabla u_1|^{p(x)} \, dx + \int_{\mathbb{R}^n} |\nabla u_2|^{p(x)} \, dx.
$$

This and an analogous for $\overline{\varrho}_{p(\cdot)}(\max\{u_1, u_2\}) + \overline{\varrho}_{p(\cdot)}(\min\{u_1, u_2\})$ yields

$$
C_{p(\cdot)}(E_1 \cup E_2) + C_{p(\cdot)}(E_1 \cap E_2) \leqslant \overline{\varrho}_{1,p(\cdot)}(u_1) + \overline{\varrho}_{1,p(\cdot)}(u_2)
$$

$$
\leqslant C_{p(\cdot)}(E_1) + C_{p(\cdot)}(E_2) + 2\varepsilon,
$$

from which (C4) follows as ε tends to zero.

To prove (C5), let $K_1 \supset K_2 \supset \cdots$ be compact sets. Since $\bigcap_{i=1}^{\infty} K_i \subset K_j$ for each $j = 1, 2, \ldots$, property (C2) gives

$$
C_{p(\cdot)}\left(\bigcap_{i=1}^{\infty} K_i\right) \leqslant \lim_{i \to \infty} C_{p(\cdot)}(K_i).
$$

To prove the opposite inequality, choose an open set U with $\bigcap_i K_i \subset U$. Because every K_i is compact (so that $\bigcap_i K_i$ is compact, as well), there is a positive integer k such that $K_i \subset U$ for all $i \geqslant k$. Thus

$$
\lim_{i \to \infty} C_{p(\cdot)}(K_i) \leqslant C_{p(\cdot)}(U),
$$

and by property (C3)

$$
\lim_{i \to \infty} C_{p(\cdot)}(K_i) \leqslant C_{p(\cdot)}\left(\bigcap_{i=1}^{\infty} K_i\right).
$$

This completes the proof.   □

A set function which has the properties (C1), (C2) and (C3) from the previous theorem is called *an outer capacity*. Theorem 10.1.2 thus yields that the Sobolev $p(\cdot)$-capacity is an outer capacity. In order to get the remaining Choquet property (that is, (C6) in the next theorem) we need extra assumptions for the variable exponent.

Theorem 10.1.3. *If $p \in \mathcal{P}(\mathbb{R}^n)$ with $1 < p^- \leqslant p^+ < \infty$, then the set function $E \mapsto C_{p(\cdot)}(E)$ has the following additional properties:*

(C6) *If $E_1 \subset E_2 \subset \cdots$ are subsets of $\mathbb{R}^n$, then*

$$\lim_{i \to \infty} C_{p(\cdot)}(E_i) = C_{p(\cdot)}\left(\bigcup_{i=1}^{\infty} E_i \right).$$

(C7) *For $E_i \subset \mathbb{R}^n$, $i \in \mathbb{N}$, we have*

$$C_{p(\cdot)}\left(\bigcup_{i=1}^{\infty} E_i \right) \leqslant \sum_{i=1}^{\infty} C_{p(\cdot)}(E_i).$$

Proof. To prove (C6), denote $E := \bigcup_{i=1}^{\infty} E_i$. Note first that (C2) implies that

$$\lim_{i \to \infty} C_{p(\cdot)}(E_i) \leqslant C_{p(\cdot)}(E). \tag{10.1.4}$$

To prove the opposite inequality we may assume that $\lim_{i \to \infty} C_{p(\cdot)}(E_i) < \infty$. Let $u_i \in S_{p(\cdot)}(E_i)$ and $\bar{\varrho}_{1,p(\cdot)}(u_i) \leqslant C_{p(\cdot)}(E_i) + 2^{-i}$ for every $i \in \mathbb{N}$. Since $W^{1,p(\cdot)}(\mathbb{R}^n)$ is reflexive (Theorem 8.1.6) and since the sequence (u_i) is bounded in $W^{1,p(\cdot)}(\mathbb{R}^n)$, there is a subsequence of (u_i) which converges weakly to a function $u \in W^{1,p(\cdot)}(\mathbb{R}^n)$. We obtain by the Banach-Saks property (Corollary 8.1.7) that $\frac{1}{m} \sum_{i=1}^{m} u_i \to u$ in the Sobolev space $W^{1,p(\cdot)}(\mathbb{R}^n)$ as $m \to \infty$. We write $v_j := \frac{1}{j(j-1)} \sum_{i=j+1}^{j^2} u_i$ and obtain

$$\left\| \frac{1}{j^2} \sum_{i=1}^{j^2} u_i - v_j \right\|_{1,p(\cdot)} \leqslant \left\| \frac{1}{j^2} \sum_{i=1}^{j} u_i \right\|_{1,p(\cdot)} + \left\| \frac{1}{j^2(j-1)} \sum_{i=j+1}^{j^2} u_i \right\|_{1,p(\cdot)}$$

$$\leqslant \frac{1}{j} \left\| \frac{1}{j} \sum_{i=1}^{j} u_i \right\|_{1,p(\cdot)} + \frac{1}{j-1} \left\| \frac{1}{j^2} \sum_{i=1}^{j^2} u_i \right\|_{1,p(\cdot)},$$

which converges to zero as j goes to infinity. Thus $v_j \to u$ in $W^{1,p(\cdot)}(\mathbb{R}^n)$. Since E_j is an increasing sequence, it follows that $u_i \geqslant 1$ in an open set containing E_j for every $j \leqslant i$. Hence $E_j \subset \text{int}\{v_j \geqslant 1\}$. By the convexity of the modular and (C2) we obtain

$$\bar{\varrho}_{1,p(\cdot)}(v_j) \leqslant \frac{1}{j(j-1)} \sum_{i=j+1}^{j^2} \int_{\mathbb{R}^n} |u_i|^{p(x)} + |\nabla u_i|^{p(x)} \, dx$$

$$\leqslant \sup_{i \geqslant j} \bar{\varrho}_{1,p(\cdot)}(u_i) \leqslant \sup_{i \geqslant j} \left(C_{p(\cdot)}(E_i) + 2^{-i} \right) \leqslant \lim_{i \to \infty} C_{p(\cdot)}(E_i) + 2^{-j}.$$

By considering a subsequence, if necessary, we may assume that $\|v_{j+1} - v_j\|_{1,p(\cdot)} \leqslant 2^{-j}$. We set

$$w_j := v_j + \sum_{i=j}^{\infty} |v_{i+1} - v_i|,$$

and observe that $w_j \in W^{1,p(\cdot)}(\mathbb{R}^n)$. Since $w_j \geqslant \sup_{i \geqslant j} v_i$, we see that $w_j \geqslant 1$ in the open set

$$\bigcup_{i=j}^{\infty} \mathrm{int}\{v_i \geqslant 1\} \supset E,$$

so $w_j \in S_{p(\cdot)}(E)$. This yields $C_{p(\cdot)}(E) \leqslant \overline{\varrho}_{1,p(\cdot)}(w_j)$ for $j = 1, 2, \ldots$ We also find that

$$\|w_j - v_j\|_{1,p(\cdot)} \leqslant \sum_{i=j}^{\infty} \|v_{i+1} - v_i\|_{1,p(\cdot)} \leqslant \sum_{i=j}^{\infty} 2^{-i} = 2^{-j+1},$$

and hence

$$\overline{\varrho}_{1,p(\cdot)}(w_j - v_j) \to 0 \quad \text{as } j \to \infty.$$

Therefore

$$C_{p(\cdot)}(E) \leqslant \lim_{j \to \infty} \overline{\varrho}_{1,p(\cdot)}(w_j) = \lim_{j \to \infty} \overline{\varrho}_{1,p(\cdot)}(v_j)$$
$$\leqslant \lim_{j \to \infty} \lim_{i \to \infty} C_{p(\cdot)}(E_i) + 2^{-j} = \lim_{i \to \infty} C_{p(\cdot)}(E_i).$$

The previous inequality and inequality (10.1.4) yield property (C6).

It remains to prove (C7). From (C4) it follows by induction that

$$C_{p(\cdot)}\left(\bigcup_{i=1}^{k} E_i\right) \leqslant \sum_{i=1}^{k} C_{p(\cdot)}(E_i) \leqslant \sum_{i=1}^{\infty} C_{p(\cdot)}(E_i)$$

for any finite family of subsets $E_1, E_2, \ldots, E_k$ in $\mathbb{R}^n$. Since $\bigcup_{i=1}^{k} E_i$ increases to $\bigcup_{i=1}^{\infty} E_i$, (C6) implies (C7). $\qquad \square$

Remark 10.1.5. The assumptions of Theorem 10.1.3 can be relaxed: we need only that $p^- > 1$ holds locally, see [195].

By the definition of outer measure (cf. [184]), properties (C1), (C2), and (C7) of the Sobolev $p(\cdot)$-capacity yield:

Corollary 10.1.6. *If $p \in \mathcal{P}(\mathbb{R}^n)$ with $1 < p^- \leqslant p^+ < \infty$, then the Sobolev $p(\cdot)$-capacity is an outer measure.*

A set function which satisfies properties (C1), (C2), (C5), and (C6) is called a *Choquet capacity*. A set $E \subset \mathbb{R}^n$ is *capacitable* if

$$\sup_{\substack{K \subset E \\ K \text{ compact}}} C_{p(\cdot)}(K) = C_{p(\cdot)}(E) = \inf_{\substack{E \subset U \\ U \text{ open}}} C_{p(\cdot)}(U).$$

For a Choquet capacity every Suslin set is capacitable [73]. The definition of Suslin sets can be found, for example, in [154, Sect. 2.2.10, p. 65]. For us it is enough to know that each Borel set is a Suslin set. Thus we obtain the following result:

Corollary 10.1.7. *Let $p \in \mathcal{P}(\mathbb{R}^n)$ with $1 < p^- \leqslant p^+ < \infty$. Then the set function $E \mapsto C_{p(\cdot)}(E)$, $E \subset \mathbb{R}^n$, is a Choquet capacity. In particular, all Borel sets $E \subset \mathbb{R}^n$ are capacitable.*

We can derive a weak form of the subadditivity (property (C7)) even if we dispense with the lower bound assumption on the variable exponent p.

Proposition 10.1.8. *Let $p \in \mathcal{P}(\mathbb{R}^n)$ be bounded. If every E_i is a subset of $\mathbb{R}^n$ with $C_{p(\cdot)}(E_i) = 0$, $i \in \mathbb{N}$, then*

$$C_{p(\cdot)}\left(\bigcup_{i=1}^{\infty} E_i \right) = 0.$$

Proof. Fix $\varepsilon \in (0,1)$. Since modular and norm convergence are equivalent (Theorem 3.4.1), we can find functions $u_i \in S_{p(\cdot)}(E_i)$ with $\|u_i\|_{1,p(\cdot)} \leqslant \varepsilon\, 2^{-i}$. Define $v_i := u_1 + \ldots + u_i$. Then (v_i) is a Cauchy sequence and since $W^{1,p(\cdot)}(\mathbb{R}^n)$ is a Banach space (Theorem 8.1.6) there exists $v \in W^{1,p(\cdot)}(\mathbb{R}^n)$ such that $v_i \to v$. By Lemma 2.3.15 we have for a subsequence (v_i) that $v_i \to v$ almost everywhere. Define $U_i := \operatorname{int}\{u_i \geqslant 1\}$. Since $v_j|_{U_i} \geqslant 1$ for $j \geqslant i$, we conclude that $v|_{U_i} \geqslant 1$ a.e. Since $\bigcup E_i \subset \bigcup U_i$, and the latter set is open, we find that $v \in S_{p(\cdot)}\left(\bigcup_{i=1}^{\infty} E_i \right)$. On the other hand

$$\|v\|_{1,p(\cdot)} \leqslant \lim_{i \to \infty} \|v_i\|_{1,p(\cdot)} \leqslant \sum_{i=1}^{\infty} \|u_i\|_{1,p(\cdot)} \leqslant \sum_{i=1}^{\infty} \varepsilon\, 2^{-i} \leqslant \varepsilon.$$

This and Lemma 3.2.4 implies

$$C_{p(\cdot)}\left(\bigcup_{i=1}^{\infty} E_i \right) \leqslant \varrho_{1,p(\cdot)}(v) \leqslant \|v\|_{1,p(\cdot)} \leqslant \varepsilon,$$

which yields the claim, as ε tends to zero.   □

Lemma 10.1.9. *Let $p \in \mathcal{P}(\mathbb{R}^n)$ be a bounded exponent and assume that $C^\infty(\mathbb{R}^n) \cap W^{1,p(\cdot)}(\mathbb{R}^n)$ is dense in $W^{1,p(\cdot)}(\mathbb{R}^n)$. If K is compact, then*

$$C_{p(\cdot)}(K) = \inf_{u \in S^\infty_{p(\cdot)}(K)} \int_{\mathbb{R}^n} |u|^{p(x)} + |\nabla u|^{p(x)}\, dx,$$

where $S^\infty_{p(\cdot)}(K) := S_{p(\cdot)}(K) \cap C^\infty(\mathbb{R}^n)$.

Proof. Let $u \in S_{p(\cdot)}(K)$ with $0 \leqslant u \leqslant 1$. We choose a sequence of functions $\varphi_j \in C^\infty(\mathbb{R}^n) \cap W^{1,p(\cdot)}(\mathbb{R}^n)$ converging to u in $W^{1,p(\cdot)}(\mathbb{R}^n)$. Choose an open bounded neighborhood U of K such that $u = 1$ in U. Let $\psi \in C^\infty(\mathbb{R}^n)$, $0 \leqslant \psi \leqslant 1$, be such that $\psi = 1$ in $\mathbb{R}^n \setminus U$ and $\psi = 0$ in an open neighborhood of K. Then the functions $\psi_j := 1 - (1 - \varphi_j)\psi$ converge to u in $W^{1,p(\cdot)}(\mathbb{R}^n)$ since $u - \psi_j = (u - \varphi_j)\psi + (1 - \psi)(u - 1) = (u - \varphi_j)\psi$. This establishes the assertion since $\psi_j \in S^\infty_{p(\cdot)}(K)$. $\square$

Proposition 10.1.10. *Let $E \subset \mathbb{R}^n$ and let $p, q \in \mathcal{P}(\mathbb{R}^n)$ be bounded with $p \geqslant q$. If $C_{p(\cdot)}(E) = 0$, then $C_{q(\cdot)}(E) = 0$.*

Proof. Let $\eta \in C_0^\infty(B(0, R+1))$ be a cut-off function with $\eta = 1$ in $B(0, R)$, $0 \leqslant \eta \leqslant 1$ and $|\nabla \eta| \leqslant 2$. Let $u \in S_{p(\cdot)}(E \cap B(0, R))$. Then $u\eta \in L^{q(\cdot)}(\mathbb{R}^n)$ since by Corollary 3.3.4 we get

$$\|u\eta\|_{L^{q(\cdot)}(\mathbb{R}^n)} = \|u\eta\|_{L^{q(\cdot)}(B(0,R+1))}$$
$$\leqslant 2\big(1 + |B(0, R+1)|\big)\|u\|_{L^{p(\cdot)}(B(0,R+1))}.$$

Since $\nabla(u\eta) = \eta\nabla u + u\nabla\eta$, we find that

$$\|\nabla(u\eta)\|_{L^{q(\cdot)}(\mathbb{R}^n)} = \|\nabla(u\eta)\|_{L^{q(\cdot)}(B(0,R+1))}$$
$$\leqslant 2\|u\|_{W^{1,q(\cdot)}(B(0,R+1))}$$
$$\leqslant 4\big(1 + |B(0, R+1)|\big)\|u\|_{W^{1,p(\cdot)}(B(0,R+1))}$$
$$\leqslant 4\big(1 + |B(0, R+1)|\big)\|u\|_{W^{1,p(\cdot)}(\mathbb{R}^n)}.$$

Since η is one in $B(0, R)$ we obtain that $u\eta$ is greater than or equal to one in an open set containing $E \cap B(0, R)$. Thus we have $u\eta \in S_{q(\cdot)}(E \cap B(0, R))$. Taking a minimizing sequence from $S_{p(\cdot)}(E \cap B(0, R))$, we get $C_{q(\cdot)}(E \cap B(0, R)) = 0$ for every $R > 0$. Since $E = \cup_{i=1}^\infty \big(E \cap B(0, i)\big)$, Proposition 10.1.8 yields the claim. $\square$

We close this section by showing that a lower estimate for the capacity of a ball can be easily derived from a suitable test function.

Remark 10.1.11. Let $B(x, r) \subset \mathbb{R}^n$ with $r \leqslant 1$. Assume that $p \in \mathcal{P}^{\log}(\mathbb{R}^n)$ is bounded. Let u be a function which equals 1 on $B(x, r)$, $3 - 2|y - x|/r$ on $B(x, 3r/2) \setminus B(x, r)$ and 0 otherwise. The function u is not a suitable test function for $C_{p(\cdot)}(\overline{B}(x, r))$ but for every $\varepsilon > 0$ the function $(1 + \varepsilon)u$ is. Thus

$$C_{p(\cdot)}(\overline{B}(x, r)) \leqslant (1 + \varepsilon)^{p^+}\overline{\varrho}_{L^{p(\cdot)}(B(x,2r))}(u) + (1 + \varepsilon)^{p^+}\overline{\varrho}_{L^{p(\cdot)}(B(x,2r))}(\nabla u).$$

For the first modular we obtain, since $r \leqslant 1$, that

$$\overline{\varrho}_{L^{p(\cdot)}(B(x,2r))}(u) \leqslant |B(x, 2r)| \leqslant c r^n \leqslant c r^{n - p(x)}.$$

For the second modular we calculate

$$\overline{\varrho}_{p(\cdot)}(\nabla u) \leqslant c \int\limits_{B(x,2r)} r^{-p(y)}\, dy$$

$$\leqslant c \int\limits_{B(x,2r)} r^{-p_{B(x,2r)}^+}\, dy = c\, r^{n-p_{B(x,2r)}^+}.$$

Since p is log-Hölder continuous we obtain $r^{n-p_{B(x,2r)}^+} \leqslant c\, r^{n-p(x)}$. Letting $\varepsilon \to 0$ we see that

$$C_{p(\cdot)}(\overline{B}(x,r)) \leqslant c\, r^{n-p(x)}.$$

10.2   Relative Capacity

In this section we introduce an alternative to the Sobolev $p(\cdot)$-capacity, in which the capacity of a set is taken relative to a surrounding open subset of $\mathbb{R}^n$. This section is based on Harjulehto, Hästö and Koskenoja [193].

Definition 10.2.1. Let $p \in \mathcal{P}(\Omega)$ and suppose that K is a compact subset of Ω. We denote

$$R_{p(\cdot)}(K,\Omega) := \{u \in W^{1,p(\cdot)}(\Omega) \cap C_0(\Omega) \colon u > 1 \text{ on } K \text{ and } u \geqslant 0\}$$

and define

$$\mathrm{cap}_{p(\cdot)}^*(K,\Omega) := \inf_{u \in R_{p(\cdot)}(K,\Omega)} \overline{\varrho}_{p(\cdot)}(\nabla u).$$

Further, if $U \subset \Omega$ is open, then we set

$$\mathrm{cap}_{p(\cdot)}(U,\Omega) := \sup_{\substack{K \subset U \\ \text{compact}}} \mathrm{cap}_{p(\cdot)}^*(K,\Omega),$$

and for an arbitrary set $E \subset \Omega$ we define

$$\mathrm{cap}_{p(\cdot)}(E,\Omega) := \inf_{\substack{E \subset U \subset \Omega \\ U \text{ open}}} \mathrm{cap}_{p(\cdot)}(U,\Omega).$$

The number $\mathrm{cap}_{p(\cdot)}(E,\Omega)$ is called the *variational $p(\cdot)$-capacity* of the *condenser* (E,Ω) or the *relative $p(\cdot)$-capacity* of E with respect to Ω.

If p is bounded we may use a different set of test functions familiar from the constant exponent case.

Proposition 10.2.2. *If $p \in \mathcal{P}(\Omega)$ is bounded, then*

$$\mathrm{cap}^*_{p(\cdot)}(K, \Omega) = \inf_{u \in \tilde{R}_{p(\cdot)}(K,\Omega)} \int_{\Omega} |\nabla u|^{p(x)} \, dx,$$

where $\tilde{R}_{p(\cdot)}(K, \Omega) := \{u \in W^{1,p(\cdot)}(\Omega) \cap C_0(\Omega) \colon u \geqslant 1 \text{ on } K\}$.

Proof. Since $R_{p(\cdot)}(K, \Omega) \subset \tilde{R}_{p(\cdot)}(K, \Omega)$ we obtain $\inf_{u \in \tilde{R}_{p(\cdot)}(K,\Omega)} \overline{\varrho}_{p(\cdot)}(\nabla u) \leqslant$ $\mathrm{cap}^*_{p(\cdot)}(K, \Omega)$. To prove the opposite inequality fix $\varepsilon > 0$ and let $u \in \tilde{R}_{p(\cdot)}(K, \Omega)$ be such that $\overline{\varrho}_{p(\cdot)}(\nabla u) \leqslant \inf_{u \in \tilde{R}_{p(\cdot)}(K,\Omega)} \overline{\varrho}_{p(\cdot)}(\nabla u) + \varepsilon$. Then $v := (1+\varepsilon)u > 1$ in K. Thus

$$\mathrm{cap}^*_{p(\cdot)}(K, \Omega) \leqslant (1 + \varepsilon)^{p^+} \overline{\varrho}_{p(\cdot)}(\nabla u)$$
$$\leqslant (1 + \varepsilon)^{p^+} \left(\inf_{u \in \tilde{R}_{p(\cdot)}(K,\Omega)} \overline{\varrho}_{p(\cdot)}(\nabla u) + \varepsilon \right),$$

from which the conclusion follows as $\varepsilon \to 0$. $\qquad\square$

It is not clear form the definition that $\mathrm{cap}^*_{p(\cdot)}(K, \Omega)$ and $\mathrm{cap}_{p(\cdot)}(K, \Omega)$ give the same value for any compact set. Next we show that they are the same i.e. we show that the relative capacity is well defined on compact sets. Note that the test function set $\tilde{R}_{p(\cdot)}$ does not work in this proof if p is unbounded.

Proposition 10.2.3. *Let $p \in \mathcal{P}(\Omega)$. Then $\mathrm{cap}^*_{p(\cdot)}(K, \Omega) = \mathrm{cap}_{p(\cdot)}(K, \Omega)$ for every compact set $K \subset \Omega$.*

Proof. The inequality $\mathrm{cap}^*_{p(\cdot)}(K, \Omega) \leqslant \mathrm{cap}_{p(\cdot)}(K, \Omega)$ follows directly from the definition. To prove the opposite inequality fix $\varepsilon > 0$ and let $u \in R_{p(\cdot)}(K, \Omega)$ be such that $\overline{\varrho}_{p(\cdot)}(\nabla u) \leqslant \mathrm{cap}^*_{p(\cdot)}(K, \Omega) + \varepsilon$. Then u is greater than one in $U = u^{-1}(1, \infty)$, which is open, since $u \in C_0(\Omega)$, and contains K. Thus u is also greater than one in every compact $K' \subset U$, and it follows that

$$\mathrm{cap}_{p(\cdot)}(U, \Omega) \leqslant \overline{\varrho}_{p(\cdot)}(\nabla u).$$

This implies that

$$\mathrm{cap}_{p(\cdot)}(K, \Omega) \leqslant \mathrm{cap}_{p(\cdot)}(U, \Omega) \leqslant \mathrm{cap}^*_{p(\cdot)}(K, \Omega) + \varepsilon,$$

from which the conclusion follows as $\varepsilon \to 0$. $\qquad\square$

We want to show that the variational capacity has the same basic properties as the Sobolev capacity even without the assumption $1 < p^- \leqslant p^+ < \infty$. Let us record the following immediate consequences of the definition:

(C1) $\mathrm{cap}_{p(\cdot)}(\emptyset, \Omega) = 0$;

(C2) If $E_1 \subset E_2 \subset \Omega_2 \subset \Omega_1$, then $\mathrm{cap}_{p(\cdot)}(E_1, \Omega_1) \leqslant \mathrm{cap}_{p(\cdot)}(E_2, \Omega_2)$; and

(C3) For an arbitrary set $E \subset \Omega$

$$\mathrm{cap}_{p(\cdot)}(E, \Omega) = \inf_{\substack{E \subset U \subset \Omega \\ U \text{ open}}} \mathrm{cap}_{p(\cdot)}(U, \Omega).$$

Thus the relative $p(\cdot)$-capacity is an outer capacity.

The proofs of the following properties (C4) and (C5) are the same as the proofs of the properties (C4) and (C5) in Theorem 10.1.2.

Theorem 10.2.4. *Suppose that $p \in \mathcal{P}(\Omega)$. The set function $E \mapsto \mathrm{cap}_{p(\cdot)}$ (E, Ω) has the following properties:*

(C4) *If K_1 and K_2 are compact subsets of Ω, then*

$$\mathrm{cap}_{p(\cdot)}(K_1 \cup K_2, \Omega) + \mathrm{cap}_{p(\cdot)}(K_1 \cap K_2, \Omega) \leqslant \mathrm{cap}_{p(\cdot)}(K_1, \Omega)$$
$$+ \mathrm{cap}_{p(\cdot)}(K_2, \Omega).$$

(C5) *If $K_1 \supset K_2 \supset \cdots$ are compact subsets of Ω, then*

$$\lim_{i \to \infty} \mathrm{cap}_{p(\cdot)}(K_i, \Omega) = \mathrm{cap}_{p(\cdot)} \left(\bigcap_{i=1}^{\infty} K_i, \Omega \right).$$

To prove the next property, we need the following lemma, which is based on iterated use of (C4).

Lemma 10.2.5. *Suppose that $p \in \mathcal{P}(\Omega)$ and $E_1, \ldots, E_k \subset \Omega$. Then*

$$\mathrm{cap}_{p(\cdot)} \left(\bigcup_{i=1}^{k} E_i, \Omega \right) - \mathrm{cap}_{p(\cdot)} \left(\bigcup_{i=1}^{k} A_i, \Omega \right)$$
$$\leqslant \sum_{i=1}^{k} \left(\mathrm{cap}_{p(\cdot)}(E_i, \Omega) - \mathrm{cap}_{p(\cdot)}(A_i, \Omega) \right)$$

whenever $A_i \subset E_i$, $i = 1, 2, \ldots, k$, and $\mathrm{cap}_{p(\cdot)}(\bigcup_{i=1}^{k} E_i, \Omega) < \infty$.

Proof. Assume first that each E_i and each A_i is compact. Note that if $K' \subset K$ and F are compact subsets of Ω, then it follows by properties (C2) and (C4) that

$$\mathrm{cap}_{p(\cdot)}(K \cup F, \Omega) + \mathrm{cap}_{p(\cdot)}(K', \Omega)$$
$$\leqslant \mathrm{cap}_{p(\cdot)}(K \cup (K' \cup F), \Omega) + \mathrm{cap}_{p(\cdot)}(K \cap (K' \cup F), \Omega)$$
$$\leqslant \mathrm{cap}_{p(\cdot)}(K, \Omega) + \mathrm{cap}_{p(\cdot)}(K' \cup F, \Omega)$$

or, equivalently,

$$\mathrm{cap}_{p(\cdot)}(K \cup F, \Omega) - \mathrm{cap}_{p(\cdot)}(K' \cup F, \Omega) \leqslant \mathrm{cap}_{p(\cdot)}(K, \Omega) - \mathrm{cap}_{p(\cdot)}(K', \Omega).$$

By choosing $E_1 = K$ and $A_1 = K' = F$ we obtain the claim for $k = 1$. We prove the claim for $k \geqslant 2$ by induction. Using the previous inequality we obtain

$$\mathrm{cap}_{p(\cdot)} \left(\bigcup_{i=1}^{k} E_i, \Omega \right) - \mathrm{cap}_{p(\cdot)} \left(\bigcup_{i=1}^{k} A_i, \Omega \right)$$

$$= \mathrm{cap}_{p(\cdot)} \left(\left(\bigcup_{i=1}^{k-1} E_i \right) \cup E_k, \Omega \right) - \mathrm{cap}_{p(\cdot)} \left(\left(\bigcup_{i=1}^{k-1} A_i \right) \cup E_k, \Omega \right)$$

$$+ \mathrm{cap}_{p(\cdot)} \left(E_k \cup \bigcup_{i=1}^{k-1} A_i, \Omega \right) - \mathrm{cap}_{p(\cdot)} \left(A_k \cup \bigcup_{i=1}^{k-1} A_i, \Omega \right)$$

$$= \mathrm{cap}_{p(\cdot)} \left(\bigcup_{i=1}^{k-1} E_i \right) - \mathrm{cap}_{p(\cdot)} \left(\bigcup_{i=1}^{k-1} A_i, \Omega \right)$$

$$+ \mathrm{cap}_{p(\cdot)} \left(E_k, \Omega \right) - \mathrm{cap}_{p(\cdot)} \left(A_k, \Omega \right).$$

By the induction assumption we obtain

$$\mathrm{cap}_{p(\cdot)} \left(\bigcup_{i=1}^{k} E_i, \Omega \right) - \mathrm{cap}_{p(\cdot)} \left(\bigcup_{i=1}^{k} A_i, \Omega \right)$$

$$\leqslant \sum_{i=1}^{k-1} \left(\mathrm{cap}_{p(\cdot)}(E_i, \Omega) - \mathrm{cap}_{p(\cdot)}(A_i, \Omega) \right)$$

$$+ \mathrm{cap}_{p(\cdot)}(E_k, \Omega) - \mathrm{cap}_{p(\cdot)}(A_k, \Omega)$$

$$= \sum_{i=1}^{k} \left(\mathrm{cap}_{p(\cdot)}(E_i, \Omega) - \mathrm{cap}_{p(\cdot)}(A_i, \Omega) \right).$$

Thus the assertion of the lemma is proved if all E_i and A_i are compact sets.

Assume now that each E_i and each A_i is open. Let $\varepsilon > 0$. We choose compact sets $F_i \subset A_i$ such that

$$\mathrm{cap}_{p(\cdot)}(F_i, \Omega) \geqslant \mathrm{cap}_{p(\cdot)}(A_i, \Omega) - \varepsilon$$

and a compact set $K \subset \bigcup_i E_i$ such that $F_i \subset K$ for every $i = 1, \ldots, k$ and

$$\mathrm{cap}_{p(\cdot)}(K, \Omega) \geqslant \mathrm{cap}_{p(\cdot)} \left(\bigcup_{i=1}^{k} E_i, \Omega \right) - \varepsilon.$$

We define
$$K_i := \{x \in K \cap E_i : \operatorname{dist}(x, \partial E_i) \geqslant \delta\}$$
where $\delta := \min\big\{\operatorname{dist}(K, \mathbb{R}^n \setminus \cup_{i=1}^{\infty} E_i), \operatorname{dist}(F_1, \partial A_1), \ldots, \operatorname{dist}(F_k, \partial A_k)\big\}$. Now we have $F_i \subset K_i$ and $K = \bigcup_{i=1}^{k} K_i$. Using the previous case we get

$$\operatorname{cap}_{p(\cdot)}\left(\bigcup_{i=1}^{k} E_i, \Omega\right) - \operatorname{cap}_{p(\cdot)}\left(\bigcup_{i=1}^{k} A_i, \Omega\right)$$

$$\leqslant \operatorname{cap}_{p(\cdot)}\left(\bigcup_{i=1}^{k} K_i, \Omega\right) - \operatorname{cap}_{p(\cdot)}\left(\bigcup_{i=1}^{k} F_i, \Omega\right) + \varepsilon$$

$$\leqslant \sum_{i=1}^{k}\left(\operatorname{cap}_{p(\cdot)}(K_i, \Omega) - \operatorname{cap}_{p(\cdot)}(F_i, \Omega)\right) + \varepsilon$$

$$\leqslant \sum_{i=1}^{k}\left(\operatorname{cap}_{p(\cdot)}(E_i, \Omega) - \operatorname{cap}_{p(\cdot)}(A_i, \Omega)\right) + (k+1)\varepsilon.$$

Letting $\varepsilon \to 0$ we obtain the claim for open sets E_i and A_i.

Assume now that E_i and A_i are arbitrary sets. Let $\varepsilon > 0$. We choose open sets U_i such that $E_i \subset U_i$ and $\operatorname{cap}_{p(\cdot)}(U_i, \Omega) \leqslant \operatorname{cap}_{p(\cdot)}(E_i, \Omega) + \varepsilon$ for every $i = 1, \ldots k$. Then we choose open sets V_i such that $A_i \subset V_i \subset U_i$ and

$$\operatorname{cap}_{p(\cdot)}\left(\bigcup_{i=1}^{k} V_i, \Omega\right) \leqslant \operatorname{cap}_{p(\cdot)}\left(\bigcup_{i=1}^{k} A_i, \Omega\right) + \varepsilon.$$

Now it follows by the previous case that

$$\operatorname{cap}_{p(\cdot)}\left(\bigcup_{i=1}^{k} E_i, \Omega\right) - \operatorname{cap}_{p(\cdot)}\left(\bigcup_{i=1}^{k} A_i, \Omega\right)$$

$$\leqslant \operatorname{cap}_{p(\cdot)}\left(\bigcup_{i=1}^{k} U_i, \Omega\right) - \operatorname{cap}_{p(\cdot)}\left(\bigcup_{i=1}^{k} V_i, \Omega\right) + \varepsilon$$

$$\leqslant \sum_{i=1}^{k}\left(\operatorname{cap}_{p(\cdot)}(U_i, \Omega) - \operatorname{cap}_{p(\cdot)}(V_i, \Omega)\right) + \varepsilon$$

$$\leqslant \sum_{i=1}^{k}\left(\operatorname{cap}_{p(\cdot)}(E_i, \Omega) - \operatorname{cap}_{p(\cdot)}(A_i, \Omega)\right) + (k+1)\varepsilon.$$

Letting $\varepsilon \to 0$ we obtain the claim for arbitrary sets E_i and A_i. This completes the proof. $\qquad\square$

Theorem 10.2.6. *Let $p \in \mathcal{P}(\Omega)$. The set function $E \mapsto \mathrm{cap}_{p(\cdot)}(E, \Omega)$ has the following properties:*

(C6) *If $E_1 \subset E_2 \subset \cdots$ are subsets of Ω, then*

$$\lim_{i \to \infty} \mathrm{cap}_{p(\cdot)}(E_i, \Omega) = \mathrm{cap}_{p(\cdot)}\left(\bigcup_{i=1}^{\infty} E_i, \Omega \right).$$

(C7) *For $E_i \subset \Omega$, $i \in \mathbb{N}$, we have*

$$\mathrm{cap}_{p(\cdot)}\left(\bigcup_{i=1}^{\infty} E_i, \Omega \right) \leqslant \sum_{i=1}^{\infty} \mathrm{cap}_{p(\cdot)}(E_i, \Omega).$$

Proof. Denote $E := \bigcup_{i=1}^{\infty} E_i$. Note first that (C2) implies

$$\lim_{i \to \infty} \mathrm{cap}_{p(\cdot)}(E_i, \Omega) \leqslant \mathrm{cap}_{p(\cdot)}(E, \Omega).$$

To prove the opposite inequality we may assume that $\lim_{i \to \infty} \mathrm{cap}_{p(\cdot)}(E_i, \Omega) < \infty$. Consequently, it follows by (C2) that $\mathrm{cap}_{p(\cdot)}(E_i, \Omega) < \infty$ for each i. Fix $\varepsilon > 0$ and choose open sets U_i such that $E_i \subset U_i \subset \Omega$ and

$$\mathrm{cap}_{p(\cdot)}(U_i, \Omega) \leqslant \mathrm{cap}_{p(\cdot)}(E_i, \Omega) + \varepsilon \, 2^{-i}.$$

Using Lemma 10.2.5 we derive from this

$$\mathrm{cap}_{p(\cdot)}\left(\bigcup_{i=1}^{k} U_i, \Omega \right) - \mathrm{cap}_{p(\cdot)}\left(\bigcup_{i=1}^{k} E_i, \Omega \right) \leqslant \sum_{i=1}^{k} \varepsilon \, 2^{-i} < \varepsilon.$$

If $K \subset \bigcup_{1=i}^{\infty} U_i$ is compact, then $K \subset \bigcup_{1=i}^{k} U_i$ for some k, and we have

$$\mathrm{cap}_{p(\cdot)}(K, \Omega) \leqslant \mathrm{cap}_{p(\cdot)}\left(\bigcup_{i=1}^{k} U_i, \Omega \right) \leqslant \mathrm{cap}_{p(\cdot)}\left(\bigcup_{i=1}^{k} E_i, \Omega \right) + \varepsilon$$
$$= \mathrm{cap}_{p(\cdot)}(E_k, \Omega) + \varepsilon \leqslant \lim_{k \to \infty} \mathrm{cap}_{p(\cdot)}(E_k, \Omega) + \varepsilon,$$

where we used that $\bigcup_{i=1}^{k} E_i = E_k$. Since $\bigcup_{i=1}^{\infty} U_i$ is open, we obtain that

$$\mathrm{cap}_{p(\cdot)}(E, \Omega) \leqslant \mathrm{cap}_{p(\cdot)}\left(\bigcup_{i=1}^{\infty} U_i, \Omega \right) = \sup_{K} \mathrm{cap}_{p(\cdot)}(K, \Omega)$$
$$\leqslant \lim_{k \to \infty} \mathrm{cap}_{p(\cdot)}(E_k, \Omega) + \varepsilon,$$

where the supremum is taken over all compact sets $K \subset \bigcup_{i=1}^{\infty} U_i$. So (C6) is proved.

The proof of property (C7) is exactly the same as the proof of property (C7) in Theorem 10.1.3.   □

In analogy with Corollaries 10.1.6 and 10.1.7 we have

Corollary 10.2.7. *If $p \in \mathcal{P}(\Omega)$, then the relative $p(\cdot)$-capacity is an outer measure and a Choquet capacity. In particular, all Borel sets $E \subset \Omega$ are capacitable, this is,*

$$\sup_{\substack{K \subset E \\ K \text{ compact}}} \text{cap}_{p(\cdot)}(K, \Omega) = \text{cap}_{p(\cdot)}(E, \Omega) = \inf_{\substack{E \subset U \subset \Omega \\ U \text{ open}}} \text{cap}_{p(\cdot)}(U, \Omega).$$

In contrast to the Sobolev capacity, where $1 < p^- \leqslant p^+ < \infty$ was needed for these properties, here we need only measurability of the exponent p.

As has been explained in Chap. 9, smooth functions are not always dense in the variable exponent Sobolev space. However, when they are, we can use the usual change of test-function set. The proof is the same as in the Sobolev capacity case, Lemma 10.1.9, and hence omitted.

Proposition 10.2.8. *Suppose that $p \in \mathcal{P}(\Omega)$ is a bounded exponent and that $C^\infty(\Omega) \cap W^{1,p(\cdot)}(\Omega)$ is dense in $W^{1,p(\cdot)}(\Omega)$. If $K \subset \Omega$ is compact, then*

$$\text{cap}_{p(\cdot)}(K, \Omega) = \inf_{u \in R_0^\infty(K,\Omega)} \int_\Omega |\nabla u(x)|^{p(x)} \, dx,$$

where $R_0^\infty(K, \Omega) := \{u \in C_0^\infty(\Omega) : u > 1 \text{ on } K\}$.

Using a suitable test function, it is easy to obtain an upper estimate for $\text{cap}_{p(\cdot)}(\overline{B}(x,r), B(x,2r))$.

Lemma 10.2.9. *Let $p \in \mathcal{P}^{\log}(\Omega)$ be bounded. For every ball $B(x, 2r) \subset \Omega$ with $r \leqslant 1$ we have*

$$\text{cap}_{p(\cdot)} \left(\overline{B}(x,r), B(x,2r) \right) \leqslant c \, r^{n-p(x)},$$

where c depends on n and the $\log$-Hölder constant of p.

Proof. Let u be a function which equals 1 on $B(x,r)$, $3 - 2|y - x|/r$ on $B(x, 3r/2) \setminus B(x,r)$ and 0 otherwise. Since u is a suitable test function for the capacity of the pair $(\overline{B}(x,r), B(x,2r))$ (Proposition 10.2.2), we find that

$$\text{cap}_{p(\cdot)}(\overline{B}(x,r), B(x,2r)) \leqslant \overline{\varrho}_{p(\cdot)}(\nabla u) \leqslant c \int_{B(x,2r)} r^{-p(y)} \, dy$$

$$\leqslant c \int_{B(x,2r)} r^{-p_{B(x,2r)}^+} \, dy = c \, r^{n-p_{B(x,2r)}^+}.$$

Since p is log-Hölder continuous we obtain $r^{n-p_{B(x,2r)}^+} \leqslant c \, r^{n-p(x)}$.   □

The following proposition gives a lower bound for $\text{cap}_{p(\cdot)}(\overline{B}(x,r), B(x,R))$. In particular we prove that

$$\text{cap}_{p(\cdot)}(\overline{B}(x,r), B(x,2r)) \geqslant c(n,p)\, r^{n-p(x)}$$

for $p(x) < n$. The result is from Alkhutov and Krasheninnikova [20, Proposition 5.2].

Proposition 10.2.10. *Let $p \in \mathcal{P}^{\log}(B(x,R))$ satisfy $1 < p^- \leqslant p^+ < \infty$. Assume that $\overline{B}(x,r) \subset B(x,R)$. If $p(x) \neq n$, then*

$$\text{cap}_{p(\cdot)}(\overline{B}(x,r), B(x,R)) \geqslant c \left| \frac{p(x)-1}{n-p(x)} \right|^{1-p(x)} \left| R^{\frac{p(x)-n}{p(x)-1}} - r^{\frac{p(x)-n}{p(x)-1}} \right|^{1-p(x)}.$$

If $p(x) = n$, then

$$\text{cap}_{p(\cdot)}(\overline{B}(x,r), B(x,R)) \geqslant c \log\left(\frac{R}{r}\right)^{1-n}.$$

In both cases the constant depends only on the dimension n, p and R.

Proof. We assume without loss of generality that $x = 0$. We change to spherical coordinates $z = (\varrho, \omega)$ with $|\omega| = 1$. Then we have

$$\text{cap}_{p(\cdot)}(\overline{B}(0,r), B(0,R)) \geqslant \inf_u \int\limits_{\partial B(0,1)} \int\limits_r^R \left| \frac{\partial u}{\partial \varrho} \right|^{p(z)} \varrho^{n-1} \, d\varrho \, d\omega$$

where the infimum is taken over continuous Sobolev functions u with compact support in $B(0,R)$ being equal to one on $\overline{B}(0,r)$. We are therefore led to minimize the integral

$$\int\limits_r^R |\psi'(\varrho)|^{p(\varrho)} \varrho^{n-1} \, d\varrho$$

among functions $\psi \in W^{1,p(\cdot)}(r,R)$ with $\psi(r) = 1$ and $\psi(R) = 0$. It is clear that we can assume $\psi' \leqslant 0$ when looking for the minimizer. Let (ψ_i) be a minimizing sequence. Then it is bounded in $W^{1,p(\cdot)}(r,R)$. Since the variable exponent Sobolev space is reflexive (Theorem 8.1.6), we find a subsequence, denoted by (ψ_i) again, converging weakly to $\psi \in W^{1,p(\cdot)}(r,R)$ and by the weak (sequential) lower semicontinuity of the modular (Theorem 3.2.9) we have

$$\int\limits_r^R |\psi'(\varrho)|^{p(\varrho)} \varrho^{n-1} \, d\varrho \leqslant \lim_{i \to \infty} \int\limits_r^R |\psi_i'(\varrho)|^{p(\varrho)} \varrho^{n-1} \, d\varrho.$$

By the Banach-Saks property, Corollary 8.1.7, we find a sequence (Ψ_i) of convex combinations of ψ_i converging to ψ in $W^{1,p(\cdot)}(r,R)$, and hence $0 \leqslant \psi \leqslant 1$ and

$$\int_r^R |\psi'(\varrho)|^{p(\varrho)} \varrho^{n-1}\, d\varrho = \lim_{i \to \infty} \int_r^R |\Psi_i'(\varrho)|^{p(\varrho)} \varrho^{n-1}\, d\varrho.$$

Since each Ψ_i is a linear combination, we find $\int_r^R \Psi_i'(\varrho)d\varrho \geqslant 1$. This yields $\int_r^R \psi'(\varrho)d\varrho \geqslant 1$ and thus ψ is the minimizer. Using $\varrho^{n-1}\, d\varrho$ as a measure, we obtain as in the proof of Lemma 13.1.3 that $p(\varrho)(\psi'(\varrho))^{p(\varrho)-1}$ is a constant almost everywhere. Thus it follows that every radial minimizer has a derivative of the form

$$\xi(\varrho) := \left(\frac{c}{p(\varrho)}\right)^{\frac{1}{p(\varrho)-1}},$$

where the constant c depends on the direction. Since $1 < p^- \leqslant p^+ < \infty$ and $\int_r^R |\xi|d\varrho = 1$, we obtain

$$c \geqslant \min\left\{1, \frac{1}{R^{p^+-1}}\right\}.$$

From now on we assume that this lower bound is used in the definition of ξ. By the log-Hölder continuity of p we obtain $|\xi|^{p(\varrho,\omega)} \geqslant c\,|\xi|^{p(0)}$. We therefore conclude that

$$\operatorname{cap}_{p(\cdot)}(\overline{B}(0,r), B(0,R)) \geqslant c \int_{\partial B(0,1)} \int_r^R |\xi(\varrho)|^{p(0)} \varrho^{n-1}\, d\varrho\, d\omega$$

$$\geqslant c \int_r^R |\xi(\varrho)|^{p(0)} \varrho^{n-1}\, d\varrho.$$

Hölder's inequality gives

$$1 \leqslant \int_r^R |\xi|\, d\varrho \leqslant \left(\int_r^R |\xi|^{p(0)} \varrho^{n-1}\, d\varrho\right)^{\frac{1}{p(0)}} \left(\int_r^R \varrho^{\frac{1-n}{p(0)-1}}\, d\varrho\right)^{\frac{1}{p(0)'}}.$$

Raising this to the power $p(0)$ and rearranging gives

$$\left(\int_r^R \varrho^{\frac{1-n}{p(0)-1}}\, d\varrho\right)^{1-p(0)} \leqslant \int_r^R |\xi|^{p(0)} \varrho^{n-1}\, d\varrho.$$

Thus we obtain

$$\text{cap}_{p(\cdot)}(\overline{B}(0,r), B(0,R)) \geqslant c \left(\int_r^R \varrho^{\frac{1-n}{p(0)-1}} \, d\varrho \right)^{1-p(0)}.$$

By a simple computation we obtain

$$\int_r^R \varrho^{\frac{1-n}{p(0)-1}} \, d\varrho = \begin{cases} \left| \frac{p(0)-1}{p(0)-n} \right| \left| R^{\frac{p(0)-n}{p(0)-1}} - r^{\frac{p(0)-n}{p(0)-1}} \right|, & \text{for } p(0) \neq n, \\ \log \left(\frac{R}{r} \right), & \text{for } p(0) = n. \end{cases}$$

Combining these estimates yields the claim.   □

10.3   The Relationship Between the Capacities

The following two theorems associate the Sobolev $p(\cdot)$-capacity and relative $p(\cdot)$-capacity. Specifically, we give sufficient conditions on the exponent p that sets of capacity zero coincide. This section is based on Harjulehto, Hästö and Koskenoja [193].

Lemma 10.3.1. *Assume that $p \in \mathcal{P}(\mathbb{R}^n)$ is bounded. If Ω is bounded and $K \subset \Omega$ is compact, then*

$$C_{p(\cdot)}(K) \leqslant c \max \left\{ \text{cap}_{p(\cdot)}(K, \Omega)^{\frac{1}{p^+}}, \text{cap}_{p(\cdot)}(K, \Omega) \right\},$$

where the constant c depends on the dimension n and $\text{diam}(\Omega)$.

Proof. We may assume that $\text{cap}_{p(\cdot)}(K, \Omega) < \infty$. Let $u \in R_{p(\cdot)}(K, \Omega)$ be a function with $\varrho_{p(\cdot)}(|\nabla u|) < \infty$. Let us extend u by zero outside of Ω and set $v := \min\{1, u\}$. Since u is greater than one in an open set containing K we obtain that $v \in S_{p(\cdot)}(K)$ and thus

$$C_{p(\cdot)}(K) \leqslant \int_{\mathbb{R}^n} |v(x)|^{p(x)} + |\nabla v(x)|^{p(x)} \, dx$$

$$\leqslant \int_{\mathbb{R}^n} |v(x)|^{p(x)} + |\nabla u(x)|^{p(x)} \, dx.$$

Since $0 \leqslant v \leqslant 1$, we have

$$\int_\Omega |v(x)|^{p(x)} \, dx \leqslant \int_\Omega |v(x)| \, dx \leqslant \int_\Omega |u(x)| \, dx.$$

Using the classical Poincaré inequality in $L^1(\Omega)$ and the embedding $L^{p(\cdot)}(\Omega)$ $\hookrightarrow L^1(\Omega)$ we obtain that

$$\|u\|_1 \leqslant c\operatorname{diam}(\Omega)\|\nabla u\|_1 \leqslant c\operatorname{diam}(\Omega)\big(1+|\Omega|\big)\|\nabla u\|_{L^{p(\cdot)}(\Omega)}.$$

Since Lemma 3.2.5 implies

$$\|\nabla u\|_{L^{p(\cdot)}(\Omega)} \leqslant \max\left\{\overline{\varrho}_{p(\cdot)}(|\nabla u|)^{\frac{1}{p^+}}, \overline{\varrho}_{p(\cdot)}(|\nabla u|)^{\frac{1}{p^-}}\right\},$$

the result follows by taking a minimizing sequence.   □

Theorem 10.3.2. *Assume that $p \in \mathcal{P}(\mathbb{R}^n)$ with $1 < p^- \leqslant p^+ < \infty$. If Ω is bounded and $E \subset \Omega$, then*

$$C_{p(\cdot)}(E) \leqslant c\max\{\operatorname{cap}_{p(\cdot)}(E,\Omega)^{\frac{1}{p^+}}, \operatorname{cap}_{p(\cdot)}(E,\Omega)\},$$

where the constant c depends on the dimension n and $\operatorname{diam}(\Omega)$.

Proof. Let $E \subset \Omega$ be a set with $\operatorname{cap}_{p(\cdot)}(E,\Omega) < \infty$. By the definition of $\operatorname{cap}_{p(\cdot)}(E,\Omega)$ there exists open sets $U_i \supset E$ with $\operatorname{cap}_{p(\cdot)}(U_i,\Omega) \to \operatorname{cap}_{p(\cdot)}(E,\Omega)$ as $i \to \infty$. Let $U := \bigcap_{i=1}^\infty U_i$. Then U is a Borel set and hence by the Choquet property (Corollary 10.1.7)

$$C_{p(\cdot)}(E) \leqslant C_{p(\cdot)}(U) = \sup_K C_{p(\cdot)}(K)$$

where the supremum is taken over all compact sets $K \subset U$. Using Lemma 10.3.1 and the Choquet property (Corollary 10.2.7) we obtain

$$C_{p(\cdot)}(E) \leqslant c\sup_K \max\{\operatorname{cap}_{p(\cdot)}(K,\Omega)^{\frac{1}{p^+}}, \operatorname{cap}_{p(\cdot)}(K,\Omega)\}$$

$$\leqslant c\max\{\operatorname{cap}_{p(\cdot)}(U,\Omega)^{\frac{1}{p^+}}, \operatorname{cap}_{p(\cdot)}(U,\Omega)\}.$$

Since $\operatorname{cap}_{p(\cdot)}(U,\Omega) = \operatorname{cap}_{p(\cdot)}(E,\Omega)$ the claim follows.   □

Remark 10.3.3. Assume that $p \in \mathcal{P}(\mathbb{R}^n)$ with $1 < p^- \leqslant p^+ < \infty$ is such that the Poincaré inequality holds, for example $p \in \mathcal{A}$ with $1 < p^- \leqslant p^+ < \infty$. We have, using the notation of Lemma 10.3.1,

$$\|v\|_{L^{p(\cdot)}(\Omega)} \leqslant \|u\|_{L^{p(\cdot)}(\Omega)} \leqslant c\|\nabla u\|_{L^{p(\cdot)}(\Omega)}.$$

Thus the proofs of Lemma 10.3.1 and Theorem 10.3.2, and Lemma 3.2.5 yield

$$C_{p(\cdot)}(E) \leqslant c\max\{\operatorname{cap}_{p(\cdot)}(E,\Omega)^{\frac{p^-}{p^+}}, \operatorname{cap}_{p(\cdot)}(E,\Omega)^{\frac{p^+}{p^-}}\},$$

where c depends on n and the constant in the Poincaré inequality.

To get the converse implication we need to assume that continuous functions are dense in the variable exponent Sobolev space.

Proposition 10.3.4. *Let $p \in \mathcal{P}(\mathbb{R}^n)$ be a bounded exponent. Suppose that $W^{1,p(\cdot)}(\mathbb{R}^n) \cap C(\mathbb{R}^n)$ is dense in $W^{1,p(\cdot)}(\mathbb{R}^n)$. If $E \subset \Omega$ with $C_{p(\cdot)}(E) = 0$, then $\mathrm{cap}_{p(\cdot)}(E, \Omega) = 0$.*

Proof. Let $K \subset \Omega$ be compact with $C_{p(\cdot)}(K) = 0$. By the density of continuous functions, it follows as in Lemma 10.1.9 that the set of admissible functions in the definition of the Sobolev $p(\cdot)$-capacity can be replaced by the subset $S^0_{p(\cdot)}(K) := S_{p(\cdot)}(K) \cap C(\mathbb{R}^n)$. Therefore we may choose a sequence (u_i) of functions belonging to $S^0_{p(\cdot)}(K)$ such that $\|u_i\|_{W^{1,p(\cdot)}(\mathbb{R}^n)} \to 0$. Let $\eta \in C^\infty_0(\Omega)$ be a cut-off function that is one in K. It is easy to show that ηu_i is in $R_{p(\cdot)}(K, \Omega)$, so we obtain $\mathrm{cap}_{p(\cdot)}(K, \Omega) = 0$.

Let $E \subset \Omega$ with $C_{p(\cdot)}(E) = 0$. Since the Sobolev capacity is an outer capacity, there exists a sequence of open sets $U_i \supset E$ with $C_{p(\cdot)}(U_i) \to 0$ as $i \to \infty$. Let $U := \bigcap_{i=1}^\infty U_i \cap \Omega$. Then U is a Borel set containing E which satisfies $C_{p(\cdot)}(U) = 0$. By the Choquet property (Corollary 10.2.7) we obtain

$$\mathrm{cap}_{p(\cdot)}(E, \Omega) \leqslant \mathrm{cap}_{p(\cdot)}(U, \Omega) = \sup_K \mathrm{cap}_{p(\cdot)}(K, \Omega)$$

where the supremum is taken over all compact sets $K \subset U$. By the first part of the proof we obtain $\mathrm{cap}_{p(\cdot)}(K, \Omega) = 0$ and hence the claim follows. $\square$

Theorem 10.3.5. *Let $p \in \mathcal{P}(\mathbb{R}^n)$ be a bounded exponent and suppose that $W^{1,p(\cdot)}(\mathbb{R}^n) \cap C(\mathbb{R}^n)$ is dense in $W^{1,p(\cdot)}(\mathbb{R}^n)$. Let B be a ball. For $E \subset B$ we have*

$$\mathrm{cap}_{p(\cdot)}(E, 2B) \leqslant c \left(1 + \max \left\{ \mathrm{diam}(B)^{-p^+_{2B}}, \mathrm{diam}(B)^{-p^-_{2B}} \right\} \right) C_{p(\cdot)}(E),$$

where the constant c depends only on p^+.

Proof. Let $K \subset B$ be compact. As in the previous proof we note that $S^0_{p(\cdot)}(K)$ can be used as admissible functions for the Sobolev capacity. Fix $u \in S^0_{p(\cdot)}(K)$. Let $\eta \in C^\infty_0(2B)$ be a cut-off function that is one in K and $|\nabla \eta| \leqslant c\, \mathrm{diam}(B)^{-1}$. It is easy to show that ηu belongs to $\tilde{R}_{p(\cdot)}(K, 2B)$. We obtain

$$\int_{2B} |\nabla(u\eta)|^{p(x)}\, dx$$

$$\leqslant \int_{2B} |\eta \nabla u|^{p(x)}\, dx + \int_{2B} |u \nabla \eta|^{p(x)}\, dx$$

$$\leqslant \int_{2B} |\nabla u|^{p(x)}\, dx + \max \left\{ \left(\frac{c}{\mathrm{diam}(B)} \right)^{p^+_{2B}}, \left(\frac{c}{\mathrm{diam}(B)} \right)^{p^-_{2B}} \right\} \int_{2B} |u|^{p(x)}\, dx.$$

By taking a minimizing sequence we obtain the claim for compact sets.

Let $E \subset B$. Since the Sobolev capacity is an outer capacity, there exists a sequence of open sets $E \subset U_i \subset 2B$ with $C_{p(\cdot)}(U_i) \to C_{p(\cdot)}(E)$ as $i \to \infty$. Let $U := \bigcap_{i=1}^{\infty} U_i$. Then U is a Borel set with $C_{p(\cdot)}(U) = C_{p(\cdot)}(E)$. By the Choquet property (Corollary 10.2.7) we obtain

$$\mathrm{cap}_{p(\cdot)}(E, 2B) \leqslant \mathrm{cap}_{p(\cdot)}(U, 2B) = \sup_K \mathrm{cap}_{p(\cdot)}(K, 2B)$$

where the supremum is taken over all compact sets $K \subset U$. By the first part of the proof and property (C2) of the Sobolev capacity we obtain

$$\mathrm{cap}_{p(\cdot)}(E, 2B) \leqslant c \sup_{K \subset U} C_{p(\cdot)}(K) \leqslant c\, C_{p(\cdot)}(U) = c\, C_{p(\cdot)}(E),$$

where the constant c is the one from the first part of the proof. $\square$

10.4 Sobolev Capacity and Hausdorff Measure

In this section we study how the Sobolev $p(\cdot)$-capacity relates to the Hausdorff measure. We start with some trivial conclusions which do not really use the variability of the exponent.

The following lemma follows easily from the definition of the capacity.

Lemma 10.4.1. *Let $p \in \mathcal{P}(\mathbb{R}^n)$. Every measurable set $E \subset \mathbb{R}^n$ satisfies $|E| \leqslant C_{p(\cdot)}(E)$.*

Proof. If $u \in S_{p(\cdot)}(E)$, then there is an open set $U \supset E$ such that $u \geqslant 1$ in U and hence

$$|E| \leqslant |U| \leqslant \overline{\varrho}_{p(\cdot)}(u) \leqslant \overline{\varrho}_{1, p(\cdot)}(u).$$

We obtain the claim by taking the infimum over all $p(\cdot)$-admissible functions for E. $\square$

The s-dimensional Hausdorff measure of a set $E \subset \mathbb{R}^n$ is denoted by $\mathcal{H}^s(E)$, see [129, Sect. 2.1] or [288] or Definition 10.4.4 which is classic when s is a constant function.

Proposition 10.4.2. *Suppose that $p \in \mathcal{P}(\mathbb{R}^n)$ is bounded, and let $E \subset \mathbb{R}^n$. If $C_{p(\cdot)}(E) = 0$, then $\mathcal{H}^s(E) = 0$ for all $s > n - p^-$. If $\mathcal{H}^{n-p^+}(E) < \infty$, then $C_{p(\cdot)}(E) = 0$.*

Proof. If $C_{p(\cdot)}(E) = 0$ we obtain by Proposition 10.1.10 that $C_{p^-}(E) = 0$. But we know from [129, Theorem 4, p. 156] that this implies the first claim.

It follows from [129, Theorem 3, p. 154] that $C_{p^+}(E) = 0$. And thus the second claim follows by Proposition 10.1.10. $\square$

Corollary 10.4.3. *Suppose that $p \in \mathcal{P}(\mathbb{R}^n)$ satisfies $n < p^- \leqslant p^+ < \infty$, and let $E \subset \mathbb{R}^n$. Then $C_{p(\cdot)}(E) = 0$ if and only if $E = \emptyset$.*

Proof. Since $\mathcal{H}^0$ is a counting measure [129, Theorem 2, p. 63], the implication '$\Rightarrow$' follows directly from Theorem 10.4.2. The implication in the other direction is (C1). □

Next we define the variable dimension Hausdorff measure that has earlier been used in Harjulehto, Hästö and Latvala [199].

Definition 10.4.4. Let $s \colon \mathbb{R}^n \to (0, n]$ be a continuous function. We define the *variable dimension Hausdorff measure* by first letting

$$\mathcal{H}^{s(\cdot)}_{\delta}(E) := \inf\Big\{\sum_i \operatorname{diam}(B_i)^{s(x_i)} : E \subset \bigcup_i B_i, \ \operatorname{diam}(B_i) \leqslant \delta\Big\},$$

where x_i is the center of the ball B_i, and then taking the limit:

$$\mathcal{H}^{s(\cdot)}(E) := \lim_{\delta \to 0} \mathcal{H}^{s(\cdot)}_{\delta}(E).$$

Note that the limit exists since for $\delta' < \delta$ we have $\mathcal{H}^{s(\cdot)}_{\delta}(E) \leqslant \mathcal{H}^{s(\cdot)}_{\delta'}(E)$. This construction is just a special case of a measure construction due to Carathéodory, hence $\mathcal{H}^{s(\cdot)}$ is a Borel regular outer measure [154, Sect. 2.10.1, p. 169].

There is some degree of arbitrariness in choosing the value of s at the center of the ball. We can similarly define lower and upper variable Hausdorff measures by taking the limit $\delta \to 0$ of

$$\underline{\mathcal{H}}^{s(\cdot)}_{\delta}(E) := \inf\Big\{\sum_i \operatorname{diam}(B_i)^{s^+_{B_i}} : E \subset \bigcup_i B_i, \ \operatorname{diam}(B_i) \leqslant \delta\Big\}$$

and

$$\overline{\mathcal{H}}^{s(\cdot)}_{\delta}(E) := \inf\Big\{\sum_i \operatorname{diam}(B_i)^{s^-_{B_i}} : E \subset \bigcup_i B_i, \ \operatorname{diam}(B_i) \leqslant \delta\Big\}.$$

However, we have the following uniqueness result:

Proposition 10.4.5. *If $s \colon \mathbb{R}^n \to (0, n]$ is log-Hölder continuous, then*

$$c_1 \, \mathcal{H}^{s(\cdot)}(E) \leqslant \underline{\mathcal{H}}^{s(\cdot)}(E) \leqslant \overline{\mathcal{H}}^{s(\cdot)}(E) \leqslant c_2 \, \mathcal{H}^{s(\cdot)}(E)$$

for every $E \subset \mathbb{R}^n$. The constants c_1 and c_2 depend only on the log-Hölder constant of p.

Proof. For any $x \in \mathbb{R}^n$ and $r < 1/2$ we have

$$1 \leqslant \operatorname{diam}(B(x,r))^{s_{B(x,r)}^- - s_{B(x,r)}^+} \leqslant c$$

by the log-Hölder continuity condition, so the claim follows. $\square$

The following lemma is a modification of [129, Chap. 2.4.3, Theorem 2, p. 77].

Lemma 10.4.6. *Let* $p \in \mathcal{P}(\mathbb{R}^n)$ *be bounded and* $s \colon \mathbb{R}^n \to (0,n]$ *be continuous. Then*

$$\mathcal{H}^{s(\cdot)}\left(\left\{x \in \mathbb{R}^n : \limsup_{r \to 0} r^{-s(x)} \int_{B(x,r)} |f|^{p(y)} dy = \infty\right\}\right) = 0$$

for every $f \in L^{p(\cdot)}(\mathbb{R}^n)$.

Proof. Let $\lambda > 0$ and
$$E_\lambda := \left\{x \in \mathbb{R}^n : \limsup_{r \to 0} r^{-s(x)} \int_{B(x,r)} |f|^{p(y)} dy > \lambda\right\}.$$

Let $\delta > 0$. For every $x \in E_\lambda$ there exists $r_x \leqslant \delta$ such that

$$\int_{B(x,r_x)} |f|^{p(y)} dy > \lambda r_x^{s(x)}.$$

By the basic covering theorem (Theorem 1.4.5) there exists a countable subfamily of pair-wise disjoint balls $B(x_i, r_{x_i})$ such that

$$E_\lambda \subset \bigcup_{i=1}^{\infty} B(x_i, 5r_{x_i}).$$

Since $E \subset E_\lambda$ for all $\lambda > 0$, we obtain

$$\mathcal{H}_\delta^s(E) \leqslant c \sum_i r_{x_i}^{s(x_i)} \leqslant c \sum_i \lambda^{-1} \int_{B(x,r_{x_i})} |f|^{p(y)} \, dy$$

$$\leqslant c \lambda^{-1} \int_{\mathbb{R}^n} |f|^{p(y)} \, dy$$

and by letting $\delta \to 0$

$$\mathcal{H}^s(E) \leqslant c \lambda^{-1} \int_{\mathbb{R}^n} |f|^{p(y)} \, dy.$$

This yields the claim, since λ can be chosen arbitrary large. $\square$

Theorem 10.4.7. *Let* $p \in \mathcal{P}^{\log}(\mathbb{R}^n)$ *be bounded and let* $s \colon \mathbb{R}^n \to (0, n]$ *be continuous and satisfy* $s > n - p$. *If* $C_{p(\cdot)}(E) = 0$, *then* $\mathcal{H}^{s(\cdot)}(E) = 0$.

Proof. Let $u_i \in W^{1,p(\cdot)}(\mathbb{R}^n)$, $i \in \mathbb{N}$, be such that $E \subset \mathrm{int}\{u_i \geqslant 1\}$ and $\|u_i\|_{1,p(\cdot)} \leqslant 2^{-i}$. We write $u := \sum_i u_i$ and note that $u \in W^{1,p(\cdot)}(\mathbb{R}^n)$. For every natural number k and $x \in E$ we find r such that $u_i(y) \geqslant 1$ for every $i = 1, \ldots, k$ and almost every $y \in B(x, r)$. Thus we conclude

$$\limsup_{r \to 0} \fint_{B(x,r)} u \, dy = \infty. \tag{10.4.8}$$

Suppose that x is such that

$$\limsup_{r \to 0} r^{-\frac{s(x)}{p(x)}} \|\nabla u\|_{L^{p(\cdot)}(B(x,r))} =: c < \infty.$$

Next we choose $R \in (0, 1)$ so small that

$$\|\nabla u\|_{L^{p(\cdot)}(B(x,r))} < c \, r^{\frac{s(x)}{p(x)}}$$

for every $0 < r \leqslant R$. Denote $B_i := B(x, 2^{-i}R)$. By the classical Poincaré inequality and Hölder's inequality, we obtain

$$|\langle u \rangle_{B_{i+1}} - \langle u \rangle_{B_i}| \leqslant \fint_{B_{i+1}} |u - \langle u \rangle_{B_i}| \, dy \leqslant c \, 2^{(i+1)n} \int_{B_i} |u - \langle u \rangle_{B_i}| \, dy$$

$$\leqslant c \, 2^{-i(1-n)} \int_{B_i} |\nabla u| \, dy \leqslant c \, 2^{-i(1-n)} \|1\|_{L^{p'(\cdot)}(B_i)} \|\nabla u\|_{L^{p(\cdot)}(B_i)}.$$

Lemma 4.5.3 and the log-Hölder continuity yield $\|1\|_{L^{p'(\cdot)}(B_i)} \leqslant c \, (2^{-i}R)^{\frac{n}{p'(x)}}$. Thus

$$|\langle u \rangle_{B_{i+1}} - \langle u \rangle_{B_i}| \leqslant c \, 2^{-i\left(1-n+\frac{n}{p'(x)}+\frac{s(x)}{p(x)}\right)} = c \, 2^{-i\frac{p(x)+s(x)-n}{p(x)}}.$$

Hence, for $k > j$ we have

$$|\langle u \rangle_{B_k} - \langle u \rangle_{B_j}| \leqslant \sum_{i=j}^{k-1} |\langle u \rangle_{B_{i+1}} - \langle u \rangle_{B_i}| \leqslant c \sum_{i=j}^{k-1} 2^{-i\frac{p(x)+s(x)-n}{p(x)}}$$

and thus $(\langle u \rangle_{B_i})$ is a Cauchy sequence since $s > n - p$. We obtain

$$\limsup_{r \to 0} \fint_{B(x,r)} u \, dy < \infty,$$

which yields by (10.4.8) that $x \notin E$. Thus we conclude that

$$E \subset \left\{ x \in \mathbb{R}^n : \limsup_{r \to 0} r^{-\frac{s(x)}{p(x)}} \|\nabla u\|_{L^{p(\cdot)}(B(x,r))} = \infty \right\}.$$

For $r \leqslant 1$ we obtain by the log-Hölder continuity of p that

$$r^{-\frac{s(x)}{p(x)}} \|\nabla u\|_{L^{p(\cdot)}(B(x,r))} \leqslant c \left\| r^{-\frac{s(x)}{p(\cdot)}} |\nabla u| \right\|_{L^{p(\cdot)}(B(x,r))}.$$

Since p is bounded the norm is finite if and only if the modular is and hence we conclude

$$E \subset \left\{ x \in \mathbb{R}^n : \limsup_{r \to 0} r^{-s(x)} \int_{B(x,r)} |\nabla u|^{p(y)} \, dy = \infty \right\},$$

and the claim follows by Lemma 10.4.6.   □

Theorem 10.4.9. *Let* $p \in \mathcal{P}^{\log}(\mathbb{R}^n)$ *with* $1 < p^- \leqslant p^+ < \infty$. *If* $E \subset \mathbb{R}^n$, *then*

$$C_{p(\cdot)}(E) \leqslant c \, \mathcal{H}^{\max\{n-p(\cdot),0\}}(E).$$

Here the constant c *depends on* n *and* p.

Proof. Fix $\delta > 0$. We cover the set E by balls $(B(x_i, r_i))_{i=1}^{\infty}$, where $x_i \in E$ and $r_i < \min\{\delta, 1\}$, for every $i \in \mathbb{N}$. Using the same test function as in the proof of Lemma 10.2.9 we obtain

$$C_{p(\cdot)}(B(x_i, r_i)) \leqslant c \, r^{n-p(x_i)}.$$

When $p(x_i) > n$ the previous estimate is bad and hence we derive a better one. Using $\max\{2 - |x - x_i|, 0\}$ as a test function we obtain

$$C_{p(\cdot)}(B(x_i, r_i)) \leqslant (2^{p^+} + 1)|B(0,2)|$$

for $r_i < 1$. Thus the subadditivity of the capacity (C7) yields

$$C_{p(\cdot)}(E) \leqslant \sum_{i=1}^{\infty} C_{p(\cdot)}(B(x_i, r_i)) \leqslant c \sum_{i=1}^{\infty} \min\{r_i^{n-p(x_i)}, 1\}.$$

The claim follows by letting $\delta \to 0$.   □

Chapter 11
Fine Properties of Sobolev Functions

In this chapter we study fine properties of Sobolev functions. By definition, Sobolev functions are defined only up to Lebesgue measure zero and thus it is not always clear how to use their point-wise properties. We pick a good representative from every equivalence class of Sobolev functions and show that this representative, called quasicontinuous, has many good properties. Our main tools are the capacities studied in Chap. 10. Our results generalize classical ones to the variable exponent case. In Sect. 11.1 we show that each Sobolev function has a quasicontinuous representatives under natural conditions on the exponent p and define a capacity based on quasicontinuous functions. In Sect. 11.2 we study different definitions of Sobolev spaces with zero boundary values. We continue by studying removable sets for Sobolev spaces in terms of capacity in Sect. 11.3. Then in Sect. 11.4 we show that quasievery point is a Lebesgue point of a Sobolev function when p is globally log-Hölder continuous. We end this chapter in Sect. 11.5 with an example which shows that for more general exponents this is not the case.

11.1 Quasicontinuity

We show that each Sobolev function has a quasicontinuous representative if p is log-Hölder continuous. Then we define the Sobolev capacity using these representatives. Since we use the Sobolev capacity we have to assume $1 < p^- \leqslant p^+ < \infty$. This section is based on [194, 196, 202] by Harjulehto, Hästö, Koskenoja, Varonen and Martio.

Definition 11.1.1. A claim is said to hold $p(\cdot)$-*quasieverywhere* if it holds everywhere except in a set of Sobolev $p(\cdot)$-capacity zero. A function $u : \Omega \to \mathbb{R}$ is said to be $p(\cdot)$-*quasicontinuous* if for every $\varepsilon > 0$ there exists an open set U with $C_{p(\cdot)}(U) < \varepsilon$ such that u restricted to $\Omega \setminus U$ is continuous.

Let u and v be quasicontinuous. It is clear that $u+v$, au $(a \in \mathbb{R})$, $\min\{u, v\}$ and $\max\{u, v\}$ are quasicontinuous.

L. Diening et al., *Lebesgue and Sobolev Spaces with Variable Exponents*, Lecture Notes in Mathematics 2017, DOI 10.1007/978-3-642-18363-8_11, © Springer-Verlag Berlin Heidelberg 2011

The following lemma and theorem give sufficient conditions for the existence of a $p(\cdot)$-quasicontinuous representative.

Lemma 11.1.2. *Let $p \in \mathcal{P}(\mathbb{R}^n)$ satisfy $1 < p^- \leqslant p^+ < \infty$. For each Cauchy sequence with respect to the $W^{1,p(\cdot)}(\mathbb{R}^n)$-norm of functions from $C(\mathbb{R}^n) \cap W^{1,p(\cdot)}(\mathbb{R}^n)$ there is a subsequence which converges pointwise $p(\cdot)$-quasievery-where in $\mathbb{R}^n$. Moreover, the convergence is uniform outside a set of arbitrary small Sobolev $p(\cdot)$-capacity.*

Proof. Let (u_i) be a Cauchy sequence in $C(\mathbb{R}^n) \cap W^{1,p(\cdot)}(\mathbb{R}^n)$. We assume without loss of generality, by considering a subsequence if necessary, that $\|u_i - u_{i+1}\|_{1,p(\cdot)} \leqslant 4^{-i}$, $i \in \mathbb{N}$. We denote

$$U_i := \{x \in \mathbb{R}^n : |u_i(x) - u_{i+1}(x)| > 2^{-i}\},$$

for $i \in \mathbb{N}$ and

$$V_j := \bigcup_{i=j}^{\infty} U_i.$$

Using Proposition 8.1.9 it is easy to show that $v := 2^i|u_i - u_{i+1}| \in W^{1,p(\cdot)}(\mathbb{R}^n)$ and by assumption we have $\|v\|_{1,p(\cdot)} \leqslant 2^{-i}$. Since $\overline{\varrho}_{p(\cdot)}(u) \leqslant \|u\|_{p(\cdot)}$ if $\|u\|_{p(\cdot)} \leqslant 1$, it follows that

$$C_{p(\cdot)}(U_i) \leqslant \overline{\varrho}_{1,p(\cdot)}(v) \leqslant 2^i\|u_i - u_{i+1}\|_{1,p(\cdot)} \leqslant 2^{-i}.$$

The subadditivity of the Sobolev $p(\cdot)$-capacity (Theorem 10.1.3 (C7)) implies that

$$C_{p(\cdot)}(V_j) \leqslant \sum_{i=j}^{\infty} C_{p(\cdot)}(U_i) \leqslant \sum_{i=j}^{\infty} 2^{-i} \leqslant 2^{1-j}.$$

Hence we obtain

$$C_{p(\cdot)}\Big(\bigcap_{j=1}^{\infty} V_j\Big) \leqslant \lim_{j\to\infty} C_{p(\cdot)}(V_j) = 0.$$

Since (u_i) converges pointwise in $\mathbb{R}^n \setminus \bigcap_{j=1}^{\infty} V_j$, we have proved the first claim of the lemma. Moreover, we have

$$|u_l(x) - u_k(x)| \leqslant \sum_{i=l}^{k-1} |u_i(x) - u_{i+1}(x)| \leqslant \sum_{i=l}^{k-1} 2^{-i} < 2^{1-l}$$

for every $x \in \mathbb{R}^n \setminus V_j$ and every $k > l > j$. Therefore the convergence is uniform in $\mathbb{R}^n \setminus V_j$. $\qquad\square$

Theorem 11.1.3. *Let $p \in \mathcal{P}(\mathbb{R}^n)$ with $1 < p^- \leqslant p^+ < \infty$ be such that $C(\mathbb{R}^n) \cap W^{1,p(\cdot)}(\mathbb{R}^n)$ is dense in $W^{1,p(\cdot)}(\mathbb{R}^n)$. Then for each $u \in W^{1,p(\cdot)}(\mathbb{R}^n)$ there exists a $p(\cdot)$-quasicontinuous function $v \in W^{1,p(\cdot)}(\mathbb{R}^n)$ such that $u = v$ almost everywhere in $\mathbb{R}^n$.*

Proof. Let $u \in W^{1,p(\cdot)}(\mathbb{R}^n)$. It follows from the density condition that there exist functions $u_i \in C(\mathbb{R}^n) \cap W^{1,p(\cdot)}(\mathbb{R}^n)$ such that $u_i \to u$ in $W^{1,p(\cdot)}(\mathbb{R}^n)$. By Lemma 11.1.2 there exists a subsequence that converges uniformly outside a set of arbitrarily small capacity. But uniform convergence implies continuity of the limit function and so we get a function continuous restricted outside a set of arbitrarily small capacity, as was to be shown. □

We recall the following lemma. For the proof of (a) we refer to [233]; (b) follows directly from (a) by using it for the function $\max\{u - v, 0\}$.

Lemma 11.1.4. *Let $p \in \mathcal{P}(\mathbb{R}^n)$ with $1 < p^- \leqslant p^+ < \infty$, and let u and v be $p(\cdot)$-quasicontinuous functions in $\mathbb{R}^n$. Suppose that $U \subset \mathbb{R}^n$ is open.*

(a) *If $u = v$ almost everywhere in U, then $u = v$ $p(\cdot)$-quasieverywhere in U.*
(b) *If $u \leqslant v$ almost everywhere in U, then $u \leqslant v$ $p(\cdot)$-quasieverywhere in U.*

Corollary 11.1.5. *Let $p \in \mathcal{P}^{\log}(\Omega)$ with $1 < p^- \leqslant p^+ < \infty$. Then for every $u \in W^{1,p(\cdot)}(\Omega)$ there exists a $p(\cdot)$-quasicontinuous function $v \in W^{1,p(\cdot)}(\Omega)$ such that $u = v$ almost everywhere in Ω.*

Proof. Let $z \in \Omega$ and let (ψ_i) be a sequence of $C_0^\infty(\Omega)$ functions that equal one in $\Omega_i := \{x \in \Omega : \operatorname{dist}(x, \partial\Omega) > \frac{1}{i}\} \cap B(z, i)$. Then $u\psi_i$ belongs to $W^{1,p(\cdot)}(\mathbb{R}^n)$ and $u\psi_i = u$ in Ω_i.

By Theorem 11.1.3 there exist quasicontinuous functions $v_i \in W^{1,p(\cdot)}(\mathbb{R}^n)$ such that v_i equals to $u\psi_i$ almost everywhere in $\mathbb{R}^n$. Let $j > i$. Since v_i and v_j coincide almost everywhere in Ω_i, they coincide $p(\cdot)$-quasieverywhere in Ω_i by Lemma 11.1.4 (a). Let $\varepsilon > 0$ and $V_i \subset \mathbb{R}^n$ be such that v_i restricted to $\mathbb{R}^n \setminus V_i$ is continuous, v_i coincides with v_{i-1} in $\Omega_{i-1} \setminus V_i$ and $C_{p(\cdot)}(V_i) \leqslant 2^{-i}\varepsilon$. Set $\tilde{u}(x) = v_i(x)$, where $i \in \mathbb{N}$ is the smallest number with $x \in B(z, i)$ and $\operatorname{dist}(x, \partial\Omega) > \frac{1}{i}$. Then $\tilde{u}$ equals to u almost everywhere in Ω, $\tilde{u}$ restricted to $\Omega \setminus \bigcup_i V_i$ is continuous and

$$C_{p(\cdot)}\left(\bigcup_i V_i\right) \leqslant \sum_i C_{p(\cdot)}(V_i) \leqslant \varepsilon. \qquad \square$$

We next consider a Sobolev $p(\cdot)$-capacity in terms of $p(\cdot)$-quasicontinuous functions.

Definition 11.1.6. *For $p \in \mathcal{P}(\mathbb{R}^n)$, $1 < p^- \leqslant p^+ < \infty$, and $E \subset \mathbb{R}^n$ we denote*

$$\widetilde{C}_{p(\cdot)}(E) := \inf_{u \in \widetilde{S}_{p(\cdot)}(E)} \int_{\mathbb{R}^n} |u|^{p(x)} + |\nabla u|^{p(x)} \, dx$$

where

$$\widetilde{S}_{p(\cdot)}(E) := \{u \in W^{1,p(\cdot)}(\mathbb{R}^n) : u \text{ is } p(\cdot)\text{-quasicontinuous},$$
$$u \geqslant 1 \ p(\cdot)\text{-quasieverywhere in } E, \text{and } u \geqslant 0\}.$$

We use the convention that $\widetilde{C}_{p(\cdot)}(E) = \infty$ if $\widetilde{S}_{p(\cdot)}(E) = \emptyset$.

Note that we can always calculate the infimum over quasicontinuous functions that are one quasieverywhere in E: this is not restrictive, since if u is quasicontinuous and greater than or equal to one in E, then $\min\{u, 1\}$ is quasicontinuous and equal to one quasieverywhere in E.

Lemma 11.1.7. *Let $p \in \mathcal{P}(\mathbb{R}^n)$ with $1 < p^- \leqslant p^+ < \infty$, and let E be a subset of $\mathbb{R}^n$. Suppose that $u \in \widetilde{S}_{p(\cdot)}(E)$. Then for every $\varepsilon \in (0, 1)$ there exists a function $v \in S_{p(\cdot)}(E)$ such that $\overline{\varrho}_{1,p(\cdot)}(u - v) < \varepsilon$.*

Proof. Let $\delta := \varepsilon/(2^{p^+}(1 + \overline{\varrho}_{1,p(\cdot)}(u)))$, and let $U \subset \mathbb{R}^n$ be an open set such that u restricted to $\mathbb{R}^n \setminus U$ is continuous and $C_{p(\cdot)}(U) < \delta$. Moreover, let $w \in S_{p(\cdot)}(U)$ be such that $\overline{\varrho}_{1,p(\cdot)}(w) < \delta$, and write $v := (1 + \delta)u + w$. It is easy to show that $v \in W^{1,p(\cdot)}(\mathbb{R}^n)$. The set

$$G := \{x \in \mathbb{R}^n \setminus U : (1 + \delta)u(x) > 1\} \cup U$$

is open and contains E. Since $v \geqslant 1$ on G, we get $v \in S_{p(\cdot)}(E)$. Moreover we obtain

$$\overline{\varrho}_{1,p(\cdot)}(u - v) = \int_{\mathbb{R}^n} |w(x) + \delta\, u(x)|^{p(x)} + |\nabla(w(x) + \delta\, u(x))|^{p(x)} \, dx$$
$$\leqslant 2^{p^+}\left(\overline{\varrho}_{1,p(\cdot)}(w) + \delta^{p^+}\overline{\varrho}_{1,p(\cdot)}(u)\right)$$
$$< 2^{p^+}\left(\delta + \delta\,\overline{\varrho}_{1,p(\cdot)}(u)\right) \leqslant \varepsilon.$$

$\square$

Next we show that if continuous functions are dense in the Sobolev space, then we can calculate the Sobolev capacity by using the quasicontinuous representatives.

Theorem 11.1.8. *Let $p \in \mathcal{P}(\mathbb{R}^n)$ satisfy $1 < p^- \leqslant p^+ < \infty$ and $E \subset \mathbb{R}^n$.*

(a) *We have $C_{p(\cdot)}(E) \leqslant \widetilde{C}_{p(\cdot)}(E)$.*
(b) *If $C(\mathbb{R}^n) \cap W^{1,p(\cdot)}(\mathbb{R}^n)$ is dense in $W^{1,p(\cdot)}(\mathbb{R}^n)$, then $C_{p(\cdot)}(E) = \widetilde{C}_{p(\cdot)}(E)$.*

Proof. The first claim follows by Lemma 11.1.7.

For the proof of the reverse inequality, assume that continuous functions are dense in $W^{1,p(\cdot)}(\mathbb{R}^n)$. Let $E \subset \mathbb{R}^n$. Take $u \in S_{p(\cdot)}(E)$ and let $U \supset E$

be an open set such that $u \geqslant 1$ on U. By Theorem 11.1.3, there exists a $p(\cdot)$-quasicontinuous function $\tilde{u}$ in $\mathbb{R}^n$ such that $u = \tilde{u}$ almost everywhere in $\mathbb{R}^n$ and thus $\tilde{u} \geqslant 1$ almost everywhere in U. Lemma 11.1.4 (b) yields $\tilde{u} \geqslant 1$ $p(\cdot)$-quasieverywhere in U. Hence $\tilde{u} \geqslant 1$ $p(\cdot)$-quasieverywhere in E and thus $\tilde{u} \in \widetilde{S}_{p(\cdot)}(E)$. This yields $\widetilde{C}_{p(\cdot)}(E) \leqslant C_{p(\cdot)}(E)$, and combining this with the first claim gives $C_{p(\cdot)}(E) = \widetilde{C}_{p(\cdot)}(E)$. □

Next we show that every quasicontinuous Sobolev function satisfies a weak type capacity inequality.

Corollary 11.1.9. *Let* $p \in \mathcal{P}(\mathbb{R}^n)$ *satisfy* $1 < p^- \leqslant p^+ < \infty$. *If* $u \in W^{1,p(\cdot)}(\mathbb{R}^n)$ *is a* $p(\cdot)$-*quasicontinuous function and* $\lambda > 0$, *then*

$$C_{p(\cdot)}(\{x \in \mathbb{R}^n : |u(x)| > \lambda\}) \leqslant \overline{\varrho}_{1,p(\cdot)}\left(\frac{u}{\lambda}\right).$$

Proof. Since $|u|$ is quasicontinuous and $|u|/\lambda$ is greater than one in the set $\{x \in \mathbb{R}^n : |u(x)| > \lambda\}$, we obtain the claim by Theorem 11.1.8 (a). □

Also the relative capacity can be calculated over quasicontinuous functions.

Theorem 11.1.10. *Let* $p \in \mathcal{P}(\mathbb{R}^n)$ *with* $1 < p^- \leqslant p^+ < \infty$ *be such that* $C_0^\infty(\Omega)$ *is dense in* $W^{1,p(\cdot)}(\Omega)$ *and let* $K \subset \Omega$ *be compact. Then*

$$\mathrm{cap}_{p(\cdot)}(K, \Omega) = \inf \int_\Omega |\nabla u|^{p(x)}\, dx,$$

where the infimum is taken over all $p(\cdot)$-*quasicontinuous functions* u *that belong to* $W^{1,p(\cdot)}(\Omega)$, *have compact support in* Ω, *and are greater than or equal to one* $p(\cdot)$-*quasieverywhere in* K.

Proof. By definition, Propositions 10.2.2 and 10.2.3

$$\inf \int_\Omega |\nabla u|^{p(x)}\, dx \leqslant \mathrm{cap}_{p(\cdot)}(K, \Omega).$$

For the opposite inequality, let $u \in W^{1,p(\cdot)}(\Omega)$ be a quasicontinuous function that has compact support in Ω and that is greater than or equal to one $p(\cdot)$-quasieverywhere in K. Let u restricted to $\Omega \setminus U$ be continuous and let w be a test function for U. Let $\eta \in C_0^\infty(\Omega)$ be a cut of function that is one in the support of u. Let $\varepsilon > 0$. We can easily calculate that if $C_{p(\cdot)}(U) \leqslant \varepsilon$ then $\overline{\varrho}_{1,p(\cdot)}(\eta w) \leqslant c\varepsilon$, where the constant c is independent of U. As in the proof of Lemma 11.1.7 we obtain that $(1 + \varepsilon)u + w\eta$ is greater than or equal to one in an open set containing K. Then using the density of smooth functions and the method presented in the proof of Lemma 10.1.9 we obtain a sequence of $C_0^\infty(\Omega)$ functions that are greater than or equal to one in an open set containing K and converging to $(1 + \varepsilon)u + w\eta$. The claim follows since $(1 + \varepsilon)u + w\eta$ converges to u as $\varepsilon \to 0$. □

For the future use we present a sharpening of Lemma 11.1.2.

Lemma 11.1.11. *Let $p \in \mathcal{P}(\mathbb{R}^n)$ satisfy $1 < p^- \leqslant p^+ < \infty$. Suppose that (u_i) is a Cauchy sequence of $p(\cdot)$-quasicontinuous functions in $W^{1,p(\cdot)}(\mathbb{R}^n)$. Then there is a subsequence of (u_i) which converges pointwise $p(\cdot)$-quasi-everywhere to a $p(\cdot)$-quasicontinuous $u \in W^{1,p(\cdot)}(\mathbb{R}^n)$.*

Proof. There is a subsequence of (u_i), denoted again by (u_i), such that

$$\sum_{i=1}^{\infty} 2^{ip^+} \|u_i - u_{i+1}\|_{1,p(\cdot)} < 1.$$

For $i \in \mathbb{N}$ denote $E_i := \{x \in \mathbb{R}^n : |u_i(x) - u_{i+1}(x)| > 2^{-i}\}$ and $F_j := \bigcup_{i=j}^{\infty} E_i$. Clearly $2^i|u_i - u_{i+1}| \in \widetilde{S}_{p(\cdot)}(E_i)$ and hence using Theorem 11.1.8 (a) we obtain

$$C_{p(\cdot)}(E_i) \leqslant \int_{\mathbb{R}^n} (2^i|u_i - u_{i+1}|)^{p(x)} + |\nabla(2^i|u_i - u_{i+1}|)|^{p(x)} \, dx$$

$$\leqslant 2^{ip^+} \overline{\varrho}_{1,p(\cdot)}(u_i - u_{i+1}).$$

Using the subadditivity of the Sobolev $p(\cdot)$-capacity (Theorem 10.1.3 (C7)) and the unit ball property Lemma 3.2.4 (a) we obtain

$$C_{p(\cdot)}(F_j) \leqslant \sum_{i=j}^{\infty} C_{p(\cdot)}(E_i) \leqslant \sum_{i=j}^{\infty} 2^{ip^+} \overline{\varrho}_{1,p(\cdot)}(u_i - u_{i+1})$$

$$\leqslant \sum_{i=j}^{\infty} 2^{ip^+} \|u_i - u_{i+1}\|_{1,p(\cdot)}.$$

Since $\bigcap_{k=1}^{\infty} F_k \subset F_j$ for each j, the monotonicity of the Sobolev $p(\cdot)$-capacity (Theorem 10.1.2 (C2)) yields

$$C_{p(\cdot)}\left(\bigcap_{k=1}^{\infty} F_k\right) \leqslant \lim_{j \to \infty} C_{p(\cdot)}(F_j) = 0.$$

Moreover, (u_i) converges pointwise in $\mathbb{R}^n \setminus \bigcap_{j=1}^{\infty} F_j$, and so the convergence $p(\cdot)$-quasieverywhere in $\mathbb{R}^n$ follows. Let u be this pointwise limit. Since $W^{1,p(\cdot)}(\Omega)$ is a Banach space (Theorem 8.1.6), we obtain $u \in W^{1,p(\cdot)}(\Omega)$.

To prove the $p(\cdot)$-quasicontinuity of u, let $\varepsilon > 0$. By the first part of this proof, there is a set $F_j \subset \mathbb{R}^n$ such that $C_{p(\cdot)}(F_j) < \frac{\varepsilon}{2}$ and that $u_i \to u$ pointwise in $\mathbb{R}^n \setminus F_j$. Since every u_i is $p(\cdot)$-quasicontinuous in $\mathbb{R}^n$, we may choose open sets $G_i \subset \mathbb{R}^n$, $i \in \mathbb{N}$, such that $C_{p(\cdot)}(G_i) < \frac{\varepsilon}{2^{i+1}}$ and $u_i|_{\mathbb{R}^n \setminus G_i}$

are continuous. Setting $G := \bigcup_i G_i$ we have

$$C_{p(\cdot)}(G) = C_{p(\cdot)}\left(\bigcup_{i=1}^{\infty} G_i\right) < \frac{\varepsilon}{2},$$

and

$$C_{p(\cdot)}(F_j \cup G) \leqslant C_{p(\cdot)}(F_j) + C_{p(\cdot)}(G) < \frac{\varepsilon}{2} + \frac{\varepsilon}{2} = \varepsilon.$$

Moreover,

$$|u_l(x) - u_k(x)| \leqslant \sum_{i=l}^{k-1} |u_i(x) - u_{i+1}(x)| \leqslant \sum_{i=l}^{k-1} 2^{-i} < 2^{1-l}$$

for every $x \in \mathbb{R}^n \setminus (F_i \cup G)$ and every $k > l > i$. Therefore the convergence is uniform in $\mathbb{R}^n \setminus (F_i \cup G)$, and it follows that $u|_{\mathbb{R}^n \setminus (F_i \cup G)}$ is continuous. This completes the proof.   □

We close this section by studying the continuity of Sobolev functions on curves. A function $u \colon \Omega \to \mathbb{R}$ is *absolutely continuous on lines*, denoted by $u \in ACL(\Omega)$, if u is absolutely continuous on almost every line segment in Ω parallel to the coordinate axes. By almost every line segment we mean that intersection of lines, that contains a line segment where u is not absolutely continuous, and $(n-1)$-dimensional hyperplane has zero Hausdorff $(n-1)$-measure for every direction. Note that an ACL function has classical partial derivatives almost everywhere.

An ACL function is said to belong to $ACL^{p(\cdot)}(\Omega)$ if $u, |\nabla u| \in L^{p(\cdot)}(\Omega)$. Since $W^{1,p(\cdot)}(\Omega) \hookrightarrow W^{1,1}_{\mathrm{loc}}(\Omega)$, we obtain the following theorem by [129, Chap. 4.9] or [399, Theorem 2.1.4].

Theorem 11.1.12. *Let $p \in \mathcal{P}(\Omega)$. If $u \in ACL^{p(\cdot)}(\Omega)$, then it has classical partial derivatives almost everywhere and these coincide with the weak partial derivatives as distributions so that $u \in W^{1,p(\cdot)}(\Omega)$. If $u \in W^{1,p(\cdot)}(\Omega)$, then there exists $v \in ACL^{p(\cdot)}(\Omega)$ such that $u = v$ almost everywhere. In short, $ACL^{p(\cdot)}(\Omega) = W^{1,p(\cdot)}(\Omega)$.*

The above theorem can be generalized to the case of curves. As in the constant exponent case, it is possible to show that every Sobolev function is absolute continuous on almost all rectifiable curves, see Harjulehto, Hästö, and Martio [202] for details.

11.2   Sobolev Spaces with Zero Boundary Values

In this section we study different definitions of variable exponent Sobolev space with zero boundary values in an open subset of $\mathbb{R}^n$. This section is based on Harjulehto [186].

Recall that $W_0^{k,p(\cdot)}(\Omega)$ is the closure of compactly supported Sobolev functions in the space $W^{k,p(\cdot)}(\Omega)$. In Theorem 8.1.13 it is proved that $W_0^{k,p(\cdot)}(\Omega)$ is a Banach space.

The following definition for another Sobolev space with zero boundary values uses C_0^∞-functions.

Definition 11.2.1. Let $p \in \mathcal{P}(\Omega)$ and $k \in \mathbb{N}$. The space $H_0^{k,p(\cdot)}(\Omega)$ is defined as the closure of $C_0^\infty(\Omega)$ in $W^{k,p(\cdot)}(\Omega)$.

Clearly $H_0^{k,p(\cdot)}(\Omega) \subset W_0^{k,p(\cdot)}(\Omega)$. The following theorem presents the basic properties of $H_0^{k,p(\cdot)}(\Omega)$.

Theorem 11.2.2. Let $p \in \mathcal{P}(\Omega)$. Then $H_0^{k,p(\cdot)}(\Omega)$ is a Banach space. If p is bounded, then $H_0^{k,p(\cdot)}(\Omega)$ is separable and if $1 < p^- \leqslant p^+ < \infty$, then it is reflexive and uniformly convex.

Proof. Since $H_0^{k,p(\cdot)}(\Omega)$ is a closed subspace of $W^{k,p(\cdot)}(\Omega)$, separability, reflexivity and uniform convexity follow from Proposition 1.4.4 and Theorem 8.1.6. $\qquad\square$

The following proposition shows that Definition 11.2.1 is natural if smooth functions are dense in the Sobolev space.

Proposition 11.2.3. Let $p \in \mathcal{P}(\Omega)$ be bounded such that $C^\infty(\Omega) \cap W^{k,p(\cdot)}(\Omega)$ is dense in $W^{k,p(\cdot)}(\Omega)$. Then $H_0^{k,p(\cdot)}(\Omega) = W_0^{k,p(\cdot)}(\Omega)$.

Proof. We prove only the case $k = 1$, the proof for the general case is similar. Clearly $H_0^{1,p(\cdot)}(\Omega) \subset W_0^{1,p(\cdot)}(\Omega)$. To show the other inclusion, fix $u \in W^{1,p(\cdot)}(\Omega)$ with a compact support in Ω. Let $\psi \in C_0^\infty(\Omega)$ be such that $0 \leqslant \psi \leqslant 1$ and $\psi = 1$ in $\mathrm{spt}(u)$. By assumption there exists a sequence $(u_i) \subset C^\infty(\Omega) \cap W^{1,p(\cdot)}(\Omega)$ converging to u in $W^{1,p(\cdot)}(\Omega)$. We show that $\psi u_i \to u$ in $W^{1,p(\cdot)}(\Omega)$. First we estimate

$$\|u - \psi u_i\|_{1,p(\cdot)} \leqslant \|u - u_i\|_{1,p(\cdot)} + \|u_i - \psi u_i\|_{1,p(\cdot)}.$$

The first term on the right-hand side converges to zero as i tends to infinity. The function in the second term is zero in $\mathrm{spt}(u)$. We obtain

$$\|u_i - \psi u_i\|_{W^{1,p(\cdot)}(\Omega)} = \|u_i - \psi u_i\|_{W^{1,p(\cdot)}(\Omega \setminus \mathrm{spt}(u))} \leqslant c(\psi)\|u_i\|_{W^{1,p(\cdot)}(\Omega \setminus \mathrm{spt}(u))}$$
$$= c(\psi)\|u_i - u\|_{W^{1,p(\cdot)}(\Omega \setminus \mathrm{spt}(u))},$$

and thus the second term also converges to zero as i tends to infinity.

Since each $u \in W_0^{1,p(\cdot)}(\Omega)$ can be approximated by compactly supported Sobolev functions, we find a sequence in $C_0^\infty(\Omega)$ converging to u. Thus we obtain $W_0^{1,p(\cdot)}(\Omega) \subset H_0^{1,p(\cdot)}(\Omega)$. $\qquad\square$

We obtain the following corollary by Proposition 11.2.3 and Theorem 9.1.8.

Corollary 11.2.4. *If $p \in \mathcal{P}^{\log}(\Omega)$ is bounded, then*

$$H_0^{k,p(\cdot)}(\Omega) = W_0^{k,p(\cdot)}(\Omega).$$

Note that $\max\{1 - |x|, 0\}$ is in $W_0^{1,\infty}(B(0,1))$ but it can not be approximated by C^1-functions and it does not belong to $H_0^{1,\infty}(B(0,1))$. Thus the restriction $p^+ < \infty$ in the previous corollary is natural. In Sect. 11.5 we give an uniformly continuous exponent p for the spaces $H_0^{1,p(\cdot)}(\Omega)$ and $W_0^{1,p(\cdot)}(\Omega)$ differ.

In Lemma 8.1.14 we showed that each $u \in W_0^{k,p(\cdot)}(\Omega)$ has a zero extension to $\mathbb{R}^n \setminus \Omega$. Next we generalize this result for $k = 1$.

Corollary 11.2.5. *Let $p \in \mathcal{P}(\Omega)$ with $1 < p^- \leqslant p^+ < \infty$ be such that $C^\infty(\Omega) \cap W^{1,p(\cdot)}(\Omega)$ is dense in $W^{1,p(\cdot)}(\Omega)$. Then for each $u \in W_0^{1,p(\cdot)}(\Omega)$ there exists a $p(\cdot)$-quasicontinuous $\tilde{u} \in W^{1,p(\cdot)}(\mathbb{R}^n)$ that equals u almost everywhere in Ω and zero $p(\cdot)$-quasieverywhere in $\mathbb{R}^n \setminus \Omega$.*

Proof. Let $u \in W_0^{1,p(\cdot)}(\Omega)$. Then by Proposition 11.2.3 we find a sequence (v_i) from $C_0^\infty(\Omega)$ converging to u in $W^{1,p(\cdot)}(\Omega)$. Clearly (v_i) is a Cauchy sequence in $W^{1,p(\cdot)}(\mathbb{R}^n)$. By Lemma 11.1.2, (v_i) has a subsequence that converges pointwise quasieverywhere and uniformly outside a set of arbitrary small capacity. Thus the limit function $\tilde{u}$ is quasicontinuous and zero quasieverywhere in $\mathbb{R}^n \setminus \Omega$. $\square$

Theorem 11.2.6. *Let $p \in \mathcal{P}(\mathbb{R}^n)$ satisfy $1 < p^- \leqslant p^+ < \infty$. If the function $u \in W^{1,p(\cdot)}(\mathbb{R}^n)$ is $p(\cdot)$-quasicontinuous and zero $p(\cdot)$-quasieverywhere in $\mathbb{R}^n \setminus \Omega$, then $u \in W_0^{1,p(\cdot)}(\Omega)$.*

Proof. We show that u can be approximated by Sobolev functions with compact support in Ω. If we can construct such a sequence for $\max\{u, 0\}$, then we can do it for $\min\{u, 0\}$, as well. Combining these results proves the assertion for $u = \max\{u, 0\} + \min\{u, 0\}$. We therefore assume that u is non-negative. By Corollary 9.1.4 we may assume that u is bounded and has compact support in $\mathbb{R}^n$.

Let $\delta > 0$ and let U be an open set such that u restricted to $\mathbb{R}^n \setminus U$ is continuous and $C_{p(\cdot)}(U) < \delta$. Let $E := \{x \in \mathbb{R}^n \setminus \Omega : u(x) \neq 0\}$. By the assumption we have $C_{p(\cdot)}(E) = 0$. Let $\omega_\delta \in S_{p(\cdot)}(U \cup E)$ be such that $0 \leqslant \omega_\delta \leqslant 1$ and $\bar{\varrho}_{1,p(\cdot)}(\omega_\delta) < \delta$. Then $\omega_\delta = 1$ in an open set V containing $U \cup E$. For $0 < \varepsilon < 1$ define $u_\varepsilon(x) := \max\{u(x) - \varepsilon, 0\}$. Since the function u is zero at $x \in \partial\Omega \setminus V$ and u restricted to $\mathbb{R}^n \setminus V$ is continuous, we find $r_x > 0$ so that u_ε vanishes in $B(x, r_x) \setminus V$. Thus the function $(1 - \omega_\delta)u_\varepsilon$ vanishes in $B(x, r_x) \cup V$ for each $x \in \partial\Omega \setminus V$, which yields that it vanishes in a neighborhood of $\mathbb{R}^n \setminus \Omega$. We have

$$\|u - (1 - \omega_\delta)u_\varepsilon\|_{1,p(\cdot)} \leqslant \|u - u_\varepsilon\|_{1,p(\cdot)} + \|\omega_\delta u_\varepsilon\|_{1,p(\cdot)}.$$

Since

$$\|u - u_\varepsilon\|_{1,p(\cdot)} \leqslant \varepsilon \|\chi_{\mathrm{spt}\, u}\|_{p(\cdot)} + \|\chi_{\{0 < u(x) \leqslant \varepsilon\}} \nabla u\|_{p(\cdot)},$$

we see that this term goes to zero with ε. Since u is bounded we find that

$$\bar{\varrho}_{1,p(\cdot)}(\omega_\delta u) \leqslant \int_{\mathbb{R}^n} |\omega_\delta(x)u(x)|^{p(x)}\, dx + 2^{p^+} \int_{\mathbb{R}^n} \omega_\delta(x)^{p(x)} |\nabla u(x)|^{p(x)}\, dx$$

$$+ 2^{p^+} \int_{\mathbb{R}^n} |\nabla \omega_\delta(x)|^{p(x)} |u(x)|^{p(x)}\, dx$$

$$\leqslant (2^{p^+} + 1)\delta \sup_{x \in \mathbb{R}^n} \{|u(x)|^{p(x)}\} + 2^{p^+} \int_{\mathbb{R}^n} \omega_\delta(x)^{p(x)} |\nabla u(x)|^{p(x)}\, dx.$$

Since $\omega_\delta \to 0$ in $L^{p(\cdot)}(\mathbb{R}^n)$, as $\delta \to 0$, we can choose a sequence $\omega_{\delta'}$ which tends to 0 pointwise almost everywhere. Then $\int_{\mathbb{R}^n} \omega_{\delta'}(x)^{p(x)} |\nabla u(x)|^{p(x)}\, dx \to 0$ by the dominated convergence theorem with $|\nabla u|^{p(x)}$ as a dominant. Therefore $\bar{\varrho}_{1,p(\cdot)}(\omega_{\delta'} u) \to 0$ and so also $\|\omega_{\delta'} u\|_{1,p(\cdot)} \to 0$ as $\delta' \to 0$. Thus we see that $(1 - \omega_{\delta'})u_\varepsilon \to u$ as $\varepsilon, \delta' \to 0$.

We have shown that u can be can be approximated by functions in $W^{1,p(\cdot)}(\Omega)$ with compact support in Ω and thus the claim follows. $\square$

Assume that $p \in \mathcal{P}(\mathbb{R}^n)$ satisfies $1 < p^- \leqslant p^+ < \infty$. A function u belongs to $Q_0^{1,p(\cdot)}(\Omega)$ if there exists a $p(\cdot)$-quasicontinuous function $\tilde{u} \in W^{1,p(\cdot)}(\mathbb{R}^n)$ such that $u = \tilde{u}$ almost everywhere in Ω and $\tilde{u} = 0$ $p(\cdot)$-quasieverywhere in $\mathbb{R}^n \setminus \Omega$. The set $Q_0^{1,p(\cdot)}(\Omega)$ is endowed with the norm

$$\|u\|_{Q_0^{1,p(\cdot)}(\Omega)} := \|\tilde{u}\|_{W^{1,p(\cdot)}(\mathbb{R}^n)}.$$

This definition of a Sobolev spaces with zero boundary values, which is common in the theory of metric measure spaces is due to Kilpeläinen, Kinnunen and Martio [234].

It easy to show that $Q_0^{1,p(\cdot)}(\Omega)$ is a closed subspace of $W^{1,p(\cdot)}(\mathbb{R}^n)$, and hence it is a separable, reflexive and uniformly convex Banach space. It follows from the definitions that $H_0^{1,p(\cdot)}(\Omega) \subset Q_0^{1,p(\cdot)}(\Omega)$. Using also Theorem 11.2.6 we obtain

$$H_0^{1,p(\cdot)}(\Omega) \subset Q_0^{1,p(\cdot)}(\Omega) \subset W_0^{1,p(\cdot)}(\Omega).$$

Under the assumptions of Corollary 11.2.5 (in particular if $p \in \mathcal{P}^{\log}(\Omega)$ with $1 < p^- \leqslant p^+ < \infty$) we have

$$H_0^{1,p(\cdot)}(\Omega) = Q_0^{1,p(\cdot)}(\Omega) = W_0^{1,p(\cdot)}(\Omega)$$

by Proposition 11.2.3, Corollary 11.2.5 and Theorem 11.2.6.

The last theorem of this section shows that if the complement is fat, then for a function from $W^{1,p(\cdot)}$ it is enough to have zero boundary values in the $W_0^{1,1}$-sense to belong to $W_0^{1,p(\cdot)}$.

Theorem 11.2.7. *Let Ω be a bounded domain with a fat complement in the sense that there exists a constant c such that*

$$|B(x,r) \setminus \Omega| \geqslant c\,|B(x,r)|$$

for every $x \in \partial\Omega$ and $r > 0$. Let $p \in \mathcal{P}^{\log}(\Omega)$ satisfy $1 < p^- \leqslant p^+ < \infty$. If $u \in W_0^{1,1}(\Omega) \cap W^{1,p(\cdot)}(\Omega)$, then $u \in W_0^{1,p(\cdot)}(\Omega)$.

Proof. We first extend p outside Ω so that the assumptions of the theorem hold in $\mathbb{R}^n$ (Proposition 4.1.7). We denote the extension still by p. Since $u \in W_0^{1,1}(\Omega)$ we get that the extension by zero belongs to $W^{1,1}(\mathbb{R}^n)$ (Lemma 8.1.14). Thus u has distributional derivatives in $\mathbb{R}^n$ and therefore $u \in W^{1,p(\cdot)}(\mathbb{R}^n)$. Let $\tilde{u}$ be the quasicontinuous representative of u (Theorem 11.1.3). By Theorem 11.4.4 the quasicontinuous representative is given by

$$\lim_{r \to 0} \fint_{B(x,r)} u\,dy.$$

Since u is zero almost everywhere in $\mathbb{R}^n \setminus \overline{\Omega}$ we find by Lemma 11.1.4 that $\tilde{u}$ is zero $p(\cdot)$-quasieverywhere in $\mathbb{R}^n \setminus \overline{\Omega}$. By Theorem 11.2.6 we obtain $u \in W_0^{1,p(\cdot)}(\Omega)$ once we have shown that

$$\lim_{r \to 0} \fint_{B(x,r)} u\,dy = 0$$

for quasievery $x \in \partial\Omega$. Clearly it is enough to verify the last statement for $|u|$.

Let $x \in \partial\Omega$. Using the fatness of the complement we obtain by [261, Lemma 3.4] or by [178, Lemma 3.8] that

$$\int_{B(x,r)} |u|\,dy \leqslant c\,r \int_{B(x,r)} |\nabla u|\,dy$$

and furthermore by Hölder's inequality and log-Hölder continuity

$$\fint_{B(x,r)} |u|\,dy \leqslant c\,r^{1-n} \|1\|_{L^{p'(\cdot)}(B(x,r))} \|\nabla u\|_{L^{p(\cdot)}(B(x,r))}$$

$$\leqslant c\,r^{1-n+\frac{n}{p'(x)}} \|\nabla u\|_{L^{p(\cdot)}(B(x,r))}$$

$$= c\,\left\| r^{\frac{p(x)-n}{p(x)}} |\nabla u| \right\|_{L^{p(\cdot)}(B(x,r))}.$$

Since p is bounded, it is enough to show that

$$\int\limits_{B(x,r)} r^{\frac{p(x)-n}{p(x)} \cdot p(y)} |\nabla u|^{p(y)} \, dy \leqslant c \, r^{p(x)} \fint\limits_{B(x,r)} |\nabla u|^{p(y)} \, dy$$

converges to zero as $r \to 0$ for quasievery $x \in \partial\Omega$ (in the inequality we used log-Hölder continuity of p) but this follows from Lemma 11.4.6. □

11.3 Exceptional Sets in Variable Exponent Sobolev Spaces

Let $E \subset \Omega$ be a relatively closed set of measure zero. By $W^{1,p(\cdot)}(\Omega \setminus E) = W^{1,p(\cdot)}(\Omega)$ we mean that the zero extension of every $u \in W^{1,p(\cdot)}(\Omega \setminus E)$ belongs to $W^{1,p(\cdot)}(\Omega)$. In fact, since $|E| = 0$, we could extend u from $\Omega \setminus E$ to Ω in an arbitrary way. The essential question for the validity of the above identity is whether a function from $W^{1,p(\cdot)}(\Omega \setminus E)$ has a gradient in the larger set Ω. For $u \in W^{1,p(\cdot)}(\Omega \setminus E)$, extended by zero to Ω, we know that

$$\int\limits_{\Omega} \psi \frac{\partial u}{\partial x_i} \, dx = - \int\limits_{\Omega} u \frac{\partial \psi}{\partial x_i} \, dx$$

for every $i = 1, \ldots, n$ and for every $\psi \in C_0^\infty(\Omega \setminus E)$. In order for u to have a gradient also in Ω we need this equation to hold also for $\psi \in C_0^\infty(\Omega)$. If $W^{1,p(\cdot)}(\Omega \setminus E) = W^{1,p(\cdot)}(\Omega)$, then E is called a *removable set* for $W^{1,p(\cdot)}(\Omega)$. This section is based on Harjulehto, Hästö, and Koskenoja [193]. We start with some results for zero boundary values Sobolev spaces.

Theorem 11.3.1. *Let $p \in \mathcal{P}(\mathbb{R}^n)$ be such that $1 < p^- \leqslant p^+ < \infty$. Suppose that $E \subset \Omega$ is a relative closed subset. Then $W_0^{1,p(\cdot)}(\Omega) = W_0^{1,p(\cdot)}(\Omega \setminus E)$ if and only if $C_{p(\cdot)}(E) = 0$.*

Proof. Suppose first that $C_{p(\cdot)}(E) = 0$. It follows from Lemma 10.4.1 that $|E| = 0$ so that the notation $W_0^{1,p(\cdot)}(\Omega) = W_0^{1,p(\cdot)}(\Omega \setminus E)$ makes sense. It is clear that $W_0^{1,p(\cdot)}(\Omega \setminus E) \subset W_0^{1,p(\cdot)}(\Omega)$. Let $u \in W_0^{1,p(\cdot)}(\Omega)$ and let $u_i \in W^{1,p(\cdot)}(\Omega)$ be bounded functions with compact support in Ω such that $u_i \to u$ in $W^{1,p(\cdot)}(\Omega)$ (we may assume that u_i are bounded by Lemma 9.1.1). Let $w_j \in S_{p(\cdot)}(E)$ be such that $0 \leqslant w_j \leqslant 1$ and $w_j \to 0$ in $W^{1,p(\cdot)}(\mathbb{R}^n)$ and also pointwise almost everywhere. We note that $u_i(1 - w_j)$ has compact support in $\Omega \setminus E$ and estimate $\|u - u_i(1 - w_j)\|_{1,p(\cdot)} \leqslant \|u - u_i\|_{1,p(\cdot)} + \|u_i w_j\|_{1,p(\cdot)}$. The first term tends to zero by the choice of u_i. For the second term we calculate $\|u_i w_j\|_{1,p(\cdot)} \leqslant \|w_j \nabla u_i\|_{p(\cdot)} + \|u_i\|_\infty \|w_j\|_{1,p(\cdot)}$. Since $|\nabla u_i| \in W^{1,p(\cdot)}(\Omega)$ and $w_j \to 0$ a.e., dominated convergence implies that $\|w_j \nabla u_i\|_{p(\cdot)} \to 0$ as

$j \to \infty$. Thus $\|u_i w_j\|_{1,p(\cdot)} \to 0$ as $j \to \infty$. Hence we have found approximations of u with compact support, so $u \in W_0^{1,p(\cdot)}(\Omega \setminus E)$.

To prove the other implication, let $x_0 \in \Omega$ and let $i_0 \in \mathbb{N}$ be such that $\mathrm{dist}(x_0, \mathbb{R}^n \setminus \Omega) > \frac{1}{i_0}$. For $i \geqslant i_0$ we define

$$\Omega_i := B(x_0, i) \cap \left\{ x \in \Omega : \mathrm{dist}(x, \mathbb{R}^n \setminus \Omega) > \frac{1}{i} \right\}$$

and $u_i : \mathbb{R}^n \to \mathbb{R}$ by

$$u_i(x) := \mathrm{dist}(x, \mathbb{R}^n \setminus \Omega_{2i}).$$

Then $u_i \in W_0^{1,p(\cdot)}(\Omega)$ is continuous and $u_i \geqslant \frac{1}{2i}$ in $E \cap \Omega_i$. By assumption, $u_i \in W_0^{1,p(\cdot)}(\Omega \setminus E)$. Fix i and let $v_j \to u_i$ be Sobolev functions with compact support in $\Omega \setminus E$. Since $3i(u_i - v_j)$ is greater than 1 in a neighborhood of $E \cap \Omega_i$, we obtain $C_{p(\cdot)}(E \cap \Omega_i) \leqslant \overline{\varrho}_{1,p(\cdot)}(3i(u_i - v_j)) \to 0$ as $j \to \infty$. Thus $C_{p(\cdot)}(E \cap \Omega_i) = 0$, and property (C7) of the Sobolev $p(\cdot)$-capacity yield

$$C_{p(\cdot)}(E) = C_{p(\cdot)}\left(\bigcup_{i=1}^{\infty} (E \cap \Omega_i) \right) \leqslant \sum_{i=1}^{\infty} C_{p(\cdot)}(E \cap \Omega_i) = 0. \qquad \square$$

Next we consider the problem of removability in the variable exponent Sobolev space $W^{1,p(\cdot)}(\Omega)$ without the zero boundary value assumption. The proof given in terms of the Sobolev $p(\cdot)$-capacity follows the proof in the fixed-exponent case from [219, Chap. 2].

Theorem 11.3.2. *Let* $p \in \mathcal{P}(\mathbb{R}^n)$ *satisfy* $1 < p^- \leqslant p^+ < \infty$. *Suppose that* $E \subset \Omega$ *is a relatively closed set. If* E *is of Sobolev* $p(\cdot)$-*capacity zero, then*

$$W^{1,p(\cdot)}(\Omega) = W^{1,p(\cdot)}(\Omega \setminus E).$$

Proof. Let $C_{p(\cdot)}(E) = 0$ and let $u \in W^{1,p(\cdot)}(\Omega \setminus E)$. Assume first that u is bounded. Choose a sequence (v_j) of functions in $S_{p(\cdot)}(E)$, $0 \leqslant v_j \leqslant 1$, such that $v_j = 1$ in an open neighborhood U_j of E, $j \in \mathbb{N}$, and $v_j \to 0$ in $W^{1,p(\cdot)}(\mathbb{R}^n)$ and also pointwise almost everywhere. Since u and $1 - v_j$ are bounded functions we find that $u_j := (1 - v_j)u \in W^{1,p(\cdot)}(\Omega \setminus E)$. Moreover $u_j = 0$ in U_j and thus $u_j \in W^{1,p(\cdot)}(\Omega)$. We easily calculate

$$\overline{\varrho}_{p(\cdot)(\Omega)}\left(|\nabla(u_i - u_j)| \right) = \int_{\Omega \setminus E} |\nabla((v_j - v_i)u)|^{p(x)} \, dx$$

$$\leqslant \int_{\Omega \setminus E} (|v_j| + |v_i|)^{p(x)} |\nabla u|^{p(x)} \, dx + \int_{\Omega \setminus E} (|\nabla v_j| + |\nabla v_i|)^{p(x)} |u|^{p(x)} \, dx$$

$$\leqslant \int_{\Omega \setminus E} (|v_j| + |v_i|)^{p(x)} |\nabla u|^{p(x)} \, dx + 2^{p^+} (\sup |u|)^{p^+} \int_{\Omega \setminus E} (|\nabla v_j| + |\nabla v_i|)^{p(x)} \, dx.$$

Clearly the second integral tends to zero as $i, j \to \infty$. Since p is bounded and $0 \leqslant v_j \leqslant 1$ for every i, we easily see that $2^{p^+} |\nabla u|^{p(x)}$ is a majorant of the first integrand which do not depend on i and j. Since $v_i(x) \to 0$ for almost every $x \in \mathbb{R}^n$, this implies, by the theorem of dominated convergence, that the second integral tends to zero as $i, j \to \infty$. Hence $\overline{\varrho}_{p(\cdot)(\Omega)}(|\nabla(u_i - u_j)|) \to 0$ as $i, j \to \infty$. The same holds for $\overline{\varrho}_{p(\cdot)(\Omega)}(u_i - u_j)$, so (u_i) is a Cauchy sequence. Since $W^{1,p(\cdot)}(\Omega)$ is a Banach space, we see that the limit u of (u_i) belongs to this space too.

The same Banach space argument also allows us to get the general case from the case of bounded functions proven above, since bounded Sobolev functions are dense in the Sobolev space (Lemma 9.1.1).   □

Remark 11.3.3. The assumption $C_{p(\cdot)}(E) = 0$ in Theorem 11.3.1 can be replaced by $\mathrm{cap}_{p(\cdot)}(E, \Omega) = 0$, see Theorem 10.3.2 and Proposition 10.3.4.

In Theorem 11.3.2 removability of a relatively closed set $E \subset \Omega$ with measure zero can be characterized also in terms of relative $p(\cdot)$-capacity. In fact, with respect to the relative $p(\cdot)$-capacity of $E \subset \Omega$, the assumption $p^- > 1$ is not needed to prove that $W^{1,p(\cdot)}(\Omega) = W^{1,p(\cdot)}(\Omega \setminus E)$. This can be shown similarly to the proof of Theorem 11.3.2 if assumed that $E \subset \Omega$ is compact with $\mathrm{cap}_{p(\cdot)}(E, \Omega) = 0$.

Corollary 11.3.4. Let $p \in \mathcal{P}(\mathbb{R}^n)$ with $1 < p^- \leqslant p^+ < \infty$. Then $W^{1,p(\cdot)}(\Omega) = W_0^{1,p(\cdot)}(\Omega)$ if and only if $\mathbb{R}^n \setminus \Omega$ has zero Sobolev $p(\cdot)$-capacity.

Proof. Suppose first that $C_{p(\cdot)}(\mathbb{R}^n \setminus \Omega) = 0$. Note that $W^{1,p(\cdot)}(\mathbb{R}^n) = W_0^{1,p(\cdot)}(\mathbb{R}^n)$ by Corollary 9.1.3. Theorems 11.3.2 and 11.3.1 now yield

$$W^{1,p(\cdot)}(\Omega) = W^{1,p(\cdot)}(\mathbb{R}^n \setminus (\mathbb{R}^n \setminus \Omega)) = W^{1,p(\cdot)}(\mathbb{R}^n)$$
$$= W_0^{1,p(\cdot)}(\mathbb{R}^n) = W_0^{1,p(\cdot)}(\mathbb{R}^n \setminus (\mathbb{R}^n \setminus \Omega)) = W_0^{1,p(\cdot)}(\Omega).$$

Suppose then that $W^{1,p(\cdot)}(\Omega) = W_0^{1,p(\cdot)}(\Omega)$. Let $u = \max\{0, 2r - |x|\}$ for $r > 0$. Then $u|_\Omega \in W^{1,p(\cdot)}(\Omega) = W_0^{1,p(\cdot)}(\Omega)$. Let $u_i \to u$ in $W^{1,p(\cdot)}(\Omega)$ have compact supports in Ω. Then $u - u_i$ is a test function for the capacity of $(\mathbb{R}^n \setminus \Omega) \cap B(0, r)$, hence $C_{p(\cdot)}((\mathbb{R}^n \setminus \Omega) \cap B(0, r)) = 0$, and subadditivity implies the claim.   □

11.4 Lebesgue Points

In this section we consider Lebesgue points of functions in Sobolev spaces. We proceed as follows: First we use a result that shows that the Hardy-Littlewood maximal function of a Sobolev function is a Sobolev function. This yields a capacity weak type estimate of the Hardy-Littlewood maximal function. Using these results we prove that

$$\lim_{r \to 0} \fint_{B(x.r)} u(y) \, dy =: u^*(x)$$

exists quasieverywhere for every $u \in W^{1,p(\cdot)}(\mathbb{R}^n)$ and u^* is the quasicontinuous representative of u. Finally, we show that

$$\lim_{r \to 0} \fint_{B(x,r)} |u(y) - u^*(x)|^{p^*(y)} \, dy = 0 \tag{11.4.1}$$

quasieverywhere in $\{x \in \mathbb{R}^n : p(x) < n\}$. Here p^* is the pointwise Sobolev conjugate exponent. A point satisfying (11.4.1) is said to be a *Lebesgue point* for the Sobolev function u. This section is based on [189] by Harjulehto and Hästö that generalizes the constant exponent proof of Kinnunen and Latvala [237] to the variable exponent case.

In this section we use the centered Hardy–Littlewood maximal operator that is calculated over balls. For every $f \in L^1_{loc}(\mathbb{R}^n)$ we denote

$$Mf(x) := \sup_{r > 0} \fint_{B(x,r)} |f(y)| \, dy.$$

We need that $M : L^{p(\cdot)}(\mathbb{R}^n) \to L^{p(\cdot)}(\mathbb{R}^n)$ is bounded and hence we first assume that $p \in \mathcal{A}$ and later strengthen it to $p \in \mathcal{P}^{\log}(\mathbb{R}^n)$.

The following proposition is an adaptation to the variable exponent case of results of Hajłasz and Onninen [183, Theorem 3], see also Kinnunen [236], and Kinnunen and Lindqvist [238].

Proposition 11.4.2. *Let* $p \in \mathcal{A}$ *be such that* $1 < p^- \leqslant p^+ < \infty$. *If* $u \in W^{1,p(\cdot)}(\mathbb{R}^n)$, *then* $Mu \in W^{1,p(\cdot)}(\mathbb{R}^n)$ *and* $|\nabla Mu| \leqslant M|\nabla u|$ *almost everywhere.*

In the remaining part of this section we will adapt the proof of [237, Theorem 4.5] by Kinnunen and Latvala to variable exponent spaces. For simplicity of exposition, we split their result into two parts, Theorems 11.4.4 and 11.4.10. The proof of the former is nearly the same as in the constant exponent case.

Proposition 11.4.3. *Let* $p \in \mathcal{A}$ *be such that* $1 < p^- \leqslant p^+ < \infty$. *Then for every* $u \in W^{1,p(\cdot)}(\mathbb{R}^n)$ *and every* $\lambda > 0$ *we have*

$$C_{p(\cdot)}\big(\{x \in \mathbb{R}^n : Mu(x) > \lambda\}\big) \leqslant c \max\left\{\left\|\frac{u}{\lambda}\right\|_{1,p(\cdot)}^{p^-}, \left\|\frac{u}{\lambda}\right\|_{1,p(\cdot)}^{p^+}\right\},$$

where the constant depends only on n, *the* $\mathcal{A}$-*constant of* p, p^- *and* p^+.

Proof. Since Mu is lower semicontinuous, the set $\{x \in \mathbb{R}^n : Mu(x) > \lambda\}$ is open for every $\lambda > 0$. By Proposition 11.4.2 we can use $\frac{Mu}{\lambda} = M\frac{u}{\lambda}$ as a test function for the capacity. This yields

$$C_{p(\cdot)}\big(\{x \in \mathbb{R}^n : Mu(x) > \lambda\}\big) \leqslant \overline{\varrho}_{1,p(\cdot)}\Big(M\frac{u}{\lambda}\Big)$$

$$\leqslant \max\Big\{\Big\|M\frac{u}{\lambda}\Big\|_{1,p(\cdot)}^{p^-}, \Big\|M\frac{u}{\lambda}\Big\|_{1,p(\cdot)}^{p^+}\Big\}.$$

Now the claim follows by Proposition 11.4.2 since M is bounded (Theorem 5.7.2). $\square$

Theorem 11.4.4. *Let $p \in \mathcal{A}$ be such that $1 < p^- \leqslant p^+ < \infty$, and let $u \in W^{1,p(\cdot)}(\mathbb{R}^n)$. Then there exists a set $E \subset \mathbb{R}^n$ of zero Sobolev $p(\cdot)$-capacity such that*

$$u^*(x) := \lim_{r \to 0} \fint_{B(x,r)} u(y)\,dy$$

exists for every $x \in \mathbb{R}^n \setminus E$. The function u^* is the $p(\cdot)$-quasicontinuous representative of u.

Proof. Since smooth functions are dense in $W^{1,p(\cdot)}(\mathbb{R}^n)$ (Theorem 9.1.6), we can choose a sequence (u_i) of continuous functions in $W^{1,p(\cdot)}(\mathbb{R}^n)$ such that $\|u - u_i\|_{1,p(\cdot)} \leqslant 2^{-2i}$. By considering a subsequence, if necessary, we may assume that $u_i \to u$ pointwise almost everywhere. For $i \in \mathbb{N}$ we denote

$$U_i := \Big\{x \in \mathbb{R}^n : M(u - u_i)(x) > 2^{-i}\Big\}, \quad V_i := \bigcup_{j=i}^{\infty} U_j, \quad \text{and} \quad E := \bigcap_{j=1}^{\infty} V_j.$$

Proposition 11.4.3 implies that $C_{p(\cdot)}(U_i) \leqslant c\,2^{-i}$, the subadditivity of the Sobolev capacity (Theorem 10.1.3 (C7)) implies that $C_{p(\cdot)}(V_i) \leqslant c\,2^{1-i}$ and Theorem 10.1.2 (C2) implies that $C_{p(\cdot)}(E) = 0$.

We next consider the relationship between u and u_i outside these sets. We have

$$|u_i(x) - u_{B(x,r)}| \leqslant \fint_{B(x,r)} |u_i(x) - u_i(y)|\,dy + \fint_{B(x,r)} |u_i(y) - u(y)|\,dy.$$

Since u_i is continuous, the first term on the right-hand side goes to zero for $r \to 0$, and so we get

$$|u_i(x) - u_j(x)| \leqslant \limsup_{r \to 0} |u_i(x) - u_{B(x,r)}| + \limsup_{r \to 0} |u_j(x) - u_{B(x,r)}|$$

$$\leqslant M(u_i - u)(x) + M(u_j - u)(x).$$

Thus we have $|u_i(x) - u_j(x)| \leqslant 2\,2^{-k}$ for $i, j \geqslant k$ and $x \in \mathbb{R}^n \setminus U_k$. It follows that (u_i) converges uniformly on $\mathbb{R}^n \setminus V_j$ for every $j > 0$. Denote the limit function, which restricted to $\mathbb{R}^n \setminus V_j$ is continuous, by $\tilde{u}$. Since $u_i \to u$ pointwise almost everywhere, we obtain $\tilde{u} = u$ almost everywhere in $\mathbb{R}^n \setminus V_j$. If $x \in \mathbb{R}^n \setminus V_j$, then

$$|\tilde{u}(x) - \lim_{r \to 0} u_{B(x,r)}| \leqslant |\tilde{u}(x) - u_i(x)| + \limsup_{r \to 0} |u_i(x) - u_{B(x,r)}|$$

$$\leqslant |\tilde{u}(x) - u_i(x)| + 2^{-i}.$$

As $i \to \infty$ the right-hand side of the previous inequality tends to zero. Since the left-hand side does not depend on i, this means that it equals zero, so that $\tilde{u}(x) = u^*(x)$ for all $x \in \mathbb{R}^n \setminus V_j$, where u^* was defined in (11.4.1). Since this holds in the complement of every V_j, it holds in the complement of E as well. Since E has capacity zero, we are done with the existence part. Since $\tilde{u}$ restricted to $\mathbb{R}^n \setminus V_j$ is continuous for every j, the claim regarding quasicontinuity is clear. □

Using the quasicontinuous representative of Sobolev functions it makes sense to study Lusin type approximations: Harjulehto, Kinnunen and Tuhkanen showed in [209] that the quasicontinuous representative of a Sobolev function coincides with a Hölder continuous Sobolev function outside a small open exceptional set. Roughly speaking the $(p(\cdot) - \varepsilon)$-capacity of the exceptional set can be chosen to be arbitrary small.

Next we move to study the Lebesgue point property, (11.4.1), using the quasicontinuous representative. We need global log-Hölder continuity of p instead of assuming $p \in \mathcal{A}$. Since $r^{p(x)-p(y)} \approx 1$ for x and y with $y \in B(x,r)$, there exists a constant c such that

$$\frac{1}{c} \leqslant \liminf_{r \to 0} r^{p(x)} \fint_{B(x,r)} r^{-p(y)} \, dy \leqslant \limsup_{r \to 0} r^{p(x)} \fint_{B(x,r)} r^{-p(y)} \, dy \leqslant c \quad (11.4.5)$$

for every $x \in \mathbb{R}^n$.

Lemma 11.4.6. *Suppose that* $p \in \mathcal{P}^{\log}(\mathbb{R}^n)$ *satisfies* $1 < p^- \leqslant p^+ < \infty$ *and let* $u \in W^{1,p(\cdot)}(\mathbb{R}^n)$. *Then*

$$C_{p(\cdot)}\left(\left\{ x \in \mathbb{R}^n : \limsup_{r \to 0} r^{p(x)} \fint_{B(x,r)} |\nabla u(y)|^{p(y)} \, dy > 0 \right\} \right) = 0.$$

Proof. Let $\delta \in (0,1)$, $\varepsilon > 0$ and set

$$E_\varepsilon := \left\{ x \in \mathbb{R}^n : \limsup_{r \to 0} r^{p(x)} \fint_{B(x,r)} |\nabla u(y)|^{p(y)} \, dy > \varepsilon \right\}.$$

For every $x \in E_\varepsilon$ there exists an arbitrarily small $r_x \in (0, \delta)$ such that

$$r_x^{p(x)} \fint_{B(x,r_x)} |\nabla u(y)|^{p(y)} \, dy > \varepsilon. \tag{11.4.7}$$

By choosing smaller r_x if necessary, we may, on account of (11.4.5) and the previous inequality, assume that

$$\int_{B(x,r_x)} |\nabla u(y)|^{p(y)} \, dy > \frac{\varepsilon}{c} \int_{B(x,r_x)} r_x^{-p(y)} \, dy, \tag{11.4.8}$$

where c does not depend on x or r_x.

By the basic covering theorem (Theorem 1.4.5) there exists a countable subfamily of pair-wise disjoint balls $B(x_i, r_{x_i})$ such that

$$E_\varepsilon \subset \bigcup_{i=1}^{\infty} B(x_i, 5r_{x_i}).$$

Denote $r_i := r_{x_i}$ and $B_i := B(x_i, r_i)$. Using Remark 10.1.11 and the log-Hölder continuity, we obtain

$$C_{p(\cdot)}(B(x_i, 5r_i)) \leqslant c\, r_i^{n-p(x)} \leqslant c \int_{B_i} r_i^{-p(y)} \, dy.$$

By subadditivity of the Sobolev capacity (Theorem 10.1.3 (C7)) we conclude that

$$C_{p(\cdot)}(E_\varepsilon) \leqslant \sum_{i=1}^{\infty} C_{p(\cdot)}(B(x_i, 5r_i)) \leqslant c \sum_{i=1}^{\infty} \int_{B_i} r_i^{-p(y)} \, dy.$$

It follows from this and (11.4.8) that

$$C_{p(\cdot)}(E_\varepsilon) \leqslant \frac{c}{\varepsilon} \sum_{i=1}^{\infty} \int_{B_i} |\nabla u(y)|^{p(y)} \, dy = \frac{c}{\varepsilon} \int_{\cup_{i=1}^{\infty} B_i} |\nabla u(y)|^{p(y)} \, dy, \tag{11.4.9}$$

where the disjointness of the balls B_i was used in the last step. We then find, by the disjointness of the balls B_i again and (11.4.7), that

$$\left| \bigcup_{i=1}^{\infty} B_i \right| = \sum_{i=1}^{\infty} |B_i| < \sum_{i=1}^{\infty} \frac{r_i^{p(x_i)}}{\varepsilon} \int_{B_i} |\nabla u(y)|^{p(y)} \, dy$$

$$\leqslant \frac{\delta^{p^-}}{\varepsilon} \int_{\mathbb{R}^n} |\nabla u(y)|^{p(y)} \, dy.$$

Hence $|\cup_{i=1}^{\infty} B_i| \to 0$ as $\delta \to 0$, which by (11.4.9) implies that $C_{p(\cdot)}(E_\varepsilon) = 0$ for every $\varepsilon > 0$. Since

$$\left\{ x \in \mathbb{R}^n : \limsup_{r \to 0} r^{p(x)} \fint_{B(x,r)} |\nabla u(y)|^{p(y)}\, dy > 0 \right\} = \bigcup_{i=1}^{\infty} E_{\frac{1}{i}},$$

the claim follows by the subadditivity of the Sobolev capacity.   □

Theorem 11.4.10. *Let $p \in \mathcal{P}^{\log}(\mathbb{R}^n)$ satisfy $1 < p^- \leqslant p^+ < \infty$ and let $u \in W^{1,p(\cdot)}(\mathbb{R}^n)$. Then there exists a set $E \subset \mathbb{R}^n$ with $C_{p(\cdot)}(E) = 0$, such that*

$$\lim_{r \to 0} \fint_{B(x,r)} |u(y) - u^*(x)|^{p^*(y)}\, dy = 0$$

for every $x \in \{ x \in \mathbb{R}^n : p(x) < n \} \setminus E$ and

$$\lim_{r \to 0} \fint_{B(x,r)} |u(y) - u^*(x)|^q\, dy = 0$$

for every $x \in \{ x \in \mathbb{R}^n : p(x) \geqslant n \} \setminus E$ and for any finite $q \geqslant 1$.

Proof. Define

$$E := \left\{ x \in \mathbb{R}^n : \limsup_{r \to 0} r^{p(x)} \fint_{B(x,r)} |\nabla u(y)|^{p(y)}\, dy > 0 \right\}.$$

Then $C_{p(\cdot)}(E) = 0$ by Lemma 11.4.6. Assume first that $p(x) < n$. We show that

$$\limsup_{r \to 0} r^{p(x)} \fint_{B(x,r)} |\nabla u(y)|^{p(y)}\, dy = 0 \quad \Rightarrow$$

$$\limsup_{r \to 0} \fint_{B(x,r)} |u(y) - u_{B(x,r)}|^{p^*(y)}\, dy = 0$$

when $p(x) < n$, from which the claim clearly follows by Theorem 11.4.4, since $p \in \mathcal{A}$ by Theorem 4.4.8.

The Sobolev-Poincaré inequality (Theorem 8.3.1 and Example 7.4.2) implies that

$$\|u - \langle u \rangle_B\|_{L^{p^*(\cdot)}(B)} \leqslant c \|\nabla u\|_{L^{p(\cdot)}(B)},$$

where we denoted $B := B(x,r)$ and where the constant c is independent of B. We may assume that r is so small the $p(\cdot)$-modular of ∇u over B is less than one, so we conclude that

$$\overline{\varrho}_{L^{p^*(\cdot)}(B)}\big(u - \langle u\rangle_B\big)^{\frac{1}{(p_B^*)^-}} \leqslant c\,\overline{\varrho}_{L^{p(\cdot)}(B)}(\nabla u)^{\frac{1}{p_B^+}}.$$

Hence we obtain

$$\fint_B |u(y) - \langle u\rangle_B|^{p^*(y)}\,dy = c\,r^{-n}\overline{\varrho}_{L^{p^*(\cdot)}(B)}\big(u - \langle u\rangle_B\big)$$

$$\leqslant c\,r^{-n}\overline{\varrho}_{L^{p(\cdot)}(B)}(\nabla u)^{\frac{(p_B^*)^-}{p_B^+}}$$

$$= c\,r^{(n-p(x))\frac{(p_B^*)^-}{p_B^+}-n}\left(r^{p(x)}\fint_B |\nabla u(y)|^{p(y)}\,dy\right)^{\frac{(p_B^*)^-}{p_B^+}}.$$

It suffices to show that $r^{(n-p(x))(p_B^*)^-/p_B^+-n} \leqslant c$ as $r \to 0$. Since $(p_B^*)^- = (p_B^-)^*$ we see that this is equivalent to

$$n\left(\frac{n-p(x)}{n-p_B^-}\frac{p_B^-}{p_B^+} - 1\right)\log r \leqslant c$$

at the same limit. By choosing the radius smaller if necessary, we may assume that $p_B^+ < n$. We have

$$0 \geqslant \frac{n-p(x)}{n-p_B^-}\frac{p_B^-}{p_B^+} - 1 \geqslant \frac{n-p_B^+}{n-p_B^-}\frac{p_B^-}{p_B^+} - 1 = \frac{n}{p_B^+(n-p_B^-)}(p_B^- - p_B^+)$$

$$\geqslant \frac{1}{n-1}(p_B^- - p_B^+).$$

Thus

$$\limsup_{r\to 0} n\left(\frac{n-p(x)}{n-p_B^-}\frac{p_B^-}{p_B^+} - 1\right)\log r \leqslant n'\limsup_{r\to 0}(p_B^- - p_B^+)\log r \leqslant c,$$

where the last inequality follows by the log-Hölder continuity of p.

Assume now that $p(x) \geqslant n$ and $q \geqslant n$. Let $\tilde{q} < n$ be such that $\tilde{q}^* = q$. Assume that $|B| \leqslant 1$ is so small that $\tilde{q} < p_B^-$. By the Sobolev-Poincaré inequality (Lemma 8.2.13), Hölder's inequality and Theorem 3.3.1 we obtain

$$\left(\fint_B |u - \langle u\rangle_B|^q \, dy\right)^{\frac{1}{q}} \leqslant c\,|B|^{-\frac{1}{q}} \|\nabla u\|_{L^{\tilde{q}}(B)} \leqslant c\,|B|^{-\frac{1}{q}+\frac{1}{q}-\frac{1}{p_B^-}} \|\nabla u\|_{L^{p_B^-}(B)}$$

$$\leqslant c\,(1+|B|)|B|^{\frac{1}{n}-\frac{1}{p_B^-}} \|\nabla u\|_{L^{p(\cdot)}(B)}$$

$$\leqslant c\,(1+|B|)|B|^{\frac{1}{p_B^+}-\frac{1}{p_B^-}} \|\nabla u\|_{L^{p(\cdot)}(B)}.$$

Since p is log-Hölder continuous $|B|^{\frac{1}{p_B^+}-\frac{1}{p_B^-}}$ is uniformly bounded. By Theorem 11.4.4 this yields the claim for $q > n$ as $r \to 0$. For $1 \leqslant q \leqslant n$ the claim follows by Hölder's inequality. $\qquad\square$

Remark 11.4.11. If $p(x) > n$ in Theorem 11.4.10, then there exists $r_x > 0$ such that

$$W^{1,p(\cdot)}(B(x,r_x)) \hookrightarrow W^{1,n+(p(x)-n)/2}(B(x,r_x)).$$

Hence u is continuous in a neighborhood of x and

$$\lim_{r \to 0} \operatorname*{ess\,sup}_{y \in B(x,r)} |u(y) - u^*(x)| = 0.$$

Theorems 11.4.4 and 11.4.10 imply the following corollary.

Corollary 11.4.12. Let $p \in \mathcal{P}^{\log}(\Omega)$ satisfy $1 < p^- \leqslant p^+ < \infty$. Assume that $u \in W^{1,p(\cdot)}(\Omega)$. Then there exists a set $E \subset \Omega$ with $\operatorname{cap}_{p(\cdot)}(E,\Omega) = 0$, such that u^* exists for every $x \in \Omega\backslash E$, u^* is the $p(\cdot)$-quasicontinuous representative of u,

$$\lim_{r \to 0} \fint_{B(x,r)} |u(y) - u^*(x)|^{p^*(y)} \, dy = 0$$

for every $x \in \big\{x \in \mathbb{R}^n : p(x) < n\big\} \setminus E$ and

$$\lim_{r \to 0} \fint_{B(x,r)} |u(y) - u^*(x)|^q dy = 0$$

for every $x \in \big\{x \in \mathbb{R}^n : p(x) \geqslant n\big\} \setminus E$ and for any finite $q \geqslant 1$.

Proof. All the claims have a local nature, so it is enough to prove them in all open balls B with $2B \subset \Omega$. Since the arguments are the same for every claim, we prove only the first one.

Fix such a ball. We first extend p as $\tilde{p}$ outside $2B$ so that the assumptions of Theorem 11.4.4 hold (Proposition 4.1.7). Let $\psi \in C_0^\infty(2B)$ be a cut off-function that is one in B. By Theorem 11.4.4 $(\psi u)^*$ exists $\tilde{p}(\cdot)$-quasieverywhere. Since $u^* = (\psi u)^*$ for every $x \in B$, we obtain by Proposition 10.3.4 that u^* exists in B outside a set of zero relative capacity. We can cover Ω by countably many balls and hence the claim follows by the subadditivity of the capacity. $\qquad\square$

11.5   Failure of Existence of Lebesgue Points*

In this section we will give an example of a uniformly continuous exponent for which not quasievery point is a Lebesgue point. The example is from Hästö [212].

We start by two technical lemmas.

Lemma 11.5.1. *Let $(a_i)_{i=1}^\infty$ be a partition of unity and $k > m - 1$. Then*

$$\sum_{i=1}^\infty a_i^m i^k \geq \left(\sum_{i=1}^\infty i^{-k/(m-1)} \right)^{1-m}.$$

Proof. Let (a_i) be a minimal sequence of the sum $\sum_{i=1}^\infty a_i^m i^k$. Fix an integer i and consider the function

$$a \mapsto (a_i + a)^m i^k + (a_{i+1} - a)^m (i+1)^k,$$

for $-a_i < a < a_{i+1}$. We find that this function has a minimum at $a = 0$ if and only if

$$i^k a_i^{m-1} = (i+1)^k a_{i+1}^{m-1}. \tag{11.5.2}$$

Thus we find that (11.5.2) holds for every a_i, $i \geq 1$. This partition is given by $a_i = i^{-k/(m-1)} a_0$ for $i > 1$ and $a_1 = (\sum_{i=1}^\infty i^{-k/(m-1)})^{-1}$ and so we easily calculate the lower bound as given in the lemma. $\qquad\square$

Lemma 11.5.3. *Let S be a subset of $\partial B(0, \frac{1}{5})$ of positive $(n-1)$-measure. Let*

$$C := \bigcup_{x \in S} [0, x].$$

Suppose that $p \in \mathcal{P}(B(0, \frac{1}{5}))$ is bounded and satisfies

$$p(x) \geq n + (n - 1 + \varepsilon) \frac{\log \log(1/|x|)}{\log(1/|x|)}$$

in C for some fixed $\varepsilon > 0$. Then

$$\inf \overline{\varrho}_{1,p(\cdot)}(u) \geq c(n, p^+),$$

where the infimum is taken over all continuous functions u that satisfy $u = 0$ in S, $u = 1$ in 0 and $|\nabla u| \in L^{p(\cdot)}(C)$.

Proof. We divide C into annuli, $A_i := \{x \in C : e^{-i} \leq |x| < e^{1-i}\}$ for $i \in \mathbb{N}$, $i \geq 3$. We set

$$q(x) := n + (n - 1 + \varepsilon) \sum_{i=3}^\infty \frac{\log i}{i} \chi_{A_i}(x).$$

Since $q \leqslant p$ we have $\|u\|_{q(\cdot)} \leqslant (1 + |C|)\|u\|_{p(\cdot)}$ and thus

$$\min\left\{\overline{\varrho}_{1,q(\cdot)}(u)^{\frac{1}{q^-}}, \overline{\varrho}_{1,q(\cdot)}(u)^{\frac{1}{q^+}}\right\} \leqslant (1 + |C|)\max\left\{\overline{\varrho}_{1,p(\cdot)}(u)^{\frac{1}{p^-}}, \overline{\varrho}_{1,p(\cdot)}(u)^{\frac{1}{p^+}}\right\}.$$

Therefore we see that it suffices to show that $\overline{\varrho}_{1,q(\cdot)}(u) > c(n)$ for every u that satisfies $u = 0$ in S, $u = 1$ in 0 and $|\nabla u| \in L^{q(\cdot)}(C)$.

The next step is crucial in making this lemma work even with possibly very irregular domains C. We estimate the gradient of u from below by the radial component of the derivative: $|\nabla u| \geqslant |\partial u/\partial r|$. (We are using that u is classically differentiable almost everywhere in C by Theorem 11.1.12.) Therefore

$$\int\limits_C |\nabla u(x)|^{q(x)}\,dx \geqslant \int\limits_C \left|\frac{\partial u(x)}{\partial r}\right|^{q(x)}\,dx.$$

It is then easy to see that the function minimizing the integral should depend only on the distance from the origin, not on the direction. If u is such a function, then

$$\int\limits_C |\nabla u(x)|^{q(x)}\,dx = \int\limits_S dm_{n-1}\int\limits_0^1 |\nabla u(rs)|^{q(rs)}\,dr,$$

where s is any fixed element in S. Thus the problem at hand is essentially a one-dimensional one.

Let e_1 be the first coordinate unit vector. We choose $r > 0$ such that

$$m_{n-1}\big(B(e_1/5, r) \cap \partial B(0, 1/5)\big) = m_{n-1}(S),$$

since $S \subset \partial B(0, 1/5)$ this is clearly possible. Define $S' := B(e_1/5, r) \cap \partial B(0, 1/5)$ and

$$C' := \bigcup_{x \in S'} [0, x].$$

Since $m_{n-1}(S) = m_{n-1}(S')$, the formula in the previous paragraph implies that

$$\int\limits_C |\nabla u(x)|^{q(x)}\,dx = \int\limits_{C'} |\nabla u(x)|^{q(x)}\,dx,$$

where u is radially symmetric.

Since the exponent q is fixed on A_i, we can use constant exponent capacity estimates for each annulus. This turns out to equal $c\, e^{i(q_i - n)}$ by Lemma 8.2.15, where q_i is the value of q on A_i. This means that if the function u increases by 1 from the inner to the outer boundary of the annulus A_i, then $\overline{\varrho}_{q_i}(|\nabla u|) \geqslant c\, e^{i(q_i - n)}$. Therefore, if the function increases by a_i on A_i, then the modular of its gradient is at least $c\, a_i^{q_i} e^{i(q_i - n)}$. Thus we seek to choose the a_i's so as to minimize

$$c \sum_{i=3}^{\infty} a_i^{q_i} e^{i(q_i - n)} = c \sum_{i=3}^{\infty} a_i^{q_i} i^{n-1+\varepsilon}. \tag{11.5.4}$$

Let N be such that

$$q_i - n = (n - 1 + \varepsilon)\frac{\log i}{i} \leqslant \frac{\varepsilon}{3}$$

for every $i \geqslant N$.

We write $\sum_{i=3}^{N-1} a_i = b$. The total increment of the function over all A_i's is at least 1 (by definition), so $\sum_{i=N}^{\infty} a_i = (1 - b)$. By the reverse Hölder's inequality for sums and Lemma 11.5.1 we obtain

$$\sum_{i=3}^{\infty} a_i^{q_i} i^{n-1+\varepsilon} \geqslant \sum_{i=3}^{N-1} a_i^{p^+} + \sum_{i=N}^{\infty} a_i^{q_i} i^{n-1+\varepsilon}$$

$$\geqslant \Big(\sum_{i=3}^{N-1} a_i \Big)^{p^+} (N - 1)^{1-p^+} + \sum_{i=N}^{\infty} a_i^{n+\varepsilon/3} i^{n-1+\varepsilon}$$

$$= \frac{b^{p^+}}{(N - 1)^{p^+-1}} + (1 - b)^{n+\varepsilon/3} \sum_{i=N}^{\infty} \Big(\frac{a_i}{1 - b} \Big)^{n+\varepsilon/3} i^{n-1+\varepsilon}$$

$$\geqslant \frac{b^{p^+}}{N^{p^+}} + (1 - b)^{n+\varepsilon/3} c > 0.$$

$\square$

We start with the unit disk D and divide it into the four quadrants. In the example of Zhikov [392, Sect. 1], the exponent is defined to be a constant $\alpha_2 > 2$ on quadrants A_1 and A_3 and a constant $\alpha_1 \in (1, 2)$ on the remaining two quadrants. In this case Zhikov proved that continuous functions are not dense in $W^{1,\alpha(\cdot)}(D)$. We will start from this and construct a uniformly continuous exponent for which the non-density still holds.

For technical reasons let's actually take $D := B(0, 1/4)$. We further partition the first and third quadrants into three parts by lines through the origin, see Fig. 11.1. For $0 < \varepsilon < 1$ we define an exponent $q \colon D \to [1, \infty)$ as follows: On A_1' and A_3' we set $p(x) := 2 + (1 + \varepsilon) \log_2(i)/i$ for $|x| \in [2^{-i}, 2^{1-i})$. On A_2 and A_4 we set $p(x) := 2 - (1 + \varepsilon) \log_2(\log_2(1/|x|))/\log_2(1/|x|)$. We extend the exponent linearly to the remaining domain. The exponent is sketched in Fig. 11.1.

Proposition 11.5.5. *For the previously defined uniformly continuous exponent p there exists a function $u \in W^{1,p(\cdot)}\big(B\big(0, \frac{1}{4}\big)\big)$ such that the origin has positive $p(\cdot)$-capacity but it is not a Lebesgue point of u. Thus not quasievery point is a Lebesgue point of u.*

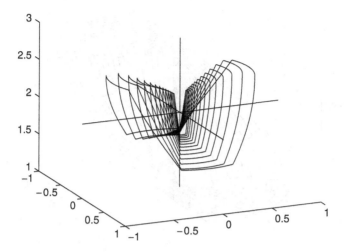

Fig. 11.1 A contour sketch of the exponent p

Proof. Define an exponent $q \in \mathcal{P}(B)$ which equals p on $A_1 \cup \ldots \cup A_4$ and 2 otherwise. Note that q is not continuous. Define the function $u \colon D \to [0,1]$ by

$$u(x) := \begin{cases} 1 & \text{for } x \in A_1 \\ x_2/|x| & \text{for } x \in A_2 \\ 0 & \text{for } x \in A_3 \\ x_1/|x| & \text{for } x \in A_4 \end{cases}$$

where $x = (x_1, x_2)$. The function is shown in Fig. 11.2. Since u is bounded, it is clear that $u \in L^{q(\cdot)}(D)$. We easily calculate $|\nabla u(x)| = |x|^{-1}$ for $x \in A_2 \cup A_4$. Using the substitution $s = \log_2(1/r)$ we calculate

$$\int\limits_D |\nabla u(x)|^{q(x)}\, dx = \pi \int\limits_0^{1/4} r^{1-q(r)}\, dr = \pi \int\limits_0^{1/4} 2^{-(1+\varepsilon)\log_2(\log_2(1/r))+\log_2(1/r)}\, dr$$

$$= \pi \log 2 \int\limits_2^{\infty} s^{-1-\varepsilon}\, ds = \pi \log(2) 2^{-\varepsilon} \varepsilon^{-1} < \infty.$$

Therefore $|\nabla u| \in L^{q(\cdot)}(D)$, and so $u \in W^{1,q(\cdot)}(D)$.

Let us next show that u cannot be approximated by a continuous function in the space $W^{1,q(\cdot)}(D)$. Let $v \in C(D) \cap W^{1,q(\cdot)}(D)$ with $v(0) = a$. We will use the estimate

$$\|u - v\|_{W^{1,q(\cdot)}(D)} \geqslant \|v\|_{W^{1,q(\cdot)}(A_3')} + \|1 - v\|_{W^{1,q(\cdot)}(A_1')}. \tag{11.5.6}$$

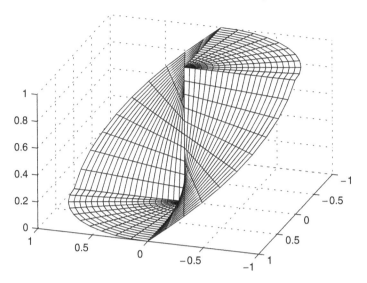

Fig. 11.2 A function u without the Lebesgue point property

By symmetry, we may then assume that $a \geqslant 1/2$ and consider only the first term on the right-hand-side of this inequality.

We can estimate $|\nabla v|$ from below by the radial derivative, which we will denote by a prime. (Note that it makes sense to speak of the radial derivative, since v is classically differentiable almost everywhere in A'_3 by [359, Theorem VIII.1.1].) Then we have

$$\int_{A'_3} |v(x)|^{q(x)} + |v'(x)|^{q(x)} \, dx \geqslant \int_{A'_3} |v_1(x)|^{q(x)} + |v'_1(x)|^{q(x)} \, dx,$$

where $v_1(x) := \min_{y \in [0,x]} v(y)$ and $[0, x]$ denotes the segment between 0 and x. We may therefore assume that v is radially decreasing. Let L be the subset of $\partial B(0, 1/5) \cap A'_3$ where $v(x) > 1/4$. Then

$$\int_{A'_3} |v(x)|^{q(x)} \, dx \geqslant 4^{-q^+} \frac{m_1(L)}{2\pi} m_2(B(1/5)),$$

where m_1 denotes the 1-dimensional Lebesgue measure. On the other hand, on $S := \left(\partial B(0, 1/5) \cap A'_3 \right) \setminus L$ the function v has value $1/4$ or less. Therefore the function $\varphi(x) := (v(x) - 1/4)/a$ is continuous, $\varphi(0) = 1$ and $\varphi(x) \leqslant 0$ for $x \in S$. Using Lemma 11.5.3, we conclude that

$$\int_{A'_3} |v'(x)|^{q(x)} \, dx \geqslant c \, m_1(S).$$

Combining these estimates we have shown that

$$\overline{\varrho}_{q(\cdot)}(v - u) + \overline{\varrho}_{q(\cdot)}(|\nabla(v - u)|) > c\, m_1(L) + c\, m_1(S) \geqslant c > 0$$

for all continuous functions v. Since $\overline{\varrho}_{p(\cdot)}(u_i) \to 0$ if and only if $\|u_i\|_{p(\cdot)} \to 0$, this means that $\|u - v\|_{1,p(\cdot)} > c > 0$ for all continuous v. Therefore continuous functions are not dense in $W^{1,q(\cdot)}(D)$. But q is not a continuous exponent, so there is still some work to be done.

We want to show that changing the exponent from q to p does not affect the properties we just proved for $W^{1,q(\cdot)}(D)$. Since $p = q$ on $A_2 \cup A_4$, it is easy to see that $u \in W^{1,p(\cdot)}(D)$ (u is as defined before). On $A_1' \cup A_3'$ we have $p \geqslant q$. We therefore have an embedding from $L^{p(\cdot)}(A_1' \cup A_3')$ to $L^{q(\cdot)}(A_1' \cup A_3')$ whose norm is at most $2(1 + |D|)$ (Theorem 3.3.1). If v is a continuous function this implies that

$$\|u - v\|_{W^{1,p(\cdot)}(B)} \geqslant \|u - v\|_{W^{1,p(\cdot)}(A_1' \cup A_3')} \geqslant \frac{\|u - v\|_{W^{1,q(\cdot)}(A_1' \cup A_3')}}{2(1 + |D|)} \geqslant c > 0,$$

so we still can not approximate by continuous functions. Thus $W^{1,p(\cdot)}(D)$ is the Sobolev space we were trying to construct. Note that in our example we have

$$|p(x) - p(0)| \approx \frac{\log_2 \log_2(1/x)}{-\log x}, \tag{11.5.7}$$

which is just barely worse than log-Hölder continuity. Incidentally, we have now proved the following result (cf. Example 9.2.6):

There exists a variable exponent Sobolev space in a bounded domain with uniformly continuous exponent such that continuous (or smooth) functions are not dense.

We continue the proof. Let $\tilde{u} \in W^{1,p(\cdot)}(D)$ be a function which equals u almost everywhere. Then

$$\lim_{r \to 0} \fint_{B(0,r)} |\tilde{u}(y) - \tilde{u}(0)|\, dy \geqslant$$

$$\lim_{r \to 0} \frac{1}{|B(0,r)|} \left(\int_{A_1 \cap B(0,r)} |1 - \tilde{u}(0)|\, dy + \int_{A_3 \cap B(0,r)} |\tilde{u}(0)|\, dy \right) > c > 0,$$

irrespective of the value of $\tilde{u}(0)$. This proves that 0 is not a Lebesgue point of any representative of u.

We will show that $C_{p(\cdot)}(\{0\}) > 0$. For this it suffices to show that

$$\inf \overline{\varrho}_{L^{p(\cdot)}(A_1')}(v) + \overline{\varrho}_{L^{p(\cdot)}(A_1')}(\nabla v) \geqslant c > 0$$

with a lower bound independent of r, where the infimum is taken over functions in $W^{1,p(\cdot)}(A_1')$ which equal 1 on the set $B(0,r) \cap A_1'$. In the set $A_1' \setminus B(0, \frac{1}{2}r)$ the exponent p is bounded away from 2, the dimension of the space. Therefore the functions we are minimizing over are continuous in $A_1' \setminus B(0, \frac{1}{2}r)$. We can then enlarge the set of functions we are considering and take the infimum over functions in $W^{1,p(\cdot)}(A_1')$ which are continuous and equal 1 at the origin. But we showed beginning with inequality (11.5.6) that precisely this infimum is positive, so we are done.   □

Remark 11.5.8. Let u be the function from Proposition 11.5.5. Let $\eta \in C_0^\infty(B(0, \frac{1}{4}))$ be a cut off -function that is one in $B(0, \frac{1}{8})$. It is easy to calculate $u\eta \in W^{1,p(\cdot)}(B(0, \frac{1}{4}))$. Moreover $u\eta$ belongs to $W_0^{1,p(\cdot)}(B(0, \frac{1}{4}))$ but not to $Q_0^{1,p(\cdot)}(B(0, \frac{1}{4}))$ and not to $H_0^{1,p(\cdot)}(B(0, \frac{1}{4}))$.

Chapter 12
Other Spaces of Differentiable Functions

We have considered spaces of measurable functions in the first part of the book and spaces of functions with a certain number of derivatives in the second part. In the final chapter of this part we look at more general spaces with other kinds of differentiability.

Although it might at first seem esoteric and perhaps unmotivated to study spaces of fractional derivatives, such spaces arise naturally in many contexts. Probably the most prominent example is the so-called "loss of $\frac{1}{p}$-th of a derivative at the boundary". To describe the precise meaning of this statement we recall the concept of trace spaces: By the trace of a function we mean its restriction to a subset of the original set of definition. For a continuous function this statement can be taken literally. For a Sobolev function some more care is needed, especially if the subset has measure zero; see the beginning of Sect. 12.1 for the exact definition.

According to classical constant exponent theory the restriction of a function $u \in W^{1,p}(\Omega)$ to the boundary behaves as if it possessed $1 - \frac{1}{p}$ derivatives on $\partial\Omega$. In the first section of this chapter we generalize this result to variable exponents in the case of the half-space $\mathbb{R}^{n+1}_{>} := \mathbb{R}^n \times (0, \infty)$. This may be thought of as a prototypical situation when the boundary is smooth or Lipschitz.

When we work with unbounded domains it is often not natural to assume that the function and the gradient belong to the same Lebesgue space. To avoid additional, superfluous assumptions, we introduce homogeneous Sobolev spaces in Sect. 12.2. These spaces are used in the study of PDEs in Chap. 14. In Sect. 12.3 we consider the dual spaces of both homogeneous and inhomogeneous Sobolev spaces, and show that the dual space can be thought of as spaces of negative smoothness.

In the constant exponent case Besov and Triebel–Lizorkin spaces have been studied both in the homogeneous and inhomogeneous case and with both positive and negative smoothness. However, in the variable exponent case only the theory of inhomogeneous Besov and Triebel–Lizorkin spaces has thus far been developed. It is presented in Sect. 12.5. These scales of

L. Diening et al., *Lebesgue and Sobolev Spaces with Variable Exponents*,
Lecture Notes in Mathematics 2017, DOI 10.1007/978-3-642-18363-8_12,
© Springer-Verlag Berlin Heidelberg 2011

function spaces include Lebesgue, Sobolev, Bessel and trace spaces, and has the advantage that it is closed under taking traces.

Let us now be a bit more precise about the meaning of a fractional derivative. Recall that the *Schwartz class* $\mathcal{S}$ consists of rapidly decaying smooth functions. If $f \in \mathcal{S}$, the *Fourier transform of* f is the function $\mathcal{F}f$ or $\widehat{f}$ defined by

$$\mathcal{F}f(\xi) = \widehat{f}(\xi) := \int_{\mathbb{R}^n} f(x)\, e^{-2\pi i \xi \cdot x} dx.$$

If f is sufficiently differentiable and β is a multi-index, then the Fourier transform turns derivatives into multiplication by powers of ξ according to the well-known formula:

$$\mathcal{F}(\partial_\beta f)(\xi) = (-2\pi i)^{|\beta|}\, \xi^\beta\, \widehat{f}(\xi).$$

For ξ^β it is not necessary that β consist only of integers. Using this idea we can define fractional derivatives and corresponding spaces. This is the basic idea behind the Bessel spaces defined in Sect. 12.4.

Let us note here that some authors have also studied Morrey spaces with variable exponents, but we will not deal with them here; see [23,303] for more information on this topic.

12.1 Trace Spaces

As mentioned above, the trace of a function is in some sense a restriction of the function to a subset of the original set of definition. Lebesgue functions are only defined almost everywhere, so it makes no sense to look at the function on a set of measure zero. A Sobolev function is, a priori, likewise only defined up to a set of measure zero. However, as we have seen in Sect. 11.1, a Sobolev function has a distinguished representative which is defined up to a set of capacity zero. Therefore it is possible to look at traces of Sobolev functions on sufficiently large measure-zero sets. Of particular interest is the trace of a function on the boundary of the set of definition, which is usually possible, since the boundary is typically of dimension at least $n-1$, hence its capacity is positive (Proposition 10.4.2).

The study of boundary trace spaces is very important in the theory of partial differential equations. Indeed, a partial differential equation is in many cases solvable if and only if the boundary values are in the corresponding trace space.

Let us start by recalling the definition of a $W^{1,1}$-trace in the half-space. By $\overline{\mathbb{R}^{n+1}_{\gtrless}}$ we denote the closure of $\mathbb{R}^{n+1}_{\gtrless}$. For $F \in W^{1,1}(\mathbb{R}^{n+1}_{\gtrless}) \cap C(\overline{\mathbb{R}^{n+1}_{\gtrless}})$ we set $\operatorname{Tr} F := F|_{\mathbb{R}^n}$. Then

$$F(x,0) = -\int_0^\infty \frac{d}{dt} F(x,t)\, dt$$

for $x \in \mathbb{R}^n$ and hence

$$|F(x,0)| \leqslant \int\limits_0^\infty |\nabla F(x,t)| \, dt.$$

Integrating this over $x \in \mathbb{R}^n$ gives $\|\operatorname{Tr} F\|_{L^1(\mathbb{R}^n)} \leqslant \|\nabla F\|_{L^1(\mathbb{R}^{n+1}_>)}$. Thus $\operatorname{Tr} \colon W^{1,1}(\mathbb{R}^{n+1}_>) \cap C^1(\mathbb{R}^{n+1}_>) \to L^1(\mathbb{R}^n)$ is a bounded, linear operator. Since the set $W^{1,1}(\mathbb{R}^{n+1}_>) \cap C^1(\mathbb{R}^{n+1}_>)$ is dense in $W^{1,1}(\mathbb{R}^{n+1}_>)$, we can extend Tr uniquely to a bounded, linear operator on $W^{1,1}(\mathbb{R}^{n+1}_>)$. The function $\operatorname{Tr} F \in L^1(\mathbb{R}^n)$ is called the *trace of F*.

Let $\Omega \subset \mathbb{R}^{n+1}$ be a Lipschitz domain and let $U \in W^{1,1}(\Omega)$. In a neighborhood of a boundary point $x_0 \in \partial\Omega$ we have a local bilipschitz chart which maps part of the boundary to $\mathbb{R}^n$. Thus the trace can be defined as in the previous paragraph and transported back to $\partial\Omega$ using the inverse chart. In this way we can define the trace $\operatorname{Tr} U$ as a function in $L^1_{\mathrm{loc}}(\partial\Omega)$. It is also possible to define traces in more general domains such as (ε, ∞)-domains, cf. [225, Chap. VIII].

Next we define the variable exponent trace space and show that it only depends on the value of the exponent on the boundary, provided the exponent is log-Hölder continuous.

Let $\Omega \subset \mathbb{R}^{n+1}$ be a Lipschitz domain and let $F \in W^1, p(\cdot)(\Omega)$ be a function. Since F is locally a $W^{1,1}$ function $\operatorname{Tr} F$ is defined as a function in $L^1_{\mathrm{loc}}(\partial\Omega)$ according to what was explained above. Note that if $F \in W^{k,p(\cdot)}(\Omega) \cap C(\overline{\Omega})$, then we still have $\operatorname{Tr} F = F|_{\partial\Omega}$. The *trace space* $\operatorname{Tr} W^{k,p(\cdot)}(\Omega)$ consists of the traces of all functions $F \in W^{k,p(\cdot)}(\Omega)$. The elements of $\operatorname{Tr} W^{k,p(\cdot)}(\Omega)$ are functions defined on $\partial\Omega$ – to emphasize this we always use lowercase letters for functions on $\partial\Omega$, whereas uppercase letters are used for functions in Ω and $\mathbb{R}^{n+1}$. The quotient norm

$$\|f\|_{\operatorname{Tr} W^{k,p(\cdot)}(\Omega)} := \inf \left\{ \|F\|_{W^{k,p(\cdot)}(\Omega)} : F \in W^{1,p(\cdot)}(\Omega) \text{ and } \operatorname{Tr} F = f \right\}$$

makes $\operatorname{Tr} W^{k,p(\cdot)}(\Omega)$ a Banach space.

The first appearance of trace spaces using this definition in the context of Sobolev spaces with variable exponents is in [104, 105], where the solvability of the Laplace equation $-\Delta u = f$ on the half-space with given boundary values is studied.

Notice that this definition allows us to treat the trace spaces as an abstract object: we simply define the boundary value function space as consisting of those functions which are the boundary values of some function. This is of course not very useful in terms of understanding the trace space. Thus the main purpose of this section is to provide an intrinsic norm for the trace space, i.e. a norm which is defined only in terms of f and not in terms of its extensions F.

Furthermore, intuitively we would expect that this intrinsic norm only depends on $p|_{\partial\Omega}$ and not on p on the whole domain Ω. Nevertheless, the definition of $\operatorname{Tr} W^{k,p(\cdot)}(\Omega)$ above is dependent on the values of p on all of Ω. Throughout the book, we have seen that log-Hölder continuity is often sufficient for variable exponent spaces to behave in a very nice way. This turns out to hold also with trace spaces:

Theorem 12.1.1. *Let* $\Omega \subset \mathbb{R}^{n+1}$ *be a Lipschitz domain, let* $p_1, p_2 \in \mathcal{P}^{\log}(\Omega)$ *with* $p_1|_{\partial\Omega} = p_2|_{\partial\Omega}$, *and let* $k \geqslant 1$. *Then* $\operatorname{Tr} W^{k,p_1(\cdot)}(\Omega) = \operatorname{Tr} W^{k,p_2(\cdot)}(\Omega)$ *with equivalent norms.*

Proof. We prove the result for $\Omega = \mathbb{R}^{n+1}_{>}$. The general result is reduced to this by the chart mappings.

Define the lower half-space $\mathbb{R}^{n+1}_{<} := \mathbb{R}^n \times (-\infty, 0)$. Set $q(x,t) := p_1(x,t)$ for $t \geqslant 0$ and $q(x,t) := p_2(x,-t)$ for $t < 0$. Then $q \in \mathcal{P}^{\log}(\mathbb{R}^{n+1})$. By Theorem 8.5.12 there exists a bounded, linear extension

$$\mathcal{E} : W^{k,p_1(\cdot)}(\mathbb{R}^{n+1}_{>}) \to W^{k,q(\cdot)}(\mathbb{R}^{n+1}),$$

Let $\mathfrak{R}$ denote the reflection $(\mathcal{R}v)(t,x) := v(-t,x)$. Then

$$W^{k,p_1(\cdot)}(\mathbb{R}^{n+1}_{>}) \xrightarrow{\mathcal{E}} W^{k,q(\cdot)}(\mathbb{R}^{n+1}) \xrightarrow{\mathcal{R}} W^{k,\mathcal{R}q(\cdot)}(\mathbb{R}^{n+1}) \xrightarrow{|_{\mathbb{R}^{n+1}_{>}}} W^{k,p_2(\cdot)}(\mathbb{R}^{n+1}_{>})$$

continuously. Since clearly $\operatorname{Tr}((\mathcal{R}\mathcal{E}F)|_{\mathbb{R}^{n+1}_{>}}) = \operatorname{Tr} F$ (e.g. by the ACL property, this proves $\operatorname{Tr} W^{k,p_1(\cdot)}(\mathbb{R}^{n+1}_{>}) \hookrightarrow \operatorname{Tr} W^{k,p_2(\cdot)}(\mathbb{R}^{n+1}_{>})$. The opposite inclusion follows by symmetry. $\qquad\square$

Proposition 4.1.7 and Theorem 12.1.1 imply that the following definition makes sense (up to equivalence of norms) for Lipschitz domains.

Definition 12.1.2. Let $p \in \mathcal{P}^{\log}(\partial\Omega)$ and let $q \in \mathcal{P}^{\log}(\Omega)$ be an arbitrary extension of p. Then we define an intrinsic trace space by

$$\left(\operatorname{Tr} W^{k,p(\cdot)} \right)(\partial\Omega) := \operatorname{Tr} W^{k,q(\cdot)}(\Omega).$$

So far we have seen that the trace space on $\mathbb{R}^n$ does not really depend on values of the exponent on $\mathbb{R}^{n+1} \setminus (\mathbb{R}^n \times \{0\})$ if p is globally log-Hölder continuous. This considerably simplifies studying the space $(\operatorname{Tr} W^{1,p(\cdot)})(\mathbb{R}^n)$. Indeed, for $x \in \mathbb{R}^n$ and $t \in [0,2]$ define $q(x,t) := p(x)$. Then q is globally log-Hölder continuous on $\mathbb{R}^n \times [0,2]$. By Proposition 4.1.7 we can extend q to the set $\mathbb{R}^{n+1}_{>}$ so that $q \in \mathcal{P}^{\log}(\mathbb{R}^{n+1}_{>})$. We have $(\operatorname{Tr} W^{1,p(\cdot)})(\mathbb{R}^n) = \operatorname{Tr} W^{1,q(\cdot)}(\mathbb{R}^{n+1}_{>})$. So we can as well assume from the beginning that the exponent $p(x,t)$ is independent of t when $t \in [0,2]$.

Since we use balls in different spaces, we denote by $B^n(x,r)$ and $B^{n+1}(x,r)$ the balls in $\mathbb{R}^n$ and $\mathbb{R}^{n+1}$, respectively. Notice the difference between the spaces $C_0^\infty(\mathbb{R}_\geqslant^{n+1})$ and $C_0^\infty(\mathbb{R}_>^{n+1})$: in the former space functions simply have bounded support, in the latter the support of the function is bounded and disjoint from the boundary $\mathbb{R}^n$ of $\mathbb{R}_>^{n+1}$.

Recall from Sect. 6.2 the definition of the sharp operator:

$$M^\sharp_{B^n(x,r)} f = \fint_{B^n(x,r)} \left| f(y) - \langle f \rangle_{B^n(x,r)} \right| dy,$$

for a function $f \in L^1_{\mathrm{loc}}(\mathbb{R}^n)$. Using the triangle inequality it is easy to show the equivalence

$$M^\sharp_{B^n(x,r)} f \leqslant \fint_{B^n(x,r)} \fint_{B^n(x,r)} \left| f(y) - f(z) \right| dy\, dz \leqslant 2\, M^\sharp_{B^n(x,r)} f. \qquad (12.1.3)$$

We define the *trace modular* $\varrho_{\mathrm{Tr},p(\cdot)}$ by

$$\varrho_{\mathrm{Tr},p(\cdot)}(f) := \int_{\mathbb{R}^n} |f(x)|^{p(x)}\, dx + \int_0^1 \int_{\mathbb{R}^n} \left(\tfrac{1}{r} M^\sharp_{B^n(x,r)} f \right)^{p(x)} dx\, dr.$$

Obviously, $\varrho_{\mathrm{Tr},p(\cdot)}$ is a modular. Thus

$$\|f\|_{\mathrm{Tr},p(\cdot)} := \inf \left\{ \lambda > 0 : \varrho_{\mathrm{Tr},p(\cdot)}(f/\lambda) \leqslant 1 \right\}$$

is a norm.

The following theorem characterizes the traces of $W^{1,p(\cdot)}(\mathbb{R}_>^{n+1})$-functions in terms of an intrinsic norm.

Theorem 12.1.4. *Let $p \in \mathcal{P}^{\log}(\mathbb{R}_\geqslant^{n+1})$ with $1 < p^- \leqslant p^+ < \infty$ and let $f \in L^1_{\mathrm{loc}}(\mathbb{R}^n)$. Then f belongs to $\operatorname{Tr} W^{1,p(\cdot)}(\mathbb{R}_>^{n+1})$ if and only if $\|f\|_{\mathrm{Tr},p(\cdot)} < \infty$, or, equivalently,*

$$\int_{\mathbb{R}^n} |f(x)|^{p(x)}\, dx + \int_0^1 \int_{\mathbb{R}^n} \left(\tfrac{1}{r} M^\sharp_{B^n(x,r)} f \right)^{p(x)} dx\, dr < \infty,$$

where $p(x) := p(x,0)$. Moreover, $\|f\|_{\mathrm{Tr},p(\cdot)}$ is equivalent to the quotient norm $\|f\|_{\operatorname{Tr} W^{1,p(\cdot)}(\mathbb{R}_>^{n+1})}$.

Before the proof of the theorem, we note that it directly generalizes to the case of Lipschitz domains:

Corollary 12.1.5. *Let $\Omega \subset \mathbb{R}^n$ be a Lipschitz domain, let $p \in \mathcal{P}^{\log}(\Omega)$ with $1 < p^- \leqslant p^+ < \infty$ and let $f \in L^1_{\mathrm{loc}}(\partial\Omega)$. Then $f \in \operatorname{Tr} W^{1,p(\cdot)}(\Omega)$ if and only if*

$$\int\limits_{\partial\Omega} |f(x)|^{p(x)}\,dx + \int\limits_{0}^{\kappa}\int\limits_{\partial\Omega}\left(\tfrac{1}{r}M^{\sharp}_{B^{n}(x,r)}f\right)^{p(x)}dx\,dr < \infty,$$

where the constant $\kappa > 0$ depends only on Ω. Moreover, $\|f\|_{\mathrm{Tr},p(\cdot)}$ is equivalent to the quotient norm $\|f\|_{\mathrm{Tr}\,W^{1,p(\cdot)}(\Omega)}$.

Proof. It follows from the definition of Lipschitz domain that the boundary of Ω can be covered by balls B_j in such a way that there exists a bilipschitz mapping $G_j\colon B_j \to \mathbb{R}^n$ with $G_j(\Omega\cap B_j) \subset \mathbb{R}^n_>$ and $G_j(\partial\Omega\cap B_j) \subset \mathbb{R}^{n-1}$. Since the boundary is covered by a finite number of balls, we can choose functions $\psi_0 \in C_0^\infty(\Omega)$ and $\psi_j \in C_0^\infty(B_j)$ for $j \geqslant 1$ such that $\sum_j \psi_j = 1$ in $\overline{\Omega}$.

Suppose first that $f \in \mathrm{Tr}\,W^{1,p(\cdot)}(\Omega)$, and let $F \in W^{1,p(\cdot)}(\Omega)$ be such that $\mathrm{Tr}\,F = f$. Then $(F\psi_j)\circ G_j^{-1} \in W^{1,p\circ G_j^{-1}(\cdot)}(\mathbb{R}^n_>)$ and $\mathrm{Tr}(F\psi_j)\circ G_j^{-1} = (f\psi_j)\circ G_j^{-1}$. Hence

$$\int\limits_{\mathbb{R}^n} |(f\psi_j)\circ G_j^{-1}(x)|^{p(G_j^{-1}(x))}\,dx + \int\limits_{0}^{1}\int\limits_{\mathbb{R}^n}\left(\tfrac{1}{r}M^{\sharp}_{B^{n}(x,r)}(f\psi_j)\circ G_j^{-1}\right)^{p(G_j^{-1}(x))}dx\,dr$$

is finite by Theorem 12.1.4. A change of variables $y = G_j^{-1}(x)$ combined with the fact that G_j is bilipschitz yields

$$\int\limits_{\partial\Omega} |f\psi_j|^{p(y)}\,dy + \int\limits_{0}^{\kappa}\int\limits_{\partial\Omega}\left(\tfrac{1}{r}M^{\sharp}_{B^{n}(y,\kappa r)}(f\psi_j)\right)^{p(y)}dy\,dr < \infty,$$

where $\kappa > 0$ is the reciprocal of the bilipschitz constant. Combining the estimates for all $j \geqslant 1$ yields the inequality on all of $\partial\Omega$, since this set is covered by the balls B_j.

The proof of the converse implication is similar, and thus skipped. $\square$

To prove Theorem 12.1.4 we proceed as follows. First, for $F \in W^{1,p(\cdot)}(\mathbb{R}^{n+1}_>)$ and $f := \mathrm{Tr}\,F$ we show that $\|f\|_{\mathrm{Tr},p(\cdot)} \leqslant c\,\|F\|_{W^{1,p(\cdot)}(\mathbb{R}^{n+1}_>)}$. Therefore, we estimate $|f|$ and $M^{\sharp}_{B^n(x,t)}f$ in terms of $|F|$ and $|\nabla F|$. Second, for $f \in \mathrm{Tr}\,W^{1,p(\cdot)}(\mathbb{R}^{n+1}_>)$ we show the existence of some $F \in W^{1,p(\cdot)}(\mathbb{R}^{n+1}_>)$ with $\mathrm{Tr}\,F = f$ and $\|F\|_{W^{1,p(\cdot)}(\mathbb{R}^{n+1}_>)} \leqslant c\,\|f\|_{\mathrm{Tr},p(\cdot)}$. We will define the extension F by $F(x,t) := (\psi_t * f)(x)$ for $x \in \mathbb{R}^n$ and $t > 0$, where (ψ_t) is a standard mollifier family in $\mathbb{R}^n$. In order to estimate $\|F\|_{W^{1,p(\cdot)}(\mathbb{R}^{n+1}_>)}$ we need to estimate $|F|$ and $|\nabla F|$ in terms of $|f|$ and $M^{\sharp}_{B^n(x,t)}f$. The following two lemmas provide these estimates.

Lemma 12.1.6. *There exists a constant $c > 0$, such that*

$$M^{\sharp}_{B^n(z,r)} \operatorname{Tr} F \leqslant cr \fint_{B^{n+1}((z,0),r) \cap \mathbb{R}^{n+1}_{>}} |\nabla F(\xi)| \, d\xi$$

for all $z \in \mathbb{R}^n$, $r > 0$ and $F \in W^{1,1}\big(B^{n+1}((z,0),r) \cap \mathbb{R}^{n+1}_{>}\big)$.

Proof. Since smooth functions are dense in $W^{1,1}(B^{n+1}(z,r))$ it suffices to prove the claim for smooth F. As usual we denote $f = \operatorname{Tr} F = F|_{\mathbb{R}^n}$. Let us estimate $|f(x) - f(y)|$ for $x, y \in \mathbb{R}^n$ by integrating the gradient over the path $\gamma_\zeta = [x, \zeta] \cup [\zeta, y]$ for $\zeta \in \mathbb{R}^{n+1}_{>}$:

$$|f(x) - f(y)| \leqslant \int_{\gamma_\zeta} |\nabla F(\xi)| \, d\xi. \tag{12.1.7}$$

Define $B_{x,y} := B^{n+1}\big(\frac{x+y}{2} + \frac{|x-y|}{4} e_{n+1}, \frac{|x-y|}{8}\big) \cap P$, where P is the mid-point normal plane of the segment $[x, y]$ and e_{n+1} is the unit vector in direction $n+1$, and let $A_{x,y} = \bigcup_{\zeta \in B_{x,y}} \gamma_\zeta$. Next we take the average integral of (12.1.7) over $\zeta \in B_{x,y}$. This so-called Riesz potential estimate (e.g. [218]) yields

$$|f(x) - f(y)| \leqslant c \int_{A_{x,y}} |\nabla F(\xi)| \big(|x - \xi|^{-n} + |y - \xi|^{-n}\big) \, d\xi.$$

Let $z \in \mathbb{R}^n$ and $r > 0$. Using the previous estimate together with (12.1.3) gives

$$M^{\sharp}_{B^n(z,r)} f \leqslant c \fint_{B^n(z,r)} \fint_{B^n(z,r)} \int_{A_{x,y}} |\nabla F(\xi)| \, t^{-n} \, d\xi \, dx \, dy, \tag{12.1.8}$$

where $\xi = (w, t) \in \mathbb{R}^n \times \mathbb{R}$ and we used that $t \leqslant \min\{|y - \xi|, |x - \xi|\}$ when $\xi \in A_{x,y}$.

The set $A_{x,y}$ consists of two cones, one emanating from y and the other from x, denoted by $A'_{x,y}$ and $A''_{x,y}$, respectively. By picking the larger integral we can replace $A_{x,y}$ by $A'_{x,y}$ or $A''_{x,y}$ by doubling the constant. By symmetry it is enough to consider in the following the case with $A_{x,y}$ replaced by $A'_{x,y}$. We want to change the order of integration. So suppose that $\xi \in A'_{x,y}$. Then certainly $\xi \in B^{n+1}(z, r)$. Also, ξ lies in a cone emanating from y whose direction depends on $x - y$. Thus we see that y lies in the cone emanating from ξ with the same base-angle but opposite direction. This means that for a fixed ξ the variable y varies in a ball $B^n(w, c't)$ and $c' > 0$ depends only on the dimension n (recall that $(w, t) = \xi$). Hence

$$M^{\sharp}_{B^n(z,r)}f \leqslant c\,r^{-2n} \int\limits_{B^n(z,r)} \int\limits_{B^n(z,r)} \int\limits_{A'_{x,y}} |\nabla F(\xi)|\, t^{-n}\, d\xi\, dx\, dy$$

$$\leqslant c\,r^{-2n} \int\limits_{B^{n+1}((z,0),r)} \chi_{\mathbb{R}^{n+1}_>}(\xi)\, |\nabla F(\xi)|\, t^{-n} \int\limits_{B^n(w,c't)} \int\limits_{B^n(z,r)} dx\, dy\, d\xi$$

$$= c\,r \fint\limits_{B^{n+1}((z,0),r)} \chi_{\mathbb{R}^{n+1}_>}(\xi)\, |\nabla F(\xi)|\, d\xi. \qquad\qquad \square$$

Recall that if $p \in \mathcal{P}^{\log}(\mathbb{R}^n)$ and $\{\psi_t\}$ is a standard mollifier family, then $\psi_t * f \to f$ in $W^{1,p(\cdot)}(\mathbb{R}^n)$ for every $f \in W^{1,p(\cdot)}(\mathbb{R}^n)$. This follows by the Lebesgue space result (Theorem 4.6.4) and $\nabla(f * \psi_t) = (\nabla f) * \psi_t$.

Lemma 12.1.9. *Let $\{\psi_t\}$ be a standard mollifier family on $\mathbb{R}^n$. Let $f \in L^1_{\mathrm{loc}}(\mathbb{R}^n)$ and define $F(x,t) := (\psi_t * f)(x)$ for $x \in \mathbb{R}^n$ and $t \in (0,\infty)$. Then there exists a constant c depending only on $\|\psi\|_{1,\infty}$ and n such that*

$$|F(x,t)| \leqslant c\, M_{B^n(x,t)}f,$$
$$|\nabla F(x,t)| \leqslant \frac{c}{t}\, M^{\sharp}_{B^n(x,t)}f,$$

for all $x \in \mathbb{R}^n$ and $t \in (0,\infty)$.

Proof. Since $|\psi_t| \leqslant \|\psi\|_\infty\, t^{-n}\chi_{B^n(0,t)}$, the first inequality is immediate.

Let $T = T_{\{2Q\}}$ denote an averaging operator where $\{Q\}$ is the family of dyadic cubes, whose side length is equal to its distance to $\mathbb{R}^n \times \{0\}$.

In the following we denote $\xi := (x,t) \in \mathbb{R}^{n+1}_>$. Since $\int_{\mathbb{R}^n} \nabla\psi\, dy = 0$, we obtain

$$\nabla_x F(\xi) = \nabla\psi_t * f(x) = \int\limits_{\mathbb{R}^n} \nabla_x \psi_t(x-y)(f(y) - \langle f\rangle_{B^n(x,t)})\, dx.$$

We use $|\nabla\psi_t| \leqslant \|\nabla\psi\|_\infty\, t^{-n-1}\chi_{B^n(0,t)}$ to derive $|\nabla_x F(x,t)| \leqslant c\,t^{-1}M^{\sharp}_{B^n(x,t)}f$. For the t-derivative we need a slightly more involved calculation: for all $a \in \mathbb{R}$ we have

$$\partial_t F(\xi) = \partial_t(f * \psi_t)(x) = \partial_t\big((f-a) * \psi_t\big)(x) = (f-a) * \frac{d}{dt}\psi_t$$

$$= \int\limits_{\mathbb{R}^n} \Big[-\frac{n}{t}\psi_t(x-y) - (\nabla_x\psi)_t(x-y)\cdot\frac{x-y}{t^2} \Big] \big(f(y) - a\big)\, dy,$$

where $(\nabla_x\psi)_t(y) := t^{-n}\nabla\psi(y/t)$. Setting $a = \langle f\rangle_{B^n(x,t)}$ we find that

$$\left| \partial_t F(\xi) \right| \leqslant \int\limits_{B^n(x,t)} \left(\frac{n}{t^{n+1}} \|\psi\|_\infty + \frac{|x-y|}{t^{n+2}} \|\nabla_x \psi\|_\infty \right) \left| f(y) - \langle f \rangle_{B^n(x,t)} \right| dy$$

$$\leqslant t^{-n-1} \left(n\|\psi\|_\infty + \|\nabla_x \psi\|_\infty \right) \int\limits_{B^n(x,t)} \left| f(y) - \langle f \rangle_{B^n(x,t)} \right| dy$$

$$\leqslant \frac{c}{|t|} \fint\limits_{B^n(x,t)} \left| f(y) - \langle f \rangle_{B^n(x,t)} \right| dy.$$

Since $|\nabla F| \leqslant |\nabla_x F| + |\partial_t F|$, this completes the proof $\square$

Thus we are ready for the proof of the main result of the section.

Proof of Theorem 12.1.4. Due to Theorem 12.1.1 and the discussion after it we can assume without loss of generality that $p(x,t) = p(x,0) = p(x)$ for $x \in \mathbb{R}^n$ and $t \in [0,2]$.

Let $\{\psi_t\}$ be a standard mollifier family on $\mathbb{R}^n$, and let $f \in L^1_{\mathrm{loc}}(\mathbb{R}^n)$ with $\|f\|_{\mathrm{Tr},p(\cdot)} \leqslant 1$, which by the unit ball property implies

$$\int\limits_{\mathbb{R}^n} |f(x)|^{p(x)}\, dx + \int\limits_0^1 \int\limits_{\mathbb{R}^n} \left(\frac{1}{r} M^\sharp_{B^n(x,r)} f \right)^{p(x)} dx\, dr \leqslant 1.$$

We have to show the existence of an extension $\widetilde{F} \in W^{1,p(\cdot)}(\mathbb{R}^{n+1}_>)$ which satisfies $\|\widetilde{F}\|_{W^{1,p(\cdot)}(\mathbb{R}^{n+1}_>)} \leqslant c$, where c is independent of f. As mentioned above, we would like to consider the extension $(x,t) \mapsto \psi_t * f(x)$. But in order to avoid difficulties as $t \to \infty$ we cut off the part for large t. Let $\eta \in C_0^\infty([0,\infty))$ with $\chi_{[0,1/2]} \leqslant \eta \leqslant \chi_{[0,1]}$. Then our extension $\widetilde{F}$ is given by $\widetilde{F}(x,t) := (\psi_t * f)(x)\,\eta(t) = F(x,t)\,\eta(t)$, where we used the notation from Lemma 12.1.9.

We now estimate the norm of $\widetilde{F}$ in $W^{1,p(\cdot)}(\mathbb{R}^{n+1}_>)$. Using Lemma 12.1.9 and noting that $M_{B^n(x,t)} f \leqslant Mf(x)$, we find that

$$\varrho_{L^{p(\cdot)}(\mathbb{R}^{n+1}_>)}(\widetilde{F}) = \int\limits_0^1 \int\limits_{\mathbb{R}^n} |\widetilde{F}(x,t)|^{p(x)}\, dx\, dt \leqslant c \int\limits_{\mathbb{R}^n} Mf(x)^{p(x)}\, dx.$$

Our assumptions on p imply that the maximal operator is bounded on $L^{p(\cdot)}(\mathbb{R}^n)$ (Theorem 4.3.8). Since $\varrho_{p(\cdot)}(f) \leqslant 1$, it follows from the previous inequality that $\varrho_{L^{p(\cdot)}(\mathbb{R}^{n+1}_>)}(\widetilde{F}) \leqslant c$. It remains to estimate the norm of the gradient of $\widetilde{F}$. Using $|\nabla \widetilde{F}| \leqslant |\nabla F|\eta + |F||\partial_t \eta|$, Lemma 12.1.9 again, $M_{B^n(x,t)} f \leqslant Mf(x)$, and the continuity of the maximal operator, we estimate

$$\varrho_{L^{p(\cdot)}(\mathbb{R}_{>}^{n+1})}(\nabla \widetilde{F}) = \int\limits_0^1 \int\limits_{\mathbb{R}^n} \left| \nabla \widetilde{F}(x,t) \right|^{p(x)} dx \, dt$$

$$\leqslant c \int\limits_0^1 \int\limits_{\mathbb{R}^n} \left| \tfrac{1}{t} M_{B^n(x,t)}^{\sharp} f \right|^{p(x)} dx \, dt + c(\eta) \int\limits_0^1 \int\limits_{\mathbb{R}^n} |Mf|^{p(x)} dx \, dt \leqslant c.$$

Thus we have shown that $F \in W^{1,p(\cdot)}(\mathbb{R}_{>}^{n+1})$. Furthermore, it follows easily that $f = \operatorname{Tr} F$, so we have proved one of the implications in the theorem.

To prove the opposite implication, we use the density of smooth functions (Theorem 9.1.7) and restrict ourselves without loss of generality to $F \in W^{1,p(\cdot)}(\mathbb{R}_{>}^{n+1}) \cap C^{\infty}(\mathbb{R}_{>}^{n+1})$. Replacing F by $F\eta$, where η is as above, we see that it suffices to consider F supported in $\mathbb{R}^n \times [0,1]$. By homogeneity, it suffices to consider the case $\|F\|_{W^{1,p(\cdot)}(\mathbb{R}_{>}^{n+1})} \leqslant 1$ and to prove $\|f\|_{\operatorname{Tr},p(\cdot)} \leqslant c$ for $f := \operatorname{Tr} F$. Since p is bounded, the latter condition is equivalent to $\varrho_{\operatorname{Tr},p(\cdot)}(f) \leqslant c$, which is what we now prove. We find that

$$|f(x)| = |F(x,0)| \leqslant \int\limits_0^1 |\partial_t F(x,t)| \, dt.$$

Hence using Hölder's inequality with respect to the variable t for the constant exponent $p(x)$ we get

$$|f(x)|^{p(x)} \leqslant \int\limits_0^1 |\partial_t F(x,t)|^{p(x)} \, dt.$$

Integrating this inequality over $x \in \mathbb{R}^n$ yields

$$\varrho_{p(\cdot)}(f) = \int\limits_{\mathbb{R}^n} |f(x)|^{p(x)} \, dx \leqslant \int\limits_{\mathbb{R}^n} \int\limits_0^1 |\partial_t F(x,t)|^{p(x)} \, dt \, dx \leqslant \varrho_{L^{p(\cdot)}(\mathbb{R}_{>}^{n+1})}(\nabla F).$$

Thus we have bounded the $L^{p(\cdot)}$ part of the trace norm.

Since $f = \operatorname{Tr} F$, we get by Lemma 12.1.6 that

$$\int\limits_0^1 \int\limits_{\mathbb{R}^n} \left(\tfrac{1}{r} M_{B^n(x,r)}^{\sharp} f \right)^{p(x)} dx \, dr$$

$$\leqslant c \int\limits_0^1 \int\limits_{\mathbb{R}^n} \left(\fint_{B^{n+1}((x,0),r) \cap \mathbb{R}_{>}^{n+1}} |\nabla F(\xi)| \, d\xi \right)^{p(x)} dx \, dr$$

$$\leqslant c \int\limits_{0}^{1}\int\limits_{\mathbb{R}^n} \Bigg(\fint\limits_{B^{n+1}((x,r),2r)\cap\mathbb{R}^{n+1}_{>}} |\nabla F(\xi)|\, d\xi \Bigg)^{p(x)} dx\, dr$$

$$\leqslant c \int\limits_{\mathbb{R}^n\times[0,1]} \Big(M\big(\chi_{\mathbb{R}^{n+1}_{>}} |\nabla F|\big)(z) \Big)^{p(z)} dz.$$

Extending the exponent to the lower half-space by reflection, we immediately see that $p \in \mathcal{P}^{\log}(\mathbb{R}^{n+1})$ and

$$\int\limits_{0}^{1}\int\limits_{\mathbb{R}^n} \Big(\tfrac{1}{r}M^{\sharp}_{B^n(x,r)}f \Big)^{p(x)} dx\, dr \leqslant c \int\limits_{\mathbb{R}^{n+1}} \Big(M\big(\chi_{\mathbb{R}^{n+1}_{>}} |\nabla F|\big)(z) \Big)^{p(z)} dz.$$

Since the maximal operator is bounded on $L^{p(\cdot)}(\mathbb{R}^{n+1})$, the right-hand-side of the previous inequality is bounded by a constant, which concludes the proof.   □

In the final part of the section we work with the weaker assumption that the exponent is such that smooth functions are dense in our Sobolev space (cf. Chap. 9). This means that we have to invoke a different machinery, in particular the capacity from Chap. 10. Recall that $H_0^{1,p(\cdot)}(\mathbb{R}^{n+1}_{>})$ is the completion of $C_0^\infty(\mathbb{R}^{n+1}_{>})$ in $W^{1,p(\cdot)}(\mathbb{R}^{n+1}_{>})$, whereas $W_0^{1,p(\cdot)}(\mathbb{R}^{n+1}_{>})$ is the completion of the set of Sobolev functions with compact support (cf. Sect. 11.2).

Lemma 12.1.10. *If $p \in \mathcal{P}(\mathbb{R}^{n+1}_{>})$ is bounded and $F \in H_0^{1,p(\cdot)}(\mathbb{R}^{n+1}_{>})$, then* $\operatorname{Tr} F = 0$.

Proof. If $F \in H_0^{1,p(\cdot)}(\mathbb{R}^{n+1}_{>})$, then there exists by definition a sequence $(\Psi_i) \subset C_0^\infty(\mathbb{R}^{n+1}_{>})$ with $F = \lim_{i\to\infty}\Psi_i$ in $W^{1,p(\cdot)}(\mathbb{R}^{n+1}_{>})$. Since $\operatorname{Tr}\Psi_i = \Psi_i|_{\mathbb{R}^n} \equiv 0$, the claim follows by continuity of $\operatorname{Tr} : W^{1,p(\cdot)}(\mathbb{R}^{n+1}_{>}) \to \operatorname{Tr} W^{1,p(\cdot)}(\mathbb{R}^{n+1}_{>})$.   □

For the converse, we need to assume the density of smooth functions, which is guaranteed e.g. for $p \in \mathcal{P}^{\log}$ (cf. Theorem 9.1.7). For a version with weaker assumptions see [98].

Theorem 12.1.11. *Let $p \in \mathcal{P}^{\log}(\mathbb{R}^{n+1}_{>})$ be bounded. Then $F \in W^{1,p(\cdot)}(\mathbb{R}^{n+1}_{>})$ with $\operatorname{Tr} F = 0$ if and only if $F \in W_0^{1,p(\cdot)}(\mathbb{R}^{n+1}_{>})$.*

Proof. Suppose first that $F \in W^{1,p(\cdot)}(\mathbb{R}^{n+1}_{>})$ with $\operatorname{Tr} F = 0$. We extend p to $\mathbb{R}^{n+1}_{<}$ by reflection. Since $W^{1,p(\cdot)}(\mathbb{R}^{n+1}_{>}) \hookrightarrow W^{1,1}(\mathbb{R}^{n+1}_{>}\cap K)$, for every compact set $K \subset \mathbb{R}^{n+1}$, it follows by classical theory that F extended by 0 to the lower half-space $\mathbb{R}^{n+1}_{<}$ is differentiable in the sense of distributions in $\mathbb{R}^{n+1}$, and hence F is in $W^{1,p(\cdot)}(\mathbb{R}^{n+1})$. Let ψ be a standard mollifier with support in $B^{n+1}(e_{n+1}/2, 1/3)$, where e_{n+1} is the unit vector in direction

$n + 1$. Then $\psi_r * F$ has support in $\mathbb{R}^{n+1}_>$ and is smooth. Since $p \in \mathcal{P}^{\log}(\mathbb{R}^{n+1})$, it follows that $\psi_r * F \to F$ in $W^{1,p(\cdot)}(\mathbb{R}^{n+1})$ as $r \to 0$ (Theorem 4.6.4). If η_R is a Lipschitz function with $\chi_{B(0,R)} \leqslant \eta_R \leqslant \chi_{B(0,2R)}$ and Lipschitz constant $2/R$, then $\eta_R \psi_r * F \to \psi_r * F$ in $W^{1,p(\cdot)}(\mathbb{R}^{n+1})$ as $R \to \infty$. Since $\eta_R \psi_r * F \in C_0^\infty(\mathbb{R}^{n+1}_>)$, it follows that $F \in H_0^{1,p(\cdot)}(\mathbb{R}^{n+1}_>)$. The converse follows from Lemma 12.1.10, Proposition 11.2.3 and Theorem 9.1.7.   □

We conclude the section by noting that the trace space inherits the density of smooth functions from the ambient space.

Theorem 12.1.12. *Suppose that $C_0^\infty(\mathbb{R}^{n+1}_\geqslant)$ is dense in $W^{1,p(\cdot)}(\mathbb{R}^{n+1}_>)$. Then $C_0^\infty(\mathbb{R}^n)$ is dense in* $\operatorname{Tr} W^{1,p(\cdot)}(\mathbb{R}^{n+1}_>)$.

Proof. Let $f \in \operatorname{Tr} W^{1,p(\cdot)}(\mathbb{R}^{n+1}_>)$, and let $F \in W^{1,p(\cdot)}(\mathbb{R}^{n+1}_>)$ be such that $\operatorname{Tr} F = f$. Then if $\Psi_i \in C_0^\infty(\mathbb{R}^{n+1}_\geqslant)$ tend to F in $W^{1,p(\cdot)}(\mathbb{R}^{n+1}_>)$, the definition of the quotient norm directly implies that also $\operatorname{Tr} \Psi_i \to f$ in $\operatorname{Tr} W^{1,p(\cdot)}(\mathbb{R}^{n+1}_>)$.   □

12.2 Homogeneous Sobolev Spaces

When working with unbounded domains it is often not natural to assume that the function and the gradient belong to the same Lebesgue space. This phenomenon is well known already in the case of constant exponents as illustrates the following example. One easily checks that $u(x) = |x|^{-1}$ is a solution of the Poisson problem

$$-\Delta u = 0 \qquad \text{in } \Omega = \mathbb{R}^3 \setminus \overline{B(0,1)} \,,$$
$$u = 1 \qquad \text{on } \partial\Omega \,,$$
$$\lim_{|x| \to \infty} u(x) = 0 \,.$$

Moreover, we see that $\nabla^2 u \in L^q(\Omega)$, $\nabla u \in L^r(\Omega)$, and $u \in L^s(\Omega)$ where $q \in (1, \infty)$, $r \in (3/2, \infty)$, and $s \in (3, \infty)$, respectively. In particular u does not belong to the Sobolev spaces $W^{2,r}(\Omega)$, $r \in (1, 3]$. This example shows that derivatives of different orders of solutions of the Poisson problem in unbounded domains belong to different Lebesgue spaces. The very same phenomenon occurs in Sobolev embedding theorems for unbounded domains. In fact from $\nabla u \in L^q(\Omega)$ one can in general not conclude that $u \in L^q(\Omega)$. These phenomena are the motivation for the introduction of *homogeneous Sobolev spaces*.

For a domain $\Omega \subset \mathbb{R}^n$, an exponent $p \in \mathcal{P}^{\log}(\Omega)$, and $k \in \mathbb{N}$ we define

$$\widetilde{D}^{k,p(\cdot)}(\Omega) := \{u \in L^1_{\mathrm{loc}}(\Omega) : \nabla^k u \in L^{p(\cdot)}(\Omega)\}.$$

The linear space $\widetilde{D}^{k,p(\cdot)}(\Omega)$ is equipped with the seminorm

$$\|u\|_{\widetilde{D}^{k,p(\cdot)}(\Omega)} := \|\nabla^k u\|_{L^{p(\cdot)}(\Omega)},$$

where $\nabla^k u$ is the tensor with entries $\partial_\alpha u$, $|\alpha| = k$. Note, that $\|u\|_{\widetilde{D}^{k,p(\cdot)}(\Omega)} = 0$ implies that u is a polynomial of degree $k-1$. Let us denote the polynomials of degree $m \in \mathbb{N}_0$ by P_m. It is evident that the seminorm $\|\cdot\|_{\widetilde{D}^{k,p(\cdot)}(\Omega)}$ becomes a norm on the equivalence classes $[u]_k$ defined for $u \in \widetilde{D}^{k,p(\cdot)}(\Omega)$ by

$$[u]_k := \{w \in \widetilde{D}^{k,p(\cdot)}(\Omega) \; : \; w - u \in P_{k-1}\}.$$

Now we can define homogeneous Sobolev spaces $D^{k,p(\cdot)}(\Omega)$.

Definition 12.2.1. Let $\Omega \subset \mathbb{R}^n$ be a domain, $p \in \mathcal{P}^{\log}(\Omega)$ a variable exponent, and let $k \in \mathbb{N}$. The *homogeneous Sobolev space* $D^{k,p(\cdot)}(\Omega)$ consists of all equivalence classes $[u]_k$ where $u \in \widetilde{D}^{k,p(\cdot)}(\Omega)$. We identify u with its equivalence class $[u]_k$ and thus write u instead of $[u]_k$. The space $D^{k,p(\cdot)}(\Omega)$ is equipped with the norm

$$\|u\|_{D^{k,p(\cdot)}(\Omega)} := \|\nabla^k u\|_{L^{p(\cdot)}(\Omega)}.$$

The natural mapping $i \colon C_0^\infty(\Omega) \to D^{k,p(\cdot)}(\Omega) \colon u \mapsto [u]_k$ implies that $C_0^\infty(\Omega)$ is isomorphic to a linear subspace of $D^{k,p(\cdot)}(\Omega)$. We define $D_0^{k,p(\cdot)}(\Omega)$ as the closure of $C_0^\infty(\Omega)$ in $D^{k,p(\cdot)}(\Omega)$.

Remark 12.2.2. If $u \in D^{k,p(\cdot)}(\Omega)$ then $\nabla^k u$ is a well defined Lebesgue function, while $\nabla^l u$, $0 \leqslant l < k$, are equivalence classes (Lebesgue function plus space of polynomials P_{k-l-1}).

Theorem 12.2.3. *Let $\Omega \subset \mathbb{R}^n$ be a domain and let $p \in \mathcal{P}^{\log}(\Omega)$. The spaces $D^{k,p(\cdot)}(\Omega)$ and $D_0^{k,p(\cdot)}(\Omega)$ are Banach spaces, which are separable if p is bounded, and reflexive and uniformly convex if $1 < p^- \leqslant p^+ < \infty$.*

Proof. We prove only the case $k = 1$, since the general case follows along the same line of arguments. Let $([u_j])$ be a Cauchy sequence in $D^{1,p(\cdot)}(\Omega)$. One easily checks that it is sufficient to show that for any $v_j \in [u_j]$ there exists $u \in \widetilde{D}^{1,p(\cdot)}(\Omega)$ such that $\nabla v_j \to \nabla u$ in $L^{p(\cdot)}(\Omega)$ as $j \to \infty$. By the completeness of $L^{p(\cdot)}(\Omega)$ we find $\mathbf{w} \in (L^{p(\cdot)}(\Omega))^n$ such that $\nabla v_j \to \mathbf{w}$ in $(L^{p(\cdot)}(\Omega))^n$ as $j \to \infty$. Now we choose an increasing sequence of bounded John domains Ω_m such that $\overline{\Omega}_m \subset \Omega$, and $\cup_{m \in \mathbb{N}} \Omega_m = \Omega$. For fixed $m \in \mathbb{N}$ we modify the sequence (v_j) by constants c_j^m such that $(v_j + c_j^m)_{\Omega_m} = 0$. Using the embedding $L^{p(\cdot)}(\Omega_m) \hookrightarrow L^{p^-}(\Omega_m)$ (Corollary 3.3.4) and Poincaré's inequality 8.2.13 for p^- we deduce that there exists a function h^m such that $v_j - c_j^m \to h^m$ in $L^{p^-}(\Omega_m)$ as $j \to \infty$. Moreover, we easily deduce that $\nabla h^m = \mathbf{w}$ a.e. in Ω_m. Thus h^m and h^l differ only by a constant. Define

$\tilde{h}^m := h^m - \langle h^m \rangle_{\Omega_l}$. Then $\tilde{h}^m = \tilde{h}^l$ on Ω_l for $m \geqslant l$. Thus the function u defined as $u(x) := \tilde{h}^m(x)$ if $x \in \Omega_m$ belongs to $L^1_{\mathrm{loc}}(\Omega)$ and satisfies $\nabla u = \mathbf{w}$ a.e. in Ω. This proves that $D^{1,p(\cdot)}(\Omega)$ is a Banach space.

By Theorem 3.4.4, $L^{p(\cdot)}(\Omega)$ is separable if $p^+ < \infty$ and by Theorems 3.4.7 and 3.4.9, it is reflexive and uniformly convex if $1 < p^- \leqslant p^+ < \infty$. Via the mapping $u \mapsto \nabla u$, the space $D^{1,p(\cdot)}(\Omega)$ is a closed subspace of $(L^{p(\cdot)}(\Omega))^n$. Thus we can prove that $D^{1,p(\cdot)}(\Omega)$ is separable, if $p^+ < \infty$, and reflexive and uniformly convex, if $1 < p^- \leqslant p^+ < \infty$, in the same way as in the proof of Theorem 8.1.6.

The statements for $D_0^{1,p(\cdot)}(\Omega)$ follow from the statements for $D^{1,p(\cdot)}(\Omega)$ since $D_0^{1,p(\cdot)}(\Omega)$ is a closed subspace of $D^{1,p(\cdot)}(\Omega)$. $\square$

The spaces $\widetilde{D}^{k,p(\cdot)}(\Omega)$ and $W^{k,p(\cdot)}(\Omega)$ essentially do not differ for bounded domains. More precisely we have:

Proposition 12.2.4. *Let Ω be a bounded John domain and let $p \in \mathcal{P}^{\log}(\Omega)$. Then we have the algebraic identity*

$$\widetilde{D}^{k,p(\cdot)}(\Omega) = W^{k,p(\cdot)}(\Omega).$$

Proof. We show the assertion only for $k = 1$, since the case $k \geqslant 2$ follows by iteration. It is sufficient to show that $\widetilde{D}^{1,p(\cdot)}(\Omega) \subset W^{1,p(\cdot)}(\Omega)$. Since Ω is bounded, Corollary 8.2.6 implies that there exists a ball $B \subset \Omega$ such that for each $u \in \widetilde{D}^{1,p(\cdot)}(\Omega)$

$$\|u - \langle u \rangle_B\|_{L^{p(\cdot)}(\Omega)} \leqslant c \operatorname{diam}(\Omega) \|\nabla u\|_{L^{p(\cdot)}(\Omega)},$$

which implies $u \in W^{1,p(\cdot)}(\Omega)$. $\square$

Remark 12.2.5. The previous lemma also implies that in the case of unbounded domains each $u \in \widetilde{D}^{k,p(\cdot)}(\Omega)$ belongs to $W^{k,p(\cdot)}_{\mathrm{loc}}(\Omega)$ if $p \in \mathcal{P}^{\log}(\Omega)$. This follows by restricting the function u to bounded subdomains.

Remark 12.2.6. As in the previous proposition, one can deduce from the density of $C_0^\infty(\Omega)$ in $D_0^{k,p(\cdot)}(\Omega)$ and $W_0^{k,p(\cdot)}(\Omega)$, and the Poincaré inequality with zero boundary values the topological identity $D_0^{k,p(\cdot)}(\Omega) = W_0^{k,p(\cdot)}(\Omega)$ for bounded John domains.

As in the case of classical Sobolev spaces we have that $D^{k,p(\cdot)}(\mathbb{R}^n)$ and $D_0^{k,p(\cdot)}(\mathbb{R}^n)$ coincide:

Proposition 12.2.7. *Let $p \in \mathcal{P}^{\log}(\mathbb{R}^n)$ satisfy $1 \leqslant p^- \leqslant p^+ < \infty$ and let $k \in \mathbb{N}$. Then $C_0^\infty(\mathbb{R}^n)$ is dense in $D^{k,p(\cdot)}(\mathbb{R}^n)$. Consequently we have $D^{k,p(\cdot)}(\mathbb{R}^n) = D_0^{k,p(\cdot)}(\mathbb{R}^n)$.*

Proof. First, consider the case $k = 1$. Let $\eta \in C_0^\infty(B(0,2))$ satisfy $0 \leqslant \eta \leqslant 1$ and $\eta|_{B(0,1)} = 1$. For $m \in \mathbb{N}$ we set $A_m := \{x \in \mathbb{R}^n : m < |x| < 2m\}$ and $\eta_{A_m}(x) := \eta(m^{-1}x)$. One easily sees that there is a constant $c = c(\eta)$ such that

$$|\nabla \eta_{A_m}(x)| \leqslant c\, m^{-1} \chi_{A_m}(x)$$

for all $j \in \mathbb{N}$. For $u \in D^{1,p(\cdot)}(\mathbb{R}^n)$ we set $u_m := \eta_{A_m}(u - c_m^0)$ where $c_m^0 := \langle u \rangle_{A_m}$, so that

$$\nabla u_m = \eta_{A_m} \nabla u + \nabla \eta_{A_m}(u - c_m^0).$$

Note that the definition of u_m is independent of the choice of the function from the equivalence class of u. In order to show that $u_m \to u$ in $D^{1,p(\cdot)}(\mathbb{R}^n)$ for $m \to \infty$ we observe that by the theorem of dominated convergence we have

$$\int_{\mathbb{R}^n} |1 - \eta_{A_m}|^{p(x)} |\nabla u|^{p(x)}\, dx \to 0.$$

Due to the Poincaré inequality (Corollary 8.2.6) and the properties of A_m we get

$$\|\nabla \eta_{A_m}(u - \langle u \rangle_{A_m})\|_{L^{p(\cdot)}(\mathbb{R}^n)} \leqslant c\, m^{-1} \|u - \langle u \rangle_{A_m}\|_{L^{p(\cdot)}(A_m)}$$
$$\leqslant c \|\nabla u\|_{L^{p(\cdot)}(A_m)} \to 0$$

for $m \to \infty$. For $m \in \mathbb{N}$ we choose $\varepsilon_m > 0$ and mollify u_m with a standard mollifier (cf. Theorem 4.6.4) such that $\|u_m - u_m * \psi_{\varepsilon_m}\|_{D^{1,p(\cdot)}(\mathbb{R}^n)} \leqslant m^{-1}$. Obviously, the sequence $(u_m * \psi_{\varepsilon_m})$ belongs to $C_0^\infty(\mathbb{R}^n)$ and converges to u in $D^{1,p(\cdot)}(\mathbb{R}^n)$.

The case $k > 1$ is treated analogously by subtracting higher order polynomials defined by mean values of higher order gradients (cf. [356, proof of Theorem 2.15, Chap. III], proof of Proposition 12.2.12). □

Remark 12.2.8. From the proof of the previous proposition it also follows that $C^\infty(\mathbb{R}^n) \cap D^{1,p(\cdot)}(\mathbb{R}^n)$ is dense in $D^{1,p(\cdot)}(\mathbb{R}^n)$.

For the treatment of the Stokes system and general elliptic problems of second order we also have to deal with trace spaces of homogeneous Sobolev spaces at least in the case of the half-space $\mathbb{R}_>^{n+1}$. Traces are defined in Sect. 12.1 for functions from $W_{\text{loc}}^{1,1}(\mathbb{R}_>^{n+1})$ and thus the notion of a trace of a function from $\tilde{D}^{1,p(\cdot)}(\mathbb{R}_>^{n+1})$ is well defined. Consequently, the *trace space* $\text{Tr}(D^{1,p(\cdot)}(\mathbb{R}_>^{n+1}))$ consists of equivalence classes modulo constants of traces of all functions $F \in \tilde{D}^{1,p(\cdot)}(\mathbb{R}_>^{n+1})$. The quotient norm

$$\|f\|_{\mathrm{Tr}(D^{1,p(\cdot)}(\mathbb{R}_>^{n+1}))} := \inf \left\{ \|F\|_{D^{1,p(\cdot)}(\mathbb{R}_>^{n+1})} : F \in D^{1,p(\cdot)}(\mathbb{R}_>^{n+1}) \text{ and } \mathrm{Tr}\, F = f \right\}$$

makes $\mathrm{Tr}(D^{1,p(\cdot)}(\mathbb{R}_>^{n+1}))$ a Banach space. The identity $\mathrm{Tr}\, F = f$ is to be understood as an identity of equivalence classes. Also for homogeneous Sobolev spaces the trace space depends only on the values of p on $\partial\mathbb{R}_>^{n+1}$. The proof is identical to that of Theorem 12.1.1 except that we use (8.5.14) instead of (8.5.13) for the extension.

Theorem 12.2.9. *Let* $p_1, p_2 \in \mathcal{P}^{\log}(\mathbb{R}_>^{n+1})$ *with* $p_1|_{\mathbb{R}^n} = p_2|_{\mathbb{R}^n}$. *Then* $\mathrm{Tr}\, D^{1,p_1(\cdot)}(\mathbb{R}_>^{n+1}) = \mathrm{Tr}\, D^{1,p_2(\cdot)}(\mathbb{R}_>^{n+1})$ *with equivalent norms.*

In application, as e.g. the Poisson problem, it happens that for a function $u \in L^1_{\mathrm{loc}}(\Omega)$ one can show that $\nabla u, \nabla^2 u \in L^{p(\cdot)}(\Omega)$. Because of the special nature of the homogeneous space this information is covered neither by $D^{1,p(\cdot)}(\Omega)$ nor by $D^{2,p(\cdot)}(\Omega)$. Thus we introduce a new space containing the full information.

Definition 12.2.10. Let $\Omega \subset \mathbb{R}^n$ be a domain and let $p \in \mathcal{P}^{\log}(\Omega)$ be a variable exponent. The space $D^{(1,2),p(\cdot)}(\Omega)$ consists of all equivalence classes $[u]_0$ with $u \in \widetilde{D}^{1,p(\cdot)}(\Omega) \cap \widetilde{D}^{2,p(\cdot)}(\Omega)$. We identify u with its equivalence class $[u]_0$ and thus write u instead of $[u]_0$. We equip $D^{(1,2),p(\cdot)}(\Omega)$ with the norm

$$\|u\|_{D^{(1,2),p(\cdot)}(\Omega)} := \|\nabla u\|_{L^{p(\cdot)}(\Omega)} + \|\nabla^2 u\|_{L^{p(\cdot)}(\Omega)}.$$

Note that $D^{(1,2),p(\cdot)}(\Omega)$ is a subspace of $D^{1,p(\cdot)}(\Omega)$ but not of $D^{2,p(\cdot)}(\Omega)$, because it consists of equivalence classes modulo constants.

The natural mapping $i \colon C_0^\infty(\Omega) \to D^{(1,2),p(\cdot)}(\Omega) \colon u \mapsto [u]_0$ implies that $C_0^\infty(\Omega)$ is isomorphic to a linear subspace of $D^{(1,2),p(\cdot)}(\Omega)$. We define $D_0^{(1,2),p(\cdot)}(\Omega)$ as the closure of $C_0^\infty(\Omega)$ in $D^{(1,2),p(\cdot)}(\Omega)$.

In the same way as in the proof of Theorem 12.2.3 one can show the following fundamental properties of $D^{(1,2),p(\cdot)}(\Omega)$ and $D_0^{(1,2),p(\cdot)}(\Omega)$.

Theorem 12.2.11. *Let* $\Omega \subset \mathbb{R}^n$ *be a domain and let* $p \in \mathcal{P}^{\log}(\Omega)$. *The spaces* $D^{(1,2),p(\cdot)}(\Omega)$, *and* $D_0^{(1,2),p(\cdot)}(\Omega)$ *are Banach spaces, which are separable if* p *is bounded, and reflexive and uniformly convex if* $1 < p^- \leqslant p^+ < \infty$.

Proposition 12.2.12. *Let* $p \in \mathcal{P}^{\log}(\mathbb{R}^n)$ *with* $1 \leqslant p^- \leqslant p^+ < \infty$, *and let* $u \in D^{(1,2),p(\cdot)}(\mathbb{R}^n)$. *Then there exists a sequence* $(u_m) \subset C_0^\infty(\mathbb{R}^n)$ *such that* $u_m \to u$ *in* $D^{(1,2),p(\cdot)}(\mathbb{R}^n)$. *Thus we have* $D^{(1,2),p(\cdot)}(\mathbb{R}^n) = D_0^{(1,2),p(\cdot)}(\mathbb{R}^n)$.

Proof. The proof of the assertion uses the arguments and notation from the proof of Proposition 12.2.7. We already know that

$$\eta_{A_m}\left(u - c_m^0\right) \to u \text{ in } D^{1,p(\cdot)}(\mathbb{R}^n),$$

$$u_m := \eta_{A_m}\left(u - c_m^0 - \sum_{i=1}^n c_m^i x_i\right) \to u \text{ in } D^{2,p(\cdot)}(\mathbb{R}^n),$$

where $c_m^0 := \langle u \rangle_{A_k}$ and $c_m^i := \langle \partial_i u \rangle_{A_k}$. Note that the definition of u_m is independent of the choice of the representative from the equivalence class of u. To prove the convergence $u_m \to u$ also in $D^{1,p(\cdot)}(\mathbb{R}^n)$ it suffices to show that

$$\partial_l(\eta_{A_m} c_m^i x_i) = (\partial_l \eta_{A_m}) c_m^i x_i + \eta_k c_m^i \delta_{il} \to 0 \qquad \text{in } L^{p(\cdot)}(\mathbb{R}^n).$$

Since $|x_i| < 2m$ in A_m, we have

$$\left\| (\partial_l \eta_{A_m}) c_m^i x_i \right\|_{L^{p(\cdot)}(\mathbb{R}^n)} \leqslant c \frac{2m}{m} \fint_{A_m} |\partial_i u| \, dx \, \|1\|_{L^{p(\cdot)}(A_m)}$$

$$\leqslant \frac{c}{|A_m|} \|\nabla u\|_{L^{p(\cdot)}(A_m)} \|1\|_{L^{p'(\cdot)}(A_m)} \|1\|_{L^{p(\cdot)}(A_m)}$$

$$\leqslant c \|\nabla u\|_{L^{p(\cdot)}(A_m)} \to 0$$

for $m \to \infty$, where we used Theorem 4.5.7. Analogously we get

$$\left\| \eta_{A_m} c_m^i \right\|_{L^{p(\cdot)}(\mathbb{R}^n)} \leqslant \fint_{A_m} |\partial_i u| \, dx \, \|1\|_{L^{p(\cdot)}(A_m)}$$

$$\leqslant c \|\nabla u\|_{L^{p(\cdot)}(A_m)} \to 0$$

for $m \to \infty$. As in the proof of Proposition 12.2.7 we choose for $m \in \mathbb{N}$ numbers $\varepsilon_m > 0$ and mollify u_m with a standard kernel (cf. Theorem 4.6.4) such that $\|u_m - u_m * \psi_{\varepsilon_m}\|_{D^{(1,2),p(\cdot)}(\mathbb{R}^n)} \leqslant m^{-1}$. Obviously, the sequence $(u_m * \psi_{\varepsilon_m})$ belongs to $C_0^\infty(\mathbb{R}^n)$ and converges to u in $D^{(1,2),p(\cdot)}(\mathbb{R}^n)$. $\square$

Remark 12.2.13. If we add to the assumptions of Proposition 12.2.12 that $u \in L^{q(\cdot)}(\mathbb{R}^n)$ with $q \in \mathcal{P}^{\log}(\mathbb{R}^n)$, $1 < q^- \leqslant q^+ < \infty$, then we obtain that the sequence $(u_m) \subset C_0^\infty(\mathbb{R}^n)$ converges also in $L_{\mathrm{loc}}^{q(\cdot)}(\mathbb{R}^n)$ to u. For that one shows that $\|\eta_{A_m} c_m^0\|_{L^{q(\cdot)}(\mathbb{R}^n)} \leqslant c \|u\|_{L^{q(\cdot)}(A_m)} \to 0$ for $m \to \infty$. Moreover, for each fixed cube $Q \subset \mathbb{R}^n$, $\|\eta_{A_m} c_m^i x_i\|_{L^{q(\cdot)}(Q)} \leqslant c(Q) \fint_{A_m} |\nabla u| \, dx \to 0$ as $m \to \infty$, where we again used Theorem 4.5.7.

12.3   Sobolev Spaces with Negative Smoothness

In this section we study the dual spaces of Sobolev spaces. We show that these spaces can be identified with Sobolev spaces of negative smoothness. We consider both the homogeneous and inhomogeneous case.

Definition 12.3.1. Let $\Omega \subset \mathbb{R}^n$ be a domain, let $p \in \mathcal{P}^{\log}(\Omega)$ satisfy $1 < p^- \leqslant p^+ \leqslant \infty$ and let $k \in \mathbb{N}$. We denote the *dual spaces* of the inhomogeneous and homogeneous Sobolev spaces as follows

$$W^{-k,p(\cdot)}(\Omega) := (W_0^{k,p'(\cdot)}(\Omega))^* \quad \text{and} \quad D^{-k,p(\cdot)}(\Omega) := (D_0^{k,p'(\cdot)}(\Omega))^*.$$

Note that under the assumptions on p in Definition 12.3.1 the space $W_0^{k,p'(\cdot)}(\Omega)$ and $H_0^{k,p'(\cdot)}(\Omega)$ coincide by Corollary 11.2.4. In particular, $C_0^\infty(\Omega)$ is dense in $W_0^{k,p'(\cdot)}(\Omega)$.

Proposition 12.3.2. *Let $\Omega \subset \mathbb{R}^n$ be a domain, let $p \in \mathcal{P}^{\log}(\Omega)$ satisfy $1 < p^- \leqslant p^+ \leqslant \infty$ and let $k \in \mathbb{N}$. For each $F \in W^{-k,p(\cdot)}(\Omega)$ there exists $f_\alpha \in L^{p(\cdot)}(\Omega)$, $|\alpha| \leqslant k$, such that*

$$\langle F, u \rangle = \sum_{|\alpha| \leqslant k} \int_\Omega f_\alpha \partial_\alpha u \, dx$$

for all $u \in W_0^{k,p'(\cdot)}(\Omega)$. Moreover,

$$\|F\|_{W^{-k,p(\cdot)}(\Omega)} \approx \sum_{|\alpha| \leqslant k} \|f_\alpha\|_{L^{p(\cdot)}(\Omega)}.$$

Proof. For $u \in W_0^{k,p'(\cdot)}(\Omega)$ we define $Pu := (\partial_\alpha u) \in (L^{p'(\cdot)}(\Omega))^{N(k)}$, where $N(k)$ is the number of multi-indexes α with $|\alpha| \leqslant k$. Clearly P maps $W_0^{k,p'(\cdot)}(\Omega)$ into a closed linear subspace W of $(L^{p'(\cdot)}(\Omega))^{N(k)}$ and we have $\|Pu\|_{(L^{p'(\cdot)}(\Omega))^{N(k)}} = \||u|\|_{W_0^{k,p'(\cdot)}(\Omega)}$ if we equip $W_0^{k,p'(\cdot)}(\Omega)$ with the equivalent norm $\||u|\|_{W_0^{k,p'(\cdot)}(\Omega)} := \sum_{|\alpha| \leqslant k} \|\partial_\alpha u\|_{L^{p'(\cdot)}(\Omega)}$. By the Hahn–Banach Theorem we can extend the bounded linear functional $F^* \in W^*$, defined by $\langle F^*, Pu \rangle := \langle F, u \rangle$ for all $u \in W_0^{k,p'(\cdot)}(\Omega)$, to a bounded linear functional $\widetilde{F} \in ((L^{p'(\cdot)}(\Omega))^{N(k)})^*$ satisfying

$$\|\widetilde{F}\|_{((L^{p'(\cdot)}(\Omega))^{N(k)})^*} = \|F^*\|_W \leqslant \|F\|_{W^{-k,p(\cdot)}(\Omega)}. \tag{12.3.3}$$

From Theorem 3.4.6 we deduce the existence of $f_\alpha \in L^{p(\cdot)}(\Omega)$, $|\alpha| \leqslant k$, such that

$$\langle \widetilde{F}, \mathbf{v} \rangle = \sum_{|\alpha| \leqslant k} \int_\Omega f_\alpha v_\alpha \, dx$$

for all $\mathbf{v} \in (L^{p'(\cdot)}(\Omega))^{N(k)}$. From this and the definition of F^* we immediately conclude that

$$\langle F, u \rangle = \langle F^*, Pu \rangle = \langle \widetilde{F}, Pu \rangle = \sum_{|\alpha| \leqslant k} \int_\Omega f_\alpha \partial_\alpha u \, dx$$

for all $u \in W_0^{k,p(\cdot)}(\Omega)$. This proves the representation for F. This representation together with (12.3.3), the equivalence of norms for $(L^{p(\cdot)}(\Omega))^*$ and $L^{p'(\cdot)}(\Omega)$ (cf. Theorem 3.4.6), and Hölder's inequality proves the equivalence of norms in the assertion. $\qquad\square$

Proposition 12.3.4. *Let $\Omega \subset \mathbb{R}^n$ be a domain, and let $p \in \mathcal{P}^{\log}(\Omega)$ satisfy $1 < p^- \leqslant p^+ < \infty$. Then $C_0^\infty(\Omega)$ is dense in $W^{-k,p(\cdot)}(\Omega)$, $k \in \mathbb{N}$.*

Proof. Let $F \in W^{-k,p(\cdot)}(\Omega)$. Then Proposition 12.3.2 implies the existence of $f_\alpha \in L^{p(\cdot)}(\Omega)$, $|\alpha| \leqslant k$, such that

$$\langle F, u \rangle = \sum_{|\alpha| \leqslant k} \int_\Omega f_\alpha \partial_\alpha u \, dx$$

for all $u \in W_0^{k,p'(\cdot)}(\Omega)$. Due to Corollary 4.6.5 there exists $f_\alpha^j \in C_0^\infty(\Omega)$ with $\|f_\alpha - f_\alpha^j\|_{L^{p(\cdot)}(\Omega)} \leqslant j^{-1}$. We set $F^j := \sum_{|\alpha| \leqslant k} (-1)^{|\alpha|} \partial_\alpha f_\alpha^j \in C_0^\infty(\Omega)$, which defines, due to the density of $C_0^\infty(\Omega)$ in $W_0^{k,p'(\cdot)}(\Omega)$ (cf. Corollary 11.2.4), an element from $W^{-k,p(\cdot)}(\Omega)$ through $\langle F^j, u \rangle := \sum_{|\alpha| \leqslant k} \int_\Omega f_\alpha^j \partial_\alpha u \, dx$, $u \in C_0^\infty(\Omega)$. One easily checks that $\|F^j - F\|_{W^{-k,p(\cdot)}(\Omega)} \leqslant c j^{-1} \to 0$ for $j \to \infty$ and the assertion follows. $\qquad\square$

We have analogous statements of Propositions 12.3.2 and 12.3.4 in the case of the dual spaces $D^{-k,p(\cdot)}(\Omega)$ of the homogeneous Sobolev spaces $D_0^{k,p(\cdot)}(\Omega)$. The proof of the next proposition is completely analogous to the proof of Proposition 12.3.2 and thus we omit it.

Proposition 12.3.5. *Let $\Omega \subset \mathbb{R}^n$ be a domain, let $p \in \mathcal{P}^{\log}(\Omega)$ satisfy $1 < p^- \leqslant p^+ \leqslant \infty$ and let $k \in \mathbb{N}$. For each $F \in D^{-k,p(\cdot)}(\Omega)$ there exists $f_\alpha \in L^{p(\cdot)}(\Omega)$, $|\alpha| = k$, such that*

$$\langle F, u \rangle = \sum_{|\alpha| = k} \int_\Omega f_\alpha \partial_\alpha u \, dx$$

for all $u \in D_0^{k,p'(\cdot)}(\Omega)$. Moreover,

$$\|F\|_{D^{-k,p(\cdot)}(\Omega)} \approx \sum_{|\alpha| = k} \|f_\alpha\|_{L^{p(\cdot)}(\Omega)}.$$

Remark 12.3.6. The last equality also implies that

$$\langle F, u \rangle_{D^{-k,p(\cdot)}(\Omega), D_0^{k,p'(\cdot)}(\Omega)} = -\sum_{|\alpha| = k} \langle \partial_\alpha f_\alpha, u \rangle_{\mathcal{D}^*(\Omega), C_0^\infty(\Omega)}$$

for all $u \in C_0^\infty(\Omega)$, i.e. the space $D^{-k,p(\cdot)}(\Omega)$ can be viewed as a subspace of the space of distributions $\mathcal{D}^*(\Omega)$.

Recall that we denote the subspace of functions $f \in C_0^\infty(\Omega)$ satisfying $\int_\Omega f \, dx = 0$ by $C_{0,0}^\infty(\Omega)$. For simplicity we restrict ourselves until the end of this section to the study of the homogeneous Sobolev space $D^{1,p(\cdot)}(\Omega)$, i.e. we consider only the case $k = 1$. For a function $f \in L^1_{\text{loc}}(\Omega)$, we define a functional $\widetilde{f}$ by $\langle \widetilde{f}, u \rangle := \int_\Omega f u \, dx$ for $u \in C_0^\infty(\Omega)$. From Proposition 3.4.14 we know that $C_{0,0}^\infty(\Omega)$ is dense in $L_0^{p(\cdot)}(\Omega)$. Next we show that this also holds for $D^{-1,p(\cdot)}(\Omega)$.

Proposition 12.3.7. *Let Ω be a domain and let $p \in \mathcal{P}^{\log}(\Omega)$ with $1 < p^- \leqslant p^+ < \infty$. Then $C_{0,0}^\infty(\Omega)$ is dense in $D^{-k,p(\cdot)}(\Omega)$, $k \in \mathbb{N}$.*

Proof. Let $F \in D^{-k,p(\cdot)}(\Omega)$. Then Proposition 12.3.5 implies the existence of $f_\alpha \in L^{p(\cdot)}(\Omega)$, $|\alpha| = k$, such that

$$\langle F, u \rangle = \sum_{|\alpha|=k} \int_\Omega f_\alpha \partial_\alpha u \, dx$$

for all $u \in D_0^{k,p'(\cdot)}(\Omega)$. Due to Proposition 3.4.14 there exists $f_\alpha^j \in C_{0,0}^\infty(\Omega)$ with $\|f_\alpha - f_\alpha^j\|_{L^{p(\cdot)}(\Omega)} \leqslant j^{-1}$. We set $F^j := \sum_{|\alpha| \leqslant k} (-1)^k \partial_\alpha f_\alpha^j \in C_{0,0}^\infty(\Omega)$, which defines through $\langle F^j, u \rangle := \sum_{|\alpha|=k} \int_\Omega f_\alpha^j \partial_\alpha u \, dx$, $u \in C_0^\infty(\Omega)$, an element from $D^{-k,p(\cdot)}(\Omega)$. One easily checks that $\|F^j - F\|_{D^{-k,p(\cdot)}(\Omega)} \leqslant c j^{-1} \to 0$ for $j \to \infty$ and the assertion follows. $\qquad\square$

Let us restrict ourselves for the remainder of this section to the case $k = 1$. Even though $C_{0,0}^\infty(\Omega)$ is dense in $D^{-1,p(\cdot)}(\Omega)$ and $L_0^{p(\cdot)}(\Omega)$ it is in general not clear that it is also dense in $D^{-1,p(\cdot)}(\Omega) \cap L_0^{p(\cdot)}(\Omega)$. However we have the following result:

Lemma 12.3.8. *Let $\Omega \subset \mathbb{R}^n$ be a domain, let $p \in \mathcal{P}^{\log}(\Omega)$ satisfy $1 < p^- \leqslant p^+ \leqslant \infty$, and let $A \subset \Omega$ be a bounded John domain.*

(a) *The space $L_0^{p(\cdot)}(A)$ embeds into $D^{-1,p(\cdot)}(A)$ which embeds into $D^{-1,p(\cdot)}(\Omega)$. We have*

$$\|f\|_{D^{-1,p(\cdot)}(\Omega)} \leqslant c \|f\|_{D^{-1,p(\cdot)}(A)} \leqslant c(\operatorname{diam} A) \|f\|_{L_0^{p(\cdot)}(A)},$$

where we have extended f by zero outside of A.

(b) *If $p^+ < \infty$ then for each $f \in L_0^{p(\cdot)}(A)$, extended f by zero outside of A, there exists a sequence $(f_k) \subset C_{0,0}^\infty(A)$ with $f_k \to f$ in $L^{p(\cdot)}(\Omega)$ and in $D^{-1,p(\cdot)}(\Omega)$.*

Proof. The embedding $D^{-1,p(\cdot)}(A) \hookrightarrow D^{-1,p(\cdot)}(\Omega)$ and first inequality in (a) follow by duality from the obvious embedding $D_0^{1,p(\cdot)}(A) \hookrightarrow D_0^{1,p(\cdot)}(\Omega)$. Using Hölder's inequality and Poincaré's inequality (Corollary 8.2.6) we find that

$$\langle \widetilde{f}, u \rangle_{D^{-1,p(\cdot)}(A), D_0^{1,p'(\cdot)}(A)} = \int_A fu \, dx = \int_A f \left(u - \langle u \rangle_A \right) dx$$

$$\leqslant c(\operatorname{diam} A) \, \|f\|_{L^{p(\cdot)}(A)} \|\nabla u\|_{L^{p'(\cdot)}(A)}$$

for each $u \in C_0^\infty(A)$. This and the density of $C_0^\infty(A)$ in $D_0^{1,p'(\cdot)}(A)$ prove that $\widetilde{f}$ defines a linear bounded functional on $D_0^{1,p'(\cdot)}(A)$ which satisfies the inequality in (a). The density of $C_{0,0}^\infty(A)$ in $L_0^{p(\cdot)}(A)$ (Proposition 3.4.14) together with the embedding in (a) proves (b). □

Remark 12.3.9. From the previous result and Proposition 3.4.14 it follows immediately that for bounded John domains $C_{0,0}^\infty(\Omega)$ is dense in $D^{-1,p(\cdot)}(\Omega) \cap L_0^{p(\cdot)}(\Omega)$ if $p \in \mathcal{P}^{\log}(\Omega)$ satisfies $1 < p^- \leqslant p^+ < \infty$.

If the domain Ω has a sufficiently large and nice boundary it is not necessary to require as in the previous results that the function f has a vanishing integral. For simplicity we formulate the results only for the half-space $\mathbb{R}_{>}^{n+1}$.

Lemma 12.3.10. *Let $p \in \mathcal{P}^{\log}(\mathbb{R}_{>}^{n+1})$ satisfy $1 < p^- \leqslant p^+ \leqslant \infty$ and let $A \subset \mathbb{R}_{\geqslant}^{n+1}$ be a bounded John domain.*

(a) *The space $L^{p(\cdot)}(A)$ embeds into $D^{-1,p(\cdot)}(\mathbb{R}_{>}^{n+1})$ and we have the estimate*

$$\|f\|_{D^{-1,p(\cdot)}(\mathbb{R}_{>}^{n+1})} \leqslant c(A)\|f\|_{L^{p(\cdot)}(A)},$$

where we have extended f by zero outside of A.

(b) *If $p^+ < \infty$ then for each $f \in L^{p(\cdot)}(A)$ there exists a sequence $(f_k) \subset C_0^\infty(A)$ with $f_k \to f$ in $L^{p(\cdot)}(\mathbb{R}_{>}^{n+1})$ and in $D^{-1,p(\cdot)}(\mathbb{R}_{>}^{n+1})$.*

Proof. We choose $x_0 \in \partial\mathbb{R}_{>}^{n+1}$ and a ball $B(x_0, R)$ such that $A \subset B(x_0, R)$. We choose an appropriate ball $B \subset \mathbb{R}_{<}^{n+1} \cap B(x_0, R)$ with $|B| \approx |A|$ and note that $u|_B = 0$ for $u \in C_0^\infty(\mathbb{R}_{>}^{n+1})$. Now we use Lemma 8.2.3 and the Poincaré inequality (Theorem 8.2.4 (b)) and obtain for $f \in L^{p(\cdot)}(A)$ that

$$\langle \widetilde{f}, u \rangle = \int_{B(x_0, R)} fu \, dx \leqslant c \, \|f\|_{L^{p(\cdot)}(B(x_0,R))} \|u - \langle u \rangle_B\|_{L^{p'(\cdot)}(B(x_0,R))}$$

$$\leqslant c \, \|f\|_{L^{p(\cdot)}(A)} \|\nabla u\|_{L^{p'(\cdot)}(\mathbb{R}_{>}^{n+1})}$$

for all $u \in C_0^\infty(\mathbb{R}_{>}^{n+1})$. This and the density of $C_0^\infty(\mathbb{R}_{>}^{n+1})$ in $D_0^{1,p'(\cdot)}(\mathbb{R}_{>}^{n+1})$ proves that $\widetilde{f}$ defines a linear bounded functional on $D_0^{1,p'(\cdot)}(\mathbb{R}_{>}^{n+1})$ which satisfies the estimate in (a). The density of $C_0^\infty(A)$ in $L^{p(\cdot)}(A)$ (Theorem 3.4.12) together with the embedding in (a) prove (b). □

Proposition 12.3.11. *Let $p \in \mathcal{P}^{\log}(\mathbb{R}_{>}^{n+1})$ with $1 < p^- \leqslant p^+ < \infty$. Then $C_0^\infty(\mathbb{R}_{\geqslant}^{n+1})$ and $C_0^\infty(\mathbb{R}_{>}^{n+1})$ are dense in $D^{-1,p(\cdot)}(\mathbb{R}_{>}^{n+1})$.*

Proof. The previous lemma implies that $C_{0,0}^\infty(\mathbb{R}_>^{n+1}) \subset C_0^\infty(\mathbb{R}_>^{n+1}) \subset C_0^\infty(\overline{\mathbb{R}_>^{n+1}}) \subset D^{-1,p(\cdot)}(\mathbb{R}_>^{n+1})$; hence the statement follows from Proposition 12.3.7. $\square$

12.4 Bessel Potential Spaces*

Almeida and Samko [26], and Gurka, Harjulehto and Nekvinda [181] have extended variable integrability Sobolev spaces to Bessel potential spaces $\mathcal{L}^{\alpha,p(\cdot)}$ for constant but potentially non-integer α. The presentation in this section follows the latter reference.

The *Bessel kernel* g_α of order $\alpha > 0$ is defined by

$$g_\alpha(x) := \frac{\pi^{n/2}}{\Gamma(\alpha/2)} \int_0^\infty e^{-s-\pi^2|x|^2/s}\, s^{(\alpha-n)/2}\,\frac{ds}{s}, \quad x \in \mathbb{R}^n.$$

The *Bessel potential space with variable exponent* $\mathcal{L}^{\alpha,p(\cdot)}(\mathbb{R}^n)$ is defined, for $p \in \mathcal{P}(\mathbb{R}^n)$ and $\alpha > 0$, by

$$\mathcal{L}^{\alpha,p(\cdot)}(\mathbb{R}^n) := \{g_\alpha * f \ : \ f \in L^{p(\cdot)}(\mathbb{R}^n)\},$$

and is equipped with the norm

$$\|g_\alpha * f\|_{\mathcal{L}^{\alpha,p(\cdot)}(\mathbb{R}^n)} := \|f\|_{p(\cdot)}.$$

If $\alpha = 0$ we put $g_0 * f := f$ and $\mathcal{L}^{0,p(\cdot)}(\mathbb{R}^n) := L^{p(\cdot)}(\mathbb{R}^n)$.

The main result of this section is the following theorem.

Theorem 12.4.1. *If $p \in \mathcal{A}$ with $1 < p^- \leqslant p^+ < \infty$ and $k \in \mathbb{N}$, then*

$$\mathcal{L}^{k,p(\cdot)}(\mathbb{R}^n) \cong W^{k,p(\cdot)}(\mathbb{R}^n).$$

Before we prove the main theorem we need a few auxiliary results. Some basic properties of the Bessel kernel g_α, $\alpha > 0$ are:

g_α is nonnegative, radially decreasing and $\|g_\alpha\|_1 = 1$,

$\widehat{g_\alpha}(\xi) = (1 + |\xi|^2)^{-\alpha/2}, \quad \xi \in \mathbb{R}^n$,

$g_\alpha * g_{\alpha'} = g_{\alpha+\alpha'}, \quad \alpha, \alpha' > 0$.

For $\alpha > 0$ we define the measure μ_α on Lebesgue measurable sets $E \subset \mathbb{R}^n$ by

$$\mu_\alpha(E) := \delta_0(E) + \sum_{k=1}^\infty b(\alpha,k) \int_E g_{2k}(y)\,dy,$$

where δ_0 is the Dirac delta measure at zero and $b(\alpha, k) := (-1)^k \binom{\alpha/2}{k} = \frac{(-1)^k}{k!} \prod_{j=0}^{k-1} (\frac{\alpha}{2} - j)$, $k \in \mathbb{N}$. Since

$$\sum_{k=1}^{\infty} |b(\alpha, k)| < \infty,$$

the measure μ_α is a finite signed Borel measure on $\mathbb{R}^n$. For $\alpha = 0$ we set $\mu_0 = \delta_0$. The origin of the coefficients is in the Taylor expansion of the function $t \mapsto (1 - t)^{\alpha/2}$, $\alpha > 0$, $t \in (0, 1]$. Indeed, for $x \in \mathbb{R}^n$ we have

$$\frac{|x|^\alpha}{(1 + |x|^2)^{\alpha/2}} = \left(1 - \frac{1}{1 + |x|^2}\right)^{\alpha/2} = 1 + \sum_{k=1}^{\infty} b(\alpha, k) (1 + |x|^2)^{-2k/2},$$

which implies that

$$\widehat{\mu_\alpha}(x) := \int_{\mathbb{R}^n} e^{-ix \cdot \xi} \, d\mu_\alpha(\xi) = \frac{|x|^\alpha}{(1 + |x|^2)^{\alpha/2}}$$

for $\alpha > 0$. Obviously, this holds for $\alpha = 0$, too. (For more details see [317, p. 32] and [359, p. 134].)

We define the *Riesz transform* $\mathcal{R}_j f$, $j = 1, \ldots, n$, of a function $f \in \mathcal{S}$ by the formula

$$\mathcal{R}_j f(x) := \frac{\Gamma((n + 1)/2)}{\pi^{(n+1)/2}} \lim_{\varepsilon \to 0+} \int_{|y| > \varepsilon} \frac{y_j}{|y|^{n+1}} f(x - y) \, dy.$$

Note that $\mathcal{R}_j$ is a Calderón–Zygmund operator. Recall that (cf. [359])

$$\mathcal{F}(\mathcal{R}_j f)(\xi) = \frac{-i\xi_j}{|\xi|} \widehat{f}(\xi).$$

Let $\beta = (\beta_1, \ldots, \beta_n) \in \mathbb{N}_0^n$ be a multi-index. Then the *multi-Riesz transform* $\mathcal{R}_\beta$ is defined by

$$\mathcal{R}_\beta f := \mathcal{R}_1^{\beta_1} \circ \cdots \circ \mathcal{R}_n^{\beta_n} f. \tag{12.4.2}$$

For $f \in \mathcal{S}$ it is easy to verify (cf. [317]) that

$$\mathcal{F}(\mathcal{R}_\beta f)(\xi) = \left(\frac{-i\xi_1}{|\xi|}\right)^{\beta_1} \cdots \left(\frac{-i\xi_n}{|\xi|}\right)^{\beta_n} \widehat{f}(\xi),$$

$$\mathcal{F}(\mathcal{R}_\beta (D^\beta f))(\xi) = \left(\frac{-2\pi\xi_1^2}{|\xi|}\right)^{\beta_1} \cdots \left(\frac{-2\pi\xi_n^2}{|\xi|}\right)^{\beta_n} \widehat{f}(\xi),$$

$$\mathcal{F}(D^\beta f)(\xi) = (-2\pi i)^{|\beta|} \, \xi^\beta \, \widehat{f}(\xi),$$

where we used the multi-index notation $\xi^\beta := \xi_1^{\beta_1} \cdots \xi_n^{\beta_n}$.

Lemma 12.4.3. *Suppose that $p \in \mathcal{A}$ with $1 < p^- \leqslant p^+ < \infty$, $\alpha \geqslant 0$ and $\beta \in \mathbb{N}_0^n$. Then there exists a positive constant c such that*

$$\|g_\alpha * f\|_{p(\cdot)} \leqslant c \|f\|_{p(\cdot)},$$
$$\|\mu_\alpha * f\|_{p(\cdot)} \leqslant c \|f\|_{p(\cdot)},$$
$$\|\mathcal{R}_\beta f\|_{p(\cdot)} \leqslant c \|f\|_{p(\cdot)}.$$

for $f \in L^{p(\cdot)}(\mathbb{R}^n)$.

Proof. We noted earlier that g_α is radially symmetric and decreasing. Furthermore, $\|g_\alpha\|_1 = \|g_1\|_1 = 1$. Hence g_α is its own radially decreasing majorant. It follows from Lemma 4.6.3 that

$$\|g_\alpha * f\|_{p(\cdot)} \leqslant c \|g_\alpha\|_1 \|f\|_{p(\cdot)},$$

which yields the first inequality.

From the definition of μ_α we obtain that

$$\mu_\alpha * f(x) = f(x) + \sum_{k=1}^{\infty} b(\alpha, k) \, g_{2k} * f(x).$$

Then the triangle inequality and the previous case complete the proof of the second claim:

$$\|\mu_\alpha * f\|_{p(\cdot)} \leqslant \|f\|_{p(\cdot)} + \sum_{k=1}^{\infty} |b(\alpha, k)| \, \|g_{2k} * f\|_{p(\cdot)}$$
$$\leqslant \left(1 + c \sum_{k=1}^{\infty} |b(\alpha, k)|\right) \|f\|_{p(\cdot)} \leqslant c \|f\|_{p(\cdot)}.$$

To prove the third inequality we note that by Corollary 6.3.13 and $p \in \mathcal{A}$ there exists a positive constant c such that

$$\|\mathcal{R}_j f\|_{p(\cdot)} \leqslant c \|f\|_{p(\cdot)}$$

for $j = 1, \ldots, n$ and $f \in L^{p(\cdot)}(\mathbb{R}^n)$. Iterating this inequality completes the proof. □

Proposition 12.4.4. *Suppose that $p \in \mathcal{P}(\mathbb{R}^n)$ is bounded. Then the Schwartz class $\mathcal{S}$ is dense in $\mathcal{L}^{\alpha, p(\cdot)}(\mathbb{R}^n)$, $\alpha \geqslant 0$.*

Proof. There is nothing to prove when $\alpha = 0$. Let $\alpha > 0$ and $u \in \mathcal{L}^{\alpha, p(\cdot)}(\mathbb{R}^n)$. Then there is a function $f \in L^{p(\cdot)}(\mathbb{R}^n)$ such that $u = g_\alpha * f$. By density of $C_0^\infty(\mathbb{R}^n)$ in $L^{p(\cdot)}(\mathbb{R}^n)$ (Theorem 3.4.12) we can find a sequence $(f_j)_{j=1}^{\infty} \subset C_0^\infty(\mathbb{R}^n) \subset \mathcal{S}$ converging to f in $L^{p(\cdot)}(\mathbb{R}^n)$. Since the mapping $f \mapsto g_\alpha * f$

maps $\mathcal{S}$ onto $\mathcal{S}$ (cf. [359]), the functions $u_j := g_\alpha * f_j$, $j \in \mathbb{N}$, belong to $\mathcal{S}$. Moreover,

$$\|u - u_j\|_{\mathcal{L}^{\alpha,p(\cdot)}(\mathbb{R}^n)} = \|f - f_j\|_{p(\cdot)} \to 0 \quad \text{as } j \to \infty$$

and the assertion follows. $\square$

Lemma 12.4.5 (cf. [317, Lemma 5.15]). *Let $f \in \mathcal{S}$ and $k \in \mathbb{N}$. Then*

$$f = g_k * \sum_{m=0}^{k} \binom{k}{m} g_{k-m} * \mu_m * (-2\pi)^{-m} \sum_{|\beta|=m} \binom{m}{\beta} \mathcal{R}_\beta(D^\beta f),$$

where $\binom{m}{\beta} = \frac{m!}{\beta_1! \beta_2! \cdots \beta_n!}$.

Proof. Using the binomial theorem for $(\xi_1^2 + \ldots + \xi_n^2)^m$ we derive

$$|\xi|^{2m} \widehat{f}(\xi) = \sum_{|\beta|=m} \binom{m}{\beta} \xi_1^{2\beta_1} \cdots \xi_n^{2\beta_n} \widehat{f}(\xi)$$

$$= \left(\frac{|\xi|}{-2\pi}\right)^m \sum_{|\beta|=m} \binom{m}{\beta} \mathcal{F}(\mathcal{R}_\beta (D^\beta f))(\xi).$$

Then we use the binomial theorem for $(1 + |\xi|^2)^k$ and the previous equation to derive

$$(1 + |\xi|^2)^k \widehat{f}(\xi) = \sum_{m=0}^{k} \binom{k}{m} \left(\frac{|\xi|}{-2\pi}\right)^m \sum_{|\beta|=m} \binom{m}{\beta} \mathcal{F}(\mathcal{R}_\beta(D^\beta f))(\xi).$$

Next we note that $\frac{|\xi|^m}{(1+|\xi|^2)^k} = \widehat{g}_k(x)\widehat{g}_{k-m}(x)\,\widehat{\mu}_m(x)$. Thus we obtain

$$\widehat{f}(\xi) = \widehat{g}_k(\xi) \sum_{m=0}^{k} \binom{k}{m} \widehat{g}_{k-m}(\xi) \widehat{\mu}_m(\xi) (-2\pi)^{-m} \sum_{|\beta|=m} \binom{m}{\beta} \mathcal{F}(\mathcal{R}_\beta(D^\beta f))(\xi)$$

by dividing the previous equation by $(1+|\xi|^2)^k$. The claim follows by applying the inverse Fourier transform. $\square$

Lemma 12.4.6 (cf. [317, Lemma 5.17]). *Let $f \in \mathcal{S}$, $k \in \mathbb{N}$ and $\beta \in \mathbb{N}_0^n$, $|\beta| \leqslant k$. Then*

$$D^\beta(g_k * f) = (2\pi)^{|\beta|} g_{k-|\beta|} * \mu_{|\beta|} * \mathcal{R}_\beta f.$$

Proof. We again work on the Fourier side:

$$\mathcal{F}(D^\beta(g_k * f))(\xi)$$
$$= (-2\pi i)^{|\beta|} \, \xi^\beta \widehat{g_k}(\xi) \, \widehat{f}(\xi)$$
$$= (2\pi)^{|\beta|} \frac{1}{(1+|\xi|^2)^{(k-|\beta|)/2}} \, \frac{|\xi|^{|\beta|}}{(1+|\xi|^2)^{|\beta|/2}} \left(\frac{-i\xi_1}{|\xi|}\right)^{\beta_1} \cdots \left(\frac{-i\xi_n}{|\xi|}\right)^{\beta_n} \widehat{f}(\xi)$$
$$= (2\pi)^{|\beta|} \, \widehat{g}_{k-|\beta|}(\xi) \, \widehat{\mu}_{|\beta|}(\xi) \, \mathcal{F}(\mathcal{R}_\beta f)(\xi).$$

The result follows by taking inverse Fourier transforms.   □

Proof of Theorem 12.4.1. By Proposition 12.4.4 it suffices to consider the case when $f \in \mathcal{S}$. Suppose first that $f \in \mathcal{L}^{k,p(\cdot)}(\mathbb{R}^n)$. Then there is a function $h \in \mathcal{S}$ such that $f = g_k * h$. By Lemmas 12.4.6 and 12.4.3, and the definition of the Bessel norm we obtain

$$\|f\|_{k,p(\cdot)} = \sum_{|\beta| \leqslant k} \|D^\beta f\|_{p(\cdot)} = \sum_{|\beta| \leqslant k} \|D^\beta(g_k * h)\|_{p(\cdot)}$$
$$= \sum_{|\beta| \leqslant k} \|(2\pi)^{|\beta|} \, g_{k-|\beta|} * \mu_{|\beta|} * \mathcal{R}_\beta h\|_{p(\cdot)} \leqslant c \, \|h\|_{p(\cdot)} = c \, \|f\|_{\mathcal{L}^{k,p(\cdot)}(\mathbb{R}^n)}.$$

We next prove the reverse inequality. Let $f \in W^{k,p(\cdot)}(\mathbb{R}^n)$. Then, by Lemmas 12.4.5 and 12.4.3,

$$\|f\|_{\mathcal{L}^{k,p(\cdot)}(\mathbb{R}^n)} = \left\| \sum_{m=0}^k g_{k-m} * \mu_m * (-2\pi)^{-m} \sum_{|\beta|=m} \binom{m}{\beta} \mathcal{R}_\beta(D^\beta f) \right\|_{p(\cdot)}$$
$$\leqslant c \sum_{|\beta| \leqslant k} \|D^\beta f\|_{p(\cdot)} = c \, \|f\|_{k,p(\cdot)}.$$ □

12.5 Besov and Triebel–Lizorkin Spaces*

From a vast array of different function spaces a well ordered superstructure appeared in the 1960s and 1970s based on two three-index spaces: the Besov space $B^\alpha_{p,q}$ and the Triebel–Lizorkin space $F^\alpha_{p,q}$. These spaces include as special cases Lebesgue spaces, Sobolev spaces, Bessel potential spaces, (real) Hardy spaces, and the trace spaces of these spaces. In this section we introduce Besov and Triebel–Lizorkin spaces with variable indices, denoted by $B^{\alpha(\cdot)}_{p(\cdot),q(\cdot)}$ and $F^{\alpha(\cdot)}_{p(\cdot),q(\cdot)}$. In the constant exponent case the Besov space was considered slightly earlier than the Triebel–Lizorkin space. However, for variable exponent spaces the order was reversed – the reason for this is the difficulty involved in having a variable secondary integrability index q, as we explain below in more detail.

Except as noted, the results on Triebel–Lizorkin spaces are based on work by Diening, Hästö and Roudenko [100] whereas the results on Besov spaces are from [24] by Almeida and Hästö. In addition to various variable exponent spaces, these scales include also variable smoothness spaces studied, e.g., by Beauzamy [41], Besov [46, 47] and Leopold [265, 266].

In order to define these spaces we need some notation. For a cube Q let $\ell(Q)$ denote the side length of Q and x_Q the "lower left corner". Let $\mathcal{D}$ be the collection of dyadic cubes in $\mathbb{R}^n$ and denote by $\mathcal{D}^+$ the subcollection of those dyadic cubes with side-length at most 1. Let $\mathcal{D}_\nu := \{Q \in \mathcal{D} : \ell(Q) = 2^{-\nu}\}$. Note that in this section φ is not used for the Φ-function of the space but instead as admissible in the sense of the following definition.

Definition 12.5.1. We say a pair (φ, Φ) is *admissible* if $\varphi, \Phi \in \mathcal{S}(\mathbb{R}^n)$ satisfy

- $\operatorname{spt} \hat{\varphi} \subset \{\xi \in \mathbb{R}^n : \frac{1}{2} \leqslant |\xi| \leqslant 2\}$ and $|\hat{\varphi}(\xi)| \geqslant c > 0$ when $\frac{3}{5} \leqslant |\xi| \leqslant \frac{5}{3}$,
- $\operatorname{spt} \hat{\Phi} \subset \{\xi \in \mathbb{R}^n : |\xi| \leqslant 2\}$ and $|\hat{\Phi}(\xi)| \geqslant c > 0$ when $|\xi| \leqslant \frac{5}{3}$.

We set $\varphi_\nu(x) := 2^{\nu n}\varphi(2^\nu x)$ for $\nu \in \mathbb{N}$ and $\varphi_0(x) := \Phi(x)$. For $Q \in \mathcal{D}_\nu$ we set

$$\varphi_Q(x) := \begin{cases} |Q|^{1/2}\varphi_\nu(x - x_Q) & \text{if } \nu \geqslant 1, \\ |Q|^{1/2}\Phi(x - x_Q) & \text{if } \nu = 0. \end{cases}$$

We define ψ_ν and ψ_Q analogously.

According to [158], given an admissible pair (φ, Φ) we can select another admissible pair (ψ, Ψ) such that

$$\widehat{\widetilde{\Phi}}(\xi)\,\widehat{\Psi}(\xi) + \sum_{\nu=1}^\infty \widehat{\widetilde{\varphi}}(2^{-\nu}\xi)\,\widehat{\psi}(2^{-\nu}\xi) = 1 \quad \text{for all } \xi \in \mathbb{R}^n.$$

Here, $\widetilde{\Phi}(x) = \overline{\Phi(-x)}$ and similarly for $\widetilde{\varphi}$.

For each $f \in \mathcal{S}'(\mathbb{R}^n)$ we define the (*inhomogeneous*) φ-transform S_φ as the map taking f to the sequence $(S_\varphi f)_{Q \in \mathcal{D}^+}$ by setting $(S_\varphi f)_Q = \langle f, \varphi_Q \rangle$. Here, $\langle \cdot, \cdot \rangle$ denotes the usual inner product on $L^2(\mathbb{R}^n; \mathbb{C})$.

Using the admissible functions (φ, Φ) we can define the norms

$$\|f\|_{F^\alpha_{p,q}} := \left\| \left\| 2^{\nu\alpha}\,\varphi_\nu * f \right\|_{l^q} \right\|_{L^p} \quad \text{and} \quad \|f\|_{B^\alpha_{p,q}} := \left\| \left\| 2^{\nu\alpha}\,\varphi_\nu * f \right\|_{L^p} \right\|_{l^q},$$

for constants $p, q \in (0, \infty)$ and $\alpha \in \mathbb{R}$. The Triebel–Lizorkin space $F^\alpha_{p,q}$ and the Besov space $B^\alpha_{p,q}$ consist of those distributions $f \in \mathcal{S}'$ for which $\|f\|_{F^\alpha_{p,q}} < \infty$ and $\|f\|_{B^\alpha_{p,q}} < \infty$, respectively. The classical theory of these spaces is presented for instance in the books of Triebel [362–364].

Throughout this section we use the following assumptions.

Assumption 12.5.2. *We assume that* $p, q \in \mathcal{P}^{\log}(\mathbb{R}^n)$ *satisfy* $0 < p^- \leqslant p^+ < \infty$ *and* $0 < q^- \leqslant q^+ < \infty$. *We also assume that* $\alpha \in L^{\infty}(\mathbb{R}^n)$ *is non-negative, locally* log-*Hölder continuous and has a limit at infinity.*

For a family of functions $f_{\nu} : \mathbb{R}^n \to \mathbb{R}$, $\nu \in \mathbb{N}_0$, we define

$$\left\|f_{\nu}(x)\right\|_{l_{\nu}^{q(x)}} := \left(\sum_{\nu=0}^{\infty} |f_{\nu}(x)|^{q(x)} \right)^{\frac{1}{q(x)}}.$$

Note that this is just an ordinary discrete Lebesgue space, since $q(x)$ does not depend on ν. The mapping $x \mapsto \|f_{\nu}(x)\|_{l_{\nu}^{q(x)}}$ is a function of x and can be measured in $L^{p(\cdot)}$. We write $L_x^{p(\cdot)}$ to indicate that the integration variable is x.

Definition 12.5.3. Let φ_{ν}, $\nu \in \mathbb{N}_0$, be as in Definition 12.5.1. The *Triebel–Lizorkin space* $F_{p(\cdot), q(\cdot)}^{\alpha(\cdot)}(\mathbb{R}^n)$ is defined to be the space of all distributions $f \in \mathcal{S}'$ with $\|f\|_{F_{p(\cdot), q(\cdot)}^{\alpha(\cdot)}} < \infty$, where

$$\|f\|_{F_{p(\cdot), q(\cdot)}^{\alpha(\cdot)}} := \left\| \left\| 2^{\nu\alpha(x)} \, \varphi_{\nu} * f(x) \right\|_{l_{\nu}^{q(x)}} \right\|_{L_x^{p(\cdot)}}.$$

Note that, *a priori*, the function space depends on the choice of admissible functions (φ, Φ). We soon show that, up to equivalence of norms, every pair of admissible functions in fact produces the same space.

In the classical case it has proved to be very useful to consider discrete Triebel–Lizorkin spaces $f_{p, q}^{\alpha}$. Intuitively, this is achieved by viewing the function as a constant on dyadic cubes. The size of the appropriate dyadic cube varies according to the level of smoothness. We now present a formulation of the Triebel–Lizorkin norm which is similar in spirit.

For a sequence of real numbers $\{s_Q\}_Q$ we define

$$\left\|\{s_Q\}_Q\right\|_{f_{p(\cdot), q(\cdot)}^{\alpha(\cdot)}} := \left\| \left\| 2^{\nu\alpha(x)} \sum_{Q \in \mathcal{D}_{\nu}} |s_Q| \, |Q|^{-\frac{1}{2}} \chi_Q \right\|_{l_{\nu}^{q(x)}} \right\|_{L_x^{p(\cdot)}}.$$

The space $f_{p(\cdot), q(\cdot)}^{\alpha(\cdot)}$ consists of all those sequences $\{s_Q\}_Q$ for which this norm is finite. The discrete representation of $F_{p, q}^{\alpha}$ as sequence spaces through the φ-transform is due to Frazier and Jawerth [158].

One of the main results of Diening, Hästö and Roudenko [100] is the following isomorphism between $F_{p(\cdot), q(\cdot)}^{\alpha(\cdot)}$ and a subspace of $f_{p(\cdot), q(\cdot)}^{\alpha(\cdot)}$ via the S_{φ} transform. For a proof we refer to [100, Corollary 3.9].

Corollary 12.5.4. *Under Assumption 12.5.2,*

$$\|f\|_{F^{\alpha(\cdot)}_{p(\cdot),\,q(\cdot)}} \approx \|S_\varphi f\|_{f^{\alpha(\cdot)}_{p(\cdot),\,q(\cdot)}}$$

for every $f \in F^{\alpha(\cdot)}_{p(\cdot),\,q(\cdot)}(\mathbb{R}^n)$.

With this kind of discrete representation, one can prove that $F^{\alpha(\cdot)}_{p(\cdot),\,q(\cdot)}(\mathbb{R}^n)$ is well-defined, see [100, Theorem 3.10] for details:

Theorem 12.5.5. *Under Assumption 12.5.2, $F^{\alpha(\cdot)}_{p(\cdot),\,q(\cdot)}(\mathbb{R}^n)$ is well-defined, i.e., the definition does not depend on the choice of the functions φ and Φ satisfying the conditions of Definition 12.5.1, up to equivalence of norms.*

We have seen in Chap. 9 that smooth functions are dense in the variable exponent space if $p^+ < \infty$ and $p \in \mathcal{P}^{\log}(\mathbb{R}^n)$. These conditions are also sufficient for density in the Triebel–Lizorkin space. The proof follows easily from the atomic decomposition of the space; we refer to [100] for this.

Corollary 12.5.6. *Under Assumption 12.5.2, the space $C_0^\infty(\mathbb{R}^n)$ is dense in $F^{\alpha(\cdot)}_{p(\cdot),\,q(\cdot)}(\mathbb{R}^n)$.*

The Triebel–Lizorkin scale $F^{\alpha(\cdot)}_{p(\cdot),\,q(\cdot)}$ includes as special cases spaces with variable differentiability or integrability which have been studied by various authors. As a first result we show that Lebesgue spaces with variable exponent are included.

Theorem 12.5.7. *Let $p \in \mathcal{P}^{\log}(\mathbb{R}^n)$ with $1 < p^- \leqslant p^+ < \infty$. Then $F^0_{p(\cdot),\,2}(\mathbb{R}^n) \cong L^{p(\cdot)}(\mathbb{R}^n)$.*

Proof. Since $C_0^\infty(\mathbb{R}^n)$ is dense in $L^{p(\cdot)}(\mathbb{R}^n)$ (Theorem 3.4.12) and also in $F^0_{p(\cdot),\,2}(\mathbb{R}^n)$ by Corollary 12.5.6, it suffices to prove the claim for all $f \in C_0^\infty(\mathbb{R}^n)$. Fix $r \in (1, p^-)$. Then

$$\big\|\|\varphi_\nu * f\|_{l^2_\nu}\big\|_{L^{r_0}(\mathbb{R}^n;\omega)} \approx \|f\|_{L^{r_0}(\mathbb{R}^n;\omega)},$$

for all $\omega \in A_1$ by [260, Theorem 1], where the constant depends only on the A_1-constant of the weight ω. Thus the assumptions of the extrapolation result, Theorem 7.2.1, are satisfied. Applying this theorem with $\mathcal{F}$ equal to either

$$\big\{\big(\|\varphi_\nu * f\|_{l^2_\nu}, f\big) : f \in C_0^\infty(\mathbb{R}^n)\big\} \quad \text{or} \quad \big\{\big(f, \|\varphi_\nu * f\|_{l^2_\nu}\big) : f \in C_0^\infty(\mathbb{R}^n)\big\}$$

completes the proof.   □

The next result states that the Bessel potential spaces studied in the previous section are also included in the scale. Xu [371–373] has studied Triebel–Lizorkin spaces with variable p, but fixed q and α. We denote these spaces by $F^{Xu,\alpha}_{p(\cdot),q}(\mathbb{R}^n)$; they are also included in the Triebel–Lizorkin scale. For the proof of the next result we refer to [100, Theorem 4.5].

Theorem 12.5.8. *Let $p \in \mathcal{P}^{\log}(\mathbb{R}^n)$ with $1 < p^- \leqslant p^+ < \infty$.*

(a) $F^{\alpha}_{p(\cdot),q}(\mathbb{R}^n) \cong F^{Xu,\alpha}_{p(\cdot),q}(\mathbb{R}^n)$ *if $\alpha \in [0,\infty)$; in particular,*

(b) $F^{\alpha}_{p(\cdot),2}(\mathbb{R}^n) \cong \mathcal{L}^{\alpha,p(\cdot)}(\mathbb{R}^n)$ *if $\alpha \in [0,\infty)$; in particular,*

(c) $F^{\alpha}_{p(\cdot),2}(\mathbb{R}^n) \cong W^{\alpha,p(\cdot)}(\mathbb{R}^n)$ *if $\alpha \in \mathbb{N}_0$.*

As an example of the power of having a unified scale of variable index function spaces we consider another trace theorem. Recall that we saw in Sect. 12.1 that the trace space of a first order variable exponent Sobolev space is not a Sobolev space. The Triebel–Lizorkin scale does not have this problem: it is, in fact, closed under taking of traces. For the proof we refer to [100, Theorem 3.13].

Theorem 12.5.9. *Let the functions p, q, and α be as in Assumption 12.5.2. If $\alpha - \frac{1}{p} - (n-1)\max\left\{\frac{1}{p}-1,0\right\} > 0$, then*

$$\operatorname{Tr} F^{\alpha(\cdot)}_{p(\cdot),q(\cdot)}(\mathbb{R}^n) = F^{\alpha(\cdot)-\frac{1}{p(\cdot)}}_{p(\cdot),p(\cdot)}(\mathbb{R}^{n-1}).$$

Note that the assumption $\alpha - \frac{1}{p} - (n-1)\max\left\{\frac{1}{p}-1,0\right\} > 0$ is optimal also in the constant smoothness and integrability case, cf. [157, Sect. 5]

Vybíral [367] has recently derived a Sobolev embedding in the variable index Triebel–Lizorkin setting. In view of the above trace result, this generalizes Fan's [134] and Liu's [273] Sobolev-type trace embeddings in the variable exponent setting. The following is his Theorem 3.5 from [367]:

Theorem 12.5.10. *Let the functions p_0, p_1, q_0, q_1 and α_0, α_1 be as in Assumption 12.5.2. If $\alpha_0 - \frac{n}{p_0} = \alpha_1 - \frac{n}{p_1}$ and $(\alpha_0 - \alpha_1)^- > 0$, then*

$$F^{\alpha_0(\cdot)}_{p_0(\cdot),q_0(\cdot)}(\mathbb{R}^n) \hookrightarrow F^{\alpha_1(\cdot)}_{p_1(\cdot),q_1(\cdot)}(\mathbb{R}^n).$$

In the special case when $q_0 = q_1 = c$, the assumption $(\alpha_0 - \alpha_1)^- > 0$ can be replaced by $\alpha_0 \geqslant \alpha_1$ [367, Theorem 3.4]. In particular, the choice $q_0 = q_1 = 2$ gives, by Theorem 12.5.8 (b), that

$$\mathcal{L}^{\alpha_0,p_0(\cdot)}(\mathbb{R}^n) \hookrightarrow \mathcal{L}^{\alpha_1,p_1(\cdot)}(\mathbb{R}^n)$$

if $\alpha_0 - \frac{n}{p_0} = \alpha_1 - \frac{n}{p_1}$ and $\alpha_0 \geqslant \alpha_1$.

In order to define Besov spaces, we introduce a generalization of the iterated function space $\ell^q(L^{p(\cdot)})$ for the case of variable q. We give a general but quite strange looking definition for the mixed Lebesgue-sequence space modular. This is not strictly an iterated function space-indeed, it cannot be, since then there would be no space variable left in the outer function space.

For all the remaining results in this section we refer to Almeida and Hästö [24].

Definition 12.5.11. Let $p, q \in \mathcal{P}(\mathbb{R}^n)$. The *mixed Lebesgue-sequence space* $\ell^{q(\cdot)}(L^{p(\cdot)})$ is defined on sequences of $L^{p(\cdot)}$-functions by the modular

$$\varrho_{\ell^{q(\cdot)}(L^{p(\cdot)})}\big((f_\nu)_\nu\big) := \sum_{\nu=0}^{\infty} \inf\Big\{\lambda_\nu > 0 \,:\, \varrho_{p(\cdot)}\Big(f_\nu/\lambda_\nu^{\frac{1}{q(\cdot)}}\Big) \leqslant 1\Big\}.$$

Here we use the convention $\lambda^{1/\infty} = 1$. The norm is defined from this as usual:

$$\big\|(f_\nu)_\nu\big\|_{\ell^{q(\cdot)}(L^{p(\cdot)})} := \inf\Big\{\mu > 0 \,:\, \varrho_{\ell^{q(\cdot)}(L^{p(\cdot)})}\big(\tfrac{1}{\mu}(f_\nu)_\nu\big) \leqslant 1\Big\}.$$

To motivate this definition, we mention that

$$\big\|(f_\nu)_\nu\big\|_{\ell^q(L^{p(\cdot)})} = \Big\|\big\|f_\nu\big\|_{p(\cdot)}\Big\|_{\ell^q}$$

if $q \in (0, \infty]$ is constant. Let $p, q \in \mathcal{P}^{\log}(\mathbb{R}^n)$ with $p^-, q^- > 0$. Then $\varrho_{\ell^{q(\cdot)}(L^{p(\cdot)})}$ is a semimodular. Additionally:

(a) It is a modular if $p^+ < \infty$.
(b) It is continuous if $p^+, q^+ < \infty$.

Of course, most importantly, $\varrho_{\ell^{q(\cdot)}(L^{p(\cdot)})}$ defines a quasinorm.

Theorem 12.5.12. *If* $p, q \in \mathcal{P}^{\log}(\mathbb{R}^n)$, *then* $\|\cdot\|_{\ell^{q(\cdot)}(L^{p(\cdot)})}$ *is a quasinorm on* $\ell^{q(\cdot)}(L^{p(\cdot)})$.

We are now able to state the definition of the Besov space without further difficulties:

Definition 12.5.13. Let φ_ν, $\nu \in \mathbb{N}_0$, be as in Definition 12.5.1. For $\alpha : \mathbb{R}^n \to \mathbb{R}$ and $p, q \in \mathcal{P}^{\log}(\mathbb{R}^n)$, the *Besov space* $B^{\alpha(\cdot)}_{p(\cdot), q(\cdot)}(\mathbb{R}^n)$ consists of all distributions $f \in \mathcal{S}'(\mathbb{R}^n)$ such that

$$\|f\|^\varphi_{B^{\alpha(\cdot)}_{p(\cdot), q(\cdot)}} := \big\|(2^{\nu\alpha(\cdot)}\varphi_\nu * f)_\nu\big\|_{\ell^{q(\cdot)}(L^{p(\cdot)})} < \infty.$$

If p and α satisfy Assumption 12.5.2, then the Besov and Triebel–Lizorkin spaces with equal primary and secondary indices agree:

$$B^{\alpha(\cdot)}_{p(\cdot), p(\cdot)}(\mathbb{R}^n) = F^{\alpha(\cdot)}_{p(\cdot), p(\cdot)}(\mathbb{R}^n).$$

In particular, the Besov spaces can also be used to describe the traces of Sobolev functions.

The following independence of the basis functions-property holds under weaker conditions than the corresponding one in the Triebel–Lizorkin case:

Theorem 12.5.14. *Let $p, q \in \mathcal{P}^{\log}(\mathbb{R}^n)$ and α be locally* log*-Hölder continuous and bounded. Then the space $B_{p(\cdot),\, q(\cdot)}^{\alpha(\cdot)}(\mathbb{R}^n)$ does not depend on the admissible basis functions φ_ν, i.e. different functions yield equivalent quasinorms.*

The following two embedding theorems generalize the constant exponent versions in the expected (and optimal) way:

Theorem 12.5.15. *Let $\alpha, \alpha_0, \alpha_1 \in L^\infty(\mathbb{R}^n)$ and $p, q_0, q_1 \in \mathcal{P}(\mathbb{R}^n)$.*

(i) If $q_0 \leqslant q_1$, then

$$B_{p(\cdot),\, q_0(\cdot)}^{\alpha(\cdot)}(\mathbb{R}^n) \hookrightarrow B_{p(\cdot),\, q_1(\cdot)}^{\alpha(\cdot)}(\mathbb{R}^n).$$

(ii) If $(\alpha_0 - \alpha_1)^- > 0$, then

$$B_{p(\cdot),\, q_0(\cdot)}^{\alpha_0(\cdot)}(\mathbb{R}^n) \hookrightarrow B_{p(\cdot),\, q_1(\cdot)}^{\alpha_1(\cdot)}(\mathbb{R}^n).$$

(iii) If $p^+, q^+ < \infty$, then

$$B_{p(\cdot),\, \min\{p(\cdot),q(\cdot)\}}^{\alpha(\cdot)}(\mathbb{R}^n) \hookrightarrow F_{p(\cdot),\, q(\cdot)}^{\alpha(\cdot)}(\mathbb{R}^n) \hookrightarrow B_{p(\cdot),\, \max\{p(\cdot),q(\cdot)\}}^{\alpha(\cdot)}(\mathbb{R}^n).$$

Theorem 12.5.16 (Sobolev inequality). *Let $p_0, p_1, q \in \mathcal{P}(\mathbb{R}^n)$ and $\alpha_0, \alpha_1 \in L^\infty(\mathbb{R}^n)$ with $\alpha_0 \geqslant \alpha_1$. If $1/q$ and*

$$\alpha_0(x) - \frac{n}{p_0(x)} = \alpha_1(x) - \frac{n}{p_1(x)}$$

are locally log*-Hölder continuous, then*

$$B_{p_0(\cdot),q(\cdot)}^{\alpha_0(\cdot)}(\mathbb{R}^n) \hookrightarrow B_{p_1(\cdot),q(\cdot)}^{\alpha_1(\cdot)}(\mathbb{R}^n).$$

Kempka [232] has studied so-called micro-local versions of both variable index Besov and Triebel–Lizorkin spaces. This setting includes also some range of weights as well as slightly more general smoothness. However, he does not include the case of Besov spaces with variable q.

Part III
Applications to Partial Differential Equations

Chapter 13
Dirichlet Energy Integral and Laplace Equation

For a constant $q \in (1, \infty)$, the Dirichlet energy integral is

$$\int_\Omega |\nabla u(x)|^q \, dx.$$

The problem is to find a minimizer for the energy integral among all Sobolev functions with a given boundary value function. The Euler–Lagrange equation of this problem is the q-Laplace equation,

$$\operatorname{div}(|\nabla u|^{q-2}\nabla u) = 0,$$

which has to be understand in the weak sense. The energy integral and q-Laplace equation have been widely studied, see for example [219, 235, 280]. The q-Laplace equation is a prototype of a non-linear elliptic equation. By non-linearity we mean that if $q \neq 2$ then the weak solutions do not form a linear space. However the set of weak solutions is closed under constant multiplication. By celebrated De Giorgi's method and Moser's iteration the minimizers and the weak solutions are locally Hölder continuous and satisfy Harnack's inequality:

$$\sup_B u \leqslant c \inf_B u,$$

where c is independent of u and the ball B.

The Dirichlet energy integral and the Laplace equation can be generalize to the variable exponent case as

$$\int_\Omega |\nabla u(x)|^{p(x)} \, dx \quad \text{and} \quad \operatorname{div}(p(x)|\nabla u(x)|^{p(x)-2}\nabla u) = 0.$$

It turns out that the minimizer exists for a given boundary value function under mild conditions of p. The minimizers and the weak solutions are locally Hölder continuous when p is log-Hölder continuous with $1 < p^- \leqslant p^+ < \infty$. Harnack's inequality does not hold by the mentioned form: in the variable

L. Diening et al., *Lebesgue and Sobolev Spaces with Variable Exponents*,
Lecture Notes in Mathematics 2017, DOI 10.1007/978-3-642-18363-8_13,
© Springer-Verlag Berlin Heidelberg 2011

exponent case the constant can not be independent of the function u. The minimizers or the weak solutions are not scalable, i.e. λu need not be a minimizer or a weak solution even if u is. These effects are visible already in the one dimensional case where the minimizers need not to be linear as in the constant exponent case.

In the first section, Sect. 13.1, we study minimizers on an interval with detailed proofs. In Sects. 13.2 and 13.3 we give a rough overview of properties of minimizers and solutions of the prototype equality. In the last section, Sect. 13.4, we generalize, with detailed proofs, Harnack's inequality to all elliptic type Laplace equations with growth conditions of a non-standard form.

The material is selected by the personal taste of the writers; it concentrates to the variable exponent Laplace equation from the potential theoretical viewpoint and all results concerning Harnack's inequality are included. In particular this chapter does not include solutions to the obstacle problems, e.g., [125, 126, 205, 331], systems, e.g., [8, 13, 49, 106, 180, 185, 272, 318, 374, 382, 383, 388], eigenvalues, e.g., [21, 88, 89, 112, 113, 130, 133, 135, 136, 145, 259, 290, 290, 291, 381], parabolic equations, e.g., [9, 36–39, 71, 257, 324, 379, 380]; regularity of solutions, e.g., [3, 5, 10, 19, 31, 72, 124, 131, 146, 147, 150, 156, 182, 276, 370, 395, 396, 398].

Existence and uniqueness of solutions has been studied in a large number of papers. For instance:

- [139, 143, 368, 385, 389, 390] deal with the one-dimensional case.
- [32, 35, 61, 80, 132, 170, 270, 282, 295, 322, 377, 384, 386, 387, 392] deal with existence of solutions to the $p(\cdot)$-Laplacian.
- [17, 30, 33, 33, 42, 55, 56, 64, 68, 78, 85–87, 110, 137, 138, 140, 144, 151, 160, 161, 170–173, 175, 220–222, 273, 274, 292–294, 320, 321, 333, 349, 354, 375–377] deal with existence related to more general equations.

A wider scope can be found from the recent surveys [201, 297].

13.1 The One Dimensional Case

Let us start by stating the Dirichlet energy integral problem on an interval. These results are mainly from [191]. We assume that the bounded interval under consideration is (r, R). Since every element in the space $W^{1,p(\cdot)}(r, R)$ has a continuous representative, we assume that every Sobolev function is continuous. We denote $u \in W_0^{1,p(\cdot)}(r, R)$ and say that u belongs to the *variable exponent Sobolev space with zero boundary values* if it can be continuously continued by 0 outside (r, R) (the extension is again denoted by u). Thus $u \in W_0^{1,p(\cdot)}(r, R)$ if and only if $u(r) = u(R) = 0$.

Definition 13.1.1. A function $u \in W^{1,p(\cdot)}(r, R)$ is a $p(\cdot)$-*minimizer* for the boundary values a and b if $u(r) = a$, $u(R) = b$ and

$$\int\limits_r^R |u'|^{p(y)} \, dy \leqslant \int\limits_r^R |v'|^{p(y)} \, dy$$

for every v with $u - v \in W_0^{1,p(\cdot)}(r, R)$.

If p is a constant, then the minimizer is linear, $u(x) = \frac{b-a}{R-r}(x - r) + a$. The next example shows that the variable exponent adds some interest to this minimization question.

Example 13.1.2. We define

$$p(x) := \begin{cases} 3, & \text{for } 0 < x \leqslant 1/2; \\ 2, & \text{for } 1/2 < x < 1. \end{cases}$$

Suppose that $u \in W^{1,p(\cdot)}(0, 1)$ is the minimizer for the boundary values 0 and $b > 0$. Denote $u(1/2) = \lambda$.

Then $u|_{(0,1/2)}$ is the solution to the classical energy integral problem with boundary values 0 and λ, and $u|_{(1/2,1)}$ is the solution with boundary values λ and b. Therefore these functions are linear, and so

$$u(x) = \begin{cases} 2\lambda x, & \text{for } 0 < x \leqslant 1/2; \\ 2\lambda + 2(b - \lambda)(x - 1/2), & \text{for } 1/2 < x < 1. \end{cases}$$

For this u we have the Dirichlet energy $4\lambda^3 + 2(b-\lambda)^2$. It is easy to see that the function $\lambda \mapsto 2\lambda^3 + (b-\lambda)^2$ has a minimum at $\lambda = (\sqrt{1 + 12b} - 1)/6$, which determines the minimizer of the variable exponent problem. The minimizing functions for some b's are shown in Fig. 13.1.

As can be seen in the figure, and confirmed by calculation, the minimizer is convex if $b > 2/3$, concave if $b < 2/3$ and linear for $b = 2/3$.

It is in fact possible to give an explicit formula for the minimizer by solving the corresponding Euler–Lagrange equation, as shown in the next theorem. The formula is not quite transparent, however, so we prove some properties of the minimizers later on. We start with preliminary results.

Lemma 13.1.3. Let $p \in \mathcal{P}(r, R)$ be bounded and strictly greater than one almost everywhere. If $u \in W^{1,p(\cdot)}(r, R)$ is a $p(\cdot)$-minimizer, then $p(x)$ $(u'(x))^{p(x)-1}$ is a constant almost everywhere.

Proof. Suppose that $p(x)(u'(x))^{p(x)-1}$ is not a constant almost everywhere. Let then $d_1 < d_2$ be such that

$$A_1 := \{x \in (r, R) : p(x)|u'(x)|^{p(x)-1} < d_1\},$$
$$A_2 := \{x \in (r, R) : p(x)|u'(x)|^{p(x)-1} > d_2\}$$

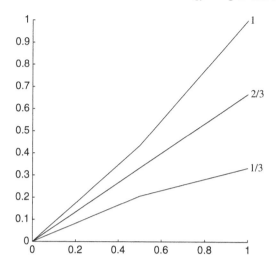

Fig. 13.1 Energy integral minimizers

have positive measure. Let $A'_1 \subset A_1$ and $A'_2 \subset A_2$ be such that $|A'_1| = |A'_2| > 0$. Define $\xi := \chi_{A'_1} - \chi_{A'_2}$.

Let $0 < \varepsilon < 1$. Using $\left||x+h|^p - |x|^p\right| \leqslant p \left||x+h| - |x|\right| \left(|x+h|^{p-1} + |x|^{p-1}\right)$ and $\varepsilon^{-1}\left||u' + \varepsilon\xi| - |u'|\right| \leqslant c$ we obtain

$$\left| \frac{|u'(x) + \varepsilon\xi(x)|^{p(x)} - |u'(x)|^{p(x)}}{\varepsilon} \right|$$

$$\leqslant \frac{p(x)\left||u'(x) + \varepsilon\xi(x)| - |u'(x)|\right| \left(|u'(x) + \varepsilon\xi(x)|^{p(x)-1} + |u'(x)|^{p(x)-1}\right)}{\varepsilon}$$

$$\leqslant c\left(|u'(x)|^{p(x)-1} + \varepsilon^{p(x)-1}\right) \leqslant c\left(|u'(x)|^{p(x)} + \varepsilon + 1\right).$$

Since p is bounded $|u'|^{p(\cdot)} \in L^1(r, R)$ and the dominated convergence theorem yields

$$\lim_{\varepsilon \to 0} \int_r^R \frac{|u' + \varepsilon\xi|^{p(x)} - |u'|^{p(x)}}{\varepsilon} \, dx = \int_r^R p(x)|u'|^{p(x)-1}\xi \, dx$$

$$\leqslant (d_1 - d_2)|A'_1| < 0.$$

Note that ξ is the classical derivative of $v(x) := \int_r^x \chi_{A'_1} - \chi_{A'_2} \, dy$. Since v' is bounded and $v(r) = v(R) = 0$ we obtain that $v \in W_0^{1,p(\cdot)}(r, R)$. Thus the previous inequality contradicts that u is the minimizer and hence $p(x)(u'(x))^{p(x)-1}$ is a constant almost everywhere. □

Lemma 13.1.4. *Let $p \in \mathcal{P}(r, R)$ be bounded and strictly greater than one almost everywhere. Let $u \in W^{1,p(\cdot)}(r, R)$. If $p(x)(u'(x))^{p(x)-1}$ is a constant almost everywhere then u is a $p(\cdot)$-minimizer for its own boundary values.*

Proof. Since $p(x)(u'(x))^{p(x)-1}$ is a constant almost everywhere we have for every v with $v - u \in W_0^{1,p(\cdot)}(r, R)$ that

$$\int_r^R p(x)(u')^{p(x)-1}(v' - u')\, dx = 0.$$

By the inequality $|b|^p \geqslant |a|^p + p|a|^{p-1}(b - a)$ we obtain

$$\int_r^R |v'|^{p(x)}\, dx \geqslant \int_r^R |u'|^{p(x)}\, dx + \int_r^R p(x)(u')^{p(x)-1}(v' - u')\, dx.$$

Since the last integral is zero we have

$$\int_r^R |u'|^{p(x)}\, dx \leqslant \int_r^R |v'|^{p(x)}\, dx. \qquad \square$$

The next theorem gives an explicit form for the minimizers. The original formulation in [191] included a mistake, so we reformulate the theorem here.

Theorem 13.1.5 ([191, Theorem 3.2]). *Let $p \in \mathcal{P}(r, R)$ be bounded and strictly greater than one almost everywhere and let $a, b \in \mathbb{R}$, $a < b$, be the boundary values at r and R. Then there exists a unique minimizer for these boundary values if and only if there exists $\tilde{m} \geqslant 0$ such that*

$$b - a \leqslant \int_r^R \left(\frac{\tilde{m}}{p(x)} \right)^{\frac{1}{p(x)-1}} dx < \infty. \tag{13.1.6}$$

In this case the minimizer is given by

$$u(x) := \int_r^x \left(\frac{m}{p(y)} \right)^{\frac{1}{p(y)-1}} dy + a,$$

for appropriate $m \in (0, \tilde{m}]$.

Note that for m we have

$$\int_r^R \left(\frac{m}{p(y)} \right)^{\frac{1}{p(y)-1}} dy = b - a.$$

Proof. Let $f_m(x) := \left(\frac{m}{p(x)}\right)^{\frac{1}{p(x)-1}}$. We first show that m can be chosen so that $\int_r^R f_m \, dx = b - a$. If $\tilde{m}$ is such that $\int_r^R f_{\tilde{m}} \, dx = b - a$, then $\tilde{m} = m$. Assume then that $\int_r^R f_{\tilde{m}} \, dx > b - a$. It is enough to show that $m \mapsto \int_r^R f_m \, dx$ is continuous on $[0, \tilde{m}]$. Fix $\varepsilon > 0$ and define

$$A_\lambda := \left\{x \in (r, R) \colon p(x) > \lambda\right\}.$$

We choose $\lambda > 1$ such that

$$\int\limits_{(r,R)\setminus A_\lambda} f_{\tilde{m}} \, dx < \varepsilon.$$

In A_λ the exponent $\frac{1}{p(x)-1}$ is bounded from above and so we can choose a small real number d, $m + |d| < \tilde{m}$, such that $\int_{A_\lambda} |f_m - f_{m+d}| \, dx < \varepsilon$. Since $m \mapsto \int_r^R f_m \, dx$ is increasing, we find that

$$\left| \int\limits_r^R f_m - f_{m+d} \, dx \right| \leqslant \int\limits_{A_\lambda} |f_m - f_{m+d}| \, dx + \int\limits_{(r,R)\setminus A_\lambda} (f_m + f_{m+d}) \, dx \leqslant 3\varepsilon.$$

Hence $m \mapsto \int_r^R f_m \, dx$ is continuous.

Let f_m be such that $\int_r^R f_m \, dx = b - a$. If $u \in W^{1,p(\cdot)}(r, R)$ is such that $f_m = u'$ and $u(r) = a$, then by Lemma 13.1.4 the function u is the minimizer we are looking for. Define therefore $u(x) := \int_r^x f_m(y) \, dy + a$ for $x \in (r, R]$. Since $a \leqslant u \leqslant b$, $u \in L^{p(\cdot)}(r, R)$. Further,

$$\bar{\varrho}_{p(\cdot)}(u') = \int\limits_r^R \left(\frac{m}{p(x)}\right)^{\frac{p(x)}{p(x)-1}} dx \leqslant \tilde{m} \int\limits_r^R \left(\frac{\tilde{m}}{p(x)}\right)^{\frac{1}{p(x)-1}} dx < \infty.$$

Therefore $u \in W^{1,p(\cdot)}(r, R)$ is a minimizer.

To prove the other direction, let u be a minimizer. Then by Lemma 13.1.3 $f_m = u'$ with $\int_r^R f_m \, dy = b - a$. So then (13.1.6) holds. Therefore the condition is both necessary and sufficient.

Finally we show that the minimizer is unique. Assume that we have two minimizers u and v with same boundary values. Since $y \mapsto y^{p(x)}$ is strictly convex when $p(x) > 1$,

$$\left| \frac{1}{2}u' + \frac{1}{2}v' \right|^{p(x)} < \frac{1}{2}|u'|^{p(x)} + \frac{1}{2}|v'|^{p(x)}$$

for almost every $x \in I$ with $u'(x) \neq v'(x)$. The function $\frac{1}{2}u + \frac{1}{2}v$ has the same boundary values than u and v. If the set $\{u'(x) \neq v'(x)\}$ has positive measure then

$$\int\limits_r^R \left| \frac{1}{2}u' + \frac{1}{2}v' \right|^{p(x)} dx < \frac{1}{2}\int\limits_r^R |u'|^{p(x)}\, dx + \frac{1}{2}\int\limits_r^R |v'|^{p(x)}\, dx = \int\limits_r^R |u'|^{p(x)}\, dx.$$

Since u is a minimizer, we have $u' = v'$ almost everywhere. We obtain $u = v$. □

The gradient of the minimizer is uniformly bounded if $1 < p^- \leqslant p^+ < \infty$. Thus we obtain the following corollary.

Corollary 13.1.7. *If $p \in \mathcal{P}(r, R)$ with $1 < p^- \leqslant p^+ < \infty$ then for every $a, b \in \mathbb{R}$, $a < b$, there exists a unique minimizer with these boundary values. The minimizer is given by*

$$u(x) := \int\limits_r^x \left(\frac{m}{p(y)} \right)^{\frac{1}{p(y)-1}} dy + a,$$

for some constant $m > 0$.

The following example shows that the Dirichlet energy integral does not always have a minimizer.

Example 13.1.8. For $p(x) := 1 + x$ in $(0, 1)$ the minimizer does not exists when the difference between the boundary values is large enough. Fix $a = 0$ and let $m > 1$. Then

$$\int\limits_0^1 m^{\frac{1}{p(x)-1}}\, dx = \int\limits_0^1 m^{\frac{1}{x}}\, dx \geqslant \max\{1, \log m\} \int\limits_0^1 \frac{dx}{x} = \infty.$$

Since $p^{1/(1-p)}$ lies between $1/e$ and 1, the condition of Theorem 13.1.5 is not satisfied for $b > \int_0^1 p(x)^{\frac{1}{1-p(x)}}\, dx$.

Example 13.1.9. Using Theorem 13.1.5 we plot some minimizers of the energy integral for $p(x) := 1.1 + x$ in $(0, 1)$ (Fig. 13.2). The left boundary value is 0 and the number on the right is again the second boundary value, b. It is easy to see that if we multiply a minimizer by a real number the result need not to be a minimizer.

Next we study regularity of minimizers.

Corollary 13.1.10. *If $p \in \mathcal{P}(r, R)$ with $1 < p^- \leqslant p^+ < \infty$, then the minimizers are bi-Lipschitz continuous.*

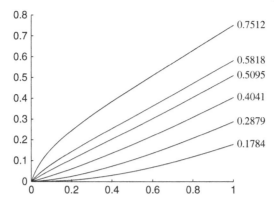

Fig. 13.2 Minimizers when $p(x) = 1.1 + x$

Proof. By Theorem 13.1.5, the minimizer has derivative $(m/p(x))^{1/(p(x)-1)}$ for some constant $m \geqslant 0$. Since

$$\left(\frac{m}{p(x)}\right)^{\frac{1}{p(x)-1}} \leqslant m^{\frac{1}{p(x)-1}} \leqslant \max\{m^{\frac{1}{p^--1}}, m^{\frac{1}{p^+-1}}\} < \infty$$

and

$$\left(\frac{m}{p(x)}\right)^{\frac{1}{p(x)-1}} \geqslant \left(\frac{m}{p^+}\right)^{\frac{1}{p(x)-1}} \geqslant \min\left\{\left(\frac{m}{p^+}\right)^{\frac{1}{p^--1}}, \left(\frac{m}{p^+}\right)^{\frac{1}{p^+-1}}\right\} > 0$$

for all $x \in (r, R)$, it follows from the mean-value theorem that the minimizer is bi-Lipschitz continuous.   □

Corollary 13.1.11. *If $p \in \mathcal{P}(r, R)$ with $1 < p^- \leqslant p^+ < \infty$, then the derivative of the minimizer is α-Hölder continuous if and only if the exponent p is α-Hölder continuous.*

Proof. Let us denote $F(y) := (m/y)^{1/(y-1)}$. Then the derivative of the minimizer equals $F(p(x))$. Since F is differentiable on $(1, \infty)$ we obtain

$$|F(p(x)) - F(p(y))| = F'(\xi)|p(x) - p(y)|,$$

where $\xi \in (p(x), p(y))$, by the mean-value theorem. It is easy to see that F' is bounded and bounded away from 0 on $[p^-, p^+]$, so that $F(p(x))$ possesses the same degree of regularity as $p(x)$.   □

The next result shows that if we relax the assumption $p^- > 1$ then we are a liable to lose a lot of the regularity of the minimizer.

Example 13.1.12. Let $p(x) := 1 + (\log(1/x))^{-1}$ in $(0, 1)$. Fix the left boundary value be 0 and let $b > 0$ be the right boundary value. We have

$$\int\limits_0^1 m^{\frac{1}{p(x)-1}}\,dx = \int\limits_0^1 x^{-\log m}\,dx = \frac{1}{1-\log m},$$

provided $m < e$, so condition (13.1.6) is satisfied for any $b > 0$. Therefore the derivative of the minimizer is $(m/p(x))^{1/(p(x)-1)}$ for some $m > 0$, by Theorem 13.1.5. Thus for $0 < y < x < 1$ we have

$$|u(x) - u(y)| = \left| \int\limits_y^x x^{-\log m} p(x)^{-\frac{1}{p(x)-1}}\,dx \right| \leqslant \frac{x^{1-\log m} - y^{1-\log m}}{1 - \log m}$$

$$\leqslant \frac{(x-y)^{1-\log m}}{1 - \log m}.$$

We see that u is $(1 - \log m)$-Hölder continuous. Moreover, if b is such that $m > 1$, then the derivative is unbounded, hence not uniformly continuous.

Some minimizers are plotted in Fig. 13.3. The number on the right is again the second boundary value, b. The lower three curves are Lipschitz continuous, the following two are 0.738- and 0.530-Hölder continuous.

Theorem 13.1.13 (Harnack's inequality, [200]). *Let $p \in \mathcal{P}(r, R)$ with $1 < p^- \leqslant p^+ < \infty$. If $u \in W^{1,p(\cdot)}(r, R)$ is a minimizer with boundary values a and b, $0 \leqslant a < b$, then*

$$\sup_{y \in B(x,r')} u(y) \leqslant c \inf_{y \in B(x,r')} u(y)$$

for every $x \in (r, R)$ and every r' with $2B(x, r') \subset (r, R)$. The constant c depends only on p^-, p^+, $b - a$, and $R - r$.

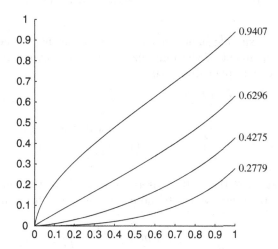

Fig. 13.3 Minimizers when $p(x) = 1 + (-\log(x))^{-1}$

Proof. We note that $t^{-1/(t-1)}$ lies between e^{-1} and 1 for all $t \geqslant 1$. By Theorem 13.1.5 we obtain

$$
\sup_{B(x,r')} u = \int_r^{x+r'} \left(\frac{m}{p(y)}\right)^{\frac{1}{p(y)-1}} dy + a \leqslant \int_r^{x+r'} m^{\frac{1}{p(y)-1}}\, dy + a
$$

$$
\leqslant (x+r'-r) \max\left\{ m^{\frac{1}{p^+-1}}, m^{\frac{1}{p^--1}} \right\} + a
$$

and

$$
\inf_{B(x,r')} u = \int_r^{x-r'} \left(\frac{m}{p(y)}\right)^{\frac{1}{p(y)-1}} dy + a \geqslant \mathrm{e}^{-1} \int_r^{x-r'} m^{\frac{1}{p(y)-1}}\, dy + a
$$

$$
\geqslant (x-r'-r) \min\left\{ m^{\frac{1}{p^+-1}}, m^{\frac{1}{p^--1}} \right\} + a.
$$

Since $2B(x,r') \subset (r,R)$, we deduce

$$
\frac{x+r'-r}{x-r'-r} \leqslant 3.
$$

Using this and the fact $\frac{c+a}{c'+a} \leqslant \max\{\frac{c}{c'}, 1\}$, the supremum and infimum estimates yield

$$
\frac{\sup_{B(x,r')} u}{\inf_{B(x,r')} u} \leqslant 3\mathrm{e} \max\left\{ m^{-\frac{1}{p^--1}}, m^{\frac{1}{p^--1}} \right\}.
$$

Here the constant m is from Theorem 13.1.5 and it depends on the boundary values of u and p. Moreover m can be estimated in terms of p^-, p^+, $b-a$ and $R-r$. $\square$

The following example shows that the constant in the Harnack inequality can not be independent of the minimizer even if the exponent is Lipschitz continuous. The example is from [208].

Example 13.1.14. We define

$$
p(x) := \begin{cases} 3 & \text{for } 0 < x \leqslant \tfrac{1}{2}; \\ 3 - 2(x - \tfrac{1}{2}) & \text{for } \tfrac{1}{2} < x < 1. \end{cases}
$$

Suppose that $u_b \in W^{1,p(\cdot)}(0,1)$ is the minimizer of the Dirichlet energy integral for the boundary values 0 and $b > 0$ given by Theorem 13.1.5:

$$
u_b(x) = \int_0^x \left(\frac{m(b)}{p(y)}\right)^{\frac{1}{p(y)-1}} dy.
$$

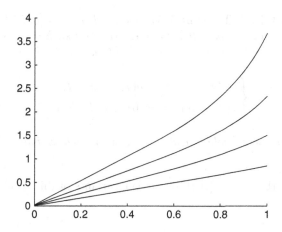

Fig. 13.4 Minimizers when p is Lipschitz continuous on the whole interval and constant on $\left(0, \frac{1}{2}\right]$

Note that if $b \to \infty$ then $m(b) \to \infty$. Three minimizers with $m(b) = 2, 4, 8$ are presented in Fig. 13.4.

In $\left(0, \frac{1}{2}\right)$ the minimizer is linear, $u_b(x) = \sqrt{\frac{m(b)}{3}} x$. In $\left(\frac{1}{2}, \frac{3}{5}\right)$ the gradient of u_b increases from $\sqrt{\frac{m(b)}{3}}$ to $\left(\frac{5m(b)}{14}\right)^{\frac{5}{9}}$. At $\frac{11}{20}$, the midpoint of $\left(\frac{1}{2}, \frac{3}{5}\right)$, the gradient of u_b equals $\left(\frac{10m(b)}{29}\right)^{\frac{10}{19}}$. Hence

$$u_b\left(\tfrac{3}{5}\right) \geqslant \sqrt{\frac{m(b)}{3}} \frac{1}{2} + \frac{1}{20}\left(\frac{10m(b)}{29}\right)^{\frac{10}{19}}.$$

Let $B := B\left(\frac{1}{2}, \frac{1}{10}\right) = \left(\frac{2}{5}, \frac{3}{5}\right)$. Then

$$\frac{\sup_{x \in B}|u_b(x)|}{\inf_{x \in B}|u_b(x)|} \geqslant \frac{\sqrt{\frac{m(b)}{3}}\frac{1}{2} + \frac{1}{20}\left(\frac{10m(b)}{29}\right)^{\frac{10}{19}}}{\sqrt{\frac{m(b)}{3}}\frac{2}{5}} = \frac{5}{4} + \frac{1}{8\sqrt{3}}\left(\frac{10}{29}\right)^{\frac{10}{19}} m(b)^{\frac{1}{38}} \to \infty$$

as $b \to \infty$.

Fan and Fan [139] have considered more complicated one-dimensional variable exponent differential equations.

Theorem 13.1.15 (Theorem 1.1, [139]). *Let $I := [0,T] \subset \mathbb{R}$ and $g \in C(I \times \mathbb{R}^N, \mathbb{R}^N)$, and suppose that there exists $r > 0$ such that $zg(x,z) \geqslant 0$ for all $x \in I$ and $z \in \mathbb{R}^N$ with $|z| = r$. If $p \in C(I)$ and $p^- > 1$, then the equation*

$$\begin{cases} \left(|u'|^{p(x)-2}u'\right)' = g(x,u), \ x \in I, \\ u(0) - u(T) = u'(0) - u'(T) = 0, \end{cases} \tag{13.1.16}$$

has at least one weak solution $u \in C^1(I, \mathbb{R}^N)$ such that $|u(x)| \leqslant r$ for all $x \in I$.

We refer to the survey [201] for further variants of this result.

13.2 Minimizers

We start this section by discussing existence of minimizers for given boundary values. Then we move to regularity of minimizers and Harnack's inequality. These results are collected from many papers and for every theorem a reference is given.

Definition 13.2.1. A function $u \in W^{1,p(\cdot)}(\Omega)$ is a minimizer for a boundary value function $w \in W^{1,p(\cdot)}(\Omega)$ if $u - w \in W_0^{1,p(\cdot)}(\Omega)$ and

$$\int_\Omega |\nabla u|^{p(x)}\, dx \leqslant \int_\Omega |\nabla v|^{p(x)}\, dx$$

for every function v with $u - v \in W_0^{1,p(\cdot)}(\Omega)$.

Theorem 13.2.2 ([196]). *Let Ω be bounded and let $p \in \mathcal{P}(\Omega)$ with $1 < p^- \leqslant p^+ < \infty$ be such that the $p(\cdot)$-Poincaré inequality holds. Assume that $w \in W^{1,p(\cdot)}(\Omega)$. Then there exists a unique minimizer for the boundary value function w.*

Let $n \geqslant 3$, $q_1 \in (1, n/(n-1))$ and $q_2 \in (q_1^*, n)$. Hästö [213] constructed a bounded domain Ω, a continuous exponent $p \in \mathcal{P}(\Omega)$ with $p^- = q_1$ and $p^+ = q_2$, and a boundary value function $w \in W^{1,p(\cdot)}(\Omega)$ such that there does not exists a minimizer for the boundary value function w.

Note that if we had $q_2 \leqslant q_1^*$ in the previous theorem, then a minimizer would always exist, by Lemma 8.2.14 and Theorems 13.2.2, so in this sense Theorem 13.2.2 is the best possible.

Theorem 13.2.3 ([213]). *Let Ω be bounded and let $p \in \mathcal{P}(\Omega)$ with $1 < p^- \leqslant p^+ < \infty$. Suppose that $w \in W^{1,p(\cdot)}(\Omega) \cap L^\infty(\Omega)$. Then there exists a unique minimizer for the boundary value function w.*

The proofs of Theorems 13.2.2 and 13.2.3 are based on a well known functional analysis result: in a reflexive Banach spaces there exists an element that minimizes every convex, lower semicontinuous and coercive operator. The space $W^{1,p(\cdot)}(\Omega)$ is a reflexive Banach space by Theorem 8.1.6. Convexity follows since $t \mapsto t^p$ is convex for every $1 < p < \infty$. Theorem 3.2.9 yields lower semicontinuity. Therefore, we need only worry about coercivity. In this setting coercivity means that $\|u\|_{p(\cdot)} \to \infty$ implies $\|\nabla u\|_{p(\cdot)} \to \infty$. Clearly this holds if the Poincaré inequality holds, see Sect. 8.2. If the boundary value function is bounded we may restrict our studied to uniformly bounded Sobolev functions and use the Poincaré inequality in the constant exponent case $p = 1$.

Assume that p is a bounded variable exponent with $p^- = 1$. For $\lambda > 1$ we set $p_\lambda := \max\{p, \lambda\}$; we can find Dirichlet $p_\lambda(\cdot)$-energy minimizers u_λ for the given bounded boundary value function $f \in W^{1,p_\delta(\cdot)}(\Omega)$ for some $\delta > 1$. The following result says that (u_λ) has a converging subsequence as $\lambda \to 1$. We denote $Y := \{x \in \Omega : p(x) = 1\}$.

Theorem 13.2.4 ([206]). *Let $p \in \mathcal{P}(\Omega)$ be bounded with $p^- = 1$ and let (λ_j) be a sequence decreasing to 1. Let (u_{λ_j}) be a sequence of Dirichlet $p_{\lambda_j}(\cdot)$-minimizers in Ω for a boundary value function $f \in W^{1,p_\delta(\cdot)}(\Omega) \cap L^\infty(\Omega)$, for some $\delta > 1$.*

Then there exists a subsequence (λ_j) and $u \in L^\infty(\Omega)$ such that:

(a) *$u_{\lambda_j} \to u$ in $L^{p_\delta(\cdot)}_{loc}(\Omega)$ for $\delta \in [1, \frac{n}{n-1})$;*
(b) *$u_{\lambda_j} \rightharpoonup u$ in $W^{1,p(\cdot)}_{loc}(\Omega \setminus Y)$;*
(c) *u is a weak solution of the $p(\cdot)$-Laplace equation in $\Omega \setminus Y$ (see the next section for the definition of weak solutions).*

If, in addition, p is log-Hölder continuous and

$$\lim_{x \to y} |p(x) - 1| \log \frac{1}{|x - y|} = 0$$

for every $y \in Y$, then the limit function u belongs to a variable exponent mixed BV-Sobolev space in Ω and it minimizes the BV-Sobolev energy among all functions with the same boundary values.

The limit function u form the previous theorem can be discontinuous as presented in Fig. 13.5 that is from [206].

At the opposite limit, when $p \to \infty$, we have the following result, where $p^\lambda := \min\{p, \lambda\}$, see also [270, 281].

Theorem 13.2.5 ([200]). *Let $p \in \mathcal{P}(\Omega)$ with $n < p^- \leqslant p^+ = \infty$. Assume that $f \in W^{1,p(\cdot)}(\Omega) \cap L^\infty(\Omega)$ with $\int_\Omega |\nabla f|^{p(x)} dx < \infty$. Let u_λ be the Dirichlet $p^\lambda(\cdot)$-energy minimizer for the boundary value function f. Then there exist a sequence (λ_i) converging to infinity and a function $u_\infty \in W^{1,p(\cdot)}(\Omega)$ such that (u_{λ_i}) converges locally uniformly to u_∞ in Ω. Moreover, $\int_\Omega |\nabla u_\infty|^{p(x)} dx$ is finite and $|\nabla u_\infty| \leqslant 1$ almost everywhere in $\{p = \infty\}$.*

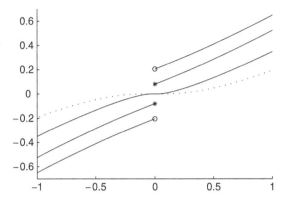

Fig. 13.5 Four BV-Sobolev-minimizers for $p(x) = 1 + |x|$ with different boundary values

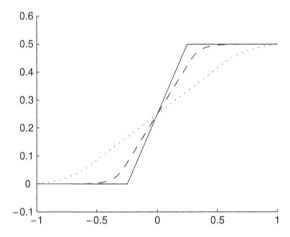

Fig. 13.6 A limit function and two solutions with $\lambda = 10, 100$

The next example shows that Harnack's inequality need not to hold for the limit function u_∞ in the form of Theorem 13.1.13. The example is from [200, Example 4.9].

Example 13.2.6. We define $p \in \mathcal{P}(0,1)$ by

$$p(x) = \begin{cases} \frac{3}{|x| - \frac{1}{4}}, & |x| > \frac{1}{4} \\ \infty, & |x| \leqslant \frac{1}{4} \end{cases}$$

and choose boundary values 0 and $\frac{1}{2}$. Figure 13.6 presents the limit function u_∞ (line) and $p^\lambda(\cdot)$-solutions with λ equal to 10 (dot) and 100 (dash). Note that the limit function u_∞ equals 0 on $(-1, -\frac{1}{4})$.

Next we study continuity of the minimizer and its gradient. A function $F : \Omega \times \mathbb{R}^n \to \mathbb{R}^n$ is a *Carathéodory function* if $x \mapsto F(x,z)$ is measurable for every $z \in \mathbb{R}^n$ and $z \mapsto F(x, z)$ is continuous for almost every $x \in \Omega$. Let $F : \Omega \times \mathbb{R}^n \to \mathbb{R}^n$ be a Carathéodory function such that

$$c^{-1}|z|^{p(x)} \leqslant F(x, z) \leqslant c(1 + |z|^{p(x)})$$

for some $c \geqslant 1$.

Definition 13.2.7. A function $u \in W^{1,1}_{\mathrm{loc}}(\Omega)$ is a local minimizer of F if $|\nabla u| \in L^{p(\cdot)}_{\mathrm{loc}}(\Omega)$ and

$$\int_{\mathrm{spt}\,\psi} F(x, \nabla u) \, dx \leqslant \int_{\mathrm{spt}\,\psi} F(x, \nabla u + \nabla \psi) \, dx$$

for every $\psi \in W^{1,1}(\Omega)$ with compact support in Ω.

Every minimizer from Definition 13.2.1 satisfies the conditions of Definition 13.2.7.

In the excellent series of paper, Acerbi, Coscia and Mingione proved the fundamental $C^{1,\alpha}$ regularity for the model equation [4, 79] and extended it to more general equations and systems [5, 7, 8]. We refer here only some their theorems and recommend reader to look the nice survey of Mingione [297].

Fan and Zhao showed [147] that if the exponent is continuous and $1 < p^- \leqslant p^+ < \infty$, then every local minimizer of F is locally bounded. Their proofs are based on De Giorgi's method.

Theorem 13.2.8 ([4, 147]). *Let Ω be bounded and $p \in \mathcal{P}^{\log}(\Omega)$ with $1 < p^- \leqslant p^+ < \infty$. Then every local minimizer of F is locally α-Hölder continuous for $\alpha \in (0, 1)$ depending the $\log$-Hölder constant of p.*

The proof of Acerbi and Mingione [4] gives also a slightly different version of the previous theorem. Namely if for every $x \in \Omega$ we have

$$|p(x) - p(y)| \log |x - y| \to 0 \text{ as } y \to x,$$

then every local minimizer of F is locally α-Hölder continuous for every $0 < \alpha < 1$.

Using the higher integrability of the gradients of local minimizers of F, Coscia and Mingione show that in some cases the gradients are continuous.

Theorem 13.2.9 ([79]). *Let Ω be bounded and $p \in \mathcal{P}(\Omega)$ be α-Hölder continuous with $1 < p^- \leqslant p^+ < \infty$ and $0 < \alpha \leqslant 1$. Then every local minimizer of F has locally β-Hölder continuous derivatives for some $\beta < \alpha$.*

Corollary 13.1.11 shows that in Theorem 13.2.9 β can not be strictly larger than α.

Definition 13.2.10. A function $u \in W_{\text{loc}}^{1,p(\cdot)}(\Omega)$ is called a quasiminimizer if there exists a constant $\kappa \geqslant 1$ such that

$$\int\limits_{\{v \neq 0\}} |\nabla u|^{p(x)} \, dx \leqslant \kappa \int\limits_{\{v \neq 0\}} |\nabla(u + v)|^{p(x)} \, dx$$

for every open set $D \subset\subset \Omega$ and for every $v \in W^{1,p(\cdot)}(D)$ with compact support in D.

In the previous section we noted that a constant multiple of a minimizers need not be a minimizer. If u is a quasiminimizer with a constant κ, then $-u$ is also a quasiminimizer with the same constant, and if $\alpha \in \mathbb{R}$ then αu is a quasiminimizer with a constant $\max\{\alpha^{p^+ - p^-}\kappa, \alpha^{p^- - p^+}\kappa\}$.

If $u \in W^{1,p(\cdot)}(\Omega)$ is a quasiminimizer and $\kappa = 1$, then u is a minimizer in a sense of Definition 13.2.1 for its own boundary values. Examples of the quasiminimizers:

- Local minimizers and minimizers with a given boundary value function of

$$\int\limits_{\Omega} \frac{|\nabla u|^{p(\cdot)}}{p(x)} \, dx.$$

These are quasiminimizers with constant $\frac{p^+}{p^-}$.
- Local minimizers of a Carathéodory function with a growing conditions

$$c^{-1}|z|^{p(x)} \leqslant F(x, \nabla u) \leqslant c|z|^{p(x)}.$$

These are quasiminimizers with constant c^2.

Fan and Zhao studied quasiminimizers in [148]. They proved higher integrability for gradients and showed that each quasiminimizer is locally Hölder continuous. Their proofs are based on De Giorgi's method.

Theorem 13.2.11 ([148]). *Let Ω be bounded and $p \in \mathcal{P}^{\log}(\Omega)$ with $1 < p^- \leqslant p^+ < \infty$. Let u be a quasiminimizer. Then u is locally Hölder continuous and $|\nabla u| \in L_{\text{loc}}^{p(\cdot)+\varepsilon}(\Omega)$ for some $\varepsilon > 0$.*

Harjulehto, Kuusi, Lukkari, Marola and Parviainen [210] extend works of Fan and Zhao [147, 148] and showed that De Giorgi's method can be fully adapted to the variable exponent case.

Theorem 13.2.12 (Harnack's inequality, [210]). *Let Ω be bounded and $p \in \mathcal{P}^{\log}(\Omega)$ with $1 < p^- \leqslant p^+ < \infty$. Let u be a nonnegative $p(\cdot)$-quasiminimizer in Ω. Further, we consider only cubes Q so small that $10Q \subset \Omega$,*

$$\int_Q |u|^{p(y)} \, dy \leqslant 1 \ and \ \int_Q |\nabla u|^{p(y)} \, dy \leqslant 1.$$

Then there exists a constant c such that

$$\operatorname*{ess\,sup}_{y \in Q} u(y) \leqslant c \left(\operatorname*{ess\,inf}_{y \in Q} u(y) + \operatorname{diam}(Q) \right).$$

The constant c depends on n, $p(\cdot)$, q, $s > p_{10Q}^+ - p_{10Q}^-$, the quasiminimizing constant and the $L^{ns}(10Q)$-norm of u.

Since p is uniformly continuous we may choose the radius of Q so that $ns \leqslant p^-$, and hence the $L^{ns}(10Q)$-norm of u is finite. Note that Harnack's inequality implies that u is continuous, for the proof see [18]. For bounded quasiminimizers the result can be write in a slightly different form:

$$\operatorname*{ess\,sup}_{y \in Q} u(y) \leqslant c \left(\operatorname*{ess\,inf}_{y \in Q} u(y) + \operatorname{diam}(Q)^{\alpha} \right)$$

for any $\alpha \geqslant 1$, where the constant c depends on n, $p(\cdot)$, q, α, the quasiminimizing constant and the L^{∞}-norm of u [210].

13.3   Harmonic and Superharmonic Functions

The Euler–Lagrange equation of the Dirichlet energy integral minimization problem is the $p(\cdot)$-*Laplace equation*

$$\operatorname{div}(p(\cdot)|\nabla u|^{p(\cdot)-2}\nabla u) = 0.$$

Next we discuss its weak solutions.

Definition 13.3.1. A function $u \in W^{1,p(\cdot)}_{loc}(\Omega)$ is a *(weak) $p(\cdot)$-supersolution* in Ω, if

$$\int_{\Omega} p(x)|\nabla u|^{p(x)-2}\nabla u \cdot \nabla \psi \, dx \geqslant 0$$

for every non-negative test function $\psi \in C_0^{\infty}(\Omega)$. A function u is a *subsolution* in Ω if $-u$ is a supersolution in Ω. A function u is a *(weak) $p(\cdot)$-solution* in Ω if u and $-u$ are supersolutions in Ω.

Note that in some paper, and also in our next section, the test functions are on from $W^{1,p(\cdot)}(\Omega)$ with a compact support in Ω. For a function $u \in W^{1,p(\cdot)}(\Omega)$ these two test classes as well $H_0^{1,p(\cdot)}(\Omega)$ and $W_0^{1,p(\cdot)}(\Omega)$ give the same result provided that smooth functions are dense in the Sobolev space.

If $u \in W^{1,p(\cdot)}(\Omega)$ is a minimizer of the Dirichlet energy integral for given boundary value function or if $u \in W^{1,p(\cdot)}_{\mathrm{loc}}(\Omega)$ is a local minimizer of $|\nabla u|^{p(\cdot)}$ then it is a solution. If u is a solution, then it is a local minimizer and if a solution belongs to $W^{1,p(\cdot)}(\Omega)$, then it is a minimizer of the Dirichlet energy integral for its own boundary values. Minimizers of the Dirichlet energy integral with an obstacle are supersolutions [205].

A more general equation which has also been considered is

$$\mathrm{div}(p(\cdot)|\nabla u|^{p(\cdot)-2}\nabla u) = B(x,u). \tag{13.3.2}$$

Weak solutions are defined analogously. There are several results on the existence of various functions B. As a preliminary result we mention that the equation with homogeneous Dirichlet boundary data has a unique weak solution if $B = B(x) \in L^{(p^*)'(1+\varepsilon)}(\Omega)$ is independent of u [144, Theorem 4.2]. Sanchón and Urbano [349] have shown that the same conclusion holds for entropy solutions even if only $B \in L^1(\Omega)$.

Theorem 13.3.3 (Theorem 4.3, [144]). *Suppose that* $p \in C(\overline{\Omega})$ *and* $|B(x,u)| \leqslant c + c|u|^{p^- - \varepsilon - 1}$. *Then* (13.3.2) *has a weak solution for Dirichlet boundary values* $g \in W^{1,p(\cdot)}(\Omega)$.

To the best of our knowledge, this is the most general result which does not require a largeness assumption on B. Newer results, by contrast, place restrictions on the growth of B at the origin or at ∞; in particular, these results do not include as a special case $B = 0$.

Theorem 13.3.4 (Theorem 4.7, [144]). *Let* $p \in C(\overline{\Omega})$ *with* $1 < p^- \leqslant p^+ < \infty$. *Suppose that the following three conditions hold.*

(a) *$|B(x,u)| \leqslant c + c|u|^{p^*(x)-1-\varepsilon}$ for some $\varepsilon > 0$.*
(b) *There exist $R > 0$ and $\theta > p^+$ such that $0 < \theta \int_0^u B(x,v)\,dv \leqslant u\,B(x,u)$ for all $u \in \mathbb{R} \setminus (-R,R)$ and $x \in \Omega$.*
(c) *$B(x,u) = o\bigl(|u|^{p^+-1}\bigr)$ as $u \to 0$ uniformly in x.*

Then (13.3.2) *has a weak solution in* $W^{1,p(\cdot)}_0(\Omega)$.

In Theorem 4.8 of the same paper it is shown that there exist infinitely many solutions if the third condition is replaced by the assumption that B is odd in the second argument, see also [220]. A variant of this result was proved in [68]: there it is assumed that $B(x,u) = -\lambda(x)|u|^{p(x)-2}u + b(x,u)$, where $\lambda \approx 1$ and b satisfies the same conditions as B in the previously stated theorem.

We now return to the $p(\cdot)$-Laplace equation. Alkhutov showed in [18] that solutions are locally bounded, locally bounded supersolutions satisfy the weak Harnack inequality and locally bounded solutions satisfy Harnack's inequality, see also [391]. Harjulehto, Kinnunen and Lukkari extended his result to unbounded supersolutions. Moser's iteration are used in both papers. The key estimate in Moser's iteration is the Caccioppoli estimate. In the proof of Harnack's inequality the Caccioppoli estimate is used for the function $u + R$, where R is a radius of a ball. The extra term R is used to handle negative powers which comes for putting together the variable exponent Caccioppoli estimate and a constant exponent modular form Sobolev inequality. Several different versions of the Caccioppoli estimate can be found from the literature, see for example [18, Lemma 1.1], [20, Proposition 6.1] and [206, Lemma 5.3].

Theorem 13.3.5 (The weak Harnack inequality, [208]). *Let Ω be bounded and $p \in \mathcal{P}^{\log}(\Omega)$ with $1 < p^- \leqslant p^+ < \infty$. Assume that u is a $p(\cdot)$-supersolution which is nonnegative in a ball $4B \subset \Omega$ and $s > p_{4B}^+ - p_{4B}^-$. Then there exists q_0 such that*

$$\left(\fint_{2B} u^{q_0} \, dx \right)^{\frac{1}{q_0}} \leqslant c \left(\operatorname*{ess\,inf}_B u(x) + \operatorname{diam}(B) \right),$$

where c depend on n, $p(\cdot)$, q and $L^{ns}(4B)$-norm of u.

Since the exponent $p(\cdot)$ is uniformly continuous, we can take for example $ns = p_{4B}^-$ by choosing $4B$ small enough. Thus the constants in the estimates are finite for all supersolutions u on a scale that depends only on $p(\cdot)$.

Combining techniques from [18] and [208] we obtain the following result. Recently Harjulehto, Hästö and Latvala noted that it can be extended to the case $p^- = 1$ [197].

Theorem 13.3.6. *Let Ω be bounded and $p \in \mathcal{P}^{\log}(\Omega)$ with $1 \leqslant p^- \leqslant p^+ < \infty$. Let B be a ball such that $4B \subset\subset \Omega$ and let u be a $p(\cdot)$-solution in Ω. Assume that $s > p_{4B}^+ - p_{4B}^-$. Then*

$$\operatorname*{ess\,sup}_B |u| \leqslant c \left(\left(\fint_{2B} |u|^t \, dx \right)^{\frac{1}{t}} + \operatorname{diam}(B) \right)$$

for every $t > 0$. The constant c depends only on n, p, t and $L^{ns}(4B)$-norm of u.

Theorems 13.3.5 and 13.3.6 yields the following full version of Harnack's inequality.

Theorem 13.3.7 (Harnack's inequality [18, 208]). *Let Ω be bounded and $p \in \mathcal{P}^{\log}(\Omega)$ with $1 < p^- \leqslant p^+ < \infty$. Assume that B is a ball such that $4B \subset\subset \Omega$, and assume that $s > p_{4B}^+ - p_{4B}^-$. Let u be a solution which is nonnegative in $4B$. Then*

$$\sup_{x \in B} u(x) \leqslant c \left(\inf_{x \in B} u(x) + \operatorname{diam}(B) \right),$$

where the constant c depends on n, p and the $L^{ns}(4B)$-norm of u.

The solutions are locally bounded, and hence the dependence of the $L^{ns}(4B)$-norm of u can be replaced by dependence of the supremum of u, as has been done in [18].

This Harnack inequality implies, as pointed out in [18], that solutions are locally Hölder continuous. Since there is the extra diameter term on the right-hand side, the inequality does not imply the strong maximum principle; by the strong maximum principle we mean that a solution can attain neither its minimum nor its maximum. Fan, Zhao and Zhang showed the strong maximum principle for weak solutions of $\operatorname{div}(|\nabla u(x)|^{p(x)-2}\nabla u) = 0$ when $p \in C^1(\overline{\Omega})$ with $1 < p^- \leqslant p^+ < \infty$ [153]. Their proof is based on choosing a suitable test function.

Since the constant in Harnack's inequality depends on the norm of u, many results that follows from Harnack's inequality have slightly different forms than in the constant exponent case even if $p \in \mathcal{P}^{\log}(\Omega)$ with $1 < p^- \leqslant p^+ < \infty$. Assume, for example, that u_i is an increasing sequence of solutions and let u be its point-wise limit. If p is a constant, then u is solution provided it is finite at some point. If p is a variable exponent, then u is solution provided that $u \in L^t_{\text{loc}}$ for some $t > 0$ [205].

Alkhutov and Krashennikova studied boundary regularity of solutions in [20]. They proved a Wiener type capacity condition for boundary regularity. Behavior of solutions up to the boundary have also been studied in [127,277].

Next we define superharmonic functions by the comparison principle.

Definition 13.3.8. We say that a function $u : \Omega \to (-\infty, \infty]$ is $p(\cdot)$-*superharmonic* in Ω if:

(a) u is lower semicontinuous;
(b) u is finite almost everywhere and;
(c) The comparison principle holds: if h is a solution in $D \subset\subset \Omega$, continuous in $\overline{D}$ and $u \geqslant h$ on ∂D, then $u \geqslant h$ in D.

Every $p(\cdot)$-supersolution in Ω which satisfies

$$u(x) = \operatorname{ess\,lim\,inf}_{y \to x} u(y)$$

for every $x \in \Omega$ is $p(\cdot)$-superharmonic in Ω. On the other hand if u is a $p(\cdot)$-superharmonic function, then $\min\{u, \lambda\}$ is a $p(\cdot)$-supersolution for every λ. For the proofs see [205]. Lukkari showed in [275] that the weak solutions of

$$-\operatorname{div}(p(x)|\nabla u(x)|^{p(x)-2}\nabla u) = \mu \quad (\mu \text{ is a finite Radon measure})$$

are $p(\cdot)$-superharmonic. See also [48] for similar equations in the case of a system. Next we list properties of $p(\cdot)$-superharmonic functions. We assume that Ω is bounded and $p \in \mathcal{P}^{\log}(\Omega)$ with $1 < p^- \leqslant p^+ < \infty$.

- Let u be a $p(\cdot)$-superharmonic in Ω and $D \subset\subset \Omega$. Then there exists an increasing sequence (u_i) of continuous $p(\cdot)$-supersolutions converging to u point-wise everywhere in D [205].
- Higher integrability properties of $p(\cdot)$-superharmonic functions and their point-wise defined "gradient" has been studied in [205].
- Superharmonic functions can be point-wise estimated by Wolff's potential [278].
- The balayage is a superharmonic function [263].
- Assume that $u \in L_{loc}^t(\Omega)$, for some $t > 0$. If u is a non-negative $p(\cdot)$-superharmonic function in Ω, then $u < \infty$ $p(\cdot)$-quasieverywhere in Ω [208]. Here the assumption $u \in L_{loc}^t(\Omega)$ is needed to adapt the weak Harnack inequality.
- If u is a $p(\cdot)$-superharmonic function, then it is $p(\cdot)$-quasicontinuous [211].

13.4   Harnack's Inequality for **A**-harmonic Functions

Harnack's inequality, as stated in Theorem 13.3.7 and in the existence literature, is formulated only for weak solutions of $p(\cdot)$-Laplace equation, although the method covers all elliptic equations with Laplace type structural conditions. Hence we prove, by Moser's iteration, Harnack's inequality here. Let $\Omega \subset \mathbb{R}^n$ be an open bounded set. We study elliptic equation of the form

$$- \operatorname{div} \mathsf{A}(x, \nabla u) = 0,$$

where the operator $\mathsf{A} : \Omega \times \mathbb{R}^n \to \mathbb{R}^n$ satisfies the following structural conditions for constants $c_1, c_2 > 0$:

$x \mapsto \mathsf{A}(x, \xi)$ is measurable for all $\xi \in \mathbb{R}^n$.

$\xi \mapsto \mathsf{A}(x, \xi)$ is continuous.

$\mathsf{A}(x, -\xi) = -\mathsf{A}(x, \xi)$ for all $x \in \Omega$ and $\xi \in \mathbb{R}^n$.

$\mathsf{A}(x, \xi) \cdot \xi \geqslant c_1 |\xi|^{p(x)}$ for all $x \in \Omega$ and $\xi \in \mathbb{R}^n$.

$|\mathsf{A}(x, \xi)| \leqslant c_2 |\xi|^{p(x)-1}$ for all $x \in \Omega$ and $\xi \in \mathbb{R}^n$.

$(\mathsf{A}(x, \eta) - \mathsf{A}(x, \xi)) \cdot (\eta - \xi) > 0$ for all $x \in \Omega$ and $\eta, \xi \in \mathbb{R}^n, \eta \neq \xi$.

By choosing c_2 larger, if necessary, we may assume that $c_2 \geqslant c_1$. For example the equations

$$- \operatorname{div}(p(x)|\nabla u|^{p(x)-1}\nabla u) = 0 \text{ and } - \operatorname{div}(|\nabla u|^{p(x)-1}\nabla u) = 0$$

satisfy the above conditions.

Following Definition 13.3.1 we define weak solutions in this case as follows.

Definition 13.4.1. A function $u \in W_{loc}^{1,p(\cdot)}(\Omega)$ is a *(weak)* A-*supersolution* in Ω, if

$$\int_\Omega A(x, \nabla u) \cdot \nabla \psi \, dx \geqslant 0$$

for every non-negative $\psi \in W^{1,p(\cdot)}(\Omega)$ with a compact support in Ω. A function u is a A-*subsolution* in Ω if $-u$ is a supersolution in Ω. A function u is a *(weak)* A-*solution* in Ω if u and $-u$ are A-supersolutions in Ω.

We start by the following technical lemma that is need later.

Lemma 13.4.2. *Let f be a positive measurable function and assume that the exponent $p \in \mathcal{P}^{\log}(\Omega)$ is bounded. Then*

$$\fint_B f^{p_B^+ - p_B^-} \, dx \leqslant c \, \|f\|_{L^s(B)}^{p_B^+ - p_B^-}$$

for any $s > p_B^+ - p_B^-$ and $B \subset \Omega$. Here the constant depends only on the dimension n and $c_{\log}(p)$.

Proof. Let $q := p_B^+ - p_B^-$ and let R be the radius of the ball B. Hölder's inequality implies that

$$\fint_B f^{p_B^+ - p_B^-} \, dx \leqslant \left(\fint_B f^s \, dx \right)^{\frac{q}{s}} = cR^{-\frac{nq}{s}} \|f\|_{L^s(B)}^q.$$

By log-Hölder continuity, $R^{-\frac{q}{s}} \leqslant R^{-q} < c < \infty$ and hence the claim follows.   □

Later we apply Lemma 13.4.2 with $f = u^{q'}$. In this case the upper bound written in terms of u is

$$c\|u\|_{L^{q's}(B)}^{q'(p_B^+ - p_B^-)}.$$

First we show that A-subsolutions are locally bounded above. We fix a ball $B := B(z, R)$ such that $R \leqslant 1$ and $4B \subset\subset \Omega$. We write

$$v := \max\{u, 0\} + R,$$

where u is a A-subsolution; also,

$$\Phi(f, q, A) := \left(\fint_A f^q \, dx \right)^{1/q}$$

for a nonnegative measurable function f and $q \neq 0$.

Lemma 13.4.3. *Let $p \in \mathcal{P}^{\log}(\Omega)$ with $1 < p^- \leqslant p^+ < \infty$. Let $1 \leqslant \tau < \kappa \leqslant 3$. Then*

$$\left(\fint_{\tau B} v^{\beta n' p_{4B}^-} \, dx \right)^{\frac{1}{n'}} \leqslant c \beta^{p_{4B}^-} \left(\frac{\kappa}{\kappa - \tau} \right)^{p_{4B}^+} \fint_{\kappa B} v^{(\beta-1)p_{4B}^- + p(x)} \, dx$$

for $\beta \geqslant 1$. The constant c depends only on n, p^-, p^+, $c_{\log}(p)$ and the structural constants c_1 and c_2.

Proof. We choose $\eta \in C_0^\infty(\kappa B)$ such that $0 \leqslant \eta \leqslant 1$. Let G be a function on $[0, \infty)$ with $G'(t) = \beta t^{\beta - 1}$. The function G_j is defined by the cut-off derivative, $G'_j(t) := \beta \min\{t, j\}^{\beta - 1}$. Fixing the origin, we see that

$$G_j(t) = \begin{cases} t^\beta, & \text{for } 0 \leqslant t \leqslant j, \\ j^\beta + \beta j^{\beta-1}(t - j), & \text{for } t \geqslant j. \end{cases}$$

We further define

$$H_j(\xi) := \int_R^\xi G'_j(t)^{p_{4B}^-} \, dt$$

for $\xi \geqslant R$.

First we show that $\psi := H_j(v)\eta^{p_{4B}^+}$ belongs to $W_0^{1, p(\cdot)}(\Omega)$. Since η has compact support in Ω, it suffices to show that $\psi \in W^{1, p(\cdot)}(\Omega)$. Note that ψ is non-negative, because η and v are. Since

$$|H_j(v)| \leqslant \frac{\beta^{p_{4B}^-}}{(\beta - 1)p_{4B}^- + 1} j^{(\beta-1)p_{4B}^- + 1} + \beta^{p_{4B}^-} j^{(\beta-1)p_{4B}^-} v,$$

we find that $\psi \in L^{p(\cdot)}(\Omega)$. For the gradient we have

$$|\nabla \psi| \leqslant p_{4B}^+ \eta^{p_{4B}^+ - 1} |\nabla \eta| H_j(v) + \eta^{p_{4B}^+} |G'_j(v)|^{p_{4B}^-} |\nabla v|$$
$$\leqslant c(\eta) p_{4B}^+ H_j(v) + \eta^{p_{4B}^+} (\beta j^{\beta-1})^{p_{4B}^-} |\nabla v|,$$

and hence $|\nabla \psi| \in L^{p(\cdot)}(\Omega)$.

Since u is a A-subsolution and ψ is an admissible test function, we have

$$\int_\Omega A(x, -\nabla u) \cdot \nabla \psi \, dx \geqslant 0$$

and furthermore

$$\int_\Omega G_j'(v)^{p_{4B}^-} \eta^{p_{4B}^+} \mathsf{A}(x, \nabla u) \cdot \nabla v \, dx \leqslant \left| \int_\Omega p_{4B}^+ \eta^{p_{4B}^+ - 1} H_j(v) \mathsf{A}(x, \nabla u) \cdot \nabla \eta \, dx \right|$$

$$\leqslant c_2 \int_\Omega p_{4B}^+ \eta^{p_{4B}^+ - 1} H_j(v) |\nabla u|^{p(x)-1} |\nabla \eta| \, dx.$$

Note that $\nabla v = 0$ and $H_j(v) = H_j(R) = 0$ whenever $u \leqslant 0$. If $u > 0$, then $\nabla v = \nabla u$. Hence we obtain

$$c_1 \int_\Omega |\nabla v|^{p(x)} |G_j'(v)|^{p_{4B}^-} \eta^{p_B^+} \, dx \leqslant c_2 \int_\Omega p_{4B}^+ |\nabla v|^{p(x)-1} H_j(v) |\nabla \eta| \eta^{p_{4B}^+ - 1} \, dx.$$

We estimate the integrand on the right-hand side by Young's inequality,

$$ab \leqslant \left(\frac{1}{\varepsilon}\right)^{p-1} a^p + \varepsilon b^{p'},$$

for the exponents $p(x)$ and $p'(x)$. For $p(x) > 1$ this yields that

$$p_{4B}^+ H_j(v) |\nabla v|^{p(x)-1} \eta^{p_{4B}^+ - 1} |\nabla \eta|$$

$$= p_{4B}^+ |G_j'(v)|^{-\frac{p_{4B}^-}{p'(x)}} H_j(v) |\nabla \eta| \eta^{p_{4B}^+ - \frac{p_{4B}^+}{p'(x)} - 1} |G_j'(v)|^{\frac{p_{4B}^-}{p'(x)}} |\nabla v|^{p(x)-1} \eta^{\frac{p_{4B}^+}{p'(x)}}$$

$$\leqslant \varepsilon^{1-p(x)} p_{4B}^{+ \, p(x)} |G_j'(v)|^{-p_{4B}^-(p(x)-1)} H_j(v)^{p(x)} |\nabla \eta|^{p(x)} \eta^{p_{4B}^+ - p(x)}$$

$$+ \varepsilon |G_j'(v)|^{p_{4B}^-} |\nabla v|^{p(x)} \eta^{p_{4B}^+}.$$

Combining this with the previous inequality and using $\eta \leqslant 1$ we obtain

$$c_1 \int_\Omega |\nabla v|^{p(x)} |G_j'(v)|^{p_{4B}^-} \eta^{p_{4B}^+} \, dx$$

$$\leqslant c_2 \varepsilon^{1-p_{4B}^+} \int_\Omega p_{4B}^{+ \, p(x)} |G_j'(v)|^{-p_{4B}^-(p(x)-1)} H_j(v)^{p(x)} |\nabla \eta|^{p(x)} \, dx$$

$$+ \varepsilon c_2 \int_\Omega |G_j'(v)|^{p_{4B}^-} |\nabla v|^{p(x)} \eta^{p_{4B}^+} \, dx.$$

Next we choose ε so small that $\varepsilon c_2 = c_1/2$. Then we can absorb the second integral on the right-hand side into the left-hand side. Thus

$$\int_\Omega |\nabla v|^{p(x)} |G_j'(v)|^{p_{4B}^-} \eta^{p_{4B}^+} \, dx \leqslant c \int_\Omega |G_j'(v)|^{-p_{4B}^-(p(x)-1)} H_j(v)^{p(x)} |\nabla \eta|^{p(x)} \, dx.$$

Then we can use the trivial estimate $|\nabla v|^{p_{4B}^-} \leqslant 1 + |\nabla v|^{p(x)}$ and the previous estimate to derive

$$
\int_\Omega |\nabla v|^{p_{4B}^-} |G_j'(v)|^{p_{4B}^-} \eta^{p_{4B}^+} \, dx
$$

$$
\leqslant \int_\Omega |G'(v)|^{p_{4B}^-} \eta^{p_{4B}^+} \, dx + c \int_\Omega |G_j'(v)|^{-p_{4B}^-(p(x)-1)} H_j(v)^{p(x)} |\nabla \eta|^{p(x)} \, dx.
$$

Since $\eta \leqslant 1$ vanishes outside κB, we get

$$
\fint_{\kappa B} |\nabla(G_j(v)\eta^{p_{4B}^+})|^{p_{4B}^-} \, dx
$$

$$
= \fint_{\kappa B} |G_j(v) p_{4B}^+ \eta^{p_{4B}^+ - 1} \nabla \eta + \eta^{p_{4B}^+} G_j'(v) \nabla v|^{p_{4B}^-} \, dx
$$

$$
\leqslant 2^{p_{4B}^-} p_{4B}^{+\,p_{4B}^-} \fint_{\kappa B} |G_j(v)|^{p_{4B}^-} |\nabla \eta|^{p_{4B}^-} \, dx + 2^{p_{4B}^-} \fint_{\kappa B} |G_j'(v)|^{p_{4B}^-} \, dx
$$

$$
+ 2^{p_{4B}^-} c \fint_{\kappa B} |G_j'(v)|^{-p_{4B}^-(p(x)-1)} H_j(v)^{p(x)} |\nabla \eta|^{p(x)} \, dx.
$$

Next we use the constant exponent Sobolev-Poincaré inequality

$$
\left(\fint_{\kappa B} \left(\frac{|w|}{R} \right)^{n' p_{4B}^-} dx \right)^{\frac{1}{n' p_{4B}^-}} \leqslant c \left(\fint_{\kappa B} |\nabla w|^{p_{4B}^-} \, dx \right)^{\frac{1}{p_{4B}^-}}
$$

with the function $w = G_j(v)\eta^{p_{4B}^+} \in W_0^{1, p_{4B}^-}(\kappa B)$. We obtain that

$$
\left(\fint_{\kappa B} \left(\frac{G_j(v)\eta^{p_{4B}^+}}{R} \right)^{n' p_{4B}^-} dx \right)^{\frac{n-1}{n}} \leqslant c \fint_{\kappa B} |\nabla(G_j(v)\eta^{p_{4B}^+})|^{p_{4B}^-} \, dx
$$

$$
\leqslant c \fint_{\kappa B} |G_j(v)|^{p_{4B}^-} |\nabla \eta|^{p_{4B}^-} \, dx + c \fint_{\kappa B} |G_j'(v)|^{p_{4B}^-} \, dx
$$

$$
+ c \fint_{\kappa B} |G_j'(v)|^{-p_{4B}^-(p(x)-1)} H_j(v)^{p(x)} |\nabla \eta|^{p(x)} \, dx.
$$

Since $G_j \leqslant G$ and $G_j' \leqslant G'$, we may replace the functions G_j on the right-hand side by the function G. Then the right-hand side does not depend on j, and we may use monotone convergence on the left-hand side to conclude that

$$\left(\fint_{\kappa B} \left(\frac{v^{\beta}\eta^{p_{4B}^+}}{R} \right)^{n'p_{4B}^-} dx \right)^{\frac{1}{n'}}$$

$$\leqslant c \fint_{\kappa B} v^{\beta p_{4B}^-} |\nabla \eta|^{p_{4B}^-} dx + c \fint_{\kappa B} \beta^{p_{4B}^-} v^{-p_{4B}^-} v^{\beta p_{4B}^-} dx + I \, ,$$

where I is given as

$$I = c \fint_{\kappa B} \beta^{(1-p(x))p_{4B}^-} v^{(\beta-1)(p_{4B}^- - p(x)p_{4B}^-)} \times$$

$$\times \left(\frac{\beta p_{4B}^-}{(\beta-1)p_{4B}^- + 1} \right)^{p(x)} v^{((\beta-1)p_{4B}^- +1)p(x)} |\nabla \eta|^{p(x)} \, dx.$$

We choose η so that $\eta = 1$ in τB and $|\nabla \eta| \leqslant \frac{c}{R(\kappa-\tau)} \leqslant \frac{c\kappa}{R(\kappa-\tau)}$. Since $v \geqslant R$, we obtain $v^{-p_{4B}^-} \leqslant R^{-p_{4B}^-}$ and

$$v^{\beta p_{4B}^-} = v^{(\beta-1)p_{4B}^- + p(x)} v^{p_{4B}^- - p(x)} \leqslant v^{(\beta-1)p_{4B}^- + p(x)} R^{p_{4B}^- - p(x)}.$$

By the log-Hölder continuity we have $R^{p_{4B}^- - p(x)} \leqslant c$. Thus there is a common integral average over $v^{(\beta-1)p_{4B}^- + p(x)}$ on the right-hand side. Since the measure of τB is comparable with the measure of κB, we can change the average on the left-hand side to the smaller ball. Multiplying both sides of the inequality by $R^{p_{4B}^-}$ now implies

$$\left(\fint_{\tau B} v^{\beta n' p_{4B}^-} dx \right)^{\frac{1}{n'}} \leqslant c \left[\left(\frac{\kappa}{\kappa-\tau} \right)^{p_{4B}^-} + \beta^{p_{4B}^-} + \beta^{p_{4B}^-} \left(\frac{\kappa}{\kappa-\tau} \right)^{p_{4B}^+} R^{p_{4B}^- - p_{4B}^+} \right]$$

$$\times \fint_{\kappa B} v^{(\beta-1)p_{4B}^- + p(x)} \, dx.$$

By the log-Hölder continuity of the exponent, the term $R^{p_{4B}^- - p_{4B}^+}$ is bounded by a constant and hence the claim follows. □

Lemma 13.4.4. *Let $p \in \mathcal{P}^{\log}(\Omega)$ with $1 < p^- \leqslant p^+ < \infty$. Let $1 \leqslant \tau < \kappa \leqslant 3$. Then*

$$\Phi(v, n'\beta, \tau B) \leqslant c^{\frac{1}{\beta}} \beta^{\frac{p_{4B}^-}{\beta}} \left(\frac{r}{r-\varrho} \right)^{\frac{p_{4B}^+}{\beta}} \Phi(v, q\beta, \kappa B)$$

for every $\beta \geqslant p_{4B}^-$, $1 < q < n'$ and $s > p_{4B}^+ - p_{4B}^-$. The constant c depends only on n, p^-, p^+, $c_{\log}(p)$, and the $L^{q's}(4B)$-norm of v and the structural constants c_1 and c_2.

Proof. Replacing β by β/p_{4B}^- in Lemma 13.4.3 we obtain

$$\left(\fint_{\tau B} \left(v^{\frac{\beta}{p_{4B}^-}} \right)^{n' p_{4B}^-} dx \right)^{\frac{1}{n'\beta}} \leqslant \left(c\beta^{p_{4B}^-} \left(\frac{\kappa}{\kappa - \tau} \right)^{p_{4B}^+} \fint_{\kappa B} v^{\left(\frac{\beta}{p_{4B}^-} - 1 \right)p_{4B}^- + p(x)} dx \right)^{\frac{1}{\beta}}.$$

This yields by Hölder's inequality and Lemma 13.4.2 that

$$\left(\fint_{\tau B} v^{\beta n'} dx \right)^{\frac{n-1}{n\beta}}$$

$$\leqslant c^{\frac{1}{\beta}} \beta^{\frac{p_{4B}^-}{\beta}} \left(\frac{\kappa}{\kappa - \tau} \right)^{\frac{p_{4B}^+}{\beta}} \left(\fint_{\kappa B} v^{q'(p(x) - p_{4B}^-)} dx \right)^{\frac{1}{\beta q'}} \left(\fint_{\kappa B} v^{\beta q} dx \right)^{\frac{1}{\beta q}}$$

$$\leqslant c^{\frac{1}{\beta}} \beta^{\frac{p_{4B}^-}{\beta}} \left(\frac{\kappa}{\kappa - \tau} \right)^{\frac{p_{4B}^+}{\beta}} \left(1 + \|v\|_{L^{q's}(4B)}^{q'(p_{4B}^+ - p_{4B}^-)} \right)^{\frac{1}{\beta q'}} \left(\fint_{\kappa B} v^{\beta q} dx \right)^{\frac{1}{\beta q}}.$$

To conclude the claim we include the term $\left(1 + \|v\|_{L^{q's}(4B)}^{q'(p_{4B}^+ - p_{4B}^-)} \right)^{\frac{1}{q'}} \leqslant 1 + \|v\|_{L^{q's}(4B)}^{q'(p_{4B}^+ - p_{4B}^-)}$ into the constant c. $\square$

Theorem 13.4.5. *Let $p \in \mathcal{P}^{\log}(\Omega)$ with $1 < p^- \leqslant p^+ < \infty$. Let B be a ball with a radius $R \leqslant 1$ such that $4B \subset\subset \Omega$ and let u be a A-subsolution in Ω. Assume that $s > p_{4B}^+ - p_{4B}^-$. Then*

$$\operatorname*{ess\,sup}_{B} u \leqslant c \left(\left(\fint_{2B} |u|^t dx \right)^{\frac{1}{t}} + R \right)$$

for every $t > 0$. The constant c depends only on n, p^-, p^+, $c_{\log}(p)$, t, $L^{ns}(4B)$-norm of u and the structural constants c_1 and c_2.

Since the exponent p is uniformly continuous, we can take for example $ns = p_\Omega^-$ by choosing B small enough. Thus the constants in the estimates are finite for all solutions u in a scale that depends only on p.

Proof. By making s slightly smaller if necessary, we may assume that there exists $q \in (1, n')$ such that $\|u\|_{L^{q's}(4B)} < \infty$. Let $1 \leqslant \tau < \kappa \leqslant 3$. For $j = 0, 1, 2, \ldots$, we write $r_j := \tau + 2^{-j}(\kappa - \tau)$ and

$$\xi_j := \left(\frac{n'}{q} \right)^j q p_{4B}^-.$$

By Lemma 13.4.4 with $\beta = \left(\frac{n'}{q}\right)^j p_{4B}^-$ we obtain

$$\Phi(v, \xi_{j+1}, r_{j+1}B) \leqslant c^{\frac{q}{\xi_j}} \xi_j^{\frac{qp_{4B}^-}{\xi_j}} \left(\frac{r_j}{r_j - r_{j+1}}\right)^{\frac{qp_{4B}^+}{\xi_j}} \Phi(v, \xi_j, r_j B).$$

Iterating and letting $j \to \infty$ we find that

$$\operatorname*{ess\,sup}_{\tau B} |v| \leqslant \prod_{j=0}^{\infty} c^{\frac{q}{\xi_j}} \xi_j^{\frac{qp_{4B}^-}{\xi_j}} \left(2^j \frac{\kappa}{\kappa - \tau}\right)^{\frac{qp_{4B}^+}{\xi_j}} \Phi(v, qp_{4B}^-, \kappa B)$$

$$\leqslant c^{\frac{qn}{p_{4B}^-}} (n')^{\frac{qp_{4B}^+}{p_{4B}^-} \sum_{j=0}^{\infty} \frac{j}{(n')^j}} 2^{\frac{qp_{4B}^+}{p_{4B}^-} \sum_{j=0}^{\infty} \frac{j}{(n')^j}} \left(\frac{\kappa}{\kappa - \tau}\right)^{\frac{qnp_{4B}^+}{p_{4B}^-}} \Phi(v, qp_{4B}^-, \kappa B).$$

By the root test the sums in the previous inequality are finite and hence

$$\operatorname*{ess\,sup}_{\tau B} |v| \leqslant c\left(1 - \frac{\tau}{\kappa}\right)^{-\frac{\lambda}{s}} \Phi(v, s, \kappa B), \qquad (13.4.6)$$

where $\lambda := \frac{p_{B_{4R}}^+ \, n'q}{(n'-q)}$ and $s := qp_{B_{4R}}^-$. By Hölder's inequality we see that $\Phi(v, s, \kappa B) \leqslant c\Phi(v, t, \kappa B)$ when $t \geqslant s$.

We then consider $t < s$. Let us show that

$$\operatorname*{ess\,sup}_{B} |v| \leqslant c\, \Phi(v, t, 2B),$$

for any $t \in (0, s)$. We adapt the argument of [280, Corollary 3.10]. Let $\sigma \in (\frac{1}{3}, 1)$. Denote

$$T(\sigma) := \operatorname*{ess\,sup}_{\sigma 2B} |v| \quad \text{and} \quad S(\sigma) := (1 - \sigma)^{\frac{\lambda}{t} - \frac{\lambda}{s}} \Phi(v, s, \sigma 2B).$$

Set $\sigma' := \frac{1+\sigma}{2}$. We rewrite the conclusion of the previous paragraph as

$$T(\sigma) \leqslant c\left(1 - \frac{\sigma}{\sigma'}\right)^{-\frac{\lambda}{s}} \Phi(v, s, \sigma'2B) \approx (1 - \sigma)^{-\frac{\lambda}{s}} \Phi(v, s, \sigma'2B).$$

Since $1 - \sigma' = \frac{1-\sigma}{2}$, we further obtain that

$$T(\sigma) \leqslant c(1 - \sigma)^{-\frac{\lambda}{t}} S(\sigma').$$

Using this in the second step, we estimate

$$\left(\fint_{\sigma 2B} v^s \, dx \right)^{\frac{1}{s}} \leqslant \left(T(\sigma)^{s-t} \fint_{\sigma 2B} v^t \, dx \right)^{\frac{1}{s}}$$

$$\leqslant c(1-\sigma)^{\frac{\lambda}{s} - \frac{\lambda}{t}} S(\sigma')^{1-\frac{t}{s}} \left(\fint_{\sigma 2B} v^t \, dx \right)^{\frac{1}{s}}.$$

Dividing both sides by $(1-\sigma)^{\frac{\lambda}{s} - \frac{\lambda}{t}}$, we obtain

$$S(\sigma) \leqslant cS(\sigma')^{1-\frac{t}{s}} \left(\fint_{\sigma 2B} v^t \, dx \right)^{\frac{1}{s}} \leqslant cS(\sigma')^{1-\frac{t}{s}} \left(\fint_{2B} v^t \, dx \right)^{\frac{1}{s}},$$

where we used $\sigma \approx 1$ in the second step. Iterating this inequality, we find that

$$S(\sigma) \leqslant c \left(\fint_{2B} v^t \, dx \right)^{\frac{1}{s} \sum_j (1-\frac{t}{s})^j} = \left(\fint_{2B} v^t \, dx \right)^{\frac{1}{t}}.$$

We choose $\tau = 1$ and $\kappa = \frac{3}{2}$ in (13.4.6) and $\sigma = \frac{3}{4}$ in the above estimate. Combining these give the claim for the function v.

The same estimate holds also for $-\min\{u, 0\}$, since $-u$ is a solution. Thus the claim follows. $\qquad\square$

If u is a solution then the above theorem holds also for $-\min\{u, 0\}$ and hence we obtain the following corollary by covering $D \subset\subset \Omega$ by a finitely many balls satisfying the conditions of the previous theorem.

Corollary 13.4.7. *Let* $p \in \mathcal{P}^{\log}(\Omega)$ *with* $1 < p^- \leqslant p^+ < \infty$. *Then every* A-*solution is locally bounded.*

Next we show that non-negative A-supersolutions satisfy the weak Harnack inequality. We write

$$v := u + R,$$

where u is a non-negative A-supersolution. Remember that $B = B(z, R)$ is fixed and $4B \subset\subset \Omega$.

We derive a suitable Caccioppoli type estimate with variable exponents.

Lemma 13.4.8 (Caccioppoli estimate). *Let* $p \in \mathcal{P}(\Omega)$ *with* $1 < p^- \leqslant p^+ < \infty$. *Let* E *be a measurable subset of* $4B \subset\subset \Omega$ *and* $\eta \in C_0^\infty(4B)$ *such that* $0 \leqslant \eta \leqslant 1$. *Then for every* $\gamma_0 < 0$ *there is a constant* c *depending only on* p^+, $c_{\log}(p)$, c_1, c_2 *and* γ_0 *such that*

$$\int_E v^{\gamma-1} |\nabla u|^{p_E^-} \eta^{p_{4B}^+} \, dx \leqslant c \int_{4B} \left(\eta^{p_{4B}^+} v^{\gamma-1} + v^{\gamma+p(x)-1} |\nabla \eta|^{p(x)} \right) dx$$

for every $\gamma < \gamma_0 < 0$.

Proof. We want to test with the function $\psi := v^\gamma \eta^{p_{4B}^+}$. Next we show that $\psi \in W_0^{1,p(\cdot)}(4B)$. Since η has compact support in $4B$ it suffices to show that $\psi \in W^{1,p(\cdot)}(\Omega)$. We observe that $\psi \in L^{p(\cdot)}(\Omega)$ since $|v^\gamma| \eta^{p_{4B}^+} \leqslant R^\gamma$. Furthermore, we have

$$|\nabla\psi| \leqslant |\gamma v^{\gamma-1} \eta^{p_{4B}^+} \nabla u + v^\gamma p_{4B}^+ \eta^{p_{4B}^+ - 1} \nabla\eta| \leqslant |\gamma| R^{\gamma-1} |\nabla u| + p_{4B}^+ R^\gamma |\nabla\eta|,$$

from which we conclude that $|\nabla\psi| \in L^{p(\cdot)}(\Omega)$.

Using the fact that u is a A-supersolution and ψ is a nonnegative test function we find that

$$0 \leqslant \int\limits_{4B} \mathsf{A}(x, \nabla u) \cdot \nabla\psi(x)\, dx$$

$$= \int\limits_{4B} \gamma \eta^{p_{4B}^+} v^{\gamma-1} \mathsf{A}(x, \nabla u) \cdot \nabla u\, dx + \int\limits_{4B} p_{4B}^+ v^\gamma \eta^{p_{4B}^+ - 1} \mathsf{A}(x, \nabla u) \cdot \nabla\eta\, dx.$$

Since γ is a negative number this implies by the structural conditions that

$$|\gamma_0| c_1 \int\limits_{4B} \eta^{p_{4B}^+} v^{\gamma-1} |\nabla u|^{p(x)}\, dx \leqslant p_{4B}^+ c_2 \int\limits_{4B} v^\gamma \eta^{p_{4B}^+ - 1} |\nabla u|^{p(x)-1} |\nabla\eta|\, dx.$$

We denote the right-hand side of the previous inequality by I. Using Young's inequality, $0 < \varepsilon \leqslant 1$, we obtain

$$I \leqslant p_{4B}^+ c_2 \int\limits_{4B} \left(\frac{1}{\varepsilon}\right)^{p(x)-1} \left(v^{\frac{\gamma+p(x)-1}{p(x)}} |\nabla\eta| \eta^{p_{4B}^+ - \frac{p_{4B}^+}{p'(x)} - 1}\right)^{p(x)}$$

$$+ \varepsilon \left(|\nabla u|^{p(x)-1} \eta^{\frac{p_{4B}^+}{p'(x)}} v^{\gamma - \frac{\gamma+p(x)-1}{p(x)}}\right)^{p'(x)}\, dx$$

$$\leqslant p_{4B}^+ c_2 \left(\frac{1}{\varepsilon}\right)^{p_{4B}^+ - 1} \int\limits_{4B} v^{\gamma+p(x)-1} |\nabla\eta|^{p(x)} \eta^{p_{4B}^+ - p(x)}\, dx$$

$$+ p_{4B}^+ c_2 \varepsilon \int\limits_{4B} |\nabla u|^{p(x)} \eta^{p_{4B}^+} v^{\gamma-1}\, dx.$$

By combining these inequalities we arrive at

$$|\gamma_0| c_1 \int\limits_{4B} |\nabla u|^{p(x)} \eta^{p_{4B}^+} v^{\gamma-1}\, dx$$

$$\leqslant p_{4B}^+ c_2 \left(\frac{1}{\varepsilon}\right)^{p_{4B}^+ - 1} \int\limits_{4B} v^{\gamma+p(x)-1} |\nabla\eta|^{p(x)} \eta^{p_{4B}^+ - p(x)}\, dx$$

$$+ p_{4B}^+ c_2 \varepsilon \int\limits_{4B} |\nabla u|^{p(x)} \eta^{p_{4B}^+} v^{\gamma-1}\, dx.$$

By choosing

$$\varepsilon = \min\left\{1, \frac{|\gamma_0|c_1}{2p_{4B}^+c_2}\right\}$$

we can absorb the last term to the left-hand side and obtain

$$\int_{4B} |\nabla u|^{p(x)}\eta^{p_{4B}^+}v^{\gamma-1}\,dx$$

$$\leqslant p_{4B}^+c_2\left(\frac{2p_{4B}^+c_2}{|\gamma_0|c_1}+1\right)^{p_{4B}^+-1}\frac{2}{|\gamma_0|c_1}\int_{4B} v^{\gamma+p(x)-1}|\nabla\eta|^{p(x)}\,dx.$$

Taking $f = v^{\gamma-1}\eta^{p_{4B}^+}$ and $g = |\nabla u|$ in the point-wise inequality

$$f(x)g(x)^{p_E} \leqslant f(x) + f(x)g(x)^{p(x)}$$

and using the previous inequality we obtain the claim. $\qquad\square$

Lemma 13.4.9. *Let $p \in \mathcal{P}^{\log}(\Omega)$ with $1 < p^- \leqslant p^+ < \infty$. Assume that u is a nonnegative A-supersolution in $4B$ and let $1 \leqslant \tau < \kappa \leqslant 3$. Then*

$$\Phi(v, q\beta, \kappa B) \leqslant c^{\frac{1}{|\beta|}}(1+|\beta|)^{\frac{p_{4B}^+}{|\beta|}}\left(\frac{\kappa}{\kappa-\tau}\right)^{\frac{p_{4B}^+}{|\beta|}}\Phi(v, n'\beta, \tau B)$$

for every $\beta < 0$ and $1 < q < n'$. The constant c depends on n, p^-, p^+, $c_{\log}(p)$, the $L^{q's}(4B)$-norm of u with $s > p_{4B}^+ - p_{4B}^-$ and the structural constants c_1 and c_2.

Proof. In the Caccioppoli estimate, Lemma 13.4.8, we take $E = 4B$ and $\gamma = \beta - p_{4B}^- + 1$. Then $\gamma < 1 - p_{4B}^-$ and thus

$$\int_{4B} v^{\beta-p_{4B}^-}|\nabla u|^{p_{4B}^-}\eta^{p_{4B}^+}\,dx \leqslant c\int_{4B}\left(\eta^{p_{4R}^+}v^{\beta-p_{4B}^-}+v^{\beta-p_{4B}^-+p(x)}|\nabla\eta|^{p(x)}\right)dx.$$

Next we take a cutoff function $\eta \in C_0^\infty(\kappa B)$ with $0 \leqslant \eta \leqslant 1$, $\eta = 1$ in τB and

$$|\nabla\eta| \leqslant \frac{c}{R(\kappa-\tau)} \leqslant \frac{c\kappa}{R(\kappa-\tau)}.$$

By the log-Hölder continuity of p we have

$$|\nabla\eta|^{-p(x)} \leqslant cR^{-p(x)}\left(\frac{\kappa}{\kappa-\tau}\right)^{p_{4B}^+} \leqslant cR^{-p_{4B}^-}\left(\frac{\kappa}{\kappa-\tau}\right)^{p_{4B}^+}. \qquad (13.4.10)$$

With this choice of η we have

$$\fint_{\kappa B} \left| \nabla \left(v^{\frac{\beta}{p_{4B}^-}} \eta^{\frac{p_{4B}^+}{p_{4B}^-}} \right) \right|^{p_{4B}^-} dx$$

$$\leqslant c \fint_{\kappa B} |\beta|^{p_{4B}^-} v^{\beta - p_{4B}^-} |\nabla u|^{p_{4B}^-} \eta^{p_{4B}^+} \, dx + c \fint_{\kappa B} v^{\beta} \eta^{p_{4B}^+ - p_{4B}^-} |\nabla \eta|^{p_{4B}^-} \, dx$$

$$\leqslant c |\beta|^{p_{4B}^-} \fint_{\kappa B} \left(\eta^{p_{4B}^+} v^{\beta - p_{4B}^-} + v^{\beta - p_{4B}^- + p(x)} |\nabla \eta|^{p(x)} \right) dx + c \fint_{\kappa B} v^{\beta} |\nabla \eta|^{p_{4B}^-} \, dx$$

$$\leqslant c(1 + |\beta|)^{p_{4B}^+} \left[\fint_{\kappa B} v^{\beta - p_{4B}^-} \, dx + \fint_{\kappa B} v^{\beta - p_{4B}^- + p(x)} |\nabla \eta|^{p(x)} \, dx + \fint_{\kappa B} v^{\beta} |\nabla \eta|^{p_{4B}^-} \, dx \right].$$

Now the goal is to estimate each integrals in the brackets by

$$\left(\fint_{\kappa B} v^{q\beta} \, dx \right)^{1/q}.$$

The first integral can be estimated with Hölder's inequality. Since $v^{-p_{4B}^-} \leqslant R^{-p_{4B}^-}$, we have

$$\fint_{\kappa B} v^{\beta - p_{4B}^-} \, dx \leqslant R^{-p_{4B}^-} \left(\fint_{\kappa B} v^{q\beta} \, dx \right)^{1/q}.$$

By (13.4.10), Hölder's inequality and Lemma 13.4.2 for the second integral we have

$$\fint_{\kappa B} v^{\beta - p_{4B}^- + p(x)} |\nabla \eta|^{p(x)} \, dx$$

$$\leqslant c R^{-p_{4B}^-} \left(\frac{\kappa}{\kappa - \tau} \right)^{p_{4B}^+} \fint_{\kappa B} v^{\beta - p_{4B}^- + p(x)} \, dx$$

$$\leqslant c R^{-p_{4B}^-} \left(\frac{\kappa}{\kappa - \tau} \right)^{p_{4B}^+} \left(\fint_{\kappa B} v^{q'(p(x) - p_{4B}^-)} \, dx \right)^{1/q'} \left(\fint_{\kappa B} v^{q\beta} \, dx \right)^{1/q}$$

$$\leqslant c R^{-p_{4B}^-} \left(\frac{\kappa}{\kappa - \tau} \right)^{p_{4B}^+} \left(1 + \|v\|_{L^{q's}(4B)}^{q'(p_{4B}^+ - p_{4B}^-)} \right)^{1/q'} \left(\fint_{\kappa B} v^{q\beta} \, dx \right)^{1/q}.$$

Finally, for the third integral we obtain the estimate by Hölder's inequality.

Now we have arrived at the inequality

$$\fint_{\kappa B} \left| \nabla \left(v^{\frac{\beta}{p_{4B}^-}} \eta^{\frac{p_{4B}^+}{p_{4B}^-}} \right) \right|^{p_{4B}^-} dx \leqslant c(1+|\beta|)^{p_{4B}^+} R^{-p_{4B}^-} \left(\frac{\kappa}{\kappa-\tau} \right)^{p_{4B}^+} \left(\fint_{\kappa B} v^{q\beta} \, dx \right)^{1/q},$$

where the term $1 + \|v\|_{L^{q's}(4B)}^{q'(p_{4B}^+ - p_{4B}^-)}$ is inside the constant c.

By the constant exponent Sobolev inequality

$$\left(\fint_{\kappa B} |u|^{n'p_{4B}^-} \, dx \right)^{\frac{1}{n'p_{4B}^-}} \leqslant cR \left(\fint_{\kappa B} |\nabla u|^{p_{4B}^-} \, dx \right)^{\frac{1}{p_{4B}^-}},$$

where $u \in W_0^{1,p_{4B}^-}(\kappa B)$, we obtain

$$\left(\fint_{\tau B} v^{\beta n'} \, dx \right)^{\frac{n-1}{n}} \leqslant \left(c \fint_{\kappa B} \left(v^{\frac{\beta}{p_{4B}^-}} \eta^{\frac{p_{4B}^+}{p_{4B}^-}} \right)^{n'p_{4B}^-} dx \right)^{\frac{n-1}{n}}$$

$$\leqslant cR^{p_{4B}^-} \fint_{\kappa B} \left| \nabla \left(v^{\frac{\beta}{p_{4B}^-}} \eta^{\frac{p_{4B}^+}{p_{4B}^-}} \right) \right|^{p_{4B}^-} dx$$

$$\leqslant c(1+|\beta|)^{p_{4B}^+} \left(\frac{\kappa}{\kappa-\tau} \right)^{p_{4B}^+} \left(\fint_{\kappa B} v^{q\beta} \, dx \right)^{1/q},$$

where the term $1 + \|v\|_{L^{q's}(4B)}^{q'(p_{4B}^+ - p_{4B}^-)}$ is inside the constant c. The claim follows from this since β is a negative number. $\square$

The next lemma is the crucial passage from positive exponents to negative exponents in the Moser iteration scheme.

Lemma 13.4.11. *Let $p \in \mathcal{P}^{\log}(\Omega)$ with $1 < p^- \leqslant p^+ < \infty$. Assume that u is a nonnegative supersolution in $4B \subset\subset \Omega$ and $s > p_{4B}^+ - p_{4B}^-$. Then there exist constants $q_0 > 0$ and c depending on n, p^-, p^+, $c_{\log}(p)$ and $L^s(4B)$-norm of u such that*

$$\Phi(v, q_0, 2B) \leqslant c\Phi(v, -q_0, 2B).$$

Proof. Choose a ball B' with a diameter r such that $2B' \subset 4B$ and a cutoff function $\eta \in C_0^\infty(2B')$ such that $\eta = 1$ in B' and $|\nabla \eta| \leqslant c/r$. Taking $E = B'$ and $\gamma = 1 - p_{B'}^-$ in Caccioppoli estimate, Lemma 13.4.8, we have

$$\fint_{B'} |\nabla \log v|^{p_{B'}^-} \, dx \leqslant c \left(\fint_{2B'} v^{-p_{B'}^-} + \fint_{2B'} v^{p(x)-p_{B'}^-} r^{-p(x)} \, dx \right).$$

Using the estimate $v^{-p_{B'}^-} \leqslant R^{-p_{B'}^-} \leqslant cr^{-p_{B'}^-}$, the log-Hölder continuity of p and Lemma 13.4.2 we find that

$$\fint_{B'} |\nabla \log v|^{p_{B'}^-}\, dx \leqslant c\Big(r^{-p_{B'}^-} + r^{-p_{2B'}^-} \fint_{2B'} v^{p(x)-p_{B'}^-}\, dx\Big)$$

$$\leqslant c\big(r^{-p_{B'}^-} + r^{-p_{2B'}^-}(1+\|v\|_{L^s(4B)}^{p_{4B}^+-p_{4B}^-})\big).$$

Let $f := \log v$. By the constant exponent Poincaré inequality, Hölder's inequality and the above estimate we obtain

$$\fint_{B'} |f - f_{B'}|\, dx \leqslant \Big(r^{p_{B'}^-} \fint_{B'} |\nabla f|^{p_{B'}^-}\, dx\Big)^{1/p_{B'}^-}$$

$$\qquad\qquad\qquad (13.4.12)$$

$$\leqslant c\big(1 + r^{p_{B'}^- - p_{2B'}^-}(1 + \|v\|_{L^s(4B)}^{p_{4B}^+-p_{4B}^-})\big)^{1/p_{B'}^-}.$$

Note that $p_{B'}^- \geqslant p_{2B'}^-$ since $B' \subset 2B'$, so that the right-hand side is bounded (and $f \in \mathrm{BMO}(2B)$).

Since the BMO-estimate (13.4.12) holds for all balls $B' \subset 4B$, the measure theoretic John-Nirenberg lemma (see for example [219, Corollary 19.10, p. 371 in Dover's edition] or [280, Theorem 1.66, p. 40]) implies that there exist positive constants c_3 and c_4 depending on the right-hand side of (13.4.12) such that

$$\fint_{2B} e^{c_3|f - f_{2B}|}\, dx \leqslant c_4.$$

Using this we can conclude that

$$\Big(\fint_{2B} e^{c_3 f}\, dx\Big)\Big(\fint_{2B} e^{-c_3 f}\, dx\Big) = \Big(\fint_{2B} e^{c_3(f - f_{2B})}\, dx\Big)\Big(\fint_{2B} e^{-c_3(f - f_{2B})}\, dx\Big)$$

$$\leqslant \Big(\fint_{2B} e^{c_3|f - f_{2B}|}\, dx\Big)^2 \leqslant c_4^2,$$

which implies that

$$\Big(\fint_{2B} v^{c_3}\, dx\Big)^{1/c_3} = \Big(\fint_{2B} e^{c_3 f}\, dx\Big)^{1/c_3} \leqslant c_4^{2/c_3}\Big(\fint_{2B} e^{-c_3 f}\, dx\Big)^{-1/c_3}$$

$$= c_4^{2/c_3}\Big(\fint_{2B} v^{-c_3}\, dx\Big)^{-1/c_3},$$

so that we can take $q_0 = c_3$. $\square$

Note that the exponent q_0 in Lemma 13.4.11 also depends on the $L^s(4B)$-norm of u. More precisely, the constant c_3 obtained from the John-Nirenberg lemma is a universal constant divided by the right-hand side of (13.4.12). Thus

$$q_0 = \frac{c}{c' + \|u\|_{L^s(4B)}^{p_{4B}^+ - p_{4B}^-}}.$$

Theorem 13.4.13 (The weak Harnack inequality). *Let $p \in \mathcal{P}^{\log}(\Omega)$ with $1 < p^- \leqslant p^+ < \infty$. Let B be a ball with a radius $R \leqslant 1$ such that $4B \subset\subset \Omega$. Assume that u is a nonnegative A-supersolution in $4B \subset\subset \Omega$ and $s > p_{4B}^+ - p_{4B}^-$. Then*

$$\left(\fint_{B_{2R}} u^{q_0}\, dx \right)^{1/q_0} \leqslant c \left(\operatorname*{ess\,inf}_{B_R} u(x) + R \right),$$

where q_0 is the exponent from Lemma 13.4.11 and c depends on n, p^-, p^+, $c_{\log}(p)$, q, $L^{ns}(4B)$-norm of u and the structural constants c_1 and c_2.

Since the exponent p is uniformly continuous, we can take for example $ns = p_\Omega^-$ by choosing R small enough. Thus the constants in the estimates are finite for all supersolutions u in a scale that depends only on p.

Proof. By making s slightly smaller if necessary, we may assume that there exists $q \in (1, n')$ such that $\|u\|_{L^{q's}(4B)} < \infty$. Let q_0 be as in the previous lemma, and assume without loss of generality that $q_0 < 1$.

Let $1 \leqslant \tau < \kappa \leqslant 3$, $r_j := \tau + 2^{-j}(\kappa - \tau)$ and

$$\xi_j := -\left(\frac{n'}{q} \right)^j q_0$$

for $j = 0, 1, 2, \ldots$ By Lemma 13.4.9 with $\beta = \frac{\xi_j}{q}$, we have

$$\Phi(v, \xi_j, r_j B) \leqslant c^{\frac{q}{|\xi_j|}} (1 + |\xi_j|)^{\frac{q p_{4B}^+}{|\xi_j|}} \left(\frac{r_j}{r_j - r_{j+1}} \right)^{\frac{q p_{4B}^+}{|\xi_j|}} \Phi(v, \xi_{j+1}, r_{j+1} B).$$

Iterating this inequality, and observing that $1 + |\xi_j| \leqslant 2(\frac{n'}{q})^j$ since $q_0 \leqslant 1$, we obtain

$$\Phi(v, -q_0, \kappa B) \leqslant \prod_{j=0}^{\infty} c^{\frac{q}{|\xi_j|}} (1 + |\xi_j|)^{\frac{q p_{4B}^+}{|\xi_j|}} \left(\frac{r_j}{r_j - r_{j+1}} \right)^{\frac{q p_{4B}^+}{|\xi_j|}} \operatorname*{ess\,inf}_{x \in \tau B} v(x)$$

$$\leqslant c^{\sum_{j=0}^{\infty} \frac{q}{|\xi_j|}} \left(\frac{2n'}{q} \right)^{\sum_{j=0}^{\infty} j \frac{q p_{4B}^+}{|\xi_j|}} \left(\frac{\kappa}{\kappa - \tau} \right)^{\sum_{j=0}^{\infty} \frac{q p_{4B}^+}{|\xi_j|}} \operatorname*{ess\,inf}_{x \in \tau B} v(x).$$

All the series in the sums converge by the root test, so

$$\Phi(v, -q_0, \kappa B) \leqslant c \operatorname*{ess\,inf}_{x \in \tau B} v(x).$$

Next we choose $\tau := 1$ and $\kappa := 2$ and use Lemma 13.4.11 to get the claim.   □

Combining Theorems 13.4.5 and 13.4.13 we obtain the following theorem.

Theorem 13.4.14 (Harnack's inequality). *Let $p \in \mathcal{P}^{\log}(\Omega)$ with $1 < p^- \leqslant p^+ < \infty$. Let B be a ball with a radius $R \leqslant 1$ such that $4B \subset\subset \Omega$. Let u be a non-negative A-solution in $4B$ and $s > p^+_{4B} - p^-_{4B}$. Then*

$$\operatorname*{ess\,sup}_{x \in B} u(x) \leqslant c \left(\operatorname*{ess\,inf}_{x \in B} u(x) + R \right),$$

where the constant c depends on n, p^-, p^+, $c_{\log}(p)$, the $L^{ns}(4B)$-norm of u and the structural constants c_1 and c_2.

Chapter 14
PDEs and Fluid Dynamics

We use the theory of Calderón–Zygmund operators to prove regularity results for the Poisson problem and the Stokes problem, to show the solvability of the divergence equation, and to prove Korn's inequality. These problems belong to the most classical problems treated in the theory of partial differential equations and fluid dynamics. It turns out that the treatment, especially of the whole space problems requires the notion of homogeneous Sobolev spaces, which have been studied in Sect. 12.2. The Poisson problem and the Stokes system are studied in the first two sections. After that we study the divergence equation and its consequences. The last section is devoted to the existence theory of electrorheological fluids. This section nicely illustrates how all the previously developed theory is used. Throughout the chapter we assume that $p \in \mathcal{P}^{\log}$.

14.1 Poisson Problem

The Poisson problem is one of the most classical problems treated in the theory of partial differential equations. Beside its importance in itself it is also very often used as an auxiliary problem in the treatment of nonlinear problems. The well established theory for the Poisson problem includes among many other results that under appropriate assumptions on the data there exists a unique strong solution $u \in W^{2,q}(\Omega)$ of the problem, provided $1 < q < \infty$. We want to generalize this result to the setting of Lebesgue spaces with variable exponents. Of course the results below apply to a much larger class of elliptic problems and we have chosen to explain the ideas only for the example of the Poisson problem. Note however, that the proof in the general case has to be modified, since the corresponding fundamental solution does not have the symmetry of the Newton potential and thus the estimates near the boundary have to be derived differently (cf. the treatment of the Stokes problem in the next section). This section is based on [103] and [264, Chap. 3], [101] where all the missing details can be found.

L. Diening et al., *Lebesgue and Sobolev Spaces with Variable Exponents*,
Lecture Notes in Mathematics 2017, DOI 10.1007/978-3-642-18363-8_14,
© Springer-Verlag Berlin Heidelberg 2011

In this section we always assume that Ω is a bounded domain in $\mathbb{R}^n$, $n \geqslant 2$, with $C^{1,1}$-boundary. We want to show that the *Poisson problem*

$$
\begin{aligned}
-\Delta u &= f && \text{in } \Omega, \\
u &= g && \text{on } \partial\Omega
\end{aligned}
\tag{14.1.1}
$$

possesses a *strong solution* u, i.e. $u \in W^{2,p(\cdot)}(\Omega)$ satisfies (14.1.1) almost everywhere, provided that the data have appropriate regularity. More precisely we prove:

Theorem 14.1.2. *Let $\Omega \subset \mathbb{R}^n$, $n \geqslant 2$, be a bounded domain with $C^{1,1}$-boundary and let $p \in \mathcal{P}^{\log}(\Omega)$ with $1 < p^- \leqslant p^+ < \infty$. For arbitrary data $f \in L^{p(\cdot)}(\Omega)$ and $g \in \mathrm{Tr}(W^{2,p(\cdot)}(\Omega))$ there exists a unique strong solution $u \in W^{2,p(\cdot)}(\Omega)$ of the Poisson problem (14.1.1) which satisfies the estimate*

$$
\|u\|_{W^{2,p(\cdot)}(\Omega)} \leqslant c \left(\|f\|_{L^{p(\cdot)}(\Omega)} + \|g\|_{\mathrm{Tr}(W^{2,p(\cdot)}(\Omega))} \right),
$$

where the constant c depends only on the domain Ω and the exponent p.

Due to the linearity of the problem and the assumption on g it is sufficient to treat the case $g = 0$. Indeed, let $v \in W^{2,p(\cdot)}(\Omega) \cap W_0^{1,p(\cdot)}(\Omega)$ be the solution of the problem $-\Delta u = f + \Delta \bar{g}$ in Ω with a corresponding estimate in terms of the right-hand side, where $\bar{g} \in W^{2,p(\cdot)}(\Omega)$ is a suitable realization of $g \in \mathrm{Tr}(W^{2,p(\cdot)}(\Omega))$ (cf. Sect. 12.1). Then we see that $u := v + \bar{g}$ satisfies the assertions of Theorem 14.1.2.

In the proof of Theorem 14.1.2 we rely on the fact that the result is well known for constant exponents (cf. [356, Theorem II.9.1] or [355]). With the usual localization technique the problem is reduced to corresponding problems in the whole space $\mathbb{R}^n$ and the half-space $\mathbb{R}^n_>$ with right-hand sides f which have bounded support. Since the structure of the Newton potential is different for $n = 2$ and $n \geqslant 3$, we restrict ourselves to the latter case. The methods presented here can be easily adapted to the case $n = 2$.

We extend the exponent p defined on Ω using Proposition 4.1.7 to an exponent defined on $\mathbb{R}^n$, which we again denote by p. Solutions of the problem

$$
-\Delta u = f \qquad \text{in } \mathbb{R}^n
\tag{14.1.3}
$$

are obtained by convolution of f with the *Newton potential*

$$
K_n(x) = K(x) := \frac{1}{(n-2)|\partial B(0,1)|} \frac{1}{|x|^{n-2}}.
\tag{14.1.4}
$$

It is well-known and easy to see that the second derivatives of the Newton potential satisfy the assumptions of Corollary 7.2.9. Consequently we get:

Proposition 14.1.5. *Let $p \in \mathcal{P}^{\log}(\mathbb{R}^n)$ with $1 < p^- \leqslant p^+ < \infty$ and let $f \in C_{0,0}^{\infty}(\mathbb{R}^n)$. Then the convolution $u := K * f$ solves the problem (14.1.3) and belongs to the space $C^{\infty}(\mathbb{R}^n)$. The first and second weak derivatives have the representations $(i, j = 1, \ldots, n)$*

$$\partial_i u(x) = \int_{\mathbb{R}^n} \partial_{x_i} K(x - y) f(y) \, dy,$$

$$\partial_i \partial_j u(x) = \lim_{\varepsilon \searrow 0} \int_{(B(x,\varepsilon))^c} \partial_{x_i} \partial_{x_j} K(x - y) f(y) \, dy - \frac{1}{n} \delta_{ij} f(x). \tag{14.1.6}$$

Moreover, $u \in D^{(1,2),p(\cdot)}(\mathbb{R}^n)$ and it satisfies the estimates

$$\|\nabla u\|_{L^{p(\cdot)}(\mathbb{R}^n)} \leqslant c \|f\|_{D^{-1,p(\cdot)}(\mathbb{R}^n)},$$

$$\|\nabla^2 u\|_{L^{p(\cdot)}(\mathbb{R}^n)} \leqslant c \|f\|_{L^{p(\cdot)}(\mathbb{R}^n)},$$

with a constant $c = c(p, n)$.

Proof. Since $f \in C_{0,0}^{\infty}$ the formulas (14.1.6), $u \in C^{\infty}(\mathbb{R}^n)$, and that u is a solution of (14.1.3) follow from the classical theory (cf. [304]). The second estimate is a direct consequence of the representation in the second line of (14.1.6) and Corollary 7.2.9. Using the norm conjugate formula (Corollary 3.4.13) and Fubini's theorem we get

$$\|\partial_i u\|_{L^{p(\cdot)}(\mathbb{R}^n)} \leqslant 2 \sup_{\substack{\xi \in C_0^{\infty}(\mathbb{R}^n), \\ \|\xi\|_{L^{p'(\cdot)}(\mathbb{R}^n)} \leqslant 1}} \int_{\mathbb{R}^n} \xi \, \partial_i u \, dx$$

$$= c_n \sup_{\substack{\xi \in C_0^{\infty}(\mathbb{R}^n), \\ \|\xi\|_{L^{p'(\cdot)}(\mathbb{R}^n)} \leqslant 1}} \int_{\mathbb{R}^n} f(y) \underbrace{\int_{\mathbb{R}^n} \partial_{x_i} \frac{1}{|x - y|^{n-2}} \xi(x) \, dx}_{=:\Phi(y)} \, dy.$$

Note that $\Phi \in C^{\infty}(\mathbb{R}^n)$ and that Corollary 7.2.9 yields

$$\|\nabla \Phi\|_{L^{p'(\cdot)}(\mathbb{R}^n)} \leqslant c \|\xi\|_{L^{p'(\cdot)}(\mathbb{R}^n)},$$

with $c = c(p, n)$. Thus $\Phi \in D^{1,p(\cdot)}(\mathbb{R}^n)$ and the first estimate follows from the above calculation. $\square$

Proposition 14.1.5 shows that the linear operator $L \colon C_{0,0}^{\infty}(\mathbb{R}^n) \to C^{\infty}(\mathbb{R}^n)$ defined through

$$L \colon f \mapsto K * f$$

is bounded as an operator from $D^{-1,p(\cdot)}(\mathbb{R}^n)$ into $D^{1,p(\cdot)}(\mathbb{R}^n)$ and from $L^{p(\cdot)}(\mathbb{R}^n)$ into $D^{2,p(\cdot)}(\mathbb{R}^n)$. Due to the density of $C_{0,0}^{\infty}(\mathbb{R}^n)$ in $L^{p(\cdot)}(\mathbb{R}^n)$ and

$D^{-1,p(\cdot)}(\mathbb{R}^n)$ (Propositions 3.4.14 and 12.3.7) we thus can extend the operator L to a bounded operator in these spaces. Moreover, for $f \in L^{p(\cdot)}(\mathbb{R}^n)$ with compact support and $\int_{\mathbb{R}^n} f \, dx = 0$ we know from Lemma 12.3.8 that f also belongs to the space $D^{-1,p(\cdot)}(\mathbb{R}^n)$ and thus both estimates of the previous proposition apply. More precisely we have:

Corollary 14.1.7. *Let $p \in \mathcal{P}^{\log}(\mathbb{R}^n)$ with $1 < p^- \leqslant p^+ < \infty$ and let L be the operator defined above.*

(a) *If $f \in L^{p(\cdot)}(\mathbb{R}^n)$, then $u = Lf \in D^{2,p(\cdot)}(\mathbb{R}^n)$ solves the problem (14.1.3) a.e. and*

$$\|Lf\|_{D^{2,p(\cdot)}(\mathbb{R}^n)} = \|\nabla^2 u\|_{L^{p(\cdot)}(\mathbb{R}^n)} \leqslant c \|f\|_{L^{p(\cdot)}(\mathbb{R}^n)} \qquad (14.1.8)$$

with a constant $c = c(p)$.

(b) *If $f \in D^{-1,p(\cdot)}(\mathbb{R}^n)$ then $u = Lf \in D^{1,p(\cdot)}(\mathbb{R}^n)$ solves the problem (14.1.3) in a distributional sense and*

$$\|Lf\|_{D^{1,p(\cdot)}(\mathbb{R}^n)} = \|\nabla u\|_{L^{p(\cdot)}(\mathbb{R}^n)} \leqslant c \|f\|_{D^{-1,p(\cdot)}(\mathbb{R}^n)}. \qquad (14.1.9)$$

with a constant $c = c(p)$.

(c) *If $f \in L^{p(\cdot)}(\mathbb{R}^n)$ has support in a bounded John domain $A \subset \mathbb{R}^n$ and satisfies $\int_{\mathbb{R}^n} f \, dx = 0$ then $u = Lf \in D^{(1,2),p(\cdot)}(\mathbb{R}^n)$ solves the problem (14.1.3) a.e. and*

$$\begin{aligned}
\|Lf\|_{D^{1,p(\cdot)}(\mathbb{R}^n)} = \|\nabla u\|_{L^{p(\cdot)}(\mathbb{R}^n)} &\leqslant c(p) \|f\|_{D^{-1,p(\cdot)}(\mathbb{R}^n)}, \\
\|Lf\|_{D^{2,p(\cdot)}(\mathbb{R}^n)} = \|\nabla^2 u\|_{L^{p(\cdot)}(\mathbb{R}^n)} &\leqslant c(\operatorname{diam} A, p) \|f\|_{L^{p(\cdot)}(\mathbb{R}^n)}.
\end{aligned} \qquad (14.1.10)$$

Proof. The first two assertions follow immediately from Proposition 14.1.5 and the discussion before the corollary. For the third statement one has to use Lemma 12.3.8 to approximate f by a sequence $(f_k) \subset C_{0,0}^\infty(\mathbb{R}^n)$ in both spaces $L^{p(\cdot)}(\mathbb{R}^n)$ and $D^{-1,p(\cdot)}(\mathbb{R}^n)$ and to observe that the corresponding sequence (Lf_k) is a Cauchy sequence in $D^{1,p(\cdot)}(\mathbb{R}^n)$ and in the closed subspace $D^{(1,2),p(\cdot)}(\mathbb{R}^n)$. The common limit satisfies the first estimate (14.1.10) due to (b), and also $\|Lf\|_{D^{(1,2),p(\cdot)}(\mathbb{R}^n)} \leqslant c (\|f\|_{L^{p(\cdot)}(\mathbb{R}^n)} + \|f\|_{D^{-1,p(\cdot)}(\mathbb{R}^n)})$ due to (a) and (b). This estimate together with Lemma 12.3.8 implies the second estimate in (14.1.10). $\qquad\square$

For $f \in L^{p(\cdot)}(\mathbb{R}^n)$ with compact support we get that f also belongs to $L^q(\mathbb{R}^n)$ for all $q \in (1, \min\{p^-, n/2\})$ and again classical theory of the Poisson problem can be used. In particular, the uniqueness of the Poisson problem follows in the classical theory from the mean value property of harmonic functions and the Theorem of Liouville (cf. [128]). More precisely, if we set $\mathcal{L}(\mathbb{R}^n) := \bigcup_{q \in [1,\infty)} L^q(\mathbb{R}^n)$ we then get that there exists at most one solution $u \in \mathcal{L}(\mathbb{R}^n)$ which satisfies (14.1.3) in a distributional sense. Using this and the previous corollary we get:

Corollary 14.1.11. *Let* $p \in \mathcal{P}^{\log}(\mathbb{R}^n)$ *with* $1 < p^- \leqslant p^+ < \infty$ *and let* $f \in L^{p(\cdot)}(\mathbb{R}^n)$ *have support in a bounded John domain* $A \subset \Omega$ *and satisfy* $\int_{\mathbb{R}^n} f \, dx = 0$. *Let* $u \in \mathcal{L}(\mathbb{R}^n)$ *be a distributional solution of* (14.1.3). *Then* $u \in D^{(1,2),p(\cdot)}(\mathbb{R}^n)$ *satisfies the estimates* (14.1.10).

From the previous results we immediately get interior estimates for the Poisson problem.

Proposition 14.1.12. *Let* Ω *be a bounded domain and let* $p \in \mathcal{P}^{\log}(\Omega)$ *with* $1 < p^- \leqslant p^+ < \infty$, $f \in L^{p(\cdot)}(\Omega)$ *and let* $\Omega_0 \subset\subset \Omega_1 \subset\subset \Omega$. *Moreover, let* $u \in W^{2,p(\cdot)}(\Omega)$ *be a solution of* $-\Delta u = f$ *in* Ω. *Then*

$$
\begin{aligned}
\|\nabla u\|_{L^{p(\cdot)}(\Omega_0)} &\leqslant c \left(\|f\|_{W^{-1,p(\cdot)}(\Omega_1)} + \|u\|_{L^{p(\cdot)}(\Omega_1 \setminus \Omega_0)} \right), \\
\|\nabla^2 u\|_{L^{p(\cdot)}(\Omega_0)} &\leqslant c \left(\|f\|_{L^{p(\cdot)}(\Omega_1)} + \|u\|_{W^{1,p(\cdot)}(\Omega_1 \setminus \Omega_0)} \right),
\end{aligned}
\tag{14.1.13}
$$

with constants $c = c(p, \operatorname{dist}(\Omega_0, \Omega_1), \operatorname{diam}(\Omega))$.

Proof. Without loss of generality we can assume that Ω_1 is a John domain. Let $\tau \in C^\infty(\mathbb{R}^n)$ with $\tau = 1$ in Ω_0 and $\operatorname{spt}(\tau) \subset\subset \Omega_1$. For $\bar{u} := u\tau$ we have

$$
-\Delta \bar{u} = f\tau - 2\nabla u \cdot \nabla \tau - u\Delta \tau =: T \quad \text{in } \mathbb{R}^n.
$$

Since $\bar{u} \in W^{2,p(\cdot)}(\mathbb{R}^n)$, $T \in L^{p(\cdot)}(\mathbb{R}^n)$ has bounded support and satisfies $\int_{\mathbb{R}^n} T \, dx = 0$, we can use Corollary 14.1.11 to get

$$
\begin{aligned}
\|\nabla^2 u\|_{L^{p(\cdot)}(\Omega_0)} &\leqslant \|\nabla^2 \bar{u}\|_{L^{p(\cdot)}(\Omega_1)} \leqslant \|\nabla^2 \bar{u}\|_{L^{p(\cdot)}(\mathbb{R}^n)} \leqslant c \|T\|_{L^{p(\cdot)}(\mathbb{R}^n)} \\
&\leqslant c \left(\|f\|_{L^{p(\cdot)}(\Omega_1)} + \|u\|_{W^{1,p(\cdot)}(\Omega_1 \setminus \Omega_0)} \right),
\end{aligned}
$$

which proves the second estimate in (14.1.13), and

$$
\|\nabla u\|_{L^{p(\cdot)}(\Omega_0)} \leqslant \|\nabla \bar{u}\|_{L^{p(\cdot)}(\Omega_1)} \leqslant \|\nabla \bar{u}\|_{L^{p(\cdot)}(\mathbb{R}^n)} \leqslant c \|T\|_{D^{-1,p(\cdot)}(\mathbb{R}^n)}.
$$

Since $\int_{\mathbb{R}^n} T \, dx = 0$, the Poincaré inequality (Theorem 8.2.4) implies that $\|\tau(\xi - \xi_{\Omega_1})\|_{W_0^{1,p'(\cdot)}(\Omega_1)} \leqslant c(\nabla \tau) \|\xi\|_{D^{1,p'(\cdot)}(\mathbb{R}^n)}$ for all $\xi \in C_0^\infty(\mathbb{R}^n)$. We use this for the first integral and partial integration in the second one to derive

$$
\|T\|_{D^{-1,p(\cdot)}(\mathbb{R}^n)} = \sup \int_{\Omega_1} T(\xi - \xi_{\Omega_1}) \, dx
\tag{14.1.14}
$$

$$
= \sup \left(\int_{\Omega_1} f\tau \, (\xi - \xi_{\Omega_1}) \, dx + \int_{\Omega_1 \setminus \Omega_0} u\Delta \tau \, (\xi - \xi_{\Omega_1}) + 2u\nabla \tau \cdot \nabla \xi \, dx \right)
$$

$$
\leqslant c \left(\|f\|_{W^{-1,p(\cdot)}(\Omega_1)} + \|u\|_{L^{p(\cdot)}(\Omega_1 \setminus \Omega_0)} \right),
$$

where the suprema are taken over all $\xi \in C_0^\infty(\mathbb{R}^n), \|\xi\|_{D^{1,p'(\cdot)}(\mathbb{R}^n)} \leqslant 1$. The last two estimates prove the first estimate in (14.1.13). $\qquad \square$

Due to the symmetry of the Newton potential we can derive estimates of the Poisson problem in the half-space with an odd reflection from Proposition 14.1.5.

Proposition 14.1.15. *Let* $p \in \mathcal{P}^{\log}(\mathbb{R}^n_{\geqslant})$ *with* $1 < p^- \leqslant p^+ < \infty$ *and let* $f \in C_0^\infty(\mathbb{R}^n_{\geqslant})$. *Let* $\tilde{p}$ *be the even reflection of* p *and* $\tilde{f}$ *be the odd reflection of* f. *Then the convolution* $u := K * \tilde{f}$ *solves the problem* $-\Delta u = f$ *in* $\mathbb{R}^n_{>}$ *and* $u = 0$ *on* $\Sigma := \partial\mathbb{R}^n_{>}$ *and belongs to the space* $C^\infty(\mathbb{R}^n_{\geqslant})$. *Furthermore,* $u \in D^{(1,2),p(\cdot)}(\mathbb{R}^n_{>})$ *satisfies the estimates*

$$
\begin{aligned}
\|\nabla u\|_{L^{p(\cdot)}(\mathbb{R}^n_{>})} &\leqslant c \|f\|_{D^{-1,p(\cdot)}(\mathbb{R}^n_{>})}, \\
\|\nabla^2 u\|_{L^{p(\cdot)}(\mathbb{R}^n_{>})} &\leqslant c \|f\|_{L^{p(\cdot)}(\mathbb{R}^n_{>})},
\end{aligned}
\tag{14.1.16}
$$

with constants $c = c(p, n)$.

Proof. Note that $\tilde{f} \in C_{0,0}^\infty(\mathbb{R}^n)$. Thus Proposition 14.1.5 yields that $u \in C^\infty(\mathbb{R}^n_{\geqslant})$ satisfies $-\Delta u = f$ in $\mathbb{R}^n_{>}$. Moreover, Proposition 14.1.5 and the estimates

$$
\begin{aligned}
\|\tilde{f}\|_{L^{\tilde{p}(\cdot)}(\mathbb{R}^n)} &\leqslant 2 \|f\|_{L^{p(\cdot)}(\mathbb{R}^n_{>})}, \\
\|\tilde{f}\|_{D^{-1,\tilde{p}(\cdot)}(\mathbb{R}^n)} &\leqslant 2 \|f\|_{D^{-1,p(\cdot)}(\mathbb{R}^n_{>})}
\end{aligned}
\tag{14.1.17}
$$

prove the estimates (14.1.16). These inequalities can be easily obtained using the definition of the norms and a decomposition of the function in an even and an odd part. Since $\tilde{f}$ is an odd function we see that $u = K * \tilde{f}$ is an odd function too and thus we have $u = 0$ on $\Sigma = \partial\mathbb{R}^n_{>}$. $\qquad \square$

Proposition 14.1.15 shows that the linear operator $H \colon C_0^\infty(\mathbb{R}^n_{>}) \to C^\infty(\mathbb{R}^n_{\geqslant})$ defined through

$$
H \colon f \mapsto K * \tilde{f},
$$

where $\tilde{f}$ is the odd reflection of f, is bounded as an operator from $D^{-1,p(\cdot)}(\mathbb{R}^n_{>})$ into $D^{1,p(\cdot)}(\mathbb{R}^n_{>})$ and from $L^{p(\cdot)}(\mathbb{R}^n_{>})$ into $D^{2,p(\cdot)}(\mathbb{R}^n_{>})$. Due to the density of $C_0^\infty(\mathbb{R}^n_{>})$ in $L^{p(\cdot)}(\mathbb{R}^n_{>})$ and $D^{-1,p(\cdot)}(\mathbb{R}^n_{>})$, respectively, (Corollary 4.6.5, Proposition 12.3.11) we thus can continue the operator L to a bounded operator in these spaces. Moreover, for $f \in L^{p(\cdot)}(\mathbb{R}^n_{>})$ with compact support in $\mathbb{R}^n_{>}$ we know from Lemma 12.3.8 that f also belongs to the space $D^{-1,p(\cdot)}(\mathbb{R}^n_{>})$ and thus both estimates of the previous proposition apply. Thus it is clear that Corollary 14.1.7 also holds for $\mathbb{R}^n_{>}$ if we replace the operator L there by the operator H defined above. Of course the corresponding $u = Hf$ solves the problem $-\Delta u = f$ in $\mathbb{R}^n_{>}$ and $u = 0$ on $\Sigma = \partial\mathbb{R}^n_{>}$. In the statement (c) of Corollary 14.1.7 it is not necessary to require that f has a vanishing integral. However the constant in the second estimate in (14.1.10) depends on A instead of $\operatorname{diam}(A)$ (cf. Lemma 12.3.10).

In analogy to the whole space case we set $\mathcal{L}(\mathbb{R}_>^n) := \bigcup_{q \in [1,\infty)} L^q(\mathbb{R}_>^n)$ and note that we also have uniqueness for the half-space problem in the appropriate class (cf. [128]). More precisely, a solution $u \in \mathcal{L}(\mathbb{R}_>^n)$ with first and second weak derivatives from $L^1_{\mathrm{loc}}(\mathbb{R}_\geqslant^n)$ of $-\Delta u = f$ in $\mathbb{R}_>^n$ and $u = 0$ on $\Sigma = \partial \mathbb{R}_>^n$ is unique. Thus we obtain in a analogous way to Corollary 14.1.11:

Corollary 14.1.18. *Let* $p \in \mathcal{P}^{\log}(\mathbb{R}_>^n)$ *with* $1 < p^- \leqslant p^+ < \infty$ *and let* $f \in L^{p(\cdot)}(\mathbb{R}_>^n)$ *have support in a bounded John domain* $A \subset \Omega$. *Let* $u \in \mathcal{L}(\mathbb{R}_>^n)$ *with first and second weak derivatives from* $L^1_{\mathrm{loc}}(\mathbb{R}_\geqslant^n)$ *be a solution of the problem* $-\Delta u = f$ *in* $\mathbb{R}_>^n$ *and* $u = 0$ *on* $\Sigma = \partial \mathbb{R}_>^n$. *Then* $u \in D^{(1,2),p(\cdot)}(\mathbb{R}_>^n)$ *satisfies the estimates*

$$\begin{aligned}
\|\nabla u\|_{L^{p(\cdot)}(\mathbb{R}_>^n)} &\leqslant c(p)\, \|f\|_{D^{-1,p(\cdot)}(\mathbb{R}_>^n)}, \\
\|\nabla^2 u\|_{L^{p(\cdot)}(\mathbb{R}_>^n)} &\leqslant c(A,p)\, \|f\|_{L^{p(\cdot)}(\mathbb{R}_>^n)}.
\end{aligned} \tag{14.1.19}$$

Using the half-space estimates and a usual localization procedure we get now estimates near the boundary for the Poisson problem. Recall that the boundary $\partial\Omega$ of a domain with boundary $\partial\Omega \in C^{1,1}$ is locally described by a function $a \in C^{1,1}([-\alpha, \alpha]^{n-1})$, $\alpha > 0$. One can always choose the function a such that $a(0) = \nabla a(0) = 0$. Moreover, the sets $V = V_+ := \{(x', x_n) \in \mathbb{R}^n : |x'| < \alpha, a(x') < x_n < a(x') + \beta\}$, $\beta > 0$, and $V' = V'_+ := \{(x', x_n) \in \mathbb{R}^n : |x'| < \alpha', a(x') < x_n < a(x') + \beta'\}$, where $0 < \beta < \beta'$, $0 < \alpha < \alpha'$, satisfy $V \subset V' \subset \Omega$.

Proposition 14.1.20. *Let* Ω *be a bounded domain with* $C^{1,1}$-*boundary, let* $p \in \mathcal{P}^{\log}(\Omega)$ *with* $1 < p^- \leqslant p^+ < \infty$, $f \in L^{p(\cdot)}(\Omega)$, *and let* $u \in W^{2,p(\cdot)}(\Omega)$ *be a strong solution of* (14.1.1) *with* $g = 0$. *Then*

$$\begin{aligned}
\|\nabla u\|_{L^{p(\cdot)}(V)} &\leqslant c\left(\|f\|_{W^{-1,p(\cdot)}(V')} + \|u\|_{L^{p(\cdot)}(V')}\right), \\
\|\nabla^2 u\|_{L^{p(\cdot)}(V)} &\leqslant c\left(\|f\|_{L^{p(\cdot)}(V')} + \|u\|_{W^{1,p(\cdot)}(V')}\right),
\end{aligned} \tag{14.1.21}$$

with constants $c = c(p, \Omega, V', V)$, *where* V *and* V' *were defined above.*

Proof. We define V'' analogously with $0 < \alpha < \alpha'' < \alpha'$ and $0 < \beta < \beta'' < \beta'$. Let $\tau \in C^\infty(\overline{\Omega})$ satisfy $\tau = 1$ in V and $\tau = 0$ outside of V''. Let us straighten the boundary with the help of the coordinate transformation $\mathbf{F} : V' \to \mathbf{F}(V') =: \widehat{V}' \subset \mathbb{R}^n$, where $(y', y_n) := \mathbf{F}(x', x_n) := (x', x_n - a(x'))$ (cf. Fig. 14.1). We set $\widehat{V} := \mathbf{F}(V)$, $\widehat{\tau} := \tau \circ \mathbf{F}^{-1}$, and analogously $\widehat{u}, \widehat{f}, \widehat{p}$. One easily computes

$$\|\widehat{f}\|_{L^{\widehat{p}(\cdot)}(\widehat{V}')} = \|f\|_{L^{p(\cdot)}(V')},$$

so $\widehat{f} \in L^{\widehat{p}(\cdot)}(\mathbb{R}_>^n)$, and $\widehat{f}$ has bounded support. Setting $\bar{u} := \widehat{u}\,\widehat{\tau}$ one checks analogously that $\bar{u} \in W^{2,\widehat{p}(\cdot)}(\mathbb{R}_>^n)$ and $\bar{u} = 0$ on Σ. A straightforward computation, using the properties of the transformation $\mathbf{F}$, yields

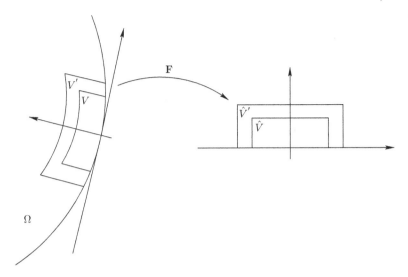

Fig. 14.1 The mapping **F**

$$\Delta\bar{u} = -\widehat{\tau}\widehat{f} + \sum_{i=1}^{n-1}\partial_{in}^2\bar{u}2\partial_i a - \partial_n^2\bar{u}\sum_{i=1}^{n-1}(\partial_i a)^2 + \sum_{i=1}^{n-1}\partial_i\widehat{u}\left(-2\partial_i a\partial_n\widehat{\tau} + 2\partial_i\widehat{\tau}\right)$$

$$+ \partial_n\widehat{u}\left(2\partial_n\widehat{\tau} + \sum_{i=1}^{n-1}\left(\widehat{\tau}\partial_i^2 a - 2\partial_i a\partial_i\widehat{\tau} + 2\partial_n\widehat{\tau}(\partial_i a)^2\right)\right)$$

$$+ \widehat{u}\sum_{i=1}^{n-1}\left(-2\partial_i a\partial_{in}^2\widehat{\tau} + \partial_n^2\widehat{\tau}(\partial_i a)^2 + \partial_i^2\widehat{\tau}\right)$$

$$=: -\widehat{\tau}\widehat{f} + \sum_{i=1}^{n}A_i\partial_{in}^2\bar{u} + \sum_{i=1}^{n}B_i\partial_i\widehat{u} + C\widehat{u} =: T \quad \text{in } \mathbb{R}^n_>.$$

Note that due to the assumptions on u, τ, and $\partial\Omega$, the function $T \in L^{\widehat{p}(\cdot)}(\mathbb{R}^n_>)$ has support in $\widehat{V}'$ and B_i and C are bounded on $\widehat{V}'$ depending on τ and a. Moreover, due to $\nabla a(0) = 0$, $A := \max\{\|A_i\|_{L^\infty(\widehat{V}')}\}$ tends to 0 for $\alpha \to 0$. Now Corollary 14.1.18 applied to $\bar{u}$ and T yields for sufficiently small A (such that the term can be absorbed in the left-hand side)

$$\|\nabla^2\bar{u}\|_{L^{\widehat{p}(\cdot)}(\widehat{V})} \lesssim \|\nabla^2\bar{u}\|_{L^{\widehat{p}(\cdot)}(\widehat{V}')} \leqslant c\left(\|\widehat{f}\|_{L^{\widehat{p}(\cdot)}(\widehat{V}')} + \|\widehat{u}\|_{W^{1,\widehat{p}(\cdot)}(\widehat{V}')}\right).$$

Transforming back with the help of $\mathbf{F}^{-1}$ we obtain

$$\|\nabla^2 u\|_{L^{p(\cdot)}(V)} \leqslant c\left(\|f\|_{L^{p(\cdot)}(V')} + \|u\|_{W^{1,p(\cdot)}(V')}\right).$$

The first inequality in (14.1.21) is proved analogously if one additionally uses ideas from the proof of (14.1.14) to estimate T in the negative norm. Moreover, one has to use partial integration to get rid of the normal derivative from the function $\bar{u}$ in the term with A_i (note that A_i depends only on y') and to choose again A sufficiently small. This finishes the proof. □

Now we have at our disposal everything what we need for the proof of the main result of this section.

Proof of Theorem 14.1.2. Since $C_0^\infty(\Omega)$ is dense in $L^{p(\cdot)}(\Omega)$ it is sufficient to construct a linear solution operator on $C_0^\infty(\Omega)$, which is bounded from $L^{p(\cdot)}(\Omega)$ into $W^{2,p(\cdot)}(\Omega)$. For $f \in C_0^\infty(\Omega)$ there exists a unique strong solution $u \in W^{2,p^+}(\Omega) \cap W_0^{1,p^+}(\Omega)$ of (14.1.1) with $g = 0$ [356, Theorem 9.1]. This defines the linear solution operator. From the definition of $\partial\Omega \in C^{1,1}$ it follows that there exists a finite covering of $\partial\Omega$ with sets Λ^i. Using the corresponding sets V^i we choose open sets $\Omega_0 \subset\subset \Omega_1 \subset\subset \Omega$ such that $\Omega = \Omega_0 \cup \bigcup_{i=1}^m V^i$. Propositions 14.1.12 and 14.1.20 now imply

$$\|\nabla u\|_{L^{p(\cdot)}(\Omega)} \leqslant \|\nabla u\|_{L^{p(\cdot)}(\Omega_0)} + \sum_i \|\nabla u\|_{L^{p(\cdot)}(V^i)}$$

$$\leqslant c\Big(\|f\|_{W^{-1,p(\cdot)}(\Omega_1)} + \sum_i \|f\|_{W^{-1,p(\cdot)}((V^i)')} + \|u\|_{L^{p(\cdot)}(\Omega)}\Big)$$

$$\leqslant c\left(\|f\|_{W^{-1,p(\cdot)}(\Omega)} + \|u\|_{L^{p(\cdot)}(\Omega)}\right),$$

and consequently

$$\|u\|_{W^{1,p(\cdot)}(\Omega)} \leqslant c\left(\|f\|_{W^{-1,p(\cdot)}(\Omega)} + \|u\|_{L^{p(\cdot)}(\Omega)}\right). \tag{14.1.22}$$

Analogously we obtain

$$\|u\|_{W^{2,p(\cdot)}(\Omega)} \leqslant c\left(\|f\|_{L^{p(\cdot)}(\Omega)} + \|u\|_{W^{1,p(\cdot)}(\Omega)}\right)$$

$$\leqslant c\left(\|f\|_{L^{p(\cdot)}(\Omega)} + \|u\|_{L^{p(\cdot)}(\Omega)}\right),$$

where we also used the previous estimate and $\|f\|_{W^{-1,p(\cdot)}(\Omega)} \leqslant c\|f\|_{L^{p(\cdot)}(\Omega)}$. From a standard contradiction argument, using also the reflexivity of the space $W^{1,p(\cdot)}(\Omega)$ and the compact embedding $W^{1,p(\cdot)}(\Omega) \hookrightarrow\hookrightarrow L^{p(\cdot)}(\Omega)$, we deduce using (14.1.22) that

$$\|u\|_{L^{p(\cdot)}(\Omega)} \leqslant c\|f\|_{W^{-1,p(\cdot)}(\Omega)}.$$

This together with the previous estimate implies the estimate in Theorem 14.1.2 in the case $g = 0$ for $f \in C_0^\infty(\Omega)$. The general result for $f \in L^{p(\cdot)}(\Omega)$, $g = 0$ now follows by continuation, since $C_0^\infty(\Omega)$ is dense in

$L^{p(\cdot)}(\Omega)$. The case of arbitrary $g \in \mathrm{Tr}(W^{2,p(\cdot)}(\Omega))$ follows by the definition of $\mathrm{Tr}(W^{2,p(\cdot)}(\Omega))$ and the linearity of the problem. $\qquad\square$

Remark 14.1.23. Using the estimates in the previous results we also have shown that a unique *weak solution* $u \in W^{1,p(\cdot)}(\Omega)$ of the Poisson problem (14.1.1) exists if the data satisfy $f \in W^{-1,p(\cdot)}(\Omega)$ and $g \in \mathrm{Tr}(W^{1,p(\cdot)}(\Omega))$. This solution satisfies (14.1.1) in the usual weak sense, i.e. for test functions from the space $W_0^{1,p'(\cdot)}(\Omega)$, and satisfies the estimate

$$\|u\|_{W^{1,p(\cdot)}(\Omega)} \leqslant c \left(\|f\|_{W^{-1,p(\cdot)}(\Omega)} + \|g\|_{\mathrm{Tr}(W^{1,p(\cdot)}(\Omega))} \right),$$

with a constant $c = c(p, \Omega)$.

14.2 Stokes Problem

In this section we treat the *Stokes problem* and prove also for this system the existence of strong solutions in spaces with variable exponents. The Stokes problem is of fundamental importance for the mathematical treatment of fluid dynamics. The treatment essentially follows along the lines in the previous section and is based on the Calderón–Zygmund theory of singular integral operators and the Agmon–Douglis–Nirenberg theory of operators in the half-space. In contrast to the previous section one cannot use in the construction of the half-space solutions the symmetry of the kernel. This problem occurs since the solution of the Stokes problem has a given divergence. It can be overcome by a generalization of the classical approach of Agmon–Douglis–Nirenberg (cf. [14, 15]) to the case of spaces with variable exponents. Moreover, one has to notice that an efficient adaptation of this method requires the usage of *homogeneous Sobolev spaces* (cf. Sect. 12.2). This section is based on [103], an improved version of [104, 105] and [264, Chap. 4], [101] where all the missing details can be found. For the classical theory we refer the reader to [67] or [169, Chap. IV].

Also in this section we always assume that Ω is a bounded domain in $\mathbb{R}^n$, $n \geqslant 2$, with $C^{1,1}$-boundary. We want to show that the *Stokes problem*

$$\begin{aligned}
\Delta \mathbf{v} - \nabla \pi &= \mathbf{f} & &\text{in } \Omega, \\
\mathrm{div}\, \mathbf{v} &= g & &\text{in } \Omega, \\
\mathbf{v} &= \mathbf{v}_0 & &\text{on } \partial\Omega,
\end{aligned} \qquad (14.2.1)$$

possesses a unique strong solution $(\mathbf{v}, \pi) \in (W^{2,p(\cdot)}(\Omega))^n \times W^{1,p(\cdot)}(\Omega)$ with $\int_\Omega \pi\, dx = 0$ provided that the data have appropriate regularity. More precisely we prove:

Theorem 14.2.2. *Let $\Omega \subset \mathbb{R}^n$, $n \geqslant 2$, be a bounded domain with $C^{1,1}$-boundary and let $p \in \mathcal{P}^{\log}(\Omega)$ with $1 < p^- \leqslant p^+ < \infty$. For arbitrary data $\mathbf{f} \in (L^{p(\cdot)}(\Omega))^n$, $g \in W^{1,p(\cdot)}(\Omega)$ and $\mathbf{v}_0 \in \mathrm{Tr}(W^{2,p(\cdot)}(\Omega))^n$ satisfying the compatibility condition $\int_\Omega g \, dx = \int_{\partial\Omega} \mathbf{v}_0 \cdot \boldsymbol{\nu} \, d\omega$, where $\boldsymbol{\nu}$ is the outer unit normal to $\partial\Omega$, there exists a unique strong solution $(\mathbf{v}, \pi) \in (W^{2,p(\cdot)}(\Omega))^n \times W^{1,p(\cdot)}(\Omega)$ with $\int_\Omega \pi \, dx = 0$ of the Stokes problem (14.2.1) which satisfies the estimate*

$$\|\mathbf{v}\|_{W^{2,p(\cdot)}(\Omega)} + \|\pi\|_{W^{1,p(\cdot)}(\Omega)}$$
$$\leqslant c \left(\|\mathbf{f}\|_{L^{p(\cdot)}(\Omega)} + \|\mathbf{v}_0\|_{\mathrm{Tr}(W^{2,p(\cdot)}(\Omega))} + \|g\|_{W^{1,p(\cdot)}(\Omega)} \right),$$

where the constant c depends only on the domain Ω and the exponent p.

As in the treatment of the Poisson problem it is sufficient to consider homogeneous boundary conditions, i.e. $\mathbf{v}_0 = \mathbf{0}$. For the general case one again uses the linearity of the problem and modifies the other data by a suitable representant of the boundary condition $\mathbf{v}_0$. The usual localization procedure reduces the problem on bounded domains to the treatment of the problem in the whole space and in the half-space with data having bounded support. For that we extend the exponent p defined on Ω by Proposition 4.1.7 to an exponent defined on $\mathbb{R}^n$, which we again denote by p. Since the structure of the fundamental solutions of the Stokes problem is different for $n = 2$ and $n \geqslant 3$, we restrict ourselves to the latter case. The methods presented here can be easily adapted to treat also the case $n = 2$. We refer the reader to [67] and [169, Chap. IV] for the assertion in the classical case of a constant exponent.

For sufficiently smooth data, the solutions of the problem

$$\begin{aligned} \Delta\mathbf{v} - \nabla\pi &= \mathbf{f} \quad \text{in } \mathbb{R}^n, \\ \operatorname{div}\mathbf{v} &= g \quad \text{in } \mathbb{R}^n, \end{aligned} \tag{14.2.3}$$

are obtained by convolution with the *fundamental solutions* of the Stokes problem $\mathbf{V} = (V^{rl})_{r,l=1,\ldots,n}$ and $\mathbf{Q} = (Q^l)_{l=1,\ldots,n}$ given by

$$V^{rl}(x) := \frac{1}{2|\partial B(0,1)|} \left(\frac{1}{n-2} \frac{\delta_{rl}}{|x|^{n-2}} + \frac{x_r x_l}{|x|^n} \right),$$

and

$$Q^l(x) := \frac{1}{|\partial B(0,1)|} \frac{x_l}{|x|^n}.$$

More precisely, set $\mathbf{F} := \mathbf{f} + \nabla g$. Then $\mathbf{v} := \mathbf{V} * \mathbf{F} + K * \nabla g$, where K is the Newton potential (14.1.4), and $\pi := \mathbf{Q} * \mathbf{F}$ solve (14.2.3). From Sect. 14.1 we already know that $K * \nabla g$ has the desired properties. Thus it is sufficient to consider the convolution of the fundamental solutions of the Stokes problem with given data from $(L^{p(\cdot)}(\mathbb{R}^n))^n$. From the classical theory it is well known that the kernels $\partial_i \partial_j V^{rl}$ and $\partial_i Q^l$ satisfy the assumptions of Corollary 7.2.9. Consequently we get:

Proposition 14.2.4. *Let* $p \in \mathcal{P}^{\log}(\mathbb{R}^n)$ *with* $1 < p^- \leqslant p^+ < \infty$ *and let* $\mathbf{f} \in (C_{0,0}^\infty(\mathbb{R}^n))^n$. *Then the convolutions* $\mathbf{u}(x) := \int_{\mathbb{R}^n} \mathbf{V}(x-y)\mathbf{f}(y)\,dy$ *and* $\varrho(x) := \int_{\mathbb{R}^n} \mathbf{Q}(x-y) \cdot \mathbf{f}(y)\,dy$ *belong to* $(C^\infty(\mathbb{R}^n))^n$. *Moreover, their first and second derivatives have the representations* $(i,j,r = 1,\ldots,n)$

$$\partial_i u_r(x) = \sum_l \int_{\mathbb{R}^n} \partial_{x_i} V^{rl}(x-y) f_l(y)\,dy,$$

$$\partial_i \partial_j u_r(x) = \sum_l \lim_{\varepsilon \searrow 0} \int_{(B(x,\varepsilon))^c} \partial_{x_i} \partial_{x_j} V^{rl}(x-y) f_l(y)\,dy$$

$$+ \frac{1}{n(n+2)}\big(-(n+1)\delta_{ij} f_r(x) + \delta_{ir} f_j(x) + \delta_{rj} f_i(x)\big),$$

$$\partial_i \varrho(x) = \sum_l \lim_{\varepsilon \searrow 0} \int_{(B(x,\varepsilon))^c} \partial_{x_i} Q^l(x-y) f_l(y)\,dy + \frac{1}{n} f_i(x),$$

(14.2.5)

and satisfy the estimates

$$\|\nabla \mathbf{u}\|_{L^{p(\cdot)}(\mathbb{R}^n)} + \|\varrho\|_{L^{p(\cdot)}(\mathbb{R}^n)} \leqslant c\,\|\mathbf{f}\|_{D^{-1,p(\cdot)}(\mathbb{R}^n)},$$

$$\|\nabla^2 \mathbf{u}\|_{L^{p(\cdot)}(\mathbb{R}^n)} + \|\nabla \varrho\|_{L^{p(\cdot)}(\mathbb{R}^n)} \leqslant c\,\|\mathbf{f}\|_{L^{p(\cdot)}(\mathbb{R}^n)},$$

(14.2.6)

with a constant $c = c(p,n)$.

Proof. The proof of the statements follows, as in the proof of Proposition 14.1.5, from the classical theory and the estimates for singular integral operators (Corollary 7.2.9). □

Proposition 14.2.4 shows that the linear operator $U \colon (C_{0,0}^\infty(\mathbb{R}^n))^n \to (C^\infty(\mathbb{R}^n))^n$ defined through

$$U \colon \mathbf{f} \mapsto \mathbf{V} * \mathbf{f}$$

is bounded from $(D^{-1,p(\cdot)}(\mathbb{R}^n))^n$ into $(D^{1,p(\cdot)}(\mathbb{R}^n))^n$ and from $(L^{p(\cdot)}(\mathbb{R}^n))^n$ into $(D^{2,p(\cdot)}(\mathbb{R}^n))^n$. Similarly we see that the operator and $P \colon (C_{0,0}^\infty(\mathbb{R}^n))^n \to (C^\infty(\mathbb{R}^n))^n$ defined through

$$P \colon \mathbf{f} \mapsto \mathbf{Q} * \mathbf{f}$$

is bounded from $(D^{-1,p(\cdot)}(\mathbb{R}^n))^n$ into $(L^{p(\cdot)}(\mathbb{R}^n))^n$ and from $(L^{p(\cdot)}(\mathbb{R}^n))^n$ into $(D^{1,p(\cdot)}(\mathbb{R}^n))^n$. Due to the density of $C_{0,0}^\infty(\mathbb{R}^n)$ in $L^{p(\cdot)}(\mathbb{R}^n)$ and $D^{-1,p(\cdot)}(\mathbb{R}^n)$ (Propositions 3.4.14 and 12.3.7) we thus can continue the operators U, P to bounded operators in these spaces. Moreover, a function $\mathbf{f} \in (L^{p(\cdot)}(\mathbb{R}^n))^n$ with compact support and $\int_{\mathbb{R}^n} \mathbf{f}\,dx = \mathbf{0}$ belongs to $(D^{-1,p(\cdot)}(\mathbb{R}^n))^n$ by Lemma 12.3.8. Thus both estimates of the previous proposition apply. More precisely like Proposition 14.1.7 we can prove:

Corollary 14.2.7. *Let $p \in \mathcal{P}^{\log}(\mathbb{R}^n)$ with $1 < p^- \leqslant p^+ < \infty$ and let U, P be the operators defined above.*

(a) *If $\mathbf{f} \in (L^{p(\cdot)}(\mathbb{R}^n))^n$ then $U\mathbf{f} \in (D^{2,p(\cdot)}(\mathbb{R}^n))^n$ and $P\mathbf{f} \in D^{1,p(\cdot)}(\mathbb{R}^n)$ satisfy the estimate*

$$\|U\mathbf{f}\|_{D^{2,p(\cdot)}(\mathbb{R}^n)} + \|P\mathbf{f}\|_{D^{1,p(\cdot)}(\mathbb{R}^n)} \leqslant c\,\|\mathbf{f}\|_{L^{p(\cdot)}(\mathbb{R}^n)}$$

with a constant $c = c(p, n)$.

(b) *If $\mathbf{f} \in (D^{-1,p(\cdot)}(\mathbb{R}^n))^n$ then $U\mathbf{f} \in (D^{1,p(\cdot)}(\mathbb{R}^n))^n$ and $P\mathbf{f} \in L^{p(\cdot)}(\mathbb{R}^n)$ satisfy the estimate*

$$\|U\mathbf{f}\|_{D^{1,p(\cdot)}(\mathbb{R}^n)} + \|P\mathbf{f}\|_{L^{p(\cdot)}(\mathbb{R}^n)} \leqslant c\,\|\mathbf{f}\|_{D^{-1,p(\cdot)}(\mathbb{R}^n)}$$

with a constant $c = c(p, n)$.

(c) *If $\mathbf{f} \in (L^{p(\cdot)}(\mathbb{R}^n))^n$ has support in a bounded John domain $A \subset \mathbb{R}^n$ and satisfies $\int_{\mathbb{R}^n} \mathbf{f}\,dx = 0$ then $U\mathbf{f} \in (D^{(1,2),p(\cdot)}(\mathbb{R}^n))^n$ and $P\mathbf{f} \in W^{1,p(\cdot)}(\mathbb{R}^n)$ satisfy the estimates*

$$\|U\mathbf{f}\|_{D^{2,p(\cdot)}(\mathbb{R}^n)} + \|P\mathbf{f}\|_{D^{1,p(\cdot)}(\mathbb{R}^n)} \leqslant c(p)\,\|\mathbf{f}\|_{L^{p(\cdot)}(\mathbb{R}^n)}\,,$$

$$\|U\mathbf{f}\|_{D^{1,p(\cdot)}(\mathbb{R}^n)} + \|P\mathbf{f}\|_{L^{p(\cdot)}(\mathbb{R}^n)} \leqslant c(p, \operatorname{diam} A)\,\|\mathbf{f}\|_{D^{-1,p(\cdot)}(\mathbb{R}^n)}\,.$$

Based on this corollary and the representation of a solution of (14.2.3) before Proposition 14.2.4 we obtain:

Corollary 14.2.8. *Let $p \in \mathcal{P}^{\log}(\mathbb{R}^n)$ with $1 < p^- \leqslant p^+ < \infty$ and let $\mathbf{f} \in (L^{p(\cdot)}(\mathbb{R}^n))^n$ and $g \in W^{1,p(\cdot)}(\mathbb{R}^n)$ have support in a bounded John domain $A \subset \mathbb{R}^n$. Moreover, let $\mathbf{f}$ have vanishing mean value and let $(\mathbf{v}, \pi) \in (W^{2,p^-}(\mathbb{R}^n))^n \times W^{1,p^-}(\mathbb{R}^n)$ be a solution of the Stokes problem (14.2.3). Then $\mathbf{v} \in (D^{(1,2),p(\cdot)}(\mathbb{R}^n))^n$ and $\pi \in W^{1,p(\cdot)}(\mathbb{R}^n)$ satisfy the estimates*

$$\|\nabla \mathbf{v}\|_{L^{p(\cdot)}(\mathbb{R}^n)} + \|\pi\|_{L^{p(\cdot)}(\mathbb{R}^n)} \leqslant c\left(\|\mathbf{f}\|_{D^{-1,p(\cdot)}(\mathbb{R}^n)} + \|g\|_{L^{p(\cdot)}(\mathbb{R}^n)}\right),$$

$$\|\nabla^2 \mathbf{v}\|_{L^{p(\cdot)}(\mathbb{R}^n)} + \|\nabla \pi\|_{L^{p(\cdot)}(\mathbb{R}^n)} \leqslant c\left(\|\mathbf{f}\|_{L^{p(\cdot)}(\mathbb{R}^n)} + \|\nabla g\|_{L^{p(\cdot)}(\mathbb{R}^n)}\right), \tag{14.2.9}$$

with a constants $c = c(p, \operatorname{diam} A)$.

Proof. From the properties of $\mathbf{f}$ and g follows that they belong to the spaces $(L^{p(\cdot)}(\mathbb{R}^n))^n \cap (L^{p^-}(\mathbb{R}^n))^n \cap (D^{-1,p(\cdot)}(\mathbb{R}^n))^n \cap (D^{-1,p^-}(\mathbb{R}^n))^n$ and $W^{1,p(\cdot)}(\mathbb{R}^n) \cap W^{1,p^-}(\mathbb{R}^n)$, respectively. This, Corollary 14.2.7(c), Corollary 14.1.7(c), and the obvious estimate $\|\nabla g\|_{D^{-1,p(\cdot)}(\mathbb{R}^n)} \leqslant c\,\|g\|_{L^{p(\cdot)}(\mathbb{R}^n)}$ imply (14.2.9) with $\mathbf{v}$ and π replaced by $\widetilde{\mathbf{v}} := U\mathbf{F} + L\nabla g$, L is the continuation of the convolution with the Newton kernel (Corollary 14.1.7), and $\widetilde{\pi} := P\mathbf{F}$, where $\mathbf{F} := \mathbf{f} + \nabla g$. Note that there holds analogous estimates with $p(\cdot)$ replaced by p^-. Using these estimates, the fact that solutions $(\mathbf{v}, \pi) \in (D^{1,q}(\mathbb{R}^n))^n \times L^q(\mathbb{R}^n)$, $1 < q < \infty$, of the Stokes problem (14.2.3) are unique up to a constant (cf. [169, Theorem IV.2.2]), and the integrability of π and $\widetilde{\pi}$ we obtain (14.2.9). $\qquad\square$

Now we are ready to prove interior estimates for solutions of the Stokes problem.

Proposition 14.2.10. *Let $p \in \mathcal{P}^{\log}(\Omega)$ with $1 < p^- \leqslant p^+ < \infty$ and let $\mathbf{f} \in (L^{p(\cdot)}(\Omega))^n$ and $g \in W^{1,p(\cdot)}(\Omega)$. Let $\Omega_0 \subset\subset \Omega_1 \subset\subset \Omega$ be open sets. Moreover, let $(\mathbf{v}, \pi) \in (W^{2,p(\cdot)}(\Omega))^n \times W^{1,p(\cdot)}(\Omega)$ be a solution of the Stokes problem (14.2.1). Then there exists a constant $c = c(p, \Omega_0, \Omega_1)$ such that $(\mathbf{v}, \pi)$ satisfy the estimates*

$$\|\nabla\mathbf{v}\|_{L^{p(\cdot)}(\Omega_0)} + \|\pi\|_{L^{p(\cdot)}(\Omega_0)}$$
$$\leqslant c\left(\|\mathbf{f}\|_{W^{-1,p(\cdot)}(\Omega_1)} + \|g\|_{L^{p(\cdot)}(\Omega_1)} + \|\mathbf{v}\|_{L^{p(\cdot)}(\Omega_1\setminus\Omega_0)} + \|\pi\|_{W^{-1,p(\cdot)}(\Omega_1\setminus\Omega_0)}\right),$$
$$\|\nabla^2\mathbf{v}\|_{L^{p(\cdot)}(\Omega_0)} + \|\nabla\pi\|_{L^{p(\cdot)}(\Omega_0)}$$
$$\leqslant c\left(\|\mathbf{f}\|_{L^{p(\cdot)}(\Omega_1)} + \|g\|_{W^{1,p(\cdot)}(\Omega_1)} + \|\mathbf{v}\|_{W^{1,p(\cdot)}(\Omega_1\setminus\Omega_0)} + \|\pi\|_{L^{p(\cdot)}(\Omega_1\setminus\Omega_0)}\right).$$

Proof. Let $\tau \in C^\infty(\mathbb{R}^n)$ with $\tau = 1$ in Ω_0 and $\mathrm{spt}(\tau) \subset\subset \Omega_1$. For $\bar{\mathbf{v}} := \mathbf{v}\tau$ and $\bar{\pi} := \pi\tau$ we have

$$\Delta\bar{\mathbf{v}} - \nabla\bar{\pi} = 2\nabla\mathbf{v}\nabla\tau + \Delta\tau\mathbf{v} - \pi\nabla\tau + \mathbf{f}\tau =: \mathbf{T} \quad \text{in } \mathbb{R}^n,$$
$$\mathrm{div}\,\bar{\mathbf{v}} = \mathbf{v}\cdot\nabla\tau + g\tau =: G \quad \text{in } \mathbb{R}^n.$$

Due to the choice of τ we see that $\bar{\mathbf{v}} \in W^{2,p(\cdot)}(\mathbb{R}^n)$, $\bar{\pi} \in W^{1,p(\cdot)}(\mathbb{R}^n)$, $G \in W^{1,p(\cdot)}(\mathbb{R}^n)$ and $\mathbf{T} \in L^{p(\cdot)}(\mathbb{R}^n)$ satisfy the assumptions of Corollary 14.2.8, which yields

$$\|\nabla\mathbf{v}\|_{L^{p(\cdot)}(\Omega_0)} + \|\pi\|_{L^{p(\cdot)}(\Omega_0)} \leqslant c\left(\|\mathbf{T}\|_{D^{-1,p(\cdot)}(\mathbb{R}^n)} + \|G\|_{L^{p(\cdot)}(\Omega_1)}\right),$$
$$\|\nabla^2\mathbf{v}\|_{L^{p(\cdot)}(\Omega_0)} + \|\nabla\pi\|_{L^{p(\cdot)}(\Omega_0)} \leqslant c\left(\|\mathbf{T}\|_{L^{p(\cdot)}(\Omega_1)} + \|\nabla G\|_{L^{p(\cdot)}(\Omega_1)}\right).$$

This estimate and the definition of $\mathbf{T}$ and G immediately imply the above estimates for $\mathbf{v}$ and π. For the estimate of $\|\mathbf{T}\|_{D^{-1,p(\cdot)}(\mathbb{R}^n)}$ we use that $\mathbf{T}$ has a vanishing mean value and proceed as in the estimate (14.1.14). $\qquad\square$

Now we turn our attention to the *Stokes problem in the half-space*

$$\Delta \mathbf{v} - \nabla \pi = \mathbf{f} \quad \text{in } \mathbb{R}^n_>,$$
$$\operatorname{div} \mathbf{v} = g \quad \text{in } \mathbb{R}^n_>, \qquad (14.2.11)$$
$$\mathbf{v} = \mathbf{0} \quad \text{on } \Sigma.$$

In order to derive estimates for this problem we cannot proceed as for the Poisson problem, since the convolution of the fundamental solutions of the Stokes problem with odd reflected data does not satisfy the divergence constraint.

Thus we proceed differently. Namely, we reflect the data in an even manner and produce by convolution a whole space solution $\mathbf{v}$ of (14.2.3). This solution does not satisfy the homogeneous boundary condition $\mathbf{v} = \mathbf{0}$ on $\Sigma = \partial \mathbb{R}^n_>$. To achieve this we add to $\mathbf{v}$ a solution of the problem in the half-space

$$\Delta \mathbf{w} - \nabla \theta = \mathbf{0} \quad \text{in } \mathbb{R}^n_>,$$
$$\operatorname{div} \mathbf{w} = 0 \quad \text{in } \mathbb{R}^n_>, \qquad (14.2.12)$$
$$\mathbf{w} = \mathbf{h} \quad \text{on } \Sigma,$$

with the special choice $\mathbf{h} = -\mathbf{v}|_\Sigma$. In order to obtain appropriate estimates of solutions of (14.2.12) we need to prove the analogue of the famous Agmon, Douglis, Nirenberg result (cf. [14]) for spaces with variable exponents. Before we proceed with the treatment of the Stokes problem in the half-space we derive this result. Recall that $C_0^\infty(\mathbb{R}^n_\geqslant)$ is the space of smooth function with compact support in $\mathbb{R}^n_\geqslant$. Moreover, we set $S := \partial B(0,1)$, $S^{n-2} := \partial B(0,1)^{n-1} \subset \mathbb{R}^{n-1}$, $S_> := S \cap \mathbb{R}^n_>$ and $S_\geqslant := S \cap \mathbb{R}^n_\geqslant$.

Theorem 14.2.13. *Let k be a kernel on $\mathbb{R}^n_\geqslant$ of the form*

$$k(x) = \frac{P(x/|x|)}{|x|^n},$$

where $P \colon S_\geqslant \to \mathbb{R}$ is continuous and satisfies

$$\int_{S^{n-2}} P(x', 0)\, dx' = 0.$$

Assume that k possesses continuous derivatives $\partial_i k$, $i = 1, \ldots, n$, and $\partial_n^2 k$ in $\mathbb{R}^n_\geqslant$, which are bounded on the hemisphere $S_>$. Let $p \in \mathcal{P}^{\log}(\mathbb{R}^n_\geqslant)$ with $1 < p^- \leqslant p^+ < \infty$. For $f \in C_0^\infty(\mathbb{R}^n_\geqslant)$ we define $Hf : \mathbb{R}^n_> \to \mathbb{R}$ through

$$(Hf|_\Sigma)(x) := \int_\Sigma k(x' - y', x_n) f(y', 0)\, dy'. \qquad (14.2.14)$$

Then

$$\|\nabla H f|_\Sigma\|_{L^{p(\cdot)}(\mathbb{R}^n_>)} \leqslant c \,\|\nabla f\|_{L^{p(\cdot)}(\mathbb{R}^n_>)}, \tag{14.2.15}$$

for all $f \in C_0^\infty(\mathbb{R}^n_>)$ with constant $c = c(p, n, P)$. In particular, H defines (by extension) a linear, bounded operator $H \colon D^{1,p(\cdot)}(\mathbb{R}^n_>) \to D^{1,p(\cdot)}(\mathbb{R}^n_>)$.

From the definition of the norm in the trace space $\operatorname{Tr} D^{1,p(\cdot)}(\mathbb{R}^n_>)$ (cf. Sect. 12.2) we immediately get from the previous theorem:

Corollary 14.2.16. *Under the assumptions of the Theorem 14.2.13 the operator H defines a linear, bounded operator $H \colon \operatorname{Tr} D^{1,p(\cdot)}(\mathbb{R}^n_>) \to D^{1,p(\cdot)}(\mathbb{R}^n_>)$.*

The proof of Theorem 14.2.13 makes use of the following consequence of the Calderón–Zygmund theorem (Corollary 7.2.9):

Lemma 14.2.17. *Let $Q \colon \mathbb{R}^n_> \to \mathbb{R}$ be a measurable function, which satisfies*

$$|Q(x)| \leqslant \frac{P(x/|x|)}{|x|^n},$$

where $P \in L^r(S_+)$, $r \in (1, \infty]$, is non-negative. Let $p \in \mathcal{P}^{\log}(\mathbb{R}^n_>)$ with $r' < p^- \leqslant p^+ < \infty$. For $f \in L^{p(\cdot)}(\mathbb{R}^n_>)$ we define $If \colon \mathbb{R}^n_> \to \mathbb{R}$ through

$$If(x) := \int\limits_{\mathbb{R}^n_>} Q(x' - y', x_n + y_n) f(y)\, dy.$$

Then I defines a the linear operator $I \colon L^{p(\cdot)}(\mathbb{R}^n_>) \to L^{p(\cdot)}(\mathbb{R}^n_>)$, which satisfies the estimate

$$\|If\|_{L^{p(\cdot)}(\mathbb{R}^n_>)} \leqslant c \,\|f\|_{L^{p(\cdot)}(\mathbb{R}^n_>)},$$

with a constant $c = c(r, p, n, P)$.

Proof. Note that for $x = (x', x_n) \in \mathbb{R}^n_>$ the kernel $Q(x' - \cdot, x_n + \cdot)$ belongs to $L^{p'(\cdot)}(\mathbb{R}^n_>)$ and thus If is well defined. We extend p by even reflection, $Q(x)$ and $\frac{P(x/|x|)}{|x|^n}$ by odd reflection, and f by 0 to $\mathbb{R}^n$. Thus $p \in \mathcal{P}^{\log}(\mathbb{R}^n)$ and $f \in L^{p(\cdot)}(\mathbb{R}^n)$. Moreover, the extended kernel $k(x) := \frac{P(x/|x|)}{|x|^n}$ on $\mathbb{R}^n$ satisfies the assumptions of Corollary 7.2.9, which due to

$$|If(x)| \leqslant \int_{\mathbb{R}^n_>} k(x' - y', x_n + y_n)|f(y)|\,dy$$

$$= \int_{\mathbb{R}^n} k(x' - y', x_n + y_n)|f(y)|\,dy$$

$$= \int_{\mathbb{R}^n} k(x' - y', x_n - y_n)|f(y', -y_n)|\,dy =: \tilde{I}f(x)$$

yields the estimate

$$\|If\|_{L^{p(\cdot)}(\mathbb{R}^n_>)} \leqslant \|\tilde{I}f\|_{L^{p(\cdot)}(\mathbb{R}^n)} \leqslant c\,\|f\|_{L^{p(\cdot)}(\mathbb{R}^n_>)},$$

which proves the lemma. $\square$

This lemma is the analogue of [14, Lemma 3.2] which is sufficient to handle the tangential derivatives in the assertion of Theorem 14.2.13. In order to handle also the normal derivative we need the following result:

Lemma 14.2.18. *Let J be a continuous kernel on $\mathbb{R}^n_>$, which is homogeneous of degree $-(n+1)$ on $\mathbb{R}^n_>$, i.e. $J(\alpha x) = \alpha^{-(n+1)} J(x)$, $\alpha > 0$, bounded on $S_>$ and satisfies for all $x_n > 0$*

$$\int_{\mathbb{R}^{n-1}} J(x', x_n)\,dx' = 0. \tag{14.2.19}$$

Let $p \in \mathcal{P}^{\log}(\mathbb{R}^n_>)$ with $1 < p^- \leqslant p^+ < \infty$ and $q \in (1, \infty)$. For $f \in C_0^\infty(\mathbb{R}^n_>)$ we define the function $If \colon \mathbb{R}^n_> \to \mathbb{R}$, through

$$If(x) := \int_{\mathbb{R}^n_>} J(x' - y', x_n + y_n)f(y)\,dy.$$

Then I satisfies

$$\|If\|_{L^{p(\cdot)}(\mathbb{R}^n_>)} \leqslant c\,\|\nabla f\|_{L^{p(\cdot)}(\mathbb{R}^n_>)}$$

for all $f \in C_0^\infty(\mathbb{R}^n_>)$ with a constant $c = c(p, n, J)$. In particular, I defines (by extension) a linear, bounded operator $I \colon D^{1,p(\cdot)}(\mathbb{R}^n_>) \to L^{p(\cdot)}(\mathbb{R}^n_>)$.

Proof. Note that for $x = (x', x_n) \in \mathbb{R}^n_>$ the kernel $J(x' - \cdot, x_n + \cdot)$ belongs to $L^r(\mathbb{R}^n_>)$, $1 \leqslant r \leqslant \infty$, and thus I is well defined. The proof of the assertion of the lemma is identical with the proof of [14, Lemma A.3.1] if one uses Lemma 14.2.17 instead of [14, Lemma 3.2]. The proof is based on the observation that one can assume without loss of generality that J additionally satisfies for $i < n$ and $x_n > 0$

$$\int_{\mathbb{R}^{n-1}} x_i J(x', x_n) \, dx' = 0.$$

Such J can be represented in the form

$$J(x', x_n) = \sum_{i=1}^{n-1} \partial_i \underbrace{\int_{\mathbb{R}^{n-1}} J(y', x_n) \partial_i K_{n-1}(x' - y') \, dy'}_{=:L_i(x', x_n)},$$

where K_{n-1} is the Newton potential in $\mathbb{R}^{n-1}$. Consequently we can write If as

$$If(x) = \sum_{i=1}^{n-1} \int_{\mathbb{R}^n_>} L_i(x' - y', x_n + y_n) \partial_i f(y) \, dy$$

and the assertion follows from Lemma 14.2.17.   □

Now we are ready to prove Theorem 14.2.13.

Proof of Theorem 14.2.13. The proof of this theorem is analogous to the proof of [14, Theorem 3.3]. We extend P to $\mathbb{R}^n_>$ through $P(x) = P(x/|x|)$. From the assumptions on the kernel one easily checks that for $i = 1, \ldots, n$

$$|k(x)| \leqslant \frac{c}{|x|^{n-1}}, \quad |\partial_i k(x)| \leqslant \frac{c}{|x|^n}, \quad |\partial_n^2 k(x)| \leqslant \frac{c}{|x|^{n+1}}. \qquad (14.2.20)$$

For $f \in C_0^\infty(\mathbb{R}^n_\geqslant)$ and $x \in \mathbb{R}^n_>$ we consider the function $g \colon \mathbb{R}^n_\geqslant \to \mathbb{R}$ defined through $g(y) := k(x' - y', x_n + y_n) f(y)$. Since the kernel $k(x' - \cdot, x_n + \cdot)$ and its first derivatives belongs to $L^r(\mathbb{R}^{n-1} \times (0, T))$, $T > 0$, $1 \leqslant r \leqslant \infty$, and due to the assumptions on f we see that $\partial_{y_n} g \in L^1(\mathbb{R}^{n-1} \times (0, T))$. This enables us to derive in the same way as in the proof of [14, Theorem 3.3] the equations

$$-\partial_i(Hf)(x) = \int_{\mathbb{R}^n_>} \big(\partial_{x_i} k(x' - y', x_n + y_n) \partial_n f(y)$$
$$+ \partial_{x_n} k(x' - y', x_n + y_n) \partial_i f(y)\big) \, dy_n dy', \qquad (14.2.21)$$

$$-\partial_n(Hf)(x) = \int_{\mathbb{R}^{n-1}} \int_0^T \big(\partial_{x_n} k(x' - y', x_n + y_n) \partial_n f(y)$$
$$+ \partial_{x_n}^2 k(x' - y', x_n + y_n) f(y)\big) \, dy_n dy' \qquad (14.2.22)$$

for $i = 1, \ldots, n - 1$. Now Lemma 14.2.17 applied to (14.2.21) and to the first term in (14.2.22) and Lemma 14.2.18 applied to second term in (14.2.22) yield the estimate (14.2.15). The remaining assertions follow from the density of $C_0^\infty(\mathbb{R}^n_\geqq)$ in $D^{1,p(\cdot)}(\mathbb{R}^n_\geqq)$. $\square$

Now we have at our disposal the tools we need for the treatment of the Stokes problem (14.2.12) in the half-space. The solutions of this problem are obtained as usual by a convolution of the *fundamental solutions* $\mathbf{Z} = (Z^{rl})_{r,l=1,\ldots,n}$ and z in the half-space, which are given by

$$Z^{rl}(x) := \frac{2}{|B(0,1)|} \frac{x_n x_r x_l}{|x|^{n+2}},$$

and

$$z := \frac{4}{|\partial B(0,1)|} \frac{x_n}{|x|^n},$$

with the boundary data $\mathbf{h}$ from (14.2.12). From the classical theory it is well known that the kernels Z^{rl} and z satisfy the assumptions of Theorem 14.2.13. This is the basis for the half-space estimates for these kernels. The result is formulated in such a way that it fits to the application in Proposition 14.2.25 showing half-space estimates for the Stokes problem. Note that the next lemma can be also shown for data $\mathbf{h} \in D^{(1,2),p(\cdot)}(\mathbb{R}^n_\geqq)$. For that one needs to use the density of $C_0^\infty(\mathbb{R}^n_\geqq)$ in $D^{1,p(\cdot)}(\mathbb{R}^n_\geqq)$ and $D^{(1,2),p(\cdot)}(\mathbb{R}^n_\geqq)$, which holds due to Propositions 12.2.7 and 12.2.12 and Theorem 8.5.12.

Lemma 14.2.23. *Let $p \in \mathcal{P}^{\log}(\mathbb{R}^n)$ with $1 < p^- \leqslant p^+ < \infty$, and $\mathbf{h} \in (D^{(1,2),p(\cdot)}(\mathbb{R}^n))^n$. Then there exists a solution*

$$(\mathbf{w}, \theta) \in (D^{(1,2),p(\cdot)}(\mathbb{R}^n_\geqq))^n \times W^{1,p(\cdot)}(\mathbb{R}^n_\geqq)$$

of the Stokes problem (14.2.12) on the half-space with boundary data $\mathbf{h}|_\Sigma$, which satisfy the estimates

$$\|\nabla \mathbf{w}\|_{L^{p(\cdot)}(\mathbb{R}^n_\geqq)} + \|\theta\|_{L^{p(\cdot)}(\mathbb{R}^n_\geqq)} \leqslant c \|\nabla \mathbf{h}\|_{L^{p(\cdot)}(\mathbb{R}^n_\geqq)}, \tag{14.2.24}$$

$$\|\nabla^2 \mathbf{w}\|_{L^{p(\cdot)}(\mathbb{R}^n_\geqq)} + \|\nabla \theta\|_{L^{p(\cdot)}(\mathbb{R}^n_\geqq)} \leqslant c \left(\|\nabla^2 \mathbf{h}\|_{L^{p(\cdot)}(\mathbb{R}^n_\geqq)} + \|\nabla \mathbf{h}\|_{L^{p(\cdot)}(\mathbb{R}^n_\geqq)} \right),$$

with a constant $c = c(p, n)$.

Proof. Since $C_0^\infty(\mathbb{R}^n)$ is dense in $D^{(1,2),p(\cdot)}(\mathbb{R}^n)$ and $D^{1,p(\cdot)}(\mathbb{R}^n)$ (Propositions 12.2.12 and 12.2.7) it suffices to show the existence of a linear solution operator satisfying the estimates (14.2.24) and then to argue by density simultaneously in $D^{(1,2),p(\cdot)}(\mathbb{R}^n)$ and $D^{1,p(\cdot)}(\mathbb{R}^n)$ (cf. the proof of Corollary 14.1.7). Thus let $\mathbf{h} \in C_0^\infty(\mathbb{R}^n)$. Then the functions $\mathbf{w} : \mathbb{R}^n_\geqq \to \mathbb{R}^n$ and $\theta : \mathbb{R}^n_\geqq \to \mathbb{R}$ defined through

$$\mathbf{w}(x) := \int_{\Sigma} \mathbf{Z}(x' - y', x_n)\mathbf{h}(y', 0)\, dy',$$

$$\theta(x) := -\sum_{i=1}^{n} \partial_i \underbrace{\int_{\Sigma} z(x' - y', x_n)h_i(y', 0)\, dy'}_{=:q_i(x)},$$

are smooth solutions of (14.2.12), which is easily shown as in the classical theory. The first estimate in (14.2.24) now follows from Theorem 14.2.13 applied to $\mathbf{w}$ and q_i, $i = 1, \ldots, n$. In order to prove the second estimate in (14.2.24) we notice that

$$\partial_k \mathbf{w}(x) = \int_{\Sigma} \mathbf{Z}(x' - y', x_n)\partial_k \mathbf{h}(y', 0)\, dy',$$

$$\partial_k \theta(x) = -\sum_i \partial_i \underbrace{\int_{\Sigma} z(x' - y', x_n)\partial_k h_i(y', 0)\, dy'}_{=:q_{ik}(x)}$$

for $1 \leqslant k < n$. Again Theorem 14.2.13 applied to $\partial_k \mathbf{w}$ and q_{ik} gives

$$\|\partial_k \nabla \mathbf{w}\|_{L^{p(\cdot)}(\mathbb{R}^n_{>})} + \|\partial_k \theta\|_{L^{p(\cdot)}(\mathbb{R}^n_{>})} \leqslant c\, \|\nabla^2 \mathbf{h}\|_{L^{p(\cdot)}(\mathbb{R}^n_{>})}.$$

Using the first two equations of (14.2.12) we compute $\partial_n^2 w_n = -\sum_{i=1}^{n-1} \partial_{ni} w_i$, $\partial_n^2 w_i = \partial_i \theta - \sum_{j=1}^{n-1} \partial_j^2 w_i$, $1 \leqslant i < n$, and $\partial_n \theta = \Delta w_n$, which together with the last estimate for $\partial_k \nabla \mathbf{w}$ and $\partial_k \theta$ gives also the missing estimate for $\partial_n^2 \mathbf{w}$ and $\partial_n \theta$. This finishes the proof of the proposition.   □

Using the whole space result Corollary 14.2.8 and the previous lemma we get half-space estimates for the Stokes problem (14.2.11).

Proposition 14.2.25. *Let* $p \in \mathcal{P}^{\log}(\mathbb{R}^n_{>})$ *with* $1 < p^- \leqslant p^+ < \infty$, *and let* $\mathbf{f} \in (L^{p(\cdot)}(\mathbb{R}^n_{>}))^n$ *and* $g \in W^{1,p(\cdot)}(\mathbb{R}^n_{>})$ *have support in a bounded John domain* $A \subset\subset \mathbb{R}^n_{>}$. *Let* $(\mathbf{v}, \pi) \in (W^{2,p^-}(\mathbb{R}^n_{>}))^n \times W^{1,p^-}(\mathbb{R}^n_{>})$ *be a solution of the Stokes problem in the half-space* (14.2.11) *corresponding to the data* $\mathbf{f}$ *and* g. *Then* $\mathbf{v}$ *belongs to the space* $D^{(1,2),p(\cdot)}(\mathbb{R}^n_{>})$ *and* π *belongs to the space* $W^{1,p(\cdot)}(\mathbb{R}^n_{>})$. *Moreover, they satisfy the estimates*

$$\|\nabla \mathbf{v}\|_{L^{p(\cdot)}(\mathbb{R}^n_{>})} + \|\pi\|_{L^{p(\cdot)}(\mathbb{R}^n_{>})} \leqslant c\left(\|\mathbf{f}\|_{D^{-1,p(\cdot)}(\mathbb{R}^n_{>})} + \|g\|_{L^{p(\cdot)}(\mathbb{R}^n_{>})}\right),$$

$$\|\nabla^2 \mathbf{v}\|_{L^{p(\cdot)}(\mathbb{R}^n_{>})} + \|\nabla \pi\|_{L^{p(\cdot)}(\mathbb{R}^n_{>})} \leqslant c\left(\|\mathbf{f}\|_{L^{p(\cdot)}(\mathbb{R}^n_{>})} + \|\nabla g\|_{L^{p(\cdot)}(\mathbb{R}^n_{>})}\right),$$
(14.2.26)

with a constant $c = c(p, A) > 0$.

Proof. From the assumptions on $\mathbf{f}$ and Lemma 12.3.10 we obtain that $\mathbf{f} \in D^{-1,p(\cdot)}(\mathbb{R}^n_>)$. We extend p and g by even reflection, and $\mathbf{f}$ by odd reflection to $\mathbb{R}^n$. Thus $p \in \mathcal{P}^{\log}(\mathbb{R}^n)$, $\mathbf{f} \in L^{p(\cdot)}(\mathbb{R}^n)$ and $g \in W^{1,p(\cdot)}(\mathbb{R}^n)$ with corresponding estimates of the whole space norms by the half-space norms. Moreover, $\mathbf{f}$ has mean value zero and $\mathbf{f}$ and g still have bounded support. Due to Lemma 12.3.8, $\mathbf{f} \in D^{-1,p(\cdot)}(\mathbb{R}^n)$ with an estimates of the whole space norm by the half-space norm (14.1.17). Let $\widetilde{\mathbf{v}} := \mathbf{V} * \mathbf{F} + K * \nabla g$, K being the Newton kernel, and $\widetilde{\pi} := \mathbf{Q} * \mathbf{F}$, where $\mathbf{F} := \mathbf{f} - \nabla g$, be whole space solutions of the Stokes problem (14.2.3) corresponding to the extended data. Corollaries 14.2.8 and 14.1.7 and the above mentioned estimates of the extended data by the original data yield the estimates

$$\|\nabla\widetilde{\mathbf{v}}\|_{L^{p(\cdot)}(\mathbb{R}^n_>)} + \|\widetilde{\pi}\|_{L^{p(\cdot)}(\mathbb{R}^n_>)} \leqslant \|\nabla\widetilde{\mathbf{v}}\|_{L^{p(\cdot)}(\mathbb{R}^n)} + \|\widetilde{\pi}\|_{L^{p(\cdot)}(\mathbb{R}^n)}$$

$$\leqslant c\left(\|\mathbf{f}\|_{D^{-1,p(\cdot)}(\mathbb{R}^n_>)} + \|\nabla g\|_{D^{-1,p(\cdot)}(\mathbb{R}^n_>)}\right),$$

$$\|\nabla^2\widetilde{\mathbf{v}}\|_{L^{p(\cdot)}(\mathbb{R}^n_>)} + \|\nabla\widetilde{\pi}\|_{L^{p(\cdot)}(\mathbb{R}^n_>)} \leqslant \|\nabla^2\widetilde{\mathbf{v}}\|_{L^{p(\cdot)}(\mathbb{R}^n)} + \|\nabla\widetilde{\pi}\|_{L^{p(\cdot)}(\mathbb{R}^n)}$$

$$\leqslant c\left(\|\mathbf{f}\|_{L^{p(\cdot)}(\mathbb{R}^n_>)} + \|\nabla g\|_{L^{p(\cdot)}(\mathbb{R}^n_>)}\right).$$

Thus $\widetilde{\mathbf{v}} \in D^{(1,2),p(\cdot)}(\mathbb{R}^n)$ and Proposition 14.2.23 yields the existence of a solution $(\mathbf{w}, \theta)$ of the Stokes problem in the half-space (14.2.12) with boundary data $-\widetilde{\mathbf{v}}|_\Sigma$ satisfying the estimates

$$\|\nabla\mathbf{w}\|_{L^{p(\cdot)}(\mathbb{R}^n_>)} + \|\theta\|_{L^{p(\cdot)}(\mathbb{R}^n_>)} \leqslant c\|\nabla\widetilde{\mathbf{v}}\|_{L^{p(\cdot)}(\mathbb{R}^n_>)},$$

$$\|\nabla^2\mathbf{w}\|_{L^{p(\cdot)}(\mathbb{R}^n_>)} + \|\nabla\theta\|_{L^{p(\cdot)}(\mathbb{R}^n_>)} \leqslant c\left(\|\nabla^2\widetilde{\mathbf{v}}\|_{L^{p(\cdot)}(\mathbb{R}^n_>)} + \|\nabla\widetilde{\mathbf{v}}\|_{L^{p(\cdot)}(\mathbb{R}^n_>)}\right).$$

These last four estimate together with $\|\nabla g\|_{D^{-1,p(\cdot)}(\mathbb{R}^n_>)} \leqslant c\|g\|_{L^{p(\cdot)}(\mathbb{R}^n_>)}$ and Lemma 12.3.10 applied to $\mathbf{f}$ and ∇g imply that $\bar{\mathbf{v}} := \widetilde{\mathbf{v}} + \mathbf{w}$ and $\bar{\pi} := \widetilde{\pi} + \theta$ satisfy the estimates (14.2.26) and solve the problem

$$\Delta\bar{\mathbf{v}} = \Delta\widetilde{\mathbf{v}} + \Delta\mathbf{w} = \mathbf{f} + \nabla\bar{\pi} \quad \text{in } \mathbb{R}^n_>,$$

$$\operatorname{div}\bar{\mathbf{v}} = g \qquad\qquad \text{in } \mathbb{R}^n_>,$$

$$\bar{\mathbf{v}} = \widetilde{\mathbf{v}} - \widetilde{\mathbf{v}} = 0 \qquad\qquad \text{on } \Sigma.$$

If we replace in the above arguments $p(\cdot)$ by p^- we get that $\bar{\mathbf{v}}$ and $\bar{\pi}$ also satisfy the corresponding estimates (14.2.26) with $p(\cdot)$ replaced by p^-. Due to the uniqueness results in classical spaces in [169, Theorem IV.3.3] and the integrability of π we get that also $\mathbf{v}$ and π satisfy estimates (14.2.26). $\square$

Now we are ready to prove estimates near the boundary for solutions of the Stokes problem (14.2.1) with homogeneous boundary data.

Proposition 14.2.27. *Let Ω be a bounded domain with $C^{1,1}$-boundary, let $p \in \mathcal{P}^{\log}(\Omega)$ with $1 < p^- \leqslant p^+ < \infty$ and let $\mathbf{f} \in (L^{p(\cdot)}(\Omega))^n$ and $g \in W^{1,p(\cdot)}(\Omega)$. Let $(\mathbf{v}, \pi) \in (W^{2,p(\cdot)}(\Omega))^n \times W^{1,p(\cdot)}(\Omega)$ be a solution of the Stokes problem (14.2.1) with $\mathbf{v}_0 = \mathbf{0}$. Let V and V' be as in Proposition 14.1.20. Then there exists a constant $c = c(p, V, V', \Omega)$ such that $(\mathbf{v}, \pi)$ satisfy the estimates*

$$\|\nabla\mathbf{v}\|_{L^{p(\cdot)}(V)} + \|\pi\|_{L^{p(\cdot)}(V)}$$
$$\leqslant c\big(\|\mathbf{f}\|_{W^{-1,p(\cdot)}(V')} + \|g\|_{L^{p(\cdot)}(V')} + \|\mathbf{v}\|_{L^{p(\cdot)}(V')} + \|\pi\|_{W^{-1,p(\cdot)}(V')}\big),$$

$$\|\nabla^2\mathbf{v}\|_{L^{p(\cdot)}(V)} + \|\nabla\pi\|_{L^{p(\cdot)}(V)}$$
$$\leqslant c\big(\|\mathbf{f}\|_{L^{p(\cdot)}(V')} + \|g\|_{W^{1,p(\cdot)}(V')} + \|\mathbf{v}\|_{W^{1,p(\cdot)}(V')} + \|\pi\|_{L^{p(\cdot)}(V')}\big).$$

Proof. The proof of this proposition is analogous to the proof of Proposition 14.1.20. We use and adapt the notations and conventions introduced there. Thus let $\widehat{\mathbf{v}}, \widehat{\pi}, \widehat{\mathbf{f}}, \widehat{g}$ and $\widehat{p}$ the transformed quantities defined on $\mathbb{R}^n_>$. With the transformed cut-off function $\widehat{\tau}$ we define $\bar{\mathbf{v}} := \widehat{\mathbf{v}}\widehat{\tau} \in W^{2,\widehat{p}(\cdot)}(\mathbb{R}^n_>)$ and $\bar{\pi} := \widehat{\pi}\widehat{\tau} \in W^{1,\widehat{p}(\cdot)}(\mathbb{R}^n_>)$. The couple $(\bar{\mathbf{v}}, \bar{\pi})$ solves the Stokes problem in the half-space with data $(\mathbf{T}, G)$ defined through

$$T_j := \widehat{\tau}\,\widehat{f}_j + \sum_{i=1}^{n} A_i \partial_{in}^2 \bar{v}_j + \sum_{i=1}^{n} B_i \partial_i \widehat{v}_j + C\widehat{v}_j + D_j \partial_n \bar{\pi} + E_j \widehat{\pi},$$

$$G := \widehat{\tau}\,\widehat{g} + \sum_{i=1}^{n} R_i \widehat{v}_i + \sum_{i=1}^{n} S_i \partial_n \bar{v}_i,$$

where the functions A_i, B_i, C are defined in the proof of Proposition 14.1.20 and where $D_j := -\partial_j a$, $E_j := \partial_n \widehat{\tau}\, \partial_j a - \partial_j \widehat{\tau}$, $R_i := \partial_i \widehat{\tau} - \partial_n \widehat{\tau}\, \partial_i a$, and $S_i = \partial_i a$. Note that D_j and S_i are as well as A_i transformations of first derivatives of the boundary description a, which can be made arbitrarily small such that they can be absorbed in the left-hand side. Since $(\mathbf{T}, G) \in (L^{\widehat{p}(\cdot)}(\mathbb{R}^n_>))^n \times W^{1,\widehat{p}(\cdot)}(\mathbb{R}^n_>)$ have bounded support we can use Proposition 14.2.25 and proceed as in the proof of Proposition 14.1.20 to obtain

$$\|\nabla^2\bar{\mathbf{v}}\|_{L^{\widehat{p}(\cdot)}(\widehat{V})} + \|\nabla\bar{\pi}\|_{L^{\widehat{p}(\cdot)}(\widehat{V})}$$
$$\leqslant c\left(\|\widehat{\mathbf{f}}\|_{L^{\widehat{p}(\cdot)}(\widehat{V}')} + \|\widehat{g}\|_{W^{1,\widehat{p}(\cdot)}(\widehat{V}')} + \|\widehat{\mathbf{v}}\|_{W^{1,\widehat{p}(\cdot)}(\widehat{V}')} + \|\widehat{\pi}\|_{L^{\widehat{p}(\cdot)}(\widehat{V}')}\right).$$

From this estimate we derive in the same way as in the proof of Proposition 14.1.20 the second estimate of the proposition. Proposition 14.2.25 also gives

$$\|\nabla\bar{\mathbf{v}}\|_{L^{\hat{p}(\cdot)}(\widehat{V})} + \|\bar{\pi}\|_{L^{\hat{p}(\cdot)}(\widehat{V})} \leqslant c\left(\|\mathbf{T}\|_{D^{-1,\hat{p}(\cdot)}(\mathbb{R}^n_>)} + \|G\|_{L^{\hat{p}(\cdot)}(\widehat{V}')}\right).$$

Proceeding as in the end of the proof of (14.1.14) one can show that

$$\|\mathbf{T}\|_{D^{-1,\hat{p}(\cdot)}(\mathbb{R}^n_>)} \leqslant c\left(\|\mathbf{f}\|_{W^{-1,p(\cdot)}(V')} + \|\pi\|_{W^{-1,p(\cdot)}(V')} + \|\mathbf{v}\|_{L^{p(\cdot)}(V')}\right)$$
$$+ A\|\nabla\bar{\mathbf{v}}\|_{L^{\hat{p}(\cdot)}(\widehat{V})} + D\|\bar{\pi}\|_{L^{\hat{p}(\cdot)}(\widehat{V})},$$

where A is defined in the proof of Proposition 14.1.20 and can be made arbitrarily small. The quantity D is defined analogously and can also be made arbitrarily small. Similarly we get

$$\|G\|_{L^{\hat{p}(\cdot)}(\widehat{V})} \leqslant c\|g\|_{L^{p(\cdot)}(V')} + c\|\mathbf{v}\|_{L^{p(\cdot)}(V')} + S\|\nabla\bar{\mathbf{v}}\|_{L^{\hat{p}(\cdot)}(\widehat{V})},$$

where S is defined analogously to A and can also be made arbitrarily small. Absorbing the appropriate terms and proceeding as in the proof of Proposition 14.1.20 we can derive from the last three estimates the first estimate of the proposition. This finishes the proof of the proposition.   □

Now the Theorem 14.2.2 follows from Propositions 14.2.10 and 14.2.27 in the same way as Theorem 14.1.2 follows from Propositions 14.1.12 and 14.1.20.

Remark 14.2.28. Using the corresponding estimates in the previous results we also have shown that a unique *weak solution* $(\mathbf{v},\pi) \in (W^{1,p(\cdot)}(\Omega))^n \times L^{p(\cdot)}(\Omega)$ of the Stokes problem (14.2.1) with $\int_\Omega \pi\, dx = 0$ exists if the data satisfy $\mathbf{f} \in (W^{-1,p(\cdot)}(\Omega))^n$, $g \in L^{p(\cdot)}(\Omega)$, $\mathbf{v}_0 \in \mathrm{Tr}(W^{1,p(\cdot)}(\Omega))^n$ and the compatibility condition $\int_{\partial\Omega} \mathbf{v}_0 \cdot \boldsymbol{\nu}\, d\omega = \int_\Omega g\, dx$. This solution satisfies the problem (14.2.1) in the usual weak sense, i.e. for test functions from the space $(W_0^{1,p'(\cdot)}(\Omega))^n$, and satisfies the estimate

$$\|\mathbf{v}\|_{W^{1,p(\cdot)}(\Omega)} + \|\pi\|_{L^{p(\cdot)}(\Omega)}$$
$$\leqslant c\left(\|\mathbf{f}\|_{W^{-1,p(\cdot)}(\Omega)} + \|g\|_{L^{p(\cdot)}(\Omega)} + \|\mathbf{v}_0\|_{\mathrm{Tr}(W^{1,p(\cdot)}(\Omega))}\right),$$

with a constant $c = c(p,\Omega)$.

14.3   Divergence Equation and Consequences

In this section we deal with the divergence equation, which is of great importance in the theory of incompressible fluids. Moreover, we will derive from the main result important consequences such as the negative norm theorem and Korn's inequality.

For given f with mean value zero we seek a solution $\mathbf{u}$ with zero boundary values of

$$\operatorname{div} \mathbf{u} = f \qquad \text{in } \Omega, \tag{14.3.1}$$

where Ω is a bounded domain. This problem has been studied by many authors. The L^s–theory in Lipschitz domains is based on an explicit representation formula is due to Bogovskiĭ [50, 51]. We generalize these results to the case of Lebesgue spaces with variable exponents and bounded John domains using the theory of Calderón–Zygmund operators, developed in Sect. 6.3 and the transfer technique from Sect. 7.4. This involves showing that in the unit ball $B(0, 1)$ the solution $\mathbf{u} \in W_0^{1, p(\cdot)}(B(0, 1))$ of (14.3.1), which is given by

$$\mathbf{u}(x) = \int\limits_{\Omega} f(y) \left(\frac{x - y}{|x - y|^n} \int\limits_{|x-y|}^{\infty} \omega \left(y + \zeta \frac{x - y}{|x - y|} \right) \zeta^{n-1} \, d\zeta \right) dy,$$

where $\omega \in C_0^\infty(B(0, 1))$, $\int_{B(0,1)} \omega \, dx = 0$, satisfies the estimate

$$\|\nabla \mathbf{u}\|_{p(\cdot)} \leqslant c \, \|f\|_{p(\cdot)}.$$

For that we need to show that the corresponding kernels satisfy the necessary assumptions. The properties of the kernel are well known and can be found e.g. in [50, 51, 103, 169].

Lemma 14.3.2. *Let* $\omega \in C_0^\infty(\mathbb{R}^n)$ *be a standard mollifier. For* $i, j = 1, \ldots, n$, *we define*

$$N_{ij}(x, z) := \frac{\delta_{ij}}{|z|^n} \int\limits_{0}^{\infty} \omega \left(x + r \frac{z}{|z|} \right) r^{n-1} \, dr + \frac{z_i}{|z|^{n+1}} \int\limits_{0}^{\infty} \partial_j \omega \left(x + r \frac{z}{|z|} \right) r^n \, dr,$$

$$k_{ij}(x, y) := N_{ij}(x, x - y) \tag{14.3.3}$$

for all $x, y, z \in \mathbb{R}^n$ *with* $z \neq 0$ *and* $x \neq y$. *Then* k_{ij} *is* C^∞ *off the diagonal and*

$$k_{ij}(x, y) = 0 \tag{14.3.4}$$

for all $x, y \in \mathbb{R}^n$ *with* $|x| > 1 + |y|$. *For* $1 \leqslant \sigma < \infty$, *there exists a constant* c, *depending on* $\|\omega\|_{W^{1,\infty}}$ *and* n, *such that*

$$N_{ij}(x, \alpha z) = \alpha^{-n} N_{ij}(x, z),$$

$$\int_{|z|=1} N_{ij}(x, z)\, dz = 0,$$

$$\left(\int_{|z|=1} |N_{ij}(x, z)|^\sigma \, dz \right)^{\frac{1}{\sigma}} \leqslant c\,(1 + |x|)^n. \tag{14.3.5}$$

Furthermore for all $x, y, z \in \mathbb{R}^n$ *with* $x \neq y$, $y \neq z$, *and* $|x - z| < \frac{1}{2}|x - y|$

$$|k_{ij}(x, y)| \leqslant c\,(1 + |x|)^n \, |x - y|^{-n},$$
$$|k_{ij}(x, y) - k_{ij}(z, y)| \leqslant c\,(1 + |y|)^{n+1} \, |x - z| |x - y|^{-n-1}, \tag{14.3.6}$$
$$|k_{ij}(y, x) - k_{ij}(y, z)| \leqslant c\,(1 + |y|)^{n+1} \, |x - z| \, |x - y|^{-n-1},$$

for all $\alpha > 0$ *and* $x \in \mathbb{R}^n$. *Especially the* k_{ij} *are kernels on* $\mathbb{R}^n \times \mathbb{R}^n$ *in the sense of Definition 6.3.1.*

We now show that the kernel k satisfies a condition similar to the condition (D) from Definition 6.3.7.

Lemma 14.3.7 (Condition (D')). *Let* k *be a kernel satisfying the second condition in* (14.3.6). *Then there exist constants* $c > 0$ *and* $N > 0$, *depending on* $\|\omega\|_{W^{1,\infty}}$ *and* n, *such that*

$$\sup_{r>0} \int_{|y-x_0|>Nr} |f(y)|\, D_{B(x_0,r)} k(y)\, dy \leqslant c\, Mf(x_0)$$

for all $f \in C_0^\infty(B(0,1))$ *and* $x_0 \in \mathbb{R}^n$.

Proof. Note that for all $y \in \Omega$, $x, z \in \mathbb{R}^n$ with $x \neq y \neq z$, and $|x - z| < \frac{1}{2}|x - y|$ there holds

$$|k_{ij}(x, y) - k_{ij}(z, y)| \leqslant c\, 2^{n+1} \, |x - z| |x - y|^{-n-1}.$$

The rest of the proof follows exactly as in the discussion after Definition 6.3.7 that a standard kernel satisfies condition (D).   □

Note that in contrast to condition (D) of Definition 6.3.7 the condition (D') is only fulfilled for smooth f with compact support in Ω not in $\mathbb{R}^n$.

Lemma 14.3.8. *Let* ω, k_{ij}, N_{ij} *be as in Lemma 14.3.2, let* $k_{ij,\varepsilon}$ *denote the truncated kernels, and let* $p \in \mathcal{P}^{\log}(\Omega)$ *satisfy* $1 < p^- \leqslant p^+ < \infty$. *Then the operators* $T_{ij,\varepsilon}$, *defined by*

$$T_{ij,\varepsilon}f(x) := \int_{B(0,1)} k_{ij,\varepsilon}(x,y)f(y)\,dy$$

for $f \in L^{p(\cdot)}(B(0,1))$, are bounded on $L^{p(\cdot)}(B(0,1))$ uniformly with respect to $\varepsilon > 0$. For all $f \in L^{p(\cdot)}(B(0,1))$

$$T_{ij}f(x) := \lim_{\varepsilon \to 0^+} T_{ij,\varepsilon}f(x) = \lim_{\varepsilon \to 0^+} \int_{B(0,1)} k_{ij,\varepsilon}(x,y)f(y)\,dy$$

exists almost everywhere and $\lim_{\varepsilon \to 0^+} T_{ij,\varepsilon}f = T_{ij}f$ in $L^{p(\cdot)}(B(0,1))$ norm. In particular the T_{ij} are bounded on $L^{p(\cdot)}(B(0,1))$.

Proof. By Theorem 5.7.2 we know that the maximal operator M is continuous from $L^{p(\cdot)}(\mathbb{R}^n)$ to $L^{p(\cdot)}(\mathbb{R}^n)$. Let $\eta \in C_0^\infty([0,\infty))$ with $\eta(t) = 1$ for all $0 \leqslant t \leqslant 3$ and $\eta(t) = 0$ for all $t \geqslant 4$. Now define $\widetilde{k}_{ij}$ and $\widetilde{N}_{ij}$ by

$$\widetilde{N}_{ij}(x,z) := \eta(x)N(x,z),$$
$$\widetilde{k}_{ij}(x,y) := \widetilde{N}_{ij}(x, x-y),$$

so that $\widetilde{k}_{ij}(x,y) = k_{ij}(x,y)$ for all $x \in B(0,1)$ and all $y \in \mathbb{R}^n$. From (14.3.5), (14.3.6) and the definition of $\widetilde{N}_{ij}$ we deduce:

(a) (14.3.5) clearly holds.
(b) For all $x,y,z \in \mathbb{R}^n$ with $x \neq y \neq z$, and $|x - z| < \frac{1}{2}|x - y|$ there holds

$$|\widetilde{k}_{ij}(x,y)| \leqslant c\,|x-y|^{-n},$$
$$|\widetilde{k}_{ij}(y,x) - \widetilde{k}_{ij}(y,z)| \leqslant c\,|x-z||x-y|^{-n-1}.$$

(c) For all $y \in B(0,1)$ and $x,z \in \mathbb{R}^n$ with $x \neq y \neq z$, and $|x - z| < \frac{1}{2}|x - y|$ there holds

$$|\widetilde{k}_{ij}(x,y) - \widetilde{k}_{ij}(z,y)| \leqslant c\,|x-z||x-y|^{-n-1}.$$

Let $\widetilde{k}_{ij,\varepsilon}$ denote the truncated kernels and define $\widetilde{T}_{ij,\varepsilon}$ by

$$\widetilde{T}_{ij,\varepsilon}f(x) := \int_{\mathbb{R}^n} \widetilde{k}_{ij,\varepsilon}(x,y)f(y)\,dy.$$

Due to the properties (a) the kernel $\widetilde{k}_{ij}$ fulfills all requirements of Proposition 6.3.4. Thus the operators $\widetilde{T}_{ij,\varepsilon}$ are uniformly bounded on $L^{\sigma'}(\mathbb{R}^n)$ with respect to $\varepsilon > 0$. Moreover

$$\widetilde{T}_{ij}f(x) = \lim_{\varepsilon \to 0^+} \widetilde{T}_{ij,\varepsilon}f(x) = \lim_{\varepsilon \to 0^+} \int_{\mathbb{R}^n} \widetilde{k}_{ij,\varepsilon}(x,y)f(y)\,dy \qquad (14.3.9)$$

exists almost everywhere and $\lim_{\varepsilon \to 0^+} \widetilde{T}_{ij,\varepsilon}f = \widetilde{T}_{ij}f$ in $L^{\sigma'}(\mathbb{R}^n)$ norm. In particular $\widetilde{T}_{ij}$ is bounded on $L^{\sigma'}(\mathbb{R}^n)$. From this and the estimates (b) one deduces as in the classical theory (cf. [360, p. 33], [103] or Remark 6.3.6)

$$\left|\{x \in \mathbb{R}^n : |\widetilde{T}_{ij}f(x)| > \alpha\}\right| \leqslant \frac{c\,\|f\|_1}{\alpha}$$

for all $\alpha > 0$ and all $f \in L^1(\mathbb{R}^n) \cap L^{\sigma'}(\mathbb{R}^n)$. Hence $\widetilde{T}_{ij}$ extends to a bounded operator from $L^1(\mathbb{R}^n)$ to w-$L^1(\mathbb{R}^n)$. Further recall that $\widetilde{T}_{ij}$ fulfills condition (D'). Thus exactly as in the proof of Proposition 6.3.8 in [29] we get

$$\left(M_1^\sharp(|\widetilde{T}_{ij}f|^s)\right)^{\frac{1}{s}}(x) \leqslant c\,Mf(x)$$

for fixed $0 < s < 1$, and all $f \in C_0^\infty(B(0,1)), x \in \mathbb{R}^n$. The restriction to smooth f with compact support in $B(0,1)$ originates from the substitution of (D) by (D'). From the previous inequality we deduce exactly as in Theorem 6.3.9 that

$$\|\widetilde{T}_{ij}f\|_{p(\cdot)} \leqslant c\,\|f\|_{p(\cdot)}$$

for all $f \in C_0^\infty(B(0,1))$. Since $C_0^\infty(B(0,1))$ is dense in $L^{p(\cdot)}(B(0,1))$ by Theorem 3.4.12, this implies

$$\|\widetilde{T}_{ij}f\|_{L^{p(\cdot)}(\mathbb{R}^n)} \leqslant c\,\|f\|_{L^{p(\cdot)}(B(0,1))} \qquad (14.3.10)$$

for all $f \in L^{p(\cdot)}(B(0,1))$. We now show that $\lim_{\varepsilon \to 0^+} \widetilde{T}_{ij,\varepsilon}f = \widetilde{T}_{ij}f$ in $L^{p(\cdot)}(\mathbb{R}^n)$ for all $f \in L^{p(\cdot)}(B(0,1))$. As an intermediate step we prove

$$\widetilde{T}_{ij}^*f(x) \leqslant c\left(M_s(\widetilde{T}_{ij}f)(x) + Mf(x)\right) \qquad (14.3.11)$$

for all $f \in L^{p(\cdot)}(B(0,1))$, where $\widetilde{T}_{ij}^*$ is the maximal truncated operator corresponding to $\widetilde{T}_{ij}$ and $0 < s \leqslant 1$. This will be done almost exactly as in the proof of Proposition 2 in [360, pp. 34–36]. The only difference in our case is that property (c) above only holds for $y \in B(0,1)$ and not all $y \in \mathbb{R}^n$. Therefore, we have to adapt the derivation of inequality (31) in [360, p. 35], i.e. we have to show the following: There exists $c > 0$ such that

$$|\widetilde{T}_{ij}f_2(z) - \widetilde{T}_{ij}f_2(x)| \leqslant c\,Mf(x)$$

for all $\varepsilon > 0$, $x \in \mathbb{R}^n$, $f_2 := \chi_{\mathbb{R}^n \setminus B(x,\varepsilon)}f$ and all $z \in \mathbb{R}^n$ with $|x - z| < c\varepsilon$. So let $f \in L^{p(\cdot)}(B(0,1))$, $\varepsilon > 0$, and set $f_2 := \chi_{\mathbb{R}^n \setminus B(x,\varepsilon)}f$. Then for all $x \in \mathbb{R}^n$ and $z \in B(x, \frac{\varepsilon}{2})$ we have with property (c) above

$$|\widetilde{T}_{ij}f_2(z) - \widetilde{T}_{ij}f_2(x)| \leqslant \int\limits_{|y-x| \geqslant \varepsilon} |\widetilde{k}_{ij}(x,y) - \widetilde{k}_{ij}(z,y)| \, |f(y)| \, dy$$

$$= \sum_{k=0}^{\infty} \int\limits_{2^{k+1}\varepsilon > |y-x| \geqslant 2^k \varepsilon} |\widetilde{k}_{ij}(x,y) - \widetilde{k}_{ij}(z,y)| \, |f(y)| \, dy$$

$$\leqslant c \sum_{k=0}^{\infty} \int\limits_{2^{k+1}\varepsilon > |y-x| \geqslant 2^k \varepsilon} |f(y)| \, \varepsilon \, |y-x|^{-n-1} \, dy$$

$$\leqslant c\varepsilon \sum_{k=0}^{\infty} |B(0,1)|(2^{k+1}\varepsilon)^n \, Mf(x) \, (2^k\varepsilon)^{-n-1}$$

$$= c \, Mf(x).$$

Using the boundedness of $\widetilde{T}_{ij}$ on $L^{\sigma'}(\mathbb{R}^n)$, the growth conditions in (b), and the previous inequality, we conclude from [360, Proposition 2, p. 34] that (14.3.11) for all $f \in L^{p(\cdot)}(B(0,1))$. The boundedness of M on $L^{p(\cdot)}(\mathbb{R}^n)$, inequalities (14.3.10), (14.3.11) (with $s = 1$), and the previous inequality imply

$$\|\widetilde{T}_{ij}^* f\|_{L^{p(\cdot)}(\mathbb{R}^n)} \leqslant c \, \|f\|_{L^{p(\cdot)}(B(0,1))}$$

for all $f \in L^{p(\cdot)}(B(0,1))$. We have already shown that $\lim_{\varepsilon \to 0+} \widetilde{T}_{ij,\varepsilon} f = \widetilde{T}_{ij} f$ almost everywhere. As in Corollary 6.3.13, the pointwise convergence, the last estimate, and the density of $C_0^\infty(B(0,1))$ in $L^{p(\cdot)}(B(0,1))$ imply $\lim_{\varepsilon \to 0+} \widetilde{T}_{ij,\varepsilon} f = \widetilde{T}_{ij} f$ also in $L^{p(\cdot)}(\mathbb{R}^n)$-norm.

Overall we have shown that Lemma 14.3.8 holds with T_{ij} replaced by $\widetilde{T}_{ij}$. But by the definition of $\widetilde{k}_{ij}$ and $\widetilde{T}_{ij}$ we have for all $x \in B(0,1)$

$$\widetilde{T}_{ij} f(x) = T_{ij} f(x),$$

whenever $\widetilde{T}_{ij} f(x)$ exists. Thus all results for $\widetilde{T}_{ij}$ on $\mathbb{R}^n$ transfer to results for T_{ij} on $B(0,1)$. This proves the lemma. $\square$

Since we are looking for solutions $\mathbf{u}$ of (14.3.1) with zero boundary values we need to assume that the right-hand side has a vanishing integral. For bounded domains we have denoted the space of such functions by

$$L_0^{p(\cdot)}(\Omega) := \left\{ f \in L^{p(\cdot)}(\Omega) : \int\limits_\Omega f(x) \, dx = 0 \right\}.$$

Due to Proposition 3.4.14 smooth function with compact support and vanishing integral are dense in $L_0^{p(\cdot)}(\Omega)$.

Proposition 14.3.12. *Let $B(0,1) \subset \mathbb{R}^n$, $n \geqslant 2$, and let $p \in \mathcal{P}^{\log}(B(0,1))$ satisfy $1 < p^- \leqslant p^+ < \infty$. Then there exists a linear, bounded operator $\mathcal{B} : L_0^{p(\cdot)}(B(0,1)) \to W_0^{1,p(\cdot)}(B(0,1))$ such that for each $f \in L_0^{p(\cdot)}(B(0,1))$ the function $\mathcal{B}f \in W_0^{1,p(\cdot)}(B(0,1))$ is a solution of the divergence equation (14.3.1), i.e. $\operatorname{div}(\mathcal{B}f) = f$, satisfying the estimate*

$$\|\nabla \mathcal{B}f\|_{L^{p(\cdot)}(B(0,1))} \leqslant c\|f\|_{L^{p(\cdot)}(B(0,1))}. \tag{14.3.13}$$

The operator $\mathcal{B}$ is independent of p, although the constant c is not. Also, $\mathcal{B}$ maps $C_{0,0}^\infty(B(0,1))$ to $C_0^\infty(B(0,1))$.

Proof. By Theorem 5.7.2 we know that M is continuous from $L^{p(\cdot)}(\mathbb{R}^n)$ to $L^{p(\cdot)}(\mathbb{R}^n)$. By Proposition 3.4.14 we know that $C_{0,0}^\infty(B(0,1))$ is dense in $L_0^{p(\cdot)}(B(0,1))$ and thus we can assume $f \in C_{0,0}^\infty(B(0,1))$. In this case we have an explicit representation (We follow the approach outlined in [169, Sect. III.3] and use the same notation.) of the solution given by

$$\mathbf{u}(x) = \mathcal{B}f(x) := \int_{B(0,1)} f(y) \left(\frac{x-y}{|x-y|^n} \int_{|x-y|}^\infty \omega\left(y + \zeta \frac{x-y}{|x-y|}\right) \zeta^{n-1} d\zeta \right) dy,$$

where $\omega \in C_0^\infty(B)$, $\int_B \omega \, dx = 0$. Note that the operator $\mathcal{B}$ itself only depends on the choice of ω. From [51] we know that $\mathbf{u} \in C_0^\infty(B(0,1))$. Due to the continuous embedding $L^{p(\cdot)}(B(0,1)) \hookrightarrow L^{p^-}(B(0,1))$ we can use the classical L^{p^-} theory and justify (cf. [51]) the formula

$$\partial_j u_i(x) = \int_{B(0,1)} k_{ij}(x, x-y) f(y) \, dy + \int_{B(0,1)} G_{ij}(x,y) f(y) \, dy$$

$$+ f(x) \int_{B(0,1)} \frac{(x-y)_i (x-y)_j}{|x-y|^2} \omega(y) \, dy \tag{14.3.14}$$

$$=: F_1(x) + F_2(x) + F_3(x),$$

where $k_{ij}(x,y)$ is given by (14.3.3), and where $G_{ij}(x,y)$ satisfies

$$|G_{ij}(x,y)| \leqslant \frac{c}{|x-y|^{n-1}}$$

for all $x, y \in B(0,1)$. By Lemma 14.3.8,

$$\|F_1\|_{L^{p(\cdot)}(B(0,1))} \leqslant c\|f\|_{L^{p(\cdot)}(B(0,1))},$$

and

$$\|F_3\|_{L^{p(\cdot)}(B(0,1))} \leqslant c \, \|f\|_{L^{p(\cdot)}(B(0,1))} \, .$$

By the estimate of G_{ij}, it is clear that $F_2 \leqslant I_1 f \leqslant c \, Mf$, where I_1 is the Riesz potential. Hence

$$\|F_2\|_{L^{p(\cdot)}(B(0,1))} \leqslant c \, \|Mf\|_{L^{p(\cdot)}(\mathbb{R}^n)} \leqslant c \, \|f\|_{L^{p(\cdot)}(\mathbb{R}^n)} = c \, \|f\|_{L^{p(\cdot)}(B(0,1))} \, .$$

From (14.3.14), and the estimates for F_1, F_2 and F_3 we immediately obtain (14.3.13). $\qquad\qquad\qquad\qquad\qquad\qquad\qquad\qquad\qquad\qquad\qquad\qquad\qquad\quad \square$

Using the previous proposition and Theorem 7.4.9 we get the following result:

Theorem 14.3.15. *Let $\Omega \subset \mathbb{R}^n$, $n \geqslant 2$, be domain satisfying the emanating chain condition and let $p \in \mathcal{P}^{\log}(\Omega)$ satisfy $1 < p^- \leqslant p^+ < \infty$. Then there exists a linear, bounded operator $\boldsymbol{\mathcal{B}} : C_0^\infty(\Omega) \to L_{\mathrm{loc}}^1(\Omega)$ which extends uniquely to an operator $\boldsymbol{\mathcal{B}} : L_0^{p(\cdot)}(\Omega) \to D_0^{1,p(\cdot)}(\Omega)$ with*

$$\mathrm{div}(\boldsymbol{\mathcal{B}}f) = f \qquad\qquad\qquad\qquad (14.3.16)$$

$$\|\nabla\boldsymbol{\mathcal{B}}f\|_{p(\cdot)} \leqslant c \, \|f\|_{p(\cdot)} \, . \qquad\qquad\qquad (14.3.17)$$

The operator $\boldsymbol{\mathcal{B}}$ is independent of p, but the constant c is not. If Ω is bounded, and $f \in C_{0,0}^\infty(\Omega)$, then $\boldsymbol{\mathcal{B}}f \in C_0^\infty(\Omega)$.

Proof. Since Ω satisfies the emanating chain condition we can find a chain-covering $\mathcal{Q}$ of Ω consisting of balls Q (cf. Sect. 7.4). Let $S_Q : L_0^{p(\cdot)}(\Omega) \to L_0^{p(\cdot)}(Q)$ be as in Theorem 7.4.9. In Proposition 14.3.12 we have shown that there exists a linear operator $\boldsymbol{\mathcal{B}}_{\mathrm{ref}}$ which maps $C_{0,0}^\infty(B(0,1))$ to $C_0^\infty(B(0,1))$ and $L_0^{p(\cdot)}(B(0,1))$ to $W_0^{1,p(\cdot)}(B_1(0))$, and satisfies $\mathrm{div}(\boldsymbol{\mathcal{B}}f) = f$ and (14.3.13). By a simple translation and scaling argument it follows that there exist linear operators $\boldsymbol{\mathcal{B}}_Q : L_0^{p(\cdot)}(Q) \to W_0^{1,p(\cdot)}(Q)$ which satisfy $\mathrm{div}\,\boldsymbol{\mathcal{B}}_Q g = g$ in Q and

$$\|\nabla\boldsymbol{\mathcal{B}}_Q g\|_{L^{p(\cdot)}(Q)} \leqslant c \, \|g\|_{L_0^{p(\cdot)}(Q)},$$

with a constant c independent of Q. Moreover, $\boldsymbol{\mathcal{B}}_Q$ maps $C_{0,0}^\infty(Q)$ to $C_0^\infty(Q)$.

Let $f \in L_0^{p(\cdot)}(\Omega)$. We extend $\boldsymbol{\mathcal{B}}_Q S_Q f$ outside of Q by zero, so that $\boldsymbol{\mathcal{B}}_Q S_Q f \in D_0^{1,p(\cdot)}(\Omega)$. We define our operator $\boldsymbol{\mathcal{B}}$ almost everywhere by

$$\boldsymbol{\mathcal{B}}f := \sum_{Q \in \mathcal{Q}} \boldsymbol{\mathcal{B}}_Q S_Q f \, .$$

Due to (B3), the fact that the cover $\mathcal{Q}$ is locally σ_1-finite (B1), and the previous estimate we see that the sum converges in $L_{\mathrm{loc}}^1(\Omega)$ and therefore in the sense of distributions. The same argument ensures that $\nabla\boldsymbol{\mathcal{B}}f = \sum_Q \nabla\boldsymbol{\mathcal{B}}_Q S_Q f$

in $L^1_{\text{loc}}(\Omega)$. Moreover, $\nabla \mathcal{B} f \in L^{p(\cdot)}_{\text{loc}}(\Omega)$. Thus Corollary 7.3.24, the previous estimate and (d) of Theorem 7.4.9 imply

$$\|\nabla \mathcal{B} f\|_{L^{p(\cdot)}(\Omega)} = \left\| \sum_{Q \in \mathcal{Q}} \nabla \mathcal{B}_Q S_Q f \chi_Q \right\|_{L^{p(\cdot)}(\Omega)}$$

$$\leqslant c \left\| \sum_{Q \in \mathcal{Q}} \chi_Q \frac{\|\nabla \mathcal{B}_Q S_Q f\|_{L^{p(\cdot)}(Q)}}{\|\chi_Q\|_{p(\cdot)}} \right\|_{L^{p(\cdot)}(\Omega)}$$

$$\leqslant c \left\| \sum_{Q \in \mathcal{Q}} \chi_Q \frac{\|S_Q f\|_{L^{p(\cdot)}_0(Q)}}{\|\chi_Q\|_{p(\cdot)}} \right\|_{L^{p(\cdot)}(\Omega)}$$

$$\leqslant c \|f\|_{L^{p(\cdot)}_0(\Omega)},$$

which proves (14.3.17). From $\nabla \mathcal{B} f = \sum_Q \nabla \mathcal{B}_Q S_Q f$ in $L^{p(\cdot)}(\Omega)$ it follows that

$$\operatorname{div} \mathcal{B} f = \sum_{Q \in \mathcal{Q}} \operatorname{div} \mathcal{B}_Q S_Q f = \sum_{Q \in \mathcal{Q}} S_Q f = f$$

for $f \in L^{p(\cdot)}_0(\Omega)$. This proves (14.3.16).

Assume now that Ω is bounded and $f \in C^\infty_0(\Omega)$. Then $S_Q f \in C^\infty_0(Q)$ for all $Q \in \mathcal{Q}$ and $S_Q f \neq 0$ for only finitely many $Q \in \mathcal{Q}$. Therefore, by the properties of $\mathcal{B}_Q$, $\mathcal{B}_Q S_Q f \in C^\infty_0(Q)$ for all $Q \in \mathcal{Q}$ and $\mathcal{B}_Q S_Q f \neq 0$ for only finitely many $Q \in \mathcal{Q}$. Thus $\mathcal{B} f \in C^\infty_0(\Omega)$. $\square$

Theorem 14.3.15 is very useful in the theory of fluid dynamics. To avoid technical problems with the pressure one uses mostly divergence free test functions $\boldsymbol{\xi}$ in the weak formulation of the fluid system. Sometimes however, these test functions have to be perturbed by functions $\mathbf{h}$ with compact support with small norm of divergence. Then Theorem 14.3.15 can be used to find a function $\mathbf{u}$ with compact support with $\operatorname{div} \mathbf{u} = \operatorname{div} \mathbf{h}$ whose full norm $\|\mathbf{u}\|_{1,p}$ is controlled only in terms of $\|\operatorname{div} \mathbf{h}\|_p$ which is small. The function $\mathbf{h} - \mathbf{u}$ is then again divergence free and the necessary correction $\mathbf{u}$ has small norm. We will use this in Sect. 14.4.

As a consequence of Theorem 14.3.15 we are able the generalize Nečas theorem on negative norms (Lions-Lemma) to variable exponent spaces and domains satisfying the emanating chain condition.

Theorem 14.3.18 (Negative norm theorem). *Let $\Omega \subset \mathbb{R}^n$ be a domain satisfying the emanating chain condition and let $p \in \mathcal{P}^{\log}(\Omega)$ satisfy $1 < p^- \leqslant p^+ < \infty$. Then*

$$\|\nabla f\|_{D^{-1,p(\cdot)}(\Omega)} \leqslant \|f\|_{L^{p(\cdot)}_0(\Omega)} \leqslant c \|\nabla f\|_{D^{-1,p(\cdot)}(\Omega)} \tag{14.3.19}$$

for all $f \in L_0^{p(\cdot)}(\Omega)$. If Ω is additionally bounded, then

$$\|f\|_{L^{p(\cdot)}(\Omega)} \leqslant c \,\|\nabla f\|_{D^{-1,p(\cdot)}(\Omega)} + \frac{c}{\operatorname{diam}(\Omega)}\|f\|_{D^{-1,p(\cdot)}(\Omega)} \qquad (14.3.20)$$

for all $f \in L^{p(\cdot)}(\Omega)$.

Proof. Let $f \in L_0^{p(\cdot)}(\Omega)$. Then

$$\|\nabla f\|_{D^{-1,p(\cdot)}(\Omega)} = \sup_{\|\mathbf{h}\|_{D_0^{1,p'(\cdot)}(\Omega)} \leqslant 1} \langle \nabla f, \mathbf{h} \rangle = \sup_{\|\mathbf{h}\|_{D_0^{1,p'(\cdot)}(\Omega)} \leqslant 1} \int_\Omega f \operatorname{div} \mathbf{h}\, dx,$$

which implies by Hölder's inequality

$$\|\nabla f\|_{D_0^{-1,p(\cdot)}(\Omega)} \leqslant \|f\|_{L_0^{p(\cdot)}(\Omega)}.$$

On the other hand the above equalities and Theorem 14.3.15 yield

$$\|f\|_{L_0^{p(\cdot)}(\Omega)} \leqslant \sup_{\|g\|_{L_0^{p'(\cdot)}(\Omega)} \leqslant 1} \int_\Omega f g\, dx \leqslant c \sup_{\|\mathcal{B}g\|_{D_0^{1,p'(\cdot)}(\Omega)} \leqslant 1} \int_\Omega f g\, dx$$

$$= c \sup_{\|\mathcal{B}g\|_{D_0^{1,p'(\cdot)}(\Omega)} \leqslant 1} \int_\Omega f \operatorname{div} \mathcal{B}g\, dx \leqslant c \,\|\nabla f\|_{D^{-1,p(\cdot)}(\Omega)}$$

where we used the norm conjugate formula. This proves (14.3.19).

Assume in the following that Ω is bounded. Let Q_0 be the central cube of Ω and choose $\eta \in C_0^\infty(Q_0)$ with $\eta \geqslant 0$, $\int_{Q_0} \eta(x)\, dx = 1$, $\|\eta\|_\infty \leqslant c/|Q_0|$, and $\|\nabla \eta\|_\infty \leqslant c/(|Q_0| \operatorname{diam}(Q_0))$, where $c = c(n)$. For $f \in L^{p(\cdot)}(\Omega)$ we estimate with Lemma 7.4.14 and (14.3.19)

$$\|f\|_{L_w^{p(\cdot)}(\Omega)} \leqslant c \,\|f - \langle f\rangle_\Omega\|_{L_0^{p(\cdot)}(\Omega)} + 4 \,\|\chi_\Omega\|_{L^{p(\cdot)}(\Omega)} \left|\int_\Omega f \eta\, dx\right|$$

$$\leqslant c \,\|\nabla f\|_{D^{-1,p(\cdot)}(\Omega)} + 4 \,\|f\|_{D^{-1,p(\cdot)}(\Omega)} \|\eta\|_{D_0^{1,p'(\cdot)}(\Omega)} \|\chi_\Omega\|_{L^{p(\cdot)}(\Omega)}.$$

By Lemma 7.4.5,

$$\|\eta\|_{D_0^{1,p'(\cdot)}(\Omega)} \|\chi_\Omega\|_{L^{p(\cdot)}(\Omega)} \leqslant \|\nabla \eta\|_\infty \|\chi_\Omega\|_{L^{p'(\cdot)}(\Omega)} \|\chi_\Omega\|_{L^{p(\cdot)}(\Omega)} \leqslant \frac{c\,\sigma_2}{\operatorname{diam}(\Omega)}.$$

This and the previous estimate yield (14.3.20). $\qquad\qquad \square$

Let us now turn our attention to *Korn's inequality*. In the context of fluid dynamics and elasticity the governing partial differential equation gives only control of the symmetric part of the gradient rather of the full gradient itself. For $\mathbf{u} \in W^{1,1}_{loc}(\mathbb{R}^n)$ we define the *symmetric gradient* $\mathbf{Du}$ by

$$\mathbf{Du} := \frac{1}{2}\big(\boldsymbol{\nabla}\mathbf{u} + (\boldsymbol{\nabla}\mathbf{u})^{\top}\big).$$

In particular the partial differential equation only ensures that the norm of $\mathbf{Du}$ can be controlled. However, from a point of view of Sobolev spaces it is desirable to have control of the gradient $\boldsymbol{\nabla}\mathbf{u}$. Although it is not possible to estimate $|\boldsymbol{\nabla}\mathbf{u}|$ point-wise by $|\mathbf{Du}|$, it is in some cases possible to bound the norm of $\boldsymbol{\nabla}\mathbf{u}$ in terms of the norm of $\mathbf{Du}$. In particular, for all $\mathbf{u} \in W^{1,q}(\mathbb{R}^n)$ with $1 < q < \infty$ we have Korn's inequality (see for example [312])

$$\|\boldsymbol{\nabla}\mathbf{u}\|_q \leqslant c\,\|\mathbf{Du}\|_q.$$

We will generalize Korn's inequality to the variable exponent spaces $L^{p(\cdot)}(\Omega)$. In the case of zero boundary value no assumption on the domain is needed, while for the general case we need a bounded John domain. The approach presented here is based on Theorem 14.3.18 (cf. [108]). A completely different proof can be found in [103].

Theorem 14.3.21 (Korn's inequality; first case). *Let $p \in \mathcal{P}^{\log}(\Omega)$ with $1 < p^- \leqslant p^+ < \infty$. Then*

$$\|\boldsymbol{\nabla}\mathbf{u}\|_{L^{p(\cdot)}(\Omega)} \leqslant c\,\|\mathbf{Du}\|_{L^{p(\cdot)}(\Omega)} \tag{14.3.22}$$

for all $\mathbf{u} \in D^{1,p(\cdot)}_0(\Omega)$.

Proof. Since every function $\mathbf{u} \in D^{1,p(\cdot)}_0(\Omega)$ can be extended by zero to a function $\mathbf{u} \in D^{1,p(\cdot)}_0(\mathbb{R}^n)$, it suffices to consider the case $\Omega = \mathbb{R}^n$.

By the vector valued version of Theorem 14.3.18 and the identity $\partial_j \partial_k u_i = \partial_j D_{ki}\mathbf{u} + \partial_k D_{ij}\mathbf{u} - \partial_i D_{jk}\mathbf{u}$ (in the sense of distributions) it follows that

$$\|\boldsymbol{\nabla}\mathbf{u}\|_{L^{p(\cdot)}(\mathbb{R}^n)} \leqslant c\|\boldsymbol{\nabla}\boldsymbol{\nabla}\mathbf{u}\|_{D^{-1,p(\cdot)}(\mathbb{R}^n)} \leqslant c\|\boldsymbol{\nabla}\mathbf{Du}\|_{D^{-1,p(\cdot)}(\mathbb{R}^n)} \leqslant c\|\mathbf{Du}\|_{L^{p(\cdot)}(\mathbb{R}^n)},$$

where we used that $L^{p(\cdot)}(\mathbb{R}^n) = L^{p(\cdot)}_0(\mathbb{R}^n)$. $\qquad\square$

Theorem 14.3.23 (Korn's inequality; second case). *Let $\Omega \subset \mathbb{R}^n$ be a bounded John domain and let $p \in \mathcal{P}^{\log}(\Omega)$ satisfy $1 < p^- \leqslant p^+ < \infty$. Then*

$$\big\|\boldsymbol{\nabla}\mathbf{u} - \langle\boldsymbol{\nabla}\mathbf{u}\rangle_\Omega\big\|_{L^{p(\cdot)}(\Omega)} \leqslant c\,\big\|\mathbf{Du} - \langle\mathbf{Du}\rangle_\Omega\big\|_{L^{p(\cdot)}(\Omega)},$$

$$\|\boldsymbol{\nabla}\mathbf{u}\|_{L^{p(\cdot)}(\Omega)} \leqslant c\,\big\|\mathbf{Du} - \langle\mathbf{Du}\rangle_\Omega\big\|_{L^{p(\cdot)}(\Omega)} + \frac{c}{\mathrm{diam}(\Omega)}\,\big\|\mathbf{u} - \langle\mathbf{u}\rangle_\Omega\big\|_{L^{p(\cdot)}(\Omega)}$$

for all $\mathbf{u} \in W^{1,p(\cdot)}(\Omega)$.

Proof. We use the identity $\partial_j \partial_k u_i = \partial_j D_{ki}\mathbf{u} + \partial_k D_{ij}\mathbf{u} - \partial_i D_{jk}\mathbf{u}$ (in the sense of distributions). Thus

$$\|\nabla\mathbf{u} - \langle\nabla\mathbf{u}\rangle_\Omega\|_{L_0^{p(\cdot)}(\Omega)} \leqslant c\,\|\nabla\nabla\mathbf{u}\|_{D^{-1,p(\cdot)}(\Omega)} \leqslant c\,\|\nabla\mathbf{Du}\|_{D^{-1,p(\cdot)}(\Omega)}$$
$$\leqslant c\,\|\mathbf{Du} - \langle\mathbf{Du}\rangle_\Omega\|_{L_0^{p(\cdot)}(\Omega)}.$$

Then the first inequality follows from Theorem 14.3.18. The proof of the second one is similar to the one of (14.3.20). Let Q_0 be the central cube of Ω and choose $\eta \in C_0^\infty(Q_0)$ with $\eta \geqslant 0$, $\int_{Q_0} \eta(x)\,dx = 1$, $\|\eta\|_\infty \leqslant c/|Q_0|$, and $\|\nabla\eta\|_\infty \leqslant c/(|Q_0|\operatorname{diam}(Q_0))$, where $c = c(n)$. For $\mathbf{u} \in W^{1,p(\cdot)}(\Omega)$ we estimate with Lemma 7.4.14

$$\|\nabla\mathbf{u}\|_{L^{p(\cdot)}(\Omega)} \leqslant c\,\|\nabla\mathbf{u} - \langle\nabla\mathbf{u}\rangle_\Omega\|_{L_0^{p(\cdot)}(\Omega)} + c\sum_{j=1}^n\sum_{k=1}^n\left|\int_\Omega \partial_k u_j\,\eta\,dx\right|\|\chi_\Omega\|_{L^{p(\cdot)}(\Omega)}.$$

Since $\int_\Omega \partial_k u_j\,\eta\,dx = \int_\Omega \partial_k(u_j - \langle u_j\rangle_\Omega)\,\eta\,dx$, we get with the already proven Korn's inequality

$$\|\nabla\mathbf{u}\|_{L^{p(\cdot)}(\Omega)}$$
$$\leqslant c\,\|\mathbf{Du} - \langle\mathbf{Du}\rangle_\Omega\|_{L_0^{p(\cdot)}(\Omega)} + c\sum_{j=1}^n\sum_{k=1}^n\left|\int_\Omega (u_j - \langle u_j\rangle_\Omega)\partial_k\eta\,dx\right|\|\chi_\Omega\|_{L^{p(\cdot)}(\Omega)}$$
$$\leqslant c\,\|\mathbf{Du} - \langle\mathbf{Du}\rangle_\Omega\|_{L_0^{p(\cdot)}(\Omega)} + c\,\|\mathbf{u} - \langle\mathbf{u}\rangle_\Omega\|_{L_0^{p(\cdot)}(\Omega)}\|\nabla\eta\|_{L^{p'(\cdot)}(\Omega)}\|\chi_\Omega\|_{L^{p(\cdot)}(\Omega)}.$$

Exactly as in the proof of Theorem 14.3.18 we get $\|\nabla\eta\|_{L^{p'(\cdot)}(\Omega)}\|\chi_\Omega\|_{L^{p(\cdot)}(\Omega)} \leqslant c/\operatorname{diam}(\Omega)$. This and the previous estimate give the second inequality. $\square$

14.4   Electrorheological Fluids

One of the driving forces for the rapid development of the theory of variable exponent function spaces has been the model of electrorheological fluids introduced by Rajagopal and Růžička [328, 329, 337]. This model leads naturally to a functional setting involving function spaces with variable exponents. Electrorheological fluids change their mechanical properties dramatically when an external electric field is applied. They are one example of smart materials, whose development is currently one of the major task in engineering sciences. Also in the mathematical community such materials are intensively investigated in the recent years [4, 6, 7, 9, 90, 94, 102, 106, 107, 296].

In the case of an isothermal, homogeneous, incompressible electrorheological fluid the governing equations read

$$\partial_t \mathbf{v} + [\nabla \mathbf{v}]\mathbf{v} - \operatorname{div} \mathbf{S} + \nabla \pi = \mathbf{g} + [\nabla \mathbf{E}]\mathbf{P} \,,$$
$$\operatorname{div} \mathbf{v} = 0 \,, \tag{14.4.1}$$

where $\mathbf{v}$ is the velocity, $[\nabla \mathbf{v}]\mathbf{v} = \left(\sum_{j=1}^{3} v_j \partial_j v_i \right)_{i=1,2,3}$ denotes the convective term, π the pressure, $\mathbf{S}$ the extra stress tensor, $\mathbf{g}$ the external body force, $\mathbf{E}$ the electric field, and $\mathbf{P}$ the electric polarization. The latter two fields are subject to the quasi-static Maxwell's equations

$$\operatorname{div}(\varepsilon_0 \mathbf{E} + \mathbf{P}) = 0 \,,$$
$$\operatorname{curl} \mathbf{E} = \mathbf{0} \,, \tag{14.4.2}$$

where ε_0 is the dielectric constant in vacuum. Equations (14.4.1) and (14.4.2) are supplemented with appropriate boundary conditions. Moreover, we have to specify constitutive relations for $\mathbf{S}$ and $\mathbf{P}$. One possibility is to assume that the polarization $\mathbf{P}$ is linear in $\mathbf{E}$, i.e., $\mathbf{P} = \chi^E \mathbf{E}$ and that the extra stress tensor $\mathbf{S}$ is given by

$$\mathbf{S} = \alpha_{21}\big((1 + |\mathbf{D}|^2)^{\frac{p-1}{2}} - 1\big)\mathbf{E} \otimes \mathbf{E} + (\alpha_{31} + \alpha_{33}|\mathbf{E}|^2)(1 + |\mathbf{D}|^2)^{\frac{p-2}{2}}\mathbf{D}$$
$$+ \alpha_{51}(1 + |\mathbf{D}|^2)^{\frac{p-2}{2}}(\mathbf{DE} \otimes \mathbf{E} + \mathbf{E} \otimes \mathbf{DE}) \,, \tag{14.4.3}$$

where α_{ij} are material constants and $p = p(|\mathbf{E}|^2)$ is a Hölder continuous function with $1 < p^- \leqslant p^+ < \infty$. The constant coefficients α_{ij} have to satisfy certain conditions which ensure the validity of the second law of thermodynamics. These requirements also ensure that the operator induced by $-\operatorname{div} \mathbf{S}(\mathbf{D}, \mathbf{E})$ is *coercive* and satisfies appropriate *growth conditions*. For the mathematical treatment we have additionally to assume that the operator induced by $-\operatorname{div} \mathbf{S}(\mathbf{D}, \mathbf{E})$ is *strictly monotone*. For simplicity and clarity we will restrict ourselves here to the treatment of the steady version of the governing equations, i.e., in (14.4.1) we omit the time derivative $\partial_t \mathbf{v}$. For results concerning the unsteady problem we refer to [336, 337], and [338].

From now on we consider the steady version of the problem. Note, that the system (14.4.1) and (14.4.2) is separated. Thus we can first solve the Maxwell's equation and obtain under appropriate assumptions on the boundary data that the solution $\mathbf{E}$ is so regular that $|\mathbf{E}|$ is Hölder continuous. Thus in the following we will assume that $\mathbf{E}$ is given and that we consider only the steady version of the equation of motion equipped with Dirichlet boundary conditions, i.e.,

$$-\operatorname{div} \mathbf{S} + \operatorname{div}(\mathbf{v} \otimes \mathbf{v}) + \nabla \pi = \mathbf{f} \qquad \text{in } \Omega \,,$$
$$\operatorname{div} \mathbf{v} = 0 \qquad \text{in } \Omega \,, \tag{14.4.4}$$
$$\mathbf{v} = \mathbf{0} \qquad \text{on } \partial\Omega \,,$$

where we have set $\mathbf{f} := \mathbf{g} + [\nabla \mathbf{E}]\mathbf{P}$ and have re-written the convective term as $\mathrm{div}(\mathbf{v} \otimes \mathbf{v})$ with $\mathbf{v} \otimes \mathbf{v}$ denoting the tensor product of the vector $\mathbf{v}$ with itself defined as $(v_i v_j)_{i,j=1,\ldots,n}$. For simplicity, we denote $p(x) := p(|\mathbf{E}(x)|^2)$ which from now on is assumed to be log-Hölder continuous with $1 < p^- \leqslant p^+ < \infty$. We refrain from considering the concrete form of the extra stress tensor in (14.4.3), but assume only that $\mathbf{S}(x, \mathbf{D})$ satisfies the following *coercivity* and *growth* conditions

$$\mathbf{S}(x, \mathbf{D}) \cdot \mathbf{D} \geqslant C_1 \left(1 + |\mathbf{D}|\right)^{p(x)-2} |\mathbf{D}|^2 , \tag{14.4.5}$$

$$|\mathbf{S}(x, \mathbf{D})| \leqslant C_2 \left(1 + |\mathbf{D}|\right)^{p(x)-2} |\mathbf{D}| \tag{14.4.6}$$

for all $\mathbf{D} \in \mathbb{R}^{n \times n}_{\mathrm{sym}} := \{\mathbf{D} \in \mathbb{R}^{n \times n} : \mathbf{D} = \mathbf{D}^\top\}$, and all $x \in \Omega$, with Ω being a bounded domain in $\mathbb{R}^n$, $n \geqslant 2$, and is *strictly monotone*, i.e.,

$$(\mathbf{S}(x, \mathbf{D}) - \mathbf{S}(x, \mathbf{C})) \cdot (\mathbf{D} - \mathbf{C}) > 0 \tag{14.4.7}$$

for all $\mathbf{D} \neq \mathbf{C} \in \mathbb{R}^{n \times n}_{\mathrm{sym}}$ and all $x \in \Omega$. In the above formulas we have used the notation $\mathbf{A} \cdot \mathbf{B}$ for the usual scalar product between two tensors. Analogously we denote the usual scalar product between two vectors by $\mathbf{v} \cdot \mathbf{w}$.

In this situation one can use the theory of monotone operators to show the existence of a weak solution. Before we do so let us introduce the relevant functions space for the treatment of problems for incompressible fluids. We denote by $\mathcal{V}$ the space of mappings $\mathbf{u}$ from $C_0^\infty(\Omega)$ which additionally satisfy $\mathrm{div}\, \mathbf{u} = 0$ and set

$$V_{p(\cdot)}(\Omega) := \text{ closure of } \mathcal{V} \text{ in } \|\mathbf{D} \cdot\|_{L^{p(\cdot)}(\Omega)} - \text{norm},$$

where $\mathbf{Du} = \frac{1}{2}(\nabla \mathbf{u} + \nabla \mathbf{u}^\top)$ is the symmetric velocity gradient. Due to Korn's inequality (Theorem 14.3.21 and Corollary 8.2.5) the space $V_{p(\cdot)}(\Omega)$ can be equivalently equipped with the $\|\cdot\|_{W^{1,p(\cdot)}(\Omega)}$-norm or the $\|\nabla \cdot\|_{L^{p(\cdot)}(\Omega)}$-norm. Consequently the space $V_{p(\cdot)}(\Omega)$ is identical with the subspace of divergence free functions from $D_0^{1,p(\cdot)}(\Omega)$.

Theorem 14.4.8. *Let* $\Omega \subset \mathbb{R}^n$ *be a bounded domain with Lipschitz boundary* $\partial\Omega$ *and let* $p : \Omega \to (1, \infty)$ *be log-Hölder continuous with* $\frac{3n}{n+2} \leqslant p^- \leqslant p^+ < \infty$. *Assume that* $\mathbf{f} \in W^{-1,p'(\cdot)}(\Omega)$ *and that* $\mathbf{S}$ *satisfies* (14.4.5)–(14.4.7). *Then there exists a weak solution* $(\mathbf{v}, \pi)$ *of the problem* (14.4.4) *such that*

$$\mathbf{v} \in V_{p(\cdot)}(\Omega) \quad and \quad \pi \in L_0^{p'(\cdot)}(\Omega),$$

satisfy the weak formulation

$$\int_\Omega \mathbf{S}(x, \mathbf{Dv}) \cdot \mathbf{D}\psi \, dx - \int_\Omega \mathbf{v} \otimes \mathbf{v} \cdot \nabla \psi \, dx - \int_\Omega \pi \, \mathrm{div}\, \psi \, dx = \langle \mathbf{f}, \psi \rangle \quad (14.4.9)$$

for all $\psi \in W_0^{1,p(\cdot)}(\Omega)$.

Proof. The proof follows from an easy adaptation of Brezis' theorem on pseudo-monotone operators (cf. [58, 378]). The main obstacle is the identification of the limit in the nonlinear elliptic operator induced by $-\mathrm{div}\,\mathbf{S}$ for a sequence of approximate solutions. In view of the divergence constraint $\mathrm{div}\,\mathbf{v} = 0$ we use first, instead of the weak formulation above, a weak formulation with divergence-free test functions, namely

$$\int_\Omega \mathbf{S}(x, \mathbf{Dv}) \cdot \mathbf{D}\psi \, dx - \int_\Omega \mathbf{v} \otimes \mathbf{v} \cdot \nabla \psi \, dx - \langle \mathbf{f}, \psi \rangle = 0 \qquad (14.4.10)$$

has to be satisfied for all $\psi \in V_{p(\cdot)}(\Omega)$. Later we will recover from this equation the pressure π.

By Theorem 12.2.3 and the above remarks the space $V_{p(\cdot)}(\Omega)$ is a reflexive, separable Banach space. The growth condition (14.4.6) and the monotonicity condition (14.4.7) imply that $-\mathrm{div}\,\mathbf{S}$ defines a strictly monotone operator from $V_{p(\cdot)}(\Omega)$ to $(V_{p(\cdot)}(\Omega))^*$ and from $W_0^{1,p(\cdot)}(\Omega)$ to $W_0^{-1,p'(\cdot)}(\Omega)$. Using the embeddings $W_0^{1,p(\cdot)}(\Omega) \hookrightarrow W_0^{1,p^-}(\Omega) \hookrightarrow\hookrightarrow L^q(\Omega)$, for all $1 \leqslant q < \frac{np^-}{n-p^-}$ one easily checks that the convective term $\mathrm{div}(\mathbf{v} \otimes \mathbf{v})$ induces a strongly continuous operator from $V_{p(\cdot)}(\Omega)$ to $(V_{p(\cdot)}(\Omega))^*$ and from $W_0^{1,p(\cdot)}(\Omega)$ to $W_0^{-1,p'(\cdot)}(\Omega)$, if $\frac{3n}{n+2} \leqslant p^- \leqslant p^+ < \infty$ (cf. [271]). Moreover one easily checks, using that $\mathbf{v}$ is divergence-free, that

$$\int_\Omega \mathbf{v} \otimes \mathbf{v} \cdot \nabla \mathbf{v} \, dx = 0 \, . \qquad (14.4.11)$$

From the assumption on $\mathbf{f}$ and the remarks above it follows that $\mathbf{f}$ can be viewed also as a linear bounded functional on the space $D_0^{1,p(\cdot)}(\Omega)$ equipped with the $\|\mathbf{D}\cdot\|_{L^{p(\cdot)}(\Omega)}$-norm. A straightforward modification of the proof of Proposition 12.3.5 implies that there exists $\mathbf{F} \in (L^{p'(\cdot)}(\Omega))^{n\times n}$ with $\mathbf{F} = \mathbf{F}^\top$, such that

$$\langle \mathbf{f}, \psi \rangle = \int_\Omega \mathbf{F} \cdot \nabla \psi \, dx = \int_\Omega \mathbf{F} \cdot \mathbf{D}\psi \, dx \qquad (14.4.12)$$

for all $\psi \in W_0^{1,p(\cdot)}(\Omega)$. This representation also implies that $\mathbf{f} \in (V_{p(\cdot)}(\Omega))^*$.

The only point which has to be modified is the derivation of the a priori estimate, due to the different behaviour of the modular and the norm,

compared to classical Sobolev spaces with constant exponents. We derive
the a priori estimate only formally, since it can be easily justified using the
Galerkin method (cf. [58, 378]). Using $\mathbf{v}$ as a test function in (14.4.10) we
obtain

$$\int_\Omega |\mathbf{Dv}|^{p(x)}\, dx \leqslant c\left(1 + \int_\Omega |\mathbf{F} \cdot \mathbf{Dv}|\, dx\right),$$

where we used the property (14.4.11) of the convective term, (14.4.12), and
(14.4.5), which implies that there exists a constant depending on p and
C_1 such that $\mathbf{S}(x, \mathbf{D}) \cdot \mathbf{D} \geqslant c(|\mathbf{D}|^{p(x)} - 1)$. Now we apply point-wise
Young's inequality to estimate $|\mathbf{F} \cdot \mathbf{Dv}| \leqslant \varepsilon |\mathbf{D}|^{p(\cdot)} + c(\varepsilon^{-1})|\mathbf{F}|^{p'(\cdot)}$. Choosing
ε sufficiently small and using $\mathbf{F} \in (L^{p'(\cdot)}(\Omega))^{n \times n}$ we thus obtain

$$\int_\Omega |\mathbf{Dv}|^{p(x)}\, dx \leqslant c\big(\|\mathbf{F}\|_{L^{p'(\cdot)}(\Omega)}\big). \tag{14.4.13}$$

Having at our disposal this a priori estimate one can now proceed exactly as
in the proof of Brezis' theorem of pseudo-monotone operators to obtain the
existence of a weak solution $\mathbf{v} \in V_{p(\cdot)}(\Omega)$ satisfying (14.4.10).

It remains to recover the pressure. Since the left-hand side of (14.4.10)
defines a bounded linear functional on $W_0^{1,p^+}(\Omega)$, which vanishes for all
$\psi \in \mathcal{V}$, we obtain from deRahm's theorem (cf. [361, Chap. 1] or [357, Chap. 2])
the existence of an associated pressure $\pi \in L_0^{(p^+)'}(\Omega)$ such that the weak formulation (14.4.9) is satisfied for all $\psi \in C_0^\infty(\Omega)$. From the properties of $\mathbf{v}$ we
deduce in the following way that $\pi \in L_0^{p'(\cdot)}(\Omega)$:

$$\|\pi\|_{L_0^{p'(\cdot)}(\Omega)} \leqslant 2 \sup_{\substack{\eta \in C_{0,0}^\infty(\Omega) \\ \|\eta\|_{L^{p(\cdot)}(\Omega)} \leqslant 1}} \left| \int_\Omega \pi \eta\, dx \right| \leqslant c \sup_{\substack{\psi \in C_0^\infty(\Omega) \\ \|\nabla\psi\|_{L^{p(\cdot)}(\Omega)} \leqslant 1}} \left| \int_\Omega \pi \operatorname{div} \psi\, dx \right|$$

$$= 2 \sup_{\substack{\psi \in C_0^\infty(\Omega) \\ \|\nabla\psi\|_{L^{p(\cdot)}(\Omega)} \leqslant c}} \left| \int_\Omega \mathbf{S}(x, \mathbf{Dv}) \cdot \mathbf{D}\psi - \mathbf{v} \otimes \mathbf{v} \cdot \nabla\psi\, dx - \langle \mathbf{f}, \psi \rangle \right|$$

$$\leqslant c\left(\|\mathbf{S}(x, \mathbf{Dv})\|_{L^{p'(\cdot)}(\Omega)} + \|\mathbf{v}\|_{L^{2(p-)'}(\Omega)}^2 + \|\mathbf{f}\|_{W^{-1,p'(\cdot)}(\Omega)} \right)$$

$$\leqslant c\left(\|\mathbf{S}(x, \mathbf{Dv})\|_{L^{p'(\cdot)}(\Omega)} + \|\mathbf{v}\|_{V_{p(\cdot)}(\Omega)}^2 + \|\mathbf{f}\|_{W^{-1,p'(\cdot)}(\Omega)} \right) < \infty,$$

where we used the norm conjugate formula in $L_0^{p'(\cdot)}(\Omega)$ (Corollary 3.4.13), the
solvability of the divergence equation (Theorem 14.3.15), the weak formula-
tion (14.4.9), Hölder's inequality, the embeddings $V_{p(\cdot)}(\Omega) \hookrightarrow (W^{1,(p^-)}(\Omega))^n$
$\hookrightarrow (L^{2(p^-)'}(\Omega))^n$ valid for $\frac{3n}{n+2} \leqslant p^- \leqslant p^+ < \infty$, and the growth condition
(14.4.6) together with (14.4.13). Since now all terms in (14.4.9) are finite for

test functions $\boldsymbol{\psi} \in (W^{1,p(\cdot)}(\Omega))^n$ we obtain (14.4.9) by continuity also for such test functions. □

The lower bound $p^- \geqslant \frac{3n}{n+2}$ in Theorem 14.4.8 comes from the fact that the convective term is well defined for test functions $\boldsymbol{\psi}$ from the energy space $V_{p(\cdot)}(\Omega)$ only for such values of the exponent. If we relax the requirements on the space of test functions, then we obtain results for lower bounds for p^-. In this case however, one cannot use the theory of monotone operators to identify the limit in the nonlinear operator induced by $-\operatorname{div} \mathbf{S}$ for a sequence of approximate solutions. This problem can be overcome by using the Theorem of Vitali (Theorem 1.4.12). The application of this theorem requires us to show the almost everywhere convergence of gradients of approximate solutions. This is possible due to the monotonicity of the operator induced by $-\operatorname{div} \mathbf{S}$ and the approximation property of Sobolev functions by Lipschitz functions from Sect. 9.5. Using these ideas one can prove the following result:

Theorem 14.4.14. *Let $\Omega \subset \mathbb{R}^n$ be a bounded domain with Lipschitz boundary $\partial\Omega$ and let p be* log-Hölder *continuous with $\frac{2n}{n+2} < p^- \leqslant p^+ < \infty$. Define the variable exponent s by $\frac{1}{s'(\cdot)} := \max\left\{1 - \frac{1}{p(\cdot)}, 2(\frac{1}{p(\cdot)} - \frac{1}{n})\right\}$. Assume that $\mathbf{f} \in W^{-1,p'(\cdot)}(\Omega)$ and that $\mathbf{S}$ satisfies (14.4.5)–(14.4.7). Then there exists a weak solution $(\mathbf{v}, \pi)$ of the problem (14.4.4) with*

$$\mathbf{v} \in V_{p(\cdot)}(\Omega) \quad and \quad \pi \in L_0^{s'(\cdot)}(\Omega).$$

The proof of this result is based on the *weak stability* of the system (14.4.4).

Lemma 14.4.15. *Let $\Omega \subset \mathbb{R}^n$ be a bounded domain with Lipschitz continuous boundary $\partial\Omega$, let p be globally* log-Hölder *continuous with $1 < p^- \leqslant p^+ < \infty$ and let $\mathbf{S}$ satisfy (14.4.6) and (14.4.7). Let $\mathbf{v}^n, \mathbf{v} \in V_{p(\cdot)}(\Omega)$ with $\mathbf{v}^k \rightharpoonup \mathbf{v}$ in $V_{p(\cdot)}(\Omega)$. Let $\mathbf{u}^k := \mathbf{v}^k - \mathbf{v}$ and let $\mathbf{u}^{k,j}$ be the approximations of $\mathbf{u}^k$ as in Theorem 9.5.2. Assume that*

$$\lim_{k\to\infty} \left| \int_\Omega \left(\mathbf{S}(x, \mathbf{Dv}^k) - \mathbf{S}(x, \mathbf{Dv})\right) \cdot \mathbf{Du}^{k,j} \, dx \right| \leqslant \delta_j \qquad (14.4.16)$$

for all $j \in \mathbb{N}$, where $\lim_{j\to\infty} \delta_j = 0$. Then, for any $0 < \theta < 1$,

$$\limsup_{k\to\infty} \int_\Omega \left(\left(\mathbf{S}(x, \mathbf{Dv}^k) - \mathbf{S}(x, \mathbf{Dv})\right) \cdot (\mathbf{Dv}^k - \mathbf{Dv})\right)^\theta dx = 0.$$

Proof. Recall that a bounded domain with Lipschitz boundary $\partial\Omega$ has a fat complement and thus Theorem 9.5.2 is applicable. The monotonicity $(\mathbf{S}(x, \mathbf{Dv}^k) - \mathbf{S}(x, \mathbf{Dv})) \cdot \mathbf{Du}^k \geqslant 0$ and Hölder's inequality yield for $0 < \theta < 1$ and all $j \in \mathbb{N}$,

$$\int_{\Omega} \left(\left(\mathbf{S}(x, \mathbf{Dv}^k) - \mathbf{S}(x, \mathbf{Dv}) \right) \cdot \mathbf{Du}^k \right)^{\theta} dx$$

$$\leqslant \left(\int_{\Omega} \left(\mathbf{S}(x, \mathbf{Dv}^k) - \mathbf{S}(x, \mathbf{Dv}) \right) \cdot \mathbf{Du}^k \chi_{\{\mathbf{u}^k = \mathbf{u}^{k,j}\}} dx \right)^{\theta} |\Omega|^{1-\theta}$$

$$+ \left(\int_{\Omega} \left(\mathbf{S}(x, \mathbf{Dv}^k) - \mathbf{S}(x, \mathbf{Dv}) \right) \cdot \mathbf{Du}^k \chi_{\{\mathbf{u}^k \neq \mathbf{u}^{k,j}\}} dx \right)^{\theta} |\{\mathbf{u}^k \neq \mathbf{u}^{k,j}\}|^{1-\theta}$$

$$=: Y_{k,j,1}^{\theta} |\Omega|^{1-\theta} + Y_{k,j,2}^{\theta} |\{\mathbf{u}^k \neq \mathbf{u}^{k,j}\}|^{1-\theta}.$$

In order to estimate the last two terms we first notice that $\mathbf{v}^k \rightharpoonup \mathbf{v}$ in $V_{p(\cdot)}(\Omega)$ implies $\sup_k \|\mathbf{v}^k\|_{V_{p(\cdot)}(\Omega)} \leqslant c$, which together with the growth condition of $\mathbf{S}$ (14.4.6) yields

$$\sup_k \|\mathbf{S}(\cdot, \mathbf{Dv}^k)\|_{L^{p'(\cdot)}(\Omega)}, \|\mathbf{S}(\cdot, \mathbf{Dv})\|_{L^{p'(\cdot)}(\Omega)} \leqslant c.$$

Hölder's inequality and the last estimate yield for all $j \in \mathbb{N}$

$$Y_{k,j,1} = \int_{\Omega} \left(\mathbf{S}(x, \mathbf{Dv}^k) - \mathbf{S}(x, \mathbf{Dv}) \right) \cdot \mathbf{Du}^{k,j} \chi_{\{\mathbf{u}^{k,j} = \mathbf{u}^k\}} dx$$

$$\leqslant \left| \int_{\Omega} \left(\mathbf{S}(x, \mathbf{Dv}^k) - \mathbf{S}(x, \mathbf{Dv}) \right) \cdot \mathbf{Du}^{k,j} dx \right|$$

$$+ \left| \int_{\Omega} \left(\mathbf{S}(x, \mathbf{Dv}^k) - \mathbf{S}(x, \mathbf{Dv}) \right) \cdot \mathbf{Du}^{k,j} \chi_{\{\mathbf{u}^{k,j} \neq \mathbf{u}^k\}} dx \right|$$

$$\leqslant \left| \int_{\Omega} \left(\mathbf{S}(x, \mathbf{Dv}^k) - \mathbf{S}(x, \mathbf{Dv}) \right) \cdot \mathbf{Du}^{k,j} dx \right|$$

$$+ c \|\nabla \mathbf{u}^{k,j} \chi_{\{\mathbf{u}^{k,j} \neq \mathbf{u}^k\}}\|_{L^{p(\cdot)}(\Omega)}.$$

Using similar estimates, we also get from Hölder's inequality and the estimates for $\mathbf{S}(\cdot, \mathbf{Dv}^k)$, $\mathbf{S}(\cdot, \mathbf{Dv})$ and $\mathbf{Dv}^k$, $\mathbf{Dv}$ in $(L^{p'(\cdot)}(\Omega))^{n \times n}$ and $(L^{p(\cdot)}(\Omega))^{n \times n}$, respectively, that

$$Y_{k,j,2} \leqslant c.$$

From the vector valued version of Corollary 9.5.4, the embedding $L^{p(\cdot)}(\Omega) \hookrightarrow L^1(\Omega)$, and $\lambda_{k,j} \geqslant 1$ we deduce

$$\limsup_{k \to \infty} |\{\mathbf{u}^k \neq \mathbf{u}^{k,j}\}| = \limsup_{k \to \infty} \|\chi_{\{\mathbf{u}^k \neq \mathbf{u}^{k,j}\}}\|_{L^1(\Omega)}$$

$$\leqslant c \limsup_{k \to \infty} \|\lambda_{k,j} \chi_{\{\mathbf{u}^k \neq \mathbf{u}^{k,j}\}}\|_{L^{p(\cdot)}(\Omega)} \leqslant \varepsilon_j$$

and

$$\|\nabla\mathbf{u}^{k,j}\chi_{\{\mathbf{u}^{k,j}\neq\mathbf{u}^{k}\}}\|_{L^{p(\cdot)}(\Omega)} \leqslant c\,\varepsilon_j\,.$$

These estimates together with (14.4.16) imply for all $j \in \mathbb{N}$

$$\limsup_{k\to\infty} \int_{\Omega} \left((\mathbf{S}(x,\mathbf{Dv}^{k}) - \mathbf{S}(x,\mathbf{Dv})) \cdot \mathbf{Du}^{k} \right)^{\theta} dx \leqslant c\left(\varepsilon_j^{\theta} + \delta_j^{\theta} + \varepsilon_j^{1-\theta}\right).$$

Since the last estimate holds for all $j \in \mathbb{N}$ and $\lim_{j\to\infty}\varepsilon_j = \lim_{j\to\infty}\delta_j = 0$, we obtain the assertion of the lemma. $\qquad\square$

Corollary 14.4.17. *Let all assumptions of Lemma 14.4.15 be satisfied. Then there exists a subsequence, labeled again* $\mathbf{v}^{k}$, *satisfying*

$$\mathbf{Dv}^{k} \to \mathbf{Dv} \qquad a.e.\ in\ \Omega$$

as $k \to \infty$.

Proof. From Lemma 14.4.15 we know that

$$\limsup_{k\to\infty} \int_{\Omega} \left((\mathbf{S}(x,\mathbf{Dv}^{k}) - \mathbf{S}(x,\mathbf{Dv})) \cdot (\mathbf{Dv}^{k} - \mathbf{Dv}) \right)^{\theta} dx = 0.$$

By monotonicity, the integrand is non-negative, hence it tends to zero almost everywhere. $\qquad\square$

Now we can prove the existence of weak solutions for the case $p^{-} > \frac{2n}{n+2}$.

Proof of Theorem 14.4.14. We will only consider the case that $p^{-} < \frac{3n}{n+2} \leqslant n$, since the other case is covered by Theorem 14.4.8. Choose $q > \frac{2p^{-}}{p^{-}-1} = 2(p^{-})'$ and let $\mathbf{v}^{k} \in V_{p(\cdot)}(\Omega) \cap L^{q}(\Omega)$ be a weak solution of the approximate problem

$$\int_{\Omega} \mathbf{S}(x,\mathbf{Dv}^{k}) \cdot \mathbf{D}\psi - \mathbf{v}^{k} \otimes \mathbf{v}^{k} \cdot \nabla\psi + \frac{1}{k}|\mathbf{v}^{k}|^{q-2}\mathbf{v}^{k} \cdot \psi\, dx = \langle \mathbf{f}, \psi \rangle \quad (14.4.18)$$

for all $\psi \in V_{p(\cdot)}(\Omega) \cap L^{q}(\Omega)$.

The existence of a weak solution to this approximate problem can be obtained in the same way as in the proof of Theorem 14.4.8, if one replaces the energy space $V_{p(\cdot)}(\Omega)$ there with the natural energy space $V_{p(\cdot)}(\Omega) \cap L^{q}(\Omega)$ of the problem (14.4.18). The choice of the value for q is due to the convective term since

$$\int_\Omega \mathbf{v}^k \otimes \mathbf{v}^k \cdot \mathbf{D}\psi\, dx \leqslant \|\mathbf{v}^k\|^2_{L^{2(p^-)'}(\Omega)} \|\nabla\psi\|_{L^{p^-}(\Omega)} \leqslant C(k)$$

for $k \in \mathbb{N}$ and $\psi \in V_{p(\cdot)}(\Omega)$ by Hölder's and Korn's inequality and the embedding $L^{p(\cdot)}(\Omega) \hookrightarrow L^{p^-}(\Omega)$.

Choosing $\psi = \mathbf{v}^k$ in (14.4.18) we easily obtain as in the derivation of (14.4.13) that

$$\int_\Omega |\mathbf{D}\mathbf{v}^k|^{p(x)}\, dx + \frac{1}{k}\|\mathbf{v}^k\|^q_{L^q(\Omega)} \leqslant c\big(\|\mathbf{F}\|_{L^{p'(\cdot)}(\Omega)}\big). \qquad (14.4.19)$$

Consequently, due to the growth condition (14.4.6), the embedding $V_{p(\cdot)}(\Omega) \hookrightarrow V_{p^-}(\Omega)$, the comments before Theorem 14.4.8, and the classical Sobolev embedding theorem

$$\begin{aligned}
\|\mathbf{S}(\cdot, \mathbf{D}\mathbf{v}^k)\|_{L^{p'(\cdot)}(\Omega)} &\leqslant c\big(\|\mathbf{F}\|_{L^{p'(\cdot)}(\Omega)}\big), \\
\|\mathbf{v}^k\|_{L^{\frac{np^-}{n-p^-}}(\Omega)} &\leqslant c\big(\|\mathbf{F}\|_{L^{p'(\cdot)}(\Omega)}\big), \\
\|\mathbf{v}^k \otimes \mathbf{v}^k\|_{L^{\frac{np^-}{2(n-p^-)}}(\Omega)} &\leqslant c\big(\|\mathbf{F}\|_{L^{p'(\cdot)}(\Omega)}\big).
\end{aligned} \qquad (14.4.20)$$

Obviously, the estimate (14.4.19) implies the existence of $\mathbf{v} \in V_{p(\cdot)}(\Omega)$, and a subsequence, which will be denoted again by $\mathbf{v}^k$ such that

$$\begin{aligned}
\mathbf{v}^k &\rightharpoonup \mathbf{v} \qquad \text{weakly in } V_{p(\cdot)}(\Omega)\,, \\
\frac{1}{k}\int_\Omega |\mathbf{v}^k|^{q-2}\mathbf{v}^k \cdot \psi\, dx &\to 0 \qquad \text{for all } \psi \in L^q(\Omega)\,,
\end{aligned} \qquad (14.4.21)$$

and due to the compact embeddings $V_{p(\cdot)}(\Omega) \hookrightarrow V_{p^-}(\Omega) \hookrightarrow\hookrightarrow L^\sigma(\Omega)$, $\sigma \in [1, \frac{np^-}{n-p^-})$, and $V_{p(\cdot)}(\Omega) \hookrightarrow\hookrightarrow L^{p(\cdot)}(\Omega)$ (Theorem 8.4.2)

$$\mathbf{v}^k \to \mathbf{v} \quad \text{strongly in } L^\sigma(\Omega) \cap L^{p(\cdot)}(\Omega)\,. \qquad (14.4.22)$$

Note that $\sigma \geqslant 2$ provided $p^- > \frac{2n}{n+2}$, which is the lower bound for p^- in Theorem 14.4.14. For these exponents p we fix some $\sigma \in (2, \frac{np^-}{n-p^-})$ and thus obtain

$$\int_\Omega \mathbf{v}^k \otimes \mathbf{v}^k \cdot \nabla\psi\, dx \to \int_\Omega \mathbf{v} \otimes \mathbf{v} \cdot \nabla\psi\, dx \qquad (14.4.23)$$

for all $\psi \in W_0^{1,(\sigma/2)'}(\Omega)$. Our next goal is to prove that also

$$\int_{\Omega} \mathbf{S}(x, \mathbf{D}\mathbf{v}^k) \cdot \mathbf{D}\psi \, dx \to \int_{\Omega} \mathbf{S}(x, \mathbf{D}\mathbf{v}) \cdot \mathbf{D}\psi \, dx \qquad (14.4.24)$$

for all $\psi \in W_0^{1,p^+ +1}(\Omega)$. By virtue of (14.4.13), (14.4.6), Hölder's inequality (Lemma 3.2.20) and Lemma 3.2.5 the integrands in (14.4.24) are equi-integrable. Thus, it suffices in view of Vitali's theorem (Theorem 1.4.12), to show at least for a subsequence that

$$\mathbf{D}\mathbf{v}^k \to \mathbf{D}\mathbf{v} \quad \text{a.e. in } \Omega. \qquad (14.4.25)$$

In view of Lemma 14.4.15, Corollary 14.4.17, and (14.4.21) we thus have to verify the assumption (14.4.16) of Lemma 14.4.15. As in that lemma we set

$$\mathbf{u}^k := \mathbf{v}^k - \mathbf{v}$$

and denote the Lipschitz approximations from Theorem 9.5.2 applied to the vector valued functions $\mathbf{u}^k$ by $(\mathbf{u}^{k,j})_{j \in \mathbb{N}}$. Note that the functions $\mathbf{u}^{k,j}$ are in general not divergence free and we have to correct them in order to use them as a test function in (14.4.18). For that we use the Bogovskiĭ operator $\mathcal{B}$, whose existence is ensured by Theorem 14.3.15. This linear, bounded operator maps $L_0^{p(\cdot)}(\Omega)$ into $W_0^{1,p(\cdot)}(\Omega)$ and simultaneously $L_0^q(\Omega)$ into $W_0^{1,q}(\Omega)$, $1 < q < \infty$. The function $\mathcal{B}h$ is a solution of the divergence equation $\mathrm{div}(\mathcal{B}h) = h$ and satisfies the estimates

$$\begin{aligned} \|\nabla \mathcal{B}h\|_{L^{p(\cdot)}(\Omega)} &\leqslant c\,\|h\|_{L^{p(\cdot)}(\Omega)}, \\ \|\nabla \mathcal{B}h\|_{L^q(\Omega)} &\leqslant c\,\|h\|_{L^q(\Omega)}, \end{aligned} \qquad (14.4.26)$$

with constants c depending only on Ω, p and Ω, q, respectively. We define

$$\boldsymbol{\xi}^{k,j} := \mathcal{B}(\mathrm{div}\,\mathbf{u}^{k,j}).$$

From Corollary 9.5.4 we know that for each fixed $j \in \mathbb{N}$, $\nabla \mathbf{u}^{k,j} \rightharpoonup \mathbf{0}$ in $L^q(\Omega)$. This together with the fact that continuous linear operators preserve weak convergence and classical compact embedding theorems implies, for each $j \in \mathbb{N}$, that

$$\begin{aligned} \boldsymbol{\zeta}^{k,j} &\rightharpoonup \mathbf{0} \quad \text{weakly in } W^{1,q}(\Omega) \quad \text{as } k \to \infty, \\ \boldsymbol{\zeta}^{k,j} &\to \mathbf{0} \quad \text{strongly in } L^q(\Omega) \quad \text{as } k \to \infty. \end{aligned} \qquad (14.4.27)$$

Due to $\nabla \mathbf{u}^k = \nabla \mathbf{u}^{k,j}$ on the set $\{\mathbf{u}^k = \mathbf{u}^{k,j}\}$ [280, Corollary 1.43] and $\mathrm{div}\,\mathbf{u}^k = 0$ we get $\mathrm{div}\,\mathbf{u}^{k,j} = \chi_{\{\mathbf{u}^k \neq \mathbf{u}^{k,j}\}}\,\mathrm{div}\,\mathbf{u}^{k,j}$. Consequently,

$$\|\boldsymbol{\xi}^{k,j}\|_{W^{1,p(\cdot)}(\Omega)} \leqslant c\,\|\mathrm{div}\,\mathbf{u}^{k,j}\,\chi_{\{\mathbf{u}^k \neq \mathbf{u}^{k,j}\}}\|_{L^{p(\cdot)}(\Omega)}$$

and Corollary 9.5.4 yields

$$\limsup_{k\to\infty} \|\boldsymbol{\xi}^{k,j}\|_{W^{1,p(\cdot)}(\Omega)} \leqslant c \limsup_{k\to\infty} \|\operatorname{div} \mathbf{u}^{k,j} \chi_{\{\mathbf{u}^k \neq \mathbf{u}^{k,j}\}}\|_{L^{p(\cdot)}(\Omega)}$$

$$\leqslant c \limsup_{k\to\infty} \|\nabla \mathbf{u}^{k,j} \chi_{\{\mathbf{u}^k \neq \mathbf{u}^{k,j}\}}\|_{L^{p(\cdot)}(\Omega)} \quad (14.4.28)$$

$$\leqslant c \varepsilon_j .$$

Now we set

$$\boldsymbol{\eta}^{k,j} := \mathbf{u}^{k,j} - \boldsymbol{\xi}^{k,j}$$

and observe that we have in view of the above construction $\boldsymbol{\eta}^{k,j} \in V_{p(\cdot)} \cap L^{2(p^-)'}(\Omega)$. Moreover, from the properties of $\mathbf{u}^{k,j}$ proved in Corollary 9.5.4, and (14.4.27) we have for each $j \in \mathbb{N}$,

$$\begin{aligned} \boldsymbol{\eta}^{k,j} \rightharpoonup \mathbf{0} \quad &\text{weakly in } W^{1,q}(\Omega) \quad \text{as } k \to \infty , \\ \boldsymbol{\eta}^{k,j} \to \mathbf{0} \quad &\text{strongly in } L^q(\Omega) \quad \text{as } k \to \infty . \end{aligned} \quad (14.4.29)$$

Thus we can use the test function $\boldsymbol{\eta}^{k,j}$ in the weak formulation (14.4.18). This equation can be re-written as

$$\int_\Omega \left(\mathbf{S}(x, \mathbf{D}\mathbf{v}^k) - \mathbf{S}(x, \mathbf{D}\mathbf{v}) \right) \cdot \mathbf{D}\mathbf{u}^{k,j} \, dx$$

$$= \int_\Omega \mathbf{S}(x, \mathbf{D}\mathbf{v}^k) \cdot \mathbf{D}\boldsymbol{\xi}^{k,j} \, dx - \int_\Omega \mathbf{S}(x, \mathbf{D}\mathbf{v}) \cdot \mathbf{D}\mathbf{u}^{k,j} \, dx$$

$$- \frac{1}{k} \int_\Omega |\mathbf{v}^k|^{q-2} \mathbf{v}^k \cdot \boldsymbol{\eta}^{k,j} \, dx + \langle \mathbf{f}, \boldsymbol{\eta}^{k,j} \rangle + \int_\Omega \mathbf{v}^k \otimes \mathbf{v}^k \cdot \nabla \boldsymbol{\eta}^{k,j} \, dx$$

$$=: J^1_{k,j} + J^2_{k,j} + J^3_{k,j} + J^4_{k,j} .$$

Using the first estimate in (14.4.20), $\nabla \mathbf{u}^{k,j} \rightharpoonup \mathbf{0}$ in $L^q(\Omega)$; (14.4.19), (14.4.29), the assumption of $\mathbf{f}$; and (14.4.22), (14.4.29) we deduce for all $j \in \mathbb{N}$

$$\lim_{k\to\infty} |J^2_{k,j}| + |J^3_{k,j}| + |J^4_{k,j}| = 0 .$$

On the other hand with Hölder's inequality, the first estimate in (14.4.20), and (14.4.28) we get

$$\limsup_{k\to\infty} |J^1_{k,j}| \leqslant c(K) \varepsilon_j.$$

Thus we have shown for all $j \in \mathbb{N}$

$$\limsup_{k\to\infty} \int_\Omega \big(\mathbf{S}(x,\mathbf{D}\mathbf{v}^k) - \mathbf{S}(x,\mathbf{D}\mathbf{v})\big) \cdot \mathbf{D}\mathbf{u}^{k,j}\, dx \leqslant c(K)\,\varepsilon_j\,,$$

which is assumption (14.4.16) of Lemma 14.4.15. Consequently, Corollary 14.4.17 yields (14.4.25), which in view of Vitali's theorem (Theorem 1.4.12) proves (14.4.24). This and (14.4.23), as well as (14.4.21) prove that the weak formulation (14.4.10) is satisfied for all sufficiently smooth divergence-free test functions $\boldsymbol{\psi}$. By continuity and the growth properties of the extra stress tensor and the convective term we than get that (14.4.10) is satisfied for all $\boldsymbol{\psi} \in V_{s(\cdot)}(\Omega)$. The existence of a pressure with the stated properties now follows as in the proof of Theorem 14.4.8. The proof of Theorem 14.4.14 is complete.   □

Remark 14.4.30. Note that in both Theorems 14.4.8 and 14.4.14 the assumption that Ω has a Lipschitz continuous boundary is not needed if one is only interested in the existence of a velocity $\mathbf{v}$ satisfying the weak formulation for divergence-free test functions only. This is due to the fact that we treat homogeneous boundary conditions and thus Korn's inequality and the properties of the function spaces also hold for arbitrary bounded domains. In Theorem 14.4.14 one has additionally to localize the Lipschitz truncation theorem (Theorem 9.5.2, Corollary 9.5.4) and the stability result (Lemma 14.4.15). Details can be found in [107].

References

1. R. Aboulaich, D. Meskine, and A. Souissi. New diffusion models in image processing. *Comput. Math. Appl.*, 56:874–882, 2008.
2. E. Acerbi and N. Fusco. An approximation lemma for $W^{1,p}$ functions. In *Material instabilities in continuum mechanics (Edinburgh, 1985–1986)*, Oxford Sci. Publ., pages 1–5. Oxford Univ. Press, New York, 1988.
3. E. Acerbi and G. Mingione. Functionals with $p(x)$ growth and regularity. *Atti Accad. Naz. Lincei Cl. Sci. Fis. Mat. Natur. Rend. Lincei (9) Mat. Appl.*, 11:169–174 (2001), 2000.
4. E. Acerbi and G. Mingione. Regularity results for a class of functionals with non-standard growth. *Arch. Ration. Mech. Anal.*, 156:121–140, 2001.
5. E. Acerbi and G. Mingione. Regularity results for a class of quasiconvex functionals with nonstandard growth. *Ann. Scuola Norm. Sup. Pisa Cl. Sci. (4)*, 30(2):311–339, 2001.
6. E. Acerbi and G. Mingione. Regularity results for electrorheological fluids: the stationary case. *C. R. Acad. Sci. Paris*, 334:817–822, 2002.
7. E. Acerbi and G. Mingione. Regularity results for stationary electro-rheological fluids. *Arch. Ration. Mech. Anal.*, 164:213–259, 2002.
8. E. Acerbi and G. Mingione. Gradient estimates for the $p(x)$–Laplacian system. *J. Reine Angew. Math.*, 584:117–148, 2005.
9. E. Acerbi, G. Mingione, and G. Seregin. Regularity results for parabolic systems related to a class of non-Newtonian fluids. *Ann. Inst. H. Poincaré Anal. Non Linéaire*, 21(1):25–60, 2004.
10. T. Adamowicz and P. Hästö. Mappings of finite distortion and $p(\cdot)$-harmonic functions. *Int. Math. Res. Not.*, pages 1940–1965, 2010.
11. R. Adams. *Sobolev spaces*. Academic Press [A subsidiary of Harcourt Brace Jovanovich, Publishers], New York-London, 1975. Pure and Applied Mathematics, Vol. 65.
12. R. Adams and J. Fournier. *Sobolev spaces*, volume 140 of *Pure and Applied Mathematics (Amsterdam)*. Elsevier/Academic Press, Amsterdam, second edition, 2003.
13. G. Afrouzi and H. Ghorbani. Existence of positive solutions for $p(x)$–Laplacian problems. *Electron. J. Differential Equations*, 177:1–9, 2007.
14. S. Agmon, A. Douglis, and L.Nirenberg. Estimates near the boundary for solutions of elliptic partial differential equations satisfying general boundary conditions. I. *Comm. Pure Appl. Math.*, 12:623–727, 1959.
15. S. Agmon, A. Douglis, and L.Nirenberg. Estimates near the boundary for solutions of elliptic partial differential equations satisfying general boundary conditions. II. *Comm. Pure Appl. Math.*, 17:35–92, 1964.

16. H. Aimar and R. A. Macías. Weighted norm inequalities for the Hardy-Littlewood maximal operator on spaces of homogeneous type. *Proc. Amer. Math. Soc.*, 91(2): 213–216, 1984.

17. K. Ali and M. Bezzarga. On a nonhomogenous quasilinear problem in Sobolev spaces with variable exponent. *Bul. Ştiinţ. Univ. Piteşti Ser. Mat. Inf.*, 14, suppl.:19–38, 2008.

18. Y. Alkhutov. The Harnack inequality and the Hölder property of solutions of nonlinear elliptic equations with nonstandard growth condition. *Differ. Uravn.*, 33:1651–1660, 1726, 1997.

19. Y. Alkhutov. On the Hölder continuity of $p(x)$-harmonic functions. *Mat. Sb.*, 196:3–28, 2005.

20. Y. Alkhutov and O. Krasheninnikova. Continuity at boundary points of solutions of quasilinear elliptic equations with a non-standard growth condition. *Izv. Ross. Akad. Nauk. Ser. Mat.*, 68(6):3–60, 2004.

21. W. Allegretto. Form estimates for the $p(x)$-Laplacean. *Proc. Amer. Math. Soc.*, 135:2177–2185, 2007.

22. A. Almeida. Inversion of Riesz potential operator on Lebesgue spaces with variable exponent. *Fract. Calc. Appl. Anal.*, 6:311–327, 2003.

23. A. Almeida, J. Hasanov, and S. Samko. Maximal and potential operators in variable exponent Morrey spaces. *Georgian Math. J.*, 15:195–208, 2008.

24. A. Almeida and P. Hästö. Besov spaces with variable smoothness and integrability. *J. Funct. Anal.*, 258(5):1628–1655, 2010.

25. A. Almeida and H. Rafeiro. Inversion of the Bessel potential operator in weighted variable Lebesgue spaces. *J. Math. Anal. Appl.*, 340:1336–1346, 2008.

26. A. Almeida and S. Samko. Characterization of Riesz and Bessel potentials on variable Lebesgue spaces. *J. Function Spaces Appl.*, 4:113–144, 2006.

27. A. Almeida and S. Samko. Pointwise inequalities in variable Sobolev spaces and applications. *Z. Anal. Anwend.*, 26:179–193, 2007.

28. A. Almeida and S. Samko. Embeddings of variable Hajłasz-Sobolev spaces into Hölder spaces of variable order. *J. Math. Anal. Appl.*, 353:489–496, 2009.

29. J. Alvarez and C. Pérez. Estimates with A_∞ weights for various singular integral operators. *Boll. Un. Mat. Ital. A (7)*, 8(1):123–133, 1994.

30. C. Alves. Existence of solution for a degenerate $p(x)$-Laplacian equation in $\mathbb{R}^N$. *J. Math. Anal. Appl.*, 345(2):731–742, 2008.

31. B. Andreianov, M. Bendahmane, and S. Ouaro. Structural stability for variable exponent elliptic problems. II. The $p(u)$-Laplacian and coupled problems. *Nonlinear Anal.*, 72(12):4649–4660, 2010.

32. S. Antontsev, M. Chipot, and Y. Xie. Uniqueness results for equations of the $p(x)-$Laplacian type. *Adv. Math. Sci. Appl.*, 17(1):287–304, 2007.

33. S. Antontsev and L. Consiglieri. Elliptic boundary value problems with nonstandard growth conditions. *Nonlinear Anal.*, 71(3–4):891–902, 2009.

34. S. Antontsev and J. Rodrigues. On stationary thermo-rheological viscous flows. *Ann. Univ. Ferrara Sez. VII Sci. Mat.*, 52(1):19–36, 2006.

35. S. Antontsev and S. Shmarev. Elliptic equations and systems with nonstandard growth conditions: Existence, uniqueness and localization properties of solutions. *Nonlinear Anal.*, 65:728–761, 2006.

36. S. Antontsev and S. Shmarev. Parabolic equations with anisotropic nonstandard growth conditions. In *Free boundary problems*, volume 154 of *Internat. Ser. Numer. Math.*, pages 33–44. Birkhäuser, Basel, 2007.

37. S. Antontsev and S. Shmarev. Anisotropic parabolic equations with variable nonlinearity. *Publ. Mat.*, 53(2):355–399, 2009.

38. S. Antontsev and S. Shmarev. Vanishing solutions of anisotropic parabolic equations with variable nonlinearity. *J. Math. Anal. Appl.*, 361(2):371–391, 2010.

39. S. Antontsev and V. Zhikov. Higher integrability for parabolic equations of $p(x,t)$-Laplacian type. *Adv. Differential Equations*, 10(9):1053–1080, 2005.

40. İ. Aydın and A. T. Gürkanlı. The inclusion $L^{p(x)}(\mu) \subseteq L^{q(x)}(\nu)$. *Int. J. Appl. Math.*, 22(7):1031–1040, 2009.

41. A. Beauzamy. Espaces de Sobolev et Besov d'ordre variable définis sur L^p. *C. R. Acad. Sci. Paris (Ser. A)*, 274:1935–1938, 1972.

42. M. Bendahmane and P. Wittbold. Renormalized solutions for nonlinear elliptic equations with variable exponents and L^1 data. *Nonlinear Anal.*, 70(2):567–583, 2009.

43. C. Bennett and R. Sharpley. *Interpolation of operators*, volume 129 of *Pure and Applied Mathematics*. Academic Press Inc., Boston, MA, 1988.

44. E. I. Berezhnoi. Two–weighted estimations for the Hardy–Littlewood maximal function in ideal Banach spaces. *Proc. Amer. Math. Soc*, 127:79–87, 1999.

45. J. Bergh and J. Löfström. *Interpolation spaces. An introduction.* Springer-Verlag, Berlin, 1976. Grundlehren der Mathematischen Wissenschaften, No. 223.

46. O. Besov. Embeddings of spaces of differentiable functions of variable smoothness. *Tr. Mat. Inst. Steklova*, 214:19–53, 1997.

47. O. Besov. Interpolation, embedding, and extension of spaces of functions of variable smoothness, (Russian). *Tr. Mat. Inst. Steklova*, 248:47–58, 2005.

48. V. Bögelein and J. Habermann. Gradient estimates via non standard potentials and continuity. *Ann. Acad. Sci. Fenn. Math.*, 35:641–678, 2010.

49. V. Bögelein and A. Zatorska-Goldstein. Higher integrability of very weak solutions of systems of $p(x)$-Laplacean type. *J. Math. Anal. Appl.*, 336:480–497, 2007.

50. M. Bogovskii. Solution of the first boundary value problem for the equation of continuity of an incompressible medium. *Dokl. Akad. Nauk SSSR*, 248:1037–1040, 1979. English transl. in Soviet Math. Dokl. **20** (1979), 1094–1098.

51. M. Bogovskii. Solution of some vector analysis problems connected with operators div and grad. *Trudy Seminar S.L. Sobolev, Akademia Nauk SSSR*, 80:5–40, 1980.

52. B. Bojarski. Remarks on Sobolev imbedding inequalities. In *Complex analysis, Joensuu 1987*, volume 1351 of *Lecture Notes in Math.*, pages 52–68. Springer, Berlin, 1988.

53. E. Bollt, R. Chartrand, S. Esedoḡlu, P. Schultz, and K. Vixie. Graduated adaptive image denoising: local compromise between total variation and isotropic diffusion. *Adv. Comput. Math.*, 31(1-3):61–85, 2009.

54. J. Boman. L_p-estimates for very strongly elliptic systems. *Department of Mathematics, University of Stockholm, Sweden*, Reports no 29, 1982.

55. M.-M. Boureanu. Existence of solutions for an elliptic equation involving the $p(x)$−Laplace operator. *Electron. J. Differential Equations*, 97:1–10, 2006.

56. M.-M. Boureanu and M. Mihăilescu. Existence and multiplicity of solutions for a Neumann problem involving variable exponent growth conditions. *Glasg. Math. J.*, 50:565–574, 2008.

57. S. Boza and J. Soria. Weighted Hardy modular inequalities in variable L^p spaces for decreasing functions. *J. Math. Anal. Appl.*, 348:383–388, 2008.

58. H. Brezis. *Analyse fonctionnelle, Théorie et applications.* Masson, Paris, 1983.

59. S. Buckley and P. Koskela. Sobolev-Poincaré implies John. *Mat. Res. Lett.*, 2:577–593, 1995.

60. S. Buckley, P. Koskela, and G. Lu. Subelliptic Poincaré inequalities: the case $p < 1$. *Publ. Mat.*, 39(2):313–334, 1995.

61. A. Cabada and R. L. Pouso. Existence theory for functional p-Laplacian equations with variable exponents. *Nonlinear Anal.*, 52:557–572, 2003.

62. A. Calderón. Intermediate spaces and interpolation, the complex method. *Studia Math.*, 24:113–190, 1964.

63. A. Calderón and A. Zygmund. On singular integrals. *Amer. J. Math.*, 78:289–309, 1956.

64. L. Calotă. On some quasilinear elliptic equations with critical Sobolev exponents and non-standard growth conditions. *Bull. Belg. Math. Soc. Simon Stevin*, 15(2):249–256, 2008.

65. S. Campanato. Proprietà di hölderianità di alcune classi di funzioni. *Ann. Scuola Norm. Sup. Pisa (3)*, 17:175–188, 1963.

66. M. Cañestro and P. Salvador. Weighted weak type inequalities with variable exponents for Hardy and maximal operators. *Proc. Japan Acad. Ser. A Math. Sci.*, 82:126–130, 2006.

67. L. Cattabriga. Su un problema al contorno relativo al sistema di equazioni di Stokes. *Rend. Sem. Mat. Padova*, 31:308–340, 1961.

68. J. Chabrowski and Y. Fu. Existence of solutions for $p(x)$-Laplacian problems on a bounded domain. *J. Math. Anal. Appl.*, 306:604–618, 2005.

69. S. Chen, Y. Cui, H. Hudzik, and T. Wang. On some solved and unsolved problems in geometry of certain classes of Banach function spaces. In *Unsolved problems on mathematics for the 21st century*, pages 239–259. IOS, Amsterdam, 2001.

70. Y. Chen, S. Levine, and M. Rao. Variable Exponent, Linear Growth Functionals in Image Restoration. *SIAM J. Appl. Math.*, 66:1383–1406, 2006.

71. Y. Chen and M. Xu. Hölder continuity of weak solutions for parabolic equations with nonstandard growth conditions. *Acta Math. Sin. (Engl. Ser.)*, 22(3):793–806, 2006.

72. V. Chiadò Piat and A. Coscia. Hölder continuity of minimizers of functionals with variable growth exponent. *Manuscripta Math.*, 93:283–299, 1997.

73. G. Choquet. Theory of capacities. *Ann. Inst. Fourier (Grenoble)*, 5:131–295, 1953.

74. M. Christ. *Lectures on singular integral operators*, volume 77 of *CBMS Regional Conference Series in Mathematics*. Published for the Conference Board of the Mathematical Sciences, Washington, DC, 1990.

75. S.-K. Chua. Extension theorems on weighted sobolev spaces. *Indiana Univ. Math. J.*, 41:1027–1076, 1992.

76. R. Coifman and Y. Meyer. *Au delà des opérateurs pseudo-différentiels*, volume 57 of *Astérisque*. Société Mathématique de France, Paris, 1978. With an English summary.

77. R. Coifman and R. Rochberg. Another characterization of BMO. *Proc. Amer. Math. Soc.*, 79(2):249–254, 1980.

78. F. Corrêa, A. Costa, and G. Figueiredo. On a singular elliptic problem involving the $p(x)$-Laplacian and generalized Lebesgue-Sobolev spaces. *Adv. Math. Sci. Appl.*, 17(2):639–650, 2007.

79. A. Coscia and G. Mingione. Hölder continuity of the gradient of $p(x)$-harmonic mappings. *C. R. Acad. Sci. Paris*, 328:363–368, 1999.

80. A. Coscia and D. Mucci. Integral representation and Γ-convergence of variational integrals with $p(x)$-growth. *ESAIM Control Optim. Calc. Var.*, 7:495–519, 2002.

81. D. Cruz-Uribe, L. Diening, and A. Fiorenza. A new proof of the boundedness of maximal operators on variable Lebesgue spaces. *Boll. Unione Mat. Ital. (9)*, 2(1):151–173, 2009.

82. D. Cruz-Uribe, L. Diening, and P. Hästö. Muckenhoupt weights in variable exponent spaces. private communications, 2010.

83. D. Cruz-Uribe, A. Fiorenza, J. Martell, and C. Pérez. The boundedness of classical operators in variable L^p spaces. *Ann. Acad. Sci. Fenn. Math.*, 31:239–264, 2006.

84. D. Cruz-Uribe, A. Fiorenza, and C. Neugebauer. The maximal function on variable L^p spaces. *Ann. Acad. Sci. Fenn. Math.*, 28:223–238; **29** (2004), 247–249, 2003.

85. G. Dai. Infinitely many non-negative solutions for a Dirichlet problem involving $p(x)$-Laplacian. *Nonlinear Anal.*, 71(11):5840–5849, 2009.

86. G. Dai. Infinitely many solutions for a hemivariational inequality involving the $p(x)$−Laplacian. *Nonlinear Anal.*, 71:186–195, 2009.

87. G. Dai. Infinitely many solutions for a $p(x)$−Laplacian equation in $\mathbb{R}^N$. *Nonlinear Anal.*, 71:1133–1139, 2009.

88. S.-G. Deng. Eigenvalues of the $p(x)$-Laplacian Steklov problem. *J. Math. Anal. Appl.*, 339:925–937, 2008.

89. S.-G. Deng. A local mountain pass theorem and applications to adouble perturbed $p(x)$−Laplacian equations. *Appl. Math.Comput.*, 211(1):234–241, 2009.

90. L. Diening. *Theoretical and Numerical Results for Electrorheological Fluids*. PhD thesis, University of Freiburg, Germany, 2002.

91. L. Diening. Maximal function on generalized Lebesgue spaces $L^{p(\cdot)}$. *Math. Inequal. Appl.*, 7:245–253, 2004.

92. L. Diening. Riesz potential and Sobolev embeddings of generalized Lebesgue and Sobolev spaces $L^{p(\cdot)}$ and $W^{k,p(\cdot)}$. *Math. Nachr.*, 263:31–43, 2004.

93. L. Diening. Maximal function on Orlicz–Musielak spaces and generalized Lebesgue spaces. *Bull. Sci. Math.*, 129:657–700, 2005.

94. L. Diening, F. Ettwein, and M. Růžička. $C^{1,\alpha}$-regularity for electrorheological fluids in two dimensions. *NoDEA Nonlinear Differential Equations Appl.*, 14(1–2)(1–2):207–217, 2007.

95. L. Diening, P. Harjulehto, P. Hästö, Y. Mizuta, and T. Shimomura. Maximal functions in variable exponent spaces: limiting cases of the exponent. *Ann. Acad. Sci. Fenn. Math.*, 34:503–522, 2009.

96. L. Diening and P. Hästö. Variable exponent trace spaces. *Studia Math.*, 183:127–141, 2007.

97. L. Diening and P. Hästö. Muckenhoupt weights in variable exponent spaces. Preprint, 2008.

98. L. Diening and P. Hästö. Further results on variable exponent trace spaces. In H. Begehr and F. Nicolosi, editors, *More progresses in analysis*, pages 101–106, Catania, Italy, 2009. World Scientific.

99. L. Diening, P. Hästö, and A. Nekvinda. Open problems in variable exponent Lebesgue and Sobolev spaces. In *FSDONA04 Proceedings*, pages 38–58. Czech Acad. Sci., Milovy, Czech Republic, 2004.

100. L. Diening, P. Hästö, and S. Roudenko. Function spaces of variable smoothness and integrability. *J. Funct. Anal.*, 256(6):1731–1768, 2009.

101. L. Diening, D. Lengeler, and M. Růžička. The Stokes and Poisson Problem in Variable Exponent Spaces. *Com. Var. Ell. Equ.*, to appear, 2011.

102. L. Diening, J. Málek, and M. Steinhauer. On Lipschitz truncations of Sobolev functions (with variable exponent) and their selected applications. *ESAIM Control Optim. Calc. Var.*, 14(2):211–232, 2008.

103. L. Diening and M. Růžička. Calderón–Zygmund operators on generalized Lebesgue spaces $L^{p(\cdot)}$ and problems related to fluid dynamics. *J. Reine Angew. Math.*, 563:197–220, 2003.

104. L. Diening and M. Růžička. Integral operators on the halfspace in generalized Lebesgue spaces $L^{p(\cdot)}$, part I. *J. Math. Anal. Appl.*, 298:559–571, 2004.

105. L. Diening and M. Růžička. Integral operators on the halfspace in generalized Lebesgue spaces $L^{p(\cdot)}$, part II. *J. Math. Anal. Appl.*, 298:572–588, 2004.

106. L. Diening and M. Růžička. Strong solutions for generalized Newtonian fluids. *J. Math. Fluid Mech.*, 7:413–450, 2005.

107. L. Diening and M. Růžička. An existence result for non-Newtonian fluids in non-regular domains. *Disc. Cont. Dyn. Sys. S (DCDS-S)*, 3:255–268, 2010.

108. L. Diening, M. Růžička, and K. Schumacher. A decomposition technique for John domains. *Ann. Acad. Sci. Fenn. Ser. A. I. Math.*, 35:87–114, 2010.

109. L. Diening and S. Samko. Hardy inequality in variable exponent Lebesgue spaces. *Fract. Calc. Appl. Anal.*, 10:1–18, 2007.

110. G. Dinca. A fixed point method for the $p(\cdot)$-Laplacian. *C. R. Math. Acad. Sci. Paris*, 347(13-14):757–762, 2009.

111. G. Dinca and P. Matei. Geometry of Sobolev spaces with variable exponent: smoothness and uniform convexity. *C. R. Math. Acad. Sci. Paris*, 347(15-16):885–889, 2009.

112. T.-L. Dinu. On a nonlinear eigenvalue problem in Sobolev spaces with variable exponent. *Sib. Elektron. Mat. Izv.*, 2:208–217, 2005.

113. T.-L. Dinu. Nonlinear eigenvalue problems in Sobolev spaces with variable exponent. *J. Funct. Spaces Appl.*, 4:225–242, 2006.

114. D. Edmunds, V. Kokilashvili, and A. Meskhi. On the boundedness and completeness of weight Hardy operators in $L^{p(x)}$ spaces. *Georgian Math. J.*, 12:27–44, 2005.

115. D. Edmunds, V. Kokilashvili, and A. Meskhi. One-sided operators in $L^{p(x)}$ spaces. *Math. Nachr.*, 281(11):1525–1548, 2008.

116. D. Edmunds, J. Lang, and A. Nekvinda. On $L^{p(x)}$ norms. *R. Soc. Lond. Proc. Ser. A Math. Phys. Eng. Sci.*, 455:219–225, 1999.

117. D. Edmunds, J. Lang, and A. Nekvinda. Some s-numbers of an integral operator of Hardy type on $L^{p(\cdot)}$ spaces. *J. Funct. Anal.*, 257(1):219–242, 2009.

118. D. Edmunds and A. Meskhi. Potential-type operators in $L^{p(x)}$ spaces. *Z. Anal. Anwendungen*, 21:681–690, 2002.

119. D. Edmunds and A. Nekvinda. Averaging operators on $\ell^{\{p_n\}}$ and $L^{p(x)}$. *Math. Inequal. Appl.*, 5:235–246, 2002.

120. D. Edmunds and J. Rákosník. Density of smooth functions in $W^{k,p(x)}(\Omega)$. *Proc. Roy. Soc. London Ser. A*, 437:229–236, 1992.

121. D. Edmunds and J. Rákosník. Sobolev embedding with variable exponent. *Studia Math.*, 143:267–293, 2000.

122. D. Edmunds and J. Rákosník. Sobolev embedding with variable exponent, II. *Math. Nachr.*, 246-247:53–67, 2002.

123. D. E. Edmunds, V. Kokilashvili, and A. Meskhi. Two-weight estimates in $l^{p(x)}$ spaces with applications to Fourier series. *Houston J. Math.*, 35(2):665–689, 2009.

124. M. Eleuteri. Hölder continuity results for a class of functionals with non standard growth. *Boll. Unione Mat. Ital.*, 7-B:129–157, 2004.

125. M. Eleuteri and J. Habermann. A Hölder continuity result for a class of obstacle problems under non standard growth conditions. to appear.

126. M. Eleuteri and J. Habermann. Regularity results for a class of obstacle problems under nonstandard growth conditions. *J. Math. Anal. Appl.*, 344(2):1120–1142, 2008.

127. M. Eleuteri, P. Harjulehto, and T. Lukkari. Global regularity and stability of solutions to elliptic equations with nonstandard growth. *Complex Var. Elliptic Equ.*, to appear.

128. L. Evans. *Partial differential equations*, volume 19 of *Graduate Studies in Mathematics*. American Mathematical Society, Providence, RI, 1998.

129. L. Evans and R. Gariepy. *Measure theory and fine properties of functions*. Studies in Advanced Mathematics. CRC Press, Boca Raton, FL, 1992.

130. X.-L. Fan. Eigenvalues of the $p(x)$-Laplacian Neumann problem. *Nonlinear Anal.*, 67:2982–2992, 2007.

131. X.-L. Fan. Global $C^{1,\alpha}$ regularity for variable exponent elliptic equations in divergence form. *J. Differential Equations*, 235(2):397–417, 2007.

132. X.-L. Fan. On the sub-supersolutions method for $p(x)$-Laplacian equations. *J. Math. Anal. Appl.*, 330:665–682, 2007.

133. X.-L. Fan. A constrained minimization problem involving the $p(x)$−Laplacian in image. *Nonlinear Anal.*, 69(10):3661–3670, 2008.

134. X.-L. Fan. Boundary trace embedding theorems for variable exponent Sobolev spaces. *J. Math. Anal. Appl.*, 339:1395–1412, 2008.

135. X.-L. Fan. A remark on Ricceri's conjecture for a class of nonlinear eigenvalue problems. *J. Math. Anal. Appl.*, 349(2):436–442, 2009.

136. X.-L. Fan. Remarks on eigenvalue problems involving the $p(x)$-Laplacian. *J. Math. Anal. Appl.*, 352(1):85–98, 2009.

137. X.-L. Fan and S.-G. Deng. Remarks on Ricceri's variational principle and applications to the $p(x)$−Laplacian equations. *Nonlinear Anal.*, 67:3064–3075, 2007.

138. X.-L. Fan and S.-G. Deng. Multiplicity of positive solutions for a class of inhomogeneous Neumann problems involving the $p(x)$-Laplacian. *NoDEA Nonlinear Differential Equations Appl.*, 16(2):255–271, 2009.

139. X.-L. Fan and X. Fan. A Knobloch-type result for $p(t)$-Laplacian systems. *J. Math. Anal. Appl.*, 282:453–464, 2003.

140. X.-L. Fan and X. Han. Existence and multiplicity of solutions for $p(x)$-Laplacian equations in R^N. *Nonlinear Anal.*, 59:173–188, 2004.

141. X.-L. Fan, J. Shen, and D. Zhao. Sobolev embedding theorems for spaces $W^{k,p(x)}$. *J. Math. Anal. Appl.*, 262:749–760, 2001.

142. X.-L. Fan, S. Wang, and D. Zhao. Density of $C^\infty(\Omega)$ in $W^{1,p(x)}(\Omega)$ with discontinuous exponent $p(x)$. *Math. Nachr.*, 279:142–149, 2006.
143. X.-L. Fan, H.-Q. Wu, and F.-Z. Wang. Hartman-type results for $p(t)$-Laplacian systems. *Nonlinear Anal.*, 52:585–594, 2003.
144. X.-L. Fan and Q.-H. Zhang. Existence of solutions for $p(x)$-Laplacian Dirichlet problem. *Nonlinear Anal.*, 52:1843–1852, 2003.
145. X.-L. Fan, Q.-H. Zhang, and D. Zhao. Eigenvalues of $p(x)$-Laplacian Dirichlet problem. *J. Math. Anal. Appl.*, 302:306–317, 2005.
146. X.-L. Fan and D. Zhao. Regularity of minimizers of variational integrals with continuous $p(x)$-growth conditions. *Chinese J. Contemp. Math.*, 17:327–336, 1996.
147. X.-L. Fan and D. Zhao. A class of De Giorgi type and Hölder continuity. *Nonlinear Anal.*, 36:295–318, 1999.
148. X.-L. Fan and D. Zhao. The quasi-minimizer of integral functionals with $m(x)$ growth conditions. *Nonlinear Anal.*, 39:807–816, 2000.
149. X.-L. Fan and D. Zhao. On the spaces spaces $L^{p(x)}(\Omega)$ and $W^{m,p(x)}(\Omega)$. *J. Math. Anal. Appl.*, 263:424–446, 2001.
150. X.-L. Fan and D. Zhao. Regularity of quasi-minimizers of integral functionals with discontinuous $p(x)$-growth conditions. *Nonlinear Anal.*, 65:1521–1531, 2006.
151. X.-L. Fan and Y. Zhao. Nodal solutions of $p(x)$−Laplacian equations. *Nonlinear Anal.*, 67:2859–2868, 2007.
152. X.-L. Fan, Y. Zhao, and D. Zhao. Compact imbedding theorems with symmetry of Strauss-Lions type for the space $W^{1,p(x)}(\Omega)$. *J. Math. Anal. Appl.*, 255:333–348, 2001.
153. X.-L. Fan, Y. Z. Zhao, and Q.-H. Zhang. A strong maximum principle for $p(x)$-Laplace equations. *Chinese J. Contemp. Math.*, 24:277–282, 2003.
154. H. Federer. *Geometric measure theory.* Die Grundlehren der mathematischen Wissenschaften, Band 153. Springer-Verlag New York Inc., New York, 1969.
155. A. Fiorenza. A mean continuity type result for certain Sobolev spaces with variable exponent. *Commun. Contemp. Math.*, 4:587–605, 2002.
156. R. Fortini, D. Mugnai, and P. Pucci. Maximum principles for anisotropic elliptic inequalities. *Nonlinear Anal.*, 70(8):2917–2929, 2009.
157. M. Frazier and B. Jawerth. Decomposition of Besov spaces. *Indiana Univ. Math. J.*, 34:777–799, 1985.
158. M. Frazier and B. Jawerth. A discrete transform and decompositions of distribution spaces. *J. Funct. Anal.*, 93:34–170, 1990.
159. S. Fröschl. Fortsetzungen in Lebesgue und Sobolevräumen mit variablen Exponenten. Master's thesis, University of Freiburg, Germany, 2007.
160. Y. Fu. The principle of concentration compactness in $L^{p(x)}$ spaces and its application. *Nonlinear Anal.*, 71(5-6):1876–1892, 2009.
161. Y. Fu and X. Zhang. A multiplicity result for $p(x)$−Laplacian problem in image. *Nonlinear Anal.*, 70(6):2261–2269, 2009.
162. T. Futamura, P. Harjulehto, P. Hästö, Y. Mizuta, and T. Shimomura. Variable exponent spaces on metric measure spaces. In *More progresses in analysis (ISAAC-5, Catania, 2005, Begehr & Nicolosi (ed.))*, pages 107–121. World Scientific, 2009.
163. T. Futamura and Y. Mizuta. Continuity properties of Riesz potentials for functions in $L^{p(\cdot)}$ of variable exponent. *Math. Inequal. Appl.*, 8(4):619–631, 2005.
164. T. Futamura and Y. Mizuta. Continuity of weakly monotone Sobolev functions of variable exponent. In *Potential theory in Matsue*, volume 44 of *Adv. Stud. Pure Math.*, pages 127–143. Math. Soc. Japan, Tokyo, 2006.
165. T. Futamura and Y. Mizuta. Maximal functions for Lebesgue spaces with variable exponent approaching 1. *Hiroshima Math. J.*, 36(1):23–28, 2006.
166. T. Futamura, Y. Mizuta, and T. Shimomura. Sobolev embeddings for Riesz potential space of variable exponent. *Math. Nachr.*, 279(13-14):1463–1473, 2006.
167. T. Futamura, Y. Mizuta, and T. Shimomura. Sobolev embeddings for variable exponent Riesz potentials on metric spaces. *Ann. Acad. Sci. Fenn. Math.*, 31:495–522, 2006.

168. T. Futamura, Y. Mizuta, and T. Shimomura. Sobolev embeddings for variable exponent Riesz potentials on metric spaces. *Ann. Acad. Sci. Fenn. Math.*, 31(2):495–522, 2006.

169. G. Galdi. *An Introduction to the Mathematical Theory of the Navier-Stokes Equations, Linearized Steady Problems*, volume 38 of *Tracts in Natural Philosophy*. Springer, New York, 1994.

170. E. Galewska and M. Galewski. On the stability of solutions for the $p(x)$-Laplacian equation and some applications to optimisation problems with state constraints. *ANZIAM J.*, 48(2):245–257, 2006.

171. M. Galewski. On a Dirichlet problem with generalized $p(x)$-Laplacian and some applications. *Numer. Funct. Anal. Optim.*, 28:1087–1111, 2007.

172. M. Galewski. On the existence and stability of solutions for Dirichlet problem with $p(x)$-Laplacian. *J. Math. Anal. Appl.*, 326:352–362, 2007.

173. M. Galewski. On a Dirichlet problem with $p(x)$–Laplacian. *J. Math. Anal. Appl.*, 337:281–291, 2008.

174. J. García-Cuerva and J. Rubio de Francia. *Weighted norm inequalities and related topics*, volume 116 of *North-Holland Mathematics Studies*. North-Holland Publishing Co., Amsterdam, 1985. Notas de Matemática [Mathematical Notes], 104.

175. J. García Melián, J. Rossi, and J. Sabina. Existence, asymptotic behavior and uniqueness for large solutions to $\Delta u = e^{q(x)u}$. *Adv. Nonlinear Stud.*, 9:395–424, 2009.

176. D. Gilbarg and N. Trudinger. *Elliptic partial differential equations of second order*. Classics in Mathematics. Springer-Verlag, Berlin, 2001. Reprint of the 1998 edition.

177. A. Ginzburg and N. Karapetyants. Fractional integro-differentiation in Hölder classes of variable order. *Dokl. Akad. Nauk*, 339(4):439–441, 1994.

178. L. Gongbao and O. Martio. Stability of solutions of varying degenerate elliptic equations. *Indiana Univ. Math. J.*, 47(3):873–891, 1998.

179. L. Grafakos. *Classical Fourier analysis*, volume 249 of *Graduate Texts in Mathematics*. Springer, New York, second edition, 2008.

180. X. Guo, M. Lu, and Q. Zhang. Infinitely many periodic solutions for variable exponent systems. *J. Inequal. Appl.*, 2009. Art. No. 714179.

181. P. Gurka, P. Harjulehto, and A. Nekvinda. Bessel potential spaces with variable exponent. *Math. Inequal. Appl.*, 10:661–676, 2007.

182. J. Habermann and A. Zatorska-Goldstein. Regularity for minimizers of functionals with nonstandard growth by $\mathcal{A}$-harmonic approximation. *NoDEA Nonlinear Differential Equations Appl.*, 15(1–2):169–194, 2008.

183. P. Hajłasz and J. Onninen. On boundedness of maximal functions in Sobolev spaces. *Ann. Acad. Sci. Fenn. Math.*, 29:167–176, 1994.

184. P. Halmos. *Measure Theory*. D. Van Nostrand Company, Inc., New York, N. Y., 1950.

185. A. E. Hamidi. Existence results to elliptic systems with nonstandard growth conditions. *J. Math. Anal. Appl.*, 300(1):30–42, 2004.

186. P. Harjulehto. Variable exponent Sobolev spaces with zero boundary values. *Math. Bohem.*, 132:125–136, 2007.

187. P. Harjulehto and P. Hästö. An overview of variable exponent Lebesgue and Sobolev spaces. In *Future trends in geometric function theory*, volume 92 of *Rep. Univ. Jyväskylä Dep. Math. Stat.*, pages 85–93. Univ. Jyväskylä, Jyväskylä, 2003.

188. P. Harjulehto and P. Hästö. A capacity approach to Poincaré inequalities and Sobolev imbedding in variable exponent Sobolev spaces. *Rev. Mat. Complut.*, 17:129–146, 2004.

189. P. Harjulehto and P. Hästö. Lebesgue points in variable exponent spaces. *Ann. Acad. Sci. Fenn. Math.*, 29:295–306, 2004.

190. P. Harjulehto and P. Hästö. Sobolev inequalities for variable exponents attaining the values 1 and n. *Publ. Mat.*, 52:347–363, 2008.

191. P. Harjulehto, P. Hästö, and M. Koskenoja. The Dirichlet energy integral on intervals in variable exponent Sobolev spaces. *Z. Anal. Anwend.*, 22:911–923, 2003.

192. P. Harjulehto, P. Hästö, and M. Koskenoja. Hardy's inequality in variable exponent Sobolev spaces. *Georgian Math. J.*, 12:431–442, 2005.

193. P. Harjulehto, P. Hästö, and M. Koskenoja. Properties of capacity in variable exponent Sobolev spaces. *J. Anal. Appl.*, 5:71–92, 2007.

194. P. Harjulehto, P. Hästö, M. Koskenoja, and S. Varonen. Sobolev capacity on the space $W^{1,p(\cdot)}(\mathbb{R}^n)$. *J. Funct. Spaces Appl.*, 1:17–33, 2003.

195. P. Harjulehto, P. Hästö, M. Koskenoja, and S. Varonen. Variable Sobolev capacity and the assumptions on the exponent. In *Orlicz centenary volume. II*, volume 68 of *Banach Center Publ.*, pages 51–59. Polish Acad. Sci., Warsaw, 2005.

196. P. Harjulehto, P. Hästö, M. Koskenoja, and S. Varonen. The Dirichlet energy integral and variable exponent Sobolev spaces with zero boundary values. *Potential Anal.*, 25:205–222, 2006.

197. P. Harjulehto, P. Hästö, and V. Latvala. Boundedness of solutions of the non-uniformly convex, non-standard growth Laplacian. *Complex Var. Elliptic Equ.*, to appear.

198. P. Harjulehto, P. Hästö, and V. Latvala. Lebesgue points in variable exponent Sobolev spaces on metric measure spaces. *Zb. Pr. Inst. Mat. NAN Ukr.*, 1(3):87–99, 2004.

199. P. Harjulehto, P. Hästö, and V. Latvala. Sobolev embeddings in metric measure spaces with variable dimension. *Math. Z.*, 254:591–609, 2006.

200. P. Harjulehto, P. Hästö, and V. Latvala. Harnack's inequality for $p(\cdot)$-harmonic functions with unbounded exponent p. *J. Math. Anal. Appl.*, 352:345–359, 2009.

201. P. Harjulehto, P. Hästö, U. V. Le, and M. Nuortio. Overview of differential equations with non-standard growth. *Nonlinear Anal.*, 72:4551–4574, 2010.

202. P. Harjulehto, P. Hästö, and O. Martio. Fuglede's theorem in variable exponent Sobolev space. *Collect. Math.*, 55:315–324, 2004.

203. P. Harjulehto, P. Hästö, and M. Pere. Variable exponent Lebesgue spaces on metric spaces: the Hardy-Littlewood maximal operator. *Real Anal. Exchange*, 30:87–103, 2004.

204. P. Harjulehto, P. Hästö, and M. Pere. Variable exponent Sobolev spaces on metric measure spaces. *Funct. Approx. Comment. Math.*, 36:79–94, 2006.

205. P. Harjulehto, P. Hästö, M. Koskenoja, T. Lukkari, and N. Marola. An obstacle problems and superharmonic functions with nonstandard growth. *Nonlinear Anal.*, 67:3424–3440, 2007.

206. P. Harjulehto, P. Hästö, and V. Latvala. Minimizers of the variable exponent, non-uniformly convex Dirichlet energy. *J. Math. Pures Appl. (9)*, 89:174–197, 2008.

207. P. Harjulehto, P. Hästö, Y. Mizuta, and T. Shimomura. Iterated maximal functions in variable exponent Lebesgue spaces. *Manuscripta Math.*, to appear.

208. P. Harjulehto, J. Kinnunen, and T. Lukkari. Unbounded supersolutions of nonlinear equations with nonstandard growth. *Bound. Value Probl.*, 2007, 2007. Art. ID 48348, 20.

209. P. Harjulehto, J. Kinnunen, and K. Tuhkanen. Hölder quasicontinuity in variable exponent Sobolev spaces. *J. Inequal. Appl.*, page 18, 2007. Art. ID 32324.

210. P. Harjulehto, T. Kuusi, T. Lukkari, N. Marola, and M. Parviainen. Harnack's inequality for quasiminimizers with non-standard growth conditions. *J. Math. Anal. Appl.*, 344:504–520, 2008.

211. P. Harjulehto and V. Latvala. Fine topology of variable exponent energy superminimizers. *Ann. Acad. Sci. Fenn. Math.*, 33:491–510, 2008.

212. P. Hästö. Counter-examples of regularity in variable exponent Sobolev spaces. In *The p-harmonic equation and recent advances in analysis*, volume 370 of *Contemp. Math.*, pages 133–143. Amer. Math. Soc., Providence, RI, 2005.

213. P. Hästö. On the variable exponent Dirichlet energy integral. *Comm. Pure Appl. Anal.*, 5:413–420, 2006.

214. P. Hästö. On the density of smooth functions in variable exponent Sobolev space. *Rev. Mat. Iberoamericana*, 23:215–237, 2007.

215. P. Hästö. The maximal operator in Lebesgue spaces with variable exponent approaching 1. *Math. Nachr.*, 280:74–82, 2007.

216. P. Hästö. Local to global results in variable exponent spaces. *Math. Res. Letters*, 16:263–278, 2009.

217. J. Heinonen. *Lectures on analysis on metric spaces*. Universitext. Springer-Verlag, New York, 2001.

218. J. Heinonen. Nonsmooth calculus. *Bull. Amer. Math. Soc. (N.S.)*, 44:163–232, 2007.

219. J. Heinonen, T. Kilpeläinen, and O. Martio. *Nonlinear potential theory of degenerate elliptic equations*. Dover Publications Inc., Mineola, NY, 2006. Unabridged republication of the 1993 original.

220. P. Ilias. Existence and multiplicity of solutions of a $p(x)$-Laplacian equation in a bounded domain. *Rev. Roumaine Math. Pures Appl.*, 52:639–653, 2007.

221. P. Ilias. Dirichlet problem with $p(x)$-Laplacian. *Math. Rep.*, 10(60):43–56, 2008.

222. C. Ji. Perturbation for a $p(x)$-Laplacian equation involving oscillating nonlinearities in $\mathbb{R}^N$. *Nonlinear Anal.*, 69:2393–2402, 2008.

223. F. John and L. Nirenberg. On functions of bounded mean oscillation. *Comm. Pure Appl. Math.*, 14:415–426, 1961.

224. P. Jones. Quasiconformal mappings and extendability of functions in Sobolev spaces. *Acta Math.*, 147(1-2):71–88, 1981.

225. A. Jonsson and H. Wallin. *Function spaces on subsets of* $\mathbf{R}^n$. Number 1 in Math. Rep. Harwood Academic Publishers, Chur, Switzerland, 1984.

226. S. Kakutani. Weak convergence in uniformly convex spaces. *Tohoku Math. J.*, 45:188–193, 1938.

227. G. Kalyabin. A nondensity criterion for $L^\infty(\mathbb{R}^n)$ in $L^{p(\cdot)}(\mathbb{R}^n)$. *Mat. Zametki*, 82(2):315–316, 2007.

228. E. Kapanadze and T. Kopaliani. A note on maximal operator on $L^{p(t)}(\Omega)$ spaces. *Georgian Math. J.*, 15(2):307–316, 2008.

229. N. Karapetyants and A. Ginsburg. Fractional integrals and singular integrals in the Hölder classes of variable order. *Integral Transform. Spec. Funct.*, 2(2):91–106, 1994.

230. A. Karlovich. Semi-Fredholm singular integral operators with piecewise continuous coefficients on weighted variable Lebesgue spaces are Fredholm. *Oper. Matrices*, 1:427–444, 2007.

231. A. Karlovich and A. Lerner. Commutators of singular integrals on generalized L^p spaces with variable exponent. *Publ. Mat.*, 49:111–125, 2005.

232. H. Kempka. 2-microlocal Besov and Triebel-Lizorkin spaces of variable integrability. *Rev. Mat. Complut.*, 22:227–251, 2009.

233. T. Kilpeläinen. A remark on the uniqueness of quasicontinuous functions. *Ann. Acad. Sci. Fenn. Math.*, 23:261–262, 1998.

234. T. Kilpeläinen, J. Kinnunen, and O. Martio. Sobolev spaces with zero boundary values on metric spaces. *Potential Anal.*, 12:233–247, 2000.

235. D. Kinderlehrer and G. Stampacchia. *An introduction to variational inequalities and their applications*, volume 88 of *Pure and Applied Mathematics*. Academic Press Inc. [Harcourt Brace Jovanovich Publishers], New York, 1980.

236. J. Kinnunen. The Hardy-Littlewood maximal function of a Sobolev function. *Israel J. Math.*, 100:117–124, 1997.

237. J. Kinnunen and V. Latvala. Lebesgue points for Sobolev functions on metric spaces. *Rev. Mat. Iberoamericana*, 18:685–700, 2002.

238. J. Kinnunen and P. Lindqvist. The derivative of the maximal function. *J. Reine Angew. Math.*, 503:161–167, 1998.

239. V. Kokilashvili and M. Krbec. *Weighted inequalities in Lorentz and Orlicz spaces*. Singapore etc.: World Scientific Publishing Co. Pte. Ltd. xii, 233 p., 1991.

240. V. Kokilashvili and A. Meskhi. Weighted criteria for generalized fractional maximal functions and potentials in Lebesgue spaces with variable exponent. *Integral Transforms Spec. Funct.*, 18:609–628, 2007.

241. V. Kokilashvili, V. Paatashvili, and S. Samko. Boundary value problems for analytic functions in the class of Cauchy-type integrals with density in $L^{p(\cdot)}(\Gamma)$. *Bound. Value Probl.*, 2005(1):43–71, 2005.

242. V. Kokilashvili, V. Paatashvili, and S. Samko. Boundedness in Lebesgue spaces with variables exponent of the Cauchy integral operator on Carleson curves. In *Modern operator theory and applications*, volume 170 of *Oper. Theory Adv. Appl.*, pages 167–186. Birkhäuser, Basel, 2006.

243. V. Kokilashvili and S. Samko. Singular integral equations in the Lebesgue spaces with variable exponent. *Proc. A. Razmadze Math. Inst.*, 131:61–78, 2003.

244. V. Kokilashvili, N. Samko, and S. Samko. Singular operators in variable spaces $L^{p(\cdot)}(\Omega, \varrho)$ with oscillating weights. *Math. Nachr.*, 280:1145–1156, 2007.

245. V. Kokilashvili, N. Samko, and S. Samko. The maximal operator in weighted variable spaces $L^{p(\cdot)}$. *J. Funct. Spaces Appl.*, 5:299–317, 2007.

246. V. Kokilashvili and S. Samko. On Sobolev Theorem for Riesz-type potentials in Lebesgue spaces with variable exponents. *Z. Anal. Anwend.*, 22:899–910, 2003.

247. V. Kokilashvili and S. Samko. Singular integrals in weighted Lebesgue spaces with variable exponent. *Georgian Math. J.*, 10:145–156, 2003.

248. V. Kokilashvili and S. Samko. Maximal and fractional operators in weighted $L^{p(x)}$ spaces. *Rev. Mat. Iberoamericana*, 20:495–517, 2004.

249. V. Kokilashvili and S. Samko. Weighted boundedness in Lebesgue spaces with variable exponents of classical operators on Carleson curves. *Proc. A. Razmadze Math. Inst.*, 138:106–110, 2005.

250. V. Kokilashvili and S. Samko. A general approach to weighted boundedness of operators of harmonic analysis in variable exponent Lebesgue spaces. *Proc. A. Razmadze Math. Inst.*, 145:109–116, 2007.

251. V. Kokilashvili and S. Samko. The maximal operator in weighted variable spaces on metric measure spaces. *Proc. A. Razmadze Math. Inst.*, 144:137–144, 2007.

252. V. Kokilashvili and S. Samko. Operators of harmonic analysis in weighted spaces with non-standard growth. *J. Math. Anal. Appl.*, 352(1):15–34, 2009.

253. A. Kolmogoroff. Zur Normierbarkeit eines allgemeinen topologischen linearen Raumes. *Studia math.*, 5:29–33, 1934.

254. T. Kopaliani. On some structural properties of Banach function spaces and boundedness of certain integral operators. *Czechoslovak Math. J.*, 54(129)(3):791–805, 2004.

255. T. Kopaliani. Infimal convolution and Muckenhoupt $A_p(\cdot)$ condition in variable L^p spaces. *Arch. Math.*, 89:185–192, 2007.

256. T. Kopaliani. On the Muckenchaupt condition in variable Lebesgue spaces. *Proc. A. Razmadze Math. Inst.*, 148:29–33, 2008.

257. O. Kováčik. Parabolic equations in generalized Sobolev spaces $W^{k,p(x)}$. *Fasc. Math.*, 25:87–94, 1995.

258. O. Kováčik and J. Rákosník. On spaces $L^{p(x)}$ and $W^{1,p(x)}$. *Czechoslovak Math. J.*, 41(116):592–618, 1991.

259. K. Kurata and N. Shioji. Compact embedding from $W_0^{1,2}(\Omega)$ to $L^{q(x)}(\Omega)$ and its application to nonlinear elliptic boundary value problem with variable critical exponent. *J. Math. Anal. Appl.*, 339:1386–1394, 2008.

260. D. Kurtz. Littlewood–Paley and multiplier theorems on weighted L^p spaces. *Trans. Amer. Math. Soc.*, 259:235–254, 1980.

261. O. Ladyzhenskaya and N. Ural'tseva. *Linear and quasilinear elliptic equations.* Translated from the Russian by Scripta Technica, Inc. Translation editor: Leon Ehrenpreis. Academic Press, New York, 1968.

262. R. Landes. Quasimonotone versus pseudomonotone. *Proc. Roy. Soc. Edinburgh Sect. A*, 126(4):705–717, 1996.

263. V. Latvala, T. Lukkari, and O. Toivanen. The fundamental convergence theorem for $p(\cdot)$-superharmonic functions. preprint 2009.

264. D. Lengeler. Regularitätstheorie in Räumen mit variablen Exponenten. Master's thesis, University Freiburg, 2008.

265. H.-G. Leopold. On Besov spaces of variable order of differentiation. *Z. Anal. Anwendungen*, 8:69–82, 1989.

266. H.-G. Leopold. Embedding of function spaces of variable order of differentiation in function spaces of variable order of integration. *Czechoslovak Math. J.*, 49(124):633–644, 1999.

267. A. Lerner. Some remarks on the Hardy–Littlewood maximal function on variable L^p spaces. *Math. Z.*, 251:509–521, 2005.

268. A. Lerner. On some questions related to the maximal operator on variable l^p spaces. *Trans. Amer. Math. Soc.*, 362(8):4229–4242, 2010.

269. F. Li, Z. Li, and L. Pi. Variable exponent functionals in image restoration. *Applied Mathematics and Computation*, 216(3):870 – 882, 2010.

270. P. Lindqvist and T. Lukkari. A curious equation involving the infinity laplacian. to appear, arXiv:0902.1771v1.

271. J. Lions. *Quelques Méthodes de Résolution des Problèmes aux Limites Non Linéaires.* Dunod, Paris, 1969.

272. J. Liu and X. Shi. Existence of three solutions for a class of quasilinear elliptic systems involving the $(p(x), q(x))$-Laplacian. *Nonlinear Anal.*, 71(1-2):550–557, 2009.

273. Q. Liu. Compact trace in weighted variable exponent Sobolev spaces $W^{1,p(x)}(\Omega; v_0, v_1)$. *J. Math. Anal. Appl.*, 348(2):760–774, 2008.

274. W. Liu and P. Zhao. Existence of positive solutions for $p(x)$-Laplacian equations in unbounded domains. *Nonlinear Anal.*, 69(10):3358–3371, 2008.

275. T. Lukkari. Elliptic equations with nonstandard growth involving measure data. *Hiroshima Math. J.*, 38:155–176, 2008.

276. T. Lukkari. Singular solutions of elliptic equations with nonstandard growth. *Math. Nachr.*, 282(12):1770–1787, 2009.

277. T. Lukkari. Boundary continuity of solutions to elliptic equations with nonstandard growth. *Manuscripta Math.*, 132(3-4):463–482, 2010. doi:10.1007/s00229-010-0355-3.

278. T. Lukkari, F.-Y. Maeda, and N. Marola. Wolff potential estimates for elliptic equations with nonstandard growth and applications. *Forum Math.*, 22(6):1061–1087, 2010. DOI: 10.1515/FORUM.2010.057.

279. F.-Y. Maeda, Y. Mizuta, and T. Ohno. Approximate identities and Young type inequalities in variable Lebesgue–Orlicz spaces $L^{p(\cdot)}(\log L)^{q(\cdot)}$. *Ann. Acad. Sci. Fenn. Math*, 35(2):405–420, 2010.

280. J. Malý and W. Ziemer. *Fine regularity of solutions of elliptic partial differential equations*, volume 51 of *Mathematical Surveys and Monographs*. American Mathematical Society, Providence, RI, 1997.

281. J. Manfredi, J. Rossi, and J. Urbano. $p(x)$-harmonic functions with unbounded exponent in a subdomain. *Ann. Inst. H. Poincaré Anal. Non Linéaire.*, 26:2581–2595, 2009.

282. J. Manfredi, J. Rossi, and J. Urbano. Limits as $p(x) \to \infty$ of $p(x)$-harmonic functions. *Nonlinear Anal.*, 72:309–315, 2010.

283. O. Martio and J. Sarvas. Injectivity theorems in plane and space. *Ann. Acad. Sci. Fenn. Math.*, 4:383–401, 1978.

284. R. Mashiyev. Some properties of variable Sobolev capacity. *Taiwanese J. Math.*, 12(3):671–678, 2008.

285. R. Mashiyev and B. Çekiç. Sobolev-type inequality for spaces $L^{p(x)}(\mathbb{R}^N)$. *Int. J. Contemp. Math. Sci.*, 2(9-12):423–429, 2007.

286. R. Mashiyev, B. Çekiç, F. Mamedov, and S. Ogras. Hardy's inequality in power-type weighted $L^{p(\cdot)}(0, \infty)$ spaces. *J. Math. Anal. Appl.*, 334(1):289–298, 2007.

287. R. Mashiyev, B. Çekiç, and S. Ogras. On Hardy's inequality in $L^{p(x)}(0, \infty)$. *JIPAM. J. Inequal. Pure Appl. Math.*, 7(3):Article 106, 5 pp. (electronic), 2006.

288. P. Mattila. *Geometry of sets and measures in Euclidean spaces*, volume 44 of *Cambridge Studies in Advanced Mathematics*. Cambridge University Press, Cambridge, 1995.

289. E. McShane. Extension of range of functions. *Bull. Amer. Math. Soc.*, 40:837–842, 1934.

290. M. Mihăilescu, P. Pucci, and V. Rădulescu. Eigenvalue problems for anisotropic quasilinear elliptic equations with variable exponent. *J. Math. Anal. Appl.*, 340(1):687–698, 2008.

291. M. Mihăilescu and V. Rădulescu. On a nonhomogeneous quasilinear eigenvalue problem in Sobolev spaces with variable exponent. *Proc. Amer. Math. Soc.*, 135:2929–2937, 2007.

292. M. Mihăilescu. Elliptic problems in variable exponent spaces. *Bull. Austral. Math. Soc.*, 74(2):197–206, 2006.

293. M. Mihăilescu. Existence and multiplicity of solutions for an elliptic equation with $p(x)$−growth conditions. *Glasgow Math. J.*, 48:411–418, 2006.

294. M. Mihăilescu. Existence and multiplicity of solutions for a Neumann problem involving the $p(x)$-Laplace operator. *Nonlinear Anal.*, 67(5):1419–1425, 2007.

295. M. Mihăilescu. On a class of nonlinear problems involving a $p(x)$-Laplace type operator. *Czechoslovak Math. J.*, 58(133):155–172, 2008.

296. M. Mihăilescu and V. Rădulescu. A multiplicity result for a nonlinear degenerate problem arising in the theory of electrorheological fluids. *Proc. R. Soc. Lond. Ser. A Math. Phys. Eng. Sci.*, 462(2073):2625–2641, 2006.

297. G. Mingione. Regularity of minima: an invitation to the dark side of the Calculus of Variations. *Appl. Math.*, 51:355–426, 2006.

298. Y. Mizuta, T. Ohno, and T. Shimomura. Integrability of maximal functions for generalized Lebesgue spaces with variable exponent. *Math. Nachr.*, 281(3):386–395, 2008.

299. Y. Mizuta, T. Ohno, and T. Shimomura. Sobolev's inequalities and vanishing integrability for Riesz potentials of functions in the generalized Lebesgue space $L^{p(\cdot)}(\log L)^{q(\cdot)}$. *J. Math. Anal. Appl.*, 345(1):70–85, 2008.

300. Y. Mizuta, T. Ohno, T. Shimomura, and N. Shioji. Compact embeddings for Sobolev spaces of variable exponents and existence of solutions for nonlinear elliptic problems involving the $p(x)$-Laplacian and its critical exponent. *Ann. Acad. Sci. Fenn. Math*, 35(1):115–130, 2010.

301. Y. Mizuta and T. Shimomura. Continuity of Sobolev functions of variable exponent on metric spaces. *Proc. Japan Acad. Ser. A Math. Sci.*, 80:96–99, 2004.

302. Y. Mizuta and T. Shimomura. Sobolev's inequality for Riesz potentials with variable exponent satisfying a log-Hölder condition at infinity. *J. Math. Anal. Appl.*, 311:268–288, 2005.

303. Y. Mizuta and T. Shimomura. Sobolev embeddings for Riesz potentials of functions in Morrey spaces of variable exponent. *J. Math. Soc. Japan*, 60:583–602, 2008.

304. C. Morrey Jr. *Multiple integrals in the calculus of variations*. Die Grundlehren der mathematischen Wissenschaften, Band 130. Springer, New York, 1966.

305. B. Muckenhoupt. Weighted norm inequalities for the Hardy maximal function. *Trans. Amer. Math. Soc.*, 165:207–226, 1972.

306. B. Muckenhoupt and R. Wheeden. Weighted norm inequalities for fractional integrals. *Trans. Amer. Math. Soc.*, 192:261–274, 1974.

307. J. Musielak. *Orlicz spaces and modular spaces*, volume 1034 of *Lecture Notes in Mathematics*. Springer-Verlag, Berlin, 1983.

308. J. Musielak and W. Orlicz. On modular spaces. *Studia Math.*, 18:49–65, 1959.

309. H. Nakano. *Modulared Semi-Ordered Linear Spaces*. Maruzen Co. Ltd., Tokyo, 1950.

310. H. Nakano. *Topology of linear topological spaces*. Maruzen Co. Ltd., Tokyo, 1951.

311. R. Näkki and J. Väisälä. John disks. *Exposition. Math.*, 9(1):3–43, 1991.

312. J. Nečas. Sur le normes équivalentes dans $w_p^k(\omega)$ et sur la coercivité des formes formellement positives. *Séminaire Equations aux Dérivées Partielles, Montreal*, 317:102–128, 1966.

313. A. Nekvinda. Equivalence of $l^{\{p_n\}}$ norms and shift operators. *Math. Inequal. Appl.*, 5:711–723, 2002.

314. A. Nekvinda. Hardy-Littlewood maximal operator on $L^{p(x)}(\mathbb{R}^n)$. *Math. Inequal. Appl.*, 7:255–266, 2004.

315. A. Nekvinda. Embeddings between discrete weighted Lebesgue spaces with variable exponents. *Math. Inequal. Appl.*, 10:165–172, 2007.

316. A. Nekvinda. Maximal operator on variable Lebesgue spaces for almost monotone radial exponent. *J. Math. Anal. Appl.*, 337:1345–1365, 2008.

317. E. Nieminen. Hausdorff measures, capacities, and Sobolev spaces with weights. *Ann. Acad. Sci. Fenn. Ser. A I Math. Dissertationes*, 81:39, 1991.

318. S. Ogras, R. Mashiyev, M. Avci, and Z. Yucedag. Existence of solutions for a class of elliptic systems in $\mathbb{R}^n$ involving the $(p(x), q(x))$−laplacian. *J. Inequal. Appl.*, 2008:Art.ID 612938, 16 pp, 2008.

319. W. Orlicz. Über konjugierte Exponentenfolgen. *Studia Math.*, 3:200–211, 1931.

320. S. Ouaro and S. Traore. Weak and entropy solutions to nonlinear elliptic problems with variable exponent. *J. Convex Anal.*, 16(2):523–541, 2009.

321. N. Papageorgiou and E. Rocha. A multiplicity theorem for a variable exponent Dirichlet problem. *Glasg. Math. J.*, 50(2):335–349, 2008.

322. M. Perez-Llanos and J. Rossi. The behaviour of the $p(x)$-Laplacian eigenvalue problem as $p(x) \to \infty$. *J. Math. Anal. Appl.*, 363(2):502–511, 2010.

323. L. Pick and M. Růžička. An example of a space $L^{p(x)}$ on which the Hardy-Littlewood maximal operator is not bounded. *Expo. Math.*, 19:369–371, 2001.

324. J. Pinasco. Blow-up for parabolic and hyperbolic problems with variable exponents. *Nonlinear Anal.*, 71(3-4):1094–1099, 2009.

325. V. Portnov. Certain properties of the Orlicz spaces generated by the functions M(x,w). *Dokl. Akad. Nauk SSSR*, 170:1269–1272, 1966.

326. V. Portnov. On the theory of Orlicz spaces which are generated by variable N-functions. *Dokl. Akad. Nauk SSSR*, 175:296–299, 1967.

327. V. Rabinovich and S. Samko. Boundedness and Fredholmness of pseudodifferential operators in variable exponent spaces. *Integral Equations Operator Theory*, 60:507–537, 2008.

328. K. Rajagopal and M. Růžička. On the modeling of electrorheological materials. *Mech. Research Comm.*, 23:401–407, 1996.

329. K. Rajagopal and M. Růžička. Mathematical modeling of electrorheological materials. *Cont. Mech. and Thermodynamics*, 13:59–78, 2001.

330. M. Rao and Z. Ren. *Theory of Orlicz spaces*, volume 146 of *Monographs and Textbooks in Pure and Applied Mathematics*. Marcel Dekker Inc., New York, 1991.

331. J. Rodrigues, M. Sanchón, and J. Urbano. The obstacle problem for nonlinear elliptic equations with variable growth and L^1-data. *Monatsh. Math.*, 154(4):303–322, 2008.

332. B. Ross and S. Samko. Fractional integration operator of variable order in the spaces H^λ. *Int. J. Math. Sci.*, 18:777–788, 1995.

333. I. Rovenţa. Boundary asymptotic and uniqueness of solution for a problem with $p(x)$-laplacian. *J. Inequal. Appl.*, 2008:Article ID 609047, 14 pp., doi:10.1155/2008/609047, 2008.

334. W. Rudin. *Real and complex analysis*. McGraw-Hill Book Co., New York, third edition, 1987.

335. W. Rudin. *Functional analysis*. McGraw-Hill Book Co., New York, second edition, 1991.

336. M. Růžička. Flow of shear dependent electrorheological fluids: unsteady space periodic case. In A. Sequeira, editor, *Applied nonlinear analysis*, pages 485–504. Kluwer/Plenum, New York, 1999.

337. M. Růžička. *Electrorheological fluids: modeling and mathematical theory*, volume 1748 of *Lecture Notes in Mathematics*. Springer-Verlag, Berlin, 2000.

338. M. Růžička. Modeling, mathematical and numerical analysis of electrorheological fluids. *Appl. Math.*, 49:565–609, 2004.

339. N. Samko, S. Samko, and B. Vakulov. Weighted Sobolev theorem in Lebesgue spaces with variable exponent. *J. Math. Anal. Appl.*, 335:560–583, 2007.

340. S. Samko. Convolution type operators in $L^{p(x)}$. *Integr. Transform. Spec. Funct.*, 7:no. 1–2, 123–144, 1998.

341. S. Samko. Convolution and potential type operators in $L^{p(x)}(\mathbb{R}^n)$. *Integr. Transform. Spec. Funct.*, 7(3–4):261–284, 1998.

342. S. Samko. Denseness of $C_0^\infty(\mathbb{R}^n)$ in the generalized Sobolev spaces $W^{m,p(x)}(\mathbb{R}^n)$ (Russian). *Dokl. Ross. Acad Nauk*, 369:451–454, 1999.

343. S. Samko. Denseness of $C_0^\infty(\mathbf{R}^N)$ in the generalized Sobolev spaces $W^{M,P(X)}(\mathbf{R}^N)$. In *Direct and inverse problems of mathematical physics (Newark, DE, 1997)*, volume 5 of *Int. Soc. Anal. Appl. Comput.*, pages 333–342. Kluwer Acad. Publ., Dordrecht, 2000.

344. S. Samko. Hardy inequality in the generalized Lebesgue spaces. *Fract. Calc. Appl. Anal.*, 6:355–362, 2003.

345. S. Samko. On a progress in the theory of Lebesgue spaces with variable exponent: maximal and singular operators. *Integral Transforms Spec. Funct.*, 16:461–482, 2005.

346. S. Samko, E. Shargorodsky, and B. Vakulov. Weighted Sobolev theorem with variable exponent for spatial and spherical potential operators. II. *J. Math. Anal. Appl.*, 325:745–751, 2007.

347. S. Samko and B. Vakulov. Weighted Sobolev theorem for spatial and spherical potentials in Lebesgue spaces with variable exponent. *Dokl. Ross. Akad. Nauk*, 403:7–10, 2005.

348. S. Samko and B. Vakulov. Weighted Sobolev theorem with variable exponent. *J. Math. Anal. Appl.*, 310:229–246, 2005.

349. M. Sanchón and J. Urbano. Entropy solutions for the $p(x)$-laplace equation. *Trans. Amer. Math. Soc.*, 361:6387–6405, 2009.

350. S. Schwarzacher. Higher integrability of elliptic differential equations with variable growth. Master's thesis, University of Freiburg, Germany, February 2010.

351. I. Sharapudinov. On the topology of the space $L^{p(t)}([0;1])$. *Math. Notes*, 26(3–4):796–806, 1979.

352. I. Sharapudinov. Approximation of functions in the metric of the space $L^{p(t)}([a,b])$ and quadrature formulas, (Russian). In *Constructive function theory '81 (Varna, 1981)*, pages 189–193. Publ. House Bulgar. Acad. Sci., Sofia, 1983.

353. I. Sharapudinov. The basis property of the Haar system in the space $L^{p(t)}([0,1])$ and the principle of localization in the mean, (Russian). *Mat. Sb. (N.S.)*, 130(172):275–283, 286, 1986.

354. X. Shi, Y. Niu, X. Ding, and Y. Jiang. Multiple solutions for elliptic equations with nonstandard growth conditions. *Int. J. Pure Appl. Math.*, 47(3):449–458, 2008.

355. C. Simader. *On Dirichlet's Boundary Value Problem*, volume 268 of *Lecture Notes in Mathematics*. Springer, Berlin, 1972.

356. C. Simader and H. Sohr. *The Dirichlet Problem for the Laplacian in Bounded and Unbounded Domains*, volume 360 of *Pitman Research Notes in Mathematics Series*. Longman, Harlow, 1996.

357. H. Sohr. *The Navier-Stokes equations*. Birkhäuser Advanced Texts: Basler Lehrbücher. [Birkhäuser Advanced Texts: Basel Textbooks]. Birkhäuser Verlag, Basel, 2001. An elementary functional analytic approach.

358. S. Spanne. Some function spaces defined using the mean oscillation over cubes. *Ann. Scuola Norm. Sup. Pisa (3)*, 19:593–608, 1965.

359. E. Stein. *Singular integrals and differentiability properties of functions*. Princeton Mathematical Series, No. 30. Princeton University Press, Princeton, N.J., 1970.

360. E. Stein. *Harmonic analysis: real-variable methods, orthogonality, and oscillatory integrals*, volume 43 of *Princeton Mathematical Series*. Princeton University Press, Princeton, NJ, 1993. With the assistance of Timothy S. Murphy, Monographs in Harmonic Analysis, III.

361. R. Temam. *Navier-Stokes Equations*. North-Holland, Amsterdam, 1977.

362. H. Triebel. *Theory of function spaces*, volume 78 of *Monographs in Mathematics*. Birkhäuser Verlag, Basel, 1983.

363. H. Triebel. *Theory of function spaces. II*, volume 84 of *Monographs in Mathematics*. Birkhäuser Verlag, Basel, 1992.

364. H. Triebel. *Theory of function spaces. III*, volume 100 of *Monographs in Mathematics*. Birkhäuser Verlag, Basel, 2006.

365. T. Tsanava. On the convergence and summability of Fourier series in weighted Lebesgue spaces with a variable exponent. *Proc. A. Razmadze Math. Inst.*, 142:143–146, 2006.

366. I. Tsenov. Generalization of the problem of best approximation of a function in the space l^s. *Uch. Zap. Dagestan Gos. Univ.*, 7:25–37, 1961.

367. J. Vybíral. Sobolev and Jawerth embeddings for spaces with variable smoothness and integrability. *Ann. Acad. Sci. Fenn. Math.*, 34:529–544, 2009.

368. X.-J. Wang and R. Yuan. Existence of periodic solutions for $p(t)$-Laplacian systems. *Nonlinear Anal.*, 70(2):866–880, 2009.

369. A. Wineman and K. Rajagopal. On constitutive equations for electrorheological materials. *Contin. Mech. Thermodyn.*, 7:1–22, 1995.

370. Q. Xiao. A strong maximum principle for $p(x)$-Laplacian equation. *Ann. Differential Equations*, 24(3):356–360, 2008.

371. J.-S. Xu. The relation between variable Bessel potential spaces and Triebel-Lizorkin spaces. *Integral Transforms Spec. Funct.*, 19:599–605, 2008.

372. J.-S. Xu. Variable Besov and Triebel-Lizorkin spaces. *Ann. Acad. Sci. Fenn. Math.*, 33:511–522, 2008.

373. J.-S. Xu. An atomic decomposition of variable Besov and Triebel-Lizorkin spaces. *Armenian J. Math.*, 2:1–12, 2009.

374. X. Xu and Y. An. Existence and multiplicity of solutions for elliptic systems with nonstandard growth condition in $\mathbb{R}^N$. *Nonlinear Anal.*, 68(4):956–968, 2008.

375. J. Yao. Solutions for Neumann boundary value problems involving $p(x)$−Laplace operators. *Nonlinear Anal.*, 68:1271–1283, 2008.

376. J. Yao and X. Wang. On an open problem involving the $p(x)$-Laplacian—A further study on the multiplicity of weak solutions to $p(x)$-Laplacian equations. *Nonlinear Anal.*, 69:1445–1453, 2008.

377. A. Zang. $p(x)$−Laplacian equations satisfying Cerami condition. *J. Math. Anal. Appl.*, 337:547–555, 2008.

378. E. Zeidler. *Nonlinear functional analysis and its applications. II/B*. Springer, New York, 1990. Nonlinear monotone operators.

379. C. Zhang and S. Zhou. A fourth-order degenerate parabolic equation with variable exponent. *J. Partial Differ. Equ.*, 22(4):376–392, 2009.

380. C. Zhang and S. Zhou. Renormalized and entropy solutions for nonlinear parabolic equations with variable exponents and L^1 data. *J. Differential Equations*, 248(6):1376–1400, 2010.

381. Q.-H. Zhang. Existence and asymptotic behavior of positive solutions to $p(x)$-Laplacian equations with singular nonlinearities. *J. Inequal. Appl.*, 2007:Article ID 19349, 9 pp., doi:10.1155/2007/19349, 2007.

382. Q.-H. Zhang. Existence of positive solutions for a class of $p(x)$−Laplacian systems. *J. Math. Anal. Appl.*, 333(2):591–603, 2007.

383. Q.-H. Zhang. Existence of positive solutions for elliptic systems with nonstandard $p(x)$−growth conditions via sub-supersolution method. *Nonlinear Anal.*, 67:1055–1067, 2007.

384. Q.-H. Zhang. Existence of solutions for weighted $p(r)$-Laplacian system boundary value problems. *J. Math. Anal. Appl.*, 327:127–141, 2007.

385. Q.-H. Zhang. Oscillatory property of solutions for $p(t)$-Laplacian equations. *J. Inequal. Appl.*, 2007:Article ID 58548, 8 pp., doi:10.1155/2007/58548, 2007.

386. Q.-H. Zhang. Boundary blow-up solutions to p(x)-Laplacian equations with exponential nonlinearities. *J. Inequal. Appl.*, 2008:Article ID 279306, 8 pp., doi:10.1155/2008/279306, 2008.

387. Q.-H. Zhang. Existence of solutions for $p(x)$-Laplacian equations with singular coefficients in $\mathbb{R}^N$. *J. Math. Anal. Appl.*, 348(1):38–50, 2008.

388. Q.-H. Zhang. Existence and asymptotic behavior of positive solutions for variable exponent elliptic systems. *Nonlinear Anal.*, 70(1):305–316, 2009.

389. Q.-H. Zhang, X. Liu, and Z. Qiu. On the boundary blow-up solutions of p(x)-Laplacian equations with singular coefficient. *Nonlinear Anal.*, 70(11):4053–4070, 2009.

390. Q.-H. Zhang, Z. Qiu, and X. Liu. Existence of multiple solutions for weighted p(r)-Laplacian equation Dirichlet problems. *Nonlinear Anal.*, 70(10):3721–3729, 2009.

391. X. Zhang and X. Liu. The local boundedness and Harnack inequality of $p(x)$-Laplace equation. *J. Math. Anal. Appl.*, 332:209–218, 2007.

392. V. Zhikov. Averaging of functionals of the calculus of variations and elasticity theory. *Math. USSR-Izv.*, 29:675–710, 877, 1987.

393. V. Zhikov. On Lavrentiev's phenomen. *Rus. J. Math. Phys.*, 3:249–269, 1995.

394. V. Zhikov. Meyer-type estimates for solving the nonlinear Stokes system. *Differ. Equ.*, 33:108–115, 1997.

395. V. Zhikov. Meyer-type estimates for solving the nonlinear Stokes system. *Differ. Equ.*, 33(1):108–115, 1997.

396. V. Zhikov. On some variational problems. *Russian J. Math. Phys.*, 5(1):105–116, 1997.

397. V. Zhikov. On the density of smooth functions in Sobolev-Orlicz spaces (Russian). *Zap. Nauchn. Sem. S.-Peterburg. Otdel. Mat. Inst. Steklov. (POMI)*, 310:285–294, 2004.

398. V. Zhikov and S. Pastukhova. On the improved integrability of the gradient of solutions of elliptic equations with a variable nonlinearity exponent. *Mat. Sb.*, 199(12):19–52, 2008.

399. W. Ziemer. *Weakly differentiable functions*, volume 120 of *Graduate Texts in Mathematics*. Springer-Verlag, New York, 1989. Sobolev spaces and functions of bounded variation.

List of Symbols

$\preccurlyeq$	Dominated by, page 156
$\ll$	Strongly dominated by, page 168
$\cong$	Equivalent in terms of domination, page 156
$\nearrow$	Limit of an increasing sequence, page 15
$\overline{A}^{\|\cdot\|_X}$	Closure of $A \subset X$ with respect to the norm $\|\cdot\|_X$, page 12
$\subset\subset$	$A \subset\subset E$ if $\overline{A} \subset E$ is compact, page 10
$\to b^+$	Tends to b from above, page 22
$\to b^-$	Tends to b from below, page 22
$2B$	Ball with same center as B but twice the diameter, page 10
$2Q$	Cubes with same center as Q but twice the diameter, page 10
A°	Annihilator of set A, page 65
$\mathcal{A}$	Class $\mathcal{A}$; generalization of Muckenhoupt class, page 117
$\mathcal{A}_\infty$	Class $\mathcal{A}_\infty$; generalization of Muckenhoupt class A_∞, page 159
$\mathcal{A}_{\mathrm{loc}}$	Class $\mathcal{A}_{\mathrm{loc}}$; generalization of Muckenhoupt class, page 117
A_q	Class of Muckenhoupt A_q weights, page 149
$(a,b), [a,b]$	Open/closed segment connecting a and b, page 10
B	Open ball, page 10
BMO	Functions with bounded mean oscillation, page 144
$B^n(x,r)$	N-dimensional ball, page 370
$B^\alpha_{p,q}$	Besov space, page 393
$B(x,r)$	Open ball with center x and radius $r > 0$, page 10
$\widetilde{C}_{p(\cdot)}$	Sobolev capacity based on quasicontinuous functions, page 342
$C_{p(\cdot)}$	Sobolev capacity, page 315
$\mathrm{cap}_{p(\cdot)}$	Relative/variational capacity, page 322
$\mathrm{cen}(Q)$	Center of Q, page 145
χ_E	Characteristic function of the set E, page 13
C^∞	Smooth functions, page 15
C_0^∞	Smooth functions with compact support, page 15
$C_{0,0}^\infty$	Compactly supported smooth functions with vanishing integral, page 91
$c_{\log}(p)$	log-Hölder constant of $\frac{1}{p}$, page 101
$D^{k,p(\cdot)}$	Homogeneous Sobolev space, page 379

$D_0^{k,p(\cdot)}$	Homogeneous Sobolev space with zero boundary values, page 379
$D^{-k,p(\cdot)}$	Dual space of $D_0^{k,p'(\cdot)}$, page 384
$\mathcal{D}'$	The space of distributions, page 17
E^φ	Set of finite elements of L^φ, page 49
e_Q	Special function on family of cubes, page 151
F	Closed set in a topological space, page 10
$\mathcal{F}f$	Fourier transform, page 368
$\widehat{f}$	Fourier transform, page 368
$F_{P(\cdot),q(\cdot)}^{\alpha(\cdot)}$	Triebel–Lizorkin space, page 394
$f_{P(\cdot),q(\cdot)}^{\alpha(\cdot)}$	Discrete Tribel–Lizorkin space, page 394
$F_{P,q}^\alpha$	Tribel–Lizorkin space, page 393
$\mathcal{G}$	Class $\mathcal{G}$; to pass from single cubes to families, page 223
g_α	Bessel kernel, page 388
$H_0^{k,p(\cdot)}$	Closure of C_0^∞-functions in $W^{k,p(\cdot)}$, page 346
$\mathbb{R}_>^n, \mathbb{R}_<^n$	Open upper/lower half-space, page 10
$\mathbb{R}_\geq^n, \mathbb{R}_\leq^n$	Closed upper/lower half-space, page 10
$\mathcal{H}^s$	Hausdorff s-measure, page 334
$\mathcal{H}^{s(\cdot)}$	Variable dimension Hausdorff measure, page 335
I_α	Riesz potential operator, page 199
$\mathrm{Im}(V)$	Image of operator V, page 65
J_g	Natural embedding from L^{φ^*} and $(L^\varphi)'$ into $(L^\varphi)^*$, page 59
$\mathbb{K}$	Either $\mathbb{R}$ or $\mathbb{C}$, page 21
K	Compact set in a topological space, page 10
$\ker(V)$	Kernel of operator V, page 65
L_{loc}^1	Space of locally integrable functions, page 14
L^0	The set of all measurable functions, page 22
L^φ	Musielak–Orlicz space, page 38
L_{OC}^φ	Musielak–Orlicz class, page 48
$\mathcal{L}^{\alpha,p(\cdot)}$	Bessel space, page 388
$L_0^{p(\cdot)}$	Lebesgue functions with vanishing integral, page 91
$l^{p(\cdot)}$	Variable exponent Lebesgue sequence space, page 84
$l^\psi(\mathcal{Q})$	Musielak–Orlicz (sequence) space with measure $|Q|$, page 151
$L_*^{p(\cdot)}$	Modified Lebesgue space, page 285
L^s	Classical Lebesgue space, page 14, 15
$M_s f$	L^s-maximal function, page 206
$M^\sharp f$	Sharp function, page 206
M_{center}	Centered maximal function, page 118
$M^\triangle$	Dyadic maximal function, page 162
Mf	Maximal function, page 111
M_φ	φ-average maximal operator, page 228
$M_{p(\cdot),Q}$	$L^{p(\cdot)}$-average operator over single cube, page 228
$M_{\varphi,Q}$	φ-average operator over Q, page 227
$M_Q\varphi$	Average of generalized Φ-function over Q, page 151

$M_Q f$	Average of $\lvert f \rvert$ over Q, page 111
$M_{s,Q}\varphi$	L^s-average of generalized Φ-function over Q, page 151
$M_{s,Q}f$	s-average of function over Q, page 111
$N(A,\mu)$	Set of N-functions, page 43
$N(\Omega)$	Set of N-functions, page 43
$\mathbb{N}_0$	Natural numbers including zero, page 237
Ω	Open set in $\mathbb{R}^n$, page 10
$\mathcal{P}(A,\mu)$	Set of variable exponents, page 69
$\mathcal{P}(\Omega)$	Set of variable exponents, page 69
$\mathcal{P}^{\log}$	Variable exponents p such that $\frac{1}{p}$ is log-Hölder continuous, page 101
$p^\sharp$	Sobolev conjugate for smoothness α, page 200
p'	Dual exponent, page 69
p^*	Sobolev conjugate $p^*(x) := \frac{np(x)}{n-p(x)}$, page 265
p^+, p^-	Essential supremum and infimum of p, page 69
p_A^+, p_A^-	Essential supremum and infimum of p in A, page 69
p_E'	Dual of harmonic mean, page 123
$\Phi(A,\mu)$	Generalized Φ-functions, page 36
$\Phi(\Omega)$	Generalized Φ-functions, page 36
$\Phi(\mathcal{Q})$	Generalized Φ-functions on locally 1-finite family of cubes, page 151
φ	Φ-function, page 34
φ^*	Conjugate Φ-function of φ, page 52
φ^{-1}	Left-continuous inverse, page 72
φ_p	Φ-function defining L^p, page 70
φ_∞	Φ-function defining L^∞, page 70
$\bar{\varphi}_{p(\cdot)}$	Generalized Φ-function defining $L^{p(\cdot)}$, page 70
$\tilde{\varphi}_{p(\cdot)}$	Generalized Φ-function defining $L^{p(\cdot)}$, page 70
p_Q	Harmonic mean, page 109
Q	Cube, and generic symbol for cube or ball, page 10
$Q_0^{1,p(\cdot)}$	Quasicontinuous Sobolev functions with zero boundary values, page 348
ϱ	Semimodular; modular, page 22
ϱ^*	Conjugate semimodular of ϱ, page 30
ϱ_φ	Semimodular induced by generalized Φ-function φ, page 37
$R_{p(\cdot)}(K,\Omega)$	Admissible test functions for the relative/variational capacity of (K,Ω), page 322
$\mathbb{R}^n$	Euclidean, n-dimensional space, page 10
$\mathbb{R}^{n+1}_>$	$\mathbb{R}^n \times (0,\infty)$, page 367
$\mathbb{R}^{n+1}_<$	$\mathbb{R}^n \times (-\infty,0)$, page 370
S	Simple functions, page 37
S_c	Simple functions with compact support, page 79
$\mathcal{S}$	Schwartz class, page 368
sgn	The sign of the argument, page 60

$\widetilde{S}_{p(\cdot)}(E)$	Admissible test functions for the quasicontinuous Sobolev capacity of E, page 342
$S_{p(\cdot)}(E)$	Admissible test functions for the Sobolev capacity of E, page 315
T_ε	Truncation of the operator T, page 209
T^*	Maximal truncated (singular integral) operator, page 212
$T_{\varphi,\mathcal{Q}}$	φ-average operator over cube family, page 227
$T_{p(\cdot),\mathcal{Q}}$	$L^{p(\cdot)}$-average operator over cube family, page 228
T_k	Averaging operator over dyadic cubes, page 200
$T_\mathcal{Q}$	Averaging operator, page 116
$\operatorname{Tr} X$	Trace of X, page 369
$\operatorname{Tr} D^{1,p(\cdot)}$	Trace space, page 381
$\operatorname{Tr} W^{k,p(\cdot)}$	Trace space, page 369
U, V	Open sets in a topological space, page 10
$W_{\mathrm{loc}}^{k,1}$	Space of local $W^{k,1}$ functions, page 15
$W^{k,p(\cdot)}$	Sobolev space, page 248
$W_0^{k,p(\cdot)}$	Sobolev space with zero boundary values, page 251
$W^{-k,p(\cdot)}$	Dual space of $W_0^{k,p'(\cdot)}$, page 384
$W^{k,s}$	Classical Sobolev space, page 14
w-L^q	Weak L^q, page 111
X_ϱ	Semimodular space or modular space, page 24
$\mathcal{X}^n$	Set of all open cubes in $\mathbb{R}^n$, page 156
$\mathcal{Y}_1^n$	Set of locally 1-finite family of cubes in $\mathbb{R}^n$, page 151

Index

$\mathcal{A}$-constant, 117
A-solution, 421
A-subsolution, 421
A-supersolution, 421
Absolutely continuity
  on lines, 345
Absolutely continuous norm, 50, 62
ACL, 345
Admissible
  for Triebel–Lizorkin space, 393
  for capacity, 315, 322
Almost everywhere, 13
Associate space, 58, 78
Averaged Taylor polynomial, 278
Averaging operator
  $M_Q f$, 111
  T_k, 200
  $T_{p(\cdot),Q}$, 228
  T_Q, 116
  $T_{\varphi,Q}$, 227

Banach couple, 213
Banach function space, 61, 78
Banach space, 11, 38, 248, 251, 346
Banach–Saks property, 250
Bell shaped, 128
Bessel
  kernel, 388
  space, 388
Biconjugate semimodular, 31
Bilipschitz, 305
Boman chain condition, 239
Bounded mapping, 11
Bounded mean oscillation, 144

Caccioppoli estimate, 429
Calderón-Zygmund decomposition, 194

Calderón–Zygmund operator, 208, 461
Capacitable set, 319
Capacity
  Choquet, 319, 320, 328
  of a ball, 328, 329
  outer, 317
  quasicontinuous, 342
  relative, 322
  Sobolev, 315
  variational, 322
Carathéodory function, 414
Circular, 41, 61, 77
Class $\mathcal{A}$, 117, 217
Class $\mathcal{A}_{\mathrm{loc}}$, 117
Class $\mathcal{G}$, 223
Closure, 12
Compact embedding, 218
Compact support, 15
Completeness, 38
Condenser, 322
Condition (D'), 461
Condition (D), 210
Cone condition, 300
Conjugate
  Φ-function, 52
  semimodular, 30
Continuous modular, 22
Convergence
  in measure, 15
  in norm, 25
  modular, 26
  strong, 12, 25
  weak, 12, 32
Convolution, 94
Covering theorem
  basic, 13
  Besicovitch, 14

Δ_2-condition
   for Φ-function, 42
   for modular, 42, 43
   weak for modular, 26
Decomposition theorem, 240
Density
   of bounded functions, 290
   of compactly supported functions, 290
   of continuous functions, 305
   of Lipschitz functions, 310
   of Schwartz functions, 390
   of smooth functions, 90, 130, 289, 395
Dini–Lipschitz, 100
Discontinuous exponent, 299
Discrete Lebesgue spaces, 114
Distribution, 17
Distributional derivative, 17
Divergence equation, 459, 466
Domain, 10
   (ε, ∞), 276
   Boman chain, 239
   emanating chain, 238
   John, 237
   Jones, 276
   Lipschitz, 237
   uniform, 276
   with $C^{k,\lambda}$-boundary, 237
   with fat complement, 310
Dominated, 156
Dominated convergence, 16, 40, 77
Doubling measure, 13
Dual exponent, 15
Dual space, 11, 383
Dyadic cube, 162, 277
Dyadic maximal function, 162

Eigenvalues, 402
Emanating chain condition, 238
Embedding between Lebesgue spaces, 82
Equi-integrable, 16
Existance of solutions, 402, 418
Existence of solutions, 405, 412
Extension domain, 276, 282, 292
Extension of exponent, 102
Extension operator, 276, 278, 282
Extrapolation
   application of, 276, 395
Extrapolation theorem, 218

Fat complement, 349
Fat complement, 310
Fatou property, 41, 61, 77

Fatou's lemma, 16
   for the modular, 40, 77
   for the norm, 41, 77
Fourier transform, 368
Fundamental solution, 455
   Poisson problem, 438
   Stokes problem, 447

$\mathcal{G}$-constant, 223
Generalized Orlicz space, 38
Generalized Φ-function, 36
   proper, 61, 78

Hardy's inequality, 247
Hardy–Littlewood maximal operator, 111
Harmonic mean, 109
Harmonic mean p_Q, 109
Harnack's inequality, 419, 436
   a counter example, 410
   for quasiminimizers, 416
   the one dimensional case, 409
   weak, 419, 435
Hausdorff measure, 334
   variable, 335
Hölder's inequality, 53, 81, 82

Initial topology, 14
Intersection of vector spaces, 85
Intrinsic trace space, 370
Inverse
   left-continuous, 72
   right-continuous, 54
Isomorphism, 11

Jensen's inequality, 17
John domain, 237
   John ball, 237
   John center, 237
   John path, 237
   unbounded, 239

Kernel, 208
   Bessel, 388
   condition (D'), 461
   condition (D), 210
   Newton potential, 438
   Riesz, 199
   standard, 208
Korn's inequality, 469

Laplace equation, 417, 421
Lattice, 250
Lavrentiev phenomenon, 289
Least bell shaped majorant, 128
Lebesgue point, 353
  for Sobolev functions, 357, 359
  non-existence for Sobolev functions, 362
Lebesgue space, 14, 25, 73
  w-L^q, 111
  w-l^2, 158
  $l^{p(\cdot)}(\mathbb{Z}^n)$, 84
  $L_*^{p(\cdot)}$, 285
  $L^{p(\cdot)}$, 73
  sequence space, 84
  weak, 17, 111
Left-continuous, 21
Left-continuous inverse, 72
Legendre transform, 52
Lipschitz boundary, 237
Lipschitz domain, 237
Lipschitz functions, 293
Lipschitz truncation, 310
Locally finite, 115
Locally integrable, 49
Locally N-finite, 115
log-Hölder continuity, 8, 100
Lower semicontinuity
  of modular, 28, 32, 40, 77
  of norm, 42, 77
Luxemburg norm, 24

Maximal inequality
  strong, 113, 114, 139, 181
  vector valued, 222
  weak, 111, 119, 120
Maximal operator
  M, 111
  Hardy–Littlewood, 111
  M_φ, 228
  M_q, 111, 228
Mean continuity, 247
Measurable functions, 22
Measure, 13
  absolutely continuous, 16
  atom-less, 13
  separable, 50
Measure space, 13, 22
Metric measure space, 114, 247
Micro-local spaces, 398
Minimizer, 412
Minkowski functional, 25, 27
Modified Lebesgue space scale, 138, 285
Modular, 22

$\varrho_{L_*^{p(\cdot)}(\Omega)}$, 285
$\varrho_{L^{p(\cdot)}}$, 73
$\varrho_{p(\cdot)}$, 74
$\varrho_{\mathrm{Tr},p(\cdot)}$, 371
$\varrho_{W^{k,p(\cdot)}}$, 248
$\varrho_{k,p(\cdot)}$, 248
Modular space, 24
Modulus of continuity, 15, 102, 293
Mollifier, 128, 129
  standard, 15
Monotone convergence, 16, 40, 77
Morrey space, 368
Muckenhoupt weight, 146, 149, 197, 218
Musielak–Orlicz
  class, 49, 87
  finite element, 49
  space, 38

N-function, 43
Natural embedding, 59
Negative norm theorem, 467
Newton potential, 438
Non-density of smooth functions, 297
Norm conjugate formula, 57
  L^φ, 61, 63
  $L^{p(\cdot)}$, 79, 91, 130
Norm-modular unit ball property, 26, 75, 88

Obstacle problem, 402
Operator
  associated to a kernel, 208
  Bogovskii, 460
  Calderón–Zygmund, 208, 461
  maximal, 111
  Riesz, 199
  Rubio de Francia, 219
  sharp, 206, 221
  singular integral, 208
Orlicz norm, 34
Orlicz space, 38
Outer measure, 319, 328, 335

Parabolic equations, 402
Partition of unity, 14, 308
Φ-function, 34
  generalized, 36
  positive, 34, 37
Poincaré inequality, 255, 256, 262
  counter example, 256
  for zero boundary values functions, 263
  in a modular form, 257

Poisson problem, 378, 437, 438
Polynomial, 379
Proper, 61, 78

Quasicontinuity, 339
Quasicontinuous representative, 341, 354
Quasieverywhere, 339
Quasiminimizer, 416
Quasinorm, 11
Quotient norm, 369, 381

Reflexive space, 12
Regularity, 402
Removable set for Sobolev space, 350
Riesz
    kernel, 199
    potential operator, 199
    transform, 389
Right-continuous inverse, 54
Rubio de Francia operator, 219

Scaling argument, 34
Schwartz class, 368
Semimodular, 22
    induced by φ, 37
    space, 24
Separable
    measure, 50
    space, 12
Sequence space, 84
Sharp operator, 206, 221, 371
Signum, 60
Simple function, 37
Singular integral, 208, 221
Smooth function, 15
    $C_{0,0}^\infty$, 91, 386
    C^∞, 15
    C_0^∞, 15
Sobolev conjugate exponent, 265
Sobolev embedding, 265, 396
    compact, 273, 274
    Hölder continuity, 271
    non-existence, 269
    Sobolev-Poincaré inequality, 265
    trace, 396
    Triebel–Lizorkin space, 396
    Trudinger type, 285
Sobolev function, 14, 248
Sobolev space, 14, 248
    $ACL^{p(\cdot)}$, 345
    $D^{k,p(\cdot)}$, 379

$D_0^{k,p(\cdot)}$, 379
$H_0^{k,p(\cdot)}$, 346
    homogeneous, 379, 446
$Q_0^{1,p(\cdot)}$, 348
    with zero boundary values, 251, 346,
        348
$W^{k,p(\cdot)}$, 248
$W_0^{k,p(\cdot)}$, 251
Sobolev-Poincaré inequality, 265
    for zero boundary values functions, 265
Sobolev-type space
    $B_{p,q}^\alpha$, 393
    $F_{p(\cdot),q(\cdot)}^{\alpha(\cdot)}$, 394
    $F_{p,q}^\alpha$, 393
    $f_{p(\cdot),q(\cdot)}^{\alpha(\cdot)}$, 394
    $\mathcal{L}^{\alpha,p(\cdot)}$, 388
    $\mathrm{Tr}\,D^{1,p(\cdot)}$, 381
    $(\mathrm{Tr}\,W^{1,p(\cdot)})$, 370
Solid, 41, 61, 77
Solution, 417, 421
Standard estimates, 208
Standard mollifier, 15
    family, 15
Stokes problem, 446
    in half-space, 451
Strong type, 115
Strongly dominated, 168
Subsolution, 417, 421
Sum of vector spaces, 85
Superharmonic function, 420
Supersolution, 417, 421
Symmetric gradient, 469
Systems of differential equations, 402

Taylor polynomial
    averaged, 278
Touch, 277
Trace, 369
    embedding, 396
    modular, 371
    space, 369, 381
Trace embeddings, 396
Truncation, 310

Uniformly convex
    N-function, 43
    norm, 12
    semimodular, 45
    space, 12, 46, 89, 249, 251
Uniqueness of solutions, 402
Unit ball property, 26, 75, 88

Variable exponent, 69
  bounded, 69
  of class $\mathcal{A}$, 117
  of class $\mathcal{A}_{\mathrm{loc}}$, 117
  of class $\mathcal{P}^{\log}$, 101
Variable smoothness, 393

Weak A-solution, 421
Weak derivative, 248
Weak gradient, 248
Weak Harnack inequality, 419, 435

Weak Lebesgue space, 111
Weak Lipschitz, 100
Weak partial derivative, 248
Weak solution, 417, 421
Weak type, 111, 115
Whitney decomposition, 277

Young's inequality, 30, 52, 55, 80

0-Hölder, 100

Lecture Notes in Mathematics

For information about earlier volumes
please contact your bookseller or Springer
LNM Online archive: springerlink.com

Vol. 1831: A. Connes, J. Cuntz, E. Guentner, N. Higson, J. E. Kaminker, Noncommutative Geometry, Martina Franca, Italy 2002. Editors: S. Doplicher, L. Longo (2004)

Vol. 1832: J. Azéma, M. Émery, M. Ledoux, M. Yor (Eds.), Séminaire de Probabilités XXXVII (2003)

Vol. 1833: D.-Q. Jiang, M. Qian, M.-P. Qian, Mathematical Theory of Nonequilibrium Steady States. On the Frontier of Probability and Dynamical Systems. IX, 280 p, 2004.

Vol. 1834: Yo. Yomdin, G. Comte, Tame Geometry with Application in Smooth Analysis. VIII, 186 p, 2004.

Vol. 1835: O.T. Izhboldin, B. Kahn, N.A. Karpenko, A. Vishik, Geometric Methods in the Algebraic Theory of Quadratic Forms. Summer School, Lens, 2000. Editor: J.-P. Tignol (2004)

Vol. 1836: C. Năstăsescu, F. Van Oystaeyen, Methods of Graded Rings. XIII, 304 p, 2004.

Vol. 1837: S. Tavaré, O. Zeitouni, Lectures on Probability Theory and Statistics. Ecole d'Été de Probabilités de Saint-Flour XXXI-2001. Editor: J. Picard (2004)

Vol. 1838: A.J. Ganesh, N.W. O'Connell, D.J. Wischik, Big Queues. XII, 254 p, 2004.

Vol. 1839: R. Gohm, Noncommutative Stationary Processes. VIII, 170 p, 2004.

Vol. 1840: B. Tsirelson, W. Werner, Lectures on Probability Theory and Statistics. Ecole d'Été de Probabilités de Saint-Flour XXXII-2002. Editor: J. Picard (2004)

Vol. 1841: W. Reichel, Uniqueness Theorems for Variational Problems by the Method of Transformation Groups (2004)

Vol. 1842: T. Johnsen, A. L. Knutsen, K_3 Projective Models in Scrolls (2004)

Vol. 1843: B. Jefferies, Spectral Properties of Noncommuting Operators (2004)

Vol. 1844: K.F. Siburg, The Principle of Least Action in Geometry and Dynamics (2004)

Vol. 1845: Min Ho Lee, Mixed Automorphic Forms, Torus Bundles, and Jacobi Forms (2004)

Vol. 1846: H. Ammari, H. Kang, Reconstruction of Small Inhomogeneities from Boundary Measurements (2004)

Vol. 1847: T.R. Bielecki, T. Björk, M. Jeanblanc, M. Rutkowski, J.A. Scheinkman, W. Xiong, Paris-Princeton Lectures on Mathematical Finance 2003 (2004)

Vol. 1848: M. Abate, J. E. Fornaess, X. Huang, J. P. Rosay, A. Tumanov, Real Methods in Complex and CR Geometry, Martina Franca, Italy 2002. Editors: D. Zaitsev, G. Zampieri (2004)

Vol. 1849: Martin L. Brown, Heegner Modules and Elliptic Curves (2004)

Vol. 1850: V. D. Milman, G. Schechtman (Eds.), Geometric Aspects of Functional Analysis. Israel Seminar 2002-2003 (2004)

Vol. 1851: O. Catoni, Statistical Learning Theory and Stochastic Optimization (2004)

Vol. 1852: A.S. Kechris, B.D. Miller, Topics in Orbit Equivalence (2004)

Vol. 1853: Ch. Favre, M. Jonsson, The Valuative Tree (2004)

Vol. 1854: O. Saeki, Topology of Singular Fibers of Differential Maps (2004)

Vol. 1855: G. Da Prato, P.C. Kunstmann, I. Lasiecka, A. Lunardi, R. Schnaubelt, L. Weis, Functional Analytic Methods for Evolution Equations. Editors: M. Iannelli, R. Nagel, S. Piazzera (2004)

Vol. 1856: K. Back, T.R. Bielecki, C. Hipp, S. Peng, W. Schachermayer, Stochastic Methods in Finance, Bressanone/Brixen, Italy, 2003. Editors: M. Fritelli, W. Runggaldier (2004)

Vol. 1857: M. Émery, M. Ledoux, M. Yor (Eds.), Séminaire de Probabilités XXXVIII (2005)

Vol. 1858: A.S. Cherny, H.-J. Engelbert, Singular Stochastic Differential Equations (2005)

Vol. 1859: E. Letellier, Fourier Transforms of Invariant Functions on Finite Reductive Lie Algebras (2005)

Vol. 1860: A. Borisyuk, G.B. Ermentrout, A. Friedman, D. Terman, Tutorials in Mathematical Biosciences I. Mathematical Neurosciences (2005)

Vol. 1861: G. Benettin, J. Henrard, S. Kuksin, Hamiltonian Dynamics – Theory and Applications, Cetraro, Italy, 1999. Editor: A. Giorgilli (2005)

Vol. 1862: B. Helffer, F. Nier, Hypoelliptic Estimates and Spectral Theory for Fokker-Planck Operators and Witten Laplacians (2005)

Vol. 1863: H. Führ, Abstract Harmonic Analysis of Continuous Wavelet Transforms (2005)

Vol. 1864: K. Efstathiou, Metamorphoses of Hamiltonian Systems with Symmetries (2005)

Vol. 1865: D. Applebaum, B.V. R. Bhat, J. Kustermans, J. M. Lindsay, Quantum Independent Increment Processes I. From Classical Probability to Quantum Stochastic Calculus. Editors: M. Schürmann, U. Franz (2005)

Vol. 1866: O.E. Barndorff-Nielsen, U. Franz, R. Gohm, B. Kümmerer, S. Thorbjønsen, Quantum Independent Increment Processes II. Structure of Quantum Lévy Processes, Classical Probability, and Physics. Editors: M. Schürmann, U. Franz, (2005)

Vol. 1867: J. Sneyd (Ed.), Tutorials in Mathematical Biosciences II. Mathematical Modeling of Calcium Dynamics and Signal Transduction. (2005)

Vol. 1868: J. Jorgenson, S. Lang, Pos_n (R) and Eisenstein Series. (2005)

Vol. 1869: A. Dembo, T. Funaki, Lectures on Probability Theory and Statistics. Ecole d'Été de Probabilités de Saint-Flour XXXIII-2003. Editor: J. Picard (2005)

Vol. 1870: V.I. Gurariy, W. Lusky, Geometry of Mntz Spaces and Related Questions. (2005)

Vol. 1871: P. Constantin, G. Gallavotti, A.V. Kazhikhov, Y. Meyer, S. Ukai, Mathematical Foundation of Turbulent Viscous Flows, Martina Franca, Italy, 2003. Editors: M. Cannone, T. Miyakawa (2006)

Vol. 1872: A. Friedman (Ed.), Tutorials in Mathematical Biosciences III. Cell Cycle, Proliferation, and Cancer (2006)

Vol. 1873: R. Mansuy, M. Yor, Random Times and Enlargements of Filtrations in a Brownian Setting (2006)

Vol. 1874: M. Yor, M. Émery (Eds.), In Memoriam Paul-André Meyer - Séminaire de Probabilités XXXIX (2006)

Vol. 1875: J. Pitman, Combinatorial Stochastic Processes. Ecole d'Été de Probabilités de Saint-Flour XXXII-2002. Editor: J. Picard (2006)

Vol. 1876: H. Herrlich, Axiom of Choice (2006)

Vol. 1877: J. Steuding, Value Distributions of L-Functions (2007)

Vol. 1878: R. Cerf, The Wulff Crystal in Ising and Percolation Models, Ecole d'Été de Probabilités de Saint-Flour XXXIV-2004. Editor: Jean Picard (2006)

Vol. 1879: G. Slade, The Lace Expansion and its Applications, Ecole d'Été de Probabilités de Saint-Flour XXXIV-2004. Editor: Jean Picard (2006)

Vol. 1880: S. Attal, A. Joye, C.-A. Pillet, Open Quantum Systems I, The Hamiltonian Approach (2006)

Vol. 1881: S. Attal, A. Joye, C.-A. Pillet, Open Quantum Systems II, The Markovian Approach (2006)

Vol. 1882: S. Attal, A. Joye, C.-A. Pillet, Open Quantum Systems III, Recent Developments (2006)

Vol. 1883: W. Van Assche, F. Marcellàn (Eds.), Orthogonal Polynomials and Special Functions, Computation and Application (2006)

Vol. 1884: N. Hayashi, E.I. Kaikina, P.I. Naumkin, I.A. Shishmarev, Asymptotics for Dissipative Nonlinear Equations (2006)

Vol. 1885: A. Telcs, The Art of Random Walks (2006)

Vol. 1886: S. Takamura, Splitting Deformations of Degenerations of Complex Curves (2006)

Vol. 1887: K. Habermann, L. Habermann, Introduction to Symplectic Dirac Operators (2006)

Vol. 1888: J. van der Hoeven, Transseries and Real Differential Algebra (2006)

Vol. 1889: G. Osipenko, Dynamical Systems, Graphs, and Algorithms (2006)

Vol. 1890: M. Bunge, J. Funk, Singular Coverings of Toposes (2006)

Vol. 1891: J.B. Friedlander, D.R. Heath-Brown, H. Iwaniec, J. Kaczorowski, Analytic Number Theory, Cetraro, Italy, 2002. Editors: A. Perelli, C. Viola (2006)

Vol. 1892: A. Baddeley, I. Bárány, R. Schneider, W. Weil, Stochastic Geometry, Martina Franca, Italy, 2004. Editor: W. Weil (2007)

Vol. 1893: H. Hanßmann, Local and Semi-Local Bifurcations in Hamiltonian Dynamical Systems, Results and Examples (2007)

Vol. 1894: C.W. Groetsch, Stable Approximate Evaluation of Unbounded Operators (2007)

Vol. 1895: L. Molnár, Selected Preserver Problems on Algebraic Structures of Linear Operators and on Function Spaces (2007)

Vol. 1896: P. Massart, Concentration Inequalities and Model Selection, Ecole d'Été de Probabilités de Saint-Flour XXXIII-2003. Editor: J. Picard (2007)

Vol. 1897: R. Doney, Fluctuation Theory for Lévy Processes, Ecole d'Été de Probabilités de Saint-Flour XXXV-2005. Editor: J. Picard (2007)

Vol. 1898: H.R. Beyer, Beyond Partial Differential Equations, On linear and Quasi-Linear Abstract Hyperbolic Evolution Equations (2007)

Vol. 1899: Séminaire de Probabilités XL. Editors: C. Donati-Martin, M. Émery, A. Rouault, C. Stricker (2007)

Vol. 1900: E. Bolthausen, A. Bovier (Eds.), Spin Glasses (2007)

Vol. 1901: O. Wittenberg, Intersections de deux quadriques et pinceaux de courbes de genre 1, Intersections of Two Quadrics and Pencils of Curves of Genus 1 (2007)

Vol. 1902: A. Isaev, Lectures on the Automorphism Groups of Kobayashi-Hyperbolic Manifolds (2007)

Vol. 1903: G. Kresin, V. Maz'ya, Sharp Real-Part Theorems (2007)

Vol. 1904: P. Giesl, Construction of Global Lyapunov Functions Using Radial Basis Functions (2007)

Vol. 1905: C. Prévôt, M. Röckner, A Concise Course on Stochastic Partial Differential Equations (2007)

Vol. 1906: T. Schuster, The Method of Approximate Inverse: Theory and Applications (2007)

Vol. 1907: M. Rasmussen, Attractivity and Bifurcation for Nonautonomous Dynamical Systems (2007)

Vol. 1908: T.J. Lyons, M. Caruana, T. Lévy, Differential Equations Driven by Rough Paths, Ecole d'Été de Probabilités de Saint-Flour XXXIV-2004 (2007)

Vol. 1909: H. Akiyoshi, M. Sakuma, M. Wada, Y. Yamashita, Punctured Torus Groups and 2-Bridge Knot Groups (I) (2007)

Vol. 1910: V.D. Milman, G. Schechtman (Eds.), Geometric Aspects of Functional Analysis. Israel Seminar 2004-2005 (2007)

Vol. 1911: A. Bressan, D. Serre, M. Williams, K. Zumbrun, Hyperbolic Systems of Balance Laws. Cetraro, Italy 2003. Editor: P. Marcati (2007)

Vol. 1912: V. Berinde, Iterative Approximation of Fixed Points (2007)

Vol. 1913: J.E. Marsden, G. Misiołek, J.-P. Ortega, M. Perlmutter, T.S. Ratiu, Hamiltonian Reduction by Stages (2007)

Vol. 1914: G. Kutyniok, Affine Density in Wavelet Analysis (2007)

Vol. 1915: T. Bıyıkoğlu, J. Leydold, P.F. Stadler, Laplacian Eigenvectors of Graphs. Perron-Frobenius and Faber-Krahn Type Theorems (2007)

Vol. 1916: C. Villani, F. Rezakhanlou, Entropy Methods for the Boltzmann Equation. Editors: F. Golse, S. Olla (2008)

Vol. 1917: I. Veselić, Existence and Regularity Properties of the Integrated Density of States of Random Schrödinger (2008)

Vol. 1918: B. Roberts, R. Schmidt, Local Newforms for GSp(4) (2007)

Vol. 1919: R.A. Carmona, I. Ekeland, A. Kohatsu-Higa, J.-M. Lasry, P.-L. Lions, H. Pham, E. Taflin, Paris-Princeton Lectures on Mathematical Finance 2004. Editors: R.A. Carmona, E. Cinlar, I. Ekeland, E. Jouini, J.A. Scheinkman, N. Touzi (2007)

Vol. 1920: S.N. Evans, Probability and Real Trees. Ecole d'Été de Probabilités de Saint-Flour XXXV-2005 (2008)

Vol. 1921: J.P. Tian, Evolution Algebras and their Applications (2008)

Vol. 1922: A. Friedman (Ed.), Tutorials in Mathematical BioSciences IV. Evolution and Ecology (2008)

Vol. 1923: J.P.N. Bishwal, Parameter Estimation in Stochastic Differential Equations (2008)

Vol. 1924: M. Wilson, Littlewood-Paley Theory and Exponential-Square Integrability (2008)

Vol. 1925: M. du Sautoy, L. Woodward, Zeta Functions of Groups and Rings (2008)

Vol. 1926: L. Barreira, V. Claudia, Stability of Nonautonomous Differential Equations (2008)

Vol. 1927: L. Ambrosio, L. Caffarelli, M.G. Crandall, L.C. Evans, N. Fusco, Calculus of Variations and Non-Linear Partial Differential Equations. Cetraro, Italy 2005. Editors: B. Dacorogna, P. Marcellini (2008)

Vol. 1928: J. Jonsson, Simplicial Complexes of Graphs (2008)

Vol. 1929: Y. Mishura, Stochastic Calculus for Fractional Brownian Motion and Related Processes (2008)

Vol. 1930: J.M. Urbano, The Method of Intrinsic Scaling. A Systematic Approach to Regularity for Degenerate and Singular PDEs (2008)

Vol. 1931: M. Cowling, E. Frenkel, M. Kashiwara, A. Valette, D.A. Vogan, Jr., N.R. Wallach, Representation Theory and Complex Analysis. Venice, Italy 2004. Editors: E.C. Tarabusi, A. D'Agnolo, M. Picardello (2008)

Vol. 1932: A.A. Agrachev, A.S. Morse, E.D. Sontag, H.J. Sussmann, V.I. Utkin, Nonlinear and Optimal Control Theory. Cetraro, Italy 2004. Editors: P. Nistri, G. Stefani (2008)

Vol. 1933: M. Petkovic, Point Estimation of Root Finding Methods (2008)

Vol. 1934: C. Donati-Martin, M. Émery, A. Rouault, C. Stricker (Eds.), Séminaire de Probabilités XLI (2008)

Vol. 1935: A. Unterberger, Alternative Pseudodifferential Analysis (2008)

Vol. 1936: P. Magal, S. Ruan (Eds.), Structured Population Models in Biology and Epidemiology (2008)

Vol. 1937: G. Capriz, P. Giovine, P.M. Mariano (Eds.), Mathematical Models of Granular Matter (2008)

Vol. 1938: D. Auroux, F. Catanese, M. Manetti, P. Seidel, B. Siebert, I. Smith, G. Tian, Symplectic 4-Manifolds and Algebraic Surfaces. Cetraro, Italy 2003. Editors: F. Catanese, G. Tian (2008)

Vol. 1939: D. Boffi, F. Brezzi, L. Demkowicz, R.G. Durán, R.S. Falk, M. Fortin, Mixed Finite Elements, Compatibility Conditions, and Applications. Cetraro, Italy 2006. Editors: D. Boffi, L. Gastaldi (2008)

Vol. 1940: J. Banasiak, V. Capasso, M.A.J. Chaplain, M. Lachowicz, J. Miękisz, Multiscale Problems in the Life Sciences. From Microscopic to Macroscopic. Będlewo, Poland 2006. Editors: V. Capasso, M. Lachowicz (2008)

Vol. 1941: S.M.J. Haran, Arithmetical Investigations. Representation Theory, Orthogonal Polynomials, and Quantum Interpolations (2008)

Vol. 1942: S. Albeverio, F. Flandoli, Y.G. Sinai, SPDE in Hydrodynamic. Recent Progress and Prospects. Cetraro, Italy 2005. Editors: G. Da Prato, M. Röckner (2008)

Vol. 1943: L.L. Bonilla (Ed.), Inverse Problems and Imaging. Martina Franca, Italy 2002 (2008)

Vol. 1944: A. Di Bartolo, G. Falcone, P. Plaumann, K. Strambach, Algebraic Groups and Lie Groups with Few Factors (2008)

Vol. 1945: F. Brauer, P. van den Driessche, J. Wu (Eds.), Mathematical Epidemiology (2008)

Vol. 1946: G. Allaire, A. Arnold, P. Degond, T.Y. Hou, Quantum Transport. Modelling, Analysis and Asymptotics. Cetraro, Italy 2006. Editors: N.B. Abdallah, G. Frosali (2008)

Vol. 1947: D. Abramovich, M. Mariño, M. Thaddeus, R. Vakil, Enumerative Invariants in Algebraic Geometry and String Theory. Cetraro, Italy 2005. Editors: K. Behrend, M. Manetti (2008)

Vol. 1948: F. Cao, J-L. Lisani, J-M. Morel, P. Musé, F. Sur, A Theory of Shape Identification (2008)

Vol. 1949: H.G. Feichtinger, B. Helffer, M.P. Lamoureux, N. Lerner, J. Toft, Pseudo-Differential Operators. Quan-

tization and Signals. Cetraro, Italy 2006. Editors: L. Rodino, M.W. Wong (2008)

Vol. 1950: M. Bramson, Stability of Queueing Networks, Ecole d'Été de Probabilités de Saint-Flour XXXVI-2006 (2008)

Vol. 1951: A. Moltó, J. Orihuela, S. Troyanski, M. Valdivia, A Non Linear Transfer Technique for Renorming (2009)

Vol. 1952: R. Mikhailov, I.B.S. Passi, Lower Central and Dimension Series of Groups (2009)

Vol. 1953: K. Arwini, C.T.J. Dodson, Information Geometry (2008)

Vol. 1954: P. Biane, L. Bouten, F. Cipriani, N. Konno, N. Privault, Q. Xu, Quantum Potential Theory. Editors: U. Franz, M. Schuermann (2008)

Vol. 1955: M. Bernot, V. Caselles, J.-M. Morel, Optimal Transportation Networks (2008)

Vol. 1956: C.H. Chu, Matrix Convolution Operators on Groups (2008)

Vol. 1957: A. Guionnet, On Random Matrices: Macroscopic Asymptotics, Ecole d'Été de Probabilités de Saint-Flour XXXVI-2006 (2009)

Vol. 1958: M.C. Olsson, Compactifying Moduli Spaces for Abelian Varieties (2008)

Vol. 1959: Y. Nakkajima, A. Shiho, Weight Filtrations on Log Crystalline Cohomologies of Families of Open Smooth Varieties (2008)

Vol. 1960: J. Lipman, M. Hashimoto, Foundations of Grothendieck Duality for Diagrams of Schemes (2009)

Vol. 1961: G. Buttazzo, A. Pratelli, S. Solimini, E. Stepanov, Optimal Urban Networks via Mass Transportation (2009)

Vol. 1962: R. Dalang, D. Khoshnevisan, C. Mueller, D. Nualart, Y. Xiao, A Minicourse on Stochastic Partial Differential Equations (2009)

Vol. 1963: W. Siegert, Local Lyapunov Exponents (2009)

Vol. 1964: W. Roth, Operator-valued Measures and Integrals for Cone-valued Functions and Integrals for Cone-valued Functions (2009)

Vol. 1965: C. Chidume, Geometric Properties of Banach Spaces and Nonlinear Iterations (2009)

Vol. 1966: D. Deng, Y. Han, Harmonic Analysis on Spaces of Homogeneous Type (2009)

Vol. 1967: B. Fresse, Modules over Operads and Functors (2009)

Vol. 1968: R. Weissauer, Endoscopy for GSP(4) and the Cohomology of Siegel Modular Threefolds (2009)

Vol. 1969: B. Roynette, M. Yor, Penalising Brownian Paths (2009)

Vol. 1970: M. Biskup, A. Bovier, F. den Hollander, D. Ioffe, F. Martinelli, K. Netočný, F. Toninelli, Methods of Contemporary Mathematical Statistical Physics. Editor: R. Kotecký (2009)

Vol. 1971: L. Saint-Raymond, Hydrodynamic Limits of the Boltzmann Equation (2009)

Vol. 1972: T. Mochizuki, Donaldson Type Invariants for Algebraic Surfaces (2009)

Vol. 1973: M.A. Berger, L.H. Kauffmann, B. Khesin, H.K. Moffatt, R.L. Ricca, De W. Sumners, Lectures on Topological Fluid Mechanics. Cetraro, Italy 2001. Editor: R.L. Ricca (2009)

Vol. 1974: F. den Hollander, Random Polymers: École d'Été de Probabilités de Saint-Flour XXXVII – 2007 (2009)

Vol. 1975: J.C. Rohde, Cyclic Coverings, Calabi-Yau Manifolds and Complex Multiplication (2009)

Vol. 1976: N. Ginoux, The Dirac Spectrum (2009)

Vol. 1977: M.J. Gursky, E. Lanconelli, A. Malchiodi, G. Tarantello, X.-J. Wang, P.C. Yang, Geometric Analysis and PDEs. Cetraro, Italy 2001. Editors: A. Ambrosetti, S.-Y.A. Chang, A. Malchiodi (2009)

Vol. 1978: M. Qian, J.-S. Xie, S. Zhu, Smooth Ergodic Theory for Endomorphisms (2009)

Vol. 1979: C. Donati-Martin, M. Émery, A. Rouault, C. Stricker (Eds.), Séminaire de Probablitiés XLII (2009)

Vol. 1980: P. Graczyk, A. Stos (Eds.), Potential Analysis of Stable Processes and its Extensions (2009)

Vol. 1981: M. Chlouveraki, Blocks and Families for Cyclotomic Hecke Algebras (2009)

Vol. 1982: N. Privault, Stochastic Analysis in Discrete and Continuous Settings. With Normal Martingales (2009)

Vol. 1983: H. Ammari (Ed.), Mathematical Modeling in Biomedical Imaging I. Electrical and Ultrasound Tomographies, Anomaly Detection, and Brain Imaging (2009)

Vol. 1984: V. Caselles, P. Monasse, Geometric Description of Images as Topographic Maps (2010)

Vol. 1985: T. Linß, Layer-Adapted Meshes for Reaction-Convection-Diffusion Problems (2010)

Vol. 1986: J.-P. Antoine, C. Trapani, Partial Inner Product Spaces. Theory and Applications (2009)

Vol. 1987: J.-P. Brasselet, J. Seade, T. Suwa, Vector Fields on Singular Varieties (2010)

Vol. 1988: M. Broué, Introduction to Complex Reflection Groups and Their Braid Groups (2010)

Vol. 1989: I.M. Bomze, V. Demyanov, Nonlinear Optimization. Cetraro, Italy 2007. Editors: G. di Pillo, F. Schoen (2010)

Vol. 1990: S. Bouc, Biset Functors for Finite Groups (2010)

Vol. 1991: F. Gazzola, H.-C. Grunau, G. Sweers, Polyharmonic Boundary Value Problems (2010)

Vol. 1992: A. Parmeggiani, Spectral Theory of Non-Commutative Harmonic Oscillators: An Introduction (2010)

Vol. 1993: P. Dodos, Banach Spaces and Descriptive Set Theory: Selected Topics (2010)

Vol. 1994: A. Baricz, Generalized Bessel Functions of the First Kind (2010)

Vol. 1995: A.Y. Khapalov, Controllability of Partial Differential Equations Governed by Multiplicative Controls (2010)

Vol. 1996: T. Lorenz, Mutational Analysis. A Joint Framework for Cauchy Problems In and Beyond Vector Spaces (2010)

Vol. 1997: M. Banagl, Intersection Spaces, Spatial Homology Truncation, and String Theory (2010)

Vol. 1998: M. Abate, E. Bedford, M. Brunella, T.-C. Dinh, D. Schleicher, N. Sibony, Holomorphic Dynamical Systems. Cetraro, Italy 2008. Editors: G. Gentili, J. Guenot, G. Patrizio (2010)

Vol. 1999: H. Schoutens, The Use of Ultraproducts in Commutative Algebra (2010)

Vol. 2000: H. Yserentant, Regularity and Approximability of Electronic Wave Functions (2010)

Vol. 2001: T. Duquesne, O. Reichmann, K.-i. Sato, C. Schwab, Lévy Matters I. Editors: O.E. Barndorff-Nielson, J. Bertoin, J. Jacod, C. Klüppelberg (2010)

Vol. 2002: C. Pötzsche, Geometric Theory of Discrete Nonautonomous Dynamical Systems (2010)

Vol. 2003: A. Cousin, S. Crépey, O. Guéant, D. Hobson, M. Jeanblanc, J.-M. Lasry, J.-P. Laurent, P.-L. Lions, P. Tankov, Paris-Princeton Lectures on Mathematical Finance 2010. Editors: R.A. Carmona, E. Cinlar, I. Ekeland, E. Jouini, J.A. Scheinkman, N. Touzi (2010)

Vol. 2004: K. Diethelm, The Analysis of Fractional Differential Equations (2010)

Vol. 2005: W. Yuan, W. Sickel, D. Yang, Morrey and Campanato Meet Besov, Lizorkin and Triebel (2011)

Vol. 2006: C. Donati-Martin, A. Lejay, W. Rouault (Eds.), Séminaire de Probabilités XLIII (2011)

Vol. 2007: E. Bujalance, F.J. Cirre, J.M. Gamboa, G. Gromadzki, Symmetries of Compact Riemann Surfaces (2010)

Vol. 2008: P.F. Baum, G. Cortiñas, R. Meyer, R. Sánchez-Garca, M. Schlichting, B. Toën, Topics in Algebraic and Topological K-Theory. Editor: G. Cortiñas (2011)

Vol. 2009: J.-L. Colliot-Thélène, P.S. Dyer, P. Vojta, Arithmetic Geometry. Cetraro, Italy 2007. Editors: P. Corvaja, C. Gasbarri (2011)

Vol. 2010: A. Farina, A. Klar, R.M.M. Mattheij, A. Mikelić, N. Siedow, Mathematical Models in the Manufacturing of Glass. Cetraro, Italy 2008. Editor: A. Fasano (2011)

Vol. 2011: B. Andrews, C. Hopper, The Ricci Flow in Riemannian Geometry (2011)

Vol. 2012: A. Etheridge, Some Mathematical Models from Population Genetics. École d'Été de Probabilités de Saint-Flour XXXIX-2009 (2011)

Vol. 2013: A. Bobenko, C.Klein, Computational Approach to Riemann Surfaces (2011)

Vol. 2014: M. Audin, Fatou, Julia, Montel. The Great Prize of Mathematical Sciences of 1918, and beyond (2011)

Vol. 2015: F. Flandoli, Random Perturbation of PDEs and Fluid Dynamic Models. École d'Été de Probabilités de Saint-Flour XL-2010 (2011)

Vol. 2016: J. Lang, D. Edmunds, Eigenvalues, Embeddings and Generalised Trigonometric Functions (2011)

Vol. 2017: L. Diening, P. Harjulehto, P. Hästö, M. Růžička, Lebesgue and Sobolev Spaces with Variable Exponents (2011)

Recent Reprints and New Editions

Vol. 1629: J.D. Moore, Lectures on Seiberg-Witten Invariants. 1997 – 2nd edition (2001)

Vol. 1638: P. Vanhaecke, Integrable Systems in the realm of Algebraic Geometry. 1996 – 2nd edition (2001)

Vol. 1702: J. Ma, J. Yong, Forward-Backward Stochastic Differential Equations and their Applications. 1999 – Corr. 3rd printing (2007)

Vol. 830: J.A. Green, Polynomial Representations of GL_n, with an Appendix on Schensted Correspondence and Littelmann Paths by K. Erdmann, J.A. Green and M. Schoker 1980 – 2nd corr. and augmented edition (2007)

Vol. 1693: S. Simons, From Hahn-Banach to Monotonicity (Minimax and Monotonicity 1998) – 2nd exp. edition (2008)

Vol. 470: R.E. Bowen, Equilibrium States and the Ergodic Theory of Anosov Diffeomorphisms. With a preface by D. Ruelle. Edited by J.-R. Chazottes. 1975 – 2nd rev. edition (2008)

Vol. 523: S.A. Albeverio, R.J. Høegh-Krohn, S. Mazzucchi, Mathematical Theory of Feynman Path Integral. 1976 – 2nd corr. and enlarged edition (2008)

Vol. 1764: A. Cannas da Silva, Lectures on Symplectic Geometry 2001 – Corr. 2nd printing (2008)

LECTURE NOTES IN MATHEMATICS

Springer

Edited by J.-M. Morel, B. Teissier, P.K. Maini

Editorial Policy (for the publication of monographs)

1. Lecture Notes aim to report new developments in all areas of mathematics and their applications - quickly, informally and at a high level. Mathematical texts analysing new developments in modelling and numerical simulation are welcome.

   Monograph manuscripts should be reasonably self-contained and rounded off. Thus they may, and often will, present not only results of the author but also related work by other people. They may be based on specialised lecture courses. Furthermore, the manuscripts should provide sufficient motivation, examples and applications. This clearly distinguishes Lecture Notes from journal articles or technical reports which normally are very concise. Articles intended for a journal but too long to be accepted by most journals, usually do not have this "lecture notes" character. For similar reasons it is unusual for doctoral theses to be accepted for the Lecture Notes series, though habilitation theses may be appropriate.

2. Manuscripts should be submitted either online at www.editorialmanager.com/lnm to Springer's mathematics editorial in Heidelberg, or to one of the series editors. In general, manuscripts will be sent out to 2 external referees for evaluation. If a decision cannot yet be reached on the basis of the first 2 reports, further referees may be contacted: The author will be informed of this. A final decision to publish can be made only on the basis of the complete manuscript, however a refereeing process leading to a preliminary decision can be based on a pre-final or incomplete manuscript. The strict minimum amount of material that will be considered should include a detailed outline describing the planned contents of each chapter, a bibliography and several sample chapters.

   Authors should be aware that incomplete or insufficiently close to final manuscripts almost always result in longer refereeing times and nevertheless unclear referees' recommendations, making further refereeing of a final draft necessary.

   Authors should also be aware that parallel submission of their manuscript to another publisher while under consideration for LNM will in general lead to immediate rejection.

3. Manuscripts should in general be submitted in English. Final manuscripts should contain at least 100 pages of mathematical text and should always include

   – a table of contents;
   – an informative introduction, with adequate motivation and perhaps some historical remarks: it should be accessible to a reader not intimately familiar with the topic treated;
   – a subject index: as a rule this is genuinely helpful for the reader.

   For evaluation purposes, manuscripts may be submitted in print or electronic form (print form is still preferred by most referees), in the latter case preferably as pdf- or zipped ps-files. Lecture Notes volumes are, as a rule, printed digitally from the authors' files. To ensure best results, authors are asked to use the LaTeX2e style files available from Springer's web-server at:

   ftp://ftp.springer.de/pub/tex/latex/svmonot1/ (for monographs) and
   ftp://ftp.springer.de/pub/tex/latex/svmultt1/ (for summer schools/tutorials).
   Additional technical instructions, if necessary, are available on request from: lnm@springer.com.

4. Careful preparation of the manuscripts will help keep production time short besides ensuring satisfactory appearance of the finished book in print and online. After acceptance of the manuscript authors will be asked to prepare the final LaTeX source files and also the corresponding dvi-, pdf- or zipped ps-file. The LaTeX source files are essential for producing the full-text online version of the book (see
http://www.springerlink.com/openurl.asp?genre=journal&issn=0075-8434 for the existing online volumes of LNM).
The actual production of a Lecture Notes volume takes approximately 12 weeks.

5. Authors receive a total of 50 free copies of their volume, but no royalties. They are entitled to a discount of 33.3% on the price of Springer books purchased for their personal use, if ordering directly from Springer.

6. Commitment to publish is made by letter of intent rather than by signing a formal contract. Springer-Verlag secures the copyright for each volume. Authors are free to reuse material contained in their LNM volumes in later publications: a brief written (or e-mail) request for formal permission is sufficient.

Addresses:
Professor J.-M. Morel, CMLA,
École Normale Supérieure de Cachan,
61 Avenue du Président Wilson, 94235 Cachan Cedex, France
E-mail: morel@cmla.ens-cachan.fr

Professor B. Teissier, Institut Mathématique de Jussieu,
UMR 7586 du CNRS, Équipe "Géométrie et Dynamique",
175 rue du Chevaleret,
75013 Paris, France
E-mail: teissier@math.jussieu.fr

For the "Mathematical Biosciences Subseries" of LNM:

Professor P.K. Maini, Center for Mathematical Biology,
Mathematical Institute, 24-29 St Giles,
Oxford OX1 3LP, UK
E-mail: maini@maths.ox.ac.uk

Springer, Mathematics Editorial, Tiergartenstr. 17,
69121 Heidelberg, Germany,
Tel.: +49 (6221) 487-259
Fax: +49 (6221) 4876-8259
E-mail: lnm@springer.com